Lecture Notes in Computer Science 3728

Commenced Publication in 1973
Founding and Former Series Editors:
Gerhard Goos, Juris Hartmanis, and Jan van Leeuwen

T0189669

Vassilis Paliouras   Johan Vounckx
Diederik Verkest (Eds.)

Integrated Circuit and System Design

Power and Timing Modeling, Optimization and Simulation

15th International Workshop, PATMOS 2005
Leuven, Belgium, September 21-23, 2005
Proceedings

 Springer

Volume Editors

Vassilis Paliouras
University of Patras
Electrical and Computer Engineering Department
26500 Patras, Greece
E-mail: paliuras@ee.upatras.gr

Johan Vounckx
Diederik Verkest
IMEC
Kapeldreef 71, 3001 Leuven, Belgium
E-mail: {Johan.Vounckx,Diederik.Verkest}@imec.be

Library of Congress Control Number: 2005931992

CR Subject Classification (1998): B.7, B.8, C.1, C.4, B.2, B.6, J.6

ISSN   0302-9743
ISBN-10   3-540-29013-3 Springer Berlin Heidelberg New York
ISBN-13   978-3-540-29013-1 Springer Berlin Heidelberg New York

Springer is a part of Springer Science+Business Media

springeronline.com

© Springer-Verlag Berlin Heidelberg 2005
Printed in Germany

Typesetting: Camera-ready by author, data conversion by Olgun Computergrafik
Printed on acid-free paper   SPIN: 11556930   06/3142   5 4 3 2 1 0

Preface

Welcome to the proceedings of PATMOS 2005, the 15th in a series of international workshops. PATMOS 2005 was organized by IMEC with technical co-sponsorship from the IEEE Circuits and Systems Society.

Over the years, PATMOS has evolved into an important European event, where researchers from both industry and academia discuss and investigate the emerging challenges in future and contemporary applications, design methodologies, and tools required for the development of upcoming generations of integrated circuits and systems. The technical program of PATMOS 2005 contained state-of-the-art technical contributions, three invited talks, a special session on hearing-aid design, and an embedded tutorial. The technical program focused on timing, performance and power consumption, as well as architectural aspects with particular emphasis on modeling, design, characterization, analysis and optimization in the nanometer era.

The Technical Program Committee, with the assistance of additional expert reviewers, selected the 74 papers to be presented at PATMOS. The papers were divided into 11 technical sessions and 3 poster sessions. As is always the case with the PATMOS workshops, the review process was anonymous, full papers were required, and several reviews were carried out per paper.

Beyond the presentations of the papers, the PATMOS technical program was enriched by a series of speeches offered by world class experts, on important emerging research issues of industrial relevance. Prof. Jan Rabaey, Berkeley, USA, gave a talk on "Traveling the Wild Frontier of Ulta Low-Power Design", Dr. Sung Bae Park, Samsung, gave a presentation on "DVL (Deep Low Voltage): Circuits and Devices", Prof. Magdy Bayoumi, Director of the Center of Advanced Computer Studies, Louisiana, USA spoke on "Wireless Sensor Networks: A New Life Paradigm", and there was a fourth presentation on "Cryptography: Circuits and Systems Approach" by Prof. Odysseas Koufopavlou.

We would like to thank all those who voluntarily worked to make this year's PATMOS possible, the expert reviewers, the members of the technical program and steering committees, and the invited speakers who offered their skill, time, and deep knowledge to make PATMOS 2005 a memorable event. Sponsorship of PATMOS 2005 by Philips is gratefully acknowledged.

September 2005

Vassilis Paliouras
Johan Vounckx
Diederik Verkest

Preface

Organization

Organizing Committee

General Co-chairs: Dr. Johan Vounckx, IMEC, Belgium
                            Dr. Diederik Verkest, IMEC, Belgium
Technical Program Chair: Assist. Prof. Vassilis Paliouras, U. of Patras, Greece
Industrial Chair: Dr. Roberto Zafalon, STMicroelectronics, Italy

PATMOS Technical Program Committee

B. Al-Hashimi, University of Southampton, UK
M. Alioto, University of Sienna, Italy
A. Alvandpour, Linkoping University, Sweden
N. Azémard, LIRMM, France
D. Bertozzi, University of Bologna, Italy
L. Bisdounis, INTRACOM, Greece
A. Bogliolo, University of Urbino, Italy
J. Bormans, IMEC, Belgium
J.A. Carballo, IBM, USA
N. Chang, Seoul National University, Korea
J. Figueras, Univ. Catalunya, Spain
E. Friedman, University of Rochester, USA
C.E. Goutis, Univ. Patras, Greece
E. Grass, IHP-GmbH, Germany
J.L. Guntzel, Universidade Federal de Pelotas, Brazil
A. Guyot, TIMA laboratory, France
R. Hartenstein, U. Kaiserslautern, Germany
J. Juan Chico, Univ. Sevilla, Spain
N. Julien, Univ. of South Brittany, France
S. Khatri, Univ. of Colorado, USA
P. Larsson-Edefors, Chalmers T. U., Sweden
V. Moshnyaga, U. Fukuoka, Japan
W. Nebel, U. Oldenburg, Germany
D. Nikolos, University of Patras, Greece
J.A. Nossek, T.U. Munich, Germany
A. Nunez, U. Las Palmas, Spain
V.G. Oklobdzija, U. California Davis, USA
M. Papaefthymiou, U. Michigan, USA
F. Pessolano, Philips, The Netherlands
H. Pfleiderer, U. Ulm, Germany
C. Piguet, CSEM, Switzerland
M. Poncino, Univ. di Verona, Italy

R. Reis, U. Porto Alegre, Brazil
M. Robert, U. Montpellier, France
A. Rubio, U. Catalunya, Spain
D. Sciuto, Politecnico di Milano, Italy
D. Soudris, U. Thrace, Greece
J. Sparsø, DTU, Denmark
A. Stauffer, EPFL, Lausanne, Switzerland
T. Stouraitis, U. Patras, Greece
A. M. Trullemans, U. Louvain-la-Neuve, Belgium
R. Zafalon, STMicroelectronics, Italy

PATMOS Steering Committee

D. Auvergne, U. Montpellier, France
R. Hartenstein, U. Kaiserslautern, Germany
W. Nebel, U. Oldenburg, Germany
C. Piguet, CSEM, Switzerland
A. Rubio, U. Catalunya, Spain
J. Figueras, U. Catalunya, Spain
B. Ricco, U. Bologna, Italy
D. Soudris, U. Thrace, Greece
J. Sparsø, DTU, Denmark
A.M. Trullemans, U. Louvain-la-Neuve, Belgium
P. Pirsch, U. Hannover, Germany
B. Hochet, EIVd, Switzerland
A.J. Acosta, U. Sevilla/IMSE-CSM, Spain
J. Juan, U. Sevilla/IMSE-CSM, Spain
E. Macii, Politecnico di Torino, Italy
R. Zafalon, STMicroelectronics, Italy
V. Paliouras, U. Patras, Greece
J. Vounckx, IMEC, Belgium

Executive Steering Sub-committee

President: Joan Figueras, U. Catalunya, Spain
Vice-president: Reiner Hartenstein, U. Kaiserslautern, Germany
Secretary: Wolfgang Nebel, U. Oldenburg, Germany

Additional Reviewers

G. Dimitrakopoulos
G. Glikiotis
P. Kalogerakis
A. Kakarountas
M. Krstic
A. Milidonis

Table of Contents

Session 3: High-Level Design

Session 4: Telecommunications and Signal Processing

Session 5: Low-Power Circuits

Session 6: System-on-Chip Design

Session 7: Busses and Interconnections

Session 8: Modeling

Session 9: Design Automation

Session 10: Low-Power Techniques

Session 11: Memory and Register Files

Poster Session 1: Applications

Poster Session 2: Digital Circuits

Poster Session 3: Analog and Physical Design

Special Session: Digital Hearing Aids: Challenges and Solutions for Ultra Low Power

Invited Talks

A Power-Efficient and Scalable Load-Store Queue Design

Fernando Castro[1], Daniel Chaver[1], Luis Pinuel[1],
Manuel Prieto[1], Michael C. Huang[2], and Francisco Tirado[1,*]

[1] ArTeCS Group, Complutense University of Madrid, Madrid, Spain
fcastror@fis.ucm.es
{dani02,mpmatias,ptirado,lpinuel}@dacya.ucm.es
[2] University of Rochester, Rochester, New York, USA
Michael.Huang@ece.rochester.edu

Abstract. The load-store queue (LQ-SQ) of modern superscalar processors is responsible for keeping the order of memory operations. As the performance gap between processing speed and memory access becomes worse, the capacity requirements for the LQ-SQ increase, and its design becomes a challenge due to its CAM structure. In this paper we propose an efficient load-store queue state filtering mechanism that provides a significant energy reduction (on average 35% in the LSQ and 3.5% in the whole processor), and only incurs a negligible performance loss of less than 0.6%.

1 Introduction

As the performance gap between processing and memory access widens, load latency becomes critical for performance. Modern microprocessors try to mitigate this problem by incorporating sophisticated techniques to allow early execution of loads without compromising program correctness.

Most out-of-order processors include two basic techniques usually denoted as *load bypassing* and *load forwarding*. The former allows a load to be executed earlier than preceding stores when the load effective address (EA) does not match with any of the preceding stores. If the EA of any preceding store is not resolved when the load is ready to execute, it must wait. When a load aliases (has the same address) with a preceding store, *load forwarding* allows the load to receive its data directly from the store.

More aggressive implementations allow *speculative execution of loads* when the EA of a preceding store is not yet resolved. Such speculative execution can be premature if a store earlier in the program order overlaps with the load and executes afterwards. When the store executes, the processor needs to detect, squash and re-execute the loads. All subsequent instructions, or at least, the dependent instructions of the load need to be re-executed as well. This is referred to as load-store replay [1]. To mitigate the performance impact of replays, some designs incorporate predictor mechanisms to control such speculation [2]. One commercial example of this technique can be found in Alpha 21264 [1]. Modern processors with support for glueless multiprocessor systems also have to include support for memory consistency in shared memory multiprocessors systems [3, 4], which makes the memory system even more complex.

* This work is supported in part by the Spanish Government research contract TIC 2002-750 and the *HiPEAC* European Network of Excellence. Manuel Prieto was also supported by the Spanish Government MECD mobility program PR2003-0306

V. Paliouras, J. Vounckx, and D. Verkest (Eds.): PATMOS 2005, LNCS 3728, pp. 1–9, 2005.

These complex mechanisms come at the expense of high energy consumption. Our investigated design accounts for around 8% of the total consumption. Furthermore, it is expected that this consumption will grow in future designs. Driven by this trend, our main objective in this paper is to design an efficient and scalable LSQ structure, which could save energy without sacrificing performance.

The rest of the paper is organized as follows. Section 2 discusses related work. Section 3 and 4 present the conventional design and our alternative LSQ mechanism respectively. Section 5 describes our experimental framework. Section 6 analyzes experimental results. Finally, Section 7 presents our conclusions.

2 Related Work

Recently, a range of schemes have been proposed that use approximate hardware hashing with Bloom filters to improve LSQ scalability [5, 9]. These schemes fall into two broad categories: The first, called *search filtering*, reduces the number of expensive, associative LSQ searches [5]. In the second category, called *state filtering*, LSQs handle memory instructions that are likely to match others in flight, and Bloom filters (denoted as EBF in [5]) encode the other operations that are unlikely to match. In [5] they report only one simple preliminary design, in which the LQ is completely eliminated and all loads are hashed in the EBF. Experimental results using such scheme suffered a significant performance loss. Our work improves upon prior art and provides an EBF-based design with a negligible performance penalty (1% on average). In addition, we have also extended early studies by evaluating the power consumption of both schemes.

In the LSQ domain, several researchers have proposed alternative solutions to the disambiguation challenges. Park *et al.* [10] proposed using the dependence predictor to filter searches to the store queue. Cain and Lipasti [11] proposed a value-based approach to memory ordering that enables scalable load queues. Roth [12] has proposed combining Bloom filters, address-partitioned store queues, and first-in, first-out retirement queues to construct a high-bandwidth load/store unit. Baugh *et al.* [13] proposed an LSQ organization that decouples the performance-critical store-bypassing logic from the rest of the LQ-SQ functionality, for which they also used address-partitioned structures.

3 Conventional Design of Load-Store Queue

Conventional implementations of the LSQ are usually based on two separate queues, the load queue (LQ) and the store queue (SQ) (see Figure 1). Each entry of these queues provides a field for the load or store EA and a field for the load or store data. For example, the Alpha 21264 microprocessor [1] incorporates 32+32 entries for the load and store queues.

Techniques such as *load forwarding and load bypassing* increase the complexity of these structures given that they require associative search in the store queue to detect aliasing between loads and preceding stores. *Speculation* increases complexity even more since associative search (AS) is also necessary in the load queue to detect premature loads: when the EA of a store is resolved, an associative search in the load queue is performed; if a match with a later load that has already been issued is found, execution from the conflicting load is squashed.

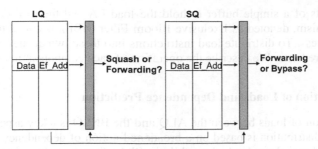

Fig. 1. Conventional LQ-SQ design. Each load instruction performs an AS of the SQ to determine if load forwarding or load bypassing is necessary. Each store also performs an AS of the LQ to forward data for a later load with the same address or to squash its execution if it is already executing

4 Efficient Load-Store Queue Management

4.1 Rationale

While the techniques outlined above such as *load forwarding* and *load bypassing* improve performance, they also require a large amount of hardware that increases energy consumption. In this paper, we propose a more efficient LSQ design that allows for a more efficient energy usage. Besides, the size of the structure scales significantly better than the conventional approach described before.

The new approach is based on the following observations, derived from Table 1:

Table 1. Percentage of bypassing loads and percentage of loads over all memory instructions

	bzip2	gap	vpr	lucas	wupwise	
Bypassing loads (%)	4,4	22,9	29,5	0,0	6,1	
Total loads (%)	75,1	69,1	68,7	59,4	69,6	
	apsi	fma3d	galgel	sixtrack	art	Average
Bypassing loads (%)	2,3	6,3	0,1	21,0	2,4	**9,5**
Total loads (%)	63,3	81,0	84,2	75,6	79,7	**72,6**

1- Memory dependencies are quite infrequent. Our experiments indicate that only around 10% of the load instructions need a *bypass*. This suggests that the complex disambiguation hardware available in modern microprocessors is often being underutilized.

2- On average, around 73% of the memory instructions that appear in a program are loads. Therefore, their contribution to the dynamic energy spent by the disambiguation hardware is much greater than that of the stores. This suggests that more attention must be paid to handling loads.

4.2 Overall Structure

As shown in Figure 2, the conventional LQ is split into two different structures: the Associative Load Queue (ALQ) and the Banked Non-associative Load Queue (*BNLQ*). ALQ is similar to a conventional LQ, but smaller. It provides fields for the EA and the load data, as well as control logic to perform associative searches. The

BNLQ consists of a simple buffer to hold the load EA and the load data. An additional mechanism, denoted as Exclusive Bloom Filter (EBF), is added to assure program correctness. To distribute load instructions into these two queues, we employ a dependence predictor. We describe the operation of each component in the following.

4.3   Distribution of Loads and Dependence Prediction

The distribution of loads between the ALQ and the BNLQ is a key aspect to be considered. Our distribution is based on a broadened notion of dependency. We classify as *dependent-loads* those loads that while in-flight, happen to have the same EA as any store from the SQ. This would include loads that overlap with younger in-flight stores as well. A more conventional definition of dependent load would only consider those loads that receive their data directly from an earlier, in-flight store – loads that overlap with older, in-flight stores. This broadened definition is due to the imprecise (although more energy-efficient) disambiguation mechanism used for handling independent loads as we will explain later. Using our alternative definition, 23% of loads (on average) are classified as dependent.

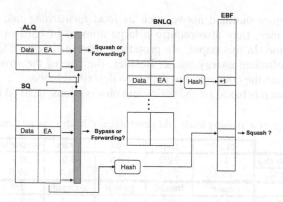

Fig. 2. Effective LQ-SQ Management. LQ is divided into two queues, an associative load queue (ALQ) for those predicted-dependent loads and a Banked Non-associative Load Queue (*BNLQ*) for those predicted-independent loads. The ALQ and the SQ work similarly to a conventional LQ-SQ. To ensure premature loads are detected even if they are sent to BNLQ, an Exclusive Bloom Filter (EBF) is used

After decoding a load instruction, it will be suggested by either a *dynamic dependence* predictor or *profiling-based* information, into which queue the load should be accommodated. According to this information and taking into account the queue occupancy, the load is allocated in the ALQ or the BNLQ. Occupancy information is used to improve the balance among both queues. In our current design, if an independent load arrives and the BNLQ is full, it is accommodated in the ALQ, since an independent load can always be allocated into the associative queue without compromising program correctness.

In a profiling-based system, every static load is tagged as dependent or independent based on profile. This way, load dependency prediction is tied to the static instructions.

Table 2. Static load distribution. We show for each benchmark the percentage of static loads that are independent for all its dynamic instances, the percentage of static loads that are dependent for all dynamic instances, and the rest of the static loads in the program

	bzip2	gap	vpr	lucas	wupwise
Static Always Independent Loads (%)	77,0	63,1	54,1	100,0	82,8
Static Always Dependent Loads (%)	2,1	2,2	0,6	0,0	0,0
Remaining Static Loads (%)	20,9	34,8	45,4	0,0	17,2

	apsi	fma3d	galgel	sixtrack	art	Average
Static Always Independent Loads (%)	93,0	88,1	93,8	73,2	97,1	82,2
Static Always Dependent Loads (%)	0,1	0,5	2,1	1,7	0,3	1,0
Remaining Static Loads (%)	6,9	11,4	4,1	25,1	2,5	16,8

Results in Table 2 show that this is a reasonable approach given that the runtime dependency behavior of static load instructions remains very stable. Only 17% (on average) of the load instructions in our simulations change their dependency behavior during program execution. Nevertheless, apart from profiling, load annotation involves some changes in the instruction set so that it is possible to distinguish between dependent and independent load instructions.

We have chosen a dynamic approach where dependency prediction is generated dynamically. To store the prediction information, we can augment the instruction cache or use a dedicated, PC-indexed table. We have opted to use a PC-indexed prediction table as in the Alpha 21264 [1]. All loads are initially considered as independent. However, as will be explained below, this initial prediction is changed if a potential dependency is detected. Once a prediction has been changed, this prediction will hold during the rest of the execution. This decision, which is based on the stable behavior observed in Table 2, simplifies the implementation and has little impact on the prediction accuracy. Unlike the profiling approach, this alternative needs neither change in the instruction set architecture nor profiling. However, as mentioned above, it requires extra storage and a prediction training phase.

Our results indicate that both strategies provide similar performance. In this paper, we have only explored the dynamic version given that its extra cost (in terms of design complexity and power) is insignificant.

4.4 Load-Store Replay and Dependence Predictor Update

Once the memory instructions are distributed to their corresponding queues, it is necessary to perform different tests to detect violations of the memory consistency.

As described above, in a conventional LQ-SQ mechanism, these violations are detected using associative searches. Our proposed mechanism follows the same strategy for those loads accommodated in the ALQ. However, for those loads accommodated in the BNLQ, an alternative mechanism needs to be incorporated to detect potential violations. As in [5], our implementation adds a small table of two-bit counters, denoted as EBF (Figure 2). When a load of the BNLQ is issued, it indexes the EBF based on its EA and increases the corresponding counter. When the load commits, the counter is decremented. When stores are issued, miss-speculated loads accommodated in the ALQ are detected by performing an associative search, as in a conventional mechanism. Miss-speculated loads in the BNLQ are detected by indexing the EBF with the store's EA. If the corresponding counter is greater than zero, a potentially truly dependent load is in-flight. In this case, our mechanism conservatively squashes execution from the subsequent load and triggers a special mode, denoted as DPU (*dependence predictor update mode*), to update the dependence prediction information. Note that any in-flight load that aliases with the store in the EBF causes a squash, independently of its relative position to the store. This is why we use a broadened notion of dependence.

Two different hashing functions (H0 and H1) have been proposed in [5] for indexing the EBF. H0 uses lower order bits of the instruction's EA to index into the hash table, whereas the H1 uses profile to index the EBF using those bits in the addresses that were the most random. We have opted to employ H0, given that H1 does not provide any significant improvement. In fact, for a large enough EBF, H0 outperforms H1 in our simulations.

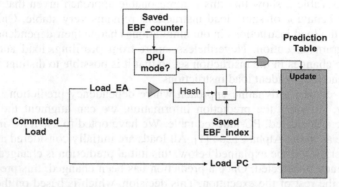

Fig. 3. Dependence predictor update in the DPU mode when loads commit

When a store triggers the DPU mode, its corresponding EBF index and EBF counter are saved in special registers (see Figure 3). During the DPU mode, when a load commits, its prediction is changed to dependent if its hashed EA matches with the EBF index value. Based on the saved EBF counter value, this mode is held until all the in-flight loads that aliased with the trailing store have committed.

5 Experimental Framework

We evaluated our proposed load-store queue design on a simulated, generic out-of-order processor. The main parameters used in the simulations are summarized in Table 3. As the evaluation tool, we employed a heavily modified version of SimpleScalar [6] that incorporates our LQ-SQ model and a tuned Wattch framework [7] that models the energy consumption in the proposed LQ-SQ mechanism. The evaluation of our proposal has been performed selecting several benchmarks from the SPEC CPU2000 suite. In selecting those applications, we tried to cover a wide variety of behavior. We simulated single sim-point regions [8] of one hundred million instructions. We have simulated three different schemes: a conventional LQ-SQ mechanism (baseline configuration), our proposed alternative and the original *state filtering* scheme proposed in [5].

The associative LQ from the conventional approach provides 80 entries, whereas the ALQ from our proposal provides only 32 entries. Our BNLQ allocates 3 banks with 16 entries in each bank, whereas the non-associative LQ (NLQ) of the original scheme has 80 entries. The EBF provides 4096 2-bit saturating counters. For this reason, if more than 4 issued loads happen to hash into the same entry, a potential issue of a fifth load would stall until any of the others commit (this situation almost never occurs in our experiments).

Table 3. System Configuration

Processor	Caches and Memory
8-issue out of order processor	**L1 data cache:**
Register File:	32-KB, 4-way, LRU, latency: 3
256 integer physical registers	cycles
256 floating-point physical registers	
Units:	**L2 data cache:**
4 integer ALUs, 2 integer Mult-Dividers, 3 floating-	2-MB, 8-way,LRU, latency: 12
point ALUs, 1 floating-point Mult-Dividers	cycles
Branch Predictor:	
Combined, bimodal: 8K entries 2level, 8K entries, 13	**L1 instruction cache:**
bits history size, meta-table: 8K entries	64-KB, 2-way, LRU, latency: 2
BTB: 4K entries	cycles
RAS: 32 entries	
Queues:	**Memory access:**
I-Queue: 128 entries	100 cycles
FP-Queue: 128 entries	

Standard LQ-SQ	Proposed LQ-SQ	EBF-based LQ-SQ
LQ: 80 entries	ALQ: 32 entries	Non-Asoc-LQ: 80 entries
SQ: 48 entries	BNLQ: 3x16 entries	SQ: 48 entries
	SQ: 48 entries	EBF: 4K entries
	EBF: 4K entries	

6 Experimental Results

Tables 4 and 5 report performance gains over the baseline configuration, as well as energy savings achieved in the LQ-SQ and in the whole processor. The original proposal of Sethumadhavan [5] manages to reduce the energy spent in the LQ-SQ mechanism (Table 4), but both overall energy and performance suffer a significant drop due to squashes caused by dependence violations.

Our scheme (Table 5) outperforms this previous proposal by reducing dramatically the number of squashes, which allows for a negligible performance loss (0.6% on average). Besides, the splitting of the LQ into a set of smaller tables (ALQ + 3xBNLQ) further improves energy consumption. On average, our implementation saves 35% energy in the LQ-SQ, which translates into an important reduction in the whole processor (around 3.5%).

We should remark that, although the end results are quite satisfactory, our current dependency predictor is far from ideal. On average, about 50% of all loads are classified as dependent, most falsely so due to conflict of hashing function: only 23% of loads will be classified as dependent loads if a hash table with an infinite size is used. This observation suggests that there is still significant room for improvement, which motivates us to include more accurate prediction and bloom filter hashing in future implementations.

Table 4. Energy savings and performance loss for the original Sethumadhavan's *state filtering* approach. A negative value for performance denotes performance loss over the conventional LQ-SQ, while a positive value for the energy means energy savings

	bzip2	gap	vpr	lucas	wupwise	
Δ*IPC (%)*	-37,3	-9,5	-21,9	-7,3	-14,3	
Δ *LQ-SQ Energy (%)*	71,7	67,3	62,8	67,0	74,7	
Δ*Energy (%)*	-19,3	-1,3	-14,2	-0,6	-1,3	
	apsi	fma3d	galgel	sixtrack	art	Average
Δ*IPC (%)*	-15,7	-6,2	-5,5	-30,4	-41,6	**-19,0**
Δ *LQ-SQ Energy (%)*	76,0	72,9	76,4	58,5	53,9	**68,1**
Δ*Energy (%)*	-0,3	1,7	3,8	-24,0	-41,8	**-9,7**

Table 5. Energy savings and performance loss of our proposed LQ-SQ design

	bzip2	gap	vpr	lucas	wupwise	
Δ*IPC (%)*	-0,02	-1,08	-0,47	-0,28	-0,01	
Δ *LQ-SQ Energy (%)*	32,9	29,4	31,4	24,6	44,7	
Δ*Energy (%)*	3,59	3,66	3,44	1,45	4,37	
	apsi	fma3d	galgel	sixtrack	art	Average
Δ*IPC (%)*	-1,92	-0,01	-0,50	-0,05	-1,32	**-0,57**
Δ *LQ-SQ Energy (%)*	45,6	49,2	29,5	32,4	33,2	**35,28**
Δ*Energy (%)*	3,72	4,23	2,77	4,26	2,24	**3,37**

7   Conclusions and Future Work

In this paper we have presented a *state filtering* scheme, which with a negligible performance penalty allows for a significant energy reduction: around 35% for the LSQ, and close to 3.5% for the whole processor. These experimental results were obtained using a moderate configuration, in terms of instruction window and LSQ sizes. When employing more aggressive configurations, the expected energy savings would be much higher. The key points of our proposal are the following:

– We have designed a simple dependence predictor, specially adapted for using with EBFs (traditional predictors are not suitable for this scheme).
– We have explored the asymmetric splitting of LQ/SQ, as well as ALQ/BNLQ.
– Since the BNLQ is not associative, banking is straightforward and provides further energy reductions. This also simplifies *gating*.

Our future research plans include applying *bank gating*, as well as dynamic structure resizing, both based on profiling information. In addition, we also envisage enhancing both the bloom filter hashing and the dependence predictor, since the actual implementation is too conservative and there is a significant room for improvement.

Acknowledgments

We want to thank Simha Sethumadhavan for his helpful and thorough comments.

References

1. R. E. Kessler. "The Alpha 21264 Microprocessor". Technical Report, Compaq Computer Corporation, 1999.
2. B. Calder and G. Reinman. "A Comparative Survey of Load Speculation Architectures". Journal of Instruction-Level Parallelism, May-2000.

3. C. Nairy and D. Soltis. "Itanium-2 Processor Microarchitecture". IEEE-Micro, 23(2):44-55, March/April, 2003.
4. J. M. Tendler, J. S. Dodson, J. S. Fields Jr., H. Le and B. Sinharoy. "Power-4 System Microarchitecture". IBM Journal of Research and Development, 46(1):5-26, 2002.
5. S. Sethumadhavan, R. Desikan, D. Burger, Charles R. Moore, Stephen W. Keckler. "Scalable Hardware Memory Disambiguation for High ILP Processors". Proceedings of MICRO-36, December-2003.
6. T. Austin, E. Larson, and D. Ernst. "SimpleScalar: An Infrastructure for Computer System Modeling". Computer, vol. 35, no. 2, Feb 2002.
7. D. Brooks, V. Tiwari, and M. Martonosi. "Wattch: A Framework for Architectural-Level Power Analysis and Optimizations". 28-ISCA, Göteborg, Sweden. July, 2001.
8. T. Sherwood, E. Perelman, G. Hamerly, B. Calder. " Automatically charecterizing large scale program behavior ". Proceedings of ASPLOS-2002, October-2002.
9. S. Sethumadhavan, R. Desikan, D. Burger, Charles R. Moore, Stephen W. Keckler. "Scalable Hardware Memory Disambiguation for High ILP Processors". IEEE-Micro, Vol. 24, Issue 6:118-127, November/December, 2004.
10. I. Park, C. Liang Ooi, T. N. Vijaykumar. "Reducing design complexity of the load-store queue". Proceedings of MICRO-36, December-2003.
11. H. W. Cain and M. H. Lipasti. "Memory Ordering: A Value-Based Approach". Proceedings of ISCA-31, June-2004.
12. A. Roth. "A high-bandwidth load-store unit for single- and multi- threaded processors". Technical Report, University of Pennsylvania, 2004.
13. L. Baugh and C. Zilles. "Decomposing the Load-Store Queue by Function for Power Reduction and Scalability". Proceedings of PAC Conference, October-2004.

Power Consumption Reduction Using Dynamic Control of Micro Processor Performance

David Rios-Arambula, Aurélien Buhrig, and Marc Renaudin

TIMA Laboratory, 46 Avenue Félix Viallet, 38031 Grenoble Cedex, France
{David.Rios,Aurelien.Buhrig,Marc.Renaudin}@imag.fr
http://tima.imag.fr

Abstract. An alternative way to reduce power consumption using dynamic voltage scaling is presented. The originality of this approach is the modeling and simulation of a system where each application indicates its performance needs (in MIPS) to the operating system, which in turn is able to know the global speed requirements of the system to meet all real time application deadlines. To achieve this level of control, a co-processor is described, that receives a set point command from the OS, and manages a DC/DC converter implemented as a charge pump, in order to have the system speed fitting this set point. This architecture is especially suited for asynchronous processors but can be adapted for synchronous ones as well.

1 Introduction

With the increased demand of embedded and portable circuits that require higher autonomy and power saving, research in low power has become of great importance [1]. Systems that use microprocessors are the perfect example of low power needs. Evolution of such systems requires a high integration of functions that demand a lot of computational power that is reflected on battery power consumption. Actually, research in batteries has achieved an excellent level but the evolution in that domain tends to be small and slow.

To bypass this problem, some solutions are viable. First it is possible to optimize the circuit design. Transistor sizing is essential for low power circuits. But this technique has been well studied and transistor sizing is no longer sufficient to obtain the power efficiency that new circuits require. Another approach is to reduce dynamic and static consumption of circuits. For this, two techniques are used.

Firstly, the Adaptive Body Biasing (ABB) technique allows to reduce the static energy that is consumed by digital circuits. In fact, a circuit that is idle has static leakage current that is becoming similar to the dynamic current used by the circuit. This means that it is essential to reduce static current to obtain the performances needed. The second technique, the Dynamic Voltage Scaling (DVS), reduces the dynamic current consumed by the circuit. Here, the power voltage of the circuit is lowered or increased to reduce the energy consumed by the circuit.

Both techniques are used in microprocessors design ([2], [3], [4]) and can be combined to obtain the maximum power saving [5]. A brief description of the DVS technique is presented in the following paragraphs.

V. Paliouras, J. Vounckx, and D. Verkest (Eds.): PATMOS 2005, LNCS 3728, pp. 10–18, 2005.
© Springer-Verlag Berlin Heidelberg 2005

1.1 DVS

As seen in [6] and [7], Dynamic voltage Scaling was used for the first time in microprocessor systems. In a CMOS microprocessor, the energy consumed can be calculated with:

$$E_{ops} \propto C * V^2 \tag{1}$$

with C the switched capacitance and V the supply voltage [8]. Therefore, to minimize the energy consumed by the circuit, we can reduce the capacitance. This is done by applying aggressive low-power design techniques [9] such as clock-gating [10]. The number of instructions executed by the application is also a parameter that has to be kept in mind to achieve a full energy optimization. This is commonly optimized by compilers. Finally, as we can see from the squared relationship between energy and voltage, a small decrement of the voltage leads to a non-negligible energy reduction factor. This voltage reduction is done by the DVS technique, and takes advantage of the square relationship to perform energy optimization. With this technique it is possible to save up to 80% of power [11].

This energy reduction doesn't come without a price. Scaling the voltage on a microprocessor changes the system speed. In fact, as seen in equation 2, with V the supply voltage and c a process dependent constant, reducing the supply voltage reduces the maximum speed at which the device can operate.

$$f_{max} \propto \frac{V - c}{V} \tag{2}$$

This is very interesting when possible to adapt the supply voltage, so that the speed is sufficient to meet the application deadline. Hence, the operating system must use the application profile to manage the energy needed by the system and control power voltage [12]. To achieve this kind of control, an interaction between hardware, software and the operating system is needed. Fig. 1 shows an application executed by the processor in a time slot T. This time slot is the maximum time allocated to the application to be executed. The total consumed energy is greater when no control is applied but the execution time increases when control is applied.

Moreover, a processor also has periods of time where it does nothing [13]. In that case the processor is still consuming energy, and could be set to an idle state reducing all the dynamic current to zero.

1.2 Asynchronous and Synchronous Processors

As described before, processors are the perfect targets for DVS technique. We have to consider that there are two kinds of technologies for processors: synchronous and asynchronous.

Synchronous processors dominate the actual market of processors as well as the research studies. They need a global clock that controls all the system. When using DVS, if the power supply of the processor is changed, the clock frequency has to be changed accordingly. This means that we have to take care of the synchronization of the PLL.

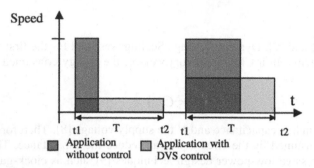

Fig. 1. Example of an application with and without DVS control

Asynchronous processors are becoming more popular on the market. They are very modular and are low-power design by themselves. In asynchronous processors, there is no global clock; all the tasks are locally controlled by a handshake protocol. This allows us to reduce the power supply without having to reduce the clock frequency. This reduces the size of the control circuit needed to control the energy consumed by the processor using a DVS technique.

Consequently, asynchronous processors are excellent candidates for DVS design [14]. The power saving obtained is higher than in synchronous processors. Although our work is focused on asynchronous processors, all the ideas and methods proposed in this work can also be applied to synchronous ones. This method has proved to be very efficient in a system with two ASPRO processors [15] and gave a 60% energy reduction in a digital image processing application [16].

1.3 Contribution

This paper explains the implementation of a DVS technique in the design of a co-processor that controls the DC/DC regulator of the power supply of a microprocessor. The co-processor dynamically controls the DC/DC to scale the supply voltage as well as the clock frequency (for synchronous processors) in order to satisfy the applications computational needs.

The idea of providing Voltage Scaling is not new, but we present in this study two main contributions:

- Modeling the power management as a process control system.
- Describing a co-processor enabling loop-back control of the DC/DC regulator to achieve the speed required by the system.

2   System Overview and Architecture

With a feedback system, it is possible to control with high accuracy the power consumed by the system. This is done respecting all the specifications and needs of the applications executed by the processor. Indeed, each application that is to be executed indicates to the coprocessor the speed it requires. The information about the speed can be inserted statically into the code at compile time or mentioned dynamically at run time. Therefore, adding the application speeds creates the speed profile of the whole system, as well as the one required by the operating system, so it is used to precisely control the speed needed. Consequently, it is possible to apply a fine-grain power management allowing real-time application deadlines to be met. Fig. 2 shows an example of application profiles and the set point applied to the system in order to operate correctly. Note that scheduling applications using this technique is an important issue that is not the gist of this paper and is not treated here.

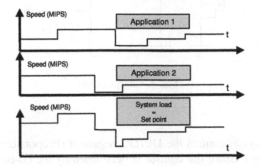

Fig. 2. Dynamic set point during a software execution

Fig. 3 shows the block diagram of the system. The co-processor integrates an instruction counter and a clock. It calculates the real speed of the processor in MIPS (Millions Instructions Per Second), averaged on a period of computation. This speed is then compared to the set point given by the application and the co-processor generates the proper digital control to the DC/DC.

2.1 DC/DC

The DC/DC chosen is a simple charge pump that is able to increase, decrease or maintain its output voltage and supplies the processor. The control of the DC/DC is a digital 2-bit code. Fig. 4 shows the schematic of the charge pump. This type of DC/DC controller is very small, very simple and can deliver a strong output current.

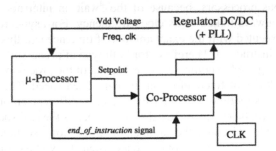

Fig. 3. System architecture

The operation of the charge pump is very simple. A 2-bit command controls the N and P transistors. When the P transistor is ON, the capacitor C is charged and the power transistor increases the output voltage. If the N transistor is ON, the capacitor C is discharged, to decrease the output voltage.

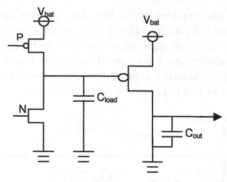

Fig. 4. Charge pump

2.2 The Control

There are many ways to control the DC/DC regulator (Proportional, Derivative, integral, fuzzy logic, etc). Since the control is performed by the co-processor, this allows us to choose the way the control is performed. In our design fuzzy logic suits very well but it has the inconvenience of needing a lot of computing resources.

We have then chosen to use a PID controller. This kind of control allows a faster response and reduces the overshoot of the output. The DC/DC regulator can be considered as an ON/OFF system while the PID gives us a digital output that cannot control the regulator by itself. Therefore the PID is translated into a time proportional control that manages the time the command is on within a time slot. This time is calculated with respect to the PID output.

2.3 Real Speed Calculation

The processor executes instructions one by one, giving us a flow of executed instructions. The quantity of instructions executed during a specific time is the speed at which the processor is running.

For synchronous processors, because of the "wait on interrupt" instructions, the real speed is not always linked to its clock frequency. For asynchronous processors, instructions are executed with the exact amount of time needed; this time is generally different for each instruction. Therefore, for both type of processors, a hardware component must be integrated to compute the real speed of the processor.

There are two kinds of fluctuations in the calculated speed. The first comes from the nature of the instructions executed that takes a different amount of time. This fluctuation is not a problem, since the calculated speed is the real speed averaged on a period of time needed to count the number of instruction executed. Therefore, this fluctuation remains small. The second occurs when the voltage (and frequency for synchronous processors) is changing. In this case, the speed of the processor starts changing and instructions are executed in a variable time. The co-processor samples the speed and gets an average number of executed instructions in the period of time considered. Consequently, as shown in Fig.5, when moving from speed S_1 to speed S_2, the speed calculated by the coprocessor, S_c, differs from the real average speed of the processor. This wrong speed disturbs the control of the DC/DC regulator.

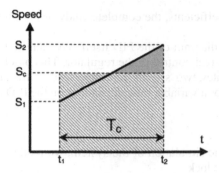

Fig. 5. Speed evolution during the transition

To solve this problem, two solutions can be used. First, it is possible to memorize the last speed of the processor. Then, when the new sample is computed, we add to the new speed S_c the difference between the previous speed S_1 and the new speed S_c and the error made is lowered.

$$S_r = S_c + (S_c - S_1) \tag{3}$$

It is also possible to consider this error like a perturbation in the PID control, and we can tune the PID to compensate it.

In the first solution, more hardware is required, while in the second one the real speed will not be defined as precisely as in the first solution and the speed control will be slower.

3 Modeling and Simulation

The system has been simulated using the ELDO simulator. The goal of simulation is to understand how the system works and to quantify the power saved.

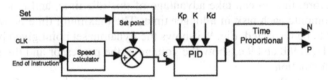

Fig. 6. Block diagram of the PID control

3.1 Control

Control is composed by two parts. The first counts the instructions at each edge of the *end_of_instruction* signal provided by the processor. The second performs the actual control. The control gets a set point and the *end_of_instruction* signal, generated by the processor each time it finishes an instruction. This way we can count the number of instructions done in a time given by an external clock signal CLK.

As we can see, the PID control receives K_p, K_i and K_d. With these coefficients the software can tune the PID as needed. In this first simulation approach, we did not try to change the coefficients dynamically, but we leave this possibility open for future work. The PID control is not the main goal of this study since each targeted processor will require its own coefficients, the complete study of the coefficients is not detailed in this paper.

Figure 6 shows the diagram of the PID itself. We can see the timed control block, which will provide the real control of the regulator. This block takes the output of the PID block and generates two signals for the P and N transistors of the regulator. Those signals are set for a variable time depending on the PID output.

3.2 Processor

To have a more realistic simulation of the system, we model the Lutonium processor [17] in a VHDL-AMS block.

3.3 Regulator

As described before, we used a charge pump to provide the supply voltage. This charge pump can scale its output voltage with a 2 bit digital command. The regulator was described using ELDO, and a low leakage transistor model from HCMOS9 of STMicroelectronics.

4 Benchmarking the Global System

To quantify the power consumption gain, an example is presented in Fig. 7 where two tasks are used to illustrate the software load profile. The first task (Fig. 7.a) is an interrupt routine. It simulates a task that has to be processed at a given frequency. This is quite common in embedded systems and generally reduces the efficiency of the traditional synchronous DVS processors, because of the synchronization time of the PLL. The second task (Fig 7.b) represents an application running in the processor. The last graph (Fig 7.c) is the resulting profile that will be used. As we can see the processor is not always active, it is processing tasks at full speed (200 MIPS). As discussed before, the OS can take advantage of this idle time, and reduce the supply voltage to compute each task in a longer time. The maximum time is set by the next task and has to be respected. Fig. 9.a shows the dynamic set point given by the operating system. Figs. 9.b and 9.c, show the voltage at the processor and the resulting real speed after simulation.

Fig. 10 presents the results of simulation for the profile shown in Fig. 7. With this DVS technique, the global system consumes about 18 µJ (Fig. 10.a), while at maximal speed the CPU would consume more than 33µJ (Fig 10.b) during the same period of time. So there is more than 45% energy saving.

Note that the last graph shows the real consumption of the processor (Fig 10.c) without the DC/DC, which is about 10 µJ. This indicates that the charge pump used has an average efficiency of 55%. Fig. 8 shows that the efficiency of this charge pump is between 45% and 70% according to the output voltage.

The global efficiency of the charge pump used is not as good as we could have expected, and could be changed or modified; nevertheless the gain obtained is very good for this study. Considering a DC/DC with an 85% efficiency for this profile, the global energy consumption would be 11.8µJ which represents a 66% energy reduction.

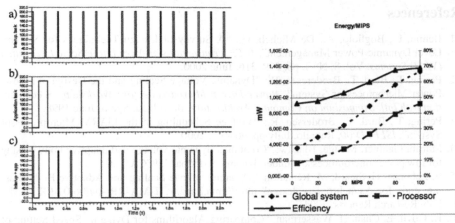

Fig. 7. Processor load versus time of the application

Fig. 8. Power consumption of the global system and power delivered to the processor as a function of speed – Efficiency of the charge pump

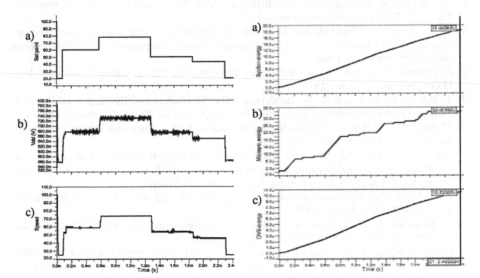

Fig. 9. Set point given by the system - Voltage supplied to the processor – Real speed of the processor

Fig. 10. Energy consumed with the profile shown before by:
- global system using DVS
- CPU alone at max power without DC/DC
- CPU alone with DVS

5   Conclusion and Future Work

This work presents simulation results on a DVS system with a feedback control for potentially unpredictable software loads. Future work will deal with increasing the efficiency of the DC/DC, and manage the voltage of the bulk to limit the static leakage.

References

1. Benini, L., Bogliolo, A., De Micheli, G.: "A Survey of Design Techniques for System-Level Dynamic Power Management", *IEEE Transactions on Very Large Scale Integration (VLSI) Systems*, Vol. 8, N° 3, pp. 299–316, June 2000.
2. Pering, T., Burd, T., Broderesen, R.: "Dynamic Voltage Scaling and the Design of a Low-Power Microprocessor System", *Power-Driven Microarchitecture Workshop, in conjunction with Intl. Symposium on Computer Architecture*, Barcelona, Spain, June 1998.
3. Pering, T., Burd, T., Brodersen, R.: "Voltage Scheduling in the lpARM Microprocessor System", *ISLPED'00*, Rapallo, Italy pp. 96–101, 2000.
4. Martin, Flautner, Mudge, Blaauw: "Combined dynamic voltage and adaptive body biasing for low power microprocessors under dynamic workload", ICCAD'02
5. E. Labonne, G. Sicard, M. Renaudin, "Dynamic voltage Scaling and Adaptive Body Biasing study for Asynchronous design" http://tima.imag.fr, TIMA-RR--04/06-01-FR, TIMA Lab. Research Reports, 2004
6. K. Govil, E. Chan, H. Wasserman, "Comparing Algorithms for Dynamic Speed-Setting of a Low-Power CPU", *Proc. 1st Int'l Conference on Mobile Computing and Networking*, Nov 1995.
7. M. Weiser, B. Welch, A. Demers, and S. Shenker, "Scheduling for reduced CPU energy," *Proc. 1st Symp. on Operating Systems Design and Implementation*, pp. 13-23, Nov. 1994.
8. T. Burd and R. Brodersen, "Energy Efficient CMOS Microprocessor Design," *Proc. 28th Hawaii Int'l Conf. on System Sciences,* 1995.
9. M. Srivastava, A. Chandrakasan, R. Brodersen. Predictive system shutdown and other architectural techniques for energy efficient programmable computation. IEEE Transactions on VLSI System, 4(1), 1996.
10. F. Emnett, M. Biegel "Power Reduction Through RTL Clock Gating" Synopsys User Group, San Jose, March 2000.
11. T. Pering, T. Burd, and R. W. Brodersen, "The Simulation and Evaluation of Dynamic Voltage Scaling Algorithms," *Proc. 1998 Int'l Symp. On Low Power Electronics Design*.
12. Flautner, K.: "Automatic Monitoring for Interactive Performance and Power Reduction", *Dissertation*, Michigan University, 2001.
13. Luca Benini, Alessandro Bogliolo, and Giovanni De Micheli, "A survey of design techniques for system-level dynamic power management", *IEEE Trans. on Very Large Scale Integration (VLSI) Systems,* vol. 8, no. 3, June 2000, pp. 299-316.
14. A.J. Martin, "An asynchronous approach to energy-efficient computing and communication" *SSGRR 2000*, August 2000.
15. M. Renaudin, P. Vivet, F. Robin, "ASPRO: an asynchronous 16-bit RISC Microprocessor with DSP capabilities", *ESSCIRC*, Duisburg, Germany, 1999.
16. Christian Piguet, "Low Power Electronics Design", *CRC Press*, ISBN 0-8493-1941-2, 2005
17. A.J. Martin, M. Nyström et al., "The Lutonium: A Sub-Nanojoule Asynchronous 8051 Microcontroller," IEEE Int. Symp. Async. Systems and Circuits, May 2003.
18. Pedram, M.: "Design Technologies for Low Power VLSI", *Encyclopedia of Computer Science and Technology*, Vol. 36. Marcel Dekker, Inc., pp. 73–96, 1997.

Low Power Techniques Applied
to a 80C51 Microcontroller
for High Temperature Applications

Philippe Manet, David Bol, Renaud Ambroise, and Jean-Didier Legat

Microelectronics Laboratory, UCL, Batiment Maxwell,
place du Levant 3, 1348 Louvain-la-Neuve, Belgium
{manet,bol,ambroise,legat}@dice.ucl.ac.be

Abstract. In this paper, we present a low power high temperature
80C51 microcontroller. The low power optimizations are applied at gate
and architectural level, by using extensive clock and data gating, and
by completely redesigning the micro-architecture. We also present original clock gating techniques: pre-computed clock gating. To validate these
techniques, extensive comparisons with other realizations of the same microcontroller are presented. It shows that gating techniques can achieve
good performances.

1 Introduction

Nowadays, the reduction of power consumption is one of the major challenges
for IC designers. Indeed, this reduction helps to solve some of the problems
met in the implementation of IC's, like the heating of chips. Furthermore, the
increasing popularity of portable devices, like personal computing devices or
wireless communication systems, still increases the need for low power systems.

The aim of our contribution is to apply some low power techniques to the
complete architecture of a microcontroller which could be used in high temperature applications, up to 225^oC. Indeed, there is a growing number of domains
where high temperature IC's are needed: oil drilling, combustion engines control
or industrial process control in extreme environments. Despite those needs, there
is only one available microcontroller running at 225^oC so far, the HT83C51 from
Honeywell [ ]. This is a 80C51 and is not power optimized at all. The microcontroller we implemented is also a 80C51. Actually, although it's an old irregular
CISC processor, it is very popular and often used in applications where the
consumption minimization is important.

High temperature applications require the use of a specific technology. The
one we used is a 1 μm Silicon-On-Insulator high temperature process, working up to 225^oC with supply voltage of 5V, which is provided by the German
manufacturer XFAB.

The need for a specific technology reduces the kind of optimizations, which
can be used. In particular, the technological methodologies to reduce consumption, like voltage scaling, cannot be used. We have then focused on optimization
at higher levels of abstraction: the gate and micro-architectural levels.

V. Paliouras, J. Vounckx, and D. Verkest (Eds.): PATMOS 2005, LNCS 3728, pp. 19–29, 2005.

In this paper, we first introduce in section 2 theoretical background: the high temperature limitations on the optimizations, a general description of low power techniques and related works. We present then in section 3 the way we implement power optimizations. Section 3.5 presents results of the experiments on these optimizations. And finally, sections 4 and 5 give the layout characterization and a comparison with the state of the art.

2   Background

2.1   High Temperature Specific Features

The targeting of high temperature applications brings some specific constraints. When working at high temperature, the performances of the devices decrease because of the drift of some technological parameters. One of the main concerns is the rise of leakage current, due to the reducing threshold voltage. This problem is solved by using the Silicon-On-Insulator technology with a high transistor threshold voltage (V_{th}). Indeed, this technology presents lower leakage current that traditional bulk technology and provides greater V_{th} stability with respect to temperature [ ].

Some other problems are solved by technological means. Metallization layers, for example, are in tungsten rather than in aluminum. The use of tungsten allows to decrease electromigration, which is more significant at high temperature and can lead to premature failure of the chip [ ]. On the other hand, tungsten is more resistive than aluminum, which means that delays rise up and the device is slower.

Furthermore, the use of high temperature technology raises some other problems because it is still experimental. The few design kits available are not mature. The cells libraries are smaller and often incomplete. The design kit we use, for example, does not contain any RAM block or specific clock gating cells. We had thus to synthesize the internal RAM with digital cells.

2.2   Low Power Generalities

Power consumption in digital circuits can be divided into static and dynamic consumption. Static consumption is mainly due to leakage current and dynamic consumption to switching activity. For any digital CMOS circuit the dynamic power is given by:

$$P_d = A_f * C_{tot} * f_{clk} * V_{dd}^2$$

where A_f is the switching activity factor, C_{tot} the total node capacitances of the circuit and V_{dd} the power supply voltage. Lowering the dynamic consumption implies reducing some of these factors. Low power techniques can be applied at several levels of abstraction: technology, circuit, gate, architecture and system levels.

Technology level involves fabrication process and devices. Optimizations are for instance the use of several V_{th} and V_{dd} scaling. In the SOI high temperature process used, transistors have only one high V_{th} and a 5 V V_{dd}. Due to the high V_{th}, leakage currents are small even at high temperature.

Circuit level is related to gates or cells design. Optimizations may imply the use of a different kind of logic like branch based logic [ ]. In the high temperature design kit we used, the few available CMOS cells are not optimized for low power design.

At the gate level, some optimizations are gating and power down techniques. Gated clock is used to reduce switching activity of latches and flip-flops. Data gating is used to decrease unwanted switching in combinational logic blocks [ ]. Power down reduces static and dynamic consumption. We will use extensively clock gating and data gating.

At the architectural level, it is possible to optimize the memory hierarchy or the data-path by retiming or scheduling. The main goal is to reduce dynamic power consumption by reducing activity and wires capacitance. For compatibility issues, we decided to keep the original architecture of the 80C51. However, a carefully redesign of the micro-architecture was done to minimize as much as possible its complexity.

Finally at the system level, possible techniques are idle modes, voltage selection and power down management. Voltage selection is not available in the library. Idle and power down modes are already implemented in the original version of the 80C51.

2.3   Related Work

The only existing high temperature microcontroller working at $225°C$ is the HT83C51 from Honeywell. It consumes 70 mA at 16 Mhz with a V_{dd} of 5V. However, in order to make further comparisons on the quality of the design (section 5), we will use some other realizations, which are able to operate only at room temperature. These realizations are presented below.

In the microcontroller area, and especially for small 8-bit microcontrollers, the main achievements were made with optimized RISC pipeline architectures and with asynchronous logic [ ] [ ].

In the CoolRisc microcontroller, a significant power reduction is achieved at architectural level [ ]. The goal is to increase the instruction throughput. This allows working at a lower operating frequency while maintaining the same performances. Optimizations are done in the instruction set and in the pipeline design to execute one instruction every clock cycle, which means a CPI (Clock Per Instruction) of 1. This is better than the usual CPI of an 8-bit microcontroller which is between 4 and 40, or 15 for the original 80C51 [ ]. Power supply voltage reduction is also used in the CoolRisc. For a $1\mu m$-process and 1.5 V supply voltage, performances are 21 000 MIPS/W at 5 MIPS with a CoolRisc81 [ ].

Another main result in low power microcontroller is achieved by asynchronous design. The 80C51 was successfully implemented with this logic. In [ ], 444 MIPS/W at 4 MIPS were obtained at 3.3 V in a $0.5\mu m$ CMOS technology. It is important to notice that the design was fully compatible with the original version. No architectural optimizations were applied to reduce the disadvantaging

CPI of 15. More recently, the Lutonium [ ], an asynchronous 80C51 microcontroller was optimized by asynchronous design and deep pipelining to reduce CPI. The microcontroller was implemented in a $0.18\mu m$ CMOS library. Up to 23 000 MIPS/W at 4 MIPS were obtained at 0.5 V and 1800 MIPS/W at 200 MIPS at 1.8 V.

3 80C51 Low Power Design

The 80C51 is an old CISC microcontroller designed in the late 70's. Nevertheless, it is still widely used in the industrial and research fields. The core is an 8-bit CISC processor with a set of 255 instructions [ ]. In the original version, the machine cycle, which is the basic time slot to execute one instruction, takes 12 clock cycles. The instructions are executed in 1, 2 or 4 machine cycles.

The target applications are small control processes with low computation loads. Therefore the power consumption needs to be reduced as much as possible and the speed is not an issue.

As mentioned in section 2.2, power techniques can be applied to a design at several levels. The high temperature SOI library we used does not allow technology and circuit level optimizations. For industrial requirements, it is necessary to be cycle accurate with the original version. Therefore, architecture modifications in order to decrease the high CPI are not possible. At system level low-power modes like idle mode and sleep mode are already defined in the original 80C51. The only fields where optimizations are relevant are thus gate and micro-architecture level.

3.1 Micro-architecture Optimization

The CISC architecture of the 80C51 is highly irregular and is difficult to model for high-level synthesis. The first design was totally made by hand [ ]. At that period, no synthesis tools were available and most of the validations were performed by human brain. For our low power version, the first step was to find a minimal micro-architecture to execute all the instructions set. This part was done by hand. A set of functions is thus obtained. The FSM of the microcontroller is then obtained by scheduling those basic functions to support the instruction set. A compromise needs to be found between the complexity of the functions performed and the complexity of the FSM. The instruction decoder and the FSM are implemented in a single logic function and automatically synthesized with Synopsys.

3.2 Pre-computed Gated Clock

The main power consumption reduction is obtained by clock gating. In figure 1a, we can see a gated-clock version of an edge-triggered flip-flop. The gating is based on a AND gate. Since the gating signal drives a clock input, it is very sensitive to glitches. For this reason a latch is inserted. This one is transparent when CLK is low and in conjunction with the AND gate ensures the filtering of the glitches. The latch and the AND are often combined in a logic cell available in most of the commercial libraries. It is not the case in the experimental library we used. One solution is to synthesize a latch but that causes testability issues. Indeed, in the final version a complete scan chain has to be generated. Therefore, a better solution was to pre-compute all gating signals when instructions are decoded. Indeed, since the architecture imposes 12 clock cycles per machine cycle, with no pipeline, the two first cycles can be dedicated exclusively to decode the instruction and compute the gating signals. Furthermore, the gating signals are used by the FSM as the only way to write data into registers and gating does not need additional control signal. Figure 1b shows the principle of the gating with pre-computation. During the two first clock cycles of the machine cycle, the logic function of the decoder pre-computes control signals for the 10 following cycles. One pre-computed signal is used per writing cycle. For example, if the register DFF of figure 1b needs to be written at cycles m and n it will need the two pre-computed signals: $enable\_@\_cycle\_n$ and $enable\_@\_cycle\_m$. At cycle m, $S2$ change to 0 and the next positive edge of the clock signal will trigger the flip-flop.

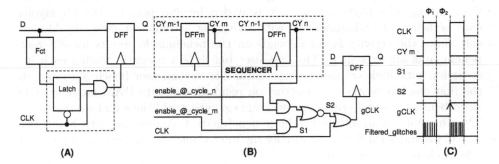

Fig. 1. Pre-computed gated clock

Two conditions are required to work properly. First, the clk inputs of the two flip-flops $DFFm$ and DFF need to be triggered at the same time. The clock tree generator of the synthesis tools supports clock gating, and thus takes into account the OR gate as part of the tree. Second, the propagation delay from the output of $DFFm$ to the clk input of DFF needs to take less than half a cycle. Indeed, the OR gate that generates the signal $gCLK$ filters glitches during the first phase of the clock (figure 1c). In order to make the design step safer, constraints are added during the synthesis and place & route steps.

3.3   Multi-level Gated Clock

Gating with pre-computation leaves some high-fanout nodes with high-switching activities. Those nodes are the clock and the CYi signals. Indeed, even if the registers are gated, the number of leafs in the clock tree remains 37 (coming

from 190). It is still high because 12 or 24 clock cycles are used to execute most
of the instructions and entire blocks of the core are unused most of the cycles.
In order to reduce further the switching activity, we add gating in the clock tree
at higher level than leaf level. The gating is done the same way it is applied at
register level. A same gating is also applied to the CYi signal.

3.4   Data Gating

Finally, data gating was added to high-fanout paths. The bigger one is the SFR
(Special Function Register) bus. The bus implementations were made by multiplexers instead of tri-state cells. This improves power consumption and testability. The gating consists of AND gates that stops the propagation of signal to
unused connections.

3.5   Gating Techniques Quantification

Figure 2 gives the normalized power consumption for 3 versions of the implemented 80C51 (80C51 UCL): a simple version with only the micro-architecture
optimization, a version with clock gating and the full version with clock and
data gating. The power reports are provided by Synopsys on the pre-layout
core netlists and back-annotated with switching activity simulated by Model-
sim. Since the activity factor depends on the benchmark, results for different
benchmarks are presented. The first user bench is a program testing each instruction with high switching of the data. The second user bench is similar to
the first one but writes each instruction result in the data RAM memory. The
third bench is a small arithmetic program executing additions and subtractions.
The dhrystone is an integer synthetic benchmark [ ],compiled with Small Device C Compiler (SDCC) [ ]. Finally the microcontroller power consumption in
idle mode is also presented.

   The application of the clock gating technique results in a power consumption
divided by a factor between 5 and 6 in normal mode and a factor of nearly 7 in
idle mode where the clock gating technique is the most efficient. The data gating
techniques allows an extra power reduction between 4 and 9 % depending on the
kind of bench. In idle mode, the data gating has no effect.

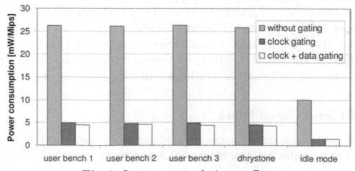

Fig. 2. Low power techniques effects

4   Layout Characterization

The layout of the chip (figure 3) was realized using the automatic place & route tool Silicon Ensemble . The core is enhanced with a scan chain and a 128-byte internal SRAM is added. The chip is ROMless, without instruction memory. The area is 23.3 mm^2 where the core is 9.5 mm^2 and the SRAM is 8.2 mm^2. The SRAM area is important because it is a synthesized RAM based on signal-hold (bus keeper) and tri-state cells.

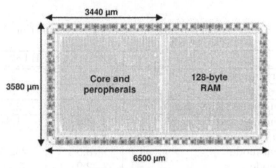

Fig. 3. Chip layout

The maximum clock frequency is 23.1 MHz (1.9 Mips) at 25^oC and 13.9 MHz (1.2 Mips) at 225^oC. The maximum power consumption is 8.52 mW/Mips for the user bench 2 at 25^oC. In idle mode, the power consumption is 2.89 mW/Mips. Figure 4 shows the distribution of this power consumption amongst the microcontroller core, the SRAM and the IO pads. Although the RAM was synthesized, its power consumption is very low because it is divided in blocks with data and address gating and because it is accessed maximum 3 times in 12 clock cycles. The major part (93%) of the IO's power consumption comes from the external program memory accesses.

5   State-of-the-Art Comparison

In this section, we first compare the 80C51 UCL with the only existing high temperature 80C51 microcontroller, the Honeywell HT83C51. Then we compare the

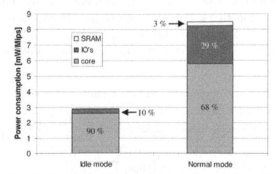

Fig. 4. Post-layout power consumption distribution

low power optimization chosen for the 80C51 UCL with the low power approach presented in section 2.3: minimization of the CPI and asynchronous design.

High temperature designs comparison. Table 1 summarizes the main features of the 80C51 UCL and the HT83C51 [ ]. The last one achieves a higher throughput but with a huge power consumption. One should take into account the presence of a 8 KB Mask ROM in the HT83C51 and an internal oscillator. Moreover, this design was not optimized for low power.

Table 1. High temperature designs comparison

	80C51 UCL	HT83C51
Power supply [V]	5	5
Process [μm]	1	?
Throughput [MIPS] @$25°C$	1.9	2.5
Throughput [MIPS] @$225°C$	1.2	1.8
Power consumption [mW/MIPS]		
Normal mode	8.5	262.5
Idle mode	2.9	56.2

Low power techniques comparison: Gating vs. pipelining. In order to make a technology-independent comparison, we synthesized the high-level behavioral description oc8051 from Open Cores [ ]. Thanks to a 2-stage pipelined architecture [ ], it offers a CPI close to 1. The oc8051 can thus achieve the same performances as the 80C51 UCL with a clock frequency divided by 12. This comparison is based on netlist level simulations before place & route, for both microcontrollers.

Table 2 reports the complexity, the maximum clock frequency calculated from the critical path reported by Synopsys, and the maximum performances of these designs. The 80C51 UCL has a 26% lower complexity. Despite its shorter critical path which allows a higher clock frequency, the 80C51 UCL theoretical performances are 6.2 times lower. Since the oc8051 is a pipelined design, its performances depend on the instruction branches and the used testbench. Simulations with different testbenches have been carried out. The ratio between the testbench execution times of both designs goes from 4.5 for the dhrystone testbench (with a lot of branches and pipeline stalls) to 7.1 for a user-defined testbench testing each instruction and writing back the results in the data RAM

Table 2. Complexity and performances comparison

	80C51 UCL	oc8051
Complexity [NAND's equ.]	9540	12944
Maximum clock frequency [MHz]	27.7	14.4
Theoretical performances [Mips]	2.3	14.4

memory. The oc8051 is able to read and write to a dual-port RAM at the same time. Therefore it achieves still higher performances when a lot of data accesses are needed. Figure 5 gives the normalized power consumption for the oc8051 and the 80C51 UCL. Thanks to the low power design, an improvement of the consumption between 39 and 56 % is achieved compared to the oc8051 which uses a 12 times lower frequency.

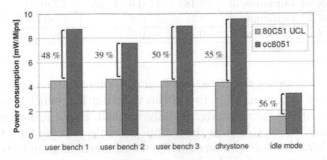

Fig. 5. Power comparison

Low power techniques comparison: Synchronous vs. asynchronous.
The 80C51 UCL has to be compared to asynchronous designs, which are the best 80C51 implementations so far. However, these designs are not able to operate at high temperature and therefore are implemented with more competitive processes than the one used for the 80C51 UCL. In order to get a fair comparison, we propose to normalize the performances relatively to V_{dd} as a quality metrics of the designs:

$$\frac{MIPS}{W} = \frac{f_{clk} * 10^{-6}}{CPI * f_{clk} * C_{tot} * A_f * V_{dd}^2}, so \quad \frac{MIPS}{W} * V_{dd}^2 = \frac{10^{-6}}{CPI * C_{tot} * A_f}.$$

Table 3 gives a comparison with asynchronous microcontrollers [ ] [ ]. We should be careful with this comparison because:

- even if the V_{dd} scaling seems to be reliable regarding the results of the asynchronous 80C51's, it does not take the process differences into account;
- the 80C51 UCL does not include an internal ROM, neither an oscillator neither a voltage regulator whereas the others do;
- however, without this internal program memory, the 80C51 UCL has a high IO's output pads consumption as shown in figure 4.

The synchronous 80C51 UCL seems to be able to deal with the asynchronous ones without raising all the special issues of this kind of implementations.

6   Conclusion

In this paper, a low power high temperature 80C51 microcontroller was presented. Low-power techniques used are mainly based on clock and data gating.

Table 3. Power consumption comparison

Microcontroller	Process $[\mu m]$	V_{dd} [V]	MIPS	MIPS/W	MIPS/W * V_{dd}^2
Synchr. 80C51 UCL	1	5	1.9	118	**2938**
Asynchr. 80C51 Philips	0.5	3.3	4	444	**4835**
Asynchr. 80C51 Lutonium 1	0.18	1.8	200	1800	**5832**
Asynchr. 80C51 Lutonium 2	0.18	1.1	100	4830	**5844**
Asynchr. 80C51 Lutonium 3	0.18	0.9	66	7200	**5832**
Asynchr. 80C51 Lutonium 4	0.18	0.8	48	10900	**6976**
Asynchr. 80C51 Lutonium 5	0.18	0.5	4	23000	**5750**

An original clock gating technique based on pre-computation of gating signals is also introduced. The main objective of this contribution, which was to realize a high temperature low power design, is fully reached. Experiments show that the results obtained are better than those from the only existing high temperature microcontroller. Measurement on the chip, which is still in manufacture, has now to be carried out, to confirm those results. Further experiments showed that the proposed microcontroller with gating achieves better results than an implementation with pipelining optimization. Finally, comparison with room temperature asynchronous designs seems encouraging. These results must nevertheless be taken carefully because of the difference of process. In this respect, an implementation in a submicron process could be very interesting and will be carried out.

References

1. "High Temperature 83C51 microcontroller product datasheet", Solid State Electronic Center, Honeywell, 1998,
   available at *www.ssec.honeywell.com/hightemp/datasheets/HT83C51.pdf*.
2. Ch. Piguet (ed.): "Low-Power Electronics Design", CRC Press, 2005.
3. D. Flandre: "Silicon-on-insulator technology for high temperature metal oxide semiconductor devices and circuits", *High-temperature electronics*, IEEE Press, Ed. R. Kirschman, pp. 303-308, 1998.
4. J. Chen and J.-P. Colinge: "Tungsten metallization for high-temperature SOI devices", J. Chen and J.P. Colinge, paper E-I.4, *European Materials Research Society*, 1994, and *Materials Science and Engineering vol. B29*, pp. 18-20, 1995.
5. I. Hassoune, A. Neve, J.-D. Legat and D. Flandre: "Investigation of Low-Power Low-Voltage Circuit Techniques for a Hybrid Full-Adder Cell", *14th International Workshop PATMOS*, pp. 189-197, 2004.
6. "MCS51 microcontroller family users's manual", Intel Corporation, 1994, available at *support.intel.com/design/mcs51/manuals/272383.htm*.
7. A.M. Volk, P.A. Stoll and P. Metrovich: "Recollections of Early Chip Development at Intel", *Intel Technology journal*, 2001.
8. H. Kapadia, L. Benini and G. De Micheli: "Reducing switching activity on datapahth busses with control-signal gating", IEEE J. Solid State-Circuits Vol.34(3), pp. 404-414, 1999.

9. L. Benini, G. De Micheli and F. Vermulen: "Transformation and synthesis of FSMs for low-power and gated-clock implementation", *ACM/SIGDA ISLP'95*, 1995.
10. oc8051 VHDL code avalaible at *www.opencores.org/cvsweb.shtml/oc8051*.
11. J. Simsic and S. Teran: "oc8051 Design Document", *www.opencores.org*, 2002.
12. Ch. Piguet et al.: "Low-Power Design of 8-b Embedded CoolRisc Microcontroller Cores", *IEEE J. of Solid-State Circuits Vol. 32(7)*, pp. 1067-1078, 1997.
13. R.P. Weicker: "Dhrystone: a synthetic systems programming benchmark", *Communications of the ACM Vol. 27(10)*, 1984.
14. H. van Gageldonk et al.: "An asynchronous low-power 80c51 microcontroller", *4th Proc Int. Symp. on Advanced Res. in Asynchronous Circuits and Syst.*, pp. 96-107, 1998.
15. A.J. Martin et al.: "The lutonium: a sub-nanojoule asynchronous 8051 microcontroller", *9th IEEE Int. Symp. on Advanced Res. in Asynchronous Circuits and Syst.*, 2003.
16. Small Device C Compiler available at *http://sdcc.sourceforge.net*.

Dynamic Instruction Cascading
on GALS Microprocessors

Hiroshi Sasaki, Masaaki Kondo, and Hiroshi Nakamura

Research Center for Advanced Science and Technology,
The University of Tokyo
{sasaki,kondo,nakamura}@hal.rcast.u-tokyo.ac.jp

Abstract. As difficulty and the costs of distributing a single global clock
throughout a processor is growing generation by generation, Globally-
Asynchronous Locally-Synchronous (GALS) designs are an alternative
approach to the conventional synchronous processors.

In this paper, we propose Dynamic Instruction Cascading (DIC). DIC
is a technique to execute two dependent instructions in one cycle by
scaling down the clock frequency. Lowering the clock frequency enables
the signal to reach farther, thereby computing two instructions in one
cycle becomes possible. DIC is effectively applied to GALS processors
because lowering only the clock frequency of the target domain is needed
and therefore unwanted performance degradation will be prevented.

The results showed average performance improvement of 7% on SPEC
CPU2000 Integer and MediaBench applications when assuming that DIC
is possible by lowering the clock frequency to 80%.

1 Introduction

Due to shrinking technologies, the difficulty and inefficiency to distribute a single global clock throughout the whole chip is growing generation by generation.
Globally-Asynchronous Locally-Synchronous (GALS) designs are an alternative
to the conventional synchronous designs [1][2][3][4][5][6]. GALS processors contain several independent synchronous blocks which operate with their own local
clocks and supply voltages. Reducing the energy of processors is becoming a major concern, and it is widely studied that GALS processors are able to achieve
low power with little performance loss compared to synchronous processors by
applying Dynamic Voltage and Frequency Scaling (DVFS) individually to each
domain [ ].

GALS designs are considerably different from the existing synchronous designs, and there should exist many microarchitectural enhancements which take
advantage of the characteristics of GALS designs. In this paper, we propose
a hardware based low power technique called Dynamic Instruction Cascading
(DIC) which work out well on GALS processors.

The main concept of our work is to execute two dependent instructions in
one cycle and improve IPC (Instructions Per Cycle). We focus on the fact that
in most general applications, ILP (Instruction Level Parallelism) is not sufficient

V. Paliouras, J. Vounckx, and D. Verkest (Eds.): PATMOS 2005, LNCS 3728, pp. 30–39, 2005.

and most of the computation resources (such as ALUs) are not in use at most cycles during the execution. In order to execute two instructions in one cycle, processor slows down the clock frequency but keep the supply voltage high. The signal will be able to reach farther in one cycle because the cycle time is extended, executing two instructions becomes possible. As this technique increases the number of instructions to be issued in each cycle, IPC will increase in compensation for high frequency. When the IPC improvement ratio is greater than the clock degradation ratio, a higher performance is expected. Applying this mechanism to a conventional synchronous processor will not work well because the whole processor core has to be slowed down. However, as each domain of GALS processors can run with there own clock frequency and supply voltage, clock frequency of the target domain can only be lowered and thus higher performance will be achieved. The end result is evident performance improvements.

The rest of this paper is organized as follows. The next section introduces previous research in related areas. Section 3 describes the proposed DIC. Section 4 details the evaluation environment and assumptions, and section 5 describes the results. Finally, we conclude in section 6.

2 Related Work

2.1 GALS Designs

Voltage Islands, a system architecture and chip implementation methodology that can be used to dramatically reduce active and also static power consumption for GALS System-on-Chip (SoC) designs is discussed recently [ ]. When turning attention to high performance microprocessors, the GALS designs applied to out-of-order superscalar processors have been studied recently by Semeraro et al. in [ ][ ] as Multiple Clock Domain (MCD) processor, and Iyer and Marculescu in [ ][ ]. Typically, these works divided the processor into four domains, comprise the front-end, integer, floating point, and load/store domains.

Most recently, YongKang Zhu et al. studied that the synchronization channels that are most responsible for the performance degradation are those involving cache access, and proposed merging the integer and load/store domain to eliminate the overhead of inter-domain communication. The result showed a significant reduction in the performance degradation and greater energy savings compared to the original MCD approach [ ].

2.2 Cascading Instructions

Intel Pentium 4 processor uses low latency integer ALU and operates "staggered add" which makes it possible to do fully dependent ALU operations at twice the main clock rate [ ].

Peter G. Sassone et al. studied by collapsing instruction strands-linear chains of dependent instructions into atomic macro-instructions, the efficiency of the issue queue and reorder buffer can be increased. The execution targets are the normal ALUs with a self-bypass mode [ ].

3   Dynamic Instruction Cascading

In this section, we describe the concept and characteristic of Dynamic Instruction Cascading (DIC).

3.1   Motivation

Recent days, processors usually apply out-of-order execution and superscalar to improve its performance. As it executes several instructions in one cycle, high performance is expected if high levels of ILP is present. However, in many general applications, the parallelism of instruction execution is not sufficient. Even if we prepare enough computation resources, they would not be in use at most cycles during the execution of the application. This is the point where we focus on and try to improve by DIC.

The aim of DIC is to effectively utilize those free computation resources. The main concept is to execute two dependent instructions in one cycle. Here, dependent instructions mean a pair of instructions that one instruction is the producer of the other consumer instruction. To execute two dependent instructions in one cycle, we lower the clock frequency with keeping high supply voltage, and thereby the signal can reach farther in an extended one cycle. Consequently, it becomes possible to execute two instructions in one cycle.

3.2   Characteristics of Dynamic Instruction Cascading

For example, assume that the instructions are carrying on as below.

1. add r5 ← r3, r2
2. add r4 ← r5, r1

As the result of instruction 1 is the input operand of the instruction 2, these two instructions cannot be executed in parallel. For example, when executing these pair of instructions on a processor running at 1 GHz, it takes two cycles (2 ns) as shown in the left-hand side of Fig. 1. In the case of DIC, however, when input operand r1 becomes ready, DIC can be applied and these two instructions will be able to complete in one cycle. It means that some pair of instructions which take two cycles to execute because of data dependency can be executed in one cycle by applying DIC. As shown in the right-hand side of Fig. 1, if the DIC can execute this pair of instructions in one cycle by lowering the clock frequency to 800 MHz, it takes one cycle (1.25 ns). It is well known that the execution of integer ALU spends half of the cycle [ ] and half on full bypass. Studies of Pentium 4 [ ] proposed double speed operation is possible, and also studies of Dynamic Strands [ ] proposed that two integer ALU execution is possible in one cycle. Thus, we conservatively assume that it is reasonable to believe that DIC can be applied on the clock frequency around 800 MHz or 900 MHz.

From the example above, it can be said that there is a trade-off between the IPC improvement and the degradation of clock frequency. That means, if the IPC improvement ratio is greater than the slow down ratio, performance will improve. This is because the product of IPC and the clock frequency means the performance itself.

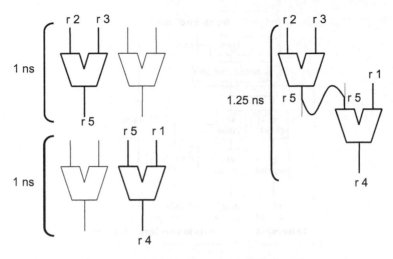

Fig. 1. Example of DIC

3.3   Dynamic Instruction Cascading on GALS Processors

When applying DIC on a conventional synchronous processor, it is necessary to lower the clock frequency of the whole processor. So even if the performance of the integer instructions improve by DIC, the performance of other pipeline stages may incredibly degrade and the total performance may not improve. However, as we are assuming a GALS processor, it is possible to lower only the clock frequency of the integer domain where DIC is applied to.

The implementation is based on a GALS processor similar to that of [ ], which has three domains, the front-end domain, floating point domain and the integer-load/store domain, as shown in Fig. 2. It differs from [ ] in detail that our base GALS processor do not have a load/store queue. Both the integer and load/store instructions are queued into the integer queue.

In this paper, we apply two types of DIC. The first is cascading integer instruction to integer instruction. Second is cascading load instruction to integer instruction. Both types can be applied at the same time.

1. To apply DIC, first, we target on the integer ALU execution which can complete in one cycle. As written above, we believe that by lowering the clock frequency, executing two dependent ALU instructions in one cycle will be possible.

   During the rename stage, the dependencies between the instructions in the queue will be checked and these pairs of instructions will be issued in the same cycle. The issue logic needs some modification, which will be similar to that of Pentium 4 [ ]. Also, we apply the original forwarding path to DIC.

   The advantage of this method is to use the redundant resources and try to issue as much instructions as possible at the same cycle. Hence, during the issue stage these pairs should be issued by priority.

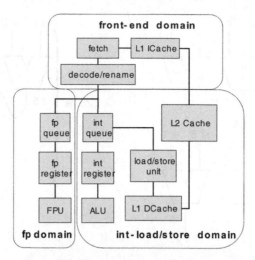

Fig. 2. The Proposed GALS Processor

2. In addition to apply DIC more aggressively, we target on cascading load instruction to integer instruction. As the hit latency of level 1 data cache is 2 cycles in most common processors, it is impossible to apply DIC without any change. To make our assumption reasonable, we assume that the processor have a line buffer [ ] in the integer-load/store domain, whose hit latency is less than 1 cycle. Line buffer is a small multi-ported fully-associative level-zero cache with a FIFO replacement policy. To implement the line buffer, the address tag and data of the returning cache access would be stored in it.

4   Experimental Setup

4.1   Evaluation Environment

We evaluated the performance and power consumption of DIC on a GALS processor with a line buffer described in section 3 compared with a GALS processor without applying DIC. We used the SimpleScalar Tool Set [ ] as our base simulation environment. We aggressively modified the SimpleScalar microarchitecture model so that it can evaluate the architecture shown in Fig. 2. Also, we modified the SimpleScalar to an fully event-driven simulator to simulate the GALS processor. To simulate the penalty of asynchronous communication, we implemented the token-ring based FIFO proposed by Chelcea and Nowick [ ]. For estimating the power consumption, we used Wattch extension [ ] to the simulation environment.

We used all the programs from the SPEC CPU2000 Integer benchmark suite [ ] and several programs from the MediaBench (adpcm, epic, g721 and mpeg2, decode and encode for each). We fast-forwarded one billion instructions and simulated 200 million instructions for the SPEC CPU2000 integer programs. The MediaBench programs were simulated from the beginning to the end.

Table 1. Processor Configuration

Fetch & Decode width	8
Branch prediction	Combined bimodal (4K-entry) gshare (4K-entry), selector(4K-entry)
BTB	1024 sets, 4-way
Miss-prediction penalty	3 cycles
Instruction queue size - integer(+load/store) - floating-point	 128 64
Issue width - integer - load/store - floating-point	 8 2 4
Commit width	8
L1 I-cache	32 KB, 32 B line, 2-way 1-cycle latency
L1 D-cache	32 KB, 32 B line, 2-way 2-cycle latency
L2 unified cache	512 KB, 64 B line, 8-way 10-cycle latency
Memory latency	80 cycles
Bus width	16 B
line buffer	4 lines
Clock frequency rate for DIC	100%, 90%, 80%, and 70%

4.2 Assumptions

Table 1 shows the assumption of the processor configurations for the evaluation. The line buffer is always used in all the evaluated processors. The size of the line buffer is 4 lines (32 B).

We evaluated the DIC technique with the GALS implementation (GALS-DIC) compared to the normal processor. We also evaluated non-GALS version of DIC (non-GALS-DIC) for comparison. In non-GALS-DIC, the clock frequency of the whole chip is scaled uniformly.

5 Evaluation Results

5.1 Performance Results

Fig. 3 shows the relative performance of DIC processors normalized to the normal processor. Each result indicates the average performance of all the evaluated programs. Because the cycle time should be extended to support DIC, we evaluated four cases of clock frequency rate (100%, 90%, 80%, and 70%) to reflect the expected cycle time degradation. Note that, the clock frequency of only the int-load/store domain is changed in GALS-DIC, whereas the clock of the whole chip is changed in non-GALS-DIC.

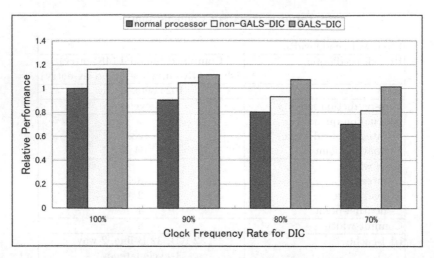

Fig. 3. Relative Performance (Average)

As seen in the figure, higher performance is obtained when the clock figure rate for DIC is 100%. About 18% performance improvement is observed in both GALS-DIC and non-GALS-DIC. However, the performance degrades almost linearly as decreasing the clock frequency rate. It should be noted that the performance degradation of GALS-DIC is less compared to non-GALS-DIC. Since non-GALS-DIC scales the clock frequency of the whole chip uniformly, the performance of the front-end, which is the most performance critical domain, is greatly degraded. On the other hand, because GALS-DIC selectively scales the clock frequency of int-load/store domain, the performance of the front-end is not degraded. Consequently, the GALS-DIC achieves higher performance than normal processor even in 80% clock frequency rate. In this case, the average performance improvement is 7%. As described in section 3, the study of Pentium 4 [9] proposed that double speed operation is possible, and the study of Dynamic Strands [11] argued that two integer ALU computation is possible in one cycle. Hereafter, we conservatively assume that the clock frequency rate is 80% to support DIC.

Fig. 4 shows the performance improvement for all the evaluated programs when the clock frequency rate is 80%. The figure shows that some programs such as adpcm, g721, 164.gzip, 197.parser, and 254.gap show a great performance improvement. On the contrary, applications such as 176.gcc, 255.vortex and epic made little performance degradation. The reason for this performance degradation is that only a small amount of instructions is selected for cascading in these programs. However, our proposed DIC has a potential of improving performance in many applications.

One future work is to invent a method in which DIC is selectively applied during the program execution. If the performance is expected to be improved by DIC in a certain time interval, the processor dynamically applies DIC for that interval. On the other hand, if the processor predicts that the performance is degraded even with DIC, the processor operates in the normal mode.

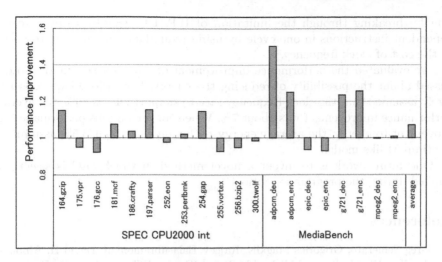

Fig. 4. Performance Improvement (Clock frequency rate for DIC = 80%)

5.2 Discussion About Power Consumption

As shown in Fig. 3, the average performance improvement of GALS-DIC is 7% when the clock frequency rate is 80%. As DIC does not scale down the supply voltage, the amount of energy consumed is the same as that of the normal processor. In other words, GALS-DIC can improve the performance without increasing the energy. This means that GALS-DISC can reduce the energy consumption by scaling the clock frequency and supply voltage for a given performance constraint. We show an brief analysis of energy reduction by GALS-DIC (clock frequency rate 80%). Table 2 shows the combinations of supply voltage and clock frequency on Intel Pentium M processor. We compared the energy consumption of the normal processor (1.6 GHZ & 1.484V) with that of GALS-DIC. Assuming GALS-DIC can choose any combination of the supply voltage and clock frequency which can be interpolated from Table 2, the maximum reduction of energy is 35% (adpcm_dec), and the average energy reduction is 6.5%. From these result, we can conclude that GALS-DIC can achieve low power without causing a performance degradation.

6 Conclusion

In this paper, we proposed a method called Dynamic Instruction Cascading (DIC) which tries to improve performance without increasing energy consump-

Table 2. The combinations of supply voltage and clock frequency on Intel Pentium M

Processor Clock	1.6 GHz	1.4 GHz	1.2 GHz	1.0 GHz	800 MHz	600 MHz
FSB Clock	400 MHz	400 MHz	400 MHz	400 MHz	400 MHz	400 MHz
Memory Bus Clock	266 MHz	266 MHz	266 MHz	266 MHz	266 MHz	266 MHz
Processor Core Vdd	1.484 V	1.420 V	1.276 V	1.164 V	1.036 V	0.956 V

tion by breaking through the limitation of ILP. DIC enables to execute two dependent instructions in one cycle by using redundant computation resources at the cost of clock frequency.

We evaluated the performance improvement of our approach, and also discussed about the possibility of reducing the energy. Even though we conservatively assumed that the clock frequency rate to support DIC is 80%, the average performance improvement was about 7%. When converting this performance improvement to energy, the average energy reduction was about 6.5% assuming a Pentium M like model.

One future work is to invent a novel method in which DIC is selectively applied during the program execution.

References

1. Greg Semeraro, Grigorios Magklis, Rajeev Balasubramonian, David H. Albonesi, Sandhya Dwarkadas, and Michael L. Scott, "Energy-Efficient Processor Design Using Multiple Clock Domains with Dynamic Voltage and Frequency Scaling," Proceedings of The Eighth International Symposium On High-Performance Computer Architecture (HPCA'02), 2002.
2. Greg Semeraro, David H. Albonesi, Steven G. Dropsho, Grigorios Magklis, Sandhya Dwarkadas and Michael L. Scott, "Dynamic Frequency and Voltage Control for a Multiple Clock Domain Microarchitecture," Proceedings of the 35th annual ACM/IEEE International Symposium on Microarchitecture (MICRO35), 2002.
3. YongKang Zhu, David H. Albonesi and Alper Buyuktosunoglu, "A High Performance, Energy Efficient GALS Processor Microarchitecture with Reduced Implementation Complexity," IEEE International Symposium on Performance Analysis of Systems and Software (ISPASS-2005), 2005.
4. Anoop Iyer, Diana Marculescu, "Power Efficiency of voltage Scaling in Multiple Clock, Multiple Voltage Cores," Proceedings of the 2002 IEEE/ACM International Conference on Computer-Aided Design (ICCAD02), 2002.
5. Anoop Iyer, Diana Marculescu, "Power and Performance Evaluation of Globally Asynchronous Locally Synchronous Processors," Proceedings of the 29th annual International Symposium on Computer Architecture (ISCA02), 2002.
6. Diana Marculescu, "Application Adaptive Energy Efficient Clustered Architectures," Proceedings of the 2004 International Symposium on low Power Electronics and Design (ISLPED 2004), 2004.
7. David Brooks, Vivek Tiwari, and Margaret Martonosi, "Wattch: A Framework for Architectural-Level Power Analysis and Optimizations," Proceedings of the 27th annual International Symposium on Computer Architecture (ISCA00), 2000.
8. Doug Burger, Todd M. Austin, and Steve Bennett, "Evaluating Future Microprocessors: The Simplescalar Tool Set," Technical Report CS-TR-1996-1308, 1996.
9. Glenn Hinton, Dave Sager, Mike Upton, Darrell Boggs, Doug Carmean, Alan Kyker, and Patrice Roussel, "The Microarchitecture Of The Pentium 4 Processor," Intel Technology Journal Q1, 2001.
10. "SPEC CPU2000 Benchmarks," http://www.spec.org.
11. Peter G. Sassone and D. Scott Wills, "Dynamic Strands: Collapsing Speculative Dependence Chains for Reducing Pipeline Communication," Proceedings of the 36th Annual IEEE/ACM International Symposium on Microarchitecture (MICRO37), 2004.

12. David E. Lackey, Paul S. Zuchowski, Thomas R. Bednar, Douglas W. Stout, Scott W. Gould, John M. Cohn, "Managing Power and Performance for System-on-Chip Designs using Voltage Islands," Proceedings of the 2002 IEEE/ACM International Conference on Computer-Aided Design (ICCAD02), 2002.
13. Tiberiu Chelcea and Steven M. Nowick, "A Low-Latency FIFO for Mixed-Clock Systems," Proceedings of the IEEE Computer Society Annual Workshop on VLSI (WVLSI'00), 2001.
14. Kenneth M. Wilson, Kunle Olukotun, and Mendel Rosenblum, "Increasing Cache Port Efficiency for Dynamic Superscalar Microprocessors," Proceedings of the 23rd annual International Symposium on Computer Architecture (ISCA96), 1996.
15. E. Fetzer and J. Orton. "A fully bypassed 6-issue integer datapath and register file on an Itanium=2 microprocessor," Proceedings of the International Solid State Circuits Conference (ISSCC 2002), 2002.

Power Reduction of Superscalar Processor Functional Units by Resizing Adder-Width

Guadalupe Miñana[1], Oscar Garnica[1], José Ignacio Hidalgo[1],
Juan Lanchares[1], and José Manuel Colmenar[2]

[1] Departamento de Arquitectura de Computadores y Automática
Universidad Complutense de Madrid
[2] Ingeniería Técnica en Informática de SistemasCES Felipe II, Aranjuez Madrid

Abstract. This paper presents a hardware technique to reduce of static and dynamic power consumption in FUs. This approach entails substituting some of the power-hungry adders of a 64-bit superscalar processor, by others with lower power-consumption, and modifying the slot protocol in order to issue as much instructions as possible to those low power consumption units incurring marginal performance penalties. Our proposal saves between a 2% and a 45% of power-performance in FUs and between a 16% and a 65% of power-consumption in adders.

1 Introduction

Many techniques have been explored to reduce power consumption on processors at multiple levels: architecture level, compiler level, operating system level and VLSI level. This paper focuses on a architecture level technique to reduce power consumption in the functional units of a Superscalar processors.

A variety of hardware techniques have been proposed for power reduction in FUs. Brooks and Martonosi proposed to use operand values to gate off portions of the execution units [1] [2]. Their scheme detects the small values of the operand and exploits them to reduce the amount of power consumed by the execution units using an aggressive form of clock gating. It disables the upper bits of the ALUs where they are not needed. A technique to exploit critical path information for power reduction was pr posed by Seng et al. [3]. A set of execution units, with different power and latency characteristics is provided and the instructions are steered to these execution units based on their criticality. Specifically, the instructions predicted as critical for performance are processed by fast and high power execution units, while the instructions predicted as not critical, are issued to slow and power-efficient execution units, thus reducing the overall power. A runtime scheme for reducing power consumption in custom logic has been proposed in [4]. It reduces power noting that operand bitwidths are often less that the maximum data width. This technique divides the functional units into two portions - the less vs. more significant portions and power reduction is seen if the data width fits within the less significant portion, in which case the more significant portion can be turned off, and its result bits computed by a simple sign-extension circuit instead. The work in [5] uses a similar intuition, but has deferent details. Haga et al. [6] presents a hardware method for functional unit assignment

V. Paliouras, J. Vounckx, and D. Verkest (Eds.): PATMOS 2005, LNCS 3728, pp. 40–48, 2005.

based on the principle that the power consumption on FUs is approximated by the switching activity of its inputs. The basic idea of this method is to reduce the number of bits whose values change between successive computations on the same functional unit.

All those methods are centred in identifying narrow-width operands. In this paper we propose a technique that reduces power consumption on functional units using narrow operands detection. However, our technique presents three important differences from those techniques. First, we analyze operation codes, in order to know if instructions involve operations with narrow operands, instead of the operand values. Second, all those techniques do not modify the structure or the number of adders in the processor. However, our technique changes the number and features of the adders with no performance penalties. Finally previous works reduce mainly the dynamic power and our technique reduces dynamic and static power-consumption because we replace high power adders by low power adders.

The rest of the paper is organized as follows. Section 2 details our new power optimization technique. Section 3 describes the simulator used to investigate our optimizations. Section 4 presents experimental results and finally section 5 describes the conclusions and future work.

2 Power Optimization Technique

2.1 Operand Characterization

In this section we provide a study that demonstrates the existence, in 64-bit processors, of a large amount of instructions that do not require a 64-bit adder. Table 1 summarizes the subset of the Alpha architecture instructions that require an adder. From this table we extract two conclusions. First, ARITH LONG instruction operands use longword data type and these instructions could be executed in a 32-bit adder. Second, the majority of memory, branch and integer arithmetic instructions, where one of the source operands is either a shift or a literal constant, could be processed in a 32-bit adder.

Initially, it looks that these instruction require a 64-bit adder because PC and source registers are 64 bit width. However, in all these instruction the other source operand is narrow (21 bit width for the BRA instruction type, 13 bit width for the JMP instruction type and so on) and for this reason it is very unlike that the operation requires a full 64-bit adder. These instructions would require a 64-bit adder provided that the length of carries would be 12 bits for the BRA instruction type, 20 bits for the JMP, 16 bits for LS and LDA and 24 bits for ARITH IMM. We call this situation "32-bit adder inhibitor" and it is illustrated in Figure 1. The key idea is that these carry chains are very unlike: $\sim 2.4 * 10^{-4}$ for BRA, $\sim 9.5 * 10^{-7}$ for JMP, $\sim 7.5 * 10^{-6}$ for LS and LDA, $\sim 3.0 * 10^{-8}$ for ARITH IMM.

In order to check the viability of this assumption we have simulated a set of benchmarks from the SPEC CPU2000 using SimpleScalar simulator to determine the percentage of addition that could be executed on a 32-bit adder. These benchmarks have been selected with the aim of covering the wider variety of behaviours. We have simulated single-sim-point regions [7] of 100M instructions, using a ref input as our default production input.

Table 1. Alpha processor instructions that need an adder. The third column shows the semantic of the operation

Instruc Type	Instruc Format	Operation
BRA	Branch	PC + Ext Sig(Desplaz(21bits))
JMP	Memory	PC + Ext Sig(Desplaz(13bits))
L/S	Memory	Rb + Ext Sig(Desplaz (16bits))
LDA	Memory	Rb + Ext Sig(Desplaz (16bits))
ARIT	Operate	Ra + Rb
ARIT_LONG	Operate	Ra(32bits)+Rb(32bits)
ARIT_IMM	Operate	Ra + Immediate(8bits)

Instruc Type	Case Study
BRA	31 21 20 0 1 1 ....1 $\boxed{X\,x\,.........x}$ PC $\boxed{X\,x\,.........x}$ Displacement
L/S y LDA	31 16 15 0 1 1 ....1 $\boxed{X\,x\,.........x}$ Rb $\boxed{X\,x\,.........x}$ Displacement
JMP	31 13 12 0 1 1 ....1 $\boxed{X\,x\,.........x}$ PC $\boxed{X\,x\,.........x}$ Displacement
ARIT_IMM	31 8 7 0 1 1 ....1 $\boxed{X\,x\,.........x}$ Rb $\boxed{X\,x\,.........x}$ Immediate

Fig. 1. Cases where the instructions cannot be issued to a 32-bit adder. The X indicates that a carry is produced in this position

Figure 2 shows the percentage of instructions that require 32-bit and 64-bit adder. On average, an 80% of instructions that require an adder can be performed using a 32-bit adder, in both FP and integer benchmarks. These percentages are respected the number of instruction that need an adder.

In summary, there are a high percentage of instructions that could be executed on a 32-bit adder instead of on a 64-bit adder. Our technique exploits this result.

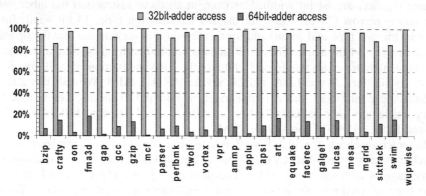

Fig. 2. Percentage of instructions on SPEC CPU2000 benchmarks that require a 32-bit and a 64-bit adder

2.2 Power Reduction Technique

In this section we present the hardware technique to reduce the power consumption in FUs of current superscalar processors. We propose to substitute some of the power-hungry adders by others with lower power-consumption. And we modify the slot protocol in order to issue as many instructions as possible to those low power-consumption units. To issue an instruction to an adder we check the operation code of the instruction, and if the "32-bit adder inhibitor" happens for the source operands.

As stated in Section 2.1 around a 80% of instructions that require an adder to be executed can use a 32-bit adder instead of the 64-bit one.

In current processors, the 64-bit adders have to perform the addition in one clock cycle. This requirement imposes an exigent design constrains during the synthesis of the adder. Due to this constraint, the synthesis tool is pushed to reach the minimum delay at the expense of any other synthesis considerations (particularly, area and power-consumption). Among the task these algorithms perform are:

- Delay-optimization techniques (i.e. logic flattening) considerably increases the circuit area and consequently the power-consumption.
- Area and power optimization techniques cannot extensively be used.
- Replacement of gates with low-driving capabilities by high-driving capabilities.

A 32-bit adder approximately dissipates half the energy than a 64-bit one (provided that the same adder architecture is used). The origin of this reduction is the fact that the former carries out half the switches, –and consequently, reduces the dynamic power-consumption– and has half the transistor –also reduces the static power-consumption–. Therefore we could use a 32-bit adder which power consumptions is approximately a 50% lesser than power consumption of the 64-bit adder. On the other hand, these 32-adders are faster than the 64-bit adders. However we do not choose to use the fastest 32-bit adder. In order to reduce the power-consumption even more, we can choose an adder whose critical path is very similar to the 64-bit adder critical path. This allows us to use smoother design constraints in the 32-bit adder synthesis. These constraints provide a margin to the synthesis to reduce the power consumption by applying two techniques:

- Reducing the area –and obviously the number of gates and transistors in the design–.
- Re-sizing transistor and/or replacement of gates by other with low-driving capabilities, so the power-consumption per gate transition is also reduced.

In summary, we have supposed a 32-bit adder whose power consumption is approximately a 25% lower than power-consumption of the 64-bit adder.

Our approach exploits this fact and proposes to substitute some of the four 64-bit adders of the Alpha 21164 for 32-bit adders, using the operation codes of the instruction to determine the FU's to send the instructions.

3 FU-Wattch Simulator

We have modified Wattch Simulator [8] in order to calculate more accurately the power values in the FUs. These are the modifications that we have implemented and its justification:

- Wattch models integer FUs just as an adder. In order to analyze a real system, we should include more components. Our simulator models the power estimation of the integer FUs as the addition of the power-consumption of an adder, a shifter unit, a specific unit for logic operations and a multiplier.
- We found the same situation in the floating point FU. Our model models the power consumption of this unit as the addition of the power-consumption of an FP adder and a FP multiplier.
- We have adapted clock gating in order to have the possibility of turn on/off each component separately.
- We have better modelled the static power-consumption. Wattch only takes into account the static power of a FU when the FU is accessed.
- Wattch supposes that static power-consumption is a 10% of the total power consumption. We have increased the static power consumption up to 25% of the total power. In this way we obtain a more realistic model regarding current and future technologies [9].

In summary, we have used the basis of Wattch Simulator and we have included those five modifications in order to improve the estimation and to obtain more realistic figures, especially as far as static power consumptions are concerned.

We have run the benchmarks presented in Section 2.1 on both simulators: Wattch and FU-Wattch. We have assumed the values used in [8] for the different components of the execution unit. The functional units in current high-end microprocessors are likely to use even more power, but detailed numbers are not yet available in the literature.

Figure 3 shows the estimated percentage of FU power consumption over total power consumption of the processors. As we can observe in the figure, the power values obtained with FU-Wattch simulation are closer to that on the literature [8] [10] [11].

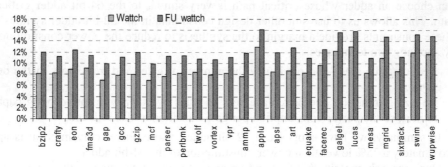

Fig. 3. Power-consumption of ALUs estimated using Wattch and our modified version of Wattch

4   Experimental Results

In order to evaluate our approach we have simulated the set of benchmarks stated in Section 2.1.

We have tested six different configurations for the execution unit in order to find which one brings a higher reduction on power-consumption, preserving the performance of the processor. The baseline configuration has four integer clusters, each of them having one 64-bit adder, a shifter unit and a specific unit for logic operations. And one of the clusters has a pipelined integer multiplier. Beside, there is one floating point cluster. It has one adder and one pipelined multiplier. So, baseline configuration uses four 64-bit integer adders (one on each cluster). In the rest of the configurations we only have changed the number and the type of the integer adders. All of the tested configurations have at least the same number of adders as the baseline configuration in order to do not modify the performance of the processor. The configurations are:

- Baseline: four 64-bit.
- Conf-1: four 64-bit and two 32-bit adders.
- Conf-2: two 64-bit and two 32-bit adders.
- Conf-3: two 64-bit and three 32-bit adders.
- Conf-4: one 64-bit and three 32-bit adders.
- Conf-5: one 64-bit and four 32-bit adders.

The protocol to choose the adder to execute an instruction is the following:

1. We check the operation code and if a "32-bit adder inhibitor" situation happens. In order to accelerate this test, we check the more significant bits of the less significant word of the PC (the number of checked bits depends on the operation code) and we detect that the addition of the PC and the displacement produces a carry just by checking the two most significant bits of the displacement and their counterparts in the PC (i.e. for the BRA instruction we check bits 20th and 19th for both operands). With this test we find more "32-bit adder inhibitor" situations that actually do but it is a quite simple mechanism.
2. If an instruction can be executed in a 32-bit adder, first we check if there is a 32-bit adder available and, in this case, the instruction is issued to it. If not, the instruction is issued to an available 64-bit adder. We apply this policy in order to preserve performance.

For all configurations, differences on the static power consumption do not depend on the benchmark, it only depends on the amount of adders. Conf-1 has more static power-consumption (1.3106 W) than the baseline configuration (1.165 W) because we have added two adders, therefore there are six adders instead of four them. The rest of configurations have less static power consumption than the baseline configuration because we have reduced the number of 64-bit adder to the half in the Conf-2 (0.7281W) and 3(0.8009W), and from four to one in the Conf-4(0.5097 W) and 5(0.5826 W). In summary, in term of static power consumption, Conf-4 is the best because it has the minimum number of 64-bit adders (just one) and the minimum number of adders permitted (just four). In this configuration, the static power has been reduced from 1.165 W to 0.5097W.

Figure 4 shows the dynamic power consumption in the adders (W/cycle) for every configuration and for all benchmarks. The baseline configuration has the higher dynamic power consumption, because all instructions that require an adder are executed in a 64-bit adder. In the rest of configurations, the dynamic power consumption is lower than the baseline configuration, because there are instructions that are executed in a 32-bits adder instead of a 64-bit one. Conf-5 is the best in terms of dynamic power consumption. The reason in that all instructions that could use a 32-bit adder find one available.

Figure 5 illustrates the total power-consumption in the adders. The best configuration depends on the benchmark. In other words, in some cases Conf-4 is the best configuration while in others Conf-5 is the optimal one. It depends on the trade-off between static consumption and the reduction of dynamic consumption.

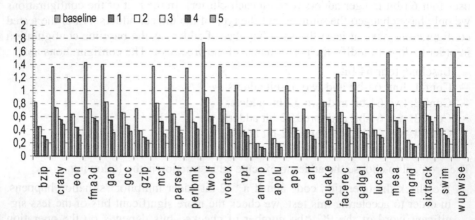

Fig. 4. Dynamic power consumption in the adders (W/cycle) for all configurations in every benchmark

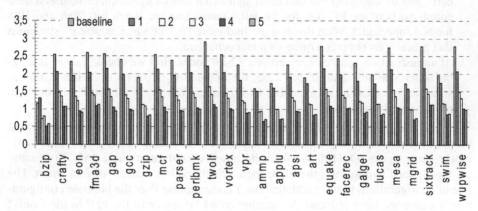

Fig. 5. Total power consumption in the adders (W/cycle) for all configurations in every benchmark

Figure 6 shows the amount of total power (in %) that is saved in the integer and floating point functional unit with our optimization. Our result shows that for every configuration and in all benchmark, our technique reduce the total power consumption of the FUs. The saving of energy is between a 2% and a 45% of power-performance in FUs and between a 16% and a 65% of power-consumption in adders.

In order to accept any change on the cluster configuration it is necessary to assure that performance it is not affected. We have obtained the execution time for all benchmarks with the six different cluster organization proposed and we have observed that this technique does not affect the processor performance. The performance penalties lie between 0.1% and 0.2%.

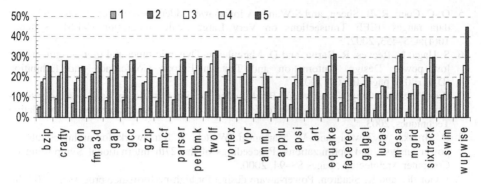

Fig. 6. Percentage of total power saved in the FU for all configurations in every benchmark

5 Conclusions

This paper presents a technique to reduce the static and dynamic power consumption in FUs of 64-bit superscalar processors. This new approach entails substituting power-hungry adders by low-power ones and modifying the slot protocol to issue as much instructions as possible in those adders.

To proof the viability of this proposal, we have shown that, on average, a 80% of instructions in the SPEC CPU2000 requiring an adder, could use a 32-bit adder instead of a 64-bit one.

We have modified Wattch in order to calculate more accurately the power consumption values in the FUs and we have used it to simulate a set of benchmarks from the SPEC CPU2000.

We have tested six different configurations for the execution unit in order to find which produces a higher reduction in the power-consumption, preserving the performance of the processor. Our results show that for every configuration, our technique reduces the total power consumption in the FUs without adding any penalty on the execution time. Our approach saves g between a 2% and a 45% of power-performance in FUs and between a 16% and a 65% of power-consumption in adders.

References

1. D. Brooks and M. Martonosi. Dynamically exploiting narrow width operands to improve processor power and performance. In Proc of the 5th Int'l Symp on High Performance Computer Architecture (HPCA), pages 13–22, 1999.
2. D. Brooks and M. Martonosi. Value-based clock gating and operation packing: dinamic strategies for improving processor power and performance. ACM Transaction on Computer Systems, (2):89, 2000.
3. J. S. Seng, E. S. Tune, and D. M. Tullsen. Reducing power with dynamic critical path information. In Proceedings of the 34th annual ACM/IEEE international symposium on Microarchitecture, page 114.
4. J. Choi, J. Jeon, and K. Choi. Power minimization of functional units partially guarded computation. In Proceedings of the 2000 international symposium on Low power electronics and design, pages 131–136, 2000.

5. O.-C. Chen, R.-B. Sheen, and S.Wang. A low-power adder operating on e ective dynamic data ranges. IEEE Transactions on Very Large Scale Integration (VLSI) Systems, 10(4):435–453, 2002.

6. S. Haga, N. Reeves, R. Barua, and D. Marculescu. Dynamic funcional unit assignment for low power. Design, Automation and Test in Europe Conference and Exhibition (DATE'03), pages 03–07, 2003.

7. Erez Perelman, Greg Hamerly, and Brad Calder. Picking statistically valid and early simulation points. In the International Conference on Parallel Architectures and Compilation Techniques, 2003.

8. D. Brooks, V. Tiwari, and M. Martonosi. Wattch: A framework for architecturallevel power analysis and optimizations. Proceedings of the 27th International Symposium on Computer Architecture, pages 83–94, 2000.

9. J. Gonzlez and K. Skadron. Power-aware design for high-performance processors. 10th International Symposium on High-Performance Computer Architecture (HPCA-10), 2004.

10. M. Gowan, L. Biro, and D. Jackson. Power considerations in the design of the alpha 21264 microprocessor. In 35th Design Automation Conference, 1998.

11. Hsien-Hsin S. Lee, Joshua B. Fryman, A. Utku Diril, and Yuvraj S. Dhillon. The elusive metric for low-power architecture research. In the Workshop on Complexity-E ective Design in conjunction with ISCA-30, 2003.

A Retargetable Environment
for Power-Aware Code Evaluation:
An Approach Based on Coloured Petri Net

Meuse N.O. Junior[1], Paulo Maciel[1], Ricardo Lima[2], Angelo Ribeiro[1], Cesar Oliveira[2], Adilson Arcoverde[1], Raimundo Barreto[1], Eduardo Tavares[1], and Leornado Amorin[1]

[1] Centro de Informática (CIn)
Universidade Federal de Pernambuco (UFPE)-Brazil
{mnoj,prmm,arnpr,aoaj,rsb,eagt,lab2}@cin.ufpe.br
[2] Departamento de Sistemas Computacionais
Universidade de Pernambuco (UPE)- Brazil
{ricardo,calo}@upe.poli.br

Abstract. This paper presents an approach for power-aware code exploration, through an analysis mechanism based on Coloured Petri Net (CPN). Given a code under interest and a CPN description of architecture, a CPN model of application (processor + code) is generated. Coloured Petri Net models allow the application of widespread analysis approaches, for instance simulation and/or state-space exploration. Additionally, this work presents a framework where a widespread CPN tool is applied to processor architecture description, model validation and analysis by simulation. A Petri net integration environment was extended in order to support specific power-aware analysis. In the present approach, such framework is focused on the Embedded Systems context, covering *pipeline-less* and *simplescalar* architectures.

1 Introduction

In many embedded computing applications the power and energy consumption are a critical characteristic, for instance: portable instruments, personal digital assistant and cellular phones. Despite of processor hardware optimization, processor consumption is affected by the dynamic behavior of software [ ], meaning that the software power analysis is crucial. This work focus on such proposition in order to implement a retargetable environment for analyzing code-consumption. Coloured Petri Nets are applied as an architecture description language (ADL) in order to generate a formal simulation mechanism. Resulting from CPN description, aspects of processor-architecture and code-structure are mapped on a formal analysis model. In the embedded system context, the designer should evaluate both code optimization and software/hardware partitioning strategies in order to guarantee the best system energy consumption. This work presents an infrastructure to perform this task, comprising an extension of previous work [ ]. This paper is organized as follows: Section 2 presents related works. Section 3 introduces Coloured Petri Nets concepts. Section 4 shows some definitions with reference to proposed approach. Section 5 presents the description model. Section 6 presents the proposed framework. Section 7 illustrates a case study and Section 8 concludes the paper.

V. Paliouras, J. Vounckx, and D. Verkest (Eds.): PATMOS 2005, LNCS 3728, pp. 49–58, 2005.
© Springer-Verlag Berlin Heidelberg 2005

2 Related Works

Works as [ ],[ ] and [ ] have shown the importance of power-aware compilation and design exploration taking into account dynamic software behavior. Such works have opened the perspectives toward a new compilers generation, that would optimize code in three axis: performance, code size and power. Even under a power-aware compilation methodology, the constraints have to be verified and software-hardware strategies options evaluated. In order to construct an evaluation infrastructure, an approach based on architecture description is an interesting option. Interesting works were proposed toward *ADLs-Architecture Description Languages*. Such languages allow an automatic generation of tools (simulators/compilers) from the processor description. LISA and EXPRESSION [ ] are interesting ADLs examples. In special, LISA 2.0 language, implemented on LISATek environment[ ], supports consumption analysis from generation of RTL simulator. SimplePower [ ] is a well-known tools-set on computer architecture community. SimplePower extends SimpleScalar implementation of micro-architecture simulator based on architectures templates, so that take into account energy consumption. In order to retarget SimpleScalar to a new processor architecture, the designer has to sequentialize the concurrency of hardware in *ad-hoc* ways, which can be time-consuming and error-prone, even though there exist works toward automatic generation of SimpleScalar models [ ]. In the Petri models context, Burns [ ] applied CPN to construct a model of a generic superscalar processor in order to perform real-time analysis. Additionally, Murugavel [ ] proposed a CPN extension, termed as Hierarchical Colored Hardware Petri Net, for dealing with gates and interconnections delays for hardware power estimation.

3 Coloured Petri Net: An Overview

Petri nets are families of formal net-based modeling techniques, that model actions and states of systems using four basics entities: places, transitions, arcs and tokens. In the majority of models, places are associated with local states and transitions with actions. Arcs represent the dependency between actions and states. An action occurs when a transition is "fired", moving tokens from incoming places to outgoing places. As a Petri net example a parallel process is described in Figure 1. Arcs describe which action (or actions) is possible from a given local state. In Figure 1 arc A1 links place P1 (state 1) to transition T1 (action 1), representing that action T1 requires local state P1. Arcs A2 and A3 connect transition T1 to places P2 and P3. There are two parallel net-paths (P3-T3-P5 and P2-T2-P4) as a representation of parallel processes. In order to accomplish the action T4 is necessary to reach the state represented by marking of places P4 and P5. In order to represent possible local states, it is used a specific mark, called token (a black ball) . This simple net is known as place-transition net [ ]. Place-transition nets are adequate to analyze some characteristics of system such as repetitiveness, liveness and reachability. In an informal approach, such definitions mean respectively: periodic behavior, absence of dead states and capability to reach specific set of states. There are various extended Petri net models, each one dealing with a specific modeling problem and distinct abstraction level. CPN is a high-level model that consider abstract data-types and hierarchy. Informally, Coloured Petri Net is a Petri net with some modeling improvements: (i) tokens express values and data structures with Types (colors); (ii) places have associated Type (color set) determining the kind of data (token) those places may contain; (iii) transitions may express complex behavior by changing token value; (iv) hierarchy can be handled at different abstraction levels. Transitions in a hierarchical net may represent more complex structures, where each transition (substitution transition) expresses another more complex net, and so on. A hierarchical net may describe complex systems by representing their behavior using a compact and expressive net. (v) Behaviors can be also described using high-level program language, specifically using CPN-ML (a Standard ML Language subset) [ ]. Hence, it is possible to model a complex system using tokens for carrying sets of internal data values and transitions to represent actions that modify internal set of data values. The entire net represents the flow of changes into the system during its states evolution. The model can be analyzed by simulation and state analysis. State analysis means to study all possible system states, or an important subset, in order to capture system patterns. There are some widespread academic tools such as Design/CPN and CPNTools [ ], for handling Coloured Petri Net and its analysis engines. CPN tools provide an environment for design, specification, validation and verification of systems [ ]. Thus, CPN provides a ruled-based declarative executable modeling. In this way, comprising an interesting instrument for processors modeling and analysis.

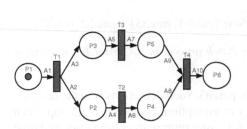

Fig. 1. Petri net example

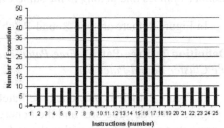

Fig. 2. Execution Profile Example

4 Nomenclatures and Definitions

In [ ] was presented some definitions that are necessary for better understanding the proposed method. These definitions are informally shown as follows.

Definition 1 (Execution Vector). *Execution Vector is a vector, enumerated by the instructions memory ordering, where each component represents the number of instruction execution.*

Definition 2 (Consumption Vector). *Consumption Vector is a vector, enumerated by the instruction memory ordering, where each component represents the respective instruction energy cost (base cost + inter-instruction cost)*[1].

Figure 2 depicts the Execution Profile for a nested-loop example. Analyzing the Execution Profile, it is possible to identify execution patterns, such as:

---

[1] Base cost is the instruction specific energy cost. Inter-instructions cost means the energy cost that appear due to *circuit overhead* when two instruction are executed consecutively [ ]

Definition 3 (Patch). *Patch is a set of instructions that are located in consecutive addresses and are executed the same number of times.*

In Figure 2, five patches are identified: from instruction 2 to instruction 6 (patch 1), from instruction 7 to instruction 10 (patch 2), from instruction 11 to instruction 14 (patch 3), from instruction 15 to instruction 18 (patch 4) and from instruction 19 to instruction 25 (patch 5).

Definition 4 (Loop-Patch). *Loop-Patch represents a single loop. It consists of a patch variant in which the incoming (first) and outgoing (last) instruction are executed only once.*

There is no Loop-Patch in Figure 2.

Definition 5 (Cluster). *Cluster is a set of Patches joined together (aggregated) in consecutive addresses, in which the incoming (first) and outgoing (last) instruction are executed only once.*

In Figure 2 there is only one cluster: {patch1, patch2, patch3, patch4, patch5}.

Definition 6 (Bound-Patch Set). *Bound-Patch Set is a set of Patches executed the same number of times and belong to the same Cluster.*

In Figure 2, there are two Bound-Patch Set: {patch1, patch5}, {patch2, patch4}.

Definition 7 (Free-Patch). *Free-Patch is a Patch present in a Cluster but not within the Cluster Bound-Patch Set.*

In Figure 2, there is only one Free-Patch: patch3. Such definitions help the designer to figure out code structures and their energy consumption. For example, a Loop-Patch represents an isolated loop within the code. A Cluster represents consumption and time cost regions. A *Cluster* with *Bound-Patch* Sets such that its *Patches* have symmetric positions in the *Execution Profile* may represent a nested-loop (see Figure 2). Inspecting *Execution Profiles* and the *Consumption Profile* graphics, the designer is able to map consumption to code structures. In the scope of embedded systems is very important to analyze code optimization in respect of total consumption and consumption profile. Under the consumption profile point of view, the designer may opt for some consumption distribution allowing best software-hardware migration from code-segments (*Patches/Clusters*), improving total consumption, as postulated in [14].

5   The Description Model

The proposed description model represents the processor in two layers: hardware resource and instruction behavior. The processor ISA (Instruction Set Architecture) is described in a Coloured Petri Net model. The hardware resources are described as CPN-ML functions embedded on such instructions model. Due to hierarchical CPN properties, the ISA description can be performed at different abstraction levels. The inferior limit is the gate-level, that can be reached by implementing descriptions as the Murugavel's hardware model [ ]. A widespread environment for CPN edition and analysis is used to generate architecture descriptions. Each instructions are constructed as an individual net so that instruction-model validation is implemented in an isolated way, for each instruction without the need of building a test program. Thus, a processor description consist of a set of instruction CPN-models. The instruction behavior is described by a CPN-ML code, present on instruction CPN-model. The CPN-ML code invokes functions to perform the instruction behavior. This functions (hardware resource) can be unfolded in order to represent more details for constructing a more accurate power-model. In the current approach, the power model is based on instruction consumption in accordance with concepts postulated in [ ]. Each instruction CPN-model computes its energy consumption during the net simulation, feeding the *Consumption Vector*. Internal context of processor (memories and registers) are described as record-structure encapsulated by the token. Additionally, the token carry *"probes variables"* such as accumulated number of cycles, energy consumption, *Execution Vector* and *Consumption Vector*. The first step on the description process comprise to define the token data-structure. The token have to carry elements that are modified due to instruction execution. In the second step, each instruction behavior is described invoking CPN-ML functions in order to modify the token. Depending on the abstraction level intended, the designer can construct such functions or to apply directly CPN-ML commands. Figure 3 shows a CJNE (Compare and Jump if Not Equal) instruction CPN-model from a 8051 architecture. The token structure of 8051 architecture is presented as follows.

```
val value={cycle=0,power=[0,0,0,0],
           Ex_Pro=inEx_Pro,patch=Inst_Preg,
           mem=write_pr(inic,129,7),Upmem=Up_inic,
           PC=0,ext_mem=ext_m_inic,
           program=programinic};
```

The variable *value* represents the token value, meaning internal context value. Fields as cycle, power, Ex_Pro , and patch are *"probes variables"* that monitors respectively: execution time in terms of clock cycles, total energy consumption, execution vector and consumption vector. Fields as mem, Upmem, ext_mem, program and PC model internal context, respectively: internal RAM memory, high memory, external memory, code memory and PC register. In the example above, such fields are being initialized. The behavior description (CPN-ML code) implements the instruction behavior by invoking functions that operates on such fields. For instance, function *write_m()*(Figure 3) receives a target memory, a target address and data, returning the target memory modified. Checking the instruction behavior is as simple as simulate its CPN-model, and to check the token value afterward. For instance, the instruction CJNE is validated executing the model presented in Figure 3. The token will flow through the instruction-model until reach place *S_Output1* or *S_Output2*, depending on *Carry* flag value. Note that the decision process is implement by a net structure (transition Jump and NotJump). Each instruction-model computes its clocks-cycle. Therefore, the CPN simulation implements an accurate simulation of processor. In order to generate automatically a simulation mechanism, a Binary-CPN compiler is applied. Given a code under interest and the processor description, a Binary-CPN compiler constructs a CPN model of application (processor + code). The CPN structure *per si* models the program possible flows, where each possible program state is modeled as a place. Each transition is a substitution-transition encapsulating instruction CPN-model. Figure 4 shows the

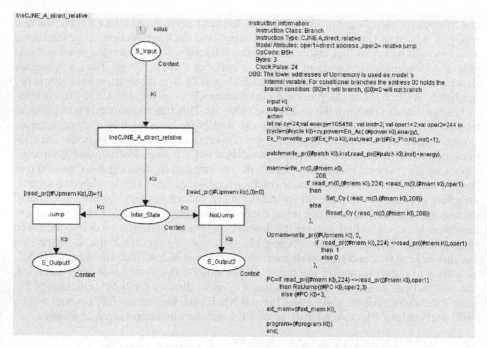

Fig. 3. coloured Petri net model for a branch instruction

CPN model at CPNTools environment. At left hand, hardware resources are modeled by functions and at right hand the application model is represented as a Coloured Petri Net.

In order to cover pipelined architecture, the model adapts the atomic concept of processor-execution (internal context modification) as a token running (*Execution-Token*) through code structure, proposed before. Such concept is replaced by an abstract entity termed as *virtual-pipeline*. The description model is extended in order to model aspects inherently related to pipeline, such as pipeline stall condition and pipeline flush cost. A novel token is introduced on the CPN-model: the *Witness-Token*. During the simulation the *Witness-Token* pass through instructions before *Execution-Token*, registering such instruction into a list (pipeline-list) and sending such list to *Execution-Token* by shared variable. The length of the list is the pipeline depth. The *Witness-Token* list shows the pipeline context and, based on it, control flags are activated on the *Execution-Token*. Such flags inform transition-code the demand for hardware resources and data conflict situations. In case of branch-instructions, a control flag instructs the next transition-code that will take the *Witness-Token* to destroy it. A novel *Witness-Token*, with an empty pipeline-list, is released and the *Execution-Token* is held until the pipeline-list be full again. In this way, the pipeline is modeled as a virtual entity having the *Witness-Token* as head and the *Execution-Token* as tail end. The Figure 5 illustrates such mechanism. The "destruction" and subsequent "reconstruction" of the *virtual-pipeline* translate the processor behavior to a CPN behavior. The behavior pattern is captured by profiling during the simulation analysis.

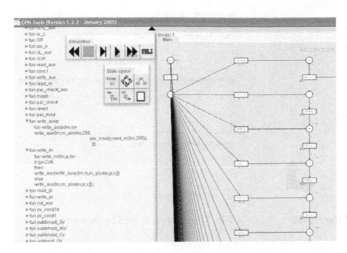

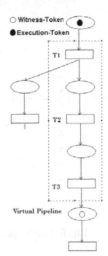

Fig. 4. CPNTools, edition and analysis interface

Fig. 5. SimpleScalar net model

6 Proposed Framework

The basic framework, illustrated in Figure 6, consists of a CPN engine, a Binary-CPN[2] compiler, that translates from machine-code to CPN model, and a set of specific functions for power and performance evaluation. These functions evaluate the *probes variables* and return metrics, among then, *Consumption Profile*, *Patch Consumption* and *Cluster Consumption*. The Binary-CPN compiler and analysis functions are integrated on EZPetri Environment [ ]. Figure 6 illustrates the framework to code evaluation, termed as EZPetri-PCAF (*EZPetri- Power Cost Analysis Framework*). EZPetri-PCAF performs Binary-CPN compilation, opens the CPN-Model in CPN-Engine, engages communication with CPN-Engine and evaluates analysis functions. Several code version can be explored. For each code version a power analysis can be performed under the EZPetri-PCAF environment. Afterward, the results can be compared with constraints that take into account code size, execution time and energy consumption. The CPNtools [ ] has been used as the CPN-Engine. The front-end is performed by the EZPetri environment, providing a specific interface for power analysis. The Binary-CPN compiler considers two entries: the machine-code and the set of instruction CPN-models. Based on such files, the Binary-CPN compiler generates a model, a Coloured Petri Net model, according to machine-code input. The model is represented in the XML file format defined by the CPNTools. Note that, due to the standardization present on the formal model (Instruction model), the Binary-CPN compiler can deal with different architectures with minimal modifications. The net entities, as places and transitions, keep the same meaning whatever the architecture. The changes will be concentrated on the CPN-instruction model, with minimal impact on the net formation rules.

---

[2] In this context Binary means executable machine-code

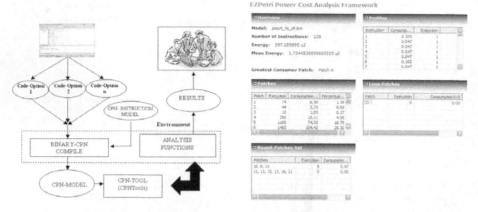

Fig. 6. The proposed framework   **Fig. 7.** Tables presented in the user interface

The EZPetri environment is based on the IBM-Eclipse platform, allowing the coupling of new features easily. It implements functions such as codes entities identification (*Patches and Clusters*) and consumption analysis. The results, generated by the analysis, are shown in a GUI as charts and tables. Figure 7 depicts the PCAF results. Additionally, these results may be presented as a portable document format (pdf) file. In order to link EZPetri-PCAF with CPNTools, a proxy application is started, opening a TCP/IP channel. Then, the proxy loads the model in the CPN Tools. Using the Comms/CPN[ ], the model establishes a TCP/IP connection in order to interchange information with the proxy during net simulation. At final, the proxy sends such information back to EZPetri-PCAF, which performs analysis and presents the results as already discussed.

7   Experiments and Results

This section presents the proposed framework implementing a power-aware evaluating of C51 Keil compiler. The evaluation comprises analysis of the set of compiler code optimization. A bubblesort algorithm was adopted as case study. It was applied on ten position vector in the reverse ordering, meaning the worst execution case. The Keil compiler allows nine optimization levels under two emphasis: *favor size* (fs) and *favor speed* (fsp). The optimization selection works in cumulative way: level 0 implements only *constant folding* technique, level 1 implement *constant folding* and *dead codes elimination*, and so on. Hence, three optimization levels was chosen for analysis: *constant folding* (cf), *register variables* (rv) and *common block subroutines* (cb). Six binary-codes was analyzed, that was generated by composing emphasis and level optimization. The estimates was performed based on the AT89S8252 consumption-model captured according to methodology proposed in [ ]. Table 1, Figure 8 and Figure 9 show the consumption pattern captured by the EZPetri-PCAF environment. The lowest consumption was achieved with the sort_fsp_cb implementation, representing 26.14% of energy savings, if compared with the highest consumption (sort_fs_cf). Note that

Table 1. Consumption Pattern

Code Options	Energy Consumption (uJ)	Energy Savings(%)	Average Power (mW)
sort_fs_rv	340.48	14.28	50.05
sort_fs_cf	397.18	0	50.42
sort_fs_cb	296.42	25.37	50.89
sort_fsp_rv	337.84	14.94	50.84
sort_fsp_cf	392.86	1.09	50.18
sort_fsp_cb	293.35	26.14	50.45

the average power is practically constant. Such savings do not impact on average device heat dissipation, but on battery autonomy, a crucial aspect for portable systems. Additionally, the environment provides identification and classification of *Patches* and *Clusters*. Such information aids designer for identifying and optimizing critical code regions. Figure 9 depicts percentile consumption of the *Highest Consumption Patch* (HCP), indicating sectors for code optimization or partitioning on hardware, a very important information in projects driven by Hardware/Software Co-Design paradigm. In this experiment, the analysis have shown that HCPs can be responsible for 79% of the total consumption. Such consumption is performed in less than 20% of total execution time (Figure 8). Therefore, representing a good reason for software-hardware migration as postulated by [14].

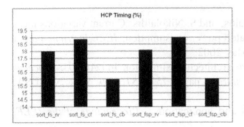

Fig. 8. Timing of Highest Consumption Patch

Fig. 9. Energy consumptions for Highest Consumption Patch

8   Conclusion

This paper presented a retargetable approach for analyzing code focusing energy consumption. Such approach is integrated in the EZPetri Environment in order to perform power-aware binary-codes analysis. Based on such analysis, the application can be *tuning*, through partitioning or code optimization so that generate the best implementation. The target processor is modeled under a formal computational model based on tokens, the Coloured Petri Nets. Due to this formal model, processor capabilities and behaviors are strictly translated into a set of rules and functions, defined under Coloured Petri Net syntax. Such modeling allows to generate automatically an analysis infrastructure from an architecture description. The resulting framework is termed EZPetri-PCAF.

References

1. Cpntools, version 1.22. *http://wiki.daimi.au.dk/cpntools/cpntools.wiki*.
2. Lisatek. *http://www.coware.com/products/lisatek_description.php*.
3. J. Ayala, A. Veidenbaum, and M. López-Vallejo. Power-aware compilation for register file energy reduction. *International Journal of Parallel Programming*, 31(6):451–467, Dec 2003.
4. F. Burns, A. Koelmans, and A. Yakovlev. Modelling of superscala processor architectures with design/CPN. In Jensen, K., editor, *Daimi PB-532: Workshop on Practical Use of Coloured Petri Nets and Design/CPN, Aarhus, Denmark, 10-12 June 1998*, pages 15–30. Aarhus University, 1998.
5. G. Gallasch and L. M. Kristensen. Comms/CPN: A communication infrastructure for external communication with design/CPN. In *3rd Workshop and Tutorial on Practical Use of Coloured Petri Nets and the CPN Tools (CPN'01) / Kurt Jensen (Ed.)*, pages 75–90. DAIMI PB-554, Aarhus University, Aug. 2001. InternalNote: Submitted by: hr.
6. A. Halambi and P. Grun. Expression: A language for architecture exploration through compiler/simulator retargetability. In *Proc. of the European Conf. on Design, Automation and Test (DATE)*, 1999.
7. A. A. Jr, G. A. Jr, R. Lima, P. Maciel, M. O. Jr, and R. Barreto. Ezpetri: A petri net interchange framework for eclipse based on pnml. In *In First Int. Symp. on Leveraging Applications of Formal Method (ISoLA'04)*, 2004.
8. M. N. O. Junior, P. Maciel, R. Barreto, and F. Carvalho. Towards a software power cost analysis framework using colored petri net. In *PATMOS 2004*, volume 3254, pages 362–371. LNCS Kluwer Academic Pubishers, September 2004.
9. L. Kristensen, S. Christensen, and K. Jensen. The practitioner's guide to coloured petri nets. *International Journal on Software Tools for Technology Transfer: Special section on coloured Petri nets*, 2(2):98–132, 1998.
10. T. Laopoulos, P. Neofotistos, C. Kosmatopoulos, and S. Nikolaidis. Current variations measurements for the estimation of software-related power consumption. *IEEE Instrumentation and Measurement Technology Conference*, May 2002.
11. W. S. Mong and J. Zhu. A retargetable micro-architecture simulator. In *DAC '03: Proceedings of the 40th conference on Design automation*, pages 752–757, New York, NY, USA, 2003. ACM Press.
12. T. Murata. Petri nets: Properties, analysis and applications. *Proceedings of the IEEE*, 77(4):541–580, April 1989.
13. A. K. Murugavel and N. Ranganathan. Petri net modeling of gate and interconnect delays for power estimation. In *Proc. of the 39th conf. on Design automation*, pages 455–460. ACM Press, 2002.
14. G. Stitt and F. Vahid. Energy advantages of microprocessor platforms with on-chip configurable logic. *IEEE Design and Test of Computers*, 19(6):36–43, Nov/Dec 2002.
15. V. Tiwari, S. Malik, and A. Wolfe. Compilation techniques for low energy: An overview. *In Proc. of Symp. Low-Power Electronics, 1994.*, 1994.
16. V. Tiwari, S. Malik, and A. Wolfe. Power analysis of embedded software: A first step towards software power minimization. *IEEE Transactions on Very Large Scale Integration Systems*, 2(4):437–445, December 1994.
17. N. Vijaykrishnan, M. Kandemir, M. J. Irwin, H. S. Kim, and W. Ye. Energy-driven integrated hardware-software optimizations using simplepower. In *Proceedings of the 27th annual international symposium on Computer architecture*, pages 95–106. ACM Press, 2000.

Designing Low-Power Embedded Software
for Mass-Produced Microprocessor
by Using a Loop Table in On-Chip Memory

Rodrigo Possamai Bastos, Fernanda Lima Kastensmidt, and Ricardo Reis

Universidade Federal do Rio Grande do Sul (UFRGS), Instituto de Informática
PO Box 15064, 91501-970, Porto Alegre, RS, Brazil
{rpbastos,fglima,reis}@inf.ufrgs.br
http://www.inf.ufrgs.br/gme

Abstract. This work presents a methodology for low-power embedded software design to mass-produced microprocessors. It is based on identifying the frequently accessed loops from the application program and to build a loop table in already present on-chip memory of standard microcontroller. By using the loop table, the loops are accessed from the on-chip memory and not any longer from a power expensive memory bus. Results based on benchmarks show a considerable reduction in power, without penalties in area or performance.

1 Introduction

Embedded systems have become a key technological component for all kinds of complex or simple technical systems. Annually many microprocessors are mass-produced for embedded systems. A large part of these embedded microprocessors are integrated in microcontrollers, which have on-chip memory or at least have compatible alternatives in the family. Some commonly used commercial microcontrollers at the industry are the Motorola M68HC11, Intel 8051 and Microchip PIC.

Embedded systems inherently claim to low-power, low cost per unit, simplicity and immediate availability to market. Low-power is one of the most important issues because embedded systems generally present a large number of integrated features supplied by light batteries or solar power. For this reason, the design elaboration should be usually dedicated to obtain low-power.

In this work we present a methodology of low-power embedded software design for mass-produced microprocessor using a loop table in on-chip memory. The method takes advantages from the frequently executed small loops, which correspond to the majority of the program execution time and consequently to many accesses to power expensive bus of the program memory. These small code loops are copied to an on-chip memory with less capacitive bus wires, thus they are accessed from there during the program execution. Since many microcontrollers have already on-chip memory utilization, it is possible to reduce the power consumption by building of the loop table in this memory, which adds some extra bytes in the original program code, but it does not insert penalties in area, performance or instructions compatibility of system.

This work is organized as follows: in section 2 we describe the problem and some related works. Section 3 analyzes the power in interconnects. The section 4 explains

V. Paliouras, J. Vounckx, and D. Verkest (Eds.): PATMOS 2005, LNCS 3728, pp. 59–68, 2005.

the proposed methodology. Then section 5 shows experimental results and finally section 6 presents some conclusions.

2 Problem Description and Related Works

The microcontrolled embedded systems usually have an on-chip volatile memory just for data (stack, context or variables) and a non-volatile program memory accessed directly by the microprocessor, i.e., without another memory level. These kind of systems typically have simple architectures and generally exclude features like multipliers, floating-point units, caches, deep pipelines and branch predictors [1].

Targeting these mass-produced systems to achieve reductions in its power consumption, our work proposes a methodology for embedded software designers to develop their programs aiming at the complete utilization of the available on-chip memory resources. The method should allow the design reusability not only on logic-functional aspect, but also on cycles-timing. The area, performance and instructions compatibility of system should be maintained. This posture foments a faster time-to-market and contributes to support small companies unable to make alterations in the microcontroller architectures.

A previous work has proposed a modification in standard architecture of microprocessor to reduce power expensive accesses to program memory [1]. It introduces a tunable component (loop table) inside the microprocessor controller, which is used to store frequently executed loops. This small memory is similar to loop-cache [2], but it differs because there is a profiling (executed in hardware just one time) to identify the most frequent loop and also because the loop stay permanently in the loop table memory. The caches need of a tag memory and comparators resulting in extra costs. There are also approaches such as ASIPs (Application-Specific Instruction-set Processors) [3], in which new low-power optimized instructions are implemented in the processor and the embedded firmware is modified to use them, i.e., the embedded software will contain incompatible instructions for the commercial standard microprocessor.

Our work uses similar principles to [4], which the small loops most frequently executed are selected at the software level and statically allocated to an on-chip memory. The program code is changed to access the loops from the on-chip memory. Another work [5] also uses similar principles, however the loops are dynamically allocated. The works [4], [5] and [6] propose, for embedded systems with hierarchy memory (DRAM as the main memory), to replace the cache or add another specific on-chip memory. These approaches differ of target systems in our work.

Our approach does not make any modification in the standard of microcontroller architecture, no extra logic is necessary. An additional specific memory is not used, the loop table is allocated within idle region of already existent standard on-chip memory of the microcontroller. The method can be applied in systems formed from off-the-shelf microcontrollers. The performance can be preserved. The identification and isolation of the frequently executed code loops are worked by the embedded software designer and the amount of different code loops in the table is not restricted or tied to a constant size of a dedicated memory (built just for it). The size and amount of different code loops are fitted in the available space of the on-chip memory, which is managed by the software designer as well.

3 Interconnects Power Analysis

In the advent of very deep sub-micron technologies, the cost of on-chip and inter-chip communication has an increasing impact on total power consumption. It is mainly due to the relative scaling of cell capacitances and the increasing of inter-wire capacitance as a result of an enlarged wire aspect ratio, aimed at reducing the wire resistance [7]. Interconnects tend to dominate the chip power consumption, due to their large total capacitance [8][9].

When a microprocessor accesses memories or devices, this communication is performed through a switching activity in the interconnects among them. The long bus wires, which go to a special chip region or even to separate chips, have capacitances larger than a typical short on-chip wire. If a small part of the switching activity was transferred from bus wires of the largest capacitance to wires of the least, there would be a power reduction. This can be observed by considering the simple power estimator [7] defined for equation:

$$P_{switch} = \tfrac{1}{2}\alpha C_L V_{dd}^2 f ,$$

(1)

whose α is the switching activity in the bus wire, C_L is the wire's capacitance, V_{dd} is the voltage and f is the switching frequency.

For a microcontroller that has an on-chip data memory and a more peripheral program memory, an access to on-chip memory is power cheaper than the same access to peripheral memory, since the on-chip capacitances are lesser. Notice that for systems, which have a program memory also completely on-chip, the access to data memory will be still power cheaper if the interconnects to this memory also have capacitances lesser than interconnects to program memory.

4 Methodology

A great part of embedded systems is compound of mass-produced microcontrollers. These sort of microcontroller generally have some on-chip memory, when do not have, almost in totality, are models from a family that offers also on-chip memory alternatives, i.e., there are other models that are identical functionally but differ just for the on-chip memory presence.

The embedded software applications that are running in mass-produced microprocessor normally reside forever in the program memory and rarely suffer any restart. This kind of software usually spends much of the execution time in small code loops. An example of program that computes checksums spends 97% of its time in 36% of the program code; another one that computes the greatest common divisor of two numbers spends 77% of its time in 31% of the code [1].

These characteristics allow us designing a low-power software when there is any available small space in on-chip memory by using this idle region to allocate, just during the initialization, those frequently accessed small code loops and thereby doing the microprocessor accesses them from power cheaper bus wires.

4.1 The Low-Power Embedded Software Design

The low-power software design to a particular application is developed identifying the program regions most frequently accessed, inserting and changing some bytes for microprocessor to copy the code loops to on-chip memory and to execute them from there. The methodology is detailed in a small software example that uses the on-chip data memory to reduce the program memory accesses and thus the power.

- First step: Identifying and isolating the frequently executed loops. Fig. 1(a) shows the startup and main loop of the program. The dark gray lines are only for the original code, which are removed in the low-power code version. The light gray lines are used only for the low-power code. The Fig. 1(b) shows an identified block as frequently executed, it is isolated in a region accessed from the main loop for a jump instruction. The identification can be performed manually or by using a software profiling tool. In this example two blocks are identified as frequently executed: the "hex_nybble" and "outstr" loops;
- Second step: Building the loop table by copying the identified code loops from program memory to on-chip memory (data memory for this example). The loop table is built inside the data memory using standard instructions. The addresses of each identified code loop in the program memory are required as well as the addresses of the chosen regions in the data memory. The Fig. 1(c) shows an example of this building. Instruction "ldd" reads each 2 bytes from the identified code loops, which one ("outstr") starts at address 0xf823 on program memory. Then instruction "std" writes the 2 bytes to the chosen region of data memory, which starts at address 0x011C. This process is repeated for the next 2 bytes and goes till the whole identified code loops is allocated within data memory;
- Third step: Inserting standard instruction on software startup, which calls the loop table building, Fig. 1(a);
- Last step: Changing the address of jump from program memory to on-chip memory. See Fig. 1(a).

Observe that in the low-power optimized code, the loop table building is executed just once after program startup as a static allocation. Since system initializations are sporadic, its power consumption does not contribute significantly for overall power.

Some mass-produced microcontrollers, especially the models from the family M68HC11 [10], the number of cycles to access off-chip blocks is the same to access on-chip blocks. It is because in these architectures the access instructions to off-chip or on-chip blocks have the same number of cycles.

Thus the performance in the main loop execution can be preserved when compared to original code. However the identified frequently executed blocks must be already isolated as a function that is accessed for a jump and return instructions. For basic blocks not isolated, the additional cost in performance is only due to the number of cycles of two extra instructions inserted in each block (6 cycles of jump and 5 cycles of return for the M68HC11). Typically it should not a great problem for designers because functions leave the software more organized and portable.

The amount of the extra bytes inserted in the original code, which represents the loop table building, is small and depends on the amount of the identified code loops, besides on the amount of bytes code of the instructions used in the loop table building. Extra bytes code in the application should not be major problem nowadays due to the facility to find low cost and high storage capacity of non-volatile memories.

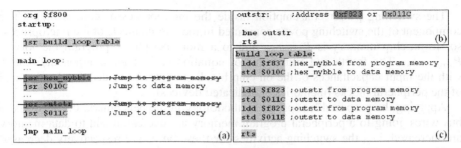

Fig. 1. The startup and main loop of the program example is presented in (a). A frequently executed loop in (b) and the building of the loop table in (c)

4.2 The Approach for Other Microcontroller Models

For those systems structured with a non-volatile program memory and a volatile data memory, which use a shared on-chip and peripheral program memory, there is a possibility to allocate the frequently executed loops in the on-chip program memory and the remainder code in the peripheral program memory. This certainly will result in power consumption decrease too and will not need extra code to build the loop table.

Our methodology is especially useful to CPUs that consider all peripheral, on-chip devices, input/output (I/O) and memory locations as addresses in the memory map (address bus). However, to attend Harvard architectures, which data and program memories are located separately, two options are also feasible to apply our approach. It is possible either to use a microcontroller model with a shared on-chip and peripheral program memory or to add some hardware mechanisms that allow an execution jump to data memory. Many commercial Harvard microcontrollers have already this hardware feature integrated in its standard model. It is the case of ROM-less Microchip PIC18C601/801 [11]. This resource is amply used in embedded systems to perform firmware updates.

5   Benchmarks Results

We performed experiments using a VHDL description of CPU M68HC11 [10] based on [12]. The microprocessor was synthesized in standard cells of the AMS 0.35 μm CMOS technology [13]. All experiments were developed using Synopsys tools [14].

The effective functional verification of our methodology was performed executing benchmarks in the original and in the low-power versions supported for a testbench, which models the embedded system formed for the microprocessor core, an on-chip data memory (volatile) and a more peripheral program memory (non-volatile). The simulation is done to obtain a count of the switching on each wire, after that we compared the power consumption, estimated for the access bus wires going to memories and devices, in the low-power and original situations.

5.1   Low-Power vs. Original Benchmark

The equation (1) can be reorganized in the following way:

$$P_{switch} = P_{inside} + P_{on-chip,bus} + P_{peripheral,bus} + P_{on-chip,input} + P_{peripheral,input} \ . \tag{2}$$

The switching power consumption inside the microprocessor core is P_{inside}. The component of the switching power consumed in the interconnects of the microprocessor to on-chip memory is $P_{on\text{-}chip,bus}$ and to a more peripheral program memory is $P_{peripheral,bus}$. The fourth and fifth terms of equation (2) are the components related with the input capacitances of the on-chip blocks (memories or dedicated units) and of the peripheral blocks (memories or dedicated devices).

Applying the proposed low-power methodology in the benchmark, the accesses to bus wires going to a peripheral program memory are decreased and to data memory are increased, i.e., the switching activities for these buses are respectively decreased and increased. Thus $P_{peripheral,bus}$ and $P_{peripheral,input}$ are decreased, while the $P_{on\text{-}chip,bus}$ and $P_{on\text{-}chip,input}$ are increased, nevertheless as the capacitances are greater for the peripheral, the total power consumption reduces.

For the embedded system experimented, the data memory was allocated in low addresses and the program memory in high addresses. When there are two successive accesses to data memory, the amount of switching in the address bus will be lesser than when there is an access to program memory followed by another access to data memory, since the most significant bits will not change. Our methodology proposes to access codes from data memory, which ones, in the original benchmarks, are accessed from program memory. Therefore there will be an overall switching activity reduction when these codes have software instructions (like stores, loads or jumps) that access the stack or registers located in the data memory.

The power consumption comparison between original and low-power benchmarks was performed using the parts of the power which represents the relevant difference between both benchmarks. The power analyzed is the switching power in the accesses up to on-chip and peripheral blocks. For simplifying, P_{inside} was not consider, however it is also reduced for the low-power benchmark, due to only the overall switching activity reduction explained above.

Summarizing the percentage of power reduction in the subsequent equation:

$$\frac{\Delta\%P_{accesses}}{100} = \frac{P_{low-power}}{P_{original}} - 1 = \frac{P_{low-power,on-chip} + P_{low-power,peripheral}}{P_{original,on-chip} + P_{original,peripheral}} - 1 \;. \tag{3}$$

It is the relation between the powers of the original benchmark $P_{original}$ and of the low-power benchmark $P_{low\text{-}power}$. Considering the on-chip components of these powers equal to sum of $P_{on\text{-}chip,bus}$ and $P_{on\text{-}chip,input}$ from equation (2) and the peripheral components equal to sum of $P_{peripheral,bus}$ and $P_{peripheral,input}$ also from equation (2).

The results strongly depend on the capacitances up to on-chip and up to peripheral blocks, which depend on the utilized technology. For evaluating the results to different technologies, the equation (3) can be rewritten based on definition of P_{switch} from equation (1) and on consideration of a numeric multiple (M) for capacitances:

$$\frac{\Delta\%P_{accesses}}{100} = \frac{M_{on-chip}C_R\alpha_{low-power,on-chip} + M_{peripheral}C_R\alpha_{low-power,peripheral}}{M_{on-chip}C_R\alpha_{original,on-chip} + M_{peripheral}C_R\alpha_{original,peripheral}} - 1 \;. \tag{4}$$

Observe that the expression $(\frac{1}{2}) \cdot f \cdot V_{dd}^2$ is canceled, because the voltage and the main loop frequency in the low-power benchmark are not changed. The capacitances are replaced for associations of different numeric multiples to a reference capacitance C_R, which represents a capacitance value for a typical short on-chip wire of a technology. Note that C_R can be also canceled and then the equation (4) only depends on the multiples and on the amounts of switching. As the switching activity (α) just depends on the software benchmark characteristics, the multiples can be factors to evaluate the results for different technologies.

The switching activities of an original benchmark $\alpha_{original,on\text{-}chip}$ and $\alpha_{original,peripheral}$ are constant elements in the equation (4). The overall switching activity of a low-power benchmark ($\alpha_{low\text{-}power,on\text{-}chip} + \alpha_{low\text{-}power,peripheral}$) must be lesser than of its original. The main effect of our methodology is the reduction of switching activity up to peripheral blocks ($\alpha_{low\text{-}power,peripheral} < \alpha_{original,peripheral}$) and the amplification up to on-chip blocks ($\alpha_{low\text{-}power,on\text{-}chip} > \alpha_{original,on\text{-}chip}$). The overall power reduction depends on how much $M_{peripheral}$ is greater than $M_{on\text{-}chip}$. Maintaining constant $M_{on\text{-}chip}$ and varying $M_{peripheral}$, the equation (4) presents an asymptote, when limited to infinite, defined for (($\alpha_{low\text{-}power,peripheral} / \alpha_{original,peripheral}$) $- 1$). It illustrates that the amount of switching up to peripheral blocks is a factor which limits the range of power reduction in the hypothesis of $M_{peripheral}$ to be so greater than $M_{on\text{-}chip}$.

5.2 Experimental Results

Benchmarks in assembly language of M68HC11 [10] were executed. One is a program of 98 bytes that converts 4 hexadecimal codes to their ASCII characters and outputs them to a terminal (Hexout benchmark). Another benchmark of 1016 bytes is a program for automatic motor control, which generates a PWM signal (Pulse Width Modulation) of constant frequency and variable duty cycle. It also outputs to a terminal the current speed and direction (Amotorcod benchmark). The Fig. 2 presents, for both benchmarks, the percentages of the total program time spend executing the code blocks most frequently accessed. Observe that in Fig. 2(a) the Hexout benchmark spends 61.79% (33.50% + 28.29%) of its time executing 35.71% (19.39% + 16.32%) of the program code (code blocks: "outstr" and "hex_nybble"). The Amotorcod benchmark in Fig. 2(b) spends 42.93% (22.67% + 20.26%) of its time in 6.99% (3.64% + 3.35%) of the program code (code blocks: "WAIT" and "OUTSCI").

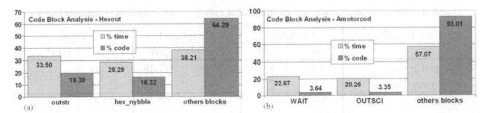

Fig. 2. Characteristics of Hexout benchmark are presented in (a) and of Amotorcod in (b)

Applying the low-power methodology in these benchmarks, those detailed code blocks are copied to on-chip data memory at the initialization and always executed from there afterwards. The size of the program code is increased in 66 bytes for both benchmarks due to the building of the loop tables. The space necessary in the on-chip data memory to allocate the code blocks was 35 bytes for Hexout benchmark (35.71% of 98) and 71 bytes for Amotorcod (6.99% of 1016).

Table 1 shows the switching activity (α) and the power in µW of the power components from equation (3). It was evaluated with the capacitance multiples $M_{on\text{-}chip} =$

10 and $M_{peripheral} = 100$, based on reference [1], and at main loop period of 130.869 µs for Hexout benchmark and of 135.069 ms for Amotorcod. Notice that since the accesses to bus wires going to on-chip data memory are increased in the low-power benchmarks, the switching activity of this bus is also increased and thus power related is grown. Nonetheless the power related to peripheral accesses, which represents the majority of the expended power (due the greater capacitances), decreases. Observe that the Amotorcod benchmark, which uses 42.93% (Fig. 2(b)) of its time executing program codes from on-chip data memory, gives an overall power reduction in the accesses of 37.47% (Table 1), whereas the Hexout, which uses 61.79% (Fig. 2(a)) of its time executing codes from on-chip memory, gives a power reduction of 57.62% (Table 1). These results are because the reductions of the switching activity (α) up to peripheral blocks (2487626 to 1462250 or –41.22% for Amotorcod and 2136 to 748 or –64.98% for Hexout) are similar to reductions of the clock cycles executed from them (for Amotorcod 42.93% from Fig. 2(b) and for Hexout 61.79% from Fig. 2(b)).

Based on equation (4), the results can be analyzed for other technological approaches. The Fig. 3(a) shows that the overall power reduction achieves its maximum when the capacitances up to peripheral blocks are about 60 times greater than capacitances up to on-chip blocks. It characterizes asymptotes in this curve defined for $((748 / 2136) - 1) \cdot 100 = -64.98\%$ for the Hexout benchmark and for $((1462250 / 2487626) - 1) \cdot 100 = -41.22\%$ for the Amotorcod. These asymptotes can be seen as the power reduction in the accesses up to peripheral blocks, as if the contribution of the power in the accesses up to on-chip blocks was negligible. The Fig. 3(b) illustrates the behavior of the power reduction when the capacitances up to on-chip blocks increase in ratio to capacitances up to peripheral blocks. Note that for ratio equal to 1, the capacitances up to on-chip blocks have the same value of the capacitances up to peripheral blocks and the power reduction associated is only due to the overall switching activity reduction (430 + 2136 = 2566 to 1755 + 748 = 2503 for Hexout and 500233 + 2487626 = 2987859 to 1245731 + 1462250 = 2707981 for Amotorcod), as explained in section 5.1.

Table 1. The switching power results in the accesses up to on-chip and peripheral blocks

Benchmarks	On-Chip		Peripheral		Total	
	α	µW	α	µW	$P_{accesses}$	$\Delta\%P_{accesses}$
Original-Hexout	430	1.92	2136	95.48	97.40	- 57.62%
Low-Power-Hexout	1755	7.84	748	33.44	41.28	
Original-Amotorcod	500233	2.17	2487626	107.74	109.91	- 37.47%
Low-Power-Amotorcod	1245731	5.40	1462250	63.33	68.73	

Nowadays, as a result of CMOS technology scaling, the leakage power has become a significant portion of the total power consumption [15]. In addition, one can say that leakage power can constitute as much as 50 % of total power consumption. The current results are based on the switching power in the accesses up to on-chip and peripheral blocks. The Fig. 3(c) presents the behavior of the total power reduction (for a system of $M_{on-chip} = 10$ and $M_{peripheral} = 100$ executing the benchmarks) with the influence of any other power components like leakage power, short circuit power, P_{inside}, power consumption during the initialization, contributions of a more sophisticated wire model as [16][7] or errors of the power estimator. If the other power components represent for example 60% of the total power consumption, the overall power reduction would be 23.05% for Hexout benchmark and 14.99% for Amotorcod.

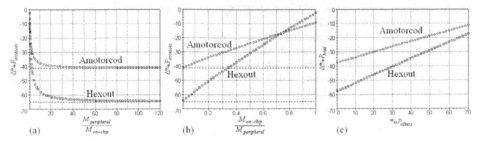

Fig. 3. Power reduction behavior for different technological approaches

6 Conclusions

This work presented a methodology for embedded software design towards reduction on power consumption of microprocessor. Some emphasized items for software development allow the planning of low-power design dedicated to target application. The implementation is simple. The memory resources already existent on-chip to standard microcontrollers are used as a loop table. It speeds up the time-to-market and reduces the cost of design, essentially for small companies.

The results indicated a significant power reduction applying our methodology. The performance to the main loop of program can be preserved. The hardware standard of system is not altered, i.e., there is no increase in area. As future work, we plan to run different sort of benchmarks to find out more generic results about the applicability of the proposed low-power software design methodology.

References

1. Vahid, F., Gordon-Ross, A.: A Self-Optimizing Embedded Microprocessor using a Loop Table for Low Power. ISLPED, ACM, pp. 219-224, 2001.
2. Lee, L. H., Moyer, B., Arends, J.: Instruction Fetch Energy Reduction Using Loop Caches for Embedded Applications with Small Tight Loops. ISLPED, ACM, pp. 267-269, 1999.
3. Kucukcakar, K.: An ASIP Design Methodology for Embedded Systems. CODES, ACM, pp. 17-21, 1999.
4. Steinke, S., Wehmeyer, L., Lee, B., Marwedel, P.: Assigning Program and Data Objects to Scratchpad for Energy Reduction. DATE, IEEE, 2002.
5. Steinke, S., et al.: Reducing Energy Consumption by Dynamic Copying of Instructions onto Onchip Memory. ISSS, ACM, pp. 213-218, 2002.
6. Panda, P. R., Dutt, N. D., Nicolau, A.: Efficient Utilization of Scratch-Pad Memory in Embedded Processor Application. ED&TC, IEEE, pp. 7-11, 1997.
7. Caputa, P., et al.: An Extended Transition Energy Cost Model for Buses in Deep Submicron Technologies. PATMOS, LNCS 3254, Springer, pp. 849-858, 2004.
8. Liu, D., Svensson, C.: Power Consumption Estimation in CMOS VLSI Chips. J. of Solid-State Circuits, Vol. 29, IEEE, pp. 663-670, 1994.
9. Chandra, G., Kapur, P., Saraswat, K. C.: Scaling Trends for the On Chip Power Dissipation. International Interconnect Technology Conference, IEEE, pp. 154-156, 2002.

10. Freescale Semiconductor, Inc.: Reference Manual and Guide M68HC11. 2002, www.freescale.com.
11. Microchip Technology, Inc.: Data Sheet PIC18C601/801. 2001, www.microchip.com.
12. Thibault, S.: GM HC11 CPU Core. Green Mountain Computing Systems, 2000.
13. Austriamicrosystem, AMS: 0.35μm CMOS Digital Standard Cell Databook. 2003, www.austriamicrosystems.com.
14. Synopsys, Inc.: Tool Manuals. 2004, www.synopsys.com.
15. Kim, N. S., et al.: Leakage Current: Moore's Law Meets Static Power. Computer Society, Vol. 36, Issue 12, IEEE, pp. 68-75, 2003.
16. Murgan, T., et al.: On Timing and Power Consumption in Inductively Coupled On-chip Interconnects. PATMOS, LNCS 3254, Springer, pp. 819-828, 2004.

Energy Characterization of Garbage Collectors for Dynamic Applications on Embedded Systems

Jose M. Velasco[1], David Atienza[1,2,*], Katzalin Olcoz[1], Francky Catthoor[2,*], Francisco Tirado[1], and J.M. Mendias[1]

[1] DACYA/U.C.M., Avenida Complutense s/n, 28040 Madrid, Spain
[2] IMEC vzw, Kapeldreef 75, 3000 Leuven, Belgium

Abstract. Modern embedded devices (e.g. PDAs, mobile phones) are now incorporating Java as a very popular implementation language in their designs. These new embedded systems include multiple complex applications (e.g. 3D rendering applications) that are dynamically launched by the user, which can produce very energy-hungry systems if they are not properly designed. Therefore, it is crucial for new embedded devices a better understanding of the interactions between the applications and the garbage collectors to reduce their energy consumption and to extend their battery life. In this paper we present a complete study, from an energy viewpoint, of the different state-of-the-art garbage collectors mechanisms (e.g. mark-and-sweep, generational garbage collectors) for embedded systems. Our results show that traditional solutions of garbage collectors for Java-based systems do not seem to produce the lowest energy consumption solutions.

1 Introduction

Currently Java is becoming one of the most popular choices for embedded/portable environments. In fact, it is suggested that Java-based systems as mobile phones, PDAs, etc. will enlarge their current market from around 150 million devices in 2000 to more than 700 millions at the end of 2005 [20]. One of the main reasons for this large growth is that the use of Java in embedded systems allows developers to design new portable services that can effectively run in almost all the available platforms without the use of special cross-compilers to port them to different platforms, as happens with other languages (e.g. C or C++). Nevertheless, the abstraction provided by Java creates an additional major problem, which is the performance degradation of the system due to the inclusion of an additional component, i.e. the Java Virtual Machine or JVM, to interpret the native Java code and to execute it onto the present architecture.

In recent years, a very important research effort has been done for Java-based systems to improve performance up to the level required in new embedded devices. This research has been mainly performed in the JVM. More specifically, it has focused on optimizing the execution time spent in the automatic object reclamation or Garbage Collector (GC) subsystem, which is one of the main sources of overall performance degradation of the system.

---

* This work is partially supported by the Spanish Government Research Grant TIC2002/0750 and E.C. Marie Curie Fellowship contract HPMT-CT-2000-00031. + Also professor at ESAT/K.U.Leuven-Belgium

However, the increasing need for efficient systems (i.e. low-power) limits very significantly the use of Java for new embedded devices since GCs are usually efficient enough in performance, but very costly in energy and power. Thus, efficient (from the energy viewpoint) automatic DM reclamation mechanisms and methodologies to define them have to be proposed for a complete integration of Java in the design of forthcoming very low-power embedded systems.

In this paper we present a detailed study of the energy consumed in current state-of-the-art GCs (i.e. generational GCs, mark-and-sweep, etc.), which is the first step to design custom energy-aware GCs for actual dynamic applications (e.g. multimedia) of embedded devices. The remainder of this paper is organized in the following way. In Section 2 we summarize some related work. In Section 3 we describe the experimental setup used to investigate the energy consumption features of GCs and the representative state-of-the-art GCs used in our study. In Section 4, we introduce our case studies and present the experimental results attained. Finally, in Section 5 we draw our conclusions.

2   Related Work

Nowadays a very wide variety of well-known techniques for uniprocessor GCs (e.g. reference counting, mark-sweep collection, copying garbage collector) are available in a general-purpose context within the software community [ ]. Recent research on GC policies has mainly focused on performance [ ]. Our work extends their research to the context of energy consumption.

Eeckout et al. [ ] investigate the microarchitectural implications of several virtual machines including Jikes. In this work, each virtual machine has a different GC, so their results are not consistent related to memory management. Similarly, Sweeney et al. [ ] conclude that GC increases the cache misses for both instruction and data. However, they do not analyze the impact of different strategies in the total energy consumed in the system as we do.

Chen et al. [ ] focus in reducing the static energy consumption in a multibanked main memory by tuning the collection frequency of a Mark&Sweep-based collector that shuts off memory banks that do not hold live data. The reduction of leakage approach is parallel to ours and can be used complementary.

Finally, a large body of research on memory optimizations and techniques exists for static data in embedded systems (see e.g. [,] for good tutorial overviews). All these techniques are complementary to our work and are applicable in the part of the Java code that accesses static data in the dynamic applications under study.

3   Experimental Setup

In this section we first describe the whole simulation environment used to obtain detailed memory access profiling of the JVM (for both the application and the collector phase). It is based on cycle-accurate simulations of the original Java code of the applications under study. Then we summarize the representative set of GCs used in our experiments. Finally we introduce the sets of applications selected as case studies.

3.1 Simulation Environment

Our simulation environment is depicted in Figure 1 and consists of three different parts. First, the detailed simulations of our case studies have been obtained after modifying significantly the code of Jikes RVM (Research Virtual Machine) from the Watson Research Center of IBM [ ]. Jikes RVM is a Java virtual machine designed for research. It is written in Java and the components of the virtual machine are Java objects [ ], which are designed as a modular system to enable the possibility of modifying extensively the source code to implement different GC strategies and custom GCs. We have used version 2.3.2 along with the recently developed memory manager JMTk (Java Memory management Toolkit) [ ].

The main modifications in Jikes have been performed to integrate in it the Dynamic SimpleScalar framework (DSS) [ ], which is an upgrade of the well known SimpleScalar simulator [ ]. DSS enables a complete Java virtual machine simulation by supporting dynamic compilation, threads scheduling and garbage collection. It is based on a PowerPC ISA and has a fully functional and accurate cache simulator. We have included a cross-compiler [ ] to be able to run our whole Jikes-DSS system onto the Pentium-based platform available for our experiments instead of the PowerPC traditionally used for DSS. In our experiments, the memory architecture consists of three different levels: an on-chip SRAM L1 memory (with separated D-cache/I-cache), an on-chip unified SRAM L2 memory of 256K and an off-chip SDRAM main memory. The L1 size is 32K and the L1 associativity has been tested between 1-way and 32-ways. The block size is 32 bytes and the cache uses and LRU blocks replacement policy.

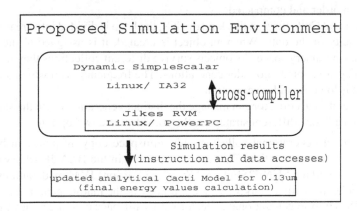

Fig. 1. Graphical overview of our whole simulation environment

Finally, after the simulation in our Jikes-DSS environment, energy figures are calculated with an updated version of the CACTI model [ ], which is a complete energy/delay/area model, scalable to different technology nodes, for embedded SRAMs. For all our results shown in Section 4, we use the .13μm technology node. In our energy results for the SDRAM main memory we also include static power values (e.g. precharging of a bank, page misses, etc.) that have been derived from a power estimation tool of Micron 16Mb mobile SDRAM [ ].

3.2   Studied State-of-the-Art Garbage Collectors

Next, we describe the main differences among the studied GCs to show how they can cover the whole state-of-the-art spectrum of choices in current GCs. We refer to [ ] for a complete overview of garbage collection techniques and for further details of the specific implementation used in our experiments with Jikes [ ].

In our study all the collectors fall into the category of GCs known as tracing *stop-the-world* [ ]. This implies that the running application (more frequently known as mutator in the GCs context) is paused during garbage collection to avoid inconsistencies in the references to dynamic memory in the system. To distinguish the live objects among the garbage, the tracing strategy relies on determining which objects are not pointed to by any living object. To this end, it needs to traverse the whole relationship graph through the memory recursively. The way of reclaiming the garbage produces the different tracing collectors of this paper. Inside this class we study the following representative GCs for embedded devices: mbdlma@hotmail.com - Mark-and-sweep (or MS): the allocation policy uses a set of different block-size *free-lists*. This produces both internal and external fragmentation. Once the tracing phase has marked the living data, the collector needs to *sweep* all the available memory to find unreacheable objects and reorganize the free-lists. The sweeping of the whole heap is very costly and to avoid it in the Jikes virtual machine, the sweep-phase is implemented as *lazy*. This means that the sweep is delayed up to the allocation phase. This is a classical collector implemented in several Java virtual machines as Kaffe [ ], JamVM [ ] or Kissme [ ].

- Copying collector (SemiSpace or SS): it divides the available space of memory in two halves, called semispaces. The objects that are found alive are copied in the other semispace in order and compacted.

Generational Collectors: in this kind of GCs, the heap is divided into areas according to the antiquity of the data. When an object is created, it is assigned to the youngest generation, the nursery space. As objects survive different collections they mature, that is to say, they are copied into older generations. The frequency with which a collection takes place is lower in older generations.

The generational collector can manage the distinct generations with the same policy or assign to each one different strategies. We consider here two options:

– GenCopy: a generational collector with semispace copying policy in both nursery and mature generation. This collector is used in the BEA JRockit virtual Machine [ ] and the SUN J2SE(Java 2 Standard Edition) JVM by default uses a very close collector with a Mark&Compact strategy in the mature generation.
– GenMS: a hybrid generational collector with semispace copying policy in the nursery and mark-and-sweep strategy in the mature generation. The Chives Virtual Machine [ ] uses a hybrid generational collector with three generations

Copying collector with Mark-and-Sweep (or CopyMS in our experiments): It is the non-generational version of the previous one. Objects that survive a collection are managed with a mark-and-sweep strategy and therefore they are not moved any more.

In Jikes, these five collectors manage objects bigger than a certain threshold (by default 16K) in a special area. Jikes also reserves space for immortal data and meta data (where the references among generations are recorded, usually known as the *remembered set*). These special memory zones are also studied in our experimental results.

Finally, it is important to mention that, even though we study all the previous GCs with the purpose of covering the whole range of options for automatic memory management, real-life Java-based embedded systems typically employ MS or SS since they are initially the GCs that possess less complex algorithms to implement; Thus, theoretically putting less pressure in the processing power of the final embedded system and achieving good overall results (e.g. performance of memory hierarchy, L1 cache behaviour, etc.)

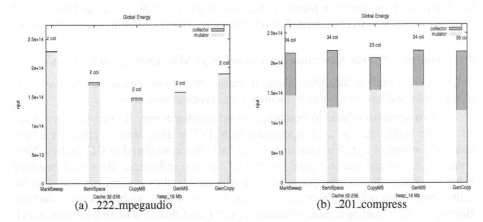

Fig. 2. Energy figures for L1-32K, direct mapped and LRU replacement policy

3.3 Case Studies

We have applied the proposed experimental setup to the GCs presented in the previous subsection running the most representative benchmarks in the suite SPECjvm98 [ ] for new embedded devices. These benchmarks could be launched as dynamic services and extensively use dynamic data allocation. The used set of applications is the following:

_222_mpegaudio: it is an MPEG audio decoder. It allocates 8 MB + 2 MB in the LOS.

_201_compress: it compresses and then uncompresses a large file. It mainly allocates objects in the LOS (18 MB) while it uses only 4MB of small objects.

_202_Jess: it is the Java version of an expert shell system using NASA CLIPS. It is compound fundamentally of structures of sentences 'if-then'. It allocates 48 MB (plus 4 MB in the LOS) and most objects are short-lived.

_205_Raytrace: raytraces a scene into a memory buffer. It allocates a lot of small data (155 MB + 1 MB in the LOS) with different lifetimes.

_213_javac: it is the java compiler. It has the highest program complexity and its data is a mixture of short and quasi-inmortal objects (35MB + 3 MB in the LOS).

The suite SPECjvm98 offers three input sets(referred as s1, s10, s100), with different data sizes. In this study we have used the medium input data size, represented as s10, as we think it is more representative of the actual input sizes of multimedia applications in embedded systems.

4   Experimental Results

This section shows the application of the previously explained experimental setup (see Section 3 for more details) to perform a complete study of automatic dynamic memory management mechanisms for embedded systems according to key metrics in embedded systems (i.e. energy, power and performance of the memory subsystem). To this end, in this section we first analyze the dynamic allocation behaviour of the different SPEC benchmarks and categorize them in dynamic behaviour scenarios. Then, we discuss how the different GC strategies respond to each scenario. Finally, we study how several key features of the memory hierarchy (i.e. associativity of the cache and the size of main memory) can affect each GC behavior from an energy consumption point of view.

4.1   Analysis of Data Allocation Scenarios for JVM in Embedded Systems

After a careful study of the different data allocation behaviours encountered in the SPEC benchmarks, we have been able to identify three types of scenarios:

1. The first scenario related to benchmarks that employ a very limited memory space within the heap, such as _222_mpegaudio in SPEC. In fact, since it allocates a very reduced amount of dynamic memory, it is usually not considered in GC studies. Nevertheless, as we can see in figure 2(a) the GC choice influences the virtual machine behavior during the mutator phase (due to allocation policy complexity and data locality) and it can achieve, if correctly selected, up to a 40% global energy reduction. Hence, we are including it as representative example of this kind of benchmarks, which in reality are not so infrequent since many traditional embedded applications (e.g. MP3 encoders/decoders, MPEG2 applications, etc.) use static data and only few data structures demand dynamic allocations.

2. The second scenario has been identified in benchmarks that mostly allocate large objects, such as _201_compress. This benchmark is the Java version of 129.compress benchmark from the SPEC CPU95 suite and it is an example of Java programming in a C-like style. The fraction of heap accesses to the TIB (Type Information Block) table is very small relative to the fraction of accesses to array elements (the Lemper-Ziv's dictionaries). This means that the application spends more time accessing static memory via C-like functions rather than generating and accessing dynamic memory using object-oriented methods as in native Java applications. Hence, similarly to the previous type of scenario, it is usually considered an exception and not included in GC studies. Nonetheless, we have included it in our study as we have verified that such kind of non-truly object-oriented Java program with dynamic behaviour is quite common in embedded systems. Moreover, we consider that compressing algorithms are very frequently present in embedded environments and the big amount of large objects allocation required in such systems demands a deep understanding.

3. The third possible scenario has been observed in benchmarks with a medium to high amount of allocated data, and with different life timespan, for instance, in SPEC: _202_Jess, _205_Raytrace and _213_javac. These benchmarks are the ones most frequently considered in performance studies of GCs and in this paper we present complementary results for them taking into consideration energy figures apart from performance values.

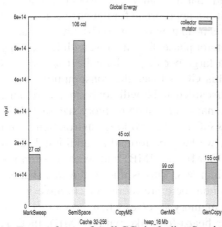

(a) Energy figures for all GCs including Semis-pace collector.

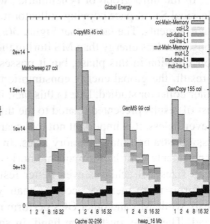

(b) Energy breakdown figures for different GCs. Associativity varies from 1-way to 32-ways.

Fig. 3. Third Scenario energy figures for all GCs with L1-32K

4.2 Comparison of GC Algorithms

Our results indicate that each GC shows a different behaviour regarding which scenario the application under study belongs to. In the first scenario, the percentage of energy wasted in the collection phase is not significative in comparison with the energy spent during the mutator phase. However, as Figure 2(a) depicts, the global JVM energy consumption varies significantly depending on the chosen GC. First, we can observe that it is very important the effect of the allocation policy associated with the GC. In fact, Figure 2(a) indicates that the simplest GC of all tested copying collectors, i.e. SS (see Section 3.2 for more details) attains better energy results than MS, which uses free-lists and lazy deletion. Second, our results indicate that since the number of accesses and L1 cache miss rates (see Table 1) of all copying-based collectors are very similar, the main factor that differentiates their final energy figures is the L2 miss rate. Thus, outlining that hybrid collectors that minimize this factor (i.e. CopyMS and GenMS) produce the best results for energy consumption. In fact, the best choice for this scenario (i.e. CopyMS) achieves an overall energy reduction of 55% compared to the classical Mark&Sweep collector employed in many embedded systems.

In the second scenario, the percentage of energy wasted in the collection phase varies from a 25% to a 50% of the final JVM energy consumption. This indicates that current GCs for embedded systems are not really tuned for this type of embedded application. Then, considering the special characteristics of _201_compress (see Section 4.1), we can observe that, contrarily to the previous scenario, the energy behavior during the mutator phase for the pure copying GCs is better than for the hybrid ones. However, the better space management that hybrids collectors exhibit produce a lesser number of collections; Thus, the energy spent in the collection step is up to 40% less than pure copying GCs. All in all, hybrid collectors attain a final energy reduction of 5% compared to copying GCs, and again CopyMS is the best election.

In the third group of benchmarks, we find that the handicaps associated with the classical SemiSpace copying collector make it not appropriate for memory constrained environments. The fact is that Figure 3(a) (for a heap memory of 16MB) shows that SS consumes less energy than MS during the mutator phase, the one next with less energy consumption in this phase, but it possesses a large penalty in the collection step. As a result, the global energy consumption of this GC is twice the consumption of any other collector studied. Due to this fact, it is discarded and it will not be included in the rest of results presented related to the third scenario for a main memory size of 16MB. Nevertheless, it is important not to discard this GC in other working environments with less constrained main memory sizes. In fact, we have performed an additional set of experiments with larger main memories (i.e. 32MB or 64MB) and, with these larger heap sizes, SS achieves even better results than MS. This is due to the fact that SS performs more memory accesses than MS when there are more objects alive in the heap, while MS and other GCs perform more memory accesses when more objects are dead. Therefore, on the one hand, in smaller heaps (e.g. like with 16 MB), there is not enough time for the objects to die and SS is more expensive than MS in energy consumption. On the other hand, in large heaps where more objects can die between memory collections, SS is favoured compared to MS and other GCs; thus, it consumes less energy.

In addition, we can observe that the best energy consumption results for this scenario are achieved by different variations of the two generational GCs studied (i.e. GenMS and GenCopy in Figure 3(a)). This occurs because for the input data size used (s10) the amount of metadata produced by the write barriers is insignificant. Thus, no performance penalties appear in generational GCs and they attain similar results to the hybrid solutions that were the best options in the other scenarios (e.g. CopyMS) during the mutator phase. Next to this, the minor collections (only in parts of the heap, see Subsection 3.2 for more details) in the generational strategy interfere less in the cache

Table 1. Summary of cache miss rates in all benchmarks with L1-32K and direct mapped

		mut-ins-L1 %	mut-data-L1 %	mut-L2 %	col-ins-L1 %	col-data-L1 %	col-L2 %
1° scenario	MS	9.9	6.6	92.5	13.3	7.3	30.8
	SS	9.9	6.2	67.3	13.3	7.8	41.1
	CopyMS	10.0	6.2	54.8	14.0	5.7	36.4
	GenMS	10.1	6.3	58.9	13.4	6.7	53.6
	GenCopy	10.0	6.3	74.0	14.0	6.8	51.2
2° scenario	MS	8.7	5.1	32.6	13.2	7.3	30.3
	SS	7.7	7.4	22.0	13.3	7.8	42.7
	CopyMS	8.6	7.0	30.5	14.0	5.6	37.1
	GenMS	8.4	7.1	33.6	13.6	5.9	45.1
	GenCopy	7.5	5.0	24.4	14.1	6.5	45.1
3° scenario	MS	13.0	8.1	56.2	13.3	7.4	31.9
	SS	13.4	8.5	52.5	13.4	7.8	44.2
	CopyMS	13.1	8.8	53.9	14.0	5.7	38.4
	GenMS	12.7	8.1	50.3	12.8	6.0	42.5
	GenCopy	12.9	7.9	49.8	13.6	6.5	43.3

behavior than the full heap collections of non-generational ones. Furthermore, GenMS does not need to reserve space for copying surviving objects in the mature generation. This produces a lesser number of both minor and mayor collections and it is the eventual reason why GenMS obtains slightly better results than GenCopy.

Finally, to test if there are important effects of the eventual memory hierarchy in the GCs, we have performed a final set of experiments varying the associativity of the L1 cache from 1-way to 32-ways and with a main memory size of 16 MB. The results accomplished are depicted in Figure 3(b), which indicates that the energy breakdown figures of the different GCs (without SS) for this third scenario, distinguishing the energy consumption during the mutator phase (mut) and collector phase (col). As these results outline, depending on the memory hierarchy (in this case simply modifying the L1 associativity), the influence of the GC algorithm choice can vary significantly. In fact, in this case the energy consumption differences between GCs can vary up to 50% in the collection phase and and its indirect effect in the mutator phase can reach an additional 20% variation in energy consumption. Hence, the global variation in energy consumption can be up to 40%. Also, this study indicates that the L1 miss rates (see Table 1) are very similar for all GCs. Finally, we can observe that Mark&Sweep has the lowest L2 miss rates. Besides, with an associativity of 8-ways, the reduction in the number of misses (Figure 3(a)) is translated in a drastic reduction of the energy spent in both main memory and L2 cache. This reduction produces a final global energy very close to the best results of the generational GCs, figure 3(b). Therefore, the Mark&Sweep handicaps can be diminished with a proper cache paremeters selection. In summary, the GC algorithm choice is a key factor for optimizing the final energy consumption of the JVM, but it should take into account the memory hierarchy to be tuned conveniently.

5   Conclusions

New embedded devices can presently execute complex dynamic applications (e.g. multimedia). These new complex applications are now including Java as one of the most popular implementation languages in their designs due to its high portability. Hence, new Java Virtual Machines (JVM) should be designed trying to minimize their energy consumption while respecting the soft real-time requirements of these embedded systems. In this paper we have presented a complete study from an energy viewpoint of the different state-of-the-art GCs mechanisms used in current JVM for embedded systems. We have shown how the GCs traditionally used in embedded devices (i.e. MS or SS) for Java-based systems do not achieve the best energy results, which are obtained with variations of generational GCs. In addition, the specific memory hierarchy selected can significantly vary the overall results for each GC scheme, thus showing the need of further research in this aspect of Java-based embedded systems.

References

1. Chives virtual machine with a jit compiler, 2005. http://chives.sunsite.dk/.
2. Kaffe is a clean room implementation of the java virtual machine, 2005.
   http://www.kaffe.org/.

3. V. Agarwal, S. Keckler, and D. Burger. The effect of technology scaling on microarchitectural structures. Technical report, Technical Report TR2000-02, University of Texas at Austin, USA, 2002.
4. T. Austin. Simple scalar llc, 2004. http://simplescalar.com/.
5. L. Benini and G. De Micheli. System level power optimization techniques and tools. In *ACM Transactions on Design Automation for Embedded Systems (TODAES)*, April 2000.
6. S. Blackburn, P. Cheng, and K. McKinley. Myths and reality: The performance impact of garbage collection. In *Proceedings of International Conference on Measurement and Modeling of Computer Systems*, SIGMETRICS, June 2004.
7. G. Chen, R. Shetty, M. Kandemir, N. Vijaykrishnan, M. J. Irwin, and M. Wolczko. Tuning garbage collection for reducing memory system energy in an embedded java environment. *ACM Transactions on Embedded Computing Systems (TECS)*, 1(1), november 2002.
8. L. Eeckhout, A. Georges, and K. D. Bosschere. How java programs interact with virtual machines at the microarchitectural level. In *Proceedings of the ACM Conference on Object-Oriented Programming, Systems, Languages, and Applications (OOPSLA)*, Anaheim, California, USA, 2003. ACM Press New York, NY, USA.
9. IBM. The jikes' research virtual machine user's guide 2.2.0., 2003. http://oss.software.ibm.com/developerworks/oss/jikesrvm/.
10. The source for java technology, 2003. http://java.sun.com.
11. R. Jones. *Garbage Collection: Algorithms for Automatic Dynamic Memory Management*. John Wiley and Sons, 4th edition, July 2000.
12. D. Kegel. Building and testing gcc/glibc cross toolchains, 2004. http://www.kegel.com/crosstool/.
13. Zbt@ sram and sdram products, 2004. http://www.micron.com/.
14. P. R. Panda, F. Catthoor, N. D. Dutt, K. Danckaert, E. Brockmeyer, and C. Kulkarni. Data and memory optimizations for embedded systems. *ACM Transactions on Design Automation for Embedded Systems (TODAES)*, 6(2):142–206, April 2001.
15. Sourceforge. Jamvm - a compact java virtual machine, 2004. http://jamvm.sourceforge.net/.
16. Sourceforge. kissme java virtual machine, 2005. http://kissme.sourceforge.net/.
17. SPEC. Specjvm98 documentation, March 1999. http://www.specbench.org/osg/jvm98/.
18. P. F. Sweeney, M. Hauswirth, B. Cahoon, P. Cheng, A. Diwan, D. Grove, and M. Hind. Using hardware performance monitors to understand the behavior of java application. In *USENIX 3rd Virtual Machine Research and Technology Symposium (VM'04)*, 2004.
19. B. Systems. Bea weblogic jrockit, 2005. http://www.bea.com/framework.jsp?CNT=index.htm&FP=/content/products/weblogic/jrockit/.
20. D. Takahashi. *Java chips make a comeback*. Red Herring, 2001.
21. The University of Massachusetts Amherst and the University of Texas. Dynamic simple scalar, 2004. http://www-ali.cs.umass.edu/DSS/index.html.

Optimizing the Configuration of Dynamic Voltage Scaling Points in Real-Time Applications*

Huizhan Yi and Xuejun Yang

Section 620, School of Computer, National University of Defense Technology,
Changsha, 410073, Hunan, P.R. China
{huizhanyi,xjyang}@nudt.edu.cn

Abstract. Compiler-directed dynamic voltage scaling (DVS) is an effective low-power technique in real-time applications, where compiler inserts voltage scaling points in a real-time application, and supply voltage and clock frequency are adjusted to the relationship between the remaining time and remaining workload at each voltage scaling point. In this paper we present the analytical energy model of proportional dynamic voltage scaling in real-time applications. Using the analytical model, we theoretically prove the optimal configuration of voltage scaling points that minimizes energy consumption. Furthermore, in order to seek the optimal configuration taking into account voltage scaling overhead in the most frequent execution case, we propose a configuration methodology, where a profile-based method constructs the abstract execution pattern of an application, voltage scaling points are inserted into the abstract execution pattern by the optimal configuration without taking into account voltage scaling overhead, and then we can find the optimal configuration considering voltage scaling overhead by deleting some voltage scaling points from the execution pattern. Finally, the remaining points are inserted into the application by compiler. The simulation results show that, when taking into account voltage scaling overhead, the configuration methodology reduces energy consumption efficiently.

1 Introduction

Embedded systems for mobile computing are developing sharply, and a crucial parameter of mobile systems is the continued time of energy supply. Although the performance in ICs has been increasing rapidly in recent years [1], battery techniques is developed very slowly [2] and it is of significant importance for battery-powered mobile systems to utilize more effective low-power techniques.

The energy consumption by the facilities from IT industry has been steadily growing year by year [3] and large quantities of energy consumption necessitate power management to improve energy efficiency. So it is very imperative not only for mobile systems but also for high-performance desktop systems to develop effective low-power techniques.

Dynamic voltage scaling (DVS) [4] is one of the low-power techniques in architecture level, and it is widely used in embedded systems for mobile computing and desk-

* Supported by the National High Technology Development 863 Program of China under Grant No. 2004AA1Z2210 and Server OS Kernel under Grant No. 2002AA1Z2101

V. Paliouras, J. Vounckx, and D. Verkest (Eds.): PATMOS 2005, LNCS 3728, pp. 79–88, 2005.

top systems. Real-time dynamic voltage scaling dynamically reduces supply voltage to the lowest possible extent that ensures a proper operation when the required performance is lower than the maximum performance. Since the dynamic energy consumption, the dominant energy consumption in ICs, is in direct proportion to the square of supply voltage V, it is possible for DVS to significantly reduce energy consumption.

For real-time applications, real-time operating system (RT-OS) makes dynamic voltage scaling by scheduling multiple tasks [5], [6], [7], where RT-OS assigns only one supply voltage to a task and does not change it in the course of task execution. Thus, OS-directed dynamic voltage scaling is called inter-task dynamic voltage scaling (InterDVS) [8].

An intra-task dynamic voltage scaling (IntraDVS) [8] assisted by compiler automatically inserts voltage scaling points in a real-time task and divides the task into some execution sections, and then supply voltage is adjusted to the relationship between the remaining time and the remaining workload. IntraDVS must properly place voltage scaling points in a real-time application, and the configuration of voltage scaling points significantly affects energy consumption. A good configuration could save more energy; however, due to voltage scaling overhead, the improper one could waste much energy. To sum up, the past algorithms have utilized two kinds of configurations of voltage scaling points. The first is to make use of fixed-length voltage scaling sections, the whole execution of a task is divided into some equal subintervals and the voltage adjustment is made at the beginning of each subinterval [9] [10]. The second is a heuristic method, the condition and loop structure in real-time applications often bring about the workload variation, and energy consumption can be reduced enormously if voltage scaling points are put at the end of the structures [8] [11]. Although both configurations can reduce energy consumption, no theoretical results pronounce which one is the optimal configuration method. Moreover, no work has comprehensively investigated the configuration method of voltage scaling points when taking into account voltage scaling overhead.

In this paper we present the analytical energy model of proportional dynamic voltage scaling in real-time applications. Utilizing the analytical model, we theoretically prove the optimal configuration of voltage scaling points that minimizes energy consumption when not taking into account voltage scaling overhead. Furthermore, in order to find out the optimal configuration taking into account voltage scaling overhead, we propose a configuration methodology, where a profile-based method constructs the abstract execution pattern of an application in the most frequent execution case, voltage scaling points are inserted into the abstract execution pattern by the optimal configuration without taking into account voltage scaling overhead, and then we can find the optimal configuration considering voltage scaling overhead by deleting some voltage scaling points from the execution pattern. Finally, the remaining points are inserted into the application by compiler. The simulation results show that, when taking into account voltage scaling overhead, the configuration methodology reduces energy consumption efficiently.

The rest of this paper is organized as follows. In Section 2 we present the analytical energy model. In Section 3, we theoretically prove the optimal solution to the configurations of voltage scaling points. In Section 4 we give the configuration methodology of voltage scaling points taking into account voltage scaling overhead. In

Section 5 we prove by simulations that the configuration methodology taking into account voltage scaling overhead saves more energy. Finally, we give the conclusions.

2 IntraDVS Models

Dynamic voltage scaling is used to exactly meet the dynamic performance requirement of real-time applications with the lowest energy consumption by dynamically adjusting supply voltage. In order to ensure a correct operation, reducing supply voltage often goes with reducing clock frequency. The switch from one voltage level to another has energy and time overhead. Nowadays, some DVS-enabled systems, such as Transmeta Crusoe, Intel Xscale, and AMD K6-IIIE+, are operated on some discrete voltage levels, and software adjusts supply voltage to performance requirement continually. In this paper, we suppose that supply voltage and clock frequency are operated in a consecutive interval $[0, f_{max}]$. With the development of microprocessor techniques, more levels are possible, and therefore the assumption doesn't affect the final result.

A real-time task has strict timing constraint and must finish before its deadline (d), missing the deadline might lead to catastrophic result. Real-time dynamic voltage scaling guarantees a correct operation of a real-time task and dynamically reduces supply voltage and clock frequency to the lowest possible extent in the execution course. Therefore, in real-time applications the worst-case execution time ($wcet$) or the worst-case execution cycle ($wcec$) must be estimated in advance [12] to ensure that the timing constraint is met, and the worst-case execution time must be less than or equal to the deadline. If the $wcet$ is less than the deadline, we can proportionally reduce clock frequency beforehand. Consequently, $wcet$ is equal to d and the obtained initial frequency is f_{static} . This is the starting point of dynamic voltage scaling.

IntraDVS divides the whole execution cycle of a task into n sections, and the worst-case execution cycle and the actual execution cycle of each section are denoted by wc_i and ac_i for $i = 1, ..., n$, respectively. It is obvious that $0 \le ac_i \le wc_i$ for $i = 1, ..., n$. At the beginning of each section, supply voltage (V_i for $i = 1, ..., n$) and clock frequency (f_i for $i = 1, ..., n$) are adjusted to the relationship between the remaining time and remaining workload, and the lowest supply voltage and clock frequency are utilized within timing constraint.

Let us simply suppose that the voltage scaling time overhead is a constant cycle o , and $v = o / wcec$. Since the proportional voltage scaling scheme is the most widely used, we utilize the scheme to analyze the IntraDVS:

$$f_i = \frac{\sum_{l=i}^{n} wc_l}{\sum_{l=1}^{n} wc_l - \sum_{l=1}^{i-1}[(ac_l + o) \cdot f_{static} / f_l]} \cdot f_{static}$$

We only take into account the dynamic power consumption of CMOS in this paper. Suppose that C is the total capacitance on the gate output node and α is the average probability of the input node changing on each clock cycle. After the partition of the whole execution interval, the dynamic energy consumption is equal to

$$E = \sum_{i=1}^{n} P_i \cdot t_i = \alpha \cdot C \cdot \sum_{i=1}^{n} V_i^2 \cdot f_i \cdot t_i .$$

The formula $f \propto (V - V_T)^2 / V$ defines the relationship between clock frequency and supply voltage of CMOS, where V_T denotes the threshold voltage of CMOS. Notice that the threshold voltage V_T is generally substantially less than the supply voltage V. Thus, after ignoring the threshold voltage, we can obtain an approximate equation $f_i = \beta \cdot V_i$ between clock frequency and supply voltage, where β is a constant relating to the technology of CMOS. The execution time t_i of each section can be computed by $t_i = ac_i / f_i$.

If considering voltage scaling overhead, we obtain the energy consumption:

$$E^w = \frac{\alpha \cdot C}{\beta^2} \cdot \left(\sum_{i=1}^{n-1} [f_i^2 \cdot (ac_i + o)] + f_n^2 \cdot ac_n \right).$$

Let $o = 0$, we have the energy consumption E^{wo} without taking into account voltage scaling overhead.

The energy consumption without making dynamic voltage scaling can be calculated by

$$E_{static} = \frac{\alpha \cdot C}{\beta^2} \cdot \sum_{i=1}^{n} f_{static}^2 \cdot ac_i = \frac{\alpha \cdot C}{\beta^2} \cdot f_{static}^2 \cdot \mu \cdot wcec,$$

where $\sum_{i=1}^{n} ac_i = \mu \cdot \sum_{i=1}^{n} wc_i = \mu \cdot wcec$.

Let $ac_i = \mu_i \cdot wc_i$ for $1 \le i \le n$. Furthermore, Let $rwc_i = \sum_{l=i}^{n} wc_l$ for $i = 1,...,n$, $rwc_{n+1} = 0$, $wcec = \sum_{i=1}^{n} wc_i$, $rwc_i' = rwc_i / wcec$ for $i = 1,...,n$. We deduce the ratio of E^w to E_{static} defined by

$$r^w = E^w / E_{static}$$

$$= \sum_{i=1}^{n} \left(\frac{rwc_i' \cdot \prod_{k=1}^{i-1} rwc_k'}{\prod_{k=1}^{i-1} \left(rwc_k' - \mu_k \cdot (rwc_k' - rwc_{k+1}') - v \right)} \right)^2 \cdot \frac{\mu_i \cdot (rwc_i' - rwc_{i+1}') + (1 - \lfloor i / n \rfloor) \cdot v}{\mu},$$

where $rwc_i' > rwc_{i+1}'$ for $i = 1,...,n$, $rwc_1' = 1$, $rwc_{n+1}' = 0$. $\lfloor \bullet \rfloor$ is a function rounding towards zero. When $v = 0$, we obtain the ratio r^{wo} of E^{wo} to E_{static}.

As a result, we obtain the analytical energy models without voltage scaling overhead and with voltage scaling overhead, respectively. For a real-time application, we can find out its most frequent execution case using a profile-guided method, and consequently, μ_i for $i = 1,...,n$ and v are fixed. By numerical methods we can find out the constrained optimal solutions of rwc_i' for $i = 1,...,n$, and finally we can calculate the minimum of the analytical models r^{wo} and r^w. The optimal solutions of rwc_i' for $i = 1,...,n$ are the optimal configuration of dynamic voltage scaling points in the most frequent execution case.

3   The Optimal Configuration

After modeling IntraDVS, we analyze the estimated execution cycle of a task, and present the theoretical optimal configuration of voltage scaling points without taking into account voltage scaling overhead.

Suppose that the execution time of the sequential code blocks in a task is estimated accurately and then only the condition and loop structure result in the workload variation that is used to reduce supply voltage and clock frequency. So after the actual execution of a task, the whole execution cycle estimated has the similar execution pattern as is shown in Fig. 1(b). The sections with the tag '1' represent the actual execution workload of the task, and since the number of loop iterations or the prediction of condition structures could be differentiated from that of the worst case, those with the tag '0' are not executed actually, i.e. the slack times (cycles).

It is simple to find the execution pattern in the most frequent execution case by a profile-guided method. For example, suppose that an application includes a condition sentence and a loop sentence, as is shown in Fig. 1(a), where the condition sentence is executed for 100 cycles if the prediction (a>b) is true, or else 50 cycles. In addition, the worst-case execution cycle and the best-case execution cycle of the loop sentence are 200 and 100 cycles, respectively. If in most cases, the prediction is false and the loop is executed for 100 cycles, then the execution pattern in the most frequent case is shown in Fig. 1(b).

It is clearly seen that the slack times are often not evenly distributed in the execution interval of an application. From the above analytical model, we can obtain the minimum of r^{wo} when the optimal configuration $OPTC_{wo}$ of voltage scaling points places voltage scaling points at the beginning of all the sections with the workload of '1' and the end of all the sections with the workload of '0'. For example, in Fig. 1(b), the optimal configuration $OPTC_{wo}$ puts voltage scaling points at the ends of both of the first section and second section marked by '0', which means that once there is slack time, we completely utilize the slack time to reduce the clock frequency at once. Only proportional voltage scaling scheme has the optimal property, and we don't include the detailed proof of the conclusion due to the space limitation.

Furthermore, suppose that a voltage scaling point set χ_w includes all points at the beginning of the sections with the workload of '1' and end of the sections with the workload of '0'. Then, in the most frequent execution case, there must be an optimal configuration $OPTC_w$ of voltage scaling points that minimizes the ratio r^w, where all the voltage scaling points belonging to $OPTC_w$ are included in the set χ_w. Due to the space limitation, we don't include the detailed proof of the conclusion.

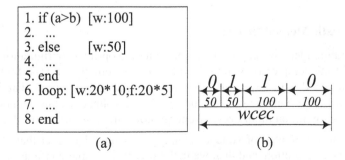

(a)   (b)

Fig. 1. An application (a) and its corresponding execution pattern (b)

4   The Optimal Configuration Methodology Taking into Account Voltage Scaling Overhead

When voltage scaling overhead is very small, in order to minimize the energy consumption in the most frequent execution case, we can insert voltage scaling points into an application by the optimal configuration without voltage scaling overhead. That is to say, we just need to find all the possible places that result in the slack times, insert voltage scaling points and totally utilize the slack time to reduce the clock frequency.

But we must consider voltage scaling overhead if it is large for an application. In the cases, it is possible that only a few voltage scaling points lead to the reduction of energy consumption, and on the contrary, the energy consumption could ascend when using large quantities of voltage scaling points [13]. So it becomes crucial to optimize the placement of voltage scaling points when taking into account voltage scaling overhead.

4.1   The Optimal Configuration Method $OPTC_w$

Using a profile-guided method, we first find out the most frequent execution pattern. Generally speaking, optimizing the most frequent execution pattern will lead to very large energy reduction. From the conclusion at the end of Section 3, we observe that all the voltage scaling points in an optimal configuration taking into account the overhead are included in χ_w, which denotes a set of all the voltage scaling points located at the beginning of the sections with the workload of '1' and end of the sections with the workload of '0' in the most frequent execution case. Suppose that χ_w includes n voltage scaling points. In order to seek the optimal configuration with the minimum energy consumption, we compute the energy consumption of the different configurations with voltage scaling points included in χ_w. In the end, we can attain the configuration with the minimum energy consumption, i.e. the optimal configuration. By a simple analysis, we need to compute 2^n times the ratio r^w, and therefore the method has $O(2^{\wedge}n)$ time complexity, where n is the number of voltage scaling points in χ_w. It is obvious that the time complexity cannot be accepted in real application.

4.2   A Heuristic Method $HOPTC$

The enumerating method can guarantee obtaining the optimal configuration, but it has exponential time complexity with the number n of sections of a task. So we attempt to utilize a method with lower time complexity. First of all, we calculate the energy consumption of the configuration including all the voltage scaling points in the set χ_w. If the configuration minimizes the energy consumption, the ratios of the configurations with fewer number of voltage scaling points must be larger than the ratio r^w of the initial configuration, and then we make use of the initial configuration to adjust supply voltage and clock frequency. Otherwise, there must be a configuration with fewer number of voltage scaling points that minimizes the energy consumption, and then, we choose a voltage scaling point and delete it from the set χ_w. The selected point should reduce voltage scaling overhead most efficiently, that is, the corresponding one-point-deleted configuration minimizes energy consumption. We have been deleting a point from the set χ_w at a time before finding out a local minimum value of the ratio r^w. Finally, we utilize the configuration with the local minimum value of the ratio r^w to set voltage scaling points of an application. The heuristic method *HOPTC* is shown in Fig. 2.

Input: the execution pattern in the most frequent execution case.
Output: a configuration of voltage scaling points.
1. Compute the ratio r^w of the configuration with n voltage
   scaling points.
2. Compute the ratio $r_{x_i}^w$ of the configurations with one point
   deleted for $i=2,...,n$.
3. Find the minimum of $r_{x_i}^w - r^w$ for $i=2,...,n$.
4. If the minimum is larger than 0, stop.
5. Or else utilize the configuration with the minimum $r_{x_i}^w$ as the
   new configuration, and update the value of n (-1).
6. Repeat the steps from 1 to 5.

Fig. 2. A heuristic method *HOPTC* seeking the optimizing configuration

First of all, observe that the ratios of the optimal configurations with different number of voltage scaling points are not decreasing with the decline of the number of voltage scaling points when not considering voltage scaling overhead. Furthermore, from the heuristic method, we are aware that every time it is most possible for us to have deleted the most ineffective point for reducing energy consumption and the most serious point for increasing the energy consumption due to voltage scaling overhead. Therefore, we believe that the first local minimum value of the ratios represents the minimum value of the ratios of all the searched configurations.

Analyzing the heuristic method, we delete all the points of the set χ_w in the worst case, and then the maximum of the calculated number of the ratio r^w is

$$(n-1)+(n-2)+,...,+1 = n \cdot (n-1)/2$$

Therefore, the time complexity of the method is $O(n^2)$.

4.3 The Configuration Methodology

Integrating the above methods, we show the methodology setting voltage scaling points.

The input is a nonpower-aware program, whereas the output is a power-aware program. Firstly, for real-time application, we make use of the *WCET* tools to evaluate the worst-case execution time of the program. Next, we utilize a profile-based method to search for the execution pattern in the most frequent execution case, and as a result we find out all the possible locations where the slack times take place in the most frequent execution case. Utilizing the execution pattern, we find out the optimal or optimizing configuration by the method of Section 4.1 or 4.2. The last step inserts voltage scaling points into the program at the selected locations. In this step, we employ the voltage scaling scheme as follows:

$$cf = \begin{cases} cf & \text{if } \dfrac{rwc_i}{d - ct - o/cf} \geq cf \\[3mm] \dfrac{rwc_i}{d - ct - o/cf} & \text{otherwise} \end{cases}$$

where cf and ct, respectively, represent current frequency and current time. The scheme guarantees the timing constrains in the worst case, and simultaneously presents the optimal or optimizing configuration in the most frequent execution case.

Apparently, compared with the past proportional voltage scheduling, we just add the profile step and the optimizing step, and therefore, all the past proportional DVS algorithms realized by compiler can be utilized.

5 Simulations

To simulate the optimizing effect in the different most frequent execution patterns, we divide the whole execution interval into λ equal subintervals and call λ simulation precision. In each subinterval there are two possible values: 0 or 1. Each subinterval with the value of '1' is the actual execution cycle, and the subintervals denoted by '0' are not executed. Then we can simulate the effect of the proposed method in the 2^λ patterns.

Let $\lambda = 16$, we calculate the energy consumption percentage using the different execution pattern. The optimal configuration $OPTC_w$ creates the minimum of energy consumption, and we compare the energy consumption percentage of $HOPTC$ with that of $OPTC_w$, as is shown in Fig. 3(a). When the overhead parameter v ranges from 0 to 0.1, the maximum of the percentage difference between $HOPTC$ and $OPTC_w$ is less than 20%.

Most of the current DVS algorithms have only taken into account meeting the timing constraints when there is voltage scaling overhead, and therefore, they do not give enough attention to the effect on energy consumption due to voltage scaling overhead. For example, in a representative DVS algorithm from Dongkun Shin [8], the clock frequency of each voltage scaling point is set to

$$cf = \begin{cases} cf & \text{if } \dfrac{rwc_{i+1} + ac_i}{rwc_{i+1} + wc_i - o} \geq 1 \\[3mm] \dfrac{rwc_{i+1} + ac_i}{rwc_{i+1} + wc_i - o} \cdot cf & \text{otherwise} \end{cases}$$

As long as there are enough slack cycles compensating the cycle loss due to voltage scaling overhead, the clock frequency is reduced. Therefore, the number of the reclaimed cycles is the decisive factor whether to adjust the clock frequency, which is a simple heuristic method to choose voltage scaling points. Using the same parameters, we present the maximum of the percentage difference between the method and $OPTC_w$ in Fig. 3(b). It is clearly seen that the simple heuristic method is so ineffective that some very large differences take place, and for example, up to more than 150% the difference is possible.

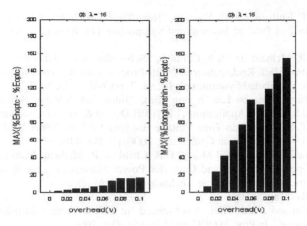

Fig. 3. The difference of the energy consumption percentage when λ=16: (a) Between *HOPTC* and *OPTC$_w$*; (b) Between the method from Dongkun Shin and *OPTC$_w$*

6 Conclusions

In this paper we present the analytical energy model of proportional dynamic voltage scaling in real-time applications. Utilizing the analytical model, we theoretically prove the optimal configuration of voltage scaling points that minimizes energy consumption. Furthermore, in order to seek the optimal configuration taking into account voltage scaling overhead in the most frequent execution case, we propose a configuration methodology, where a profile-based method constructs the abstract execution pattern of an application, voltage scaling points are inserted into the abstract execution pattern by the optimal configuration without taking into account voltage scaling overhead, and then we find out the optimal configuration taking into account overhead by deleting some voltage scaling points from the execution pattern. Finally, the remaining points are inserted into the application by compiler. The simulation results show that when taking into account voltage scaling overhead, the configuration methodology reduces energy consumption efficiently.

References

1. ITRS, "International Technology Roadmap for Semiconductors 2003 Edition," Can get from http://public.itrs.net
2. Kanishka Lahiri, "Battery-Driven System Design: A New Frontier in Low Power Design, " ASP-DAC/VLSI Design 2002, January 07 - 11, 2002, Bangalore, India.
3. Trevor Mudge, "Power: A First Class Design Constraint for Future Architectures," HiPC 2000: 215-224.
4. T. Burd, T. Pering, A. Stratakos, and R. Brodersen, "A Dynamic Voltage Scaled Micro-process- or System," in Proc. of IEEE International Solid-State Circuits Conference, 2000, pp. 294–295.
5. H. Aydin, R. Melhem, D. Mosse, and P. M. Alvarez, "Power-Aware Scheduling for Periodic Real-Time Tasks," IEEE TRANSACTIONS ON COMPUTERS, VOL. 53, NO. 5, MAY 2004.

6. T. Okuma, T. Ishihara, and H. Yasuura, "Real-Time Task Scheduling for a Variable Voltage Processor," in Proc. of International Symposium On System Synthesis, 1999, pp. 24–29.
7. Dakai Zhu, R. Melhem, and B.R. Childers, "Scheduling with Dynamic Voltage/Speed Adjustment Using Slack Reclamation in Multi-Processor Real-Time Systems," IEEE Trans. on Parallel & Distributed Systems, vol. 14, no. 7, pp. 686 - 700, 2003.
8. Dongkun Shin, Seongsoo Lee, Jihong Kim, "Intra-Task Voltage Scheduling for Low-Energy Hard Real-Time Applications," In IEEE Design & Test of Computers, Mar. 2001.
9. S. Lee and T. Sakurai, "Run-Time Voltage Hopping for Low-Power Real-Time Systems," in Proc. of Design Automation Conference, 2000, pp. 806–809.
10. Nevine AbouGhazaleh, Daniel Mosse, B.R. Childers, R. Melhem, Matthew Craven, "Collaborative Operating System and Compiler Power Management for Real-Time Applications," in Proc. of The Real-time Technology and Application Symposium, RTAS, Toronto, Canada (May 2003).
11. Dongkun Shin and Jihong Kim, "Look-ahead Intra-Task Voltage Scheduling Using Data Flow Information," In Proc. ISOCC, pp. 148-151, Oct. 2004.
12. Peter Puscher, Alan Burns, "A Review of Worst-Case Execution-Time Analysis (Editorial)," Kluwer Academic Pubilishers, September 24, 1999.
13. Nevine AbouGhazaleh, Daniel Mosse, B.R. Childers, R. Melhem, "Toward the placement of power management points in real-time applications," Compilers and operating systems for low power, Pages:37-52, 2003. ISBN: 1-4020-7573-1, Kluwer Academic Publishers Norwell, MA, USA.

Systematic Preprocessing of Data Dependent Constructs for Embedded Systems

Martin Palkovic[1], Erik Brockmeyer[1], P. Vanbroekhoven[2],
Henk Corporaal[3], and Francky Catthoor[1]

[1] IMEC Lab., Kapeldreef 75, 3001 Leuven, Belgium
[2] KU Leuven, Celestijnenlaan 200A, 3001 Leuven, Belgium
[3] TU Eindhoven, Den Dolech 2, 5612 AZ Eindhoven, The Netherlands

Abstract. Data transfers and storage are dominating contributors to the area and power consumption for all modern multimedia applications. A cost-efficient realisation of these systems can be obtained by using high-level memory optimisations. This paper demonstrates that the state-of-the-art memory optimisation techniques only partly can deal with code from real-life multimedia applications. We propose a systematic preprocessing methodology that can be applied on top of the existing work. This opens more opportunities for existing memory optimisation techniques. Our methodology is complemented with a postprocessing step, which eliminates the negative effects of preprocessing and may further improve the code quality [,]. Our methodology has been applied on several real-life multimedia applications. Results show a decrease in the number of main memory accesses up to 45.8% compared to applying only state-of-the-art techniques.

1 Introduction

Modern multimedia systems are characterised as applications with huge amount of data transfers and large memories. Thus, the memory subsystem in these applications consumes a major part of the overall area and energy. The utilisation of the memory subsystem by optimisation of global memory accesses usually brings significant energy savings [ ]. Improving the temporal locality of the memory accesses also decreases the life-time of data elements and hence the required memory footprint.

Every recent high-level low-power design methodology contains a global loop transformation (**GLT**) stage [, ,]. This stage either improves the parallelisation opportunities at the instruction or data level, or it improves the locality of data accesses so that data can be stored in lower levels of the memory hierarchy, resulting in significant area and power gains.

The loop transformations are nearly always performed on a geometrical model [,] which is very effective in dealing with generic loop transformations [, , ,]. However, the geometrical model imposes strict limitations on the input code, namely it considers only the static control parts (SCoP) [ ]. The static control part is a maximal set of consecutive statements without *while* loops, where loop bounds and conditionals may only depend on invariants within this set of statements. These invariants include symbolic constants, formal function parameters and surrounding loop counters. Also, the geometrical model requires pointer-free single-assignment code in one function [ ]. Clearly, this is not acceptable from a user and application point of view.

V. Paliouras, J. Vounckx, and D. Verkest (Eds.): PATMOS 2005, LNCS 3728, pp. 89–98, 2005.

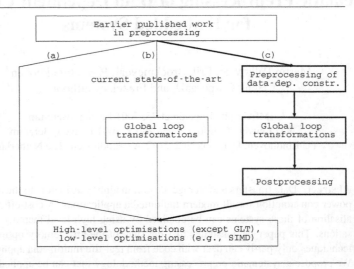

Fig. 1. Optimisation approaches compared in the paper: (a) without GLT, (b) with GLT and existing preprocessing, (c) with GLT and our systematic preprocessing on top of existing design flow

To alleviate this situation we propose a systematic preprocessing methodology dealing with data dependent constructs which are the major limitation for geometrical model. We do not consider dynamic data types and dynamic memory allocation because this is out of the scope of GLT and high-level optimisations we are using as a backend. For dynamic memory management methodology see [ ]. The steps of our methodology are applied on top of the earlier published work in preprocessing (see Section 2) as shown in Figure 1. All of our steps described in this paper are performed manually, however in a systematic way. The automation process is ongoing and it is out of the scope of the paper. For details about the current automation status see [,].

The rest of the paper is structured as follows. Section 2 surveys related work in preprocessing. In Section 3 we describe MP3 audio decoder which is used as an example in this paper. Section 4 explains our preprocessing methodology. Section 5 describes our results on three real-life multimedia benchmarks and compares the results to the state-of-the-art work. Finally, in Section 6 we draw several conclusions.

2   Related Work

The state-of-the art in application preprocessing which overcomes the strict limitations of the geometrical model used by most loop transformation approaches includes selective function inlining [ ] (SFI), pointer analysis and conversion [,   ] (PA&C), dynamic single assignment (DSA) conversion [ ] and hierarchical rewriting [ ]. Hierarchical rewriting partitions an application in three function layers. The first layer contains process control flow, the second layer contains loop hierarchy and indexed signals and the third layer contains arithmetic, logic and data-dependent operations. Only the second layer containing memory management is the target for high-level low-power optimisation.

As mentioned in the introduction loop transformations performed on geometrical model do not deal with data dependent behaviour [ ]. Still, some related work can be found in the high level hardware synthesis area. To deal with data-dependent control flow changes, a transformation technique to convert an iterative algorithm whose loop bounds are data dependent to an equivalent data independent regular algorithm has been proposed in [ ]. However, the authors use the worst case assumption to define the structure of a given situation and lose information about data-dependent iterator bounds. In [ ] a special internal representation is used to treat both data-flow and control-flow designs. This representation is meant for developing scheduling heuristics and it does not enable the GLT.

3   Benchmark Description and Profiling

We use one of our applications, the MP3 decoder [ ], as example during the rest of the paper. The MP3 decoder is a frame-based algorithm for decoding a bitstream from the perceptual audio coder. A frame is coding 1152 mono or stereo samples and is divided into two granules of 576 samples. Each granule consists of 32 subband blocks of 18 frequency lines and has two channels. One channel in one granule is called a block.

After receiving a frame, the frame is Huffman decoded (see decoder frontend in Figure 2). Then, on the Huffman decoded frame, several kernels are applied (see the decoder backend in Figure 2). For details about the different kernels, see [ ]. In the rest of the paper, we will concentrate only on the decoder backend. The decoder frontend (Huffman decoder) is not data access intensive compared to the rest of the whole

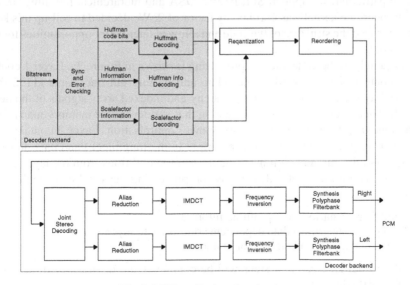

Fig. 2. MP3 audio decoder structure

decoder and is also not suited for high-level memory optimisation because of irregular memory accesses and a lot of small look-up tables.

For the decoding of the MP3 blocks (a block is one channel in one granule), different methods are used depending of the initial coding of the block on the encoder site. Each block can be coded using short window, mixed window or one of three types of long window. Also, each block can be coded using stereo decoding or joint stereo decoding (middle side and/or intensity stereo decoding). Combination of these options introduces different modes of decoding, which are represented with data dependent conditions in the benchmark code. However, the occurrence of different modes is not the same. To identify the most probable ones, we have profiled our application with dozens realistic mp3 bitstreams (see Table 1).

Table 1. Frequency of use (in %) of different modes in MP3 audio decoder for real-life bitstreams. This information will be used in Section 4, Subsection 4.1

Mode	Frequency	Scenario
Long block (type 0)/Joint Middle Side Stereo Decoding	90.1%	1
Long block (type 0)/Stereo Decoding	3.2%	1
Long block (type 1&3)/Joint Middle Side Stereo Decoding	3.1%	1
Long block (type 1&3)/Stereo Decoding	0.9%	1
Short block (type 2)/Joint Middle Side Stereo Decoding	1.4%	2
Short block (type 2)/Stereo Decoding	1.2%	2
Other	<0.1%	3

4   Systematic Preprocessing of Data Dependent Constructs

We consider that the preprocessing steps listed in Section 2 have been already completed on the initial source code before applying the remaining steps of our methodology. In particular, we applied SFI, PA&C, DSA and hierarchical rewriting. In MP3 audio decoder one kernel was unrolled by factor of 2. We decided to roll-up this kernel which improved the structure of the code and allowed later better opportunities for GLT (loop fusion).

An example of the code after preprocessing is shown in Figure 3a. Layer 1 contains process control flow, i.e. reading, decoding and writing out one frame (1152 stereo samples). Layer 2 describes the loop hierarchy and indexed array signals of the decoding itself. This layer contains all information that is relevant for memory management which is the target for the high-level low-power optimisation. Layer 3 contains arithmetic, logic (e.g. requantization of one sample in the long block) and data-dependent operations (so this already eliminates the innermost conditions from the layer 2 code). The second layer is the input code for 3 new preprocessing steps addressed in this paper, namely:

1. Identifying scenarios from profiling information
2. Moving non-innermost data dependent conditions to layer 3
3. Extending data dependent iterator bounds

In the sequel all three new steps will be explained and demonstrated on different parts of an MP3 decoder.

```
/* Layer 1 source code */              /* Layer 1 source code */
while (!EOF) {                         ...
   ... read_frame_from_bitstream ...
   ... huffman_decoder ...
   ... mp3_decoder ...
   ... write_raw_PCM_samples ...
}

/* Layer 2 source code */              /* Layer 2 source code */
if(g_side_info.block_type[gr][ch]==2)  /* Scenario partition 1 */
   for(i=0;...)                        if(g_side_info.block_type[gr][ch]!=2 &&
      ... requantize_long_block(is[gr][ch][i]);   ...
if(g_side_info.block_type[gr][ch]!=2)  g_frame_header.mode_extension & 0x2) {
   ...                                    ... requantize_long_block ...
...                                       ... joint_m/s_stereo_decode ...
if(g_frame_header.mode_extension & 0x2)   ... antialias ...
   for(i=0;...)                           ... IMDCT ...
      ... joint_m/s_stereo_decode ...     ... frequency_inversion ...
...                                       ... synthesis_polyphase_filterbank ...
                                       }
/* Layer 3 source code */
int requantize_long_block(int in) {    /* Layer 3 source code */
   tmp1 = pow(2.0, ...);               ...
   tmp2 = pow(2.0, ...);
   tmp3 = in <0.0 ? -pow_43(in) : pow_43(in);
   return tmp1*tmp2*tmp3;
}
```

(a)   (b)

Fig. 3. (a) MP3 code after hierarchical rewriting. (b) Scenarios for an MP3 decoder

4.1 Identifying Scenarios from Profiling Information

In real-life multimedia applications, various parts of the bitstream are coded with different methods depending on audio and video characteristics of the input signal (see Section 3 for an MP3 audio decoder). These different options introduce different modes of decoding as we discussed in previous Section (see Table 1), with possibly different computation and memory requirements. A set of modes is called a scenario. Scenarios are interesting because they can remove the outermost data dependent *if* conditions which are preventing from performing the GLT.

Identifying scenarios means first collecting the profiling information of all possible modes. Similar modes are then grouped together and create a scenario. Similarity is measured as ratio of the number of identical statements covered by execution of two decoding modes and the total number of statements covered by these modes, i.e.

$$Similarity(mode_i, mode_j) = \frac{\|statements(mode_i) \cup statements(mode_j)\|}{\|statements(mode_i) \cap statements(mode_j)\|}$$

Note, that if $i = j$ all statements are identical (we compute the similarity of the same mode) and the *Similarity* is *1*. When there are no common statements in $mode_i$ and $mode_j$ the *Similarity* is *0*. The similarity is computed only for modes, that are relevant, i.e. first n most frequently used modes that cover 90% of the whole execution time of the application. For an MP3 we observed that only 6 modes cover 99.9% of the whole execution time so we took all of them. The 6 modes cover approximately 26% of all possible modes.

After profiling and computing the similarity for the $n=6$ most frequently used modes we grouped together modes that had similarity measure close to 1. Using this approach we created 2 scenarios for the MP3 audio decoder. The first scenario groups long blocks type 0,1,3 coded with stereo or middle side coding (see Table 1). The second scenario contains short blocks type 2 coded with stereo or middle side coding. For the rest, i.e. the 18 uncovered modes we create one backup scenario.

By creating scenarios, we eliminate most of the outermost data dependent conditions in the code targeted for optimisation. This is done by separation of modes to different groups we call scenarios (compare layer 2 in Figure 3a and Figure 3b). On the other hand, this leads to significant code size overhead due to the duplication of the code. The cost of scenario creation is in code duplication and thus the code size increase. The gain is in the reduction of number of memory accesses to expensive off-chip memories after high-level GLT. These transformations can be also beneficial for code size as we will see in the sequel. Without scenario creation these transformations will not be feasible.

To illustrate this on our benchmark, with the grouping mentioned above, we have initially increased the code size by 50%. However, this grouping has enabled GLT that have not been feasible before. After these transformations the overall code size increase is only 2%, while the reduction in the number of main memory accesses is significant as we will see in Section 5. In general, we group paths with the highest similarity measure, thus the code size increase is kept under control. The remaining steps of the methodology do not increase the code size at all.

4.2   Moving Non-innermost Data Dependent Conditions to Layer 3

In this step we move data dependent *if* conditions down in the loop hierarchy.

We have eliminated most of the outermost data dependent conditions by creating scenarios as we have described in Subsection 4.1. However, one scenario contains several modes, i.e. some outermost data dependent conditions remain. E.g. in scenario 1 of our MP3 audio decoder example (long blocks type 0,1,3 and stereo/joint middle side stereo decoding) the data dependent condition in stereo decoding kernel is present (see Figure 4a). Note, that the data dependent condition which decides the active one among long blocks 0,1,3 is innermost and is already hidden in layer 3 of the IMDCT kernel. Also, the code can contain data dependent conditions in the middle of the loop hierarchy.

To eliminate the rest of the outermost and middle data dependent conditions we move them down in the loop hierarchy to hide them also in layer 3. When moving down in the loop hierarchy, the condition is evaluated many more times. If these conditions

```
if ( g_frame_header . mode_extension & 0x2)
  for ( i =0; i <576; i ++)
    os [ gr ][ ch ][ i ] = ...
```

(a)

```
for ( i =0; i <576; i ++)
  os [ gr ][ ch ][ i ] =
  ( g_frame_header . mode_extension & 0x2)
  ? ... : os [ gr ][ ch ][ i ];
```

(b)

Fig. 4. Stereo kernel before (a) / after (b) moving data dependent condition to layer 3

create too much overhead for the control-flow, they can be optimised using techniques similar to Falk et. al. [ ] after GLT. Also, for our example, the stereo decoding kernel is the only place where the data dependent condition is preventing the loop fusion of all kernels. So to hide his condition in layer 3 may have a huge positive impact on data transfers to main memory and thus on overall power consumption. This stresses the importance of moving the condition down.

Note, that the step 4.2 is complementary with the step 4.1 and together they solve the whole problem of data dependent conditions in the application. I.e. the outermost and middle data dependent conditions that have not been handled in step 4.1 are moved to layer 3 in step 4.2. The overhead of moving data-dependent conditions to layer 3 will be eliminated in our postprocessing step by applying the hoisting techniques of Gupta et. al. [ ] and the loop nest splitting techniques by Falk et. al. [ ].

4.3 Extending Data Dependent Iterator Bounds

In this step we replace the data dependent iterator bound by a worst case bound and data dependent *if* condition inside the loop.

In MP3 decoder the decoded data are divided in 3 regions: big_value, count1 and rzero region. The borders of the count1 region are data dependent and only first two regions are decoded in the requantization kernel. This is resulting in data dependent iteration bound (see Figure 5a) which is not acceptable for GLT. Thus, we have to extend the data dependent iterator bound to the worst case and data dependent condition inside the loop (see Figure 5b). The extension to worst case increases the number of iterations of that particular loop, however could be beneficial for energy after applying GLT.

If we use the worst case for the upper bound (576 samples) in the requantization kernel of the MP3 decoder, we increase the average number of iterations for this loop from 473 to 576. This results in 22% more iterations. The benefit is in the opportunity to merge this kernel with the remaining kernels. Thus we eliminate large intermediate buffers and reduce the number of main memory accesses as we will see in Section 5.

Note, that if the introduced condition is non-innermost, we have to move *if* condition to layer 3. This means to repeat step 4.2 for this condition.

5   Results

In this Section we report our results for all 3 real-life multimedia applications: an MP3 audio decoder [ ], a DivX5.x video decoder [ ] and a QSDPCM audio coder [ ].

```
for(i=0;i<g_side_info.count1[gr][ch];i++)
   ... = requantize_long_block (...);
```

(a)

```
for(i=0;i<576;i++)                          for(i=0;i<3476;i++)
   if(i<g_side_info.count1[gr][ch])            if(i<g_side_info.count1[gr][ch])
     ... = requantize_long_block (...);          ... = requantize_long_block (...);
```

(b)   (c)

Fig. 5. Requantization kernel before (a) / after (b) data dependent iterator bounds extension

We have applied our preprocessing methodology on top of existing preprocessing work as explained in Section 4. After that, the GLT step has been applied (mainly strip mining, loop interchange and loop fusion) to improve the temporal locality. First strip mining and loop interchange have been applied to be able to fuse the kernels optimally. The script to apply such loop transformations is not the focus of this paper though. A description of the main principles can be found in related work for loop transformations [, 2, 5, 4]. After loop fusion, the advanced signal propagation [2] and in-place [ ] have been used to eliminate large intermediate buffers.

Although beneficial for memory optimisations, the preprocessing introduces usually instruction and control-flow overhead (e.g., code duplication, extending the loop bounds, moving *if* conditions down in the loop hierarchy and thus execute it more often). However, the postprocessing techniques and tools like the one of Falk et al. [ ] for optimising control flow or code hoisting techniques and common subexpression elimination techniques for elimination of code duplication [ ] can be used to remove the negative preprocessing effects and further improve the code quality.

After these high-level optimisations, we have used the ATOMIUM profiling tool together with Memory Hierarchy Layer Assignment (MHLA) [ ] tool. We have observed the traffic from/to main memory and we have determined the number of main memory accesses for an optimised memory organisation.

Table 2. Comparison of # main memory (MM) accesses of initial application (initial), loop transformed application after existing preprocessing (GLT) and loop transformed application after our preprocessing for data-dependent constructs on top (PRE+GLT). The last two columns show improvement using our techniques comparing PRE+GLT and GLT results

Application	Initial	GLT		PRE+GLT		Δ	
	#MM access	#MM acc.	reduction(%)	#MM acc.	red.(%)	#MM acc.	red.(%)
MP3 [$\times 10^6$]	714.2	126.9	82.2%	68.8	90.4%	58.1	45.8%
DivX5.x [$\times 10^6$]	9.29	9.29	0%	8.98	3.3%	0.31	3.3%
QSDPCM [$\times 10^3$]	542.1	445.5	17.8%	306.1	43.5%	139.4	31.3%

In Table 2 we can see the results in the number of main memory accesses for our three benchmarks. Column two shows the results for initial applications. Columns three and four show the results for state-of-the-art preprocessed and loop transformed applications. Column five and six show the results for loop transformed applications after both, state-of-the-art and our preprocessing methodology. It is clear that our methodology enables better utilisation of GLT, leading to overall better results when compared to the state-of-the-art approaches without our technique. This will be substantiated in the experiments described below.

We have showed our technique on an MP3 audio decoder through this paper (see Section 3). Using only state-of-the-art work, we were able to optimise 25% lines of source code of the application, namely frequency inversion and polyphase synthesis filterbank kernels. This leads to 82.2% reduction in the number of main memory accesses compared to the initial version (see column 4 in Table 2). This reduction is mainly due to transformations in the synthesis polyphase filterbank kernel. The remaining kernels cannot be preprocessed and optimised. However, if we apply also our methodology, it enables to optimise 87.5% lines of source code of the application and leads to a further improvement of 45.8% (see column 7 and 8 in Table 2) in the number of main memory accesses.

The DivX5.x video decoder is taken from the ffmpeg library of mplayer application [6] and profiled using dozens realistic DivX movies. This application was already heavily optimised. Still we have been able to gain 3.3% (see column 7 and 8 in Table 2) in the number of main memory accesses. Without our preprocessing it is infeasible to do any loop transformation on this application. In this application, also a lot of postprocessing (e.g., control-flow optimisation, elimination of code duplication) is needed.

The QSDPCM application does not contain all important features for our methodology. It has only one mode resulting in 1 scenario. So the first step of our methodology cannot be applied for this application. Still a lot of data dependent conditions are hidden by applying step 2 of our methodology resulting in 31.3% gain (see column 7 and 8 in Table 2) in the number of main memory accesses compared to existing techniques.

Our methodology does not only improve the opportunity for high-level low-power optimisations, particularly GLT. It can also improve the opportunity for low-level optimisations, like the introduction of fine-grain data level parallelism. By enabling loop transformations for our example benchmark, the two channels can be decoded in parallel using SIMD instructions. This can halve the number of remaining main memory accesses and significantly reduce the area and energy cost on top of GLT improvements.

6 Conclusions

Current multimedia programs are characterised as applications with a lot of data dependent constructs. These constructs may considerably hamper optimisation opportunities. Therefore, novel preprocessing methodologies are needed.

This paper proposes a systematic preprocessing methodology dealing with data dependent constructs in real-life multimedia applications. The methodology has been explained and demonstrated on the MP3 audio decoder. Finally, results from three real-life multimedia applications have been discussed showing the large potential of this methodology giving reduction in the number of main memory accesses up to 45.8%.

References

1. J.Absar, F.Catthoor, K.Das, "Call-instance based function inlining for increasing data access related optimisation opportunities", *Technical report*, IMEC, Leuven, Belgium, 2003.
2. D.Atienza et al., "Dynamic memory management methodology for reduced memory footprint in multi-media and wireless network applications", *Proc. 7th ACM/IEEE Design and Test in Europe Conf.* (DATE), Paris, France, pp.532-537, Feb. 2004.
3. C.Bastoul et al., "Putting polyhedral loop transformations to work", *LCPC'16 International Workshop on Languages and Compilers for Parallel Computers*, LNCS 2958, pp. 209-225, College Station, Sept. 2003.
4. E.Brockmeyer, M.Miranda, F.Catthoor, H.Corporaal, "Layer Assignment Techniques for Low Energy in Multi-layered Memory Organisations", *Proc. 6th ACM/IEEE Design and Test in Europe Conf.* (DATE), Munich, Germany, pp.1070-1075, March 2003.

5. F.Catthoor, K.Danckaert, C.Kulkarni, E.Brockmeyer, P.G.Kjeldsberg, T.Van Achteren, T.Omnes, "Data access and storage management for embedded programmable processors", ISBN 0-7923-7689-7, Kluwer Acad. Publ., Boston, 2002.

6. A.Darte, Y.Robert, "Affine-by-statement scheduling of uniform and affine loop nests over parametric domains", *Journal of Parallel and Distributed Computing 29(1)*, 43-59, 1995.

7. H.Falk, P.Marwedel, "Control Flow Optimization by Loop Nest Splitting at the Source Code Level", *Proc. 6th ACM/IEEE Design and Test in Europe Conf.* (DATE), Munich, Germany, pp.410-415, March 2003.

8. B.Franke, M.O'Boyle, "Array Recovery and High-Level Transformations for DSP Applications", *ACM Transactions on Embedded Computing Systems (TECS)*, Vol. 2, Issue 2, pp 132-162, May 2003.

9. E.De Greef, F.Catthoor, H.De Man, "Program transformation strategies for reduced power and memory size in pseudo-regular multimedia applications", *IEEE Trans. on Circuits and Systems for Video Technology*, Vol.8, No.6, pp.719-733, Oct. 1998.

10. S.Gupta, M.Miranda, F.Catthoor, R.Gupta, "Analysis of high-level address code transformations for programmable processors", *Proc. 3rd ACM/IEEE Design and Test in Europe Conf.* (DATE), Paris, France, pp.9-13, April 2000.

11. B.Jung, Y.Jeong, W.P.Burleson, "Distributed Control Synthesis for Data-Dependent Iterative Algorithms", *Proc. on IEEE Int. Conf. on Application Specific Array Processors*, San Francisco, pp. 57-68, August 1994.

12. M.Kandemir, J.Ramanujam, A.Choudhary, P.Banerjee, "A layout-conscious iteration space transformation technique", *IEEE Transactions on Computers*, 50(12):1321-1335, 2001.

13. A.A.Kountouris, Ch.Wolinski, "Hierarchical Conditional Dependency Graphs as a Unifying Design Representation in the CODESIS High-Level Synthesis System", *Proc. of 13th Intnl. Symposium on System Synthesis (ISSS'00)*, Madrid, Spain, pp.66-71, Sept. 2000.

14. K.Lagerström, "Design and Implementation of an MP3 Decoder", M.Sc. thesis, Chalmers University of Technology, Sweden, http://www.kmlager.com/mp3/, May 2001.

15. K.McKinley, S.Carr, C-W.Tseng, "Improving data locality with loop transformations", *ACM Trans. on Programming Languages and Systems*, Vol.18, No.4, pp.424-453, July 1996.

16. *http://www.mplayerhq.hu/*

17. M.Palkovic, E.Brockmeyer, F.Catthoor, "Hierarchical rewriting and hiding of data dependent conditions to enable global loop transformations", *Proc. 2nd Wsh. on Optim. for DSP and Embedded Systems (ODES)*, Palo Alta CA, March 2004.

18. M.Palkovic, H.Corporaal, F.Catthoor, "Global Memory Optimisation for Embedded Systems allowed by Code Duplication", *Proc 9th Intnl. Wsh. on Software and Compilers for Embedded Systems (SCOPES)*, Dallas TX, Sep. 2005.

19. W.Pugh, "The Omega Test: a fast and practical integer programming algorithm for dependence analysis", *Communications of the. ACM*, Vol.35, No.8, Aug. 1992.

20. L.Semeria, G.De Micheli, "SpC: synthesis of pointers in C", *Proc. IEEE Intnl. Conf. on Comp. Aided Design*, Santa Clara CA, pp.340-346, Nov. 1998.

21. P.Strobach, "QSDPCM – A New Technique in Scene Adaptive Coding", *Proc. 4th Eur. Signal Processing Conf.*, EUSIPCO-88, Grenoble, France, pp.1141–1144, Sep. 1988.

22. P.Vanbroekhoven, G.Janssens, M.Bruynooghe, H.Corporaal, F.Catthoor, "Advanced copy propagation for arrays", *Proc. of the SIGPLAN Conf. on Languages, Compilers, and Tools for Embedded Systems (LCTES'03)*, San Diego CA, pp.24-33, June 2003.

23. D.Wilde, "A Library for Doing Polyhedral Operations", M.Sc. thesis, Oregon State Univ., Dec. 1993. In co-operation with IRISA/INRIA, Rennes, France.

24. M.Wolf, M.Lam, "A data locality optimizing algorithm", *Proc. of the SIGPLAN'91 Conf. on Programming Language Design and Implementation*, Toronto, Canada, pp.30-43, June 1991.

Temperature Aware Datapath Scheduling

Ali Manzak

Suleyman Demirel University, Isparta 32260, Turkey
manzak@mmf.sdu.edu.tr

Abstract. This paper presents temperature aware low power scheduling under resource and latency constraints. We assume resources with different energy delay values are available. These resources are optimized in terms of energy for a certain delay, using variable supply voltage, multiple threshold voltages and sizing techniques. The proposed algorithms are based on temperature and power efficient distribution of slack among the nodes in the data-flow graph. The distribution procedure tries to implement the minimum energy scheduling when there is no temperature critical points. If a functional unit reaches a critical temperature, algorithm tries not to schedule any nodes in the data flow graph to high temperature resources, thus decrease the chip temperature. Experiments with some HLS benchmark examples show that the proposed algorithms achieve significant power/energy reduction. For instance, when the latency constraint is 2 times the critical path delay and one of the resource temperature is critical the average power reduction is 50.8% and utilization of the hot resource is average 1%.

1 Introduction

Power consumption of VLSI circuits have increased substantially with the increase of speed and number of transistors per unit chip area. High power consumption has increased the chip temperature to undesirable levels. Power and temperature have become new design constraints along with area and speed in VLSI design. Low power design techniques have been developed in all levels of design abstractions. However these techniques not necessarily decreased the circuit temperature. Instead of average power, peak power is more accountable in generating hot spots in the chip. New design methodologies and tools are required to overcome this new design obstical.

Dynamic power still dominates the total power consumption of CMOS chips. An effective way to reduce dynamic power consumption is to lower the supply voltage level of a circuit. Reducing the supply voltage, however, increases the circuit delay and reduces the throughput. To maintain the throughput, parallelism and/or pipelining has to be incorporated [ ]. The resulting circuit consumes lower average power while meeting the global throughput constraint at the cost of increased circuit area. Another way of maintaining the throughput is to use resources operating at multiple voltages [ ], [ ], [ ], [ ], [ ], [ ]. This has the advantage of allowing modules on the critical paths to be assigned to the highest voltage levels (thus meeting the required timing constraints) while allowing

V. Paliouras, J. Vounckx, and D. Verkest (Eds.): PATMOS 2005, LNCS 3728, pp. 99–106, 2005.

modules on noncritical paths to be assigned to the lower voltages (thus reducing the power consumption).

Recently leakage power consumption increased the percentage in total power consumption. Using low threshold devices with the reduced channel width increased leakage power exponantially. Now power optimization techniques include leakage power consumption into account. Since leakage power is exponentially related to temperature, increased temperature increases the leakage power exponentialy. This results in even more increment in the temperature. Therefore, temperature now should be observed more carefully in low power design methods.

In this paper, we address the problem of scheduling a data-flow graph (DFG) under resource and latency constraints for the case when the resources have different energy/delay values. The scheduling algorithm eliminates the hot spots in the chip and minimizes power/energy consumption for the case when different energy/delay resources are available. We have no restriction on supply and threshold voltages of functional units as long as energy/delay profile is known. With the development of microsensors technogy, it is now possible to get temperature information of the different parts of the chip [2].

The proposed algorithm operates in two passes. In the first pass, minimum-time resource-constrained scheduling is done. In the second pass, the difference between the given latency and the time needed by the resource-constrained schedule (obtained in the first pass) is distributed among the nodes in a way that the total power/energy consumption is minimum. The distribution procedure (derived using the Lagrange multiplier method) uses the energy/delay (E/D) ratio of the nodes to distribute the slack. The procedure is implemented by an iterative algorithm, where in each iteration, increasing number of resources with high E/D ratio are disabled. The iterations continue until there is a timing violation. Experiments with some HLS benchmarks (HAL, AR Lattice, EW Filter, LMS, FIR Filter, DCT and Band Pass Filter) show the effectiveness of these approaches in reducing power/energy. When the latency constraint is 2 times the critical path delay and one of the resource temperature is critical the average power reduction is 50.8% and utilization of the resource at critical temperature is average 1%. When the latency constraint is 1.5 times the critical path delay and one of the resource temperature is critical the average power reduction is 35.5% and utilization of the resource at critical temperature is average 5.1%.

There are several scheduling algorithms for multiple-voltage resources in the literature today [4], [ ], [ ], [12], [ ], [ ]. These algorithms can be classified into i) only latency-constrained (i.e., latency is a hard constraint and resources are minimized) [ ], [ ], ii) only resource-constrained (i.e., resource is a hard constraint) [ ], and iii) latency and resource constrained (i.e., both latency and resource are hard constraints) [ ], [ ]. While the (only) latency-constrained and (only) resource-constrained algorithms have polynomial or pseudopolynomial time complexity, the latency and resource-constrained algorithms are based on integer linear programming and have (worst case) exponential time complexity. In this paper, we propose a heuristic algorithm for temperature aware low power latency and resource-constrained scheduling with only polynomial time complexity. As of our knowledge the proposed algorithm is the first temperature-aware low power scheduling algorithm in the literature.

The rest of the paper is organized as follows. Section II presents the definitions and the Lagrange formulation. Section III describes the algorithms and illustrates them with examples. Section IV includes the results on some HLS benchmark examples. Section V concludes the paper.

2   Preliminaries

2.1   Definitions

The input to our algorithm is a data flow graph, a timing constraint, and a resource constraint.

Timing Constraint: This is the time available to execute the operations in the data flow graph. It is also referred to as the latency constraint.

Resource Constraint: This is specified by the number of resources for each type, where each type of resource can be operated at a specific supply and threshold voltages and corresponding energy and delay values are also given. These resources are power optimized for different delay values. Examples of resources include adder/subtractor, multiplier, etc.

We have also microsensors available associated with each functional units and capable of supplying online temperature data.

2.2   Slack Distribution Using the Lagrange Multiplier Method

In the proposed algorithm, the Lagrange multiplier method is used to distribute the slack among the nodes in the critical path. The relation between the voltages of the nodes in the critical path is derived in the following way.

We have resources available with different power delay combinations and the power of each functional unit is optimized for a given delay. The energy of functional unit is decreased with the increase of delay constraint. Let

$$E_i = f(T_i)$$

where E_i is the energy and T_i is the delay of functional unit i. Our aim is to minimize E_{total} subject to the T_{total}.

$$minimize \ \ E_{total} = \sum_{i=1}^{n} E_i, \quad subject \ \ to, \ \ T_{total} = \sum_{i=1}^{n} T_i = constant,$$

We use Lagrange multiplier method to determine optimum energy/delay combination of each node.

$$\frac{\partial E_{total}}{\partial T_i} = \frac{\partial f(T_i)}{\partial T_i} = \lambda$$

If df/dT is invertible, we find that,

$$E_{total} \quad is \quad minimum, \quad when \quad T_1 = T_2 = ... = T_n \tag{1}$$

Therefore algorithm tries to use equal delay (equal voltage) functional units as much as possible on the datapath. When different functional units (multiplier, adder etc.) are present, algorithm gives the priority to the functional units with high E/D ratio [ ].

3   Resource and Latency-Constrained Scheduling

Algorithm Overview: In a nutshell, the proposed temperature aware algorithm first schedules the functional units with resource constrained algorithm such that computation time is minimum. Next the slack between the given latency and the computation time obtained by the resource-constrained algorithm is distributed to the nodes such that first hot spots are eliminated and the next the total power/energy consumption is minimum.

3.1   Minimum-Time Resource-Constrained Algorithm

The minimum-time resource-constrained algorithm schedules the nodes such that the computation time is minimum. The nodes in a ready set are prioritized based on the freedom: the nodes with the lowest freedom are chosen among the ready nodes. Then the nodes are scheduled such that if the freedom of a node is low, it is assigned to a high-power (low-delay) resource and if the freedom of a node is high, it is assigned to a low-power (high-delay) resource [ ].

The computation time obtained by application of the minimum-time resource-constrained algorithm is referred to as L_{low}. Thus if the latency constraint is $L < L_{low}$, a feasible solution cannot be obtained.

3.2   Temperature Aware Low-Power Algorithm

At the end of the first pass, if $L > L_{low}$, each node has a nonzero slack. The objective of the low-power algorithm is to distribute the available slack between the nodes (i.e., determine the resource assignment of the nodes) such that the latency constraint is satisfied, hot spots are eliminated and the minimum energy condition (eqn. 1) is satisfied as much as possible.

The procedure is iterative; in each iteration, high temperature resources and increasing number of resources with high E/D values are disabled respectively and then the nodes are scheduled. After each iteration, resources with high E/D values are disabled and more nodes are assigned to the resources with similar delay (voltage) values and the minimum energy condition (eqn. 1) is better satisfied. The iterations continue till there is a timing violation.

Priority Assignment: The priority of which resources to disable first is the resources with high temperature. Next is the resources with high energy delay ratio. The priorities are given in Table 2 for the modules in Table 1. According to this table, the multipliers operating at 5 V have the highest chance of being

Table 1. Energy E (in pJ) and normalized delay for modules in [ ]

	5V		3.3V		2.4V		1.5V	
Module	Delay	Energy	Delay	Energy	Delay	Energy	Delay	Energy
mult16	5	2504	9	1090.7	15	576.9	36	225.3
add16	1	118	2	51.4	3	27.2	8	10.6
sub16	1	118	2	51.4	3	27.2	8	10.6

Table 2. Order in which the resources are disabled

Disabling Priority	Resource
1	Hot resources
2	5V multiplier
3	3.3V multiplier
4	5V adder
5	2.4V multiplier
6	3.3V adder
7	2.4V adder
8	1.5V multiplier
9	1.5V adder

disabled followed by multipliers operating at 3.3 V, followed by adders operating at 5 V, followed by multipliers operating at 2.4 V, etc.

The proposed algorithm disables resources in the order of their priority. For each configuration, it schedules the nodes and checks if its computation time $L_{alg} < L$. If it is true, then it disables the resources with the next highest priority and reschedules the nodes. If it is not true ($L_{alg} > L$), then the specific resource cannot be disabled for all the control cycles.

Temperature Aware Low Power Schedule:

Step 1: Calculate the freedom of the nodes using ASAP and ALAP.
Step 2: Apply Minimum-Time Resource-Constrained Algorithm
If $L > L_{low}$,
    Step 3: Make/update the priority table for resource disabling.
    Step 4: Disable resources with the highest priority.
    Step 5: Make/update the ready table.
    Step 6: Schedule the ready nodes with respect to their freedom.
    Step 7: If not feasible go to step 4.
    Step 8: Check temperature sensors, go to step 3.

3.3 Complexity Analysis

The complexity of the Minimum-Time Resource-Constrained Algorithm is dominated by the step where the nodes are ordered with respect to their freedom.

The complexity of ordering is $O(nlogn)$ where n is number of nodes (multiplication, addition etc.). In the temperature-energy aware latency and resource constrained algorithm, some resources are disabled with respect to their priorities and nodes are again scheduled to resources with respect to their freedom. The complexity is $O(Rnlogn)$, where R is number of different resources. Since $R << n$ the complexity of overall algorithm is $O(nlogn)$

4   Results

In this section, we present the results obtained by running our algorithm on some high-level synthesis benchmarks (HAL, AR Lattice, EW Filter, LMS, FIR Filter, DCT and Band Pass Filter). We present the results when actual energy consumption values in [ ] are used. The switching activity of the nodes is assumed to be the same. The results for (i.e., 1 multiplier and 1 adder operating at 5 V, 3.3 V, 2.4 V, and 1.5 V) have been tabulated in Table 3. In this table, E_5 is the energy dissipation corresponding to the supply voltage of 5V. Timing constraints are given for two different values: $1.5L_{crit}$ and $2L_{crit}$, where L_{crit} is the critical path delay. R_{hot} is the resource with the critical temperature. R_{hot} is chosen 5V, 3.3V, 2.4V, 1.5V and randomly respectively.

Energy savings (E_{redn}) for each latency and critical temperature resource has been tabulated in Table 3. If $L = 1.5L_{crit}$, then the average energy reduction is 35.5% and for $L = 2L_{crit}$, then the average energy reduction is 50.8% compared to E_5. Energy savings are higher when $L = 2L_{crit}$ than $L = 1.5L_{crit}$, since an increase in latency facilities more nodes to being assigned to lower voltages. Energy reduction when 5V resources reach the critical temperature ($R_{hot} = 5V$) is the highest since we have sufficient latency to schedule nodes to the remaining

Table 3. % Energy reduction for the set of benchmarks when $res = 1$

Benchmark	Latency	$R_{hot} = 5V$ % E_{redn}	$R_{hot} = 3.3V$ % E_{redn}	$R_{hot} = 2.4V$ % E_{redn}	$R_{hot} = 1.5V$ % E_{redn}	$R_{hot} = rand.$ % E_{redn}
Diffeq	$1.5L_{crit}$	41	14.3	28.7	40.9	23.2
$E_5 = 15614pJ$	$2L_{crit}$	63.5	26.7	38	62.5	35.5
AR-Lattice	$1.5L_{crit}$	49.2	27.3	33.9	47.3	35.8
$E_5 = 41480pJ$	$2L_{crit}$	66.3	39.7	42.1	62.4	46.6
Elliptic Wave	$1.5L_{crit}$	55.6	32.2	39.2	52.5	40.1
$E_5 = 23100pJ$	$2L_{crit}$	64.2	37.4	46	61	47.8
LMS	$1.5L_{crit}$	49.1	18.1	34.5	48	24.1
$E_5 = 23480pJ$	$2L_{crit}$	67.1	44.2	53.4	62.3	64.3
FIR	$1.5L_{crit}$	47.2	20.3	27	45.1	27.1
$E_5 = 21802pJ$	$2L_{crit}$	68.1	38.3	44.2	60.4	42.7
DCT	$1.5L_{crit}$	44.9	28.2	31	41.3	32.3
$E_5 = 35974pJ$	$2L_{crit}$	66.8	41	46.5	61	46.5
BPF	$1.5L_{crit}$	43.6	25.8	30.1	41.2	24.1
$E_5 = 32054pJ$	$2L_{crit}$	66	32.1	39.3	56.2	39.7
Average	$1.5L_{crit}$	47.2	23.7	32.1	45.2	29.5
	$2L_{crit}$	66	37.1	44.2	60.8	46.1

Table 4. % Utilization of hot resources

Benchmark	Latency	$R_{hot} = 5V$ % U_{5V}	$R_{hot} = rand$ % $U_{rand.}$
Diffeq	$1.5L_{crit}$	20.6	0
	$2L_{crit}$	0	13
AR-Lattice	$1.5L_{crit}$	23.5	11.5
	$2L_{crit}$	0	4.6
Elliptic Wave	$1.5L_{crit}$	15.1	9.7
	$2L_{crit}$	0	2.1
LMS	$1.5L_{crit}$	8.3	7.5
	$2L_{crit}$	0	3
FIR	$1.5L_{crit}$	18.3	15.8
	$2L_{crit}$	0	3.8
DCT	$1.5L_{crit}$	12.4	5.4
	$2L_{crit}$	0	2.6
BPF	$1.5L_{crit}$	25	2.8
	$2L_{crit}$	0	6.2
Average	$1.5L_{crit}$	17.6	7.5
	$2L_{crit}$	0	5

lower voltage resources (3.3V, 2.4V, 1.5V). On the other hand, disabling 3.3V resource ($R_{hot} = 3.3V$) causes more nodes to be assigned to 5V and reduces power savings significantly.

Proposed algorithm tries not to assign any nodes to the hot resources. However that might not be always possible. Table 4 shows utilization of hot resources when $R_{hot} = 5V$ and $R_{hot} = random$. Algorithm forces to use 5V resources when the latency is tight. The average hot resource usage when $L = 1.5L_{crit}$ is 17.6%. Please note that algorithm did not use hot resource when temperature critical resource is 3.3V, 2.4V or 1.5V. When hot resource is randomly changed the average hot resource utilizations are 7.5% and 5% for $L = 1.5L_{crit}$ and $L = 2L_{crit}$ respectively. Hot resources usage occurs if 5V resources reach the critical temperature and latency is tight or temperature of resource is changed before execution of task in that specific resource is not completed. Similarly when the latency is tight we will have less flexibility to use low voltage or cold resources. Therefore energy consumption and usage of hot resources increase with the decrease of given latency.

5   Conclusion

In this paper, we present a new scheduling scheme under resource and latency constraint that eliminates hot spots and minimizes power/energy consumption for the case when different energy/delay ratio resources are available and temperature information of the resources is provided online by microsensors. The proposed scheme minimizes the power/energy consumption by distributing the slack among the nodes according to the condition derived using the Lagrange multiplier method. The scheme is implemented using an iterative algorithm, where in each iteration, hot temperature resources and increasing number of resources with high-energy-delay ratio are disabled and the nodes in data flow graph are scheduled using a list-based algorithm. The average reduction obtained by the low power algorithm is 50.8% when the latency constraint is 2 times the critical-path delay and one of the resource temperature is critical. Utilization of the hot resource is average 1% for this case.

References

1. Augsburger, S., Nikolij, B.: Reducing power with dual supply, dual threshold, and transistor sizing. Proc of Int. Conf. on Comp. Design. (2002) 316-321
2. Bakker A., Huijsing. J.: High-Accuracy CMOS Smart Temperature Sensors. Kluwer Academic Publishers. (2000) 113–118
3. Chandrakasan A., et al.: Design considerations and tools for low-voltage digital system design. Proc. of the Design Automation Conference. (1996) 113–118
4. Chang, J. -M., Pedram, M.: Energy minimization using multiple supply voltages. IEEE Trans. on VLSI Systems. vol. 10. no. 1. (2002) 6-14
5. Hung, W., et al.: Total power optimization through simultaneously multiple-Vdd multiple-Vth assignment and device sizing with stack forcing. Proc. of the Int. Symp. on Low Power Electronics and Design. (2004)
6. Linden, H. D.: Handbook of Batteries. 2nd edition. McGraw-Hill, New York. (1995)
7. Johnson, M. C., Roy, K.: Datapath scheduling with multiple supply voltages and level converters. ACM Trans. Design Automation Electronic Syst. (1997) 227-248
8. Lin, Y.-R., et al.: Scheduling techniques for variable voltage low power design. ACM Trans. Design Automation Electronic Syst. (1997) 81-97
9. Manzak, A., Chakrabarti, A.: A low power scheduling scheme with resources operating at multiple voltages. IEEE Trans. on VLSI Systems. vol. 10. (2002) 6-14
10. Nguyen, D. et al.: Minimization of dynamic and static power through joint assignment of threshold voltages and sizing optimization. Proc. of the Int. Symp. on Low Power Electronics and Design. (2003)
11. Pant, P., Roy, R., and Chatterj, A.: Dual-Threshold voltage assignment with transistor sizing for low power CMOS circuits. IEEE Trans. on VLSI, vol. 9, (2001) 390-394
12. Raje, S., Sarrafzadeh, M.: Scheduling with multiple voltages. Integr. VLSI J. (1997) 37-60
13. Shiue, W.-T., Chakrabarti, C.: Low power scheduling with resources operating at multiple voltages. IEEE Trans. Circuits Syst. II. vol. 47. (2000) 536-543
14. Stojanovic, V., et al.: Energy-delay tradeoffs in combinational logic using gate sizing and supply voltage optimization. Proc. European Solid-State Circuits Conf. (2002)
15. Usami, K., Horowitz, M.: Clustered voltage scaling technique for low power. Proc. Int. Symp. Low Power Electronics Design. (1995) 3-8.

Memory Hierarchy Energy Cost of a Direct Filtering Implementation of the Wavelet Transform

Bert Geelen[1,2], Gauthier Lafruit[1], V. Ferentinos[1,3],
R. Lauwereins[1,2], and Diederik Verkest[1,2,4]

[1] IMEC vzw, Kapeldreef 75, B-3001 Leuven, Belgium
{bgeelen,lafruit,ferentin,lauwerei,verkest}@imec.be
[2] Department of Electrical Engineering, Katholieke Universiteit Leuven, Belgium
[3] Department of Electrical and Computer Engineering, University of Patras, Greece
[4] Department of Electrical Engineering, Vrije Universiteit Brussel, Belgium

Abstract. A new implementation of the Wavelet Transform (WT), foregoing the traditional recursive filtering operations and with promising memory requirement properties is described: the Direct Filtering implementation. Its memory hierarchy energy performance is compared to the traditional Level-By-Level implementation and the memory optimized Block-based approach. This comparison is performed using the Memory Hierarchy Layer Assignment tool (MHLA). The results indicate the Direct Filtering implementation described here as such is not a likely candidate to replace the other implementations, but it has improvement possibilities and characteristics that can make it useful in certain contexts.

1 Introduction

Portable multimedia applications impose difficult requirements on the platforms they run on, because they should be energy efficient for extended battery life but also provide high performance. Many multimedia applications have huge data storage requirements in addition to their pressing computational requirements. Previous studies [,] show that high off-chip memory latencies and energy consumptions are likely to be the limiting factor for future embedded systems.

The subject of this paper is the Wavelet Transform (WT). The WT produces a space/frequency decomposition of a signal, with inherent multi-resolution characteristics. These characteristics are useful to achieve scalability. Currently the WT is already the basis of the JPEG2000 [ ] and the MPEG4-VTC [ ] compression standards. It is also the subject of an exploration activity as an alternative to the upcoming MPEG Scalable Video Coding (SVC) standard [ ]. SVC – which from the multi-resolution filtering perspective is very similar to the WT – intends to permit video coding/decoding under (rapidly) varying conditions such as channel variations, bandwidth, terminal resources,... Therefore it will be useful for video coding applications in wireless devices. A complete video

V. Paliouras, J. Vounckx, and D. Verkest (Eds.): PATMOS 2005, LNCS 3728, pp. 107–116, 2005.

coding chain will combine advanced motion estimation/compensation variants, the WT's multi-resolution filtering and entropy coding and will put tremendous performance constraints on the executing platform. Therefore attention should be paid to an efficient memory hierarchy study for the WT.

Reference [ ] presents an extended Data Transfer and Storage Exploration methodology (DTSE) developed at IMEC. Of special interest for these experiments is the Data Reuse Exploration step. [ ] presents the Memory Hierarchy Layer Assignment (MHLA) tool, which analyzes code to find possible data reuse and explores how this reuse can be optimally exploited by introducing small arrays containing *copies* of the larger arrays that exhibit the data reuse [ ]. Most accesses to these larger arrays are then done through the smaller copies, which can be stored in smaller levels of the data memory hierarchy, leading to significant power savings and performance gains.

MHLA is further used to determine the actual energy cost related to the memory transfers in three promising software implementations of the WT: level-by-level, block-based and direct-filtering. Level-by-level is the most straightforward and simple implementation of the WT algorithm, but it requires large amounts of intermediate memory and has a corresponding high energy cost. The block-based approach reschedules the WT calculations to produce so-called parent-children-trees ASAP. This allows smaller and more efficient intermediate memories, but it also leads to quite complex code and bigger foreground memory demands to exploit all data reuse. In direct-filtering, intermediate calculations and the corresponding memory accesses are avoided altogether. This leads to simple code again, but also to more complex filtering operations. In [ ] a high-level theoretical comparison was made of the energy dissipation and throughput of the level-by-level and block-based approaches. This paper adds a practical evaluation of the energy costs using MHLA, including an optimized mapping to a memory hierarchy, confirming the conclusions of [ ] that in most cases the best results are achieved with the block-based approach.

The paper is organized as follows: in Section 3 an overview is given of the Wavelet Transform (WT) and its possible implementation methods. Section 2 gives a brief explanation of the Memory Hierarchy Layer Assignment tool (MHLA), used to optimize and compare the different implementations. Section 4 presents the results of these comparisons, while Section 5 shows some possibilities for improvements to the Direct Filtering implementation. Finally, conclusions are drawn in Section 6.

2   Memory Hierarchy Layer Assignment

In these experiments, the studied WT implementations were realized in software. Since a large part of the energy consumption in multimedia applications is caused by the transfer and storage of data, memory optimizations should be a prime focus in their design. Traditionally management of the memory hierarchy is a task for a hardware cache, but these can often lead to suboptimal results. Applying the "Memory Hierarchy Layer Assignment" tool (MHLA) to explicitly manage the memory hierarchy in software can lead to much better results. MHLA is based on the principles of IMEC's DTSE-methodology (Data Transfer and Storage [ ]), which focuses on the different aspects of data transfer and storage and indicates how this should be performed efficiently. MHLA allows optimal exploitation of the memory hierarchy through the study of the data reuse and memory hierarchy mapping possibilities and it evaluates the energy-cost and execution time when executing (optimally) on a certain candidate memory hierarchy.

Figure 1 illustrates the principles on which MHLA is based. Multimedia applications consume immense amounts of data, frequently stored in large SDRAM memories. These large background memories are also rather slow and energy-inefficient. The upper configuration in Fig. 1 represents an architecture without a memory hierarchy, where 100% of the necessary data is always accessed from this expensive background memory. Energy-savings can be realized by accessing frequently used data from cheaper foreground memory. This optimization requires adding layers of smaller memory to the architecture, to which the frequently used data can then be copied. In the second configuration of Fig. 1 a smaller, three times more energy-efficient layer of memory A' has been added to the system. The processor still consumes the same 100% of data, but it accesses this data more frequently from the cheaper foreground memory, while accesses to the background memory are reduced to 5% of the original amount. This would lower the power consumption to a factor of $P = 100\% \times 0.3 + 5\% \times 1 = 0.35$ of the upper configuration.

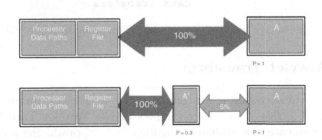

Fig. 1. MHLA Memory Hierarchy Optimization principles

These principles are the same as those on which caches are based: exploitation of spatial and temporal locality of data accesses. Based on these principles a hardware cache automatically copies data from the background memory to the cache memory, assuming the processor will require it in the near future. In contrast to hardware caches, the MHLA approach performs these transfers explicitly in software, to foreground memory configured as scratchpad memory. In this way the energy (and area) overhead corresponding to automatic hardware caching can be avoided. Moreover MHLA studies the data reuse possibilities of the application globally at design time, whereas a hardware cache only relies on very general principles of high probability of reusing recently used data. MHLA can therefore evaluate the trade-offs resulting from the constraints of the limited foreground memory sizes and different data reuse copies with specific data lifetimes, mapped in different ways to memory hierarchy, resulting in an optimal exploitation of a memory hierarchy and hence high energy efficiency.

The results of an earlier paper [ ] demonstrate what gains can be made using MHLA. In this paper, a level-by-level implementation of a 2D inverse WT was mapped to a ADSP-BF533 Blackfin DSP [ ]. After general and shared low-level optimizations (such as data type refinements and data layout transformations), both a traditional hardware-cache-assisted implementation and a MHLA-assisted implementation were realized. Figure 2 illustrates that in this case, using the MHLA-assisted, software-controlled memory transfers results in a twice as fast implementation (or equivalently, lower energy consumption at fixed execution time performances). Moreover, since MHLA can also be used for evaluating the energy and timing costs of an application given a certain memory hierarchy, it can also be used to perform a comparison of the real energy dissipation of the three implementations, as opposed to the high-level estimations of [ ].

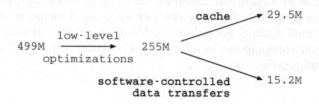

Fig. 2. Comparison of traditional and MHLA mapping to ADSP-BF533 DSP

3   The Wavelet Transform

Wavelet-based video coding is a promising new technology for use in scalable video coding. It is based on the principles of the Wavelet Transform (WT), which is a special case of a subband transform [,   ] producing a type of localized time-frequency analysis [ ]. The transform coefficients reflect the energy distribution of the source signal in both the space and the frequency domain. Figure 3 shows an example of the transformed Lena image, as a hierarchy of subimages grouped in levels. Multiresolution coding can be easily achieved by selective decoding of transform coefficients related to a certain frequency range. Schemes coded by wavelet-based video coding have proven that they can preserve excellent rate-distortion behavior, while offering full scalability [,   ].

3.1   Level-by-Level

In digital compression applications, the WT is traditionally computed as an iterated filter bank [ ]. Figure 4 shows a schematic representation of the multilevel 1D Forward Wavelet Transform (FWT) procedure.

Fig. 3. Lena Input Image, 2 Level Wavelet Transformed Lena and WT Output Subband Organization

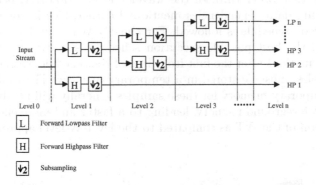

Fig. 4. Forward Wavelet Transform Filtering Procedure

Each level of the FWT consists of a Lowpass (L) and a Highpass (H) filtering followed by a subsampling operation. The output of the lowpass filtering, after subsampling, represents the input of the next level of the transformation, while the highpass samples are not involved in the calculation of the successive FWT levels. The final output of an n level FWT is the result of the nth lowpass filtering (n-LP), completed with all levels highpass filtering results (1-HP, 2-HP, 3-HP, ... , n-HP). The filtering procedure consists of a number of multiply-accumulate operations. The efficiency of this implementation procedure as a set of iterated filter banks has contributed to the success of the WT in many applications, due to its simplicity: only one filter pair needed to be designed and the only necessary operations were (rather short) filtering operations.

The filter creates strong dependencies between the data in the different levels of the transformation. The simplest scheduling order to respect these dependencies is the traditional level-by-level implementation. Here, the WT is calculated *level by level* starting from the original input signal. However, this schedule requires large amounts of memory to store the results of the intermediate levels. More precisely, in the level-by-level algorithm it is only after all the data of an intermediate level has been calculated (and temporarily stored in memory) that the calculation of the next (higher) level starts. Consequently, the required memory size to store the temporal data is equal to the input length. These large amounts of intermediate data are unlikely to fit in small, efficient memories and will need to be stored in SDRAM.

3.2 Block-Based

The block-based implementation analyzed in [ ] is an "as soon as possible" scheduling algorithm (ASAP), which consumes input data and intermediate results as soon as possible after their creation, so their storage in SDRAM can be avoided. The block-based schedule takes advantage of the fact that due to certain application constraints (Motion Estimation/Compensation, Entropy Coding) the WT has to be performed in a block-by-block basis. Ideally, for each calculation step (guided by each input block), the execution schedule would prioritize the vertical order through the wavelet levels. However, because of the wavelet filtering data dependencies mentioned earlier, this is not feasible, and thus a "skewed" schedule is followed. For the 1D WT, the difference between the ideal and the best possible execution schedule is drawn in Fig. 5. To resume the calculations after the first front line of the schedule, the last processed lowpass samples, must be stored in a temporary memory. The total amount of necessary temporary memory for these samples generally will be able to fit in a more efficient foreground memory, leading to a faster and more energy efficient implementation of the WT as compared to the level-by-level approach.

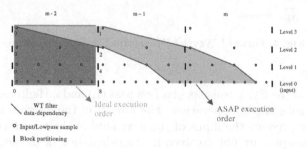

Fig. 5. 1D WT, the optimal As-Soon-As-Possible (ASAP) execution schedule compared to the ideal (but not feasible) schedule

3.3 Direct Filtering

The block-based implementation in section 3.2 tried to minimize intermediate memory demands, by changing the execution order. The outputs were generated "as soon as possible", but essentially the identical calculations were performed as for the level-by-level implementation in section 3.1. More specifically, lowpass samples of intermediate levels, not used in the final output, were still calculated. In the proposed implementation method, *Direct Filtering*, this is not the case. In this implementation method, only the values actually used in the output are calculated, directly from the input samples as shown in Fig. 6. It also shows one of the immediately apparent disadvantages of the direct-filtering method: the filter sizes can get extremely large, exponentially increasing with the number of levels. This is presumably also the reason why this rather straightforward calculation method has never been considered. It can be argued though that, since in modern multimedia applications the largest bottleneck is not actually the

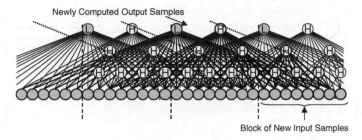

Fig. 6. Direct Filtering

arithmetic processing cost but the data transfer and storage cost, this increase in filter widths can be compensated by a decrease in memory accesses, as all transfers corresponding to intermediate lowpass values are avoided. Moreover the analysis filter data dependencies throughout the levels are not present here, allowing more scheduling freedom.

For the comparisons in these experiments an "as soon as possible" schedule was more-or-less also used for the direct-filtering implementation: every time a new block of input samples was read all output values that could be fully calculated were calculated.

4   Comparison

Since MHLA can be used not only for memory hierarchy optimizations, but also for evaluating the energy and timing costs of an application given a certain memory hierarchy, it could also be used to perform a comparison between the energy efficiency of the traditional WT implementations and the proposed Direct Filtering approach. For this comparison a fixed size background memory of 1Mb was used, and a variable sized foreground memory, to compare the energy necessary for one execution of a 3 level 1D-WT.

Figure 7 gives the results of these experiments. It confirms the conclusions of [ ]: in most cases the block-based approach achieves the highest energy efficiency. Generally speaking the curves for the three implementations exhibit the same characteristics: initially, for small foreground memories, there is not enough foreground capacity to exploit much data reuse. While increasing the foreground memory size, the energy consumption gradually decreases, as more and more useful data can be stored in the foreground. At a certain point, the energy consumption slowly increases again due to the decreasing energy efficiency of the growing foreground memory, which is then no longer compensated by extra energy gains through storing more data in the foreground.

We can conclude that, for sufficient foreground memory, the block-based implementation achieves the highest energy efficiency: it consumes about 25% less energy than the direct-filtering implementation, while the direct-filtering itself also consumes about 25% less then the level-by-level implementation. The good block-based performance can be explained by its low arithmetic cost due

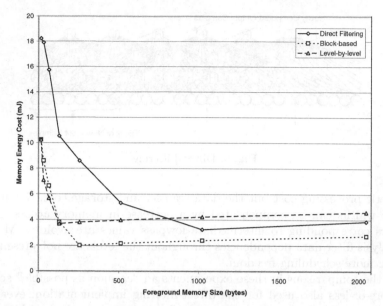

Fig. 7. Comparison of results

to short filter lengths and its ability to access intermediate data from cheap foreground memory. Direct-filtering on the other hand has some very large filter lengths (29 taps for the highest level, compared to 5 taps for block-based in a 5/3 tap 3 level WT), also leading to quite some memory accesses and long data lifetimes compared to block-based. It does have the advantage that it doesn't have to access intermediate lowpass data from expensive background memory which level-by-level is required to do, explaining its better performance compared to level-by-level.

Another conclusion that can be made is that for the optimal performance, block-based and level-by-level require much less foreground memory space than direct-filtering: 256 and 128 bytes respectively, compared to 1024 bytes for direct-filtering. This can be partially explained by the longer data lifetimes corresponding to the long filter lengths: to perform one of these filterings at least 29 samples should fit in the foreground, while for level-by-level this would be around 5 samples and three times 5 samples for the block-based (since all three levels are alive simultaneously). Another reason is that these longer filters should themselves also be stored in memory (62 filter taps in total for all filters for direct-filtering), which corresponds to increased memory requirements.

A last conclusion is that for underdimensioned foreground memories, the best performance is surprisingly achieved by level-by-level. This can be motivated by the simultaneous lifetimes of all filtering levels for block-based: the foreground should then be large enough to simultaneously accommodate copies for all levels to exploit all possible reuse, while for level-by-level, the filtering of the levels occurs sequentially, so data reuse for all levels can already be exploited when there is enough foreground space to store copies for one level.

5 Future Work

The experiments show that the direct-filtering approach doesn't achieve the lowest energy consumption anywhere. It should be remarked though that the block-based approach has a quite complex control flow, caused by following the skewed bands of Fig. 5. This is not the case for direct-filtering and level-by-level. On specific, less control flow oriented architectures this can make direct-filtering the recommended implementation. Moreover, there are still quite some improvements possible for direct-filtering. One option is to perform direct-filtering up to a certain level, and processing the remaining levels in a level-by-level or block-based manner. This way, the largest filters of the upper levels (see Fig. 6) are avoided, thereby discarding the largest bottleneck for direct-filtering. Another possible improvement is exploiting the symmetry of the wavelet filters. When working with typical filter sizes of about 5 taps, as for level-by-level and block-based, this might not be very relevant, but for the increased direct-filtering filter sizes around halve the memory necessary for the taps can be saved. Finally, since there is more scheduling freedom for direct-filtering, the input data lifetimes can be slightly shortened by reordering the filtering operations.

6 Conclusion

We have shown the application of MHLA is very useful in the evaluation of the energy efficiency of different WT implementations. Specifically, we have observed that the newly proposed direct-filtering approach does not immediately offer huge advantages over existing implementations. Given sufficient attention to possible optimizations, it might in the future become a likely candidate under certain circumstances. Moreover, we have confirmed the conclusion of [ ] that in most cases a block-based approach has the highest energy efficiency.

Acknowledgements

This work was supported in part by the Institute for the Promotion of Innovation through Science and Technology in Flanders (IWT-Vlaanderen, PhD bursary Bert Geelen).

References

1. Catthoor, F., Wuytack, S., De Greef, E., Balasha, S., Nachtergaele, L., Vandecappelle, A.: Custom Memory Management Methodology. Kluwer (1998)
2. Vijaykrishnan, N., Kandemir, M., Irwin, M.J., Kim, H.Y., Ye, W.: Energy-driven integrated hardware software optimizations using simplepower. In: Proc. Int. Symp. on Computer Architecture. (2000)
3. Skodras, A., Christopoulos, C., Ebrahimi, T.: The jpeg2000 still image compression standard. IEEE Signal Processing Magazine **9** (2001) 36–58

4. Sodagar, I., Lee, H., Hatrack, P., Zhang, Y.: Scalable wavelet coding for synthetic/natural hybrid images. IEEE Trans. on Circuits and Systems for Video Technology **9** (1999) 244–254
5. MPEG: Call for proposals on scalable video coding technology. In: http://www.chiariglione.org/MPEG/working_documents/mpeg-04/svc/cfp.zip. (2003)
6. Masselos, K., Catthoor, F., Kakarudas, A., Goutis, C., De Man, H.: Memory hierarchy layer assignment for data re-use exploitation in multimedia algorithms realized on predefined processor architectures. In: The 8th IEEE Int. Conf. on Electronics, Circuits and Systems - ICECS. Volume 1. (2001) 281–287
7. Van Achteren, T., Lauwereins, R., Catthoor, F.: Data reuse exploration techniques for loop-dominated applications. In: Proc. 5th ACM/IEEE Design and Test in Europe Conf. - DATE. (2002) 428–435
8. Zervas, N.D., Anagnostopoulos, G.P., Spiliotopoulos, V., Andreopoulos, Y., Goutis, C.: Evaluation of design alternatives for the 2d-discrete wavelet transform. IEEE Trans. on Circuits and Systems for Video Technology **11** (2001) 1246–1262
9. Geelen, B., Brockmeyer, E., Durinck, B., Lafruit, G., Lauwereins, R.: Alleviating memory bottlenecks by software-controlled data transfers in a data-parallel wavelet transfor on a multicore dsp. Proc. IEEE Sig. Proc. Symp. SPS-DARTS (2005)
10. AnalogDevices: Analog devices blackfin processors. In: http://www.analog.com/processors/processors/blackfin. (2005)
11. Woods, J.W., O'Neil, S.: Subband coding of images. IEEE Trans. on Acoust., Speech and Signal Processing **34** (1986) 1278–1288
12. Woods, J.W.: Subband Image Coding. Kluwer Academic Publishers (1991)
13. Vetterli, M., Kovacevic, J.: Wavelets and Subband Coding. Prentice-Hall (1995)
14. Van der Auwera, G., Munteanu, A., Schelkens, P., Cornelis, J.: Bottom-up motion compensated prediction in the wavelet domain for spatially scalable video coding. IEEE Electronics Letters **38** (2002) 1251–1253
15. Woods, J.W.: A resolution and frame-rate scalable subband/wavelet video coder. IEEE Trans. on Circuits and Systems for Video Technology **11** (2001) 1035–1044
16. Mallat, S.G.: A theory for multiresolution signal decomposition: The wavelet representation. IEEE Trans. on Pattern Analysis and Machine Intelligence **11** (1989) 674–693
17. Lafruit, G., Nachtergaele, L., Bormans, J., Engels, M., Bolsens, I.: Optimal memory organization for scalable texture codecs in mpeg-4. IEEE Trans. on Circuits and Systems for Video Technology **9** (1999) 218–243

Improving the Memory Bandwidth Utilization Using Loop Transformations*

Minas Dasygenis[1], Erik Brockmeyer[2], Francky Catthoor[2,**],
Dimitrios Soudris[1], and Antonios Thanailakis[1]

[1] VLSI Design and Testing Center,
Department of Electrical and Computer Engineering,
Democritus University of Thrace, 67 100 Xanthi, Greece
{mdasyg,dsoudris,thanail}@ee.duth.gr
[2] DESICS, IMEC, Kapeldreef 75,
Leuven, Belgium
{Erik.Brockmeyer,Francky.Catthoor}@imec.be

Abstract. Embedded devices designed for various real-time multimedia and telecom applications, have a bottleneck in energy consumption and performance that becomes day by day more crucial. This is imposed by the increasing gap between processor and memory speed. Many authors have addressed this problem, but all existing techniques either consider only performance without any other trade-off, or they operate at the level of individual loops. We fill this gap, by presenting a technique which achieves parallelization in the memory accesses through four loop transformations. Our estimations from two real-life applications from the multimedia and telecom domain, reveal that using our technique, we can either increase the performance (up to 35%) or lower the energy consumption (up to 20%) for the same cost.

1 Introduction

The design of embedded or integrated systems has become more and more complex, especially due to their particular characteristics and specific usage (for instance mobile computing), which require stringent energy and area constrains. On the other hand, in data dominated applications, like interactive multi-media and telecom applications, data storage and transfers are the most important factors in terms of meeting the real time constrains that are imposed. The main bottleneck of these applications is the huge amount of transfers and storage requirements from and to (on-chip and off-chip) memory that results in extreme energy consumption and performance degradation [ ]. This bottleneck becomes gradually more crucial, as the gap between processor and memory speeds continues to grow. Hence, techniques are required that on the one hand allow to

---

* This work was partially sponsored by a scholarship from Public Benefit Foundation of Alexander S. Onasis and from Marie Curie Host Fellowship project HPMT-CT-2000-00031
** Also Professor at the Katholieke Universiteit Leuven

V. Paliouras, J. Vounckx, and D. Verkest (Eds.): PATMOS 2005, LNCS 3728, pp. 117–126, 2005.

meet the real time constrains and on the other hand achieve efficient embedded realizations of the memory architecture.

Various authors have suggested different techniques to increase the utilization of the memory hierarchy and gap the mismatch evolution in bandwidth between the faster processors and slower memories, but most of them do not take into consideration the various trade-offs possibilities that can be found in the applications. A thorough survey of the state-of-the-art techniques that are used in performing data and memory related optimizations is presented by Panda *et al* [ ], which describes a broad spectrum of optimization techniques at varying levels of granularity. Chen Ding *et al* [ ], also, addressed the memory bandwidth constrain and the lack of compiler support for an efficient usage of it, and proposed a group of compiler optimizations combined with a graph model, for optimizing it. Zhong Wang *et al* [ ], also contributed to this domain, by suggesting the use of prefetch combined with three levels of memory hierarchy and special hardware memory units using a multidimensional data flow graph. Furthermore, Antoine Fraboulet *et al* [ ] proposed a framework for performing loop alignment transformations for optimizing the memory accesses using a number of ILP equations, but their use is limited to applications without any conditionals. Kandemir *et al* [ ], presents a unified strategy to optimize out-of-core programs for locality, parallelism and communication on distributed memory message passing systems. Other authors, also, published techniques on scheduling using pipelining [ ], but their techniques are oriented at speed optimization only and do not consider any trade-off. Finally, a formalized methodology called Data Transfer and Storage Exploration (DTSE) [ ] has been presented, which allows to systematically optimize data-dominated applications for energy and system bus load reduction. Our work can fit in there as a new sub-step.

In this work we address the problem of low utilization of the bandwidth of the memory hierarchy by employing a set of well known loop transformations, namely loop pipelining, loop pealing, loop merging and loop unrolling, on two realistic algorithms belonging to two distinct application domains: The QSDPCM application from the multimedia processing domain and the DWRR application from the telecom domain. Our results demonstrate that by increasing the parallelism, significant benefits in energy and performance are achieved.

2   Target Memory Architecture

To estimate the effect of applying the transformations, we used two tools to analyze the C source code of every application: (i) Storage Bandwidth Optimization (SBO), and (ii) Memory Allocation and Assignment (MAA) [ ]. The first tool produces a set of constrains for the specific user-defined memory architecture. The set of constrains is used to compute a cost factor. The second tool is used to evaluate the energy consumption of the algorithm given the set of constrains and a description of the memory architecture. These tools are part of the ATOMIUM (**A** **T**oolbox for **O**ptimizing **M**emory **I**/**O** **U**sing geometrical **M**odels) tool suite [ ], which supports the DTSE (Data Transfer and Storage Exploration) methodology. Concerning the target memory architecture, the tools of ATOMIUM are flexible; various memory architectures can be used.

We evaluated the effect of our transformation on a generic memory architecture, that is often used in embedded systems (Figure 1). It is a multi-level memory hierarchy that has one or more instances of various storage types per layer. The platform can be fully predefined, fully custom or partial flexible. The template consists also of a (multi) processor programmable (software) or application specific (hardware) core(s).The memory subsystem is divided into layers. These layers consist out of one or multiple storage partitions (denoted as M1, M2, C1, SD1 in Fig. 1). Typically the storage type of a partition is either a hardware controlled cache or a local SRAM. Also, memory layers (on-chip and off-chip) support pipeline accesses, a characteristic which is especially beneficial when a large number of write or read accesses must occur.

Even though, the ATOMIUM tools can support a variety of memory architectures, in this research we configured the memory architecture to have two memory layers: an on-chip and off-chip layer. The on-chip layer can have 4 or 8 single-port scratchpad (SRAM) memories, while the off-chip layer has one single-port DRAM memory. Layer 1 (on-chip) has one partition M1 and Layer 2 (off-chip) has one partition SD1. The SBO and MAA tools estimate both performance and power using memory libraries. In our example we used an on-chip Mietec memory library of 0.35 μm and an off-chip Siemens pipelined SDRAM.

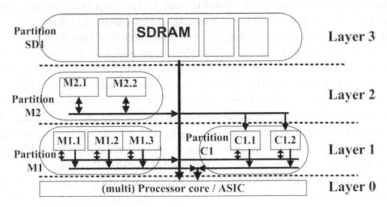

Fig. 1. In our technique we use a multilayered memory architecture. Every memory layer consists of one or more partitions (memory types)

3   Loop Transformations for Increasing Utilization in Memory Bandwidth

The basic loop transformations proposed in this section, are by themselves not new. Various authors have proposed them in the past to improve locality of data accesses and thus reduce cost, and on the other hand to increase processing speed. However, the combination of these loop transformations is important and is used to achieve a trade-off between speed, energy consumption and cost. We study the effect of these transformation on two real-life algorithms from the video and telecom domain (Section 4).

Our transformations belong to two groups: (i) enabling group, and (ii) parallelizing group transformations. In the first group belong transformations such as loop pipelining and loop pealing which are used as enabling transformations. The former is used to break dependencies (which is differentiated with scalar level pipeline), to enable the loop merging transformation and to increase the access scheduling freedom. The later is used to produce similar loops. In the second group belong transformations such as loop merging and loop (partial) unrolling that are essential for an efficient utilization of the memory bandwidth. Loop merging gives a boost on the parallelization of memory accesses with perhaps a drawback on an increase on the on-chip memory. Loop (partial) unrolling reduces the trip count of a given loop by putting more work inside the nest with the aim of promoting register re-use. This is a costly high level transformation, because it multiplies the code size and produces many self-conflicts. In this case, we can reduce the incurred cost of this transformation by combining it with a data group structuring or using an interleaved memory.

Loop pipelining is a well known transformation for lifting dependencies from different loop bodies. Specifically, the freedom for parallelizing memory accesses is limited by dependencies between the memory accesses. Since most of the compilers treats the accesses of every loop independently, these dependencies result in a minimum number of cycles in which this loop body is executed: this is called the critical path. However, in most of the cases, this does not take into account the possibility of parallelizing the last cycles of one iteration with the first cycles of the next iteration. Thus, the loops are usually pipelined in order to increase the execution time, where the execution periods of several iterations are overlapped. Pipelining the loop bodies means that memory transfers will happen in different iteration than the one that they suppose to happen. This also means that the lifetime of some arrays will be increased. The lifetime of every data structure affects the allocation and assignment to the different layers of the memory hierarchy, resulting in some cases in an increased layer size. Fortunately, this is not always the case, as the experimental results illustrate.

Loop pealing is another enabling loop transformation. This transformation is used when loop nests have different bounds, making loop merging impossible. Loop pealing is employed to produce loop nest with similar bounds, by removing a number of iterations and making them new loop nests (Figure 2).

Loop merging or loop fusion combines two loops without or with very limited dependencies, into one loop. This transformation is employed to increase the potential for increasing the parallelizing in memory accesses and may lead to an increase in cost in terms of memory size, similar to loop pipelining. Loop fusion may also introduce an overhead due to increased memory size, increased number of variables and increased code size. What typically happens is that two arrays, which had non-overlapping lifetimes before the merging, and therefore could use the same memory space, will be alive at the same time after the merging. The

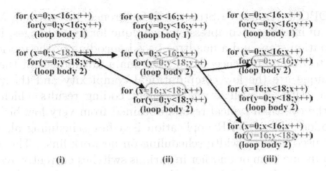

Fig. 2. Loop Pealing Example: (a) Original Loop structure, (b) Intermediate Loop Structure, and (c) Final Loop Structure

number of variables (registers) is increased for the same reason. Finally, the code size and control-flow complexity may increase due to the enabling transformations before the loop merging. After loop merging, the merged loop has only one body, which is executed as often as the minimum common number of times the two original loops were executed. However, the critical path of the merged loops can be larger than the maximum critical path of the two original loops, for instance due to new read-after-write dependencies. Also, new loops may have been introduced by enabling transformations (e.g. loop pealing).

Loop (partial) unrolling is the last transformation considered in this research work. Usually, compilers do not exploit parallelism between different iterations of a loop nest. To allow such parallelism, the loop has to be unrolled with a certain factor, making different iterations of the original loop nest to be executed in one iteration of the new loop nest. Loop unrolling may introduce extra cost in terms of memory size and certainly in terms of code size. Memory size may be increased when an outer loop is unrolled and this is combined with loop merging. On the other hand, the code size is always increased: the size of the unrolled loop body is multiplied by the unrolled factor. Even for hardwired, custom platforms this is an important consideration, since all unrolled iterations will be executed in parallel and resource sharing between them will therefore not be possible. In the best case, all unrolled iterations of the loop can be executed fully in parallel. However, when loop-carried dependencies are present, these limit the parallelization possibilities; in the worst case, even with unrolling, no further parallelization is possible. Because this transformation produces many self conflicts, a data group structuring step or interleaved memory may be required to relax this constrain.

4   Demonstrators and Experimental Results

Our demonstrator applications were selected to be two real-life representative applications of their respective domains: (*i*) one application of the multimedia domain: Quadtree Structured Difference Pulse Code Modulation (QSDPCM) [ ], and (*ii*) one application of the telecom domain, which is Deficit Weighted Round Robin (DWRR) [ ].

The QSDPCM (Quadtree Structured Difference Pulse Code Modulation) technique is an interframe compression technique for video images. It involves a motion estimation step, and a quadtree based encoding of the motion compensated frame-to-frame difference signal. The main advantages of the QSDPCM coding technique are the low computational complexity and the very regular algorithm structure, as well as the excellent coding results which look more pleasant to the eye than typical results obtained from very low bit-rate hybrid transform coders. The DWRR application is a fair scheduling algorithm that is commonly used for bandwidth scheduling on network links. The algorithm is implemented in one form or another in various switches currently available (e.g., Cisco 12000 series). Its format categorization is queue maintenance and packet scheduling for fair resource utilization. We selected two applications belonging to different domains in order to analyze and evaluate more objective the proposed transformations.

In order to evaluate the impact of the transformations, two prototype tools Storage Bandwidth Optimization (SBO) and Memory Allocation and Assignment (MAA) [ ], which are part of ATOMIUM [ ] tool-suite, were used. The SBO tool analyzes the application source file and trades off the memory bandwidth with the system real-time constrains. This is achieved by partly ordering the memory accesses such that the maximally required memory bandwidth is minimized. This can be done by a (more) efficient use of the memory ports. Typically, an overall target storage cycle budget is imposed, corresponding to the overall throughput. In addition, other real time constrains can be present which restrict the ordering freedom. The result of SBO is a set of constrains for the memory architecture. After that, the MAA tool tries to fit the memory architecture to the application's needs (Memory Allocation and Assignment), taking into account the constrains generated into the previous step. To carefully evaluate the effect of the previous step, a detailed custom memory architecture has to be decided. This objective is tackled in the Memory Allocation and Assignment (MAA) step, for which a systematic technique has been published in [ ].

4.1 Performance Estimations

In our experiments the measured time in cycles for the QSDPCM involves the cost of reading the luminance component of two sequential QCIF (176x144) frames, sub-sampling them by 2 and 4, performing all the three motion estimation steps (me-4, me-2, me) and computing the motion vectors.

In the QSDPCM code, transformations of loop merging and loop unrolling were applied for the sub-sampling previous frame by 2 and sub-sampling current by 2, as well as for the sub-sampling previous by 4 and sub-sampling current by 4. The next step was to perform loop pipelining of the different stages of the application. Thus, after breaking dependencies between me-2 and me-4 with this transformation, loop merging of these two computations was performed. Of course, together with loop merging, also enabling transformations like loop unrolling and loop pealing had to be done in order to bring the loops into a form that is easy to merge. Figure 3(i) depicts the optimized and original performance versus trade-off curves for the original implementation and the implementation after all the transformations proposed in this research work were performed. In this figure, the cost axis is computed by SBO, and corresponds to a specific assignment for every array into the memory hierarchy. The cost factor is related with the number of memories, the number of ports and the size of every memory. Costly implementations mean that in order to achieve the specific execution time, multi-port memories or bigger memories should be used. This curve presents all the possible implementations for the given application in terms of a relative cost and performance. After performing the transformations, the optimized trade-off curve is shifted towards lower execution times, as expected. It is evident that for the same cost value, greater performance (e.g. lower execution time) is achieved which proves that the memory hierarchy for the same application is utilized better. For example, in Figure 3 our results indicate that for cost 1 the optimized QSDPCM has execution time 2.4×10^6 cycles compared with the original QSDPCM, which has 3.7×10^6 (35% improvement).

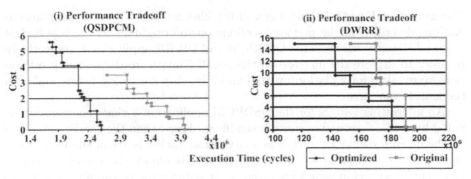

Fig. 3. Our technique allows the system to perform better for the same cost, illustrated for (i) QSDPCM and (ii) DWRR

Figure 3(ii) presents the performance estimations for the DWRR application. This code consists of two major functions. The fill_next_request which takes the packets as they arrive into the system and the schedule_packet which checks all the active queues and if there are enough "credits" on that queue the packet is transmitted. Software pipelining was performed both to these functions independently, making them access the next packet when operations were performed on the current packet. Loop merging was performed also on the second function which had more than one loops. This transformation was not able to be performed on the first function, because it only consists of a single loop. Finally, loop unrolling was performed by a factor of two, which gave a boost on the performance of the application. These transformations reduced the execution time of the application for the same cost factor (for example for cost 5 the optimized DWRR requires 165×10^6 cycles, while the original DWRR requires 192×10^6 cycles (14% improvement)), but on the other hand they increased the complexity, which is affiliated with the number of cycles per packet. In the original version, the complexity was 730 CPU cycles/packet, while in the optimized code the complexity was 1090 CPU cycles/packet. Although our transformations have increased the complexity per packet the total performance of the system is increased, because the memory communication is the bottleneck in this application. The transformations help to exploit the memory bandwidth much better (with some minor penalty in CPU cycles). Furthermore, the increase of CPU cycles per packet is not alarming, because CPUs have many cycles to spare, given the fact that they spend a lot of time stalling, waiting for memory reads/writes.

Finally, a common remark on both applications is that by performing transformations on the codes, the new trade-off curves have more points (e.g. more feasible implementations), because new memory access scheduling possibilities have arise to the surface. The designer has a higher flexibility in implementing his application(s) by either increasing the bandwidth for a given cost, or decreasing the cost for the same performance.

4.2   Energy Consumption Estimations

The graphs of Fig. 4(i-ii) and Fig.4(iii-iv) illustrate the effect of the transformations described in the previous sections on the original energy-cycle budget trade-off (Pareto) curves of the QSDPCM and DWRR applications, respectively. In order to explore the increase on the parallelization freedom, energy estimations were taken for a platform with 4 and 8 on-chip memories, before and after the transformations.

An interesting remark for the QSDPCM application is that the energy dissipated by the background memory is significant larger than the energy dissipated by the on-chip memory (on-chip memory consumption is 1% of the total memory energy consumption for QSDPCM). For this reason the use of a memory hierarchy of 8 on-chip memories, although it reduces the consuming energy, has a much lighter improvement on the overall energy consumption. But our transformations also have a very positive effect on the performance of the off-chip accesses.

On the DWRR application the use of 8 on-chip memories has a much better effect, because on-chip memory energy consumption is near 20% of the total energy consumption of the memory hierarchy (DWRR has higher on-chip/off-chip memory energy consumption than QSDPCM). Figures 4(ii) and 4(iv) illustrate the effect on the transformations on the energy dissipated on the on-chip memory hierarchy only. It is evident, that the memory bandwidth transformations resulted in gains to both QSDPCM and DWRR applications. In Figure 4(iii-iv), the curves of 4 and 8 on-chip memories for the original DWRR application almost coincide, which means that the flexibility of using a large number of memories is reduced and no parallelization memory accesses possibilities exist. Thus, using a memory hierarchy with a higher degree of parallelism has no effect. On the other hand, the optimized DWRR can utilize efficiently the 8 on-chip memories. For this reason the optimized DWRR (Figure 4(iii-iv)) for the 8 on-chip memories has lower energy consumption compared with the original DWRR of 8 on-chip memories.

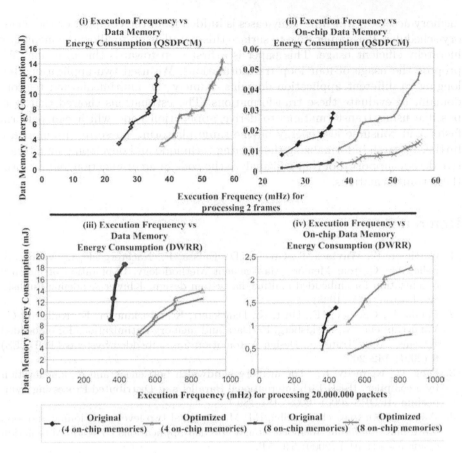

Fig. 4. The estimations of energy consumption of the on-chip and off-chip memory, and on-chip memory only, indicate that for the same cost we can have lower energy consumption. This is true for both applications

Both applications show energy improvement. This is caused by the following reasons: (i) loop transformations have broken dependencies and thus the system executes faster, (ii) the number of memory accesses to the off-chip memory is decreased due to loop merging, (iii) the lifetimes of some arrays are changed and thus, arrays can share on-chip memory space at non-overlapping periods, (iv) the size of some arrays change; the MAA tool can use in the same on-chip partition, memories of different sizes that match the size of some arrays. Smaller memories have lower power consumption than bigger memories.

5   Conclusions

The memory hierarchy plays an important role and should be used efficiently in embedded systems. Most of the applications have a potential parallelism in memory accesses, which in many cases is hidden from the compiler or the memory scheduler. It is imperative to surface this parallelism, to allow the memory's hierarchy efficient usage. This paper is a first step towards this direction. We propose the usage of four loop transformations. We used two applications belonging to different application domains, namely the multimedia and telecom domain, to evaluate these transformations. The estimations showed that it is possible using transformations to derive application code which can perform faster and consume less energy on the same platform. These results motivate further research in formalizing the steering technique and developing an interactive design support tool that will analyze the code of an application and perform these transformations.

References

1. Catthoor, F., Wuytack, S., Greef, E.D., Balasa, F., Nachtergaele, L., Vandecappelle, A.: Custom Memory Management Methodology, Exploration of memory organization for embedded multimedia system design. Kluwer Academic Publishers, Boston, MA (1998)
2. Panda, P., Catthoor, F., Dutt, N., Danckaert, K., Brockmeyer, E., Kulkarni, C., Vandecappelle, A., Kjeldsberg, P.: Data and memory optimizations for embedded systems. ACM Trans. on Design Automation for Embedded Systems (TODAES) 6 (2001) 142–206
3. Ding, C., Kennedy, K.: The memory bandwidth bottleneck and its amelioration by a compiler. IEEE 14th International Parallel and Distributed Processing Symposium (IPDPS 2000) (2000) 181–189
4. Wand, Z., Kirkpatrick, M., Sha, E.H.M.: Optimal two level partitioning and loop scheduling for hiding memory latency for dsp applications. Design Automation Conference (DAC) (2000) 540–545
5. Fraboulet, A., Huard, G., Mignotte, A.: Loop alignment for memory accesses optimization. 12th International Symposium on System Synthesis (ISSS) (1999) 71–77
6. Kandemir, M., Choudhary, A.: Compiler-directed scratch pad memory hierarchy design and management. 39th ACM/IEEE DAC (2002) 690–695
7. Altman, E., Gao, G.: Optimal software pipelining through enumeration of schedules. In: Proceedings of EuroPar Conf. Lecture Notes in computer science, Lyon, France, Springer Verlag (1996) 833–840
8. Wuytack, S., Catthoor, F., Jong, G.D., Man, H.J.D.: Minimizing the required memory bandwidth in VLSI system realizations. IEEE Trans. VLSI Systems 7 (1999) 433–441
9. IMEC: http://www.imec.be/design/atomium/ (2005)
10. Strobach, P.: QSDPCM - a new technique in scene adaptive coding. Proceedings 4th Eur. Signal Processing Conf. (EUSIPCO) (1988) 1141–1144
11. Shreedhar, M., Varghese, G.: Efficient fair queuing using deficit round robin. Proceedings of ACM SIGCOMM (1995) 231–242

Power-Aware Scheduling for Hard Real-Time Embedded Systems Using Voltage-Scaling Enabled Architectures

Amjad Mohsen and Richard Hofmann

Department of Computer Science 7, University of Erlangen-Nürnberg
91058 Erlangen, Germany
amjad.muhsen@informatik.uni-erlangen.de, rhofmann@cs.fau.de

Abstract. In this paper, we present a power-aware scheduling scheme for hard real-time embedded systems design. Our new approach can enhance the efficiency of both dynamic voltage scaling (DVS) and dynamic V_{th} scaling (DVTS). While optimizing the schedule in the time domain, the priorities of the tasks are modified dynamically based on their contribution to the overall power/energy reduction. The scheduling scheme leads to better "distribution" and "utilization" of slack intervals in the system which in return improves the efficiency of voltage scaling techniques. The voltage schedule is generated based on a global view of the components' energy profile when executing different tasks. The experimental results prove the applicability of our approach.

1 Introduction

Reducing power/energy consumption is a major design interest for mobile as well as for stationary digital systems. The constraint on the consumed power/energy override now other design constraints such as performance, especially in mobile environments. Hard real-time systems in mobile communication as well as in other application areas bring this topic to a real challenge: Delay constraints must be satisfied to guarantee correct operation while the consumed power/energy must be kept below a certain level for safety and reliability reasons.

To address this challenge, the design is optimized at different abstraction levels. Low-level approaches and tools can not deal with future chip complexity. Based on the fact that at high abstraction levels design tradeoffs are better understood, design decisions made at these levels cause drastic reduction in the consumed power and can also shorten the design cycle. Algorithm optimization, components' selection as well as application partitioning are critical issues in low power/energy design.

A widely used technique which can reduce the consumed power/energy is dynamic voltage scaling (DVS). It trades performance for power during system operation when peak performance is not essentially required [1], [2]: The supply voltage and the operational frequency are adapted under the control of the operating system based on the required performance. For applications with a pre-defined performance limits, we suggest to plan the required voltage levels to execute different tasks during system synthesis. As a result, the overhead of calculating or predicting the required voltage level(s) at run-time is reduced to a minimum.

Basically, more energy reduction can be achieved when longer slack intervals are utilized by DVS. Additionally, the way in which the available slack intervals are exploited directly influences the efficiency of DVS. The higher the energy profile the

V. Paliouras, J. Vounckx, and D. Verkest (Eds.): PATMOS 2005, LNCS 3728, pp. 127–136, 2005.

more energy reduction a task can achieve when applying DVS. Therefore, adapting the time schedule of the tasks such that tasks which are major energy consumers are enabled to utilize longer slack intervals can remarkably maximizes the saved energy when applying DVS.

In future technologies (*70 nm* and less), the leakage power becomes more dominant and will account for about 50% of the total consumed power or even more. Dynamic V_{th} scaling (DVTH) is suggested to reduce the leakage power by adaptively changing the body bias voltage [3]. The availability and the distribution of slack intervals are key issue also for this technique. Additionally, the amount of saved leakage power depends on the time duration the system spends in each operation mode (active, inactive). Modifying the time schedule of the tasks to take these issues into consideration can be a source of extra power reduction.

This paper is organized as follows: The next section presents a summary of selected related work. Section 3 overviews our low power/energy system-level co-synthesis approach. In section 4, the power-aware scheduling scheme is explained. Experimental results are presented and analyzed in section 5. Section 6 concludes this paper.

2 Related Work

Dynamic voltage scaling is used to trade performance for power without impacting the peak performance of the system. In [4], algorithms were presented to determine the needed operating speed of a processor at run-time under software control. The operating voltage was adjusted based on the expected needed performance.

A task scheduling heuristic based on list-scheduling was introduced in [5]. The objective was reducing the consumed energy in systems by applying DVS. The scheme dynamically calculated the tasks' priorities and chose the supply voltage levels that could minimize the consumed energy. The presented priority function was aware of energy but might fail under tight deadline. A priority function tuning mechanism was used to compensate for this drawback

A hybrid global/local search optimization approach in a multi-processor system for dynamic voltage scaling was presented in [6]. Although this approach could yield the voltage levels that could minimize energy, the optimization itself was time consuming. Also, the authors assumed that the schedule was computed and the order of tasks' execution was known in advance. Hence, the influence of scheduling on power reduction was not tackled.

A static voltage scheduling problem was proposed and formulated as an integer linear programming (ILP) problem in [7]. It was assumed that the tasks have different average switching capacitance per cycle to consider energy profiles of processing elements. The study showed that considering energy profiles when scaling the voltage is a source of extra power reductions. Other studies also reached a similar conclusion [8]. However, these studies were done for single processor systems.

Grajcar suggested a genetic list scheduling algorithm without tackling the power problem [9]. Schmitz et al. used this scheduling approach to achieve energy–efficient scheduling when using DVS-enabled architectures [10]. The authors suggested the genetic list-based scheduling algorithm inside a mapping optimization loop. The latter optimization loop was based on genetic algorithms.

A low power co-synthesis tool (LOPOCOS) was presented in [11]. The objective was to help the designer in identifying energy-efficient application partitioning for distributed embedded systems. This approach assumed heterogeneous and DVS-enabled architectures. Although it performs better than previously suggested approaches, for applications with stringent delay constraints, the amount of saved power/energy could be moderate.

Our approach for low power/energy co-design is comprehensive and targets real-time systems. Starting at the FDT level (formal description techniques), the design space is explored to find low power/energy implementation alternatives that satisfy the constraints. The core issue which is handled in this paper is the scheduling problem. When applying voltage scaling mechanisms, the scheduling problem becomes even more involved, because the voltage scaling mechanism performs well only if the time schedule is optimized for this purpose.

Due to this correlation, the time schedule is iteratively adapted in order to enhance the efficiency of voltage scaling mechanisms: Tasks' priorities are calculated dynamically based on their contribution to power/energy reduction. Tasks that cause higher power/energy reductions are given higher priority to execute in an attempt to increase the available slack intervals for these tasks. DVS and DVTS can benefit from this approach. Available slack intervals are efficiently exploited by optimizing the voltage schedule based on a global view of energy profiles of all tasks and their mappings.

3 System Co-synthesis and Voltage Scaling Aspects

System-level synthesis can be seen as mapping a behavioural description onto a structural specification. Functional objects have the granularity of algorithms, tasks, procedures, etc, while structural objects are processors, ASICs, buses, etc. This section briefly presents our system-level low power/energy co-synthesis approach. Then, some basic issues in voltage scaling are introduced.

3.1 System Co-synthesis

In our approach, system-level synthesis is considered as a multiobjective optimization problem: Performance, power, and cost are considered while optimizing allocation, binding, and scheduling. Further refinement steps are proposed in our approach to optimize the design under stringent performance constraints. The optimization process searches the design space to find implementations that satisfy the design constraints and performs a rating on them with respect to the optimization goal(s). Evolutionary-based design space exploration is adopted in our approach. For this purpose, we have integrated to our automated co-synthesis tool the well-known evolutionary multi-objective optimizer SPEA2 [12]. SPEA2 is responsible for assigning fitness values to individuals. Each individual encodes a possible implementation alternative. Fittest individuals are selected to the mating process.

System synthesis is performed automatically using an internal system model which consists of two graphs: A task graph (TG), and an architecture graph (AG). Both are automatically generated from the specification. The TG is a directed acyclic graph $F_p(\Psi, \Omega)$, where Ψ represents the set of vertices in the graph ($\psi_i \in \Psi$) and Ω is the set of directed edges representing the precedence constraints and data dependencies ($\omega_i \in \Omega$). The AG is $F_A(\Theta, \mathfrak{R})$, where Θ represents the available architectures ($\theta_i \in \Theta$) and ($\rho_i \in \mathfrak{R}$) represents the available connections between the hardware components. For each hardware component ($\theta_i \in \Theta$), a finite set of resource types (S) is defined. For each resource type ($s_i \in S$) there is a set of associated ratios (R_s) that specify the scaling of power, delay, and cost when using this type for the selected component. Hard real-time constraint(s) T_i are forced on a node or on a set of nodes as well as an absolute time constraint T_T.

3.2 Applying Voltage Scaling

DVS and DVTS are efficient techniques which can reduce the overall consumed power/energy. But some factors such as the availability of enough slack intervals, their distribution among tasks, and the way in which these intervals are exploited have crucial influence on the efficiency of these techniques.

In systems which employ DVS technology, the supply voltage and the operational frequency can be adapted based on the required performance when executing a task. Reducing the supply voltage by a factor of two reduces the consumed energy by a factor of four. The dynamic energy ($E'(\psi_i)$) consumed by task ψ_i when executed at the reduced voltage level V_{level} [1] can be related to its nominal energy consumption, $E(\psi_i)$, as given below:

$$E'(\psi_i) = \left(\frac{V_{level}^2}{V_{supply}^2} \right) E(\psi_i) \Big|_{V = V_{supply}} \tag{1}$$

In this equation, it is seen that the larger the consumed energy by a task the larger the achieved energy reduction when scaling the supply voltage. So, the time schedule has to be modified to increase the available slack for tasks that have major influence on power/energy.

Reducing the body bias voltage yields an exponential leakage current/power reduction. This can be inferred from the leakage power equation given below:

$$P_{static} = V_{dd} K_3 e^{K_4 V_{dd}} e^{K_5 V_{bs}} + |V_{bs}| I_j \tag{2}$$

where V_{dd} is the supply voltage, V_{bs} is the body bias voltage, K_3, K_4, K_5, K_6, are constant fitting parameters which are technology dependent, and I_j is the current due to junction leakage [3].

As demonstrated above, voltage scaling causes significant power reduction, but scaling the voltage increases the circuit delay. The relation between path delay and voltage can be modeled similar to the alpha-model of an inverter as [3]:

$$d = \frac{L_d K_6}{(V_{dd} - V_{th})^\alpha} \tag{3}$$

In the above equation, α is a measure of the velocity saturation with typical value between 1 and 2. L_d is the logic depth of the path. Reducing the supply voltage V_{dd} or

---

[1] V_{level} is less than (or equal to) the maximum voltage level V_{supply}

increasing the threshold voltage V_{th} increases the circuit delay. Therefore, available slack intervals directly influence the level to which the voltage can be reduced. Without any loss of generality, only DVS is applied here. Applying DVTS requires slight modifications, but the proposed algorithms are valid for both schemes.

4 Solving the Scheduling Problem

Scheduling in our case can be defined as optimizing the start time and the required voltage level(s) needed to execute each task. Therefore, the problem can be seen as a two dimensional (2-D) optimization problem: The start time $\tau_i(t)$ of each task ψ_i, and the voltage level(s) $V(\tau_i) = \{v_1, v_2, ..., v_n\}$ should be optimized to achieve maximum power/energy reduction. The time schedule is optimized in our approach with an eye on energy. Based on the time schedule the voltage schedule is optimized.

4.1 Time Scheduling

The time schedule in our approach has not only to guarantee time feasibility, but also to improve the availability and the distribution of the slack. A list based scheduler determines the start time of each task such that the time constraints are fulfilled and maximum possible energy reduction is achieved when applying voltage scaling. The fitness of each individual is evaluated after the 2-D scheduling process. The general time scheduling algorithm is depicted in Fig. (1).

```
Input: Tasks' Mapping, Technology Library
Output: Power-Aware Time Schedule

Get_PT(K_C);
Get_PP(K_C, Step);
C = 0;
REPEAT {
    FOR (i = 1) to N {
        Get(ST_C,i);
        Get(SP_C,i);
        IF(ST_C,i ≠ SP_C,i) {
            IF (ST_C,i can be delayed)
                ST_C,i = SP_C,i;
        }
        Schedule_task (S, ST_C,i, C);
        Update (K_C,i);
        Get_PT(K_C,i);
        Get_PP(K_C,i);
    }
    C = C + 1;
    UNTIL (All tasks are scheduled)
}
```

Fig. 1. Time scheduling algorithm

The time scheduling algorithm computes the time priority for ready tasks, set K_C, by calling the function Get_PT(). The priority is determined initially based on ALAP and ASAP. The function Get_PP() ranks ready tasks according to their potential influence on energy reduction when applying voltage scaling. *Step* is the used time step as shown in Fig. 2. N is the number of allocated hardware components. Get($ST_{C,i}$) determines the set of ready tasks, $ST_{C,i} \subseteq K_{C,i}$, which has the maximum priority. Get($SP_{C,i}$) determines the set of ready tasks, $SP_{C,i} \subseteq K_{C,i}$, which has the highest power reduction effects when scheduled earlier in time. If the set $ST_{C,i}$ is different from the set $SP_{C,i}$, the priority of tasks with higher power rank is changed to be scheduled earlier if higher priority tasks can be delayed. The chosen tasks are scheduled at the given scheduling step using the function Schedule_task(). Subsequently, the set of ready tasks is updated for each hardware component by Update($K_{C,i}$).

The *power rank* (priority) is assigned to a tasks based on its potential influence on energy reduction. A task is scheduled in the scheduling step that leads to higher energy reduction according to equation 4. This in return increases the opportunity of tasks with high *power ranks* to utilize longer slack(s). Since scheduling tasks based on information related to the "critical path" only may cause other tasks to be scheduled near their deadlines. This may prevent additional energy saving when scaling the voltage. For instance, consider the case when tasks on the non-critical path have higher energy reduction effects. Delaying the execution of these tasks may lead to small or even no slack intervals available for these tasks.

We have experimentally observed that permitting ready tasks born at the k_{th} scheduling step to compete with tasks delayed from previous scheduling steps when ranking the tasks according to their energy reduction abilities causes inefficiency problems in some of the studied cases. To handle this deficiency, tasks can be delayed only n scheduling steps based on their power rank. This means that tasks with low power rank are scheduled after n scheduling steps depending on their time priority even if they can be delayed and their power rank is low. This is relatively significant to compensate for the lack of a lock-ahead mechanism in our fast heuristic in an attempt to avoid time-consuming methodologies. Due to this fast heuristics which avoids time-consuming look-ahead schemes, the resulting algorithm may not be optimal. But these cases where it may fail to yield remarkable improvements compared to general list scheduling are automatically eliminated by the general genetic optimization loop.

We further improved the efficiency of the proposed scheduling algorithm by executing it iteratively and one task is rescheduled at a time. The approach is based on the paradigm originally presented by Kernighan and Lin for graph partitioning [14]. According to ASAP and ALAP, a task is scheduled earlier or later as long as no data dependency or design constraint is violated. Each move leads to a new schedule. A task that is moved is then locked and is not allowed to move later unless all tasks are locked. The new (temporal) schedule is evaluated after scaling the voltage. After a set of m such moves, a subsequence of $q \leq m$ that maximizes the total power reduction can be reached. The schedule is then changed to include these q moves. The process is repeated until no further improvement is achieved.

A basic advantage in our algorithm is that it does not fail when the performance constraints are tight such as the one in [5]. The reason is that power and time aspects are considered at the same time in our scheduling approach.

4.2 Voltage Scheduling

The time schedule is adapted to increase the efficiency of voltage scaling techniques. The voltage levels required to execute the tasks should then be planned carefully in order to exploit the available slack intervals efficiently. This can achieve additional energy reduction when applying voltage scaling. Therefore, planning the voltage levels in our approach is based on a global view of all tasks and their energy profiles. The voltage scheduling algorithm is depicted in Fig. 2.

In this algorithm, y refers to the number of tasks. ΔEN_i refers to the energy saving for task (ψ_i) when extending its execution time (by one time step (Step, $n=1$)) by scaling its operating voltage:

$$\Delta EN_i = E(\psi_i) - E'(\psi_i)\big|_{n=1} \tag{4}$$

The achieved energy reduction is related to ΔEN_i, according to equation (1). So, tasks with larger energy profiles are given more preference for extending their execution and reducing their energy. The power priority ($P_{priority}$) for task ψ_i is proportional to the calculated ΔEN_i multiplied by sl_i which is defined as:

$$sl_i = \begin{cases} 1, & slack_i \neq 0 \\ 0, & otherwise \end{cases} \tag{5}$$

The voltage scheduling algorithm terminates when one of two conditions is satisfied: 1) when the list LS is empty. This occurs if the voltage level is reduced to a value around $2V_{th}$ for all tasks. Actually, this voltage limit is chosen to avoid drastic increase in delay when voltage is reduced below this limit. 2) when $P_{priority} = 0$ for all tasks which means that there is no available slack to be exploited by any of the tasks.

Input: $F_p(\Psi,\Omega)$, $F_A(\Theta, \beta)$, *Mapping, Time Schedule, Step.*
Output: *Voltage Schedule* $V_{ss}(t)$

Step 0:
- Calculate ΔEN_i of all tasks $\psi_i \in \Psi$
- Assign $P_{priority}$ to all tasks $\psi_i \in \Psi$
- Create empty list LS of size y

Step 1:
    Arrange the tasks in LS in a descending order of $P_{priority}$.

Step 2:
    Get a (ψ_j) with the highest non-zero $P_{priority}$ from LS.
- If (V_{dd} is no longer $\geq 2V_T$) → remove ψ_j from LS.
- Else, extend the task (ψ_j) in steps of (n*step).
- Update the tasks profile.

Step 3:
    Return if LS is empty OR all tasks have $P_{priority} = 0$

Step 4:
- Calculate ΔEN of all tasks in LS.
- Assign $P_{priority}$ to all tasks.
- Go to step 1.

Fig. 2. Planning the voltage levels planning

5 Experimental Results

A set of publicly available benchmarks was used to justify the applicability of our approach. The benchmarks include a set of 15 hypothetical examples generated originally by using the "Task Graphs For Free" TGFF [13]. Our 2-D scheduling methodology has been integrated to our automated co-synthesis tool. As a preliminary step, the general mapping and optimization approach presented in [11] have been implemented. In this preliminary step, we concentrated on optimizing the time schedule to satisfy performance constraints without tackling issues related to power. We call this step the 1-D scheduling.

The obtained energy reductions in percentage are presented in columns 2 and 3 of Table (1) for the 1- and 2-D scheduling, respectively. The last column presents the additional energy reductions achieved by the suggested 2-D scheduling algorithm over 1-D scheduling. The presented results show that all benchmarks made benefit of our 2-D scheduling algorithm, but in varying amounts.

The results can be categorized into two groups: The first group includes these benchmarks which made high benefit from the 2-D scheduling algorithm. This group of benchmarks indicates that for some applications the 1-D schedule might prevent the voltage scaling from achieving effective energy reductions. This reduces the efficiency of voltage scaling algorithms. The second group appears to make less benefit from our approach. This might partially be related to the performance constraints as well as to the mapping itself. At the same time, the lack of optimality in the algorithm might have an additional influence. But, in both cases, the 2-D scheduling algorithm is able to achieve additional energy reductions. Up to about 18% additional energy reduction was achieved by the 2-D scheduling algorithm over the 1-D. Additionally, the voltage planning algorithm which optimizes the voltage schedule partially hides the influence of time scheduling. Since the voltage scaling algorithm operates based on a global view of all tasks and their mappings.

Table 1. Achieved energy reductions by 1-D and 2-D scheduling

Benchmark	1-D Scheduling	2-D Scheduling	Extra Improvement
tgff1	68.1	83.5	15.4
tgff2	36.4	49.1	12.7
tgff3	64.6	75.9	11.3
tgff4	82.6	90.9	8.3
tgff5	60.1	61.1	1.0
tgff6	83.5	90.1	6.6
tgff7	30.2	45.0	14.8
tgff8	76.6	77.0	0.4
tgff9	37.3	44.9	7.6
tgff10	19.6	33.7	14.1
tgff11	24.3	33.6	9.3
tgff12	62.6	76.0	13.4
tgff13	60.9	73.2	12.3
tgff14	10.0	27.6	17.6
tgff15	14.1	27.9	13.8

6 Conclusions

An integrated methodology for low power/energy co-synthesis of real-time embedded systems is presented. The emphasis is on tasks' scheduling in time domain and scheduling the voltage level(s) required to execute each task. In this 2-D scheduling approach, the time schedule is prepared to increase the efficiency of voltage scaling algorithms while satisfying all performance constraints.

The presented results are encouraging and demonstrate that optimizing the time schedule increases the efficiency of voltage scaling algorithms. Additional energy reduction factor of up to about 18% is achieved when applying our 2-D scheduling methodology. The work also indicates the need to consider both power/energy consumption and performance constraints at the same time. This avoids time schedules which perform well in the time domain but causes low power/energy reductions when applying voltage scaling. It also avoids schedules that have high power/energy reduction effects but fail to satisfy performance constraints.

Acknowledgement

We are grateful indeed that DAAD (German Academic Exchange Service) supports this research since 2002.

References

1. Pering, T., Burd, T., Broderson, R.: Dynamic Voltage Scaling and the Design of a Low-Power Microprocessor System. In Power Driven Micro-Architectures Workshop, attached to ISCA'98, Barcelona, Spain, June, (1998).
2. Burd, T., Pering, T., Stratakos, A., Broderson, R.: A Dynamic Voltage Scaled Microprocessor System. IEEE Journal of Solid-State Circuits, vol. 35, no. 11, Nov. (2000) 1571-1580.
3. Martin, S., Flautner, K., Mudge, T., Blaauw, D.: Combined Dynamic Voltage Scaling and Adaptive Body Biasing for Lower Power Microprocessors under Dynamic Workloads. In International Conference on Computer-Aided Design, ICCAD'02, (2002) 721-725.
4. Pering, T., Burd, T., Broderson, R.: The Simulation and Evaluation of Dynamic Voltage Scaling. In International Symposium on Low Power Electronics and Design, ISLPED, (1998) 76-81.
5. Gruian, F., Kuchcinski, K.: LEnS: Task Scheduling for Low-Energy Systems Using Variable Supply Voltage Processors. In Asia and South Pacific Design Automation Conference, ASP-DAC, Jan. (2001) 449-455.
6. Bambha, N., Bhattacharyya, S., Teich, J., Zitzler. E.: Hybrid Global/Local Search for Dynamic Voltage Scaling in Embedded Multiprocessor. In the 1st International Symposium on Hardware/Software Co-Design, CODES'01, (2001) 243-248.
7. Ishihara, T., Yasuura, H.: Voltage Scheduling Problem for Dynamically Variable Voltage Processors. In International Symposium on Low Power Electronics and Design, ISLPED, (1998) 197-202.
8. Manzak, A., Chakrabarti, C.: Variable Voltage Task Scheduling for Minimizing Energy or Minimizing Power. In International Conference on Acoustics, Speech, and Signal Processing, Nov. (2000) 3239-3242.
9. Grajcar, M.: Genetic List Scheduling Algorithm for Scheduling and Allocation on a Loosely Coupled Heterogeneous Multiprocessor System. In the 36th ACM/IEEE Conference on Design Automation, (1999) 280-285.

10. Schmitz, M., Al-hashimi, B., Eles, P.: Energy-Efficient Mapping and Scheduling for DVS Enabled Distributed Embedded Systems. In Design, Automation and Test in Europe Conference and Exhibition, DATE, March, (2002) 514-521.
11. Schmitz, M., Al-Hashimi, B., Eles, P.: Synthesizing Energy-Efficient Embedded Systems with LOPOCOS. In Design Automation for Embedded Systems, (2002) 401-424.
12. Zitzler, E., Laumanns, M., Thiele, L.: SPEA2: Improving the Strength Pareto Evolutionary Algorithm for Multiobjective Optimization. Evolutionary Methods for Design, Optimization, and Control, CIMNE, Barcelona, Spain, (2002) 95-100.
13. Dick, R., Rhodes, D., Wolf, W.: TGFF: Tasks Graphs for Free. In International Workshop on Hardware/Software Codesign, March, (1998).
14. Kernighan, K., Lin, S.: An Efficient Heuristic Procedure for Partitioning Graph. Bell System Technical Journal, vol. 49, no. 2, (1970) 291-307.

Design of Digital Filters for Low Power Applications Using Integer Quadratic Programming

Mustafa Aktan and Günhan Dündar

Department of Electrical & Electronics Engineering, Bogazici University
34342 Bebek, Istanbul, Turkiye
{aktanmus,dundar}@boun.edu.tr

Abstract. An integer quadratic programming based formulation is proposed for the design of FIR filters implemented on Digital Signal Processors (DSP). The method unifies the cost of switching activity and number of ones in coefficients and is applicable to DSPs having multiple multiply accumulate units. Four FIR filter examples are designed with the proposed method. Power simulation results show that up to 38% power reduction can be achieved in the multiply accumulate unit of a DSP using the optimized coefficients. The resulting coefficients show better performance than coefficients optimized with previously proposed methods such as reordering coefficients.

1 Introduction

Digital filters are the most frequently used elements in signal processing applications. They are realized with ASICs or can be implemented by programming of digital signal processors. They require sequential arithmetic calculations and thus consume large power and require dedicated fast hardware resources. Therefore, power aware design of digital filters is essential. In this work, we focus on FIR filters since they are the basic building block of most digital filtering structures.

FIR filtering operation of an N tap filter can be expressed by the equation

$$y[n] = \sum_{k=0}^{N-1} h[n]x[n-k] \tag{1}$$

where x represents the input data stream, h the coefficients of the filter, and y the output data stream. The low power FIR filter design problem can be divided into two categories depending on the choice of implementation: Constant coefficient and variable coefficient FIR filter synthesis.

Constant coefficient applications are also referred to as multiplierless implementation of FIR filters. Since multiplication is a combination of addition operations, the less the number of additions, the less the number of adders, which then translates to a reduction in area and power of the final implementation. The number of addition operations is determined by the number of nonzero bits in the coefficients. Many researchers have focused on this problem and proposed several optimal and suboptimal algorithms to solve the problem [1], [2], [3].

Variable coefficient implementations of FIR filters are generally realized on digital signal processors (DSP) where the filtering algorithm is translated into a series of multiply accumulate operations. The basic source of power consumption is the multiplication operation, which is performed on a dedicated multiplier unit. A filtering operation on a single multiply-accumulate (MAC) unit is shown in figure 1. The

V. Paliouras, J. Vounckx, and D. Verkest (Eds.): PATMOS 2005, LNCS 3728, pp. 137–145, 2005.

power dissipated in a multiplier is related to the switching activity in the multiplier, which in turn is directly affected by the switching activity (Hamming Distance) at the inputs [4].

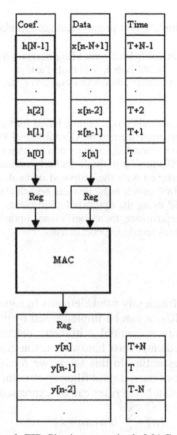

Fig. 1. FIR filtering on a single MAC unit

To reduce the power consumed in the MAC unit, the coefficients can be reordered so as to minimize the Hamming distance between successively applied coefficients [5], [6], [7]. However, reordering of coefficients requires reordering of data. One should keep in mind that data is usually correlated and thus there are very few sudden jumps between consecutive data. This then may cancel out the reduced Hamming distance for the coefficients by increasing the Hamming distance in the data stream. One can alleviate this problem by both considering the Hamming distance of the data and coefficient stream simultaneously. In this case, the possible reordering of data, especially in real-time systems or even systems where data is stored in consecutive addresses in memory, may offset expected the gains in power. Thus this approach should be restricted to problems where both data and the coefficients are readily available and reordering does not bring much power overhead.

In [4], a method that only reduces the switching activity between filter coefficients is proposed. The method formulates the coefficient optimization problem as a local search problem to find low switching activity coefficients, thus resulting in suboptimal solutions. In [9] the same problem is formulated as an integer linear programming problem targeting low Hamming distance coefficients thus reducing the power consumed. However it lacks the contribution of the number of ones in the coefficients thus resulting filters are optimum for Hamming distance but not necessarily for power.

In our work, we converted the low power FIR filter coefficient synthesis problem to a problem to find low switching activity (Hamming distance) and number of ones coefficients which is then formulated as a quadratic integer programming problem. Moreover since today's processors may possess multiple MAC units the formulation proposed handles this situation. The resulting coefficients are optimum in terms of switching activity and number of ones for the desired number of multiplier units. A couple of example filters are designed and the power performances are tested on a pre-designed multiply-accumulate (MAC) unit. The effectiveness of our approach is also shown on a processor having multiple MAC units.

2 The Formulation

The frequency response of a linear phase FIR filter having N taps is given by:

$$H(\omega) = \sum_{i=0}^{M-1} h_i T_i(\omega) \tag{2}$$

$$M = \left\lfloor \frac{N+1}{2} \right\rfloor \tag{3}$$

where $H(\omega)$ is the magnitude response without phase, T is a trigonometric function determined by the number of taps (even or odd) and the type of symmetry (symmetric or anti-symmetric) the coefficients have. The two's complement representation of a coefficient can be expressed as:

$$h_i = -x_{i,0} + \sum_{j=1}^{B-1} x_{i,j} 2^{-j} \ , \ x_{i,j} \in \{0,1\} \tag{4}$$

where B is the quantization word length, and $x_{i,j}$ corresponds to the j'th bit of coefficient h_i.

Now suppose we want to design an FIR filter having a magnitude response $H(\omega)$ for which the desired magnitude response at any frequency is given with $H_d(\omega)$, and the maximum magnitude deviation allowed at any frequency is given as $\delta(\omega)$. Then the magnitude constraints for the filter at any frequency is

$$\left| H(\omega) - H_d(\omega) \right| \le \delta(\omega), \text{ for } 0 \le \omega \le \pi \tag{5}$$

2.1 Formulation of the Cost of Switching Activity (Hamming Distance)

Formulation of the switching activity between successively applied coefficients is done as follows: A switching between coefficients h_i and h_{i+1}, which are quantized according to (4), at bit j occurs when Boolean XOR of the two bits $x_{i,j} \oplus x_{i+1,j}$ evaluates to a one. The arithmetic expression for the Boolean XOR operation is

$$x_{i,j} \oplus x_{i+1,j} \equiv x_{i,j}^2 + x_{i+1,j}^2 - 2x_{i,j}x_{i+1,j}$$
$$\equiv (x_{i,j} - x_{i+1,j})^2 \tag{6}$$

Having defined the cost function for the switching activity between two bits, the cost of switching from coefficient h_i to h_{i+1} is given by

$$\sum_{j=0}^{B-1}(x_{i,j} - x_{i+1,j})^2 \tag{7}$$

where B is the coefficient wordlength. The total cost of switching of an FIR filter having N taps is

$$C_{swa} = \sum_{i=0}^{N-2}\sum_{j=0}^{B-1}(x_{i,j} - x_{i+1,j})^2 \tag{8}$$

Since for a linear phase filter the coefficients are symmetric the above cost function reduces to

$$C_{swa} = 2\sum_{i=0}^{M-2}\sum_{j=0}^{B-1}(x_{i,j} - x_{i+1,j})^2 \tag{9}$$

where M is calculated using (3).

When there are more than one MAC unit then coefficients are assumed to be applied in the following sequence: Assuming N number of taps and P MAC units then coefficients applied to a MAC unit are h_i, h_{i+P}, h_{i+2P}, ... An example is shown in figure 2 for four MAC units and an FIR filter having ten coefficients. The new formulation of switching activity of successively applied coefficients for P MAC units is

$$C_{swa} = \sum_{p=0}^{P-1}\sum_{i=0}^{K-1}\sum_{j=0}^{B-1}(x_{p+iP,j} - x_{p+(i+1)P,j})^2, \quad K = \left\lfloor \frac{N-1-p}{P} \right\rfloor \tag{10}$$

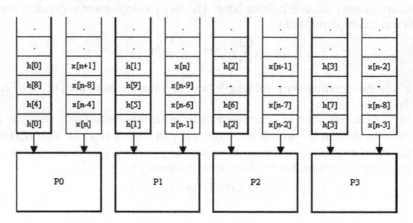

Fig. 2. FIR filtering on multiple MAC units

2.2 Formulation of the Cost of Number of Ones

Formulation of the number of ones in the coefficients of an FIR filter is straightforward. For a linear phase FIR filter having N taps and for which the coefficients are represented in two's complement representation using (4) the cost of number of ones in the coefficients can be expressed as

$$C_{one} = \sum_{i=0}^{N-1} \sum_{j=0}^{B-1} x_{i,j} , \quad x_{i,j} \in \{0,1\} \tag{11}$$

Since for a linear phase filter the coefficients are symmetric the above cost function reduces to

$$C_{one} = 2\sum_{i=0}^{M-1} \sum_{j=0}^{B-1} x_{i,j} , \quad x_{i,j} \in \{0,1\} \tag{12}$$

where M is calculated using (3).

2.3 Formulation of the Problem

Having defined the cost functions of switching activity and number of ones three sets of optimized coefficients can be obtained: minimum number of ones (MONE), minimum switching activity (MSWA) filters, and minimum switching activity and ones (MSWO) filters.

The optimization problem for MONE filters can be formulated as

$$\begin{aligned}
\text{Minimize} \quad & 2\sum_{i=0}^{M-1}\sum_{j=0}^{B-1} x_{i,j} \\
\text{Subject to} \quad & |H(\omega) - H_d(\omega)| \le \delta(\omega), \ 0 \le \omega \le \pi \\
\text{Where} \quad & H(\omega) = \sum_{i=0}^{M-1} h_i T_i(\omega) \\
& h_i = -x_{i,0} + \sum_{j=1}^{B-1} x_{i,j} 2^{-j} , \quad x_{i,j} \in \{0,1\}
\end{aligned} \tag{13}$$

This problem can be solved optimally using integer programming.

The optimization problem for MSWA filters for a filter core having P MAC units can be formulated as

$$\begin{aligned}
\text{Minimize} \quad & \sum_{p=0}^{P-1}\sum_{i=0}^{K-1}\sum_{j=0}^{B-1} \left(x_{p+iP,j} - x_{p+(i+1)P,j} \right)^2 \\
\text{Subject to} \quad & |H(\omega) - H_d(\omega)| \le \delta(\omega) \quad , \qquad 0 \le \omega \le \pi \\
\text{Where} \quad & H(\omega) = \sum_{i=0}^{M-1} h_i T_i(\omega) \\
& h_i = -x_{i,0} + \sum_{j=1}^{B-1} x_{i,j} 2^{-j} \quad , \qquad x_{i,j} \in \{0,1\} \\
& K = \left\lfloor \frac{N-1-p}{P} \right\rfloor
\end{aligned} \tag{14}$$

Due to the quadratic term in the objective function this problem can be formulated as an integer quadratic problem. By introducing new variables this problem can be converted to an integer linear programming problem as described in [9].

By combining the switching activity cost and number of ones cost the optimization problem for MSWO filters for a filter core having P MAC units can be formulated as

$$\text{Minimize} \quad \sum_{p=0}^{P-1}\sum_{i=0}^{K-1}\sum_{j=0}^{B-1}\left(x_{i,j}-x_{i+1,j}\right)^2 + \sum_{i=0}^{N-1}\sum_{j=0}^{B-1}x_{i,j}$$

$$\text{Subject to} \quad \left|H(\omega)-H_d(\omega)\right| \le \delta(\omega) \quad , \qquad 0 \le \omega \le \pi$$

$$\text{Where} \quad H(\omega)=\sum_{i=0}^{M-1}h_i T_i(\omega) \tag{15}$$

$$h_i = -x_{i,0} + \sum_{j=1}^{B-1}x_{i,j}2^{-j} \ , \ x_{i,j}\in\{0,1\}$$

$$K = \left\lfloor\frac{N-1-p}{P}\right\rfloor$$

where the cost of switching activity is given the same weight as the cost for number of ones.

3   Design Examples

In this section results for four filters taken from [4] are given. The problems are solved using ILOG CPLEX quadratic programming tool. The characteristics of the filters are shown in table 1.

Table 1. Filter characteristics

Filter	Number of Taps	Sampling Freq.	Passband		Stopband	
			Cutoff Freq.	Ripple (dB)	Cutoff Freq.	Ripple (dB)
B	24	16kHz	3kHz	0.20	4.5kHz	42
D	28	12kHz	2kHz	0.12	3kHz	45
E	34	12kHz	2.2kHz	0.16	3.1kHz	49
F	29	10kHz	2kHz	0.05	3kHz	40

Table 2 shows the results for switching activity and number of ones counts in the coefficients generated using five methods, namely NOPT, RORD, MONE, MSWA, and MSWO. NOPT coefficients are the coefficients generated using MATLAB's Remez function and quantized to 16 bits. Thus they are the reference coefficients for which no optimization is done. RORD coefficients are actually NOPT coefficients reordered by the method given in [5] for minimum switching activity. MSWA coefficients are the coefficients optimized for minimum switching activity using the formulation in (14) with ILOG CPLEX integer programming tool. MONE coefficients are the coefficients optimized for minimum number of ones using the formulation in (13) with ILOG CPLEX integer programming tool. MSWO coefficients are the coefficients optimized for both minimum number of ones and minimum switching activity using the formulation in (15) with ILOG CPLEX integer programming tool. The optimization problems were solved on a PC having INTEL P4 1.7GHz processor with 256 MB of RAM. Since solution time of integer programming problems depend on the number of variables the method proposed in [3] is used to determine the boundary values of the coefficients that satisfy the filter constraints. Using these boundary values the most significant bits of the coefficients can be determined. Hence a reduction in the number of variables is achieved.

The power performance of the generated coefficients are tested on a single MAC unit having a 16 bit Booth encoded Wallace tree multiplier and 40 bit accumulator. The MAC unit is synthesized with AMS 0.6μ technology cell library. FIR filtering is performed on 15625 samples of voice data quantized to 16 bits. Power simulations were done with an event driven gate-level simulator using a variable delay model which accounts for glitches. The operating frequency was taken to be 1MHz and supply voltage to be 5V. The resulting power dissipations are given in table 3.

The percentage power reduction is calculated by taking the NOPT coefficients' power as reference. The results indicate that by just reducing the switching activity between coefficients the best power performance cannot be achieved. By reordering coefficients one can get 19% reduction in power. Minimum switching activity (MSWA) coefficients could achieve a power reduction of 22% on average. The best power performance is obtained from MSWO coefficients having a power reduction of 35% on average. MONE coefficients have a comparable performance to MSWO coefficients with 30% power reduction on average. When design time is important, which might be the case for filters having large number of coefficients, MONE coefficients are preferable to MSWO coefficients.

Table 2. Switching activity counts and number of ones in synthesized filters for one MAC unit

Filter	Optimization method	Number of Switching	Number of Ones	Time (sec)
B	NOPT	178	186	-
	RORD	59	189	<1
	MONE	94	88	20
	MSWA	82	150	800
	MSWO	90	90	340
D	NOPT	188	244	-
	RORD	68	244	<1
	MONE	116	140	24
	MSWA	100	174	1200
	MSWO	102	144	80
E	NOPT	246	272	-
	RORD	75	272	<1
	MONE	148	140	300
	MSWA	122	290	13742
	MSWO	140	146	5328
F	NOPT	216	245	-
	RORD	59	245	<1
	MONE	122	61	4
	MSWA	108	328	2240
	MSWO	118	63	66

Another set of coefficients were generated targeting a filter core having four MAC units. The coefficients are generated for filter B. The optimization method used is MSWO but now targeting 4 MAC units, i.e. P=4 in (15). The resulting coefficients' switching activity counts for each MAC unit are given in table 4. The switching activity counts are compared to those coefficients generated using methods NOPT, and MSWO targeting one MAC unit.

The power performances of the coefficients are tested using the same MAC unit mentioned above. The operating frequency is 1MHz and supply voltage 5V. The resulting average power dissipation in each MAC unit is given in table 5. The performance of the coefficients generated by the method MSWO targeting 4 units is the best, as expected. However there is a little performance increase (3%) over MSWO coefficients targeting one MAC unit.

Table 3. Power simulation results using one MAC unit

Filter	Optimization	Power (uW)	Reduction (%)
B	NOPT	1317	-
	RORD	1077	18.2
	MONE	890	32.4
	MSWA	982	25.4
	MSWO	850	35.4
D	NOPT	1324	-
	RORD	1083	18.2
	MONE	913	31.0
	MSWA	957	27.7
	MSWO	874	34.0
E	NOPT	1321	-
	RORD	1065	19.4
	MONE	978	26.0
	MSWA	1137	13.9
	MSWO	899	31.9
F	NOPT	1249	-
	RORD	993	20.5
	MONE	787	37.0
	MSWA	1001	19.9
	MSWO	775	38.0

Table 4. Switching activity counts and number of ones in synthesized filters for four MAC units

Filter	Optimization method	Number of Switching				Number of Ones
		MAC 0	MAC 1	MAC 2	MAC 3	
B	NOPT	47	38	37	46	186
	MSWO for 1 MAC	36	22	24	36	90
	MSWO for 4 MAC	22	16	16	22	92

Table 5. Power simulation results using four MAC units

Filter	Optimization method	Power (uW)				Average Reduction (%)
		MAC 0	MAC 1	MAC 2	MAC 3	
B	NOPT	1523	1357	1341	1525	-
	MSWO for 1 MAC	994	937	906	981	33.4
	MSWO for 4 MAC	944	875	855	972	36.5

4   Conclusion

In this paper we have demonstrated the formulation of finding power optimum coefficients for realization of FIR on programmable DSPs. The coefficient optimization results for 4 low pass FIR filters were shown. The results indicate that when minimization of switching activity is our goal the most effective method is to reorder coefficients. However when it comes to power performance coefficients optimized with our method, which minimizes the number of ones in addition to switching activity, outperformed reordered coefficients in all cases. The effectiveness of the formulation on DSPs having multiple units is also shown on a design example.

References

1. Samueli, H.: An Improved Search Algorithm for the Design of Multiplierless FIR Filters with Powers of Two Coefficients. In IEEE Transactions on Circuits and Systems, Vol. 36, pp. 1044-1047, July 1989.
2. Fox, T. W., Turner, L. E.: The Design of Peak Constrained Least Squares FIR Filters with Finite Precision Coefficients. In IEEE Transactions on Circuits and Systems II, Vol. 49, No. 2, pp. 151-154, Feb. 2002.
3. Yli-Kaakinen, J., Saramäki, T.: A systematic approach for the design of multiplierless FIR filters. In Proc. IEEE Int. Symp. Circuits Syst., Sydney, Australia, vol. II, pp. 185-188, May 2001.
4. Mehendale, M., Sherlekar, S.D., Venkatesh, G.,: Coefficient optimization for low power realization of FIR filters. In Proc. IEEE Workshop on VLSI Signal Processing, pp. 352 - 361, 1995
5. Masselos, K., Merakos, P., Theoharis, S., Stouraitis, T., Goutis, C. E.: Power Efficient Data Path Synthesis of Sum-of-Products Computations. In IEEE Transactions on VLSI Systems, Vol. 11, No. 3, pp. 446-450, June 2003.
6. Mehendale, M., Sherlekar, SD., Venkatesh, G.,: Low Power Realization of FIR Filters on Programmable DSP' s. In IEEE Transactions on VLSI Systems, Vol. 6, pp. 546-553, Dec. 1998.
7. Erdogan A. T., Arslan, T.: Low Power Implementation of linear Phase FIR Filters for single Multiplier CMOS Based DSPs. In IEEE ISCAS, California, USA, pp. D425-D428, May 1998.
8. Erdogan, A. T., Hasan, M., Arslan, T.: A low power FIR filtering core. In 14th Annual IEEE International ASIC/SOC Conference, Washington D.C., pp. 271-275, 12-15 September 2001.
9. Gustafsson, O., Wanhammar, L.: Design of linear-phase FIR filters with minimum Hamming distance. In IEEE Nordic Signal Processing Symp., Hutigruten, Norway, Oct. 4-7, 2002.

A High Level Constant Coefficient Multiplier Power Model for Power Estimation on High Levels of Abstraction

Arne Schulz, Andreas Schallenberg, Domenik Helms, Milan Schulte, Axel Reimer, and Wolfgang Nebel

CvO University of Oldenburg, Faculty II - Department of Computing Science, Escherweg 2, 26121 Oldenburg, Germany
Arne.Schulz@uni-oldenburg.de

Abstract. Early power estimation in current designflows becomes more important nowadays. To meet this need, power estimation even on the algorithmic level has become an important step in the typical design flow. This helps the designer to choose the right algorithm right from the start and much optimisation potential can be used due to the focus on the crucial parts. In particular, algorithms for digital signal processing as applied in mobile communication systems are very power sensitive. Such algorithms massively contain multiplications with constants on parts of digital filters. In this paper we propose on the one hand our new decomposition algorithm for (nearly) optimal synthesis of constant coefficient multipliers which we use for the evaluation of our new power model. On the other hand we propose a new power model based on the canonical signed digit (CSD) approach which can be used very fast and where the deviation of the power compared to the time consuming decomposition is 4.9 %[1].

1 Introduction

In current digital signal processing algorithms often digital filters (like FIR, IIR) are used. These filters consist of many multiplications with constant multiplicands. So it's inevitable to estimate the dynamic power of these functional units as well as it is reasonable to use the already established estimation of standard arithmetic functions. This has to be performed quickly due to the fact that one advantage of high level power analysis is the fast estimation. Therefore time consuming approaches like our decomposition algorithm (DecAlg) do not apply for this purpose.

The problem of the optimization of multiplication by constants has been studied for a long time. For instance the famous recoding presented by Booth [ ] can simplify the multiplication by constant operation as well as the full multiplication. It is well known that the so-called canonic-signed-digit (CSD) code is the best signed-power-two (SPT) code because the number of non-zero digits of a

---

[1] This work is partly funded by the German DFG as part of the project AVSy

V. Paliouras, J. Vounckx, and D. Verkest (Eds.): PATMOS 2005, LNCS 3728, pp. 146–155, 2005.

CSD code is minimal. Many publications offer lots of improvements for this problem. To name some of the most important solutions in the wide field of optimal constant coefficient multiplier implementations there are some solutions for the CSD code with reuse of internal results [ ]. Other authors propose to consider all coefficients of transposed-form FIR filters as a whole and replace coefficient multiplications by a multiplier block. Their methods find the redundancy across the SPT coefficients in the multiplier block of a transposed-form FIR filter, for example the common-subexpression-elimination (CSE) methods [, −,] and the graph-dependence algorithms [−].

Offering power models for constant coefficient multipliers enables a big chance for further improvements in terms of power consumptions because filter coefficients can be easily adapted. This can reduce significantly the necessary adders (and / or subtractors) in constant coefficient multipliers. The modification of filter constants is often applicable in the field of speech processing when the speech intelligibility isn't affected by the adaption.

In this paper we present a new power model for high level power estimation of power-optimal implemented constant coefficient multipliers. The aim is not another optimal synthesis of constant coefficient multipliers but the fast and efficient power estimation of such a synthesized function is the goal. Therefore we describe on the one hand the model itself and on the other hand our Decomposition Algorithm (DecAlg) which we use for comparison and evaluation reasons. The paper is organized as follows. The Decomposition algorithm is presented in Section 2, the power model itself is then presented in Section 3. In Section 4 we evaluate our model in comparison to our Decomposition Algorithm and also in comparison to the results retrieved from a commercial synthesis tool. Finally, some conclusions are drawn in Section 5.

2   Decomposition Algorithm

In this section we present our Decomposition Algorithm (DecAlg) which initially was intended to be our power estimation model. Due to the runtime problem this approach is only used for evaluation reasons. The next paragraphs describe the working flow, starting with building up an initial construction table, followed by the description of the modification algorithm. Finally we give an overview of the table size and usage.

To be able to assess the quality of a solution for multiplication with a constant factor in terms of delay, area and power it is necessary to have a reference which is close to the optimal solution. It is very unlikely to find such optimal solution in terms of all those three criteria. Furthermore, to guarantee that there is no better one would force to test all available solutions using a brute force algorithm.

It is obvious that this is infeasible so we have chosen a heuristic approach. It works on an algorithmic level and performs strength reduction for the multiplication into a term which uses shifts, additions and subtractions.

2.1   Building a Construction Table

The input is the constant c to be decomposed. The algorithm then sets up a table of size $2c$ with the following fields for each entry:

- *Op* $\in$ *init, shift, add, sub, notYet*: operation which constructs the entry
- *Left* $\in \{1, \ldots, 2c\}$: left operand
- *Right* $\in \{1, \ldots, 2c\}$: right operand

We use $Dec(n)$ to indicate the decomposition for the constant n. The entry n is constructed the following way:

Op	formula	costs
init	value is given from start	0
shift	$Dec(n) = Left(n) << Right(n)$	$Cost(Left(n))$
add	$Dec(n) = Left(n) + Right(n)$	$Cost(Left(n)) + Cost(Right(n)) + AddCosts$
sub	$Dec(n) = Left(n) - Right(n)$	$Cost(Left(n)) + Cost(Right(n)) + SubCosts$
notYet	no way known yet	∞

Initially all *Op* fields are set to *notYet* except for the entry 1, which is set to init. The rightmost column shows adustable example costs for each construction.

If an entry a is constructed using an entry b the entry a is dependent on b. The dependency relation is transitive. The table can be seen as a graph of nodes where the dependency relation represents the directed edges. Selecting one a and all nodes which are reachable from a will give a subgraph containing all operation nodes needed to calculate the costs for a.

This subgraph is processed with a recursive tagging algorithm. Whenever a node is being visited a tag on it is being checked. If it was not tagged before then the tag is being set, its costs (AddCosts or SubCosts)[2] are being counted and the dependent nodes are being visited. Visiting a node which is already tagged does not lead to any operation. This way diamond-like dependency relations are taken into account and costs are counted only once for each node. Such diamond like structures represent the sharing of subexpressions in the final construction of a.

2.2   Modifying the Construction Table

The table needs to be modified after its inital construction. This is done in three phases. In these phases it is necessary to take care that no recursive dependencies would exist.

In the first phase all known values are tried to be shifted left. One rule is that if b can be constructed by shifting a to the left, than this is to be done using the smallest a possible. Using this algorithm it is guaranteed that shifting this way cannot introduce recursive dependencies. The phase stops when no further modifications are done to the table.

After this the add operations are applied. The left operand has always to be smaller than the right one and both operands must not depend on the sum of the operation. An add may be introduced when the resulting cost function would

---

[2] We used AddCosts of 10 and SubCosts of 11 which may be adjusted if the cost relation of adders and subtracters is known. Realistic values would consider the bit width of the variable to be multiplied since this is one of the effects which modify the balance of AddCosts to SubCosts

show lower costs than before or the cost would be the same but the old construction was a subtraction. This is done once with each entry. If there was a modification we go back to the first phase. Otherwise we proceed with the third one.

The third phase is similar to the second and introduces subtractions. Replacing an operation with a newly found subtraction is only allowed if the costs would really improve in this phase. Each entry is inspected once. If modifications are done, the shift phase will be next, otherwise the algorithm ends here.

2.3   Construction Table Size

The subtract operation is the reason why the table is created up to $2c$. The upper half is not processed completely since there may be a $i > c$ which would be constructed best by calculating it $i = a - b$ with $a \geq 2c$. This a would be outside the table. Therefore only shift and add operations are tested to construct the upper half of the table.

2.4   Using the Table

The construction of an entry is done in a way which is very similar to the cost calculation. Decomposing a multiplication $x * 71$ would show:

Index	Op	Left	Right
1	init	-	-
...	...	...	...
8	shift	[1]	3
9	add	[8]	[2]
...	...	...	...
71	sub	[72]	[1]
72	shift	[9]	3
...	...	...	...

The interpretation would be $x * 71 == (((x << 3) + x) << 3) - x$.

3   Power Model

In this section we present our power model which we use for a fast and efficient high level power analysis. For this evaluation we used a prototypic implementation and the estimation tool ORINOCO [ ]. Firstly we give a brief introduction into the CSD code computation used in the model. Secondly we show the limitations and the extension for the reuse of intermediate results using an approach for common subexpression elemination.

The results of this CSD decomposition is then used to drive the power models for arithmetic components described in [-].

3.1   CSD

The original Booth [ ] encoding to realize 2's complement multiplication was improved by Reitwiesner [ ]. This method is the so called canonical-signed-digit (CSD) code. The algorithm works as follows:

Reitwiesener's Algorithm: let number x to recode be given 2's complement format[3] $x = x_{m-1}, \ldots, x_0$; $x_i \in \{0, 1\}$.
The new coded number $r = r_m, \ldots, r_0$ has one more bit and each $r_i \in \{-1, 0, 1\}$, since $15_{10} = 1111_2$ will be recoded to $16_{10} - 1_{10} = 100\overline{1}_{SD}$[4].

Now the algorithm performs the following loop:

1. Set initial $\text{carry}_{old} = 0$, $r_m = r_{m-1} = 0$.
2. For each $i = 0, \ldots, n$ do
   (a) Set $\text{carry}_{new} := \lfloor (\text{carry}_{old} + x_i + x_{i+1}) \rfloor / 2$
   (b) Set $r_i := \text{carry}_{old} + x_i - 2 \cdot \text{carry}_{new}$
   (c) $\text{carry}_{old} = \text{carry}_{new}$

As easily can be seen, the algorithm goes linearly through the number of bits (representing the constant), not with the constant itself. Another advantage is that the canonical signed digit number representation has a minimal Hamming-weight. A CSD number representation has not more than $(m + 1)/2$ nonzeros and that tends to $m/3 + 1/9$ at the average [ ].

The disadvantage of a number representation with minimal Hamming-weight is that this representation is not unique, as shown in the following example:

$$3_{10} = 11_2 = 10\overline{1}_{SD}.$$

Both representations have the same weigth but with different costs.

This initial algorithm can efficiently be extended for the reuse of intermediate results (common subexpression elemination, CSE) at nearly no computational cost. According to Hartley [, p. 683], 101_{SD} and $10\overline{1}_{SD}$ are statistically the most common subexpressions found over all CSD-recoded numbers. He shows that an average m-bit CSD number can be broken down into about $n/18$ terms of type 101_{SD}, the same amount of type $10\overline{1}_{SD}$ and about $n/9$ isolated 1 or $\overline{1}$. So that at the average it can be broken down into $2(n + 1) + 1/9$ nonzero terms which gives another 33% saving to Reitwiesner's algorithm.

In our approach we use Reitwiesner's algorithm to construct our architecture which is then used for dynamic power estimation using the dynamic power models for functional units proposed in [–]. The results are now presented in section 4.

4   Results

In this section we present the evaluation results on the one hand for the Dec-Alg presented in Section 2 in comparison with the results for the CSD-based approach. The CSD-based approach is then improved as described in Section 3 and again compared against the results of the DecAlg.

4.1   Evaluation of DecAlg

Firstly we evaluated the DecAlg in comparison with the built-in capabilities of a commercial synthesis tool (Synopsys DesignCompiler) for all constants from

---

[3]   signed magnitude: Add a 0 digit at the left, increase bitwidth from m-1 to m and start the algorithm with the new number
[4] $\overline{1} := -1$

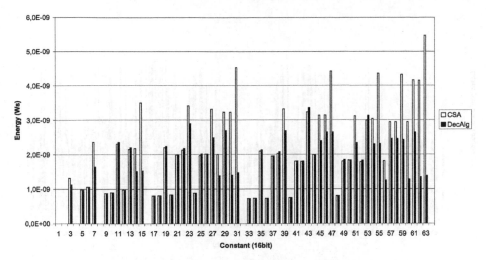

Fig. 1. Comparison of architecture generated with DecAlg vs. Synopsys DC (CSA-architecture) (in Ws)

1 to 64. Therefore we ensured that the necessary variables for optimizations (`"hlo_share_common_subexpressions"`) were set. We then implemented the results given from DecAlg automatically in Verilog. The power estimation was performed with Synopsys PowerCompiler after incremental compile (to make sure maximal optimizations were done by Synopsys) with a $0.18\mu m$ low-power technology at 2.5V. For the Synopsys solution we instantiated a CSA multiplier. The results shown in Figure 1 make clear that the synthesis tool often provides worse solutions than the DecAlg. Especially for constants slightly smaller than power-of-two values Synopsys does not find the solution with a subtractor.

Therefore we assume for the following evaluation that the results given from the DecAlg are adequate and better reference values than Synopsys' generated estimation.

4.2 Evaluation of the CSD-Model

The evaluation of the CSD-model is structured in two steps: The first step evaluates the model without the CSE-approach whereas the second one uses the final model which includes the reuse of intermediate results.

For better good comparison we also used the normalized costs as described in Section 2, which are set so that shifts have costs of 0, additions costs of 10 and subtractions costs of 11.

We computed all constants from 0 up to 4095 and compared the results of the DecAlg to the plain CSD (without CSE) prediction. The results are depicted in Figure 2 where only the difference is shown. It is obvious that the plain CSD algorithm overestimates in many cases the solution provided by the DecAlg. Table 1 shows the related numbers.

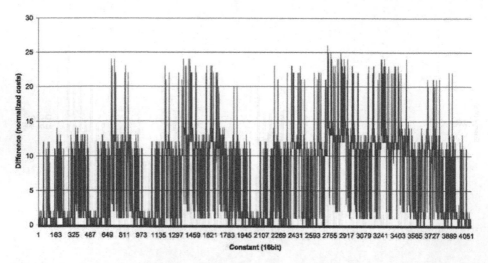

Fig. 2. Overestimation of initial CSD architecture vs. DecAlg architecture (in normalized "costs")

Table 1. Results: DecAlg vs. CSD without CSE for constants from 0 to 4095 (in normalized costs)

	DecAlg	CSD
Cost total	121249	148827
Equivalent estimations	721	
Maximum difference at same constant	0	26
Average of the difference	6.73291 %	
Variance of the difference	39.6712 %	
Standard deviation	6.2985 %	

To summarize this evaluation it can be said that:

- peek difference was 26
- average difference was 6.73291
- sum of total difference was 27578
- sum of total cost (DecAlg) was 121249 and (CSD) 148827, in percent 18.5302
- average cost (DecAlg) was 29.6018 and (CSD) 36.3347

The results obtained from this evaluation showed that this approach was insufficient in terms of accuracy. Therefore we extended the CSD approach with the proposed CSE approach.

For this evaluation we used the power estimation tool ORINOCO from ChipVision. In both cases(DecAlg and CSD+CSE) C language code was used as input. For the simulation (to achieve the dynamic switching activity) we used randomly generated patterns with well defined settings. The normalized Hamming-distance and the normalized signal distance were set to 0.5. We used the same input stimuli patterns for the DecAlg as well as for the CSD+CSE ap-

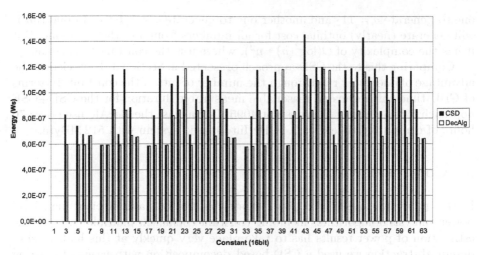

Fig. 3. Comparison of architecture generated with DecAlg vs. estimated power consumption with enhanced CSD-model (in Ws)

proach. The results for the first constants are depicted in Figure 3. The results for constants 0 to 4095 are shown in Table 2. The used technology in ORINOCO was a 0.25μ technology with 2.5V.

The results presented in this section and summarized in Table 2 show the applicability of the model. There is an additional error of ca. 4 % one has to take into account for the dynamic power estimation (the error of the power model for the adders and subtractors are unchanged) of constant coefficient multipliers. This is reasonable enough keeping in mind that the estimation takes place in nearly no computation time compared to the DecAlg algorithm (cf. 4.3) or other efficient approaches described in the literature.

4.3 Complexity

The DecAlg algorithm works table-based. It recognizes subpatterns in numbers. Therefore it will reuse 11_2 in 110011_2 and produce only the cost for two adders,

Table 2. Results: DecAlg vs. CSD with CSE for constant from 0 to 4095 (in normalized costs)

	DecAlg	CSD + CSE
Cost total	121249	137749
Equal cost in [cases]		831
Maximum difference at same constant	-1	24
Average of the difference		4,02832 %
Variance of the difference		23,99 %
Standard deviation		4,89796 %

one to generate $a = 11_2$ and another one to generate $a << 4 + a$. In one run it will generate (nearly) optimal cost for all numbers from 0 to the given one. But it has the complexity of $O(\log^2(n) * n^2)$, where n is the constant to calculate.

Comparing this to the CSD-approach presented in Section 3 shows clearly the advantage of one linear run through the number of bits of the constant. In terms of CPU time this means no reasonable measurable duration for the CSD+CSE approach for computing the constant 4095. In contrast the DecAlg used over 70 hours of computing time on a 2.4 GHz Intel-based system with 512 MByte.

5   Conclusion

In this paper we presented a power model for the fast and efficient high level power analysis of algorithms including constant coefficient multipliers. Since the estimation of power results has to take place very quickly at this high level of design abstraction we used a CSD-based decomposition with usage of existing adder and subtractor power models – knowing very well that the estimation might slightly overestimate the finally implemented optimal solution for a specific constant coefficient multiplier. Nevertheless the evaluation has shown that this approach is valid. For comparison reasons we also presented an heuristic approach for a nearly optimal decomposition of a constant coefficient multiplier into shifts, additions and subtractions. This approach produces for specific constants up to 200 % better results than a commercial synthesis tool. The power estimation error is approximately 4 % and the standard deviation of the model compared to our heuristic approach is approx. 4.9 %.

References

1. A. D. Booth: A signed binary multiplication technique. Quart. J. Mech. App. Math., IV(2) (1951) pp.236–240
2. G.W.Reitwiesner, "'Binary Arithmetic"', in *Advances in Computers*, Vol. 1 (1960) pp. 231–308
3. Richard I. Hartley, Subexpression Sharing in Filters Using Canonical Signed Digit Multipliers, IEEE Transactions on Circuits and Systems-II: Analog and Digital Signal Processing Vol. 43 (1996) pp. 677–688
4. A. Stammermann, D. Helms, M. Schulte, A. Schulz, W. Nebel, Binding, Allocation and Floorplanning in Low Power High-Level Synthesis, International Conference on Computer Aided Design (ICCAD), 2003
5. M. Potkonjak, M. B. Srivastava, and A. P. Chandrakasan, Multiple constant multiplications: efficient and versatile framework and algorithms for exploring common subexpression elemination, IEEE Transactions CAD, vol. 15, 1996, pp. 151–165
6. R. Paško, P. Schaumont, V. Derudder, S. Vernalde, and D. Ďuracková, A new algorithm for elimination of common subexpressions, IEEE Transactions CAD, vol. 18, 1999, pp. 58–68
7. M. Martínez-Peiró, E. I. Boemo, and L. Wahammar, Design of highspeed multiplierless filters using a nonrecursive signed common subexpression algorithm, IEEE Transactions on Circuits and Systems-II, vol. 49, 2002, pp. 196–203

8. D. R. Bull, and D. H. Horrcks, Primitive operator digital filter, Proceedings Inst. Elect. Eng. Circuits Devices Syst., vol. 138, 1991, pp. 401–412

9. A. G. Dempster and M. D. Macleod, Use of minimum adder multiplier blocks in FIR digital filters, IEEE Transactions on Circuits and Systems-II, vol. 42, 1995, pp. 569–577

10. H.-J. Kang, and I.-C. Park, FIR filter synthesis algorithms for minimizing the delay and the number of adders, IEEE Transactions on Circuits and Systems-II, vol. 48, 2001, pp. 770–777

11. A. G. Dempster, S. S. Dimirsoy, and I. Kale, Designing multiplier blocks with low logic design, Proceedings IEEE Int. Symp. Circuits and Systems, vol. 5, 2002, pp 773–776

12. C.-Y. Yao, H.-H. Chen, T.-F. Lin, C.-J. Chien, and C.-T. Hsu, A novel common-subexpression-elmination method for synthesizing fixed-point FIR filters, IEEE Transactions on Circuits and Systems-I, vol. 51, 2004, pp 2215–2221

13. G. von Cölln (Jochens), L. Kruse, E. Schmidt, A. Stammermann, and W. Nebel, Power macro-modelling for firm-macros, Proceedings of PATMOS'00, 2000, pp. 24–35

14. E. Schmidt, A. Schulz, L. Kruse, G. von Cölln (Jochens), and W. Nebel, Automatic Generation of Complexity Functions for High Level Power Analysis, Proceedings of PATMOS'01, 2001, pp. 26–35

15. D. Helms, E. Schmidt, A. Schulz, A Stammermann, and W. Nebel, An Improved Power Macro-Model for Arithmetic Datapath Components, Proceedings of PATMOS'02, 2002, pp. 16–24

An Energy-Tree Based Routing Algorithm in Wireless Ad-Hoc Network Environments⋆

Hyun Ho Kim, Jung Hee Kim, Yong-hyeog Kang, and Young Ik Eom

School of Information and Communication Engineering, Sungkyunkwan Univ.
300 Cheoncheon-dong, Jangan-gu, Suwon, Gyeonggi-do 440-746, Korea
{must,kimjh,yhkang1,yieom}@ece.skku.ac.kr

Abstract. It is important to provide energy-balanced routing protocols, since a critical limiting factor for a mobile host is its operation
time, which is restricted by its battery capacity. We propose a scheme in
which each mobile host maintains an energy tree and uses it to evaluate
the amount of energy remaining in other mobile hosts. Our proposed
algorithm extends the lifetime and upgrades the performance of the system, in that the amount of energy consumed by each mobile host in
wireless ad-hoc network is balanced, thereby prolonging the lifetime of
those mobile hosts whose energy capacity is limited.

1 Introduction

An ad-hoc network is a collection of wireless mobile nodes dynamically forming
a temporary network, without the use of any existing network infrastructure or
centralized administration. This type of network does not have any fixed control
devices connecting to the backbone host or other MHs(Mobile Hosts). Each MH
plays the role of the router in order to transmit the packets it receives to other
MHs[1, 2]. The study of wireless ad-hoc network is being actively pursued in the
MANET working group of IETF[3].

The amount of energy available to the nodes constituting this type of network is limited. In this paper, we propose a new algorithm, whose purpose is
to determine how the energy consumption of the MHs participating in wireless
ad-hoc network should be balanced, in order to solve some of the problems[4,
5, 6], namely the lowering of the system lifetime and the system performance
in this wireless network, due to the concentration of the energy consumption at
certain MHs. Each MH maintains an energy tree and broadcasts the information of energy tree into the directly neighboring MHs by means of message tree
packets. When a MH sets up a routing path to forward a packet, it chooses the
most appropriate path in terms of the energy consumption, by making use of its
energy tree and the breadth first search. In this way, the energy consumption
of each MH is balanced and its operating time is lengthened, by optimizing its

---

⋆ This work was supported (in part) by the Ministry of Information & Communications, Korea, under the Information Technology Research Center (ITRC) Support
Program

V. Paliouras, J. Vounckx, and D. Verkest (Eds.): PATMOS 2005, LNCS 3728, pp. 156–165, 2005.

limited battery capacity so that it improves the system lifetime and the system performance.

This paper consists of six sections. In section 2, we discuss previous studies concerned with the energy consumption of MHs. In section 3, we discuss some of the considerations involved in achieving balanced energy consumption. In section 4, we describe the new routing algorithm based on the proposed energy tree. In section 5, we explain the performance evaluation model used to compare the proposed algorithm with the AODV protocol[7], and analyze the results of the simulation. We conclude our paper in section 6.

2 Related Works

In the ad-hoc network environments, various routing algorithms related to the energy consumption of the nodes have been proposed[8]. The Energy Conserving Routing is a mechanism, which balances the energy consumption ratio. By comparing the amount of energy left in the different nodes, this mechanism causes those nodes with higher energy consume more energy and those nodes with lower energy consume less energy[9]. This mechanism sets up a routing path by means of the flow augmentation algorithm and the flow redirection algorithm, in order to balance the energy consumption ratio between the nodes involved in routing.

The Power Aware Routing protocol is not a new routing protocol, but suggests the use of different metrics when determining a routing path. The following energy-related metrics have been suggested instead of the shortest routing path between a source and a destination: minimizing the energy consumed per packet, maximizing the time to network partition, minimizing the variance in the node power levels, minimizing the cost per packet, and minimizing the maximum node cost[10].

The Localized Energy Aware Routing protocol directly controls the energy consumption. In particular, it achieves balanced energy consumption among all participating mobiles nodes. When a route path is searched for, each mobile node relies on local information on the remaining battery level to decide whether or not to participate in the selection process of a routing path. An energy-hungry node can conserve its battery power by not forwarding data packets on behalf of others. The decision-making process in LEAR is distributed to all relevant nodes, and the destination node does not need to wait or block itself in order to find the most energy-efficient path[11].

3 Considerations for Balanced Energy Consumption in Setting up the Routing Path

3.1 Outline

In this paper, the energy tree-based routing algorithm for the purpose of balancing the energy consumption is designed according to the following principles:

1) In the wireless ad-hoc network, a MH having a relatively large amount of energy is forced to consume more energy than a MH having less energy.
2) Each MH can reduce the amount of network traffic by locally broadcasting the message tree packet only to its neighboring MHs.
3) When a MH sets up a routing path, it sets up a path avoiding the neighboring MH, which has the least amount of energy.

3.2 The Energy Tree and the Message Tree Packet

In this paper, each MH maintains an energy tree that contains the energy information for both its own and the other MHs. The MH constructs the message tree packet which it forwards to its neighboring MHs by means of its current energy tree. A message tree is an energy tree received from a neighboring MH, and a message tree packet contains the message tree information which is forwarded to the neighboring MHs.

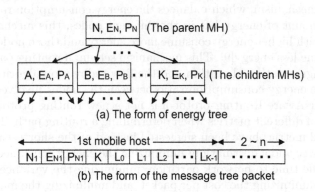

(a) The form of energy tree

(b) The form of the message tree packet

Fig. 1. The form of energy tree and the message tree packet

Figure 1a shows an energy tree maintained by MH N, and each MH entry in this tree is composed of the MH id(N), the energy information of the MH(E_N), the parent of the MH (P_N), and pointers pointing to the children of the MH. The form of the message tree is the same that of the energy tree.

Figure 1b shows the general form of the message tree packet, and the first MH contains the following information: the MH id(N_1), the energy level of the MH(E_{N_1}), the parent of the MH(P_{N_1}), the total number of pointers pointing to children MHs(K), and the links related to the pointers pointing to the children MHs(L_0, L_1, L_2, ..., L_{K-1}). For example, if the value of K is 3, three links are needed, viz. L_0, L_1, and L_2. The second formation of the MH follows the first formation.

4 Our Proposed Routing Algorithm Based on the Energy Tree

The A.1 of figure 2 shows the initialization process that the MH performs the algorithm when it connects to the wireless ad-hoc network. The MH creates the

---

A.1 When the mobile host N connects to the wireless ad-hoc network

    Create the root entry $(N, E_N, null)$ by using the mobile host information pertaining to itself in the energy tree

        Invoke a broadcast event for the message tree packet

A.2 When the broadcast event of the message tree packet is invoked in mobile host N

    Update the current energy information of mobile host to the E_N of the energy tree

    Create the message tree packet using the energy tree

    Broadcast this message-tree packet to all neighboring mobile hosts

    Reserve the broadcast event of the message tree packet after t seconds

A.3 When mobile host receives a message tree packet from a neighboring mobile host

    Create the message tree using the message tree packet

    Insert the root mobile host entry of the message tree into the bfs queue

    /* the entry type is $(N, E_N, null)$ */

    While (the queue is not empty)

    {

        entry (V, E_V, P_V) = dequeue from bfs queue

        /* V is any mobile host, E_V is the energy level of V, P_V is the parent mobile host of V */

        Insert the entries of all of the children of V into the bfs queue

        /* the entry type is (N, E_N, P_N) */

        if (V does not exist in the energy tree)

          if $P_V = null$

            append V as the child mobile host of the root mobile host in the energy tree

          else

            append V as the child mobile host of P_V in the energy tree

        else /* V exists in the energy tree */

        {

          update E_V of the energy tree as E_V of the message tree

          if the parent of V in the message tree is not the parent of V in the energy tree

            if the candidate path is a more balanced energy path than the old path

            /* the candidate path is V + the path from P_V to root in the energy tree */

            /* the old path is the path from V to root in the energy tree */

              change the parent of V in the energy tree to P_V on the candidate path

        }

    }

Fig. 2. The algorithm that each MH executes

root entry by using the MH information pertaining to itself in the energy tree. Then, the MH invokes a broadcast event for the message tree packet.

The A.2 of figure 2 shows the algorithm that each MH broadcasts the message tree packet to the neighboring MHs in order to exchange energy information with them every t seconds. This algorithm updates the current energy information of the MH into energy information of energy tree (E_N), and creates the message tree packet by using the energy tree. Then, this algorithm broadcasts the newly created message tree packet to all neighboring MHs, and reserves the broadcast event of the message tree packet, t seconds later.

The A.3 of figure 2 shows the algorithm that each MH executes when it receives a message tree packet from a neighboring MH. The detailed sequence of this algorithm is as follows. First, when the MH receives a message tree packet, it creates the message tree using this packet. Then, it creates the entry for the root MH of the message tree as a form of $(N, E_N, null)$, and inserts this entry into the bfs(breadth first search) queue. While the queue is not empty, the MH extracts MH V, which is the first entry in the queue and becomes the root MH of the message tree. The MH then creates the information for all of the children of MH V as a form of (N, E_N, P_N) and inserts this information into the bfs queue. If the MH V does not exist in the energy tree maintained by the MH and the parent of MH V is null, the MH appends MH V as the child MH of the root MH in the energy tree. If the parent of the MH V is not null, the MH appends MH V to the child of the parent of MH V, which already exists in the energy tree. If MH V exists in the energy tree, the MH updates the E_V of the energy tree as the E_V of the message tree. If P_V of the message tree is not the P_V of the energy tree and the candidate path is a more balanced than the old path in terms of energy consumption, then the MH changes the parent of MH V in the energy tree into P_V of the candidate path. The candidate MHs are the MHs that are MH V and MHs on the path from the root MH to P_V in the energy tree. The old MHs are those MHs that are situated on the path from the root MH to MH V in the energy tree. In this paper, we consider the average amount of energy left in the MHs as the factor to use to balance the energy consumption.

5   Performance Evaluation and Analysis

The purpose of this performance evaluation is to evaluate whether the proposed algorithm can continuously balance the energy consumption of each MH better than the AODV protocol. And we check the number of MHs below the threshold value and evaluate the network lifetime of the wireless ad-hoc networks.

5.1   The Performance Evaluation Model

In this paper, we used the AODV protocol to evaluate the performance of the proposed algorithm, using the following performance evaluation condition. 50 moving MHs communicate with each other within a rectangle (1500m by 900m). Direct communication between one MH and other MHs was limited to a maximum distance of 250m and the proposed algorithm was evaluated until 100,000 packets were completely transmitted to the destination.

Firstly, we describe the mobility model used in this paper. In this model, each MH constituting the wireless ad-hoc network moves, according to the random waypoint model[12]. We evaluated the performance using a uniform distribution in which the minimum moving speed is 0m/s and the maximum 2m/s.

Secondly, We modeled the source MH of the traffic to transmit as many packets as it transmits in one session to the destination MH with the constant bit rate (CBR). The number of CBR Source MHs is the half of the total number of MHs and each CBR source chooses an arbitrary destination. After this, it transmits 4, 8, 16, 32, and 64 packets per second to this destination. The think time has a uniform distribution with a minimum value of 0 seconds and a maximum value of 100 seconds. And the pause time has a uniform distribution with a minimum value of 0 seconds and a maximum value of 250 seconds.

Lastly, we describe the energy model used in this paper. The initial amount of energy in each MH is 10,000mWatt, 180mWatt is consumed when a packet is transmitted, and 130mWatt is consumed when a packet is received. Also, the threshold value is limited to the 33% of the initial energy.

5.2   The Results of the Performance Evaluation

We used the SIMLIB, which is an event-driven simulation tool, in order to evaluate the performance of the proposed algorithm.

Figure 3a shows the result of the performance evaluation concerning the average amount of energy left in the MHs changing by the traffic. The average amount of energy left in the MHs with the proposed algorithm is less than that with the AODV protocol. This is because, in the case of the proposed algorithm, the MH transmits the message tree packet to all of the neighboring MHs, in order to exchange the energy information with them. If we suppose that the total average of energy left in all of the MHs in the case of the AODV protocol is 100%, regardless of the number of the packets transmitted per second, the total average of energy left in all of the MHs in the case of the proposed algorithm is 96%.

Figure 3b shows the result of the performance evaluation concerning the variance of energy left in the MHs changing by the traffic. The more the number of packets transmitted per second gets, the smaller the variance of energy in the case of the proposed algorithm becomes in comparison with that of energy in the case of the AODV protocol. This is because, when the source MH sets up the routing path in the proposed algorithm, it chooses the routing path that avoids the neighboring MH that has the least energy. Therefore, the network lifetime of the proposed algorithm is longer than that of the AODV protocol. If we suppose that the total energy variance left in all of the MHs in the case of the AODV protocol is 100%, regardless of the number of packets transmitted per second, the total energy variance left in all of the MHs in the case of the proposed algorithm is 86%. Therefore, the performance of the proposed algorithm is excellent.

The result of the comparison between figures 3a and 3b is as follows. Regardless of the number of packets transmitted per second, the total average of energy left in all of the MHs was 4% smaller in the case of the proposed algorithm than in the case of the AODV protocol. However, the total variance of energy left in all of the MHs in the case of the proposed algorithm was 14% less than that in the case of the AODV protocol. Therefore, the energy consumption of the MHs is more balanced in the case of the proposed algorithm.

Figure 3c shows the result of the performance evaluation concerning the number of MHs whose energy is below the threshold value changing by the traffic. The more the number of packets transmitted per second gets, the less

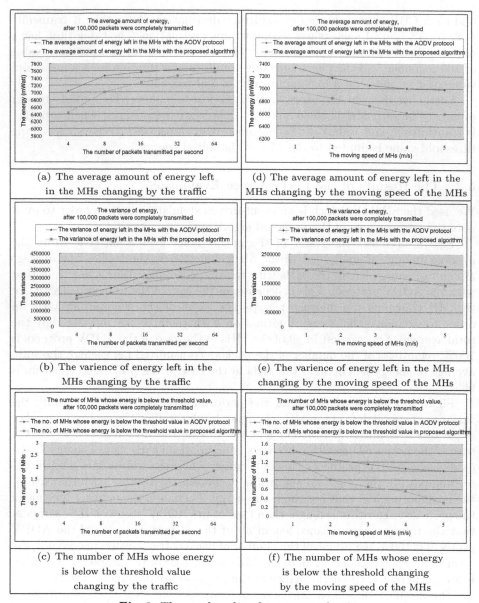

(a) The average amount of energy left in the MHs changing by the traffic

(d) The average amount of energy left in the MHs changing by the moving speed of the MHs

(b) The varience of energy left in the MHs changing by the traffic

(e) The varience of energy left in the MHs changing by the moving speed of the MHs

(c) The number of MHs whose energy is below the threshold value changing by the traffic

(f) The number of MHs whose energy is below the threshold changing by the moving speed of the MHs

Fig. 3. The results of performance evaluation

the number of MHs whose energy is below the threshold value in the case of the proposed algorithm becomes little by little in comparison with that in the case of the AODV protocol. This is because, when the source MH sets up the routing path in the case of the proposed algorithm, it chooses the routing path that avoids the neighboring MH that has the least energy. Therefore, the network lifetime of the proposed algorithm is longer than that of the AODV protocol.

Figure 3d shows the result of the performance evaluation concerning the average amount of energy left in the MHs changing by the moving speed of the MHs. On the whole, the average amount of energy left in the MHs in the case of the proposed algorithm is less than that in the case of the AODV protocol. This is because, in the case of the proposed algorithm, the MH transmits the message tree packet to all of the neighboring MHs, in order to exchange the energy information with them. If we suppose that the total average of energy left in all of the MHs in the case of the AODV protocol is 100%, regardless of the moving speed of the MHs, the total average of energy left in all of the MHs in the case of the proposed algorithm is 95%.

Figure 3e shows the result of the performance evaluation concerning the variance of energy left in the MHs changing by the moving speed of the MHs. The faster the moving speed of the MHs gets, the smaller the variance of energy of the proposed algorithm becomes in comparison with that of energy of the AODV protocol. This is because, when the source MH sets up the routing path in the case of the proposed algorithm, it chooses the routing path that avoids the neighboring MH that has the least energy. Therefore, the network lifetime of the proposed algorithm is longer than that of the AODV protocol. If we suppose that the total variance of energy left in all of the MHs in the case of the AODV protocol is 100%, regardless of the moving speed of the MHs, the total variance of energy left in all of the MHs in the case of the proposed algorithm is 78%. Therefore, the performance of the proposed algorithm is excellent.

The result of the comparison between figures 3d and 3e is as follows. Regardless of the moving speed of the MHs, the total average of energy left in all of the MHs in the case of the proposed algorithm is 5% smaller than that in the case of the AODV protocol. However, the total variance of energy left in all of the MHs in the case of the proposed algorithm is 22% less than that in the case of the AODV protocol. Therefore, the energy consumption of the MHs is more balanced in the case of the proposed algorithm.

Figure 3f shows the result of the performance evaluation concerning the number of MHs whose energy is below the threshold value changing by the moving speed of the MHs. The faster the moving speed of the MHs gets, the less the number of MHs whose energy is below the threshold value in the case of the proposed algorithm gets in comparison with that in the case of the AODV protocol. This is because, when the source MH sets up the routing path in the proposed algorithm, it chooses the routing path that avoids the neighboring MH that has the least energy. Therefore, the network lifetime of the proposed algorithm is longer than that of the AODV protocol.

Figure 4 shows the results of the performance evaluation, according to the broadcast periods, concerning the average amount of energy left in the MHs, the variance of energy left in the MHs, and the number of MHs whose energy is below the threshold value, after 100,000 packets were completely transmitted to the destination. These results have no relationship with the broadcast period in the AODV protocol, so that these factors are always regular.

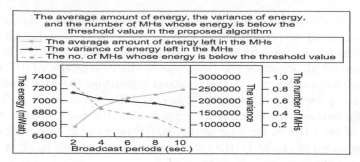

Fig. 4. The average amount of energy left in the MHs, the variance of energy left in the MHs, and the number of MHs whose energy is below the threshold values, according to the broadcast periods

As the broadcast period of the proposed algorithm increases, the number of the message tree packets to be sent decreases and the energy consumption also decreases. Because the energy consumption decrease, the average energy left in the MHs gradually increases, and the variance of energy left in the MHs and the number of MHs whose energy is below the threshold value gradually decrease.

6   Conclusion

It is important to maximize the durability of the batteries in the MHs in wireless ad-hoc network environments, because this extends the system lifetime and performance. Since the MHs play the role of the router, the network structure and location of the MHs affect their energy consumption. In this paper, we propose a new routing algorithm based on the energy tree by means of various principles. By balancing the energy consumption of the MHs, the proposed algorithm extends the system lifetime.

We compared the proposed algorithm with the AODV protocol through simulation. The results of the performance evaluation are as follows. Although the average energy left in each MH in the case of the proposed algorithm is less than that in the case of the AODV protocol, the variance of energy left in each MH in the case of the proposed algorithm is much less than that in the case of the AODV protocol. Therefore, the energy consumption of each MH is more balanced in the case of the proposed algorithm. Also, the number of MHs whose energy is below the threshold value in the case of the proposed algorithm is less than that in the case of the AODV protocol. Therefore, the system lifetime and the operating time of the MHs can be extended, because the proposed algorithm balances the energy consumption of the MHs better than the AODV protocol. In this paper, the criterion used to evaluate the balance in the energy consumption is the average energy left in each MH. Our future work is to evaluate the performance in various environments by applying different methods to the proposed algorithm.

References

1. C. S. R. Murthy and B. S. Manoj, Ad Hoc Wireless Networks:Architectures and Protocols, Prentice Hall PTR, pp. 193-196, 2004.
2. D. A. Maltz, "Resource Management in Multi-hop Ad hoc Networks", Technical Report CMU-CS-00-150, November, 1999.
3. IETF MANET WG, http://www.ietf.org/html.charters/manet-charter.html
4. D. J. Baker and A. Ephremides, "The architectural organization of a mobile radio network via a distributed algorithm", IEEE Transactions on Communications, vol. COM-29, no. 11, pp. 56-73, January, 1981.
5. A. Ephremides, J. E. Wieselthier, and D. J. Baker, "A design concept for reliable mobile radio networks with frequency hopping signaling", Proceedings of IEEE, vol. 75, no. 1, pp. 56-73, January, 1987.
6. V. Rodoplu and T. H. Meng, "Minimum energy mobile wireless networks", Proceedings of the 1998 IEEE Int'l Conf. on Communications, ICC'98, vol. 3, pp. 1633-1639, June, 1998.
7. C. Perkins and E. Royer, "Ad-hoc On-Demand Distance Vector Routing", 2nd IEEE Workshop on Mobile Computing Systems and Applications, February, 1999.
8. Mohammad Ilyas, The Handbook of Ad Hoc Wireless Networks, CRC PRESS, 2003.
9. J. H. Chang and L. Tassiulas, "Energy Conserving Routing in Wireless Ad-hoc Networks", Proceedings of IEEE INFOCOM2000, March, 2000.
10. S. Singh, M. Woo, and C. Raghavendra, "Power-Aware Routing in Mobile Ad Hoc Networks", Int'l Conf. on Mobile Computing and Networking (MobiCom '98), Dallas, TX, October, 1998.
11. K. Woo, C. Yu, H. Y. Youn, and B. Lee, "Non-Blocking Localized Routing Algorithm for Balanced Energy Consumption in Mobile Ad Hoc Networks", Int'l Symp. on Modeling, Analysis and Simulation of Computer and Telecommunication Systems (MASCOTS 2001), pp. 117-124, August, 2001.
12. J. Broch, D. B. Johnson, D. A. Maltz, Y. C. hu, and J. Jetcheva, "A Performance Comparison of Multi-Hop Wireless Ad Hoc Network Routing Protocols", Proceeding of the Forth Annual ACM/IEEE Int'l Conf. on Mobile Computing and Networking, October, 1998.

Energy-Aware System-on-Chip
for 5 GHz Wireless LANs*

Labros Bisdounis[1], Spyros Blionas[1], Enrico Macii[2],
Spiridon Nikolaidis[3], and Roberto Zafalon[4]

[1] INTRACOM S.A., Markopoulo Ave., GR-19002 Peania, Athens, Greece
[2] Dipartimento di Automatica e Informatica, Politecnico di Torino, I-10129 Torino, Italy
[3] Department of Physics, Aristotle University of Thessaloniki, GR-54124 Thessaloniki, Greece
[4] STMicroelectronics, Advanced System Technology, I-20041 Agrate Brianza, Milan, Italy

Abstract. This paper presents the realization of an energy-aware system-on-chip that implements the baseband processing as well as the medium access control and data link control functionalities of a 5 GHz wireless system. It is compliant with the HIPERLAN/2 standard, but it also covers critical functionality of the IEEE 802.11a standard. Two embedded processor cores, dedicated hardware, on-chip memory elements, as well as advanced bus architectures and peripheral inter-faces were carefully combined and optimized for the targeted application, leading to a proper trade-off of silicon area, flexibility and power consumption. A system-level low-power design methodology has been used, due to the fact that power consumption is the most critical parameter in electronic portable system design. The 17.5 million-transistor solution was implemented in a 0.18 m CMOS pro-cess and performs baseband processing at data rates up to 54 Mbit/s, with average power consumption of about 550 mW.

1 Introduction

In wireless data communications there have been many standardization efforts in order to meet the increased needs of users and applications. In the 5 GHz band, there were the IEEE 802.1a [1] and the HIPERLAN/2 [2-3], both specified to provide data rates up to 54 Mbps for wireless LAN applications in indoor and outdoor environments. Both aforementioned standards operate in the same frequency band, and utilize orthogonal frequency division multiplexing (OFDM) for multicarrier transmission [4].

The purpose of this paper is to present the realization of a low-power system-on-chip that implements the baseband processing as well as the medium access control and data link control functionalities of a 5 GHz wireless LAN system. Wireless communication systems require optimization of different factors including real time performance, area, power, flexibility and time-to-market. In order to optimize the combination of the above factors, instruction-set processors, custom hardware blocks as well as low-power memory and bus interface synthesis and mapping techniques are followed, offering a good balance between flexibility and implementation efficiency. The evolving scenario has serious consequences for any SoC development to be used in wireless systems. The protocol processor (running the upper protocol's layers) is included to the implemented SoC on the contrary to previous ASIC implementations

---

* This work was partially supported by the IST-2000-30093 EASY project

V. Paliouras, J. Vounckx, and D. Verkest (Eds.): PATMOS 2005, LNCS 3728, pp. 166–176, 2005.

[5], [6] in which an external protocol processor should be used. Additional advantages of the implemented architecture are the adopted low-power design methodology, and the flexibility inserted by using an additional embedded processor core for controlling the baseband modem and implementing the lower-MAC processing of the HIPER-LAN/2 standard, and a custom processor (MAC hardware accelerator) for implementing the lower MAC processing of the IEEE 802.11a standard [7]. During the development of the SoC, power optimization techniques were applied in the whole design range of the system, starting from the embedded software and the hardware-software mapping until the memory and bus interface synthesis [7].

The system-level design, the architectural choices of the implemented SoC, as well as the validation methodology have been presented in detail in [7-9]. In this paper we present an overview of the used design methodology, while our emphasis is on the presentation of the implementation results.

2 System-Level Design and SoC Architecture

The major problem in the design of a complicated system, such as a wireless LAN SoC, is the verification of the design and the early detection of design faults. For this reason a UML-based flow for the system design was followed in order to produce an executable system specification (virtual prototype) for early verification. This virtual prototype is based on a UML model, which uses the UML-RT profile [10]. The structure of the UML-based system model is presented in Figure 1.

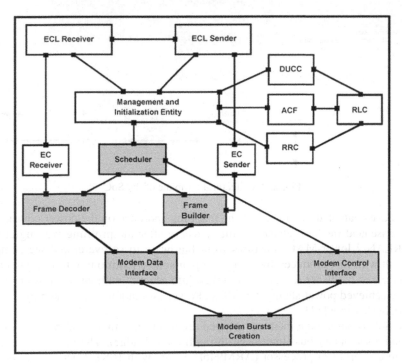

Fig. 1. Object structure of the HIPERLAN/2 high-level model

Within the structure of the model there is a group of three objects (modem data interface, modem control interface, modem bursts creation) that model the interface of the modem with the rest system, as well as the behavior of the modem's control parts. The remaining objects regard the upper layers of the HIPERLAN/2 protocol stack [2] and a custom implementation of the lower-MAC layer.

In order to support the hardware-software mapping of the system, a profiling and energy estimation methodology and tool were developed. The tool is based on a commercial instruction-set simulator, augmented with functional, timing and power models for peripherals and memory systems. It takes, as input, the executable specification (source code) of the target application, as well as power and timing models for various system components and it generates detailed profiling information, with aggregate performance and energy estimation for all functions in the source code. The profiling data are then used to identify the most energy and performance critical kernels in the application, in order to enable efficient mapping of the functional specification of the chip onto the target architectural template (Figure 2).

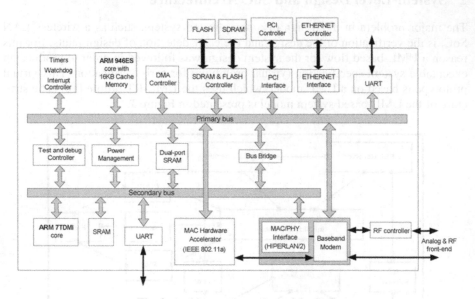

Fig. 2. Architectural template of the SoC

We have evaluated the impact of the mapping approach on the power consumption of the baseband modem-related circuitry, where after the profiling we migrated the HIPERLAN/2 lower-MAC functions from hardware to software, and we compared the result with that produced by using an open-source partitioning tool in which performance and hardware resources are taken into account without any power parameter. The achieved power saving was 19 %. Power estimations were taken by using the PowerChecker tool [11].

The SoC is based on a memory-mapped architecture with a primary (AMBA AHB bus) and a secondary bus (interfaced through a bus bridge). The main macrocells of the chip are: protocol processor (ARM946E-S core with 16KB Cache memory along with ARM-related blocks, PCI and Ethernet interfaces for the mobile terminal and access point devices respectively, SDRAM and Flash controller, internal SRAMs (single and dual port), DMA controller, test and debug controller, power management unit (accounting for the special circuitry for clock gating and supply shutdown), two UART I/O serial interfaces, dedicated ARM7TDMI core (for controlling the baseband modem and support critical lower-MAC processes), MAC hardware accelerator (supporting critical lower-MAC functionality of the IEEE 802.11a standard such as encryption/decryption, fragmentation, timing control, protocol medium access) [1], dedicated block for the interface of the MAC and PHY processing elements (for the case of the HIPERLAN/2 standard), and baseband modem implementing the digital part of the physical layer of both HIPERLAN/2 and IEEE 802.11a standards [1-3].

3 Low-Power Design Methodology

The first stage of the applied low-power design methodology regards the preprocesssing and pruning of the executable specification in order to extract useful information and produce a pruned version of the code, which will be used in the next optimization stage that regards data transfer and storage energy optimization. At the preprocessing stage, the information concerning the algorithmic behavior was extracted, as well as the different tasks and their communication way, the used data types and the manner in which data are manipulated. During pruning some parts of the code were removed as they are consider unworthy of being optimized. In the second stage, a transformation-based methodology was applied to the pruned code guided by power and performance cost functions. The output is an optimized code in terms of memory accesses, which contributed in the reduction of the total power consumption of the system. In order to evaluate the impact of the applied software optimizations we performed an experiment, in which we estimated the power consumed by the execution of the software specification on the ARM946-ES processor. The achieved power saving in comparison with the original executable specification was 18 %.

In addition, a methodology and an instrumentation setup for measuring the instruction-level power consumption of an embedded processor were developed [12]. They are based on the measurement of the instantaneous current drawn by the processor in each clock cycle. The instrumentation setup for the accurate measurement of the processor current includes a current sensing circuit, a digital storage oscilloscope and a PC with data processing software. After continuous monitoring and measurement of the instantaneous current of the processor, we derived instruction-level power models. Based on these models a framework was developed for the estimation of the energy consumed by the software running on the ARM processors. The framework calculates the base and inter-instruction energy costs of a given program and takes into account all other factors influencing the total power consumed by the software. Estimations and analysis that were performed using the developed framework were used for the optimization of the lower-MAC processes running on the ARM7TDMI processor.

Based on existing memory partitioning techniques [13], an automatic optimization methodology for on-chip memories was applied to the main memories of the SoC. According to the used methodology, after the mapping of the memory addresses range onto an on-chip SRAM and the analysis of its dynamic access profile, a multi-banked memory architecture is synthesized, optimally fitted to such profile [13]. The profile (obtained by instruction-level simulation) gives for each address in the range, the number of reads and writes to the address during the execution of the target application. Assume for instance the profile of Figure 3, where a small subset of the addresses is accessed very frequently. A power-optimal partitioned memory organization is shown in the right part of Figure 3 that consists of three memory cuts and a memory selection block. The larger cuts contain the top and bottom part of the range, while 'hot' addresses are stored into a small memory. The average power in accessing the memory is decreased, because a large fraction of accesses is concentrated on a small memory, and the memory banks that are not accessed are disabled through chip select (CS). According to our system architecture (Figure 2), there are in principle two internal memory structures on which memory partitioning was applied: a single-port 16KB SRAM and a dual-port 120KB SRAM. Table 1 gives the memory components of each memory after their partition made according to the analysis of the access profile. The power savings were about 22% for the case of the single-port SRAM, and 62% for the case of the dual-port SRAM. Power estimations were taken by using the PowerChecker tool [11].

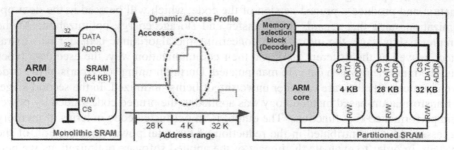

Fig. 3. Memory partitioning for power optimization

Table 1. On-chip memories partitioning

Memory components	Components after partition
Single-port SRAM: 16 KB	Block 1: 3.4 KB Block 2: 12.6 KB
Dual-port SRAM: 120 KB	Block 1: 1 KB Block 2: 119 KB

Another method that was used to reduce the power consumption in the SoC was to apply existing data encoding techniques on the information transmitted on the buses. These techniques consist of modifying the way the binary words are represented. In order to apply bus-encoding techniques on the secondary bus of the SoC (see Figure 2), an exploration tool has been developed, in which several of the existing encoding methods (bus invert [14], gray [15], zone [16], adaptive [17] etc.) have been imple-mented in software. Data and address traces that have been obtained by profiling of the embedded application were used as input to the exploration tool. The output of the exploration tool is the savings in terms of power-consuming transitions number. Given a set of input bus traces, regarding addresses and data, the first step we have performed was to identify the most convenient encoding scheme using our exploration tool. After the exploration regarding the address secondary bus we found that significant power savings (33%) are achieved by applying gray coding [15], while regarding the data bus we found that power savings up to 31% are achieved by applying bus-invert [14] coding. The derivation of the above savings took into account the energy consumed by the required encoders and decoders.

4 Embedded Software Development

The software parts of the system include high-level protocol oriented processes, low-level protocol processes and system specific hardware drivers. The different parts of the system software can be categorized into two different types: software processes and driver parts (Figure 4). The development of the software processes has been performed at a high-level of abstraction, using UML as a modelling language [10], while code generation techniques were utilised for the production of the executable code.

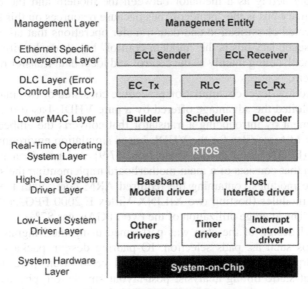

Fig. 4. Software architecture

Based on the results of the mapping procedure (section 2), as well as on exhaustive simulation using the UML-based high-level model of the system, a software-mapping scheme was derived and presented in the following table.

Table 2. Mapping of the software processes

ARM946ES	ARM7TDMI
ECL Sender	Scheduler
ECL Receiver	Frame builder
Management entity	Frame Decoder
EC Transmit	Modem driver
EC Receive	Interrupt Controller driver
RLC	Timer driver
Host Interface driver	UART driver
Interrupt Controller driver	
Timer driver	
DMA driver	

5 SoC Implementation and Validation

The chip is built on a dual-bus architecture (Figure 2), each bus dealing with different layers of the protocol stack. This allows each layer to have all the necessary resources for data and control transfer without having to compete for a single-bus with other layers. Furthermore, as each bus has a processor core attached, the layers become independent in what regards processing and control resources. The two buses communicate through a dual-port RAM, where large data-blocks are placed by one bus for the other to read and through a bridge, which allows the upper bus to push (or pull) small mounts of control/data information to (or from) the lower bus. Another device that connects to both buses is the baseband modem. Since we want to transfer data between layers, the modem has ports to both buses. This relieves the secondary bus from simply acting as a mediator between the modem and the upper protocol layers. The primary bus is the "protocol" bus because resources on this bus implement all the complicated classification and data transfer operations that are inherent to the protocol. The secondary bus is the "modem-control" bus. It deals with the lower layers of the protocol, acting mainly as the control and local data transfer resource for the baseband modem.

The followed low-level co-simulation procedure consisted of two main phases. The first phase exploited the advantage offered by a pure VHDL-based design and simulation environment by introducing a translator that converts the embedded core program into a set of force files for the VHDL-based simulation environment [8-9]. The second phase (FPGA-based design evaluation) [7-9] was based on the ARM Integrator platform that consists of a main motherboard implementing the system architecture, two core modules containing the required ARM cores with their peripherals, and two logic modules (hosting two XILINX Virtex E 2000 FPGAs that implement the rest logic). The average utilization of the two FPGAs was 87%.

The physical design of the chip was performed using the Magma flow [18] and con-tained steps such as: pads selection, IO-padring design, package selection and bon-ding, floorplanning, place and route, formal verification, parasitics extraction, pre/ post-layout static timing analysis, post-layout simulation, physical verification, DRC, and mask preparation. Implementation data for the chip are given in the following table. Figure 5 shows the placement of the chip's macrocells and a die photograph.

Table 3. Chip implementation data

Process technology	0.18 μm CMOS
Supply voltage	1.8 V (core), 3.3 V (I/O pads)
Operating frequency	80 MHz (some modem's blocks in 40 MHz)
Average power consumption	554 mW
Equivalent gates count	4,400,000
Transistors count	> 17,500,000
Pins count	456 (200 of them are for testing/debugging purposes)
Packaging	BGA 35mm x 35mm
Chip area	9.408mm x 9.408mm ≈ 88.5 sq. mm
Core area	8.576 mm x 8.576 mm ≈ 73.5 sq. mm
Area occupied by logic	43 sq. mm
Area occupied by memories	30.5 sq. mm
Total length of interconnections	47 m (six metal layers)

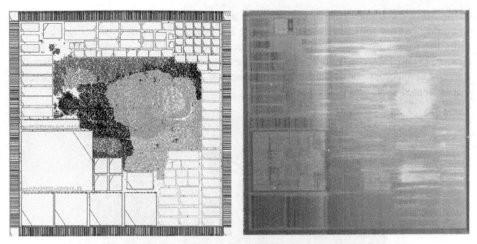

Fig. 5. Placement of the chip's macrocells and photograph of the die

The power consumption of the chip was estimated by using the PowerChecker tool [11]. The estimation was based on a testbench that includes an access point (AP) and a mobile terminal (MT) instance (side) of the chip that exchanging data. The total estimated power for the AP side was 460 mW, while for the MT side was 480 mW. For the embedded processors, we used the nominal power consumption (mW/MHz) for the target technology. In the following figure, we present the contribution of the main macrocells to the total chip power consumption.

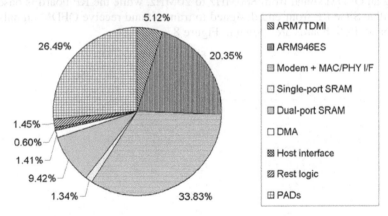

Fig. 6. Contribution of the system parts to the total power consumption

Power measurements were taken after the placement of the chip in two reference boards (operating as AP and MT, respectively). The average power consumption of the chip hosted by the AP board was 545 mW, while the average power consumption of the chip hasted by the MT board was 563 mW. In comparison with the power estimations given above, there was a deviation of about 18%.

In order to demonstrate and validate the SoC, we developed a system of reference boards. The system contained three boards: the main board hosting the SoC and implementing the digital part of the protocol as well as the analog to digital and digital to analog conversions, the IF board and the RF board. The main board (Figure 7) contains the following major items: Ethernet controller, SDRAM memory, Flash memories, A/D and D/A conversion and clock generation circuitry, power supply circuitry, various connectors for system's interconnection and test purposes, and of course the implemented system-on-chip (SoC).

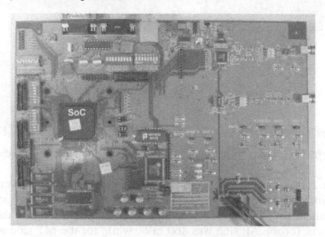

Fig. 7. Reference board hosting the implemented SoC

The IF board is based on a monolithic CMOS fully differential IF transceiver con-ver-ting an OFDM signal from 880MHz to 20MHz, while the RF board is based on a mo-nolithic SiGe RF front-end designed to transmit and receive OFDM signals in the 5GHz band. Both boards are shown in Figure 8.

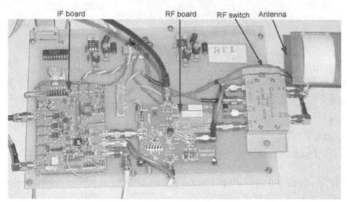

Fig. 8. IF and RF boards

The target area of the final demonstration setup of the EASY project includes two reference board systems (one operating as AP and a second operating as MT) exchanging Ethernet data through 5GHz RF connection. Operations such as association (between MT and AP) and connection setup, as well as network applications such as ping and FTP were tested successfully.

6 Conclusion

In this paper, the realization of an energy-aware system-on-chip for 5 GHz wireless LANs has been presented. The SoC implements the baseband processing as well as the medium access control and data link control functionalities of a 5 GHz wireless system. It is compliant with the HIPERLAN/2 standard, but it also covers critical functionality of the IEEE 802.11a standard. The embedded software development, the hardware implementation and the application of power optimization techniques have been described. Finally, implementation data as well as power measurements and the followed validation strategy were presented. The implemented SoC meets the functional and timing requirements of the WLAN system, and offers a power-efficient and flexible solution.

References

1. IEEE: 'Supplement to IEEE standard for information technology - Telecommunications and informa-tion exchange between systems - Local and metropolitan area networks - Specific requirements - Part 11: WLAN Medium Access Control & Physical Layer in the 5 GHz band', IEEE Std. 802.11a, 1999.
2. ETSI BRAN: HIPERLAN Type 2: 'System overview', TR 101 683, February 2000.
3. ETSI BRAN: HIPERLAN Type 2: 'Physical (PHY) layer', TS 101 475, April 2000.
4. Van Nee, R., Prasad, R.: 'OFDM for mobile multimedia communications', Artech House, MA (1999).
5. Kneip, J., Weiss, M., Drescher, W., Aue, V., Strobel, J., Oberthur, T., Bole, M., Fettweis, G.: 'Single chip programmable baseband ASSP for 5 GHz wireless LAN applications', IEICE Trans. Electronics, vol. 85 (2002) 359–367.
6. Eberle, W., Derudder, V., Vanwijnsberghe, G., Vergara, M., Deneire, L., Van der Perre, L., Engels, M.G.E., Bolsens, I., De Man, H.: '80-Mb/s QPSK and 72-Mb/s 64-QAM flexible and scalable digital OFDM transceiver ASICs for wireless local area networks in the 5-GHz band', IEEE J. Solid-State Circuits, vol. 36 (2001) 1829–1838.
7. Bisdounis, L., Dre, C., Blionas, S., Metafas, D., Tatsaki, A., Ieromnimon, F, Macii, E., Rouzet, Ph., Zafalon, R., Benini, L.: 'Low-power system-on-chip architecture for wireless LANs', IEE Proc. Com-puters and Digital Techniques, vol. 151 (2004) 2-15.
8. Drosos, S., Bisdounis, L., Metafas, D., Blionas, S., Tatsaki, A., Papadopoulos, G.: 'Hardware-software design and validation framework for wireless LAN modems', IEE Proc. Computers and Digital Tech-niques, vol. 151 (2004) 173-182.
9. Drosos, S., Bisdounis, L., Metafas, D., Blionas, S., Tatsaki, A.: 'A multi-level validation methodology for wireless network applications', pp. 332-341, in Integrated Circuit and System Design: Power and Timing Modeling, Optimization and Simulation, edited by E. Macii, V. Paliouras, O. Koufopavlou, Lecture Notes in Computer Science Series, No. 3254, Springer-Verlag, Berlin, September 2004.
10. Martin, G.: 'UML for embedded systems specification and design: Motivation and overview', Proc. DATE, Paris, France, 4-8 March (2002) 773-775.
11. BullDAST s.r.l., "PowerChecker", User Manual, Version 4.0, 2004.
12. Nikolaidis, S., Kavvadias, N., Neofotistos, P., Kosmatopoulos, K., Laopoulos, T., Bisdounis, L.: 'In-strumentation set-up for instruction-level power modeling', pp. 71-80, in Integrated Circuit Design: Power and Timing Modeling, Optimization and Simulation, edited by B. Hochet, A.J. Acosta, M.J. Bellido, Lecture Notes in Computer Science Series, No. 2451, Springer-Verlag, Berlin, Sept. 2002.

13. Benini L., Macchiarulo, L., Macii, A., Poncino, M.: 'Layout-driven memory synthesis for embedded systems-on-chip', IEEE Trans. on VLSI Systems, vol. 10 (2002) 96-105.
14. Stan, M.R., Burleson, W.P.: 'Bus-invert coding for low-power I/O', IEEE Trans. on VLSI Systems, vol. 3 (1995) 49-58.
15. Su, C.L., Tsui, C.Y., Despain, A.M.: 'Saving power in the control path of embedded processors', IEEE Design and Test of Computers, vol. 11 (1994) 24-30.
16. Musoll, E., Lang, T., Cortadella, J.: 'Working-zone encoding for reducing the energy in microproces-sor address buses', IEEE Trans. on VLSI Systems, vol. 6 (1998) 568-572.
17. Benini, L., Macii, A., Macii, E., Poncino, M., Scarsi, R.: 'Architectures and synthesis algorithms for power-efficient bus interfaces', IEEE Trans. on Computer-Aided Design, vol. 19 (2000) 969-980.
18. Magma Design Automation, "Gain-based synthesis: Speeding RTL to silicon", White paper, 2002.

Low-Power VLSI Architectures for OFDM Transmitters Based on PAPR Reduction

Th. Giannopoulos and Vassilis Paliouras

Electrical & Computer Engineering Department,
University of Patras,
25600 Patras, Greece

Abstract. This paper introduces a quantitative approach to the reduction of system-level power dissipation reducing the Peak-to-Average
Power Ratio (PAPR) in multicarrier systems. In particular introduces
a VLSI implementation of Partial Transmit Sequences (PTS) approach.
PTS is a distortionless Peak-to-Average Power Ratio (PAPR) reduction scheme suitable for Orthogonal Frequency Division Multiplexing
(OFDM) which imposes low additional complexity to the overall system.
We show that the application of this method reduces the power consumption of the complete digital-analog system by even 12.6%. Furthermore,
this paper examines theoretically the relationship between the achieved
PAPR reduction and the corresponding PA efficiency. Subsequently the
achieved PAPR reduction and the corresponding power saving are evaluated via simulation.

1 Introduction

New generation wireless technology standards, like 802.11a used in Wireless
Local Area Network (WLAN) and 802.16a suited for Wireless Metropolitan Area
Networks (WMAN), use OFDM due to its inherent error resistance and high bit-
rate capacity in a multipath environment. OFDM systems are sensitive to time
and frequency synchronization errors. Furthermore, OFDM symbols suffer from
high PAPR and thus can be distorted easily due to the nonlinearity of power
amplifiers in transmitters, leading to significant performance loss. Nonlinearities
cause imperfect reproduction of the amplified signal resulting in distortion and
out-of-band noise. Therefore the use of highly linear Power Amplifiers (PA) is
required. High linearity normally implies low efficiency, since large back off needs
to be applied. Therefore, the use of a PAPR reduction method is essential.

Several alternative solutions have been proposed to reduce PAPR. A simple
and effective approach is clipping the OFDM signal [ ]. However clipping may
cause significant in-band distortion, and out-of-band noise. Another solution is
to use appropriate block coding [ ]. However, the particular coding-based PAPR
reduction schemes require large look-up tables both at the transmitter and the
receiver, limiting their usefulness to applications with a small number of sub-
channels and small constellation sizes. Other PAPR reduction schemes introduce
a nonlinear predistortion of the transmit signal to combat signal peaks prior to

V. Paliouras, J. Vounckx, and D. Verkest (Eds.): PATMOS 2005, LNCS 3728, pp. 177–186, 2005.

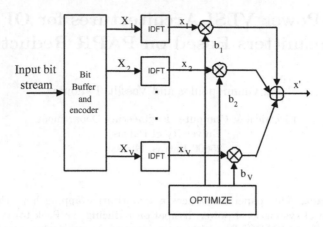

Fig. 1. General Architecture of PTS

amplification [ ]. An efficient and distortionless PAPR reduction scheme utilizes the so-called Partial Transmit Sequences (PTS) [ ][ ]. All PAPR reduction techniques impose additional processing. Therefore, the area requirements and the power consumption, of the digital part are increased. However, a benefit is anticipated at the complete digital-analog system level in terms of performance and power dissipation.

This paper explores the trade-off between the additional processing, required to reduce PAPR, and the corresponding power savings in the analog part of the transmitter. To quantify digital realization costs, a VLSI implementation of PTS algorithm is presented. The cost of PTS application is evaluated in terms of area, latency, and power consumption. Complexity trade-offs in the digital part are investigated, as follows: Two different ways of estimating PAPR are comparatively explored and the effect of the required data wordlength, is studied. As case study, for an OFDM system with 64 carriers, the effect of both the PAPR estimation method and the wordlength on the resulting PAPR reduction are quantified.

To quantify the impact of the PAPR reduction architecture at the analog-digital system level, the efficiency of a class-A PA is studied. In particular a relationship between the achieved PAPR reduction and the PA efficiency is derived, so that the designer can relate PAPR reduction figures with the corresponding power savings at the PA. Finally the relationship between the power dissipation increase in the digital part and the corresponding power reduction in the analog part of the transmitter is determined.

The remainder of the paper is as follows: Section 2 discusses the basics of OFDM transmission, defines PAPR and Crest Factor and outlines the PTS approach. In section 3, PA efficiency is defined and its relationship to PAPR reduction is depicted. In section 4 the proposed architecture is presented and the alternative strategies for estimating PAPR are explored. Section 5 presents the simulation results, while section 6 discusses conclusions.

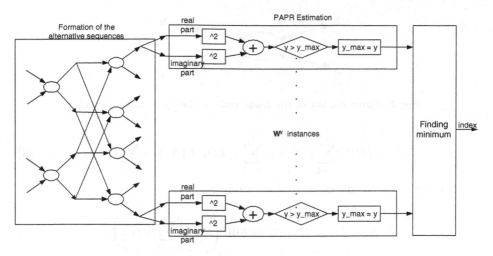

Fig. 2. Block Diagram of the optimize block

2   Basic Theory

Initially, the binary input data are mapped onto QPSK or QAM symbols. An IFFT/FFT pair is used as a modulator/demodulator. The N-point IFFT output sequence is $x_k = \frac{1}{\sqrt{N}} \sum_{n=0}^{N-1} X_n e^{j\frac{2\pi nk}{N}}$, where X_n is the transmitted symbol sequence and N is the block size.

The PAPR of the signal x_k is defined as square of the ratio of the peak power magnitude and the square root of the average power of the signal; i.e.,

$$\text{PAPR} = \frac{(\max |x_k|)^2}{E[|x_k|^2]}, \tag{1}$$

where $E[\cdot]$ is the expected value operator.

In the PTS approach [ ][ ], the input data block is partitioned into disjoint subblocks of equal size, each one consisting of a contiguous set of subcarriers. These subblocks are properly combined to minimize the PAPR (Fig. 1). Let the data block, $\{X_n, n = 0, 1, \cdots, N-1\}$, be represented as a vector, $X = [X_0\ X_1\ \ldots\ X_{N-1}]^T$. Then X is partitioned into V disjoint sets, represented by the vectors $\{X_v, v = 1, 2, \ldots, V\}$, such as $X = \sum_{v=1}^{V} X_v$. The objective of the PTS approach is to form a weighted combination of the V subblocks,

$$X' = \sum_{v=1}^{V} b_v X_v, \tag{2}$$

where $\{b_v, v = 1, 2, \cdots, V\}$ are the weighting factors such that the PAPR corresponding to $x' = IDFT\{X'\}$ is minimized. In order to calculate x' the linearity of the IDFT is exploited. Accordingly, the subblocks are transformed by V separate and parallel IDFTs yielding

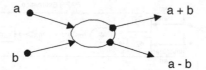

Fig. 3. Functionality of the basic unit of the optimization block

$$x' = IDFT\{\sum_{v=1}^{V} b_v X_v\} = \sum_{v=1}^{V} b_v IDFT\{X_v\} = \sum_{v=1}^{V} b_v x_v. \tag{3}$$

The optimum weighting factors are such that [ ]:

$$\mathbf{b} = [b_1, b_2, \ldots, b_V] = \arg\min\left(\max\left\|\sum_{u=1}^{V} b_v x_v\right\|\right), \tag{4}$$

resulting in the optimum transmitting sequence x' in terms of low PAPR. $\|\cdot\|$ denotes a norm for PAPR estimation of each alternative sequence. The calculation of the exact symbol PAPR requires the computation of symbol's average power. However, PTS algorithm does not seek the actual PAPR value; instead the vector x' with the lowest PAPR is sought. Therefore, assuming that the average power remains almost the same for all symbols, (4) is a very good approximation of the optimal weighting factors [ ]. Subsequently, the sequence with the lowest PAPR is chosen to be transmitted.

3   Amplifier Efficiency

Efficiency is a critical factor in PA design. PA's power consumption is evaluated by drain efficiency, defined as the ratio of RF output power to DC input power [ ],

$$n = P_{out}/P_{in}. \tag{5}$$

The instantaneous efficiency is the efficiency at one specific output level. Therefore, signals with time-varying amplitudes (amplitude modulation) produce time-varying efficiencies. In this case, a useful measure of performance is the average output power to the average DC-input power ratio:

$$n = P_{outAVG}/P_{inAVG}. \tag{6}$$

Signals with constant envelopes are always at peak output. In multicarrier communications systems, the average efficiency of a class-A amplifier, which is the most linear amplifier, is

$$n_{AVG} = n_{PEP}/\xi, \tag{7}$$

where ξ is the corresponding PAPR value, and n_{PEP} is the average efficiency if the signal had constant envelope and equal with the peak value [ ]. Assume

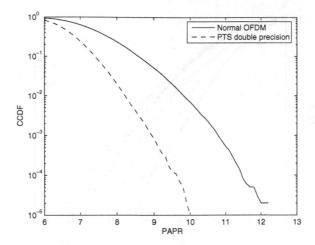

Fig. 4. PAPR reduction using PTS for $N = 64$

that a PAPR reduction method is employed. The achieved average efficiency is, $n'_{AVG} = n_{PEP}/\xi'$, where ξ' is the reduced PAPR value and in comparison with n_{AVG}, is

$$n'_{AVG} = \frac{\xi}{\xi'} n_{AVG}. \tag{8}$$

According to (6) and (8) the average DC-input power, required to give the same output power, is

$$P'_{inAVG} = \frac{\xi'}{\xi} P_{inAVG}. \tag{9}$$

The application of any PAPR reduction method, imposes additional processing onto the digital part of transmitter, and extra power, P_{dig}, is consumed. In order that the power consumption for the complete analog/digital system decreases, P_{dig} should be less than the power gain in the analog part;i.e.,

$$P_{dig} \leq P_{inAVG} - P'_{inAVG} \tag{10}$$

$$P_{dig} \leq \frac{\xi - \xi'}{\xi} P_{inAVG}. \tag{11}$$

4   Proposed Architecture

Several symbol partitioning strategies and many ways of choosing weighting factors b_v have been proposed in the literature [ ][ ][ ]. This paper presents the VLSI implementation of the optimization block (Fig. 2); i.e., a hardware unit which calculates the PAPR of all possible alternatives sequences, obtained of various values for **b**, and chooses among them the one to be transmitted. For the straightforward implementation, W^V alternative sequences should be examined, where W is the number of different values of weighting factors. The alternative

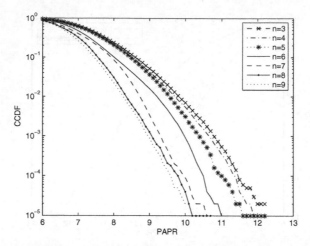

Fig. 5. CCDF for the first estimation method

sequences are the result of the weighted combination of the PTS according to (3). Each alternative sequence includes N samples. For every sample, the corresponding magnitude is calculated, and the maximum of them denotes PAPR of the corresponding alternative sequence. Finally, the sequence with the minimum PAPR value is identified and chosen to be transmitted.

In the proposed architecture the input block is partitioned into $V = 4$ subblocks $X_v(i)$,

$$X_v(i) = \begin{cases} X[4i + v], & i = 0, 1, 2, \ldots, N/4 - 1 \\ 0, & \text{otherwise} \end{cases}, \tag{12}$$

with $v = 0, 1, 2, 3$. The use of the particular partitioning allows the computation of the 4 $N/4$-point IFFTs with the same number of arithmetic equations as a N-point IFFT [ ]. Weighting factors are restricted to ± 1, thus no multiplication is required, when combining the partial sequences x_v to the peak-optimized transmit sequence x'.

A substantial reduction of the number of alternative sequences is achieved by exploiting the following property,

$$|x'| = \left| \sum_{v=1}^{4} b_v x_v \right| = |-1| \cdot \left| \sum_{v=1}^{4} b_v x_v \right| = \left| \sum_{v=1}^{4} (-b_v) x_v \right|. \tag{13}$$

The number of alternative sequences is reduced by a factor of 2, without loss of PAPR reduction capability. According to (13) the apply of weighting factors $[b_1 \ b_2 \ b_3 \ b_4] \ [-b_1 - b_2 - b_3 - b_4]$ on the PTSs, gives two sequences with exactly the same PAPR value. Therefore, only one of them needs to be computed. Hence, the alternative sequences that need to be computed in terms of their PAPR are just $W^V/2$. The corresponding sets of b_i, are that for which the value of the first weighting factor is fixed [ ].

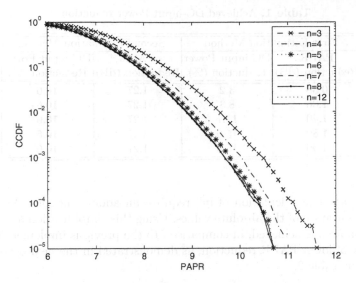

Fig. 6. CCDF for the second estimation method

According to (4), in order to combine the alternative sequences, N sums of 4 complex numbers are required, one per sample. For the addition of 4 numbers, 3 adders are required. Hence, the formation of all alternative sequences requires $3 \cdot W^V \cdot N$ additions. In the proposed architecture, the sums for the combination of the partial sequences are not calculated directly. The 4 common partial sums, $x_1 + x_2$, $x_1 - x_2$, $x_3 + x_4$, $x_3 - x_4$ are initially calculated and subsequently combined to derive the elements of the 8 alternative sequences. In that case, for all the alternative sequences only $12N$ complex additions are required (Fig. 2). The functionality of the basic implementation unit for the formation of the alternative sequences is depicted in Fig. 3.

A major task performed by the unit which derives the transmitted sequence, is PAPR estimation. PAPR is estimated by means of (4) and a suitable choice of a norm $\| \cdot \|$ applied on the complex sum $\sum_{u=1}^{V} b_v x_v$. This paper examines two alternative ways of estimating PAPR, distinguished by the choice of norm. The first choice calculates the magnitude of a complex number $y = y_1 + \jmath \cdot y_2$; i.e., $|y| = \sqrt{y_1^2 + y_2^2}$. Since it holds that $|y_1| > |y_2|$ then $|y_1|^2 > |y_2|^2$, for every real number, the computation of

$$|y|^2 = y_1^2 + y_2^2, \tag{14}$$

suffices, as a norm by means of which two complex numbers can be compared. The particular estimation does not impose any performance loss. The straightforward way to compute (14) in hardware, requires a multiplier. In order to calculate (14) in a more efficient way, an application-specific squarer [ ] has been implemented, taking into account the binary representation of the input.

The alternative way of estimating PAPR is by using a simple norm,

$$|y'| = |y_1| + |y_2| \tag{15}$$

Table 1. Achieved DC-input Power reduction

	First Estimation Method		Second Estimation Method	
n	PAPR Reduction (dB)	DC-input Power Reduction (%)	PAPR Reduction (dB)	DC-input Power Reduction (%)
5	0.38	3.2	1.27	10.6
6	1.00	8.3	1.27	10.6
7	1.40	11.7	1.27	10.6
8	1.80	15.0	1.27	10.6
9	1.90	15.8	1.27	10.6

instead of $|y|^2$. The calculation of $|y'|$ requires an adder and some XOR gates for the computation of the absolute values. Using this simplification a significant reduction in area is achieved, in comparison to the previous implementation, at the penalty of lower PAPR reduction, as demonstrated in the next section, and quantified in Table 2.

5   Simulation Results

In an OFDM symbol the large peaks occur with very low probability, so for characterizing the PAPR of OFDM signals, the statistical distribution of the PAPR should be taken into account [ ]. For that reason, the complementary cumulative distribution function $CCDF = Pr(PAPR > PAPR_o)$ is used. In the following results, 100000 random OFDM symbols were generated to obtain the CCDF. Each symbol includes 64 carriers, each one QPSK modulated. Cyclic Prefix is restricted to the 25% of the OFDM symbol. The transmitted signal is oversampled by a factor of 4 [ ], in order to better approximate the continuous-time PAPR.

Throughout this section all PAPR values, $PAPRo$, refer to $Pr(PAPR > PAPRo) = 10^{-5}$. This means that less than one symbol out of 100000 have PAPR value bigger than $PAPRo$ dB. Fig. 4 depicts the achieved PAPR reduction, resulting by the application of PTS algorithm and assuming that arithmetic operations are performed with double precision. The application of PTS leads to 2 dB PAPR reduction in comparison to normal (without PAPR reduction) OFDM.

A VLSI implementation of the PTS algorithm based on double-precision arithmetic (64-bit floating-point operations) is not viable, because of the extended area and power consumption of the corresponding circuit. According to simulation results, a fixed-point representation with 7 bits for the real and imaginary part, respectively, is sufficient. In the remainder of this section, the achieved PAPR reduction is examined, for an OFDM system with 64 carriers, using the two different norms for PAPR estimation and for various values of bits word length. It is assumed that two's complement numbers are used. In the complex-valued input of the optimization block, the length of real and imaginary part

Table 2. Synthesis results

Estimation Method	$|y|^2 = y_1^2 + y_2^2$ (using multiplier)			$|y|^2 = y_1^2 + y_2^2$ (using squarer)			$|y| = |y_1| + |y_2|$		
n	area (μm^2)	power (mW)	delay (ns)	area	power	delay	area	power	delay
5	120873	31.50	3.5	104588	43.25	2.4	76418	62.73	1.5
6	147205	36.43	3.6	130769	48.35	2.6	88988	74.63	1.5
7	178824	41.30	3.8	154263	43.69	3.3	98221	67.98	1.8
8	214332	45.20	4.0	182869	48.45	3.4	111109	77.50	1.8
9	248612	54.10	4.0	216474	53.20	3.5	127530	90.93	1.8

is n bits, respectively. During optimization, the word length increases by 1 bit following each addition and remains constant at the output of the squarer.

In the case of the first estimation method (Fig. 5) when the wordlength is restricted to 3 or 4 bits there is no PAPR reduction. For $n \geq 8$, the PAPR reduction is almost the same with that achieved when double precision arithmetic is used. But even for $n = 7$ or $n = 6$ there is a significant reduction of 1.4 dB and 1 dB, respectively. Using the second estimation method even for $n = 4$ bits there is a significant reduction of 1 dB. But for $n \geq 5$ the achieved PAPR reduction is the same and equal with 1.27 dB. Table 1 tabulates the achieved PAPR reduction per case, the power delay product, and the corresponding reduction of the average DC-input power, of PA, required to give the same output power, according to (9).

Table 2 tabulates the area, the latency and the total, including dynamic and leakage, power dissipation of the optimization block for different word lengths and for the two alternative estimation methods. Furthermore, Table 2 compares the case of implementing a general-purpose multiplier or an application-specific squarer. The corresponding results are obtained using Synopsys Design Compiler using an $0.18\mu m$ ASIC library. In order to estimate dynamic power, it is assumed that the switching activity of all the nets of the optimize circuit is 50%.

A commercial PA [ ] has DC-input power 1.4 W. From Table 1, when PTS is applied, with the first estimation method and for data wordlength $n = 5$, the average power dissipation of the transmitter reduces by $44,08$ mW. For that case the power consumption of the optimize circuit is 31.50 mW, hence the total power saving is 1%. Furthermore, it should be noted that reduced PAPR improves the linearity of the PA, which has a straightforward impact on the Bit-Error-Rate of the communication system. For wordlength of $n = 9$, the corresponding total power saving is 12.6%.

6   Conclusions

PA efficiency is a crucial matter for every wireless application, because of PA's power dissipation. This paper quantifies the impact of a PAPR reduction scheme adopted at the digital part of the system, versus the corresponding efficiency increase expected at the analog part of the transmitter, firstly theoretically and subsequently via extended simulation. The introduced analysis focuses on the

PTS PAPR reduction algorithm, for two different estimation methods and for various data wordlengths. Applying the corresponding results on an commercial PA, a total power saving of 1% to 12,6% is expected.

References

1. Li, X., Cimini Jr., L.J.: Effects of clipping and filtering on the performance of OFDM. IEEE Comm. Letts. **2** (1998) 131–133
2. Jones, A.E., Wilkinson, T.A., Barton, S.K.: Block coding scheme for reduction of peak to mean envelope power ratio of multicarrier transmission scheme. Elec. Letts. **30** (1994) 2098–2099
3. Kang, H.W., Cho, Y.S., Youn, D.H.: On compensating nonlinear distortions of an OFDM system using an efficient adaptive predistorter. IEEE Trans. on Comm. **47** (1999)
4. Müller, S.H., Bäuml, R.W., Fischer, R.F.H., Huber, J.B.: OFDM with reduced peak-to-average power ratio by multiple signal representation. In: In Annals of Telecommunications. Volume 52. (1997) 58–67
5. Cimini Jr., L., Sollenberger, N.R.: Peak-to-average power ratio by optimum combination of partial transmit sequences. In: Proc. of ICC'99. (1999) 511–515
6. Raab, F., Asbeck, P., Cripps, S., Kennington, P., Popovic, Z., Pothecary, N., Sevic, J., Sokal, N.: Power amplifiers and transmitters for RF and microwave. IEEE Transactions on Microwave Theory **50** (2002) 814–826
7. Kang, S.G., Kim, J.G., Joo, E.K.: A novel subblock partition scheme for partial transmit sequence OFDM. IEEE Transactions on Broadcasting **45** (1999) 333–338
8. Tellambura, C.: Improved phase factor computation for the PAR reduction of an OFDM signal using PTS. IEEE Communications Letters **5** (2001) 135 – 137
9. Giannopoulos, T., Paliouras, V.: An efficient architecture for peak-to-average power ratio reduction in OFDM systems in the presence of pulse-shaping filtering. In: Proc. of ISCAS'04. Volume 4. (2004) 85–88
10. Müller, S.H., Hüber, J.B.: OFDM with reduced peak-to-average power ratio by optimum combination of partial transmit sequences. Elec. Letts. **33** (1997) 368–369
11. K.E. Wires, M.J. Schulte, L.P. Marquette, P.I.Balzola: Combined unsigned and two's complement squarers. In: 33 Asilomar Conference on Signals, Systems, and Computers. (1999) 1215–1219
12. Ochiai, H., Imai, H.: On the distribution of the peak-to-average power ratio in OFDM signals. IEEE Trans. on Comm. **49** (2001)
13. STA-6033 datasheet. (In: www.sirenza.com)

An Activity Monitor for Power/Performance Tuning of CMOS Digital Circuits

Josep Rius[1], José Pineda[2], and Maurice Meijer[2]

[1] Departament d'Enginyeria Electrònica, Universitat Politècnica de Catalunya
Diagonal 647, 9th floor, 08028 Barcelona, Spain
rius@eel.upc.edu
[2] Philips Research Laboratories, Digital Design and Test Group, Building WAY4.81,
Post-box WAY41, Prof. Holstlaan 4, 5656 AA, Eindhoven, The Netherlands
{jose.pineda.de.gyvez,maurice.meijer}@philips.com

Abstract. The requirement to control each possible degree of freedom of digital circuits becomes a necessity in deep submicron technologies. This requires getting a set of monitors to measure each one of the parameters of interest. This paper describes a monitor fabricated in a 90nm CMOS technology which is able to estimate the circuit activity. The output of such monitor can be used as a tool to decide how to adjust the circuit working conditions to get the best power/performance circuit response. The paper presents the implementation and experimental results of a test chip including such monitor.

1 Introduction

The techniques to control power consumption of an IC are receiving more attention because of the increasing power consumption of ICs. In this way, many research efforts have been oriented to discover techniques to reach the desired performance at acceptable power levels.

Design techniques (we call them *static* techniques) to optimize power/performance, as scheduling, behavioral transformations, pipelining, bus encoding or optimization of FSMs [1] are good solutions to obtain optimized designs according to algorithms and/or the average expected data. However, after the design is finished, they cannot be changed to adapt the processor to different conditions. There are other techniques (we call them *dynamic* techniques), that try to optimize the power consumed [1] by adapting the system/circuit behavior to the current algorithms and/or data. System-level power management techniques with predictive power-management strategies exploits the past history of the active and idle intervals of each part of the system to shutdown the system resources during its periods of inactivity. Other RTL/gate-level techniques, as pre-computation or gated-clocks, take advantage of the same idea by shutting down or reducing the activity of the circuit portions when they are idle. Other methods try to dynamically find the optimum Energy Delay Product (EDP) of a given circuit by changing the supply voltage V_{DD} and/or the transistor threshold voltage V_{TH} [2], or by dynamically changing the transistor threshold voltage to obtain the desired performance under leakage power constraints taking into account in the control scheme the die-to-die and within-die process variations [3] [4]. Other proposal, [5] consists on dynamically change V_{TH} (by forward and reverse well biasing) to obtain

V. Paliouras, J. Vounckx, and D. Verkest (Eds.): PATMOS 2005, LNCS 3728, pp. 187–196, 2005.
© Springer-Verlag Berlin Heidelberg 2005

the best performance in the periods of time the circuit is computing and the lowest leakage in the periods the circuit is idle.

In the context of these dynamic techniques to optimize power/performance, monitoring in real time the circuit activity appears as an interesting way to fit the circuit to the best power/performance place. An activity monitor scheme is presented in [7] which has the drawback to require calibration and to know the characteristics of the circuit where the digital monitor is implanted. Also, it do not takes into account the data dependencies. Other analog activity monitors have presented in [8][9] for self-timed circuits. What we present in this paper is an activity monitor conceived as a fine grain monitor that collects information on the present internal activity of any part of a circuit. This information is useful in the following cases:

- Where it is not convenient or possible to fix the circuit working conditions for an average power consumption (which can be usually known from simulation or statistical analysis). In this case some type of control scheme to adapt these conditions to the changing consumption becomes useful. To do so, it is necessary to have a knowledge of the internal activity.
- Where the power consumption and the computational needs strongly depends on the input data or on the algorithm being executed. In this case, some type of trade between speed and power is also useful. Thus, an estimation of the activity (power) is a necessity.

This paper is organized as follows. Next Section describes the monitor. Section 3 is devoted to show an implementation of such monitor in a test chip and describes the experimental results. Section 4 concludes the paper.

2 Description

Consider the generic synchronous system depicted in Figure 1. If the content of the flip-flops does not change, the dynamic power consumption of the whole circuit, except the clock, is zero because the internal nodes remain in the same state. If, as response of a clock edge, the flip-flop's state changes, these changes propagate through the circuit generating a given amount of dynamic power consumption.

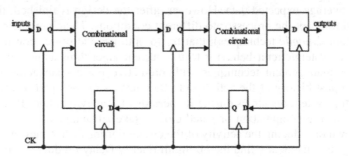

Fig. 1. A generic synchronous sequential system

After some time before the next clock edge, the new values at the flip-flop inputs are settled, thus preparing they for a new activity cycle. According to this description, the circuit power consumption indirectly depends on the number of switching flip-flops in each clock cycle. *As average*, the greater the number of switching flip-flops, the greater the power consumption is [6]. Thus, if we count the number of switching flip-flops in each clock cycle, we obtain an approximate image of the circuit power consumption. By counting the number of switching flip-flops in successive clock cycles, this image becomes more and more exact.

Assuming the circuit is designed using D flip-flops, we know if a FF is ready to switch in the next clock cycle by adding a two input XOR gate with one input connected to the D terminal and the other input to the Q terminal. It is clear that the flip-flop will change its state only if $D \neq Q$. In this case, the XOR output is "1". Thus, by counting the number of XOR outputs at "1" in each clock cycle we have the required information. As we need to obtain the result in one clock cycle, the counter has to be an adder. For a circuit with N flip-flops, the digital implementation of the previous scheme needs N two input XOR gates and a digital adder with N one bit inputs and $\log_2 (N)$ outputs. It is clear that with practical circuits (where N may be many thousands), the area overhead is too large and delay is too long to make this solution feasible.

2.1 Monitor Description

The proposed circuit consists on an analog solution to monitor the activity of the whole or a part of a digital circuit. Instead of the digital solution previously described, we propose to use an analog circuit working in current mode. The monitor adds four extra transistors to each flip-flop (Figure 2A) that work as a controlled current source. If $D \neq Q$, a pair of transistors are in conducting state supplying a small current to the monitor's output. If $D = Q$, the output remains in high impedance and no current is supplied. We call such monitor Flip-Flop Activity Monitor (FFAM).

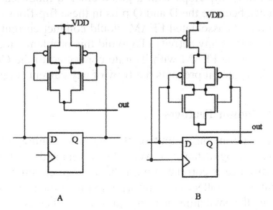

Fig. 2. Proposed monitor

The problem of summing the output of many of such monitors has a solution in the analog current mode here proposed: by connecting a wire to the output of each monitor the summation is accomplished, see Figure 3A. The current passing through the output wire is the sum of all the currents supplied by the FFAMs connected to flip-flops that will change their state at the next clock edge. By connecting this wire to a resistor, the total current is converted to a measurable voltage, if desired. On the other hand, introducing a PMOS transistor between V_{DD} and the output wire (not shown in the figure), we can disable the FFAMs and suppress any current.

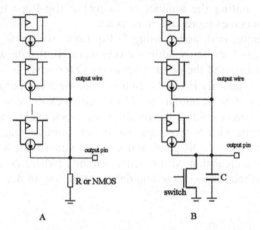

Fig. 3. Summing the monitor's output

Other possibility is to use a capacitor instead of a resistor as an I/V transducer, thus integrating the current supplied by the FFAMs in one or more clock cycles (Figure 3B). In this case, the voltage waveform at the output wire would reflect the energy spent by the circuit during the integration time. This capacitor has to be discharged (by the switch of the Figure 3B) at the beginning of the integration time in order to initialize its charge.

In circuits which have flip-flops with a gated clock a modification in the FFAM needs to be introduced because the D and Q pins in these flip-flops can be at different values for a long time. Its associated FFAM would conduct current continuously but this current does not reflect any activity. To avoid this problem, an additional transistor is put in series with the FFAM with its gate connected to the CK input in such a way that if the clock is gated it prevents the flow of any current (Figure 2B).

2.2 Discussion of Monitor Features

In the previous description, the output of the activity monitor performs an analog calculation of the Hamming distance between the present and the next circuit state. As is known [6], this distance is correlated with the average circuit power consumption. However, the monitor can collect more information than the state Hamming distance because in addition, the switching activity due to glitches is also captured if the glitches propagate until the flip-flop inputs and activate the monitor. Thus, the current waveform in the output wire reflects more accurately than the Hamming distance the transient power consumption in cycle-by-cycle basis.

The following paragraphs discuss the main concerns in the monitor operation:

- **Monitor speed and extra circuit delay.** The monitor speed is limited by the time required by the currents supplied by the conducting FFAMs to charge the output wire capacitance. This capacitance may be large due to the length of the output wire. If this were the case, "current repeaters" made using current mirrors can be distributed in the circuit to improve the speed. Some extra delay is introduced to the monitored circuit due to the increase in the load capacitance of the output of the flip-flop. This extra load capacitance is equivalent to the input capacitance of two small inverters.

- **Area overhead.** Each flip-flop including a FFAM needs four (or five) small size transistors. As each flip-flop has about 30 transistors, the overhead for flip-flop Ov_{FF} is 15-20 %. If the number of transistors in flip-flops is N_{FF} and the total number of transistors in the circuit is N_C, the total overhead is Ov_{FF} times N_{FF} / N_C. Also, the output wire connecting the output of all the FFAMs, and the I/V converter introduce additional area overhead. For large designs with many thousands of flip-flops this area overhead may be unacceptable if each FF has its own FFAM. If it is the case, some wise strategy is necessary to include the FFAMs only in the FF of proper parts of the circuit or activate them just in the proper times. However, in other smaller designs the area overhead may be acceptable.
- **Extra power consumption.** The amount of current delivered by the activity monitor depends on the design of the FFAM and on the circuit activity and it is maximum for the maximum activity. It is worth to note that this current is independent of clock frequency. The monitor current is taken from the same power supply as the monitored circuit, thus increasing its power consumption. Like the increase in area overhead, if the extra power consumption is unacceptable in large digital designs which include a FFAM in each FF, wise strategies to include or activate the FFAMs would be needed. At high clock frequencies, however, the relative weight of the extra consumption decreases.

3 Monitor Implementation and Experiments

In order to prove the feasibility of the proposed activity monitor and to check its dependence with process, voltage and temperature, a test chip has been designed and fabricated in a 90nm CMOS technology with nominal V_{DD} voltage of 1.2V. Figure 4 shows a view of such chip, which has about 8000 FF and 50000 gates. Two instances of the test circuit have been implemented. One with normal flip-flops and other with modified FFs including FFAM (5 additional transistors in each FF). All outputs of the FFAMs are connected together to an analog output pin.

The area overhead of the modified FF is 26 %. At room temperature and nominal conditions, the simulated delay time from a change in the D input to the FFAM output is lesser than 300 ps and from the rising edge of CK is lesser than 200 ps. The monitor delivers a current around 3.7 μA when it is excited. Total area overhead, including routing of the FFAMs output signals is 33 %. The test chip was fabricated in a 90nm CMOS technology with low power V_{TH} features. The modules of the test chip are basically circular shift registers including several levels of logic between each FF, and with programmable working frequencies. One can emulate the activity of a core with this circular shift register by shifting in a sequence of zeros and ones. The more swinging of bit patterns, the higher the activity of the shift register will be. Bit patterns like '111...111' or '000...000' correspond to no activity and bit pattern '0101...0101' correspond to 100% of activity. The clock frequency range is between 10MHz to 100MHz, voltage supply ranges from 0.7V to 1.4V and temperature range is from -40 to 125 degrees Celsius. In this first series of experiments only the average core current and activity monitor current is measured for a set of seven IC samples at different voltage and temperature conditions.

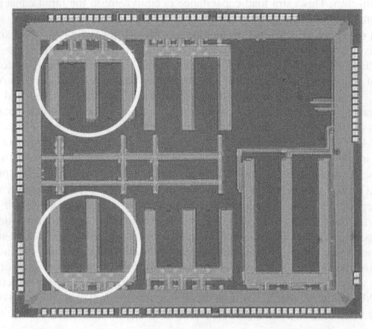

Fig. 4. Activity Monitor test chip. Upper circle: test circuit with normal flip-flops. Lower circle: test circuit with modified flip-flops including FFAMs.

3.1 Experiments

Figure 5 shows the correlation between the monitor's and the core's current for 0 to 100% activity. Each cloud in the graph represents a data activity, and each point in the cloud corresponds to a physical test chip sample. As can be seen a linear relationship exists between the monitor and core current. The current due to clock activity is subtracted from the core current and is measured with the input pattern '000...000' as shown in Figure 6. Figure 7 shows the output current for a given data activity as a function of the clock period. Basically, the current supplied by the monitor at a given activity should be only data and not clock dependent. That is, the monitor current remains always the same, whereas the core current increases with clock frequency. In this way, at high clock frequencies the monitor current becomes a small fraction of the total power of the circuit.

The dependency of the monitor response with supply voltage and temperature have also been measured. Figure 8 shows how the linear dependency between core and monitor current holds for a range of supply voltages from 0.7V to 1.2V.

In its turn, Figures 9 and 10 show the dependency of core and monitor current with temperature. In Figure 9, core current includes the current due to the clock network. As can be seen, at high activity, the monitor current shows a significant dependence with temperature.

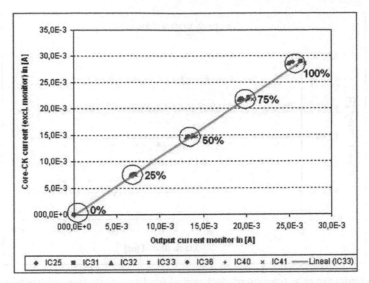

Fig. 5. Correlation of core's and monitor's current. Core current do not include the current of the clock network

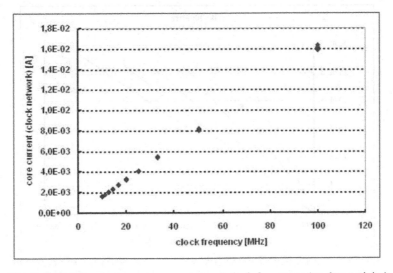

Fig. 6. Correlation between core current and clock frequency (no data activity)

4 Conclusion

As summary, a monitor of the activity of a digital circuit has been presented. The monitor's output gives an estimation of the power consumption in every clock cycle or in N clock cycles. This estimation contains useful information on: (a), how many changes in the flip-flop inputs have been produced in the present clock cycle (or in the last N clock cycles). And (b), how many changes in the flip-flop outputs will be produced in the next clock period. This information may be used to take decisions to

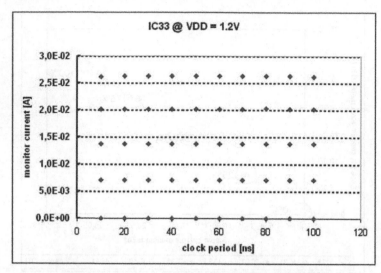

Fig. 7. Monitor current as function of clock period for data activity from 0% (bottom) to 100% (top)

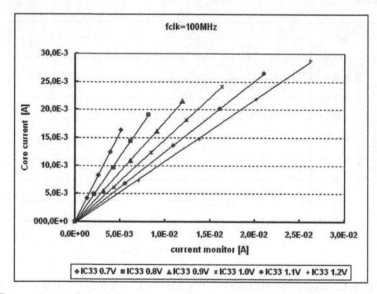

Fig. 8. Core current and monitor current as a function of supply voltage. Core current do not include the current of the clock network

improve the circuit power, performance or both. A test chip including such monitor has been fabricated in a 90nm CMOS technology and the experiments measuring the correlation between the average activity and average monitor current show that the monitor behaves as predicted. Also the independency of the monitor's output with clock frequency has been demonstrated. Further experimental work is necessary to collect the monitor response in a cycle-by-cycle basis.

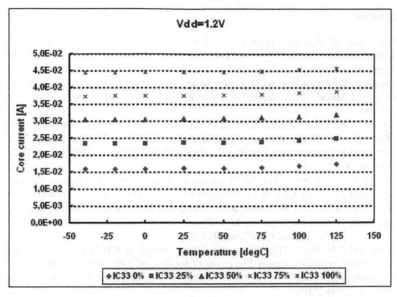

Fig. 9. Dependency of core current with temperature

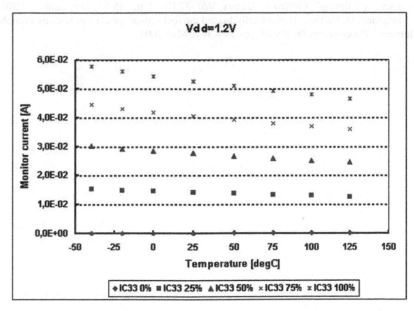

Fig. 10. Dependency of monitor current with temperature

Acknowledgements

The work of J. Rius was partially supported by the Comisión Interministerial para la Ciencia y Tecnología under Project TIC2004-02625, and the Secretaría de Estado de Educación, Universidades, Investigación y Desarrollo in Spain.

References

1. E. Macii and M. Pedram, "High-Level Power Modeling, Estimation and Optimization", *IEEE Transactions on CAD of Integrated Circuits and Systems*, Vol 17, No 11, November 1998, pp. 1061-1079.
2. R. Gonzalez, B.M. Gordon and M.A. Horowitz, "Supply and Threshold Voltage Scaling for Low Power CMOS", *IEEE Journal of Solid-State Circuits*, Vol 32, No 8, August 1997, pp. 1210-1216.
3. J. Tschanz, J. Kao, S. Narendra, R. Nair, D. Anoniadis, A. Chandrakasan and V. De, "Adaptive Body Bias for Reducing Impacts of Die-to-Die and Within-Die Parameter Variations on Microprocessor Frequency and Leakage", *Proceedings of ISSCC 2002*, paper 25.7. pp. 422 and 344-345 and 538.
4. M. Meijer, F. Pessolano, J. Pineda, "Technology Exploration for Adaptive Power and Frequency Scaling in 90nm CMOS", *Proceedings of ISLPED'04*, August 2004, pp. 14-19.
5. K. Nose, M. Hirabayashi, H. Kawaguchi, S. Lee and T. Sakurai, "VTH-Hopping Scheme to Reduce Subthreshold Leakage for Low-Power Processors", *IEEE Journal of Solid-State Circuits*, Vol 37, No 3, March 2002, pp. 413-419.
6. S. Gupta and F.N. Najm, "Energy-per-Cycle Estimation at RTL", *Proceedings of International Symposium on Low Power Electronics and Design*, 1999, pp. 121-126.
7. Intel Corporation, "Microprocessor with Digital Power Throttle", *International Patent* WO 01/48584 A1, July 2001.
8. E. Grass, S. Jones, "Activity-Monitoring Completion-Detection (AMCD): a new approach to achieve self-timing", *Electronic Letters*, Vol. 32 No. 2, pp. 86-88, 18th January 1996.
9. H. Lampinen, O. Vainio, "Dynamically biased current sensor for current-sensing completion detection", *Proceedings ISCAS'01*, pp. 394-397, May 2001.

Power Management for Low-Power Battery Operated Portable Systems Using Current-Mode Techniques

Jean-Félix Perotto and Stefan Cserveny

CSEM SA, Neuchâtel, CH
{jean-felix.perotto,stefan.cserveny}@csem.ch

Abstract. High performance voltage reducers were developed for ultra low-power battery operated systems in which several parts need a supply voltage adapted dynamically in order to optimize the energy consumption. Above 90% efficiency has been obtained for a 6 mA output current with a sliding mode Buck converter using a miniature external inductor; a very simple current mode approach has been proposed for this converter. Linear dissipative voltage regulators are used to supply very low duty cycle parts; their dynamic behavior for large and fast load variation has been improved adding the derivative of the regulated voltage to the control signal.

1 Introduction

The power management has become a critical part for the realization of complex Systems-on-Chips for which the optimization of the power consumption has become the most important issue. Efficient DC-DC converters should satisfy the specific requirements of a large range of very different applications. Even more challenging, the same system often needs several different power supplies drawing their energy from a common source such as a battery; each of these multiple supplies should satisfy its own requirements as efficiently as possible. Complex application specific power management systems [1-8] were developed with renewed solutions to satisfy such more and more demanding requirements.

Portable electronic equipments are generally powered by a battery whose voltage is often larger than the optimum supply voltage of the integrated circuits. Moreover, the optimization of the total energy consumption requires an adaptive power supply changing the regulated voltage according to the lowest value required to satisfy the desired performance at different operation modes; in some case, this adaptation is performed dynamically (DVS = Dynamic Voltage Scaling [2, 9]).

Reduced stand-by power consumption has also become an essential issue for very large systems in deep sub-micron technologies, especially for applications with very long idle modes [6, 10, 11].

The two voltage reducers presented in this paper were specifically designed for a 2.4 GHz transceiver whose requirements are briefly presented in the next section; nevertheless, different requirements for other foreseen applications were also considered in these developments.

The following sections present the novel solutions proposed to improve the performance of the inductive step-down and the linear dissipative regulators. Finally a few details concerning the test circuit which has been integrated are given before the conclusive review of what has been obtained.

V. Paliouras, J. Vounckx, and D. Verkest (Eds.): PATMOS 2005, LNCS 3728, pp. 197–206, 2005.

2 Power Control Blocks for a 2.4 GHz Transceiver

The considered 2.4 GHz transceiver has several blocks with very different power supply requirements. Fig. 1 presents schematically how its power management is organized. The corresponding power control blocks are considered in this paper.

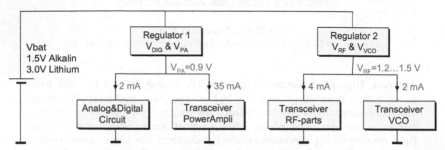

Fig. 1. Power control schematic for the considered receiver

The power supply for this system is either a 1.5 V alkaline or a 3 V lithium battery.

The analog and digital circuits and the transceiver power amplifier need a V_{PA} = 0.9 V supply. Due to the very low duty cycle for these blocks, a dissipative regulation is acceptable for this regulator (Regulator 1) without significant efficiency penalty. When the digital circuit and the power amplifier are inactive, the remaining blocks biased all the time with the same 0.9 V voltage consume about 5 µA.

On the contrary, it is very important to have a high efficiency regulator for the V_{RF} = 1.2 to 1.5 V supply, especially when a lithium battery is used. Therefore an inductive adiabatic step-down converter has been designed for this regulator (Regulator 2).

These two regulators, which have been integrated, are described in this paper.

3 Sliding Mode Inductive Step-Down Converter

A high efficiency regulator has to be adiabatic. In order to obtain the desired continuous range of regulated voltage, the inductive Buck approach had to be chosen because it is more flexible than the capacitive approach.

Most inductive converters use a classical PWM regulation loop to stabilize the output voltage to the specified value [7, 8, 12]. In this approach the regulation loop is very sensitive to the values of any element. Another approach, the sliding mode, which is a part of the Robust Control Technique, is much less sensitive [13]. The considered Buck converter is itself part of the VSS (Variable Structure System) class of circuits for which the sliding mode regulation turns out to be very efficient.

In the sliding mode regulation the system to be regulated is constrained to follow a particular trajectory in its state space, the sliding line, forcing the state variable that is regulated to converge toward the predefined target value. This technique applied to a typical Buck converter regulating V_{OUT} to the V_{REF} target is shown in Fig.2.

This schematic corresponds to the usual straight sliding line implementation, according to the simple equation

$$\tau \frac{dV_{OUT}}{dt} - (V_{REF} - V_{OUT}) = 0. \tag{1}$$

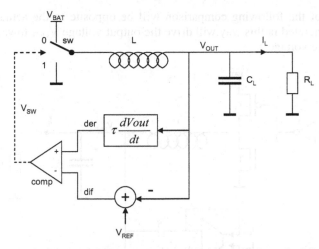

Fig. 2. Schematic of the sliding mode Buck regulation principle

The sliding line given by the equation (1) is represented in the left part of Fig. 3 state diagram. At power-on, the system evolves from the starting point A until it reaches the sliding line Δ at the point B; at this point it is captured by this line and guided to the target point C (V_{REF} in our case).

However, it is much easier to implement a sliding trajectory as shown on the right part of Fig. 3, which corresponds to the behavior of simple OTA output current characteristic. In this case the system, after being captured in the point B, evolves horizontally while the OTA output is saturated. At the point B', where the OTA leaves the saturation mode, the system is then guided to the target point C.

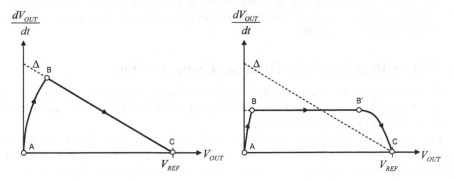

Fig. 3. Sliding mode state diagram for the Buck regulation principle (left) and for the proposed current mode (right)

The current mode approach proposed in Fig. 4 implements this last approach.

In this circuit the current I_{der} corresponds to the derivative of the output voltage V_{OUT}. This current is compared with the current I_{dif} corresponding to the difference between V_{OUT} and the reference voltage V_{REF}. Depending on the sign of the result of this comparison, the V_{err} node voltage will be much higher or lower than the V_{REF} reference. Therefore the inductor L is switched via the T1-T2 inverter in such a way that the sign of the following comparison will be opposite to the actual. The clock signal V_{SW} generated in this way will drive the output voltage V_{OUT} towards the value of the reference voltage.

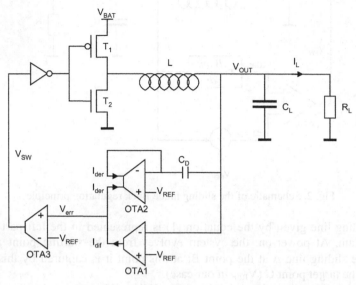

Fig. 4. Schematic of the sliding mode step-down converter using the proposed current mode

It is important to notice that this system is self-clocked therefore does not need any external clock. The switching frequency is defined by the delay in the loop depending primarily on the slew rate of OTA 3; it is about 2 MHz in the present design.

Fig. 5 shows the good V_{REF} tracking obtained without any over- or under-shoot. The efficiency which has been obtained is larger than 90% for a nominal 6 mA output current. The PSRR is -45 dB.

4 Linear Regulator with Dynamic Compensation

In the linear regulator there is a ballast transistor whose gate is controlled in a closed loop with an OTA as shown in Fig. 6. High voltage PMOS ballast is needed to satisfy the specified voltage range requirements.

Fig. 6 also shows the proposed addition on the ballast transistor gate node of a current proportional to the derivative of the output voltage V_{OUT} in order to improve the dynamic behavior of this regulator. The formulation of this current expresses the fact that the derivative is obtained with a derivation capacitance C_D while by other means its effect is thereafter multiplied k times.

Including this additional current, the closed loop transfer function $H(s) = V_{out}/V_{ref}$ of the regulator becomes

$$H(s) = \frac{\dfrac{g_{mB}g_{m1}}{C_L C_G}}{s^2 + s\,\dfrac{1}{R_L C_L}\left(1 + g_{mB} R_L\,\dfrac{kC_D}{C_G}\right) + \dfrac{g_{mB}g_{m1}}{C_L C_G}} . \tag{2}$$

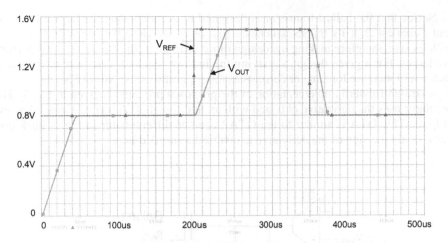

Fig. 5. V_{OUT} response to a V_{REF} change in the step-down converter

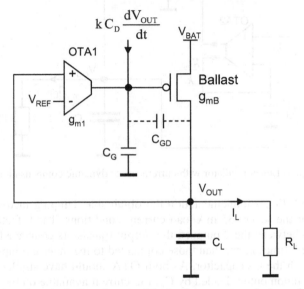

Fig. 6. Linear regulator including the principle for the current mode dynamic compensation

For the sake of clarity, this formula has been obtained assuming an ideal OTA with the transconductance g_{m1} and neglecting the very small effect of the C_{GD} feedback capacitance; in this formula g_{mB} is the transconductance of the ballast transistor and C_G the total capacitance on its gate.

The transfer function (2) corresponds to a second order system, characterized by its oscillation frequency and dumping factor. The parenthesis in the denominator of this function shows the multiplying factor by which the dumping increases due to the additional current proportional to the derivative of V_{OUT}. This increase becomes significant only if kC_D is large enough relatively to the C_G gate capacitance of the very large ballast PMOS (about 10 pF in our design).

Notice that the ballast transistor is designed to accommodate the largest load current (smallest R_L); its transconductance g_{mB} and gate capacitance C_G both increase with its size, therefore making the design more difficult as the maximum load currents are larger. For a given design, satisfying the highest required load current, the $g_{mB}R_L$ factor will increase for smaller currents as the g_m/I ratio of the ballast PMOS.

A straightforward implementation of the proposed principle is given in Fig. 7. This regulator uses the derivative obtained with an additional OTA (OTA2) which has two outputs. This OTA is similar to the OTA2 used as differentiator in Fig. 4 buck converter, only the sign of the derivative is inverted as needed here.

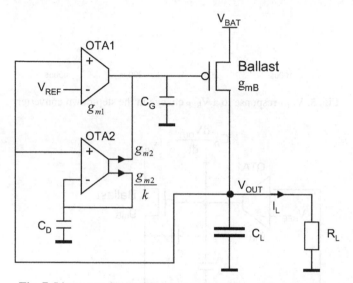

Fig. 7. Linear regulator with current mode dynamic compensation

However, with Fig. 7 implementation the obtainable damping factor increase was quite limited for the severe 35 mA load current conditions. The k factor is given by the mirror ratio between the 2 outputs, the output transistors connected to the ballast gate node being k times larger than those connected to the inverting input to make the derivative through the C_D capacitor. As both OTA should have similar driving capability on the common output loaded by C_G, the current available on the output loaded by C_D will be k times less. In the same time, the maximum C_D value is limited because, for an efficient dynamic compensation, this smaller current flowing into C_D should reach fast enough the value proportional to dV_{OUT}/dt. Therefore it is not possible to reach a very high kC_D value.

Larger kC_D values can be obtained with the two-stage circuit proposed in Fig. 8.

Here the derivative signal is not implemented directly as a current added to the ballast PMOS gate; it is transformed in the first stage (OTA2, C_D and R_D) into a voltage signal. As far a g_{m2} is much larger than $1/R_D$, the equivalent k becomes $g_{m1}R_D$ and there is more freedom to increase C_D. However this approach needs a careful design of OTA2, C_D and R_D taking into account the large spread of the poly resistor used in R_D and the parasitic capacitance C_P proportional to its value; a too large R_D can lead to oscillations.

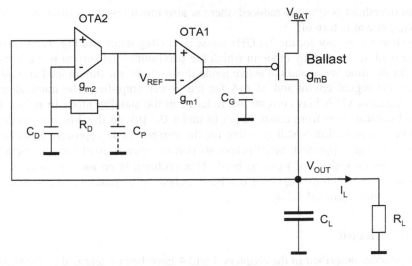

Fig. 8. Two-stages dynamically compensated linear regulator

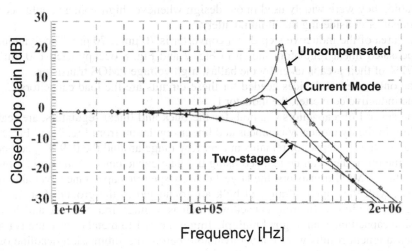

Fig. 9. Closed loop Bode plots for the uncompensated and for current mode and two-stages compensated linear regulators at the maximum $I_L = 35$ mA

The effectiveness of the proposed dynamic compensation on the closed loop frequency characteristics is demonstrated in Fig. 9.

This comparison shows the improvement obtained with the current mode compensation (Fig. 7); however full spike elimination is obtained with the two-stage compensation (Fig. 8) which has a much stronger dumping. Similar behavior is obtained for the PSRR, which is better than 40 dB at all frequencies for the two-stage compensation.

The transient responses in Fig. 10 show the improvement obtained with the two-stage dynamic compensation for both rising and falling 35 mA load current steps. At the I_L on switching there is no more visible V_{OUT} ringing and the quite harmful battery current overshoot is strongly reduced; there is also much less V_{OUT} drop at I_L turn on and V_{OUT} rise at I_L turn off.

The linear regulator for the 2.4 GHz transceiver (Regulator 1 in Fig. 1) has two operating modes, the standby mode in which the maximum load current is 5 µA, essentially for the time basis, and the active mode in which the maximum load current is 2 mA for the digital circuit and 35 mA for the power amplifier. The nominal active mode regulator OTA bias currents are reduced in the standby mode to reduce their internal consumption from about 30 µA to under 0.5 µA. In this low bias mode, the dynamic compensation is still effective for the current mode approach, however not for the two-stage approach; nevertheless, its performance is good enough, especially if a C_L high enough (2 µF) can be used. This problem is no more present in other applications in which the regulator can be switched off in standby, accepting a start-up time for the regulated voltage.

5 Test Circuit

The regulators described in the chapters 3 and 4 have been integrated in the tsmc018 standard digital process. This is a 1.8 V process, however 3.3 V transistors are also available; they were widely used in our design whenever high voltage could occur on the transistor considering the lithium battery.

The area of the inductive step-down converter is 70 µm x 70 µm. The dynamically compensated linear regulators shown in Fig. 7 and Fig. 8 occupy each 130 µm x 75 µm; 72% of their area is taken by the ballast high voltage PMOS transistor.

The only external devices needed for these circuits are the load capacitors C_L and, for the inductive converter, the inductor L.

The several pF capacitors C_D, both for the buck and linear regulators, are realized on chip. In the dynamically compensated linear regulators from Fig. 7 and Fig. 8, the voltage on the C_D capacitor remains in a narrow range around the 0. 9 V specified for V_{OUT}; therefore they can be realized using the gate capacitance of the normal threshold voltage NMOS transistors, the largest density specific capacitance available in the process, with the source, drain and bulk all connected to the ground. For the Buck converter, needing a floating capacitor which is voltage independent and has low parasitic capacitance, an original multiple finger metal 1 to metal 6 structure is used.

Measurement results were not yet available before the submission deadline of this final draft for the conference proceedings; however, they are expected to be presented at the conference.

6 Conclusions

A current mode approach has been proposed for a sliding mode inductive step-down converter. Relying on the normal behavior of a simple OTA, this approach is very stable because it behaves like a 1st-order system; it is also very robust and particularly easy to be implemented in comparison with the widely used PWM technique. It keeps the same high efficiency, above 90%, in a very large load current range due to its small own internal consumption (less than 100 µA).

Another development concerns a linear dissipative regulator with enhanced dumping characteristics. It uses a dynamic compensation by adding a current proportional

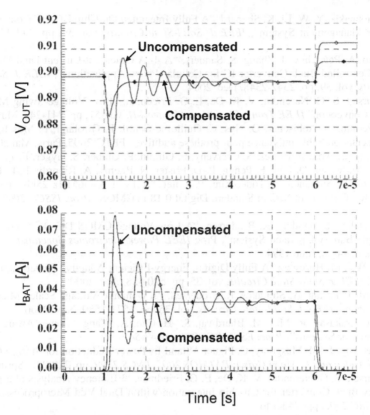

Fig. 10. Output voltage (top) and battery current (bottom) responses for a 35 mA I_L pulse for the two-stage dynamically compensated regulator compared with the uncompensated linear regulator

to the derivative of the regulated voltage on the ballast transistor gate node. A straightforward implementation of this current mode compensation shows good results as far as the load currents are not very large; however, for the required extreme conditions, a two stage approach has been used to reach the desired compensation factor. With an internal consumption about 25 µA, the regulated voltage over/undershoot is kept in a ± 10 mV range for a 35 mA load current step.

These circuits have been designed for the power management system of a battery operated 2.4 GHz transceiver. They have been integrated on a test circuit using the 0.18 µm TSMC standard digital process; the simulated performances presented now will be completed at the conference by measurement results that are expected soon.

References

1. D. Ma, W.-H. Ki, C.-Y. Tsui, "An Integrated One-Cycle Control Buck Converter with Adaptive Output and Dual Loops for Output Error Correction", *IEEE J. Solid-State Circuits*, vol. 39, pp. 140-149, Jan. 2004

2. G. Patounakis, Y. W. Li, K. Shepard, "A Fully Integrated On-Chip DC-DC Conversion and Power Management System", *IEEE J. Solid-State Circuits*, vol. 39, pp. 443-451, March 2004
3. J. Xiao, A. Peterchev, J. Zhang, S. Sanders, "A 4- A Quiescent-Current Dual-Mode Digitally Controlled Buck Converter IC for Cellular Phone Applications", *IEEE J. Solid-State Circuits*, vol. 39, pp. 2342-2348, Dec. 2004
4. V. Kursun, S. G. Narendra, V. K. De, E. G. Friedman, "Low-Voltage-Swing Monolithic dc-dc Conversion" *IEEE Trans. Circuits and Systems-II,* vol. 51, pp. 241-248, May 2004
5. J-F. Perotto, D. Aebischer, O. Nys, C. Guillome-Gentil, P. Girolami, P. Mosch, "Circuit d'alimentation faible puissance pour prothèse auditive", FFTC'2005, Paris, Mai 2005
6. V. Peiris, C. Arm, S. Bories, S. Cserveny, F. Giroud, P. Graber, S. Gyger, E. Le Roux, T. Melly, M. Moser, O. Nys, F. Pengg, P.-D. Pfister, N. Raemy, A. Ribordy, P.-F. Ruedi, D. Ruffieux, L. Sumanen, S. Todeschini, P. Volet, "A 1V 433/866MHz 25kb/s-FSK 2kb/s-OOK RF Transceiver SoC in Standard Digital 0.18 m CMOS" *Proc. ISSCC 2005*, pp. 258-259
7. A. Stratakos, S. Sanders, R. Brodersen, "A Low-Voltage CMOS DC-DC Converter for a Portable Battery-Operated System", *Proc IEEE Power Electronics Specialists Conf.*, vol. 35, pp. 619-626, June 1994
8. G.-Y. Wei, M. Horowitz "A Fully Digital, Energy-Efficient Adaptive Power-Supply Regulator", *IEEE J. Solid-State Circuits*, vol. 35, pp. 520-528, Apr. 2000
9. T. Burd, T. Pering, A. Stratakos, R. Brodersen, "A Dynamic Voltage Scaled Microprocessor System", *IEEE J. Solid-State Circuits*, vol. 35, pp. 1571-1580, Nov. 2000
10. A. Chandrakasan, R. Min, M. Bhardwaj, S. -H. Cho, A. Wang, "Power Aware Wireless Microsensor Systems", *Proc. ESSCIRC 2002*, pp. 47-54
11. S. Cserveny, J. -M. Masgonty, C. Piguet, "Stand-by Power Reduction for Storage Circuits", in *J. J. Chico and E. Macii (Eds.): PATMOS 2003, LNCS 2799*, pp. 229-238, Springer 2003
12. V. Kursun, S. G. Narendra, V. K. De, E. G. Friedman, "Efficiency Analysis of a High Frequency Buck Converter for On-chip Integration with a Dual-Vdd Microprocessor", *Proc. ESSCIRC 2002*, pp. 743-746
13. J-F Perotto, C. Condemine, "Convertisseurs DC-DC inductifs: vers un contrôle optimale", *Proc. TAISA 2004*, EPFL Lausanne, Sep.-Oct. 2004

Power Consumption in Reversible Logic Addressed by a Ramp Voltage

Alexis De Vos and Yvan Van Rentergem

Imec v.z.w. and Universiteit Gent, B-9000 Gent, Belgium

Abstract. Reversible MOS or r-MOS is a logic family that inherently promises asymptotically-zero power consumption. We deduce a simple formula for calculating the power consumption. It rightly highlights the unfortunate influence of the threshold voltages of the MOS transistors.

1 Introduction

Pass-transistor logic families offer different advantages with respect to conventional (so-called static) c-MOS. Less power consumption and smaller power-delay product are possible [1] [2] [3]. If we combine pass-transistor circuits with (quasi)-adiabatic addressing [4] [5] and with reversible logic [6] [7] [8], particularly low power consumption becomes possible. In the resulting circuits, there are neither power nor clock signals. All information, power, and clocking propagates from the input pins to the output pins, or the other way around.

2 Theory

Figure 1a shows the basic circuit: a source voltage $v(t)$ charges a capacitor (with capacitance C) to a voltage $u(t)$. Between voltage source and capacitor, we have a switch. Its off-resistance is infinite; its on-resistance is R. In practice, the switch is a transmission gate, i.e. the parallel connection of an n-MOS transistor and a p-MOS transistor. We call $w(t)$ the control voltage of the switch. In fact, the voltage w is applied to the gate of the n-MOS and the voltage $-w$ is applied to the gate of the p-MOS.

The analog input signals $v(t)$ and $w(t)$ as well as the analog output signal $u(t)$ represent binary digital signals: V, W, and U, respectively. Thus, we denote logic values with capital letters. E.g. A is either 0 or 1. The analog signal that represents the logic variable A is denoted by the lower-case letter a. In the ideal case, $A = 0$ is materialized by $a = -V_{dd}/2$ and $A = 1$ by $a = V_{dd}/2$. Non-ideal analog signals are interpreted as follows: if $a < 0$, it is interpreted as $A = 0$ and if $a > 0$, it is interpreted as $A = 1$.

An ideal switch is open whenever $w < 0$ and is closed whenever $w > 0$. Unfortunately, a transmission gate is not ideal. We assume it works as follows:

- whenever the gate voltage w exceeds $v + V_{tn}$, the switch is closed, because the n-MOS transistor is on;
- whenever the gate voltage $-w$ sinks below $v + V_{tp}$, the switch is closed, because the p-MOS transistor is on.

V. Paliouras, J. Vounckx, and D. Verkest (Eds.): PATMOS 2005, LNCS 3728, pp. 207–216, 2005.

The parameters V_{tn} and V_{tp} are called the threshold voltages of the transistors. Note that, in standard technologies, V_{tn} is positive and V_{tp} is negative. Thus the switch is always closed, except if both transistors are off, i.e. if

$$w < \min\left(v + V_{tn}, -v - V_{tp}\right).$$

We compare this rule with the law of the ideal switch: an ideal switch is always closed, except if $w < 0$.

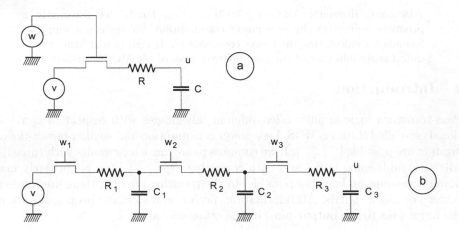

Fig. 1. Basic RC model

Let V be an input signal, which changes from $V = 0$ to $V = 1$ at time $t = 0$. For this purpose, let $v(t)$ be a ramp voltage:

$$v(t) = -\frac{V_{dd}}{2} \quad \text{for} \quad t \le -\frac{T}{2}$$
$$= at \quad \text{for} \quad -\frac{T}{2} \le t \le \frac{T}{2}$$
$$= \frac{V_{dd}}{2} \quad \text{for} \quad t \ge \frac{T}{2}.$$

Thus V_{dd} is the height of the ramp, T is the rise time, and $a = V_{dd}/T$ is the slope of the ramp. See Figure 2a. We stress that such addressing strategy is not optimal from an energy-consumption point-of-view [9] [10]. However, among all possible functions $v(t)$ which rise from the value $-V_{dd}/2$ to the value $V_{dd}/2$ in a finite time T, its performance is close to optimal. Because the above v profile is a simple time function and independent of circuit parameters C, R, and V_t, it is very often applied as the pseudo-optimal addressing of adiabatic logic [11] [12].

If the switch stays permanently closed, then Svensson [13] and Alioto et al. [14] show that the energy E dissipated during the transient phenomenon is

$$E = CV_{dd}^2 f(\frac{T}{\tau}),$$

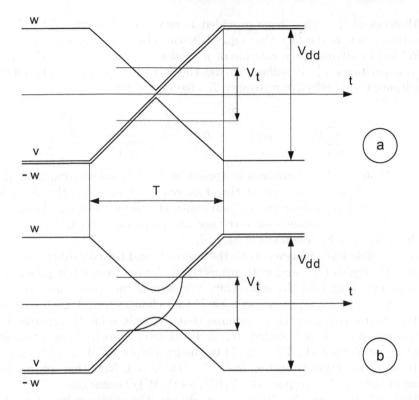

Fig. 2. Input signals $v(t)$, $w(t)$, and $-w(t)$

where $\tau = RC$ and

$$f(x) = \frac{1}{x} \left[\frac{1}{x} (e^{-x} - 1) + 1 \right] .$$

Alioto et al. [13] have proposed the following very useful approximation:

$$f(x) \approx \frac{1}{2 + x} ,$$

leading to the result

$$E \approx CV_{dd}^2 \frac{1}{2 + x} \tag{1}$$

$$\text{with} \quad x = \frac{T}{\tau} . \tag{2}$$

In the limit of short rise time, we find the conventional energy consumption per c-MOS computational step:

$$\lim_{T \to 0} E = \frac{1}{2} CV_{dd}^2 . \tag{3}$$

For infinitely slow addressing, we find the zero-dissipation **adiabatic limit**:

$$\lim_{T \to +\infty} E = 0 . \tag{4}$$

Alioto et al. [ ] thus have provided a very concise result. What is even more important, is the fact that eqn (1-2) can also be applied in the case of an RC ladder circuit, i.e. a cascade of n sections, each having a resistance R_i and a capacitance C_i. It suffices to use the following values for the effective capacitance C and effective resistance R (See Figure 1b) :

$$C = \sum_{i=1}^{n} C_i \qquad \text{and} \qquad R = \frac{\sum_{i=1}^{n} R_i \left(\sum_{j=i}^{n} C_j\right)^2}{\left(\sum_{i=1}^{n} C_i\right)^2} .$$

The Alioto et al. [ ] analysis is applicable to the transmission gate, if the latter is closed the whole time. It therefore can be regarded as the analysis of the case 'rising ramp source voltage and constant control voltage'. The opposite case, i.e. the case 'constant source voltage and rising ramp control voltage' has also been studied: by Nikolaidis et al. [ ].

In reversible logic, however, both the source V and the control W are driven by ramping signals ($v(t)$ and $w(t)$, respectively), either from input pins or from previous logic stages on the same chip. We assume the same ramping source voltage $v(t)$ as above, representing a bit V that changes from logic 0 to logic 1. For the control voltage $w(t)$, we assume that the logic value W remains 1, i.e. the transmission gate is intended to remain 'closed'. Thus we have a transition from the initial input $(V, W)_i = (0, 1)$ to the final input $(V, W)_f = (1, 1)$, which will result in an output transition from $U = 0$ to $U = 1$. Note that we thus focus on one of the $4 \times 4 = 16$ possible $(V, W)_i \to (V, W)_f$ transitions [ ].

Although its logic value W does not change, the analog value of w of the gate does change in a ramping way:

$$w(t) = \tfrac{V_{dd}}{2} \quad \text{for } t \le -\frac{T}{2}$$

$$= -at \quad \text{for } -\frac{T}{2} \le t \le 0$$

$$= at \quad \text{for } 0 \le t \le \frac{T}{2}$$

$$= \tfrac{V_{dd}}{2} \quad \text{for } t \ge \frac{T}{2} .$$

See Figure 2a. This is in contrast with the constant $w(t)$ scenario in Figure 2(b) of Chatzigeorgiou et al. [ ]. It is thus no surprise that conclusions will be different. As a result of the particular $v(t)$ and $w(t)$ waveforms, the transmission gate, intended to remain closed all the time, unfortunately is open for a short period

$$-\alpha_n \frac{T}{2} < t < \alpha_p \frac{T}{2} .$$

Here, the αs are dimensionless (and signless) threshold voltages: $\alpha_n = V_{tn}/V_{dd}$ and $\alpha_p = |V_{tp}|/V_{dd} = -V_{tp}/V_{dd}$. We will restrict ourselves to the common case $V_{tp} = -V_{tn}$. We will denote $V_{tn} = |V_{tp}|$ by V_t and $\alpha_n = \alpha_p$ by α. After laborious (but straightforward) calculations (see Appendix), we find:

$$E = \frac{1}{2}\, CV_{dd}^2 \left[\, \alpha(2-\alpha) - \alpha(1-\alpha)\, z\, f(\tfrac{z}{2}) + 2(1-\alpha)^2 f(z) \,\right] ,$$

where $z = (1-\alpha)\frac{T}{\tau} = (1-\alpha)x$. If we twice apply the Alioto–Palumbo–Poli approximation for f, we get:

$$E \approx \frac{1}{2}\, CV_{dd}^2 \, \frac{8 + 2(1+2\alpha)z + \alpha^2 z^2}{(4+z)(2+z)} \tag{5}$$

$$\text{with} \qquad z = (1-\alpha)x = \frac{(1-\alpha)T}{\tau}. \tag{6}$$

We thus find that $E(x)$ is a second-degree polynomial in x divided by another second-degree polynomial in x, thus a simple rational function of the x variable[1].

Note that, for $\alpha = 0$, we recover the Alioto–Palumbo–Poli formula (1-2). If α is not zero, i.e. if V_t is not zero, eqn (3) is still valid, but not eqn (4). We find the following **quasi-adiabatic limit** for slow addressing:

$$\lim_{T \to +\infty} E = \frac{1}{2}\, CV_t^2 . \tag{7}$$

Thus, unfortunately, the existence of a threshold voltage excludes asymptotically zero-energy switching. Note that, in this limit, we have $E \approx E_3$: the energy is mainly dissipated after the transmission gate conducts again and signal $u_3(t)$ subsequently catches up with the signal $v(t)$. Here, the symbols E_3 and u_3 have the meanings presented in the Appendix. Eqn (7) is well-known in literature: see e.g. [9] [16] [17] [18] [19] [20] [21]. New however is eqn (5-6), which expresses **how** the limit (7) is reached for large T.

We now assume that, just like result (1-2) holds for arbitrary RC-ladders, also result (5-6) holds for arbitrary chains of logic gates. This assumption is justified, because intermediate source signals $v_2(t)$, $v_3(t)$, ... and intermediate control signals $w_2(t)$, $w_3(t)$, ... (Figure 1b) are allowed to be themselves derived from intermediate result signals $u_1(t)$, $u_2(t)$, ... Indeed, source signals $v_i(t)$ and control signals $w_i(t)$ are allowed to be highly distorted, without affecting the value of E. In Figure 2b, the input voltages $v(t)$ and $w(t)$ are distorted during the second interval, however satisfying $w \leq v + V_{tn}$ and $w \leq -v - V_{tp}$ during that interval. Our simplifying assumption is supported by the numerical waveform simulations of transmission-gate ripple adders by Alioto and Palumbo [22].

3   Experiment

The above theory has been checked by means of a small-integration chip in the *AMI Semiconductor* standard n-well c-MOS 0.35 μm technology. The layout is designed with **Cadence** full-custom software. All transistors have minimum length: $L = 0.35$ μm. The n-MOS transistors have width W equal to 0.5 μm, whereas the p-MOS transistors have $W = 1.5$ μm. The treshold voltages are

---

[1] If α_n and α_p are different, i.e. if $V_{tp} \neq -V_{tn}$, then we obtain a third-degree polynomial in x divided by another third-degree polynomial in x

$V_{tn} = 0.6$ V and $V_{tp} = -0.6$ V. The circuit is a full adder consisting of four reversible logic gates: three CONTROLLED NOT gates and one FREDKIN gate, using a total of 20 switches [23] [24]. The circuit thus consists of 40 transistors. It measures about 65 μm $\times$ 30 μm. See Figure 3.

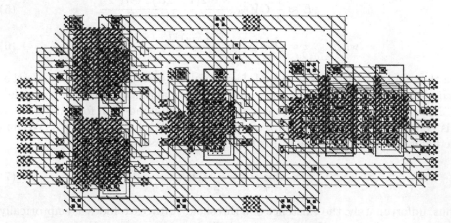

Fig. 3. Computer layout of full adder

Figure 4 shows the results of Spectre simulations for the full adder, in case of a ramp addressing of the circuit. The energy consumption E per computational cycle is calculated for 19 different values of T, ranging from 100 ps to 10 μs. In one cycle time, the input (A, B, C_i) switches from $(1, 0, 0)$ to $(1, 1, 0)$ in a rise time T and back in a second (fall) time T. Here, A, B, and C_i are addend, augend, and carry-in bits of the addition.

Figure 4 shows the curve fitting by eqn (5-6) with following parameter set:

$$\frac{1}{2} CV_{dd}^2 = 1.6 \text{ pJ}$$

$$\tau = 310 \text{ ps}$$

$$\alpha = 0.21 \ .$$

With $V_{dd} = 3.6$ V, this yields $C = 250$ fF, $R = 1.2$ kΩ, and $V_t = 0.76$ V. The threshold voltage is higher than the nominal one because of body effect.

We note that such $E(T)$ curve resembles a Bode diagram, thanks to the fact that eqn (5-6) has a rational form. The bending points of the Bode diagram are at $T = \frac{2\tau}{1-\alpha}$ and at $T = \frac{2\tau}{\alpha^2(1-\alpha)}$, because of the pole at $z = -2$ and the zero approximately at $z = -2/\alpha^2$. The second pole (at $z = -4$) and the second zero (at $z \approx -4 + 8\alpha$) are too close to one another to cause any bends in the curve.

Figure 5a shows experimental signals $v(t)$ and $u(t)$. Here $v(t)$ is the input signal applied to input bit B, whereas $u(t)$ is the voltage of the carry-out bit C_o. Figure 5b shows the chip current. Note the non-adiabatic spike during interval 3.

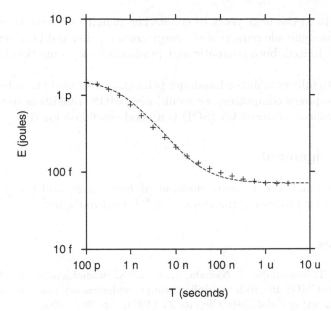

Fig. 4. Energy dissipation per computational step in a ramp-addressed reversible full adder: crosses are `Spectre` simulations; smooth curve is the analytical expression (5-6)

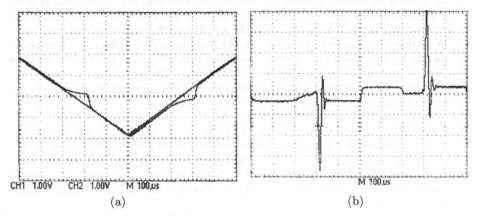

Fig. 5. Experimental oscilloscope view of (a) input and output voltages (1 V/div. vert.) and (b) output current (approx. 5 nA/div. vert.)

4 Conclusion

Static c-MOS logic has a $\frac{1}{2}CV_{dd}^2$ power consumption per computational step; adiabatic r-MOS logic consumes only $\frac{1}{2}CV_t^2$. Thus the power reduction corresponds to $(V_t/V_{dd})^2$. Now, the scaling approach of the *International Technology Roadmap for Semiconductors* [25] not only foresees a continuous shrinking of V_{dd}, but also prescribes a V_t reduced along with V_{dd}, with V_{dd} between $2 \times V_t$ and $6 \times V_t$. We therefore can assume that $(V_t/V_{dd})^2$ will remain of the order of

$(1/4)^2 = 1/16$ in the next years to come. The penalty we have to pay for this one-order-of-magnitude reduction of energy consumption and heat generation, is loss of speed. Indeed, both adiabatic and quasi-adiabatic computing is inherently slow.

In order to fully exploit the Landauer principle, i.e. in order to achieve asymptotically zero-power computing, we would need MOS transistors from a zero-V_t technology. Silicon-on-insulator (SOI) is a good candidate for this [26] [27].

Acknowledgement

The authors thank the *Invomec* division of *Imec v.z.w.* and the *Europractice* organization, for processing the chips at *AMI Semiconductor*.

References

1. K. Yano, T. Yamanaka, T. Nishida, M. Saito, K. Shimohigashi, and A. Shimizu : " A 3.8-ns CMOS 16×16-b multiplier using complementary pass-transistor logic ", *IEEE Journal of Solid-State Circuits* **25** (1990), pp. 388 - 395.
2. A. Chandrakasan, S. Shengl, and R. Brodersen : " Low-power CMOS digital design ", *IEEE Journal of Solid-State Circuits* **27** (1992), pp. 473 - 483.
3. R. Jiménez, A. Acosta, J. Juan, M. Bellido, and M. Valencia : " Study and analysis of low-voltage/low-power CMOS logic families for low switching noise ", *Proc. 9 th Patmos Workshop*, Kos, 6 - 8 Oct. 1999, pp. 377 - 386.
4. W. Athas, L. Svensson, J. Koller, N. Tzartzanis, and E. Chou : " Low-power digital systems based on adiabatic-switching principles ", *IEEE Transactions on Very Large Scale Integration Systems* **2** (1994), pp. 398 - 407.
5. M. Alioto and G. Palumbo : " Modeling of power consumption of adiabatic gates versus fan in and comparison with conventional gates ", *Proc. 10 th Patmos Workshop*, Göttingen, 13 - 15 Sept. 2000, pp. 265 - 275.
6. A. De Vos, B. Desoete, A. Adamski, P. Pietrzak, M. Sibiński, and T. Widerski : " Design of reversible logic circuits by means of control gates ", *Proc. 10 th Patmos Workshop*, Göttingen, 13 - 15 Sept. 2000, pp. 255 - 264.
7. A. De Vos, B. Desoete, F. Janiak, and A. Nogawski : " Control gates for reversible computers ", *Proc. 11 th Patmos Workshop*, Yverdon, 26 - 28 Sept. 2001, pp. 9.2.1 - 9.2.10.
8. A. De Vos : " Lossless computing ", *Proc. 7 th IEEE Workshop on Signal Processing*, Poznań, 10 Oct. 2003, pp. 7 - 14.
9. B. Desoete and A. De Vos : " Optimal charging of capacitors ", *Proc. 8 th Patmos Workshop*, Lyngby, 7 - 9 Oct. 1998, pp. 335 - 344.
10. A. De Vos and B. Desoete : " Equipartition principles in finite-time thermodynamics ", *Journal of Non-Equilibrium Thermodynamics* **25** (2000), pp. 1 - 13.
11. L. Svensson : " Energy-recovery CMOS in the deep-submicron domain ", *Proc. 9 th Patmos Workshop*, Kos, 6 - 8 Oct. 1999, pp. 83 - 92.
12. A. Schlaffer and J. Nossek : " Register design for adiabatic circuits ", *Proc. 9 th Patmos Workshop*, Kos, 6 - 8 Oct. 1999, pp. 103 - 111.
13. M. Alioto, G. Palumbo, and M. Poli : " Evaluation of energy consumption in *RC* ladder circuits driven by a ramp input ", *IEEE Transactions on Very Large Scale Integration Systems* **12** (2004), pp. 1094 - 1107.

14. S. Nikolaidis, H. Pournara, and A. Chatzigeorgiou : " Output waveform evaluation of basic pass transistor structure ", *Proc. 12 th Patmos Workshop*, Sevilla, 11 - 13 Sept. 2002, pp. 229 - 238.
15. A. Chatzigeorgiou, S. Nikolaidis, and I. Tsoukalas : " Timing analysis of pass transistor and CPL gates ", *Proc. 9 th Patmos Workshop*, Kos, 6 - 8 Oct. 1999, pp. 367 - 376.
16. Y. Moon and D. Song : " An efficient charge recovery logic circuit ", *IEEE Journal of Solid-State Circuits* **31** (1996), pp. 514 - 522.
17. Y. Ye and K. Roy : " Energy recovery circuits using reversible and partially reversible logic ", *IEEE Transactions on Circuits and Systems – I: Fundamental Theory and Applications* **43** (1996), pp. 769 - 778.
18. D. Mateo and A. Rubio : " Design and implementation of a 5 × 5 trits multiplier in a quasi-adiabatic ternary CMOS logic ", *IEEE Journal of Solid-State Circuits* **33** (1998), pp. 1111 - 1116.
19. J. Lim, D. Kim, and S. Chae : " A 16-bit carry-lookahead adder using reversible energy recovery logic for ultra-low-energy systems ", *IEEE Journal of Solid-State Circuits* **34** (1999), pp. 898 - 903.
20. C. Lo and P. Chan : " An adiabatic differential logic for low-power digital systems ", *IEEE Transactions on Circuits and Systems – II: Analog and Digital Signal Processing* **46** (1999), pp. 1245 - 1250.
21. J. Fisher, E. Amirante, F. Randazzo, G. Iannaccone, and D. Smith–Landiedel : " Reduction of the energy consumption in adiabatic gates by optimal transistor sizing ", *Proc. 13 th Patmos Workshop*, Torino, 10 - 12 Sept. 2003, pp. 309 - 318.
22. M. Alioto and G. Palumbo : " Analysis and comparison on full adder block in sub-micron technology ", *IEEE Transactions on Very Large Scale Integration Systems* **10** (2002), pp. 806 - 823.
23. Y. Van Rentergem and A. De Vos : " Optimal design of a reversible full adder ", *International Journal of Unconventional Computing*, to be published.
24. Y. Van Rentergem and A. De Vos : " Reversible full adders applying Fredkin gates ", *Proc. 12 th Mixdes Conference*, Kraków, 23 - 25 June 2005, pp. 179 - 184.
25. P. Zeitzoff and J. Chung : " A perspective from the 2003 ITRS ", *IEEE Circuits & Devices Magazine* **21** (January 2005), pp. 4 - 15.
26. M. Belleville and O. Faynot : " Low-power SOI design ", *Proc. 11 th Patmos Workshop*, Yverdon, 26 - 28 Sept. 2001, pp. 8.1.1 - 8.1.10.
27. M. Nagaya : " Fully-depleted type SOI device enabling an ultra low-power solar radio wristwatch ", *O.K.I. Technical Review* **70** (2003), pp. 48 - 51.

Appendix: Calculations

For $t \leq -\frac{T}{2}$, we assume the circuit is in equilibrium: $v = u = -V_{dd}/2$ (Figures 2a and 6). No current is flowing and thus no energy is dissipated in the resistor R (Figure 1a). For $t \geq -\frac{T}{2}$, we distinguish four time intervals (See Figure 6) :
(1) $-\frac{T}{2} \leq t \leq -\frac{\alpha_n T}{2}$, (2) $-\frac{\alpha_n T}{2} \leq t \leq \frac{\alpha_p T}{2}$, (3) $\frac{\alpha_p T}{2} \leq t \leq \frac{T}{2}$, (4) $\frac{T}{2} \leq t < +\infty$.
During interval 1, we have:

$$u_1(t) = at - a\tau + b_1 \exp(-t/\tau)$$

$$E_1 = \int_{-T/2}^{-\alpha_n T/2} \frac{1}{R}(at - u_1)^2 \, dt \ .$$

During the second interval, the switch is open, because w is both lower than $v + V_{tn}$ and lower than $-v - V_{tp}$. Therefore, no current flows and thus we have:

$$u_2(t) = b_2$$
$$E_2 = 0 .$$

During intervals 3 and 4, we have:

$$u_3(t) = at - a\tau + b_3 \exp(-t/\tau) \qquad \text{and} \qquad u_4(t) = aT/2 + b_4 \exp(-t/\tau)$$

$$E_3 = \int_{\alpha_p T/2}^{T/2} \frac{1}{R}(at - u_3)^2 \, dt \qquad \text{and} \qquad E_4 = \int_{T/2}^{\infty} \frac{1}{R}(aT/2 - u_4)^2 \, dt .$$

In the above, the four constants b_j are found from initial conditions. These are consequences of the continuity of the capacitor voltage u: $u_1(-\frac{T}{2}) = -\frac{aT}{2}$, $u_2(-\frac{\alpha_n T}{2}) = u_1(-\frac{\alpha_n T}{2})$, $u_3(\frac{\alpha_p T}{2}) = u_2(\frac{\alpha_p T}{2})$, and $u_4(\frac{T}{2}) = u_3(\frac{T}{2})$. Finally, the total energy dissipated in the resistor R is

$$E = E_1 + E_2 + E_3 + E_4 .$$

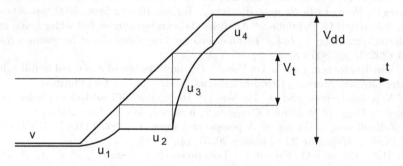

Fig. 6. Input voltage $v(t)$ and output voltage $u(t)$

Leakage and Dynamic Glitch Power Minimization Using Integer Linear Programming for V_{th} Assignment and Path Balancing

Yuanlin Lu and Vishwani D. Agrawal

Auburn University, Department of ECE, Auburn, AL 36849, USA
luyuanl@auburn.edu, vagrawal@eng.auburn.edu

Abstract. This paper presents a novel technique, which uses integer linear programming (ILP) to minimize the leakage power in a dual-threshold static CMOS circuit by optimally placing high-threshold devices and simultaneously reduces the glitch power using the smallest number of delay elements to balance path delays. The constraint set size for the ILP model is linear in the circuit size. Experimental results show 96%, 40% and 70% reduction of leakage, dynamic and total power, respectively, for the benchmark circuit C7552 implemented in the 70nm BPTM CMOS technology.

1 Introduction

In the past, the dynamic power has dominated total power dissipation of a CMOS device. Since dynamic power is proportional to the square of the power supply voltage, lowering the voltage reduces the power dissipation. However, to maintain or increase the performance of a circuit, its threshold voltage should be decreased by the same factor, which increases the subthreshold (leakage) current of transistors exponentially [1]. Therefore, with the trend of CMOS technology scaling, leakage power is becoming a dominant contributor to the total power consumption. To reduce leakage power, a large number of techniques have been proposed, including transistor sizing, multi-Vth, dual-Vth, optimal standby input vector selection, stacking transistors, etc.

Dual-V_{th} assignment [2-6] is an efficient technique to decrease leakage power. Wei *et al.* [3] describe an algorithm to find the optimal high V_{th} for different circuit structure. However, in reality, the available threshold voltages in a process are predetermined and a designer does not have the choice of arbitrary V_{th}. The back trace algorithm [3] used to determine the dual-V_{th} assignment also has a disadvantage. Because the back trace search direction for non-critical paths is always from primary outputs to primary inputs, the gates close to the primary outputs always have the priority to be assigned high V_{th}, even though their leakage power reduction due to V_{th} increase may be smaller than that of gates close to the primary inputs. This algorithm only gives a possible solution, not an optimal one. On the contrary, using ILP, a global optimization solution can be easily achieved. Nguyen *et al.* [6] use linear programming (LP) to minimize the leakage and dynamic power by gate sizing and dual-threshold voltage devices assignment. However, they have not considered the glitch power, which can account for 20%-40% of the dynamic switching power [7]. To eliminate these unnecessary transitions, a designer can adopt techniques of hazard filter [7] or path balance

V. Paliouras, J. Vounckx, and D. Verkest (Eds.): PATMOS 2005, LNCS 3728, pp. 217–226, 2005.

[8]. Raja *et al.* [8] have proposed a technique which uses a reduced constraint set linear program (LP) to eliminate dynamic glitch power.

The present work was motivated by the above research. A new ILP model is proposed to minimize leakage power by dual-V_{th} assignment and simultaneously eliminate dynamic glitch power by inserting zero-subthreshold delay elements to balance path delays. To our knowledge, no previous work on optimizing dynamic and static power has adopted such a combined approach.

This ILP method is specifically devised with a set of constraints whose size is linear in the number of gates. Thus, large circuits can be handled. The ILP either holds the critical path delay corresponding to the all-low V_{th} gates, or allows an increase by a user-specified amount. As a result, a tradeoff between power saving and performance degradation can be allowed.

To deal with the complexities of delay models and leakage calculation, two look up tables for the delay and leakage current are constructed in advance for each cell. This greatly simplifies the optimization procedure.

To further reduce power, other approaches such as gate sizing can be easily implemented by extending our cell library and look up tables.

2 Leakage and Delay

The leakage current of a transistor is mainly the result of gate leakage, reverse bias PN junction leakage and subthreshold leakage. Compared to the subthreshold leakage, the reverse bias PN junction leakage can be ignored. The subthreshold leakage current is the weak inversion current between source and drain of an MOS transistor when the gate voltage is less than the threshold voltage [1]. Subthreshold current is given by [2]:

$$I_{sub} = u_0 C_{ox} \frac{W_{eff}}{L_{eff}} V_T^2 e^{1.8} \exp\left(\frac{V_{gs} - V_{th}}{nV_T}\right) \cdot \left(1 - \exp\left(\frac{-V_{ds}}{V_T}\right)\right) \tag{1}$$

where u_0 is the zero bias electron mobility, and n is the subthreshold slope coefficient. Due to the exponential relation between V_{th} and I_{sub}, we can increase the V_{th} to reduce the subthreshold current sharply.

Our SPICE simulation results on the leakage current of a two-input NAND gate show that, for 70nm CMOS technology (Vdd=1V, Low V_{th}=0.20V, High V_{th}=0.32V), the leakage current in a high V_{th} gate is only about 2% of the leakage current in a low V_{th} gate. If all gates in a CMOS circuit could be assigned the high threshold voltage, the total leakage power consumed in the active and standby modes can be reduced by 98%, which is a significant improvement. However, the gate delay increases with the increase of V_{th}. From SPICE simulation result for a NAND gate delay when the output fans out to a specified number of inverters, we observe that the gate delay increases 30%-40% by increasing V_{th} form 0.20V to 0.32V.

Thus, we can make tradeoffs between leakage power and performance, leading to a significant reduction in the leakage power while sacrificing only some or no circuit performance. Such a tradeoff is made in the ILP. Results in Section 4.1 show that the leakage power of all ISCAS85 benchmark circuits can be reduced by over 90% if the delay of the critical path is allowed to increase by 25%.

3 Integer Linear Programming

To minimize the leakage power, we use an ILP model to determine the optimal assignment of V_{th} while controlling any sacrifice in performance. Due to the constraints on the maximum path delay, all the gates on the critical path are assigned low V_{th}. The V_{th} assignments of gates on the non-critical path are determined jointly by their delay increases and leakage reductions if high V_{th} were assigned to them. To eliminate the glitch power, additional ILP constraints determine the positions and values of the delay elements to be inserted to balance path delays. Unlike the heuristic algorithms [2-5], this ILP gives us a globally optimal solution.

We can easily make a tradeoff between power reduction and performance degradation by changing the constraint for the maximum path delay in the ILP model.

3.1 ILP for Leakage Power Reduction

Raja *et al.* [8] proposed a LP formulation to reduce dynamic glitch power by a reduced constraint set linear program whose number of constraints is proportional to the total number of gates. We first modify their formulation into an integer linear program (ILP) to reduce subthreshold leakage power as described below.

3.1.1 Variables
Each gate has two variables.

T_i: the latest time at which the output of gate i can produce an event after the occurrence of an input event at primary inputs of the circuit.

X_i: the assignment of low or high V_{th} to gate i; X_i is an integer which can only be 0 or 1. A value 1 means that gate i is assigned low V_{th}, and 0 means that gate i is assigned high V_{th}.

3.1.2 Objective Function
In a CMOS static circuit, the leakage power is

$$P_{leak} = V_{dd} \sum_i I_{leaki} \tag{2}$$

If we know the leakage currents of all gates, the leakage power can be easily obtained. Therefore, the objective function for this ILP is to minimize the sum of all gate leakage currents, which is given by

$$Min \sum_i \left(X_i \cdot I_{Li} + (1 - X_i) \cdot I_{Hi} \right) \tag{3}$$

I_{Li} is the leakage current of gate i with low V_{th};
I_{Hi} is the leakage current of gate i with high V_{th};

The leakage current of a gate depends on the input vector. Therefore, we make a leakage current look up table, which is indexed by the gate type and the input vector. I_{Li} and I_{Hi} can both be searched from this look-up table. The values in this lookup table, as found by Smart-SPICE simulation, are the total leakage currents including subthreshold and gate leakages of a cell under specific input vector conditions.

3.1.3 Constraints
Constraints for Each Gate

$$T_i \geq T_j + X_i \cdot D_{Li} + (1 - X_i) \cdot D_{Hi} \tag{4}$$

$$0 \leq X_i \leq 1 \tag{5}$$

Constraint (5) assigns either low V_{th} ($X_i = 1$) or high V_{th} ($X_i = 0$) to gate i ;

D_{Hi} is the delay of gate i with high V_{th};
D_{Li} is the delay of gate i with low V_{th}.

With the increase of the fanout, the delay of the gate also increases proportionately. Therefore, a second look-up table is constructed and specifies the delay for given gate type and fanout number. D_{Hi} and D_{Hi} can be searched from this look-up table indexed by the gate type and the number of fanout of gate i.

We explain constraint (4) using the circuit of Figure 1. Let us assume that all primary input (PI) signals on the left arrive at the same time. For gate 2, one input is from gate 0 and the other input is directly from a PI. Its constraints corresponding to inequality (4) are given by

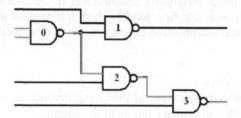

Fig. 1. Circuit for explaining ILP constraints

$$T_2 \geq T_0 + X_2 \cdot D_{L2} + (1 - X_2) \cdot D_{H2} \tag{6}$$

$$T_2 \geq 0 + X_2 \cdot D_{L2} + (1 - X_2) \cdot D_{H2} \tag{7}$$

T_2 that satisfies both inequalities is the latest time at which an event (signal change) may occur at the output of gate 2.

Max Delay Constraints for Primary Outputs (PO)

$$T_i \leq T_{max} \tag{8}$$

T_{max} can be the critical path delay when all the gates are assigned low V_{th} or the maximum delay specified by the designer. We use a simplified ILP model, whose description is omitted here, to find the delay T_c of the critical path. If T_{max} equals T_c the objective function of our ILP model will be to minimize the total leakage current without affecting the circuit performance. By making T_{max} larger than T_c, we can further reduce leakage power with some performance compromise, and thus make a tradeoff between leakage power consumption and performance

3.2 ILP for Leakage Power and Dynamic Glitch Power Reduction

Glitches can account for 20%-40% dynamic power [7]. To eliminate these unnecessary transitions, a designer can adopt techniques of hazard filter [7] or path balance

[8]. Combined with the method of path balance, the technique of Section 3.1 can be extended to reduce leakage power and dynamic glitch power simultaneously. Such an extended ILP model is developed below.

3.2.1 Variables
Each gate has four variables:

X_i: the assignment of low or high V_{th} to gate i; X_i is an integer which can only be 0 or 1. A value 1 means that gate i is assigned low V_{th}, and 0 means that gate i is assigned high V_{th}.

T_i: the latest time at which the output of gate i can produce an event after the occurrence of an input event at primary inputs of the circuit.

t_i: the earliest time at which the output of gate i can produce an event after the occurrence of an input event at primary inputs of the circuit.

$\Delta d_{i,j}$: the delay of the inserted buffer at the j_{th} input path of gate i.

3.2.2 Objective Function
The objective function for this ILP is to minimize the sum of all gate leakage currents and the sum of all inserted delays:

$$Min\left(\sum_i I_{leaki} + \sum_i \sum_j \Delta d_{i,j}\right) = Min\left(\sum_i (X_i I_{Li} + (1-X_i)I_{Hi}) + \sum_i \sum_j \Delta d_{i,j}\right) \tag{9}$$

Besides the objective to minimize the leakage power consumption which is the same as Equation (3), an additional objective function is to minimize the glitch power. We insert minimal delays to balance path delays and eliminate glitches. This leads to another objective function:

$$Min \sum_i \sum_j \Delta d_{i,j} \tag{10}$$

Our objective function (9) combines objectives (3) and (10).

When implementing these delay elements, we use transmission gates with only the gate leakage, which is much smaller than the subthreshold leakage and can be ignored.

3.2.3 Constraints
Constraints for Each Gate

$$0 \le X_i \le 1 \tag{11}$$

$$T_i >= T_j + \Delta d_{i,j} + (X_i \cdot D_{Li} + (1-X_i) \cdot D_{Hi}) \tag{12}$$

$$t_i <= t_j + \Delta d_{i,j} + (X_i \cdot D_{Li} + (1-X_i) \cdot D_{Hi}) \tag{13}$$

$$X_i \cdot D_{Li} + (1-X_i) \cdot D_{Hi} >= T_i - t_i \tag{14}$$

where, i is the the gate on which constraints are set, and j is the the gate whose output is gate i's fanin. Constraints (12-14) ensure that gate i's inertial delay is always larger than the delay difference of its input paths by inserting some delays on its faster input paths. Therefore, glitches can be eliminated.

We explain constraints (12-14) using the circuit shown in Figure 1. Let us assume that all primary input (PI) signals on the left arrive at the same time. For gate 2, one input is from gate 0 and the other input is directly from a PI. Its constraints corresponding to inequality (12-14) are:

$$T_2 \geq T_0 + X_2 \cdot D_{L2} + (1 - X_2) \cdot D_{H2} \tag{15}$$

$$T_2 \geq 0 + X_2 \cdot D_{L2} + (1 - X_2) \cdot D_{H2} \tag{16}$$

$$t_2 \leq t_0 + X_2 \cdot D_{L2} + (1 - X_2) \cdot D_{H2} \tag{17}$$

$$t_2 \leq 0 + X_2 \cdot D_{L2} + (1 - X_2) \cdot D_{H2} \tag{18}$$

$$X_2 \cdot D_{L2} + (1 - X_2) \cdot D_{H2} \geq T_2 - t_2 \tag{19}$$

Time T_2 that satisfies both inequalities (15) and (16) is the latest time at which an event (signal change) may occur at the output of gate 2.

Time t_2 is the earliest time at which an event may occur at the output of gate 2, if it satisfies both inequalities (17) and (18).

Constraint (19) means that the difference of T_2 and t_2, which equals the delay difference between two input paths, is smaller than gate 2's inertial delay, which may be either low V_{th} gate delay, D_{L2}, or high V_{th} gate delay, D_{H2}.

Max Delay Constraints for Primary Outputs (PO)

$$T_i \leq T_{max} \tag{20}$$

As in Section 3.1, T_{max} can be the maximum delay specified by the circuit designer or the critical path delay.

When we use the ILP model to simultaneously minimize leakage power with dual-V_{th} assignments and reduce dynamic power by balancing path delays with inserted delay elements, the optimized version for the circuit in Figure 2 is shown in Figure 3. The label in or near a gate is its inertial delay.

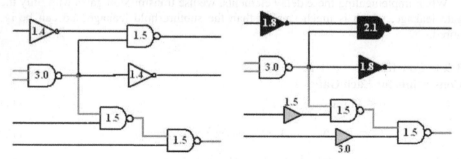

Fig. 2. Unoptimized circuit with potential glitches

Fig. 3. Optimized circuit

Three black shaded gates are assigned high V_{th} since they are not on the critical path and their delay increases do not affect the critical path delay. Two delay elements (grey shaded) are inserted to eliminate glitches. Although delay elements, if implemented as buffer gates, may consume additional leakage power, we may assign high V_{th} to them. Therefore, the three low V_{th} gates (without shading) on the critical path still dominate the total leakage power. Actually, in our design, delay elements are implemented by CMOS transmission gates that have no subthreshold leakage. Transmission gates also consume very little dynamic power since they are not driven by any supply rails [9].

4 Results

To study the increasingly dominant effect of leakage power dissipation, we use the BPTM 70nm CMOS technology. Low V_{th} for NMOS and PMOS are 0.20V and -0.22V. High V_{th} for NMOS and PMOS are 0.32V and -0.34V, respectively.

We regenerate the netlists of all ISCAS'85 benchmark circuits using a cell library in which the maximum gate fanin is 5. Two look-up tables for gate delays and leakage currents, respectively, of each type of cell are constructed using SPICE simulation. A C program parses the netlist and generates the constraint set (see Section 3) for the CPLEX ILP solver in the AMPL software package [10]. CPLEX then give the optimal V_{th} assignment as well as the value and position of every delay element.

4.1 Leakage Power Reduction

The results of the leakage power reduction for ISCAS'85 benchmark circuits are shown in Table 1. The numbers of gates in column 2 are for the gate library used and differ from those for original benchmark netlists. T_c in column 3 is the minimum delay of the critical path when all gates have low V_{th}. Column 4 shows the leakage reduction (%) for optimization without sacrificing any performance. Column 6 shows the leakage reduction with 25% performance sacrifice. The CPU times shown are for the ILP runs and are, as expected, linear in circuit size since both number of variables and number of constraints are linear in circuit size. From Table 1, we see that by V_{th} reassignment the leakage current of most benchmark circuits is reduced by more than 60% without any performance sacrifice (column 4). For several large benchmarks leakage is reduced by 90% due to a smaller percentage of gates being on the critical path. However, for some highly symmetrical circuits, which have many critical paths, such as C499 and C1355, the leakage reduction is less. Column 6 shows that the leakage reduction reaches the highest level, around 98%, with some performance sacrifice.

The curves in Fig. 4. show the relation between normalized leakage power and normalized critical path delay in a dual-V_{th} process. Unoptimized circuits with all low V_{th} gates are at point $(1,1)$ and have the largest leakage power and smallest delay. With optimal V_{th} assignment, leakage power can be reduced sharply by 60% (from point$(1,1)$ to point$(1,0.4)$) to 90% (from point$(1,1)$ to point$(1,0.1)$), depending on the circuit, without sacrificing any performance. When normalized T_{max} becomes greater than 1, i.e., we sacrifice some performance, leakage power further decreases in a slower reduction trend. When the delay increase is more than 30%, the leakage reduction saturates at about 98%. Therefore, Figure 4 provides a guide for making a trade-off between leakage and power.

Table 1. Leakage reduction due to dual-V_{th} reassignment (@ 27°C)

Ckt.	Gates #	T_c (ns)	Leakage Red. (%) $(T_{max}=T_c)$	Sun OS 5.7 CPU secs.	Leakage Red. (%) $(T_{max}=1.25T_c)$	Sun OS 5.7 CPU secs.
C432	160	0.75	61.0	0.25	95.0	0.25
C499	182	0.39	19.3	0.31	94.8	0.30
C880	328	0.67	88.1	0.54	96.5	0.53
C1355	214	0.40	25.0	0.33	93.3	0.36
C1908	319	0.57	66.4	0.57	96.6	0.56
C2670	362	1.26	90.4	0.68	97.9	0.53
C3540	1097	1.75	93.8	1.71	98.0	1.70
C5315	1165	1.59	87.1	1.82	98.0	1.83
C6288	1177	2.18	73.8	2.07	97.1	2.00
C7552	1046	1.92	96.0	1.59	98.0	1.68

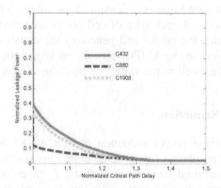

Fig. 4. Tradeoff between leakage power and performance

4.2 Leakage, Dynamic Glitch and Total Power Reduction

The leakage current strongly depends on the temperature. Our SPICE simulation shows that for a 2-input NAND gate with low V_{th}, when temperature increases from 27°C to 90°C, the leakage current increases by a factor 10X. For a 2-input NAND gate with high V_{th}, this factor is 20X. The leakage in the look-up table is from simulation for a 27°C operation. To manifest the dominant effect of leakage power, we estimate the leakage currents at 90°C by multiplying the total leakage current obtained from CPLEX [10] by a factor between 10X and 20X as determined by the proportion of low to high threshold transistors.

The dynamic power is estimated by a glitch filter event driven simulator, and is given by

$$P_{dyn} = \frac{E_{dyn}}{T} = \frac{0.5 \cdot C_{inv} \cdot V_{dd}^2 \cdot \sum_i T_i FO_i}{1000(1.2 \cdot T_c)}$$

(21)

where C_{inv} is the gate capacitance of one inverter, T_i is the number of transitions at gate i's output when 1,000 random test vectors are applied at PIs, and FO_i is the number of fanouts. Vector period is assumed to be 20% greater than the critical path delay, T_c. By simulating each gate's transition number, we can estimate the glitch power reduction.

Table 2. Leakage, dynamic and total power reduction comparison for unoptimized and optimized circuits (@ 90℃)

Ckt.	$P_{leak}1$ (uW)	$P_{leak}2$ (uW)	Leak. Red. (%)	$P_{dyn}1$ (uW)	$P_{dyn}2$ (uW)	Dyn. Red. (%)	$P_{total}1$ (uW)	$P_{total}2$ (uW)	Total Red. (%)
C432	35.8	11.9	66.8	101	73	27.4	137	85	37.7
C499	50.4	39.9	20.7	226	160	29.0	276	200	27.5
C880	85.2	11.1	87.0	177	128	27.8	263	139	47.0
C1355	54.1	40.0	26.3	293	166	43.5	347	206	40.8
C1908	92.2	29.7	67.8	255	198	22.4	347	227	34.5
C2670	116	11.3	90.2	129	101	21.6	244	112	54.1
C3540	303	18.0	94.1	333	228	31.5	636	246	61.3
C5315	421	9.80	88.2	466	304	34.6	887	354	60.1
C6288	389	97.2	75.0	1691	406	76.0	2080	503	75.8
C7552	444	18.8	95.8	381	228	40.2	825	247	70.1

We compare the leakage power and dynamic power at 90℃ in Table 2. The suffix-1 means the unoptimized circuit which has all the low threshold gates and the largest glitch power, and suffix-2 means the optimized circuit whose V_{th} has already been optimimally assigned and most of the glitches have been eliminated. We observe that for 70nm BPTM CMOS technology at 90℃, unoptimized leakage power (column 2) of some large ISCAS'85 benchmark circuits can account for about one half or more of the total power consumption (column 8). With V_{th} reassignment, the optimized leakage power of most benchmark circuits is reduced to less than 10%. With further glitch (dynamic) power reduction, total power reductions for most circuits are more than 50%. Some have a total reduction of up to 70%.

5   Conclusion

A new technique to reduce the leakage and glitch dynamic power simultaneously in a dual-V_{th} process is proposed in this paper. An integer linear programming (ILP) model is generated from the circuit netlist and the AMPL CPLEX [10] solver determines the optimal V_{th} assignments for leakage power minimization and the delays and positions of inserted delay elements for glitch power reduction. The experimental results for ISCAS'85 benchmark show reductions of 20%-96% in leakage, 28%-76% in dynamic (glitch) and 27%-76% in total power. We believe some of the other techniques, such as gate sizing and dual power supply can also be incorporated in the ILP formulation.

References

1. L. Wei, K. Roy and V. K. De, "Low Voltage Low Power CMOS Design Techniques for Deep Submicron ICs," *Proc. 13th International Conf. VLSI Design*, 2000, pp. 24-29.
2. M. Ketkar and S. S. Sapatnekar, "Standby Power Optimization via Transistor Sizing and Dual Threshold Voltage Assignment," *Proc. ICCAD*, 2002, pp. 375-378.
3. L. Wei, Z. Chen, M. Johnson and K. Roy, "Design and Optimization of Low Voltage High Performance Dual Threshold CMOS Circuits," *Proc. DAC*, 1998, pp. 489-494.
4. L. Wei, Z. Chen, K. Roy, Y. Ye and V. De, "Mixed-Vth (MVT) CMOS Circuit Design Methodology for Low Power Applications," *Proc. DAC*, 1999, pp.430-435.

5. Q. Wang, and S. B. K. Vrudhula, "Static Power Optimization of Deep Submicron CMOS Circuits for Dual V_T Technology," *Proc, ICCAD*, 1998, pp490-496.
6. D. Nguyen, A. Davare, M. Orshansky, D. Chinney, B. Thompson, and K. Keutzer, "Minimization of Dynamic and Static Power Through Joint Assignment of Threshold Voltages and Sizing Optimization," *Proc. ISLPED*, 2003, pp. 158-163.
7. V. D. Agrawal, "Low Power Design by Hazard Filtering," *Proc. 10th International Conference on VLSI Design*, 1997, pp. 193-197.
8. T. Raja, V. D. Agrawal and M. L. Bushnell, "Minimum Dynamic Power CMOS Circuit Design by a Reduced Constraint Set Linear Program," *Proc. 16th International Conference on VLSI Design*, 2003, pp. 527-532.
9. N. R. Mahapatra, S. V. Garimella. A. Tarbeen, "An Empirical and Analytical Comparison of Delay Elements and a New Delay Element Design," *Proc. IEEE Computer Society workshop on VLSI, 2000,* pp. 81 – 86.
10. R. Fourer, D. M. Gay, and B. W. Kernighan, *AMPL: A Modeling Language for Mathematical Programming*. South San Francisco, California: The Scientific Press, 1993

Back Annotation in High Speed Asynchronous Design

Pankaj Golani and Peter A. Beerel

Department of Electrical Engineering – Systems
Andrew & Erna Viterbi School of Engineering
University of Southern California, Los Angeles CA 90089, USA
{pgolani,pabeerel}@usc.edu

Abstract. This paper presents the next step in an evolving back-end design flow for high performance asynchronous ASICs using single-track full-buffer (STFB) standard cells and industry standard CAD tools. This paper demonstrates that these cells can be efficiently modeled in Verilog, effectively characterized in the standard Liberty format, and support accurate Verilog back-annotation using the standard-delay-format (SDF) flow, thereby enabling digital simulation-based performance and timing verification. Experimental results on several test designs including a 260K transistor parallel prefix 64 bit adder design demonstrate the proposed back-annotation flow yields less than 5% error compared with much more time-consuming analog-level simulation using a circuit simulator.

1 Introduction

Semi-custom standard-cell based design methodologies offers good performance with typically 12-month design times [1]. They are supported by a large array of evolving CAD tools that support simulation, synthesis, verification, and test. A large library of standard-cell components that have been carefully designed, verified, and characterized supports the synthesis task. This library is generally limited to static CMOS gates because they are robust to different environmental loads and have high noise margins, thus requiring little block-level analog verification. But the time-to-market advantage of standard-cell based design is being attacked by the increasing difficult task of estimating wire delay and increasing process variability making worst-case design overly conservative. These limitations have created an opportunity for alternative circuit styles, such as asynchronous design (e.g., [5] [6] [8] [9]).

In particular, Fulcrum Microsystems has demonstrated the commercial viability of high-speed asynchronous design. Their chips have asynchronous cores with over 2X the performance of standard-cell-based synchronous design using overly conservative design style called quasi-delay-insensitive (QDI) [7][14] that costs in area and limits performance. Moreover, Fulcrum Microsystems relies on a semi-automated full-custom flow that includes the development of a liquid library and time-consuming analog simulation for performance verification. In conjunction with a larger effort at Columbia University, USC has been for past five years exploring a semi-custom approach to designing such high-speed asynchronous designs as well as alternative circuit styles that tradeoff robustness with performance [3][4]. We have recently developed single-track full-buffers (STFB), which have 3X the performance of quasi-delay-insensitive circuits and are 50% smaller [16]. We recently demonstrated the

V. Paliouras, J. Vounckx, and D. Verkest (Eds.): PATMOS 2005, LNCS 3728, pp. 227–236, 2005.
© Springer-Verlag Berlin Heidelberg 2005

advantages of our STFB library and the standard-cell flow on a 260K-transistor test chip that includes a 64-bit adder and test circuitry in **TSMC 0.25u** technology [3]. The chips worked flawlessly at a performance of over 1.4 GHz over a wide range of voltages and temperatures, giving more than 3X faster than most standard-cell based ASIC designs in this process. At that time, however, our design process was immature and required extensive and time-consuming block-level analog performance and timing verification.

The next step in the evolution of this standard-cell-based design flow is to determine if we can properly model these cells in a standard HDL, effectively characterize them in a standard library format, and enable accurate back-annotation using a standard back-annotation flow, thereby enabling digital simulation-based performance and timing verification. These questions are particularly interesting for STFB designs that use a non-standard single-track protocol that involves complicated tri-state logic and several atypical timing constraints that must be modeled. This paper answers this question in the affirmative. In particular, it describes how STFB cells can be modeled in Verilog, what timing arcs must be characterized to enable accurate performance and timing verification, and how these timing arcs can be represented in the commercially-standard Liberty format. Moreover, it successfully shows how this characterization enables performance verification and simulation based validation of the timing constraints using back-annotated Verilog on several designs including our test chip [3]. Experimental results show that the modeling error is less than 5%.

The remainder of this paper is organized as follows. Section 2 reviews asynchronous channels and STFB templates and semi-custom design flow. Section 3 describes modeling STFB cells in Verilog, timing arc definition in the Liberty format, and back-annotation using SDF. Section 4 presents the experimental results and Section 5 draws some conclusions.

2   Background

This section first describes asynchronous channels, the basic structure and operation of single-track full-buffers (STFB). It then introduces the Standard Delay Format (SDF) and Liberty Format (.lib).

2.1   Asynchronous Channels

In general asynchronous designs are composed of blocks communicating using handshaking via asynchronous communication channels. An asynchronous communication channel is a bundle of wires and a protocol to communicate between blocks. The encoding scheme in which one wire per bit is used to transmit the data and an associated request is sent to identify when data is valid is called single rail encoding [7]. If the data is sent using two wires for each bit of information, the encoding is called a dual-rail channel and extensions to 1-of-N encoding also exist. In 1-of-N channels, the receiver detects the presence of the token from the data itself and, once the data is no longer needed, it acknowledges the sender as shown in Fig. 1(a). In the 1-of-N single-track channel, the receiver detects the presence of the token, as in the 1-of-N channel, but is also responsible for consuming it. The sender detects that the token was consumed before sending another token as shown in Fig. 1(b) [2].

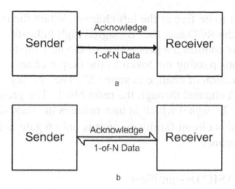

Fig. 1. Asynchronous channels, (a) 1-of-N Channel. (b) 1-of-N single track channel

Single Track Full Buffer (STFB) Template

Fig. 2 shows a typical STFB cell's block-diagram and more detailed transistor-level implementation of a STFB buffer with a single 1-of-N input and output channel.

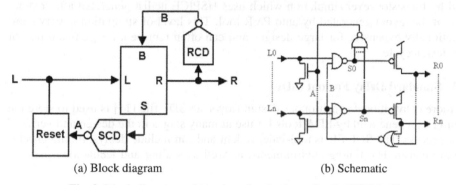

(a) Block diagram   (b) Schematic

Fig. 2. Block diagram and transistor level schematic of a STFB buffer

In the STFB buffer, the NOR gate with inputs R0 and R1 is the Right Completion Detector (RCD) and the NAND gate with inputs S0 and S1 is the State Completion Detector (SCD). The operation of a STFB cell when a token arrives at the input can be partitioned into two cases depending on the state of the output channels:

1. **Output channel is free:** The cell is waiting for a token from left environment. The RCD sets the "B" signal high indicating that the output channel is free and enabling the processing of the new input token. The token arrives at the left channel and then the state signal "S" is lowered thus creating an output token for the right environment. The SCD then asserts the "A" signal on detection of the change in state "S". The "A" signal in turn removes the token from left channel through reset block. The presence of output token on the right channel resets the "B" signal which will restore the state signal "S" and disable the stage from firing while the output channel is busy even if a new token arrives at the left channel.

2. **Output channel is busy:** In this case, when the token arrives at the left environment, the RCD sets the "B" signal low indicating that the output channel is busy and thus disables the processing of the new input token. The new token waits for

the output channel to be free at the left channel. When the right environment consumes the token, the RCD sets the 'B" signal high indicating that the output channel is free and thus enables the processing of the new input token. The state signal "S" is lowered, thus passing the token to the output channel. The SCD asserts the "A" signal on detection of change in state "S". The "A" signal in turn removes the token from the left channel through the reset block. The presence of the output token will reset the "B' signal which in turn restores the state signal "S".

The cycle time of this circuit family amount is only 6 gate delays, which enable ultra-high speed performance.

2.2 Asynchronous ASIC Design Flow

The following ASIC design flow was proposed by USC Asynchronous Group [3, 4]. After creating a standard cell library, the largely conventional standard-cell ASIC back-end design flow using conventional place and route tools is used to create the final layout. One important difference, however, is that unlike conventional standard-cell designs that rely on static timing analysis, the performance of the design is verified by transistor level simulation which uses HSPICE netlist generated after extraction of the layout generated by auto P&R tool. This level of simulation is very computationally expensive for large designs and can often represent a significant part of the design cycle.

2.3 Standard Delay Format (SDF)

In more conventional synchronous design flows, an SDF file [11] is used to store the timing data generated by EDA tools for use at many stages in the design and verification process. The SDF file is tool-independent and can include delays, timing checks, timing constraints, timing environments, as well as scaling and technology parameters.

2.4 Liberty Format

The Liberty Format [10] is a standard format used to characterize timing and power consumption properties of a standard cell library. Liberty format supports many timing models including the linear delay model, table-lookup model, scalable polynomial delay model, and piecewise linear delay model. In our work, we adopted the three term linear model for characterizing delay, which provides a reasonable tradeoff between accuracy and simplicity. Of course, more accurate models can also be used and are expected to provide more accurate estimate of the delays at the expense of a somewhat more time-consuming library characterization process. In the linear model the delay of the cell is modeled as the sum of the intrinsic delay, delay due to the load capacitance and delay due to the input slope.

3 Proposed Design Flow

Fig. 3 shows the proposed enhanced back-end design flow that includes SDF back annotation. In particular, the SDF file is generated after place and routing using a characterized cell library and is used to simulate the design in Verilog.

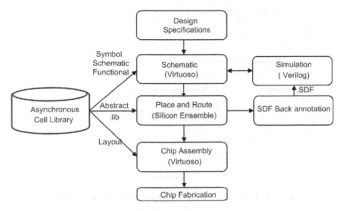

Fig. 3. SDF back annotation design flow

3.1 Identifying the Timing Arcs

For characterizing the timing information of cells in STFB library we need to identify a set of timing arcs that can completely model the performance and timing assumptions of STFB cells. As explained earlier the behavior of a STFB cell can be partitioned into two cases.

1. The right channels are free and a token arrives at the left environment. Since all the channels are bidirectional (tri-state) wires, all the transitions are either from high impedance to logic state or logic state to high impedance. In this case, for the right channel to go high the timing arcs corresponding to it will be from L to R and the transition is L going from high impedance (Z) to 1. Similarly after R goes high the left channel resets itself through signal A, so the timing arc corresponding to it will be L to L with the transition at L going from high impedance to 0. After L goes low, the right channel tri-states itself and the left channel using signal B and A respectively. So in this case the timing arcs are R to R (1 => Z) and R to L (0 => Z) respectively as shown in Fig. 4.
2. The output channels are busy and a token arrives at the input channels. In this case the token at the input channels will have to wait for the output channels to be free. So in this case the triggering event is output channels getting free. So this case can be characterized by two new timing arcs and two timing arcs that are in common with case 1. For processing of the input token, the new timing arc is R => R (Z => 1), for resetting of the input channels the new timing arc is R => L (Z => 0) and, for tri-stating of the input and output channels the common timing arcs are R => R (1 => Z) and R => L (0 => Z).

So the timing arcs for both the cases are as shown in Fig. 4:

$$L => R\ (Z => 1) \quad R => R\ (Z => 1)$$
$$L => L\ (Z => 0) \quad R => L\ (Z => 0)$$
$$R => R\ (1 => Z) \quad R => L\ (0 => Z)$$

These same arcs exist for all N data lines associated with the 1-of-N channel. Thus, for a dual rail STFB Buffer, there are 6 unique timing arcs per rail (shown in Fig. 4) and 12 timing arcs in total.

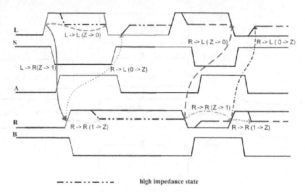

---··-··—··---   high impedance state

Fig. 4. Timing arcs on a single rail of STFB buffer

3.2 Characterization Methodology

Because the input and output ports of the cell under consideration are bidirectional and must tri-state, in our simulation environment the left and right environment are controlled using similar STFB cells. The load for the cell under consideration is varied by changing the length of the interconnect wire thus varying the interconnect capacitance. The delay values are then fitted into a plane of form as follows.

$$Delay = A * C_L + B * Slew_{in} + D_{int} \tag{1}$$

C_L, $Slew_{in}$ and D_{int} indicate the load capacitance, the input slew, and the intrinsic delay of the cell, respectively. For the transition going to high impedance, we measure the delay from 50% of L to the delay of signal "A" going to the threshold voltage when transition L goes from 0 to Z. Note that determining if and how commercially-available library characterization tools can be used to automate this process is still an open question.

3.3 Functional Description of Cells

For SDF back annotation to be possible, we have to specify the pin to pin delays in the functional view of the cells. In Verilog-HDL we can describe the pin to pin delays in the specify block [12]. For example the functional description of a STFB Buffer will look like as shown in Figure 5, where (L0 => R0) means that a timing path exist from L0 to R0 and the six delay values in parenthesis correspond to 0 -> 1, 1 ->0, 0 -> Z, Z -> 1, 1 -> Z, Z -> 0 transitions respectively. Because each *specify* line handles all transitions between unique input and output pins, there are fewer lines than timing arcs. In particular, for a STFB Buffer we need 8 total lines ignoring global reset. It is also important to note that this timing specification is independent of the description of the behavior of the cell, which simplifies the Verilog specifications.

There are many ways to model the behavior of STFB cells in Verilog. We developed a consistent approach in which the Verilog models all have three *always* blocks for each cell. The first always block is activated whenever a new input comes at the input channels and the output channels are free, the second always block is activated when the output channels consumes the tokens present at the input channels and then drive both the input and output channels to high impedance state. The third always block is used for the global reset signal. Since the input and output channels are bidirectional we have to declare them as trireg variables. Unfortunately, trireg variable cannot be assigned inside an always block. To solve this problem we have to declare a temporary register variable and assigned it in the always block, and then using standard Verilog primitives like *bufif* which model a tri-state buffer to transfer the value to the trireg variables [12]. To tri-state a *trireg* variable, we just assign the enable input of the *bufif* gates to 0. Using these standard primitives solves the bidirectional problem and it also does not affect the functionality or timing of the cells in any case.

```
module STFBBuffer(L0, L1, R0, R1, NReset);
inout L0, L1, R0, R1;
input NReset;
trireg L0, L1, R0, R1;
reg a,b,c,d,enable1, enable2 ;

bufif b1(L0,a, enable1);
bufif b2(L1,b, enable1);
bufif b3(R0,c, enable2);
bufif b4(R1,d, enable2);

specify
  (L0 => R0) = (0, 0, 0, 2, 0, 0);
  (L0 => L0) = (0, 0, 0, 0, 0, 3);
  (R0 => L0) = (0, 0, 3, 0, 0, 4);
  (R0 => R0) = (0, 0, 0, 3, 3, 0);
  (L1 => R1) = (0, 0, 0, 2, 0, 0);
  (L1 => L1) = (0, 0, 0, 0, 0, 3);
  (R1 => L1) = (0, 0, 3, 0, 0, 4);
  (R1 => R1) = (0, 0, 0, 3, 3, 0);
endspecify

always @(L0 or L1)
wait( (L0 | L1) && (R0 == 0 && R1 == 0) && NReset == 1))
begin c <= L0; d <= L1; a <= 0; b <= 0; enable1 <= 1; enable2 <= 1; end

always @(R0 or R1)
wait (R0 == 1 | R1 == 1) begin enable1 <= 0; enable2 <= 0; end

always @(NReset)
begin
if (NReset == 0) begin a <= 0; b <= 0; enable1 <= 1; end
else enable1 <= 0;
end

endmodule
```

```
library(STFB) { // Start Library
technology (cmos) ;
delay_model : generic_cmos ;
..
cell(STFBBuffer) { // Start Cell
  pin(L0) {
  ..
  }
  ..
  pin(R0) {
  direction : inout;
  capacitance : 0.034;
  timing(){
    related_pin : "L0";
    timing_type : three_state_enable;
    intrinsic_rise : 0.080;    // Factor C in equation 1
    rise résistance : 0.70;    // Factor A in equation 1
    slope_rise : 0.11;         // Factor B in equation 1
  }
  }
  ..
} // End Cell
..
} // End Library
```

(a) Functional description   (b) Liberty model

Fig. 5. Functional and Liberty model for STFB buffer

3.4 Timing Arcs in Liberty Format

We are using standard Liberty format to represent the characterized asynchronous cell library. We are using *generic_cmos* model which means the linear delay model to represent the delay values. Since all the transitions in STFB templates are from tristate to logic or from logic to tri-state, we have to use *three_state_enable* and *three_state_disable* constructs, where *three_state_disable* is a construct used to model the delay leading to the transitions (0 -> Z and 1 -> Z) and *three_state_enable* is used to model the delay leading to the transitions (Z -> 1 and Z -> 0). A timing model for timing arc L->R (Z->1) is shown in Fig. 5(b).

4 Simulation Results

The proposed back-annotation flow was evaluated on asynchronous linear pipelines with variable number of stages and a 260 K transistor 64-bit parallel prefix asynchronous adder with the input and output circuitry to feed the adder and sample the results [3]. The SDF back annotation flow is also used to verify if these designs meet the timing constraints of STFB [2].

4.1 Performance of Pipelines

We analyzed the throughput of the 9-stage pipeline by varying number of tokens in the pipeline as shown in Fig. 6. Our pipeline structure consists of 6 buffers, a fork, a merge, an exclusive OR and a controlled bit-generator (BG), and the experimental result using HSPICE and Verilog simulation with back-annotation is shown in Fig. 6 (b). In addition, the maximum throughput was measured on pipelines of different lengths with the results in Table 1 which shows that the back annotation flow yielded a maximum error of 4.4%.

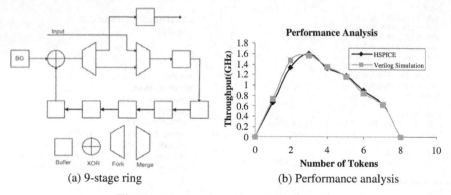

(a) 9-stage ring   (b) Performance analysis

Fig. 6. Performance analysis of a 9 stage ring

Table 1. Throughput comparison of pipelines of different length

Design	HSPICE Simulation	Verilog simulation with back annotation	Error (%)
9 stage ring	1.6 GHz	1.58GHz	1.8
15 stage ring	1.7 GHz	1.78 GHz	4.4
30 stage ring	1.6 GHz	1.58 GHz	0.6

4.2 Prefix Adder

Fig. 7 shows the test circuitry used to demonstrate the ASIC standard design flow on a 64 bit asynchronous prefix adder which uses cells from STFB standard cell library. The input bit generator is made up of input rings which generate input fast enough to continuously feed the adder block so that the adder can work at its peak throughput. The sampler block has the responsibility of sampling the addition results by a variable sampling ratio and will set the ReqCout and ReqSum high [3]. In our simulations we have set the sampling ratio to be 1:1000 that means out of every 1000 additions the sampler block will set the ReqSum and the ReqCout variable high for just one addition. In this case the throughput will be the throughput of ReqCout field multiplied by the sampling ratio i.e. 1000. The Adder design has around 5000 STFB standard cells and the layout of this design occupies an area of 4.1 mm^2.

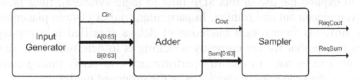

Fig. 7. Block diagram of Prefix Adder and its test circuitry

The throughput for the adder design after simulation using transistor level simulation scheme like Nanosim is 1.01 GHz[1]. After performing the same experiment using Verilog Simulation with SDF Back annotation we get the throughput to be 1.06 GHz. The error that we get is 4% which is within acceptable limits and it increases the speed of simulation over two orders of magnitude.

4.3   Simulation-Based Verification of Timing Constraints

STFB cells are based on the timing assumption that one stage will tri-state the channels associated with it before the next stage can drive the wire. Using Verilog Simulation with SDF back annotation we can verify the validation of this timing assumption after P&R. Fig. 8 shows the verilog simulation of two STFB Buffers I1 followed by I2 in cascade. In Fig. 8 we can see that there is a fight[2] between logic state '1' and logic state '0' marked by grey area showing the violation of the timing assumption that I1 should tri-state R before I2 starts to drive it.

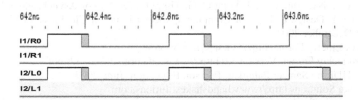

Fig. 8. Verilog simulations showing timing constraint has been violated

5   Conclusions

This paper presents a STFB cell characterization and back-end timing flow using SDF back-annotation. The experimental result indicates that the flow yields over two

---

[1]   The difference in the throughput of the design reported here and that reported in [3] is because we used more conservative spice model card to simulate the design

[2]   Normally for such small designs the timing assumption are satisfied. But just for the demonstration purposes a fight has been created by editing the delays of the wires

orders of magnitude increase in simulation speed with less than 5% error. Our future work involves testing and improving this flow on smaller nanometer processes in which the introduced error may be larger for a variety of reasons, including increased crosstalk noise. In addition to back-annotation of timing data for analysis, SDF supports the forward annotation of timing constraints to design synthesis tools. Thus, we also hope to expand the use of this SDF flow to logic synthesis, floor planning, and timing-driven placement and routing. In particular, trial or custom placements can be verified against maximum legal interconnect delays specified in a pre-layout SDF file. A key hurdle that we are currently attacking is to determine how static timing analysis tools can be used to quantify performance and verify timing constraints in the presence of the numerous timing loops in the proposed model.

References

1. International Technology Roadmap for Semiconductors. http://public.itrs.net/, June 2004.
2. Ferretti M., Beerel P. A. Single-Track Asynchronous Pipeline Templates Using 1-of-N Encoding, DATE'02, Mar. 2002.
3. Ferretti M, Ozdag R. O., Beerel P. A. High Performance Asynchronous ASIC Back-End Design Flow Using Single-Track-Full-Buffer Standard Cells, ASYNC'04, April 2004.
4. Ozdag R. O., Beerel P. A., A Channel based Asynchronous low power High Performance Standard Cell Based Sequential Decoder Implemented with QDI Templates, ASYNC'04, April 2004.
5. Beerel P. A. Asynchronous Circuits: An Increasingly Practical Design Solution, ISQED'02, pp. 367 -372, March. 2002.
6. Martin A. J., Nyström M., Penzes P., Wong C. G. Speed and Energy Performance of an Asynchronous MIPS R3000 Microprocessor. Technical Report Caltech, June 2001.
7. Lines A. M. Pipelined Asynchronous Circuits, Master Thesis, California institute of Technology, June 1998.
8. Singh M. and Nowick S. M. High throughput Asynchronous Pipelines for Fine-Grain Dynamic paths, ASYNC'00, April 2000.
9. Martin A. J., Nyström M. and Wong C. G. Three Generation of Asynchronous Microprocessors, IEEE Design and test of Computers, special issue on Clockless VLSI design November/December 2003.
10. Synopsys Liberty Format http://www.synopsys.com
11. Standard Delay Format Version 3.0 http://www.eda.org/sdf/sdf_3.0.pdf
12. Verilog-HDL by Sameer Palnitkar, Pearson Education June 2003
13. Handshake Solutions http://www.handshakesolutions.com.
14. Cummings U. Terabit Crossbar Switch Core for Multi-Clock-Domain SoCs. In the Proceedings of 2003 Hot Chips, August 2003
15. Beerel P.A., Cortadella J., Kondratyev A., Bridging the Gap between Asynchronous Design and Designers. VLSI Design, 2004,
16. Ferretti M., Single-Track Asynchronous Pipeline Templates, PhD Thesis, Submitted to USC, June 2004.

Optimization of Reliability and Power Consumption in Systems on a Chip

Tajana Simunic[1], Kresimir Mihic[2], and Giovanni De Micheli[2]

[1] CSE Department, UCSD, 9500 Gilman Drive, La Jolla, CA 92093
tajana@ucsd.edu
[2] CSL, Stanford U., 353 Serra Mall, Stanford, CA 94305
kmihic@stanford.edu

Abstract. Aggressive transistor scaling, decreased voltage margins and increased processor power and temperature, have made reliability assessment a much more significant issue in design. Although reliability of devices and interconnect has been broadly studied, here we characterize reliability at the system level. Thus we consider component-based System on Chip designs. Reliability is strongly affected by system temperature, which is in turn driven by power consumption. Thus, component reliability and their power management should be addressed jointly. We present here a joint reliability and power management optimization problem whose solution is an optimal management policy. When careful joint policy optimization is performed, we obtain a significant improvement in energy consumption (40%) in tandem with meeting reliability constraint for all operating temperatures.

1 Introduction

Advances in technology lead to higher device density and operating frequency, and consequently to higher power dissipation and operating temperatures. To deal with such problems, *dynamic power management* (DPM) has been applied in various forms, to both single and networked on-chip components [3],[5]. Reducing energy consumption to the required levels ensures correct and useful operation of the integrated systems. DPM also affects the reliability of the system components. Curbing power dissipation helps lowering the device temperatures and reducing the effect of temperature-driven failure mechanisms, thus making components more reliable. On the other hand, aggressive power management policies can decrease the overall component reliability because of the degradation effect that temperature cycles have on modern IC materials [9],[12]. As a result, there is a need to evaluate the *System on Chip* (SoC) reliability along with power consumption and performance. There are several interesting problems that can be considered.

The first problem is to determine whether or not, for a given system topology, DPM affects reliability and to find if such effect is beneficial or not. The second problem is to include reliability as an objective or constraint in the policy optimization. The third problem is the combined search for system topologies and joint DPM policies to achieve reliable low-energy design. All problems involve both run-time strategies as well as design issues. In this paper we focus on the first two problems. The first one enables us to understand the relationship between run-time power management and reliability analysis. We evaluate reliability, performance and power

V. Paliouras, J. Vounckx, and D. Verkest (Eds.): PATMOS 2005, LNCS 3728, pp. 237–246, 2005.

consumption of computational elements (cores) in SoCs by modeling system-level reliability as a function of failure rates, system configuration and management policies. The overall objective is to introduce design constraints, such as *mean time to failure* (MTTF), in the design space spanned by performance and energy consumption. The major novelty and contribution of this paper is the definition of a joint dynamic power management (DPM) and reliability (DRM) optimization method, that yields optimal system-level run-time policies. In addition, we evaluate the effect of the policy on single core and multi-core systems. Experimental results show that with careful joint optimization we can save energy by 40% while meeting both reliability and performance constraints.

The rest of the paper begins with a discussion of related work. Our approach for assessing and optimizing reliability and power is presented in Sections 3 and 4. Optimization results for a typical SoC design are presented in Section 5.

2   Related Work

Integrated systems have been in production for a while in the form of Systems on Chips (SoCs). A number of issues related to SoC design have been discussed to date ranging from managing power consumption, to addressing problems with interconnect design. Previous work for energy management of networked SoCs mainly focused on controlling the power consumption of interconnects, while neglecting managing power of the cores. A stochastic optimization methodology for core-level dynamic voltage and power management of multi-processor SoCs with using a closed-loop control model has been presented in [3].

Reliability of SoCs is another area of increasing concern. A good summary of research contributions that combine performance and reliability measures is given in [1]. An approach to improve system reliability and increase processor lifetime by implementing redundancy at the architecture level is discussed in [2]. A number of fault-tolerant micro-architectures have been proposed that can handle hard failures at performance cost [6]. Minimizing energy and performance by exploiting architecture and application-level adaptability has been presented in [8]. The RAMP methodology models chip MTTF as a function of the failure rates of individual structures on chip due to different failure mechanisms [7]. Soft (or transient) failure mechanisms and their effect on power consumption have been studied by a number of researchers (e.g. [22], [23]). Incorrect signal levels due to cross talk is an example of a soft failure. In this work we address hard failure mechanisms which cause irrecoverable component failures. Open interconnect line due to electromigration is an example of a hard failure. An overview of most commonly observed hard failure mechanisms that affect the current semiconductor technologies is given in [11]. The effect of a temperature gradient on the electromigration failure mechanism is investigated in [12]. The description of the connection between fast thermal cycling and thin film cracking (interlayer dielectric, interconnections) is presented in [13] and a model is given in [32]. A model for Time-Dependent Dielectric Breakdown is developed in [14]. Our work presents the first unified methodology for optimization of reliability, power and performance in SoCs.

3 Reliability Modeling

Our objective is to optimize system-level power consumption under reliability and performance constraints. The *reliability* of a component (or system) is a probability function $R(t)$, defined on the interval $[0,\infty]$, that the component (system) operates correctly with no repair up to time t. The *failure rate* of a component (or system) is the conditional probability that the component (or system) fails in the interval $[t, t+ \Delta t]$ while assuming correct operation up to time t. The *mean time to failure* (MTTF) is the expected time at which a component fails, i.e. $MTTF = \int R(t)dt$. In the particular case that the failure rate λ_f is constant with time, then MTTFF is $1/\lambda_f$ and the reliability is $R(t) = e^{-\lambda_f t}$.

In general, failure rates depend on time because of material aging and on temperature because of thermodynamic issues. In this work we focus on the reliability of components during their useful life and thus we neglect aging but we do consider temperature dependence. We assume that components can be in different operational states (e.g., *active, idle, sleep*) characterized by parameters such as voltage and frequency, which determine the component temperature. Thus failure rates can be considered constant within any given operational state. We consider three failure mechanisms most commonly used by semiconductor industry: *Electromigration* (EM), *Time Dependant Dielectric Breakdown* (TDDB) and *Thermal Cycles* (TC).

Electromigration is a result of momentum transfer from electrons to the ions which make interconnect lattice. It leads to opening of metal lines/contacts, shortening between adjacent metal lines, shortening between metal levels, increased resistance of metal lines/contacts or junction shortening. The MTTF due EM process is commonly described by Black's model:

$$MTTF_{EM} = A_o(J - J_{crit})^{-n} e^{\frac{Ea}{kT}} \tag{1}$$

where A_o is an empirically determined constant, J is the current density in the interconnect, J_{crit} is the threshold current density and k is the Boltzmann's constant, $8.62*10^{-5}$. For aluminum alloys E_a and n are 0.7 and 2 respectively. The value of MTTF for EM can also be obtained by silicon measurements for the cores. We model the EM failure rate for idle and active states only, because leakage current present in the sleep state is not of large enough to cause the migration:

$$\lambda_{core,s}^{EM} = A_o^{'}(J_s - J_{crit})^n e^{\frac{-Ea}{kT_s}} \; ; \; \forall s = active, idle \tag{2}$$

Time Dependent Dielectric Breakdown is a wear out mechanism of dielectric due electric field and temperature. The mechanism causes the formation of conductive paths through dielectrics shortening the anode and cathode. In this work we use the field-driven model:

$$MTTF_{TDDB} = A_o e^{-\gamma E_{ox}} e^{\frac{Ea}{kT}} \tag{3}$$

where A_o is an empirically determined constant, γ is the field acceleration parameter and E_{ox} is the electric field across the dielectric. The activation energy, E_a, for intrinsic failures in SiO_2 is found to be 0.6-0.9 and for extrinsic failures about 0.3 [11]. The failure rate due to TDDM mechanism can be defined as follows:

$$\lambda_{core,s}^{TDDB} = A_o'e^{\gamma E_{ox,s}}e^{\frac{-Ea}{kT_s}} \; ; \forall s = active, idle, sleep \tag{4}$$

Temperature cycling induces plastic deformations of materials that accumulate every time the cycle is experienced. This eventually leads to creation of cracks, fractures, short circuits and other failures of metal films and interlayer dielectrics as well as fatigue at the package and die interface. The effect of low-frequency thermal cycles, such as turning a device on/off during normal operation, has been well studied by the packaging community [11]. Thermal cycles that occur with higher frequencies and on chip, instead of just at the interface with package, are gaining in importance as power management gets more aggressive, the features sizes get smaller and low-k dielectric becomes more prevalent in the fabrication process [9]. Recent work [12] showed that such cycles play the major role in cracking of thin film metallic interconnects and dielectrics. Expected number of thermal cycles before core failure is given in Equation below. It does not only depend on the temperature range between power states (T_{max}-T_{min}) but is also strongly influenced by the average temperature in the sleep state, $T_{avg,s}$ and the molding temperature of the package process, T_{mold}. The exponent q ranges from 6-9, and $C_{1,2}$ are fitting constants defined in [12] for on chip structures. Mechanical properties of the interlayer dielectric layers are very dependant on the nature of the processing steps. As a result, when $T_{avg,s}$ increases, the stress buildup on the silicon due the package decreases resulting in a longer lifetime.

$$N_f = C_o \left[C_1(T_{max} - T_{min}) - C_2(T_{avg} - T_{mold}) \right]^{-q} \tag{5}$$

$$\lambda_{core,s}^{TC} = C_o' \left[(T_{active} - T_s) - (T_{avg,s} - T_{mold}) \right]^{q} t^{-1} \quad \forall s = sleep \tag{6}$$

Since a component fails when at least one of the failure mechanisms occurs, we express the overall component failure rate as a sum of failure rates due to all three mechanisms.

System Reliability
Systems are interconnections of components. We use the term *core* to refer to one of the SoC components that perform computing, storage or communication function. From a reliability analysis standpoint, components can be viewed as in *series (parallel)* if the overall correct operation hinges upon the *conjunction (disjunction)* of the correct operation of components. For example, a system consisting of a processor, a bus and a memory is seen as the series interconnection of three components. A system is therefore characterized by its topology, i.e., by a *reliability graph* [15]. In this work we use reducible graphs, as they display the conjunctive and disjunctive relations among components. Thus, system reliability can be computed bottom-up, by considering series/parallel compositions of sub-systems. When failure rates are constant, the failure rate of a series composition is the sum of the failure rates of each component:

$$\lambda_{system} = \sum_i \lambda_{core_i}$$

Systems with parallel structures offer built-in redundancy. Such systems can either have all components concurrently operating (*active parallel*) or only one component active while the rest are in low power mode (*standby parallel*). Active parallel combination has higher power consumption and lower reliability than standby parallel, but also faster response time to failure of any one component. The combined failure rate of M active components, λ_{fap}, is defined using binomial coefficient, C_i^M, and active reliability rate, λ_f [15]: $\lambda_{fap} = \sum_{i=1}^{M}(-1)^{i-1}\frac{C_i^M}{i\lambda_f}$. Since our goal is to minimize power consumption while improving system reliability, in this work we focus on standby parallel configurations with only one active component. In this case, the failure rate is: $\lambda_{fsp} = \lambda_{fs}/M$ [15].

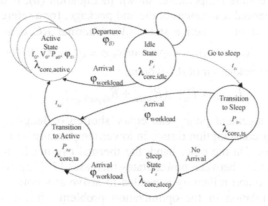

Fig. 1. System Model

4   Joint Policy Optimization

We can define an optimization problem given a system topology and a set of component operational states characterized by failure rate, power consumption and performance. Cores are modeled with a *power and reliability state machine* (PRSM) as shown in Figure 1, a state diagram relating service levels to the allowable transitions among them. Multiple cores are a reliability network of series and parallel combinations of single core PRSM models.

Single core PRSM characterizes each state by its failure rate, $\lambda_{core,state}$, and power consumption, P_{state}. Thus, active state i is characterized by the failure rate $\lambda_{core,activei}$, frequency and voltage of operation, f_i, V_i, which is equivalent to the core processing rate φ_{fi}, and power consumption P_{ai}. We assume for simplicity that workload and core's data processing times follow exponential distribution with rates $\varphi_{workload}$ and φ_{core_fi}. More complex distributions can also be used [4],[5]. In idle state a core is active but not currently processing data. Sleep state represents one or more low power states a core can enter. *TransitionToSleep* and *TransitionToActive* states model the time and power consumption required to enter and exit each sleep state. Transition times to/from low-power states follow uniform distribution with average transition times t_{ts}, t_{ta} [10]. The arcs represent transitions between the states with the associated transition times and rates. The transitions can occur due to normal operation of the system, or because of a command (action) is given as a part of the management policy. We define two actions, *"go to sleep"*, which causes a transition to sleep state, and *"continue"*, which allows the system to continue normal operation.

In order to obtain the failure rate for each state we need to evaluate failure rates of each of the three mechanisms described in Section 0 as functions of the component temperature. Expected temperature in a state is estimated using reference active state temperature T_{active}, the expected time spent in a state s, due to an action a, $y(s,a)$, and state's steady state temperature $T_{state,ss}=T_{active}P_{state}/P_{active}$:

$$T_{state} = (T_{active} - T_{state,ss})e^{\frac{y(s,a)}{\tau}} + T_{state,ss} \qquad (7)$$

Reference active state temperature, shown in Equation (8), is defined using R_{thdie} and $R_{thpackage}$ the thermal resistances of die and package, for a reference frequency and voltage of operation in the active state, f_0V_0. Thermal RC constant, $\tau \approx c\rho a^2$ has $c=10^6 J/m^3K$ for silicon thermal capacitance, $\rho = 10^{-2} mK/W$ for thermal resistivity [10] and the wafer thickness a of 0.1-$0.6mm$.

$$T_{active} \propto P_{active}(R_{th\,die} + R_{th\,package}) \qquad (8)$$

The power management policy can either shorten or lengthen the lifetime of a core. Lower power consumption results in lower temperature and thus lower EM and TDDB failure rates. On the other hand, the thermal cycling failure rate rises as the frequency of switching between power states increases [7],[12]. Joint optimization of power, performance and reliability is needed to arrive at a policy that meets all constraints. The formulation of the optimization problem, shown in Equation (9), is based on the Semi-Markov Decision Process model [4],[5].

$$
\begin{aligned}
\min \quad & \sum_{c=1}^{N} cost_{energy,\,c} \\
s.t. \quad & \sum_{a \in A} f(s,a) - \sum_{a \in A} \sum_{s' \in S} m(s'|\,s,a) f(s',a) = 0; \quad \forall s, \forall c_s \\
& \sum_{a \in A} \sum_{s \in S} y(s,a) f(s,a) = 1; \quad \forall c_s \\
& \sum_{c=1}^{N} cost_{energy,\,c} < Perf_{const}; \quad \forall c \\
& Tpl\,(\lambda_c) \le Rel_{const}; \quad \forall c_s \\
& \lambda_c = \sum_{i \in F} \sum_{a \in A} \sum_{s \in S} \lambda_{core}^i(s,a) y(s,a) f(s,a)
\end{aligned}
\qquad (9)
$$

This linear program minimizes the cost in energy over all cores, $cost_{energy,c}$, under a set of constraints. As such it can be solved with a linear program solver. The unknowns are state-action frequencies $f(s,a)$ which represent the expected number of times that the system is in state s when command a is issued. The management policy is derived for each core that has a low power state where "go to sleep" command can be given. The policy is in form of a table of probabilities for entry into each low-power state a: $p(s,a)=f(s,a) / \sum f(s,a')$.

The first constraint shown in Equation 9 is a "balance equation" which specifies that for each core c the number of entries to any state has to equal the number of exits. Here $m(s'|s,a)$ is the probability of arriving to state s' given that the action a was taken in state s. The second constraint specifies that the sum of probabilities over each core states and actions has to equal one. Third constraint specifies that each core's expected performance penalty for transitioning into low power states has to be lower than the specified limit, $Perf_{constr,c}$. We next describe the reliability constraint, represented by the last two lines in Equation (9), since definitions of other constraints are in [5].

The reliability constraint, Tpl is a function of the system topology, i.e. $Tpl=f(series,\ parallel\ combinations)$. For example, with series combinations $Tpl = \sum \lambda_{core,s}$, and with parallel standby $Tpl=\lambda_{core,standby}/N_{standby}$. Cleary, a reliability network normally has a number of series and parallel combinations of cores. Each core's failure rate, λ_c, as shown in the last line of the Equation (9), is in turn a sum of failure mechanisms, $i \in \{ EM,\ TDDB,\ TC\}$, when the core is in the state s and the action a is given. For example, the reliability constraint is given in Equation (10) for a core that has one *active* (A), *idle* (I) and *sleep* (S) state and two actions: *go to sleep* (S) and *continue* (C). Failure rate in each state, $\lambda_{core,state}$, is a sum of failure rates due to failure mechanisms active for that state as described in Section 0.

$$\lambda_A y(A,C)f(A,C)+$$
$$\lambda_I y(I,C)f(I,C)+\lambda_I y(I,S)f(I,S)+ \qquad (10)$$
$$\lambda_S y(S,C)f(S,C) \le \text{Rel}_{const}$$

We have thus far shown how to perform synthesis of optimal power, reliability and performance policy. We next present the optimization results for SoCs.

Table 1. SoC Parameters

IP block	P_{active} [W]	P_{idle} [W]	P_{sleep} [W]	t_{ts} [s]	t_{ta} [s]
DSP [17]	1.1	0.5	0.01	250u	100n
Video [18]	0.44	N/A	0.07	110m	0.9
Audio [19]	0.11	0.03	3e-3	6u	0.13
I/O [20]	1e-3	N/A	6e-6	100n	6u
DRAM [21]	1.58	0.37	1e-2	16n	16n

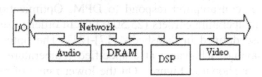

Fig. 2. System on a Chip

5   Results

The methodology presented in this work has been tested on an SoC shown in Figure 2. Input to the optimizer are power, reliability and performance characteristics of each core, along with a reliability network topology. The output is a set of management policies obtained from state-action frequencies $f(s,a)$ which are the unknowns in Equation (9). The policies determine when each core can enter any one of its low-power states.

Power and performance characteristics of cores come from the datasheets [17]-[21] and are summarized in Table 1. Each core supports multiple power modes (*active, idle, sleep* and *off*). *Off* state is supported by all cores with zero power consumption. Transition times between active and sleep state are defined by t_{ts} and t_{ta}. Reliability rates for each failure mechanism (EM, TDDB, TC) are based on actual silicon measurements obtained for 95nm technology. Due to confidentiality reasons we are unable to provide their exact values. Each of the cores in the system is designed to meet MTTF of 10 years. Core's workload and data consumption rates ($\varphi_{workload}$ and φ_{core_fi}) are obtained from cycle-accurate simulation of algorithms running on the cores (e.g. MPEG video, MP3 audio). The optimization results have been successfully validated against analytical models [15] for simpler reliability networks. We first present results of single core optimization followed by a discussion on design changes to the core that influence reliability. Then we show system level optimization results for the whole SoC.

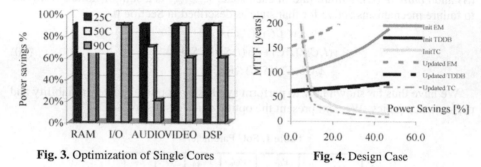

Fig. 3. Optimization of Single Cores   **Fig. 4.** Design Case

5.1 Single Core Optimization

We optimize the power consumption of each core presented in Table 1 while keeping the minimum lifetime requirement of 10 years. The objective is to observe how cores based on the same technology of 95nm feature size and comparative dimensions but with different power consumption respond to DPM. Optimization is performed at three internal chip temperature corners (25,50,90°C) in order to set the die operating points close to those defined in datasheets [17]-[21]. The optimization results for maximum power savings achievable at a specified temperature given MTTF constraint of 10 years are shown in Figure . On the lower range of temperatures (25°C-50°C) most of the cores react positively to DPM and allow the maximum power savings to be achieved. Figure shows that maximum power savings decrease for DSP, Video and Audio cores working at 90°C. This decrease is due to thermal cycles failure mechanism. Thus, a system designed to meet a specific MTTF requirement without power management may fail sooner once DPM is introduced. One way to try to address this problem is by redesigning the core.

Influencing the lifetime of power managed core by means of changing the design is a matter of finding the equilibrium between related physical parameters. In Figure 4 we show results of design updates done to the RAM core. EM failure rate is lowered by widening critical metal lines. Core area expanded by 5%, current density dropped by 20% and the core temperature dropped by 2%. Although both EM and TDDB gain from design change, the TC failure rate increase sufficiently to worsen the net reliability by 10%.

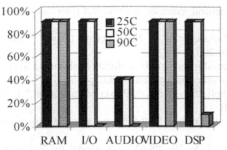
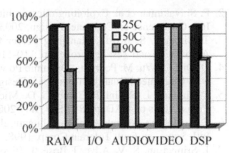

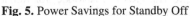

Fig. 5. Power Savings for Standby Off   **Fig. 6.** Power Savings for Standby Sleep

5.2 SoC Optimization

Here we examine the influence of redundant components to the overall system reliability. We use the SoC shown in Figure 2 with the core parameters given in Table 1 and the operating characteristics described in the previous section. Since all cores are essential to the correct SoC operation, the initial reliability network is their series combination. Unfortunately, although each core meets MTTF requirement, the overall system does not. Therefore, we add to the SoC redundant components at the cost of increased area. Two redundancy models described in Section 3 are studied: standby sleep, with redundant cores in sleep state until needed, and standby off, with redundant cores turned off. Figures 5 and 6 show that the best power savings are with standby off model. However, this model has the largest wakeup delay for redundant components. The standby sleep model shown in Figure 6 gives more moderate power savings with faster activation time. Results for both models show that not all cores can operate reliably at the highest temperature (e.g. no power savings for AUDIO core at 90°C show that the reliability constraint is not met). Thus, we expand the reliability network to have DSP, AUDIO and I/O with one redundant component in standby sleep and the other in standby off model, while VIDEO and RAM remain with a single redundant component in standby sleep. The new system meets MTTF of ten years at the cost of die area increase while getting power savings of 40% and a faster response time to component failure.

6 Conclusion

In this work we show that a functional and highly reliable core may fail to meet the lifetime requirement once power management is enabled due to thermal cycle failure mechanism. As technology scales down, limitations set by thermal cycling are going to be an even more important factor in system design. Thus the methodology we presented in this work for joint optimization of reliability, power consumption and performance is going to be even more crucial. With our optimizer we show that we can obtain large power savings on SoCs while meeting the reliability constraint.

References

1. M. D. Beaudry. Performance-related reliability measures for computing systems. IEEE Trans. on Comp., c-27:540(6), June 1978.

2. P. Shivakumar et al. Exploiting microarchitectural redundancy for defect tolerance. In 21st Intl. Conference on Computer Design, 2003.
3. T. Simunic, S. Boyd, "Managing Power Consumption in Networks on Chips," Design, Automation and Test in Europe, pp. 110-116, 2002.
4. Q. Qiu, Q. Wu, M. Pedram, "Dynamic Power Management in a Mobile Multimedia System with Guaranteed Quality-of-Service," Design Automation Conference, pp. 701-707, 2001.
5. T. Simunic, L. Benini, P. Glynn, G. De Micheli, "Event-driven Power Management," IEEE Transactions on CAD, pp.840-857, July 2001
6. E. Rotenberg. Ar/smt: A microarchitectural approach to fault tolerance in microprocessors, Intl. Symp. on Fault Tolerant Comp, 1998.
7. J. Srinivasan, S. V. Adve, P.Bose, J. Rivers, C.K. Hu, "RAMP: A Model for Reliability Aware MicroProcessor Design,"IBM Research Report, December 29, 2003
8. J. Srinivasan and S. V. Adve. Predictive dynamic thermal management for multimedia applications. In Proc. of the 2003 Intl Conf. on Supercomputing, 2003.
9. J. Srinivasan, P. Bose, J.Rivers, "The impact of Technology Scaling on Processor Lifetime Reliability," UIUC CS Technical Report, December 2003
10. K. Skadron, T. Abdelzaher, M.R. Stan, "Control-Theoretic Techniques and Thermal-RC Modeling for Accurate Localized Dynamic Thermal ManagementProceedings", Proc. of the 8th Intl. Symp. on High-Performance Computer Architecture (HPCA'02)
11. "Semiconductor Device Reliability Failure Models", International Sematech Technology Transfer document 00053955A-XFR, 2000
12. H.V. Nguyen, "Multilevel Interconnect Reliability on the effects of electro-thremomechanical stresses", Ph.D. dissertation, Univ. of Twente, Netherland, March 2004.
13. M. Huang, Z. Suo, "Thin film cracking and rathcheting caused by temperature cycling", J. Mater. Res. v.15, n..6, pp. 1239 (4), Jun 2000.
14. R. Degraeve, J.L. Ogier, et. al, "A New Model for the Field Dependence of Intristic and Extrinsic Time-Dependent Dielectric Breakdown", IEEE Trans. on Elect. Devices, 472(10),v.45,n.2, Feb. 1998
15. E.E. Lewis, Introduction to Reliability Engineering, Wiley 1996.
16. T. Simunic, K. Mihic, G. De Micheli, "Reliability and Power Management of Integrated Systems," DSD 2004.
17. "TMS320C6211, TMS320C6211B, Fixed-Point Digital Signal Processors", Texas Instruments, 2002.
18. "SAF7113H datasheet", Philips Semiconductors, March 2004
19. "SST-Melody-DAP Audio Processor", Analog Devices, 2002.
20. "MSP430x11x2, MSP430x12x2 Mixed Signal Microcontroller", Texas Instruments, August 2004.
21. "RDRAM 512Mb", Rambus, July 2003.
22. A. Maheshwari, W. Burleson, R. Tessier, "Trading off Reliability and Power-Consumption in Ultra-Low Power Systems", ISQED 2002.
23. P. Stanley-Marbell, D. Marculescu, "Dynamic Fault-Tolerance and Metrics for Battery Powered, Failure-Prone Systems," ICCAD 2003.

Performance Gains from Partitioning Embedded Applications in Processor-FPGA SoCs

Michalis D. Galanis, Gregory Dimitroulakos, and Costas E. Goutis

VLSI Design Laboratory, ECE Department, University of Patras
{mgalanis,dhmhgre,goutis}@ee.upatras.gr

Abstract. In this paper, we propose a hardware/software partitioning method for improving performance in single-chip embedded systems comprised by processor and Field Programmable Gate Array reconfigurable logic. Speedups are achieved by executing critical software parts on the reconfigurable logic. A generic hybrid System-on-Chip platform, which can model existing processor-FPGA systems, is considered. The partitioning flow utilizes an automated analysis procedure at the basic-block level for detecting kernels in software. Three different instances of the considered generic platform and two sets of benchmarks are used in the experiments. For the systems composed by 32-bit processors the speedup of five applications ranges from 1.3 to 3.7 relative to an all software solution. For an 8-bit platform, the speedups of eight DSP algorithms are considerably greater, since they range from 3.2 to 68.4.

1 Introduction

In past few years, academic [1], [2] and commercial [3], [4], [5] single-chip platforms emerged that employ processor(s) with Field Programmable Gate Array (FPGA) logic. These System-on-Chip (SoC) platforms are mainly composed by 8-bit micro-controllers, as in the ATMEL's Field Programmable System-Level Integrated Circuit (FPSLIC) [5], and 32-bit processors as in the Altera's Excalibur [4], in Xilinx's Virtex-II Pro [3], and in Garp architecture [1]. A significant advantage of using FPGA logic is that the functionality of custom-made coprocessors or peripherals implemented in this logic, can be changed due to the reconfiguration capabilities of such devices. This is not the case in the implementation in Application Specific Integrated Circuits (ASIC), where a small change in an application or in a standard requires the re-design of the ASIC component. The benefits of on-chip FPGA logic become larger in mass-produced devices where a low-cost implementation is sought. The processor-FPGA SoCs are expected to become more widespread in the future due to emergence of standards, like telecom ones, that their specification changes over time to meet the contemporary demands.

It is important to efficiently utilize the reconfigurable logic in single-chip microprocessor-FPGA systems. A hardware/software partitioning methodology that divides the application into software running on the microprocessor and on the FPGA logic is essential for such systems. Partitioning can improve performance [6] and in some cases even reduce power consumption [7]. More recently, hardware/software partitioning techniques for SoC platforms composed by a processor and FPGA [8], [9], [10], [11], were developed. The FPGA unit is treated as an extension of the processor. Critical parts of the application, called *kernels*, are moved for execution on the FPGA for improved performance and usually reduced energy consumption. This design choice stems from the observation that most embedded DSP and multimedia applica-

V. Paliouras, J. Vounckx, and D. Verkest (Eds.): PATMOS 2005, LNCS 3728, pp. 247–256, 2005.

tions spend the majority of their execution time in few small code segments (usually loops), the kernels. This means that an extensive solution space search, as in past hardware/partitioning works [6], [7], is not necessary.

In this work, we propose a hardware/software partitioning flow for accelerating software kernels of an embedded application on the reconfigurable logic of a generic processor-FPGA system. The processor executes the non-critical part of the application's software. This type of partitioning is possible in embedded systems, where the application is usually invariant during the lifetime of the system or of the specification. The considered generic processor-FPGA SoC can model a variety of existing systems, like the ones considered in [3], [4], [5]. Furthermore, the proposed method considers the communication time for exchanging data values between the FPGA and the processor, which was not the case in past works in hardware/software partitioning for processor-FPGA systems [8], [9], [10], [11]. An analysis tool at the basic block (BB) level has been developed. The term basic block expresses a sequence of instructions (operations) with no branches into or out of the middle. At the end of each basic block there is a branch instruction that controls which basic block executes next. This tool identifies kernels in the input software and targets RISC processor based SoCs, which are the case in both academia and in industry [1], [3], [4], [5].

For validating the hardware/software partitioning flow, we have used three different instances of the considered processor-FPGA platform: (i) four embedded 32-bit processors coupled with two devices from the Xilinx's Virtex FPGA family, (ii) an 32-bit processor with two devices from the Altera's APEX FPGAs [4], and (iii) an 8-bit microcontroller coupled with an ATMEL's AT40K FPGA device [5]. The (ii) and (iii) platform instances correspond to the processor and the FPGA units used in the Altera's Excalibur family [4] and the ATMEL's FPSLIC [5], respectively.

We have used two set of benchmarks for the experimentation: (a) five real-life applications coded in C language: an IEEE 802.11a OFDM transmitter, a video compression technique, a medical imaging application, a wavelet-based image compressor and a JPEG image encoder. This set of benchmarks is used for the experimentation with the 32-bit systems. (b) Eight DSP and multimedia algorithms, coded in C, from the Texas Instruments (TI) benchmark suite. This set of benchmarks is executed on the FPSLIC-simulated platform. The extensive experiments show that the kernels in the five real-world applications contribute an average of 69% of the total instruction count, while their size is 11% on average of the total code size. For the Virtex-based platform the speedups range from 1.3 to 3.7, while for the Excalibur-simulated SoC the speedups are from 1.3 to 3.2. The average speedup of the TI's algorithms on the FPSLIC-simulated platform is 28.1.

The rest of the paper is organized as follows: section 2 presents the proposed partitioning method. Section 3 describes the extensive experiments for the three different platforms. Finally, section 4 concludes this paper and outlines future activities.

2   Hardware/Software Design Flow

2.1   Target SoC Architecture

A general diagram of the considered hybrid SoC architecture is shown in Fig. 1. The platform includes: (a) FPGA logic for executing software kernels, (b) shared system data memory, (c) instruction and configuration memories, and (d) an embedded processor. The processor is typically a RISC processor, like an ARM7. Communication between the FPGA and the processor takes place via the system's shared data memory. Direct communication is also present between the FPGA and the processor. Part of the direct signals is used by the processor for controlling the FPGA by writing values to configuration registers located in the FPGA. The rest direct signals are used by the FPGA for informing the processor. Local data memories exist in the FPGA for quickly loading data, as in modern FPGAs [3], [4], [5]. The main configuration memory of Fig. 1 is used to store the whole configuration bitstream for programming the execution of the application's kernels on the FPGA. This generic system architecture can model the majority of the contemporary processor-FPGA SoCs, like the ones considered in [3], [4], [5].

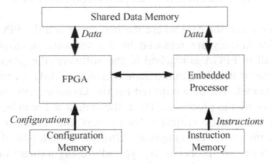

Fig. 1. Target hybrid embedded SoC

2.2 Flow Description

The proposed hardware/software partitioning method for processor-FPGA systems interests in increasing application's performance by mapping critical software parts on the reconfigurable hardware. The flow is illustrated in Fig. 2. The input is a software description of the application in a high-level language, like C/C++. Firstly, the Control Data Flow Graph (CDFG) Intermediate Representation (IR) is created from the input source code. The CDFG is the input to the analysis step. In the analysis, an ordering of the basic blocks in terms of the computational complexity is performed. The computational complexity is represented by the instruction count, which is the number of instruction executed in running the application on the processor. The instruction count is found by a combination of dynamic (profiling) and static analysis. A threshold, set by the designer, is used to characterize specific basic blocks as kernels. The rest of the basic blocks are executed on the processor.

The kernels are synthesized on the FPGA architecture for acceleration. The non-critical application's parts are converted from the CDFG IR back to the source code representation. Then, the source code is compiled using a compiler for the specific processor and it is run on the processor. The separation of the application to critical and non-critical parts, defines the data communication requirements between the processor and the FPGA. The proposed design flow considers the data exchange time through the shared memory for calculating the application's execution time, which is not the case in previous works for processor-FPGA systems [9], [10], [11].

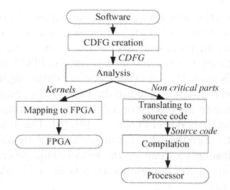

Fig. 2. Hardware/software partitioning design flow

Currently, we consider the case where the processor and the FPGA execute in mutual exclusion. The kernels are replaced in the software description with calls to FPGA. When a call to FPGA is reached in the software, the processor activates the FPGA and the proper state of the Finite State Machine (FSM) is enabled on the FPGA for executing the kernel. The data required for the kernel execution are written to the shared data memory by the processor. Then, these data are read by the FPGA. When the FPGA executes a specific critical software part, the processor usually enters an idle state for reducing power consumption. After the completion of the kernel execution, the FPGA informs the processor by typically using a direct interrupt signal and writes the data required for executing the remaining software. Then, the execution of the software is continued on the processor and the FPGA remains idle. Since the partitioning method interests in accelerating a sequential software program, which is often the case in implementing embedded applications in a high-level language like C, the performance gains from concurrent execution of the FPGA and the processor could be likely small. We mention that works in single-chip processor-FPGA systems [9], [10], [11] also assumed a mutual exclusive operation. However, the parallel execution on the processor and on the FPGA is a topic of our future research activities.

Due to the mutual exclusive operation of the processor and the FPGA, the total execution cycles after hardware/software partitioning are:

$$Cycles_{hw/sw} = Cycles_{sw} + Cycles_{FPGA} + Cycles_{comm} \qquad (1)$$

where $Cycles_{sw}$ represents the number of cycles needed for executing non-critical parts on the processor, $Cycles_{FPGA}$ corresponds to the cycles that are required for executing the kernels on the FPGA, and $Cycles_{comm}$ is the time required for transferring data, through the shared data memory of Fig. 1, between the processor and the FPGA. The $Cycles_{hw/sw}$ are multiplied with the clock period of the processor for calculating the total execution time $t_{hw/sw}$ after the partitioning.

For estimating the $Cycles_{FPGA}$ of the application's kernels on the FPGA, we realize the following procedure. We describe each kernel in a synthesizable Register-Transfer Level (RTL) description using VHDL language. Loop unrolling and pipelining transformations are used for achieving better performance when each kernel is synthesized on the FPGA. Each kernel is a state of an FSM (controller), so that when the kernels are synthesized they could share the same hardware. This sharing is doable because the kernels are not executed concurrently since they are belonging to a sequential software description. For executing a specific kernel on the FPGA, the proper state of the controller is selected. The reconfigurable logic runs at the maximum possible clock frequency after synthesizing all the kernels of an application. For synthesis, placing and routing of the RTL descriptions of the kernels, standard commercial tools can be used. In this work, we have utilized the Synplify Pro ver. 7.3.1 of the Synplicity Inc. [13].

Parts of the hardware/software partitioning method have been automated for a software description in C language. In particular, for the CDFG creation from the C code, we have used the SUIF2 [14] and MachineSUIF compiler infrastructures [15]. The automation of the analysis step is described in section 2.2.3. For the translation from the CDFG format to the C source code, the *m2c* compiler pass from the Machine-SUIF distribution is utilized.

2.3 Analysis

The analysis step of the partitioning flow outputs the kernel and non-critical parts of the software description. The inherent computational complexity of basic blocks, represented by the dynamic instruction count, is a meaningful measure to detect dominant kernels. The number of instructions executed when an application runs on the processor is obtained by a combination of profiling and static analysis within basic blocks. Fig. 3 shows the flow of the analysis. The input to the analysis process is the CDFG IR of the source code. For the CDFG representation, we have chosen the SUIF Virtual Machine (SUIFvm) representation for the instruction opcodes inside basic blocks [15]. The SUIFvm instruction set assumes a generic RISC machine, not biased to any existing architecture. Thus, the information obtained from the analysis, could stand for any RISC processor architecture. This means that the detected critical software parts are kernels for various types of RISC processors. This was justified by experimentation, using the profiling utilities of the compilation tools of the processors considered in the experiments.

We have used the HALT library [15] of the Machine-SUIF distribution for performing *profiling* at the basic block level. The profiling step reports the execution frequency of the basic blocks. For the *static analysis*, a MachineSUIF pass has been developed that identifies the type of instructions inside each basic block. Afterwards, a custom developed compiler pass calculates the static size of the basic block using the SUIFvm opcodes. The static size and the execution frequency of the basic blocks are inputs to a developed instruction mix pass that outputs the dynamic instruction count. After the instruction count calculation for each basic block, an ordering of the basic blocks is performed. We consider kernels, the basic blocks which have an instruction count over a user-defined threshold. This threshold represents the percentage of the contribution of the basic block's instruction count in the application's overall instruction count. For example, basic blocks contributing more than 10% in the total instruction count can be considered as kernels.

3 Experiments

Two sets of benchmarks are used for validating the proposed hardware/software partitioning flow. The first one consists of five applications and it is used for the systems

composed by 32-bit RISC processors. The first application is an IEEE 802.11a
OFDM transmitter. The second one is a cavity detector which is a medical image
processing application. The third one is a video compression technique, called Quadtree Structured Difference Pulse Code Modulation (QSDPCM), while the fourth application is a still-image JPEG encoder. Finally, the fifth application is a wavelet-
based image compressor [12].

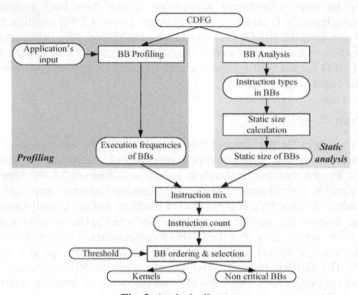

Fig. 3. Analysis diagram

The second set of benchmarks is used for the considered 8-bit platform. It is composed by smaller applications, actually algorithmic kernels, that can be handled by the
computational capabilities of a low-cost and low-performance (compared to the 32-bit
cores) 8-bit RISC processor core. These algorithms are derived from the TI's DSP
and imaging benchmark suite. Eight representative DSP and multimedia algorithms
are used: an 3x3 convolution kernel, a 2-Dimensional (2D) 8x8 Forward Discrete
Cosine Transform (FDCT), a 2D 8x8 Inverse DCT (IDCT), a 64-point Fast Fourier
Transform (FFT), a complex FIR filter, a 16x16 point Minimum Absolute Differences
(MAD) unit, a 168 by 256 matrix multiplication, and the vertical pass of a 2D Discrete Wavelet Transform (DWT).

For calculating the $Cycles_{comm}$ in eq. (1), we assume that the shared data memory is
modelled as a multi-port and single clock cycle accessible Static RAM (SRAM),
which is a reasonable assumption for modern configurable SoCs [3], [4].

3.1 Analysis Results

The results using the proposed analysis flow for the five real-life applications of the
first benchmark set are shown in Table 1. The contributions of the kernels to the total
static size (in instruction bytes) and to the total instructions are reported. The threshold for the kernel detection was set to the 10% of the total dynamic instructions of the

application. The number of kernels detected in each application is also given. It is noted that there is a small number of critical basic blocks in the considered applications, at maximum equal to 4. The detected kernels are loop bodies without conditional statements inside them. From the analysis results, it is inferred that an average of 10.9% of the code size, representing the kernels' size, contributes 68.5% on average to the total executed instructions.

Table 1. Results from the analysis process

Application	Total Size	Kernels size	% size	% total instructions	# of kernels
Cavity	12,039	910	7.6	79.8	4
OFDM	15,579	1,440	9.2	61.5	4
Compressor	12,835	602	4.7	78.8	4
JPEG	10,995	2,534	23.0	71.3	4
QSDPCM	24,767	2,477	10.0	51.0	3
Average			10.9	68.5	

3.2 Virtex-Based Systems

In this section, we present the results from partitioning the five applications of the first set of benchmarks in a SoC that has a Virtex FPGA device as its reconfigurable logic. These results show the speedups after executing the kernels on the FPGA. We have used four different types of 32-bit embedded RISC processors: an ARM7, an ARM9, and two SimpleScalar processors [16]. The SimpleScalar processor is an extension of the MIPS32 IV core. These processors are widely used in embedded SoCs. The first type of the MIPS processor (*MIPSa*) uses one integer ALU unit, while the second one (*MIPSb*) has two integer ALU units. We have used instruction-set simulators for the considered embedded processors for estimating the number of execution cycles. More specifically, for the ARM processors, the ARM Developer Suite (version 1.2) was utilized, while the performance for the MIPS-based processors is estimated using the SimpleScalar simulator tool (version 3.0) [16]. Typical clock frequencies are considered for the four processors: the ARM7 runs at 100 MHz, the ARM9 at 250 MHz, and the MIPS processors at 200 MHz. The five applications were optimized for best performance when compiled for the considered processors.

The performance gains from applying the partitioning flow in the five applications are presented in Fig. 4. For each application, the four aforementioned processor architectures co-exist with the FPGA in the hybrid SoC. We have assumed two different Virtex FPGA devices: (a) the smallest available Virtex device, the XCV50 FPGA, and (b) the medium size device XCV400. The clock frequencies after synthesizing, placing and routing the designs using the Synplify Pro toolset [13], range from 45 to 77 MHz for the XCV50 device and from 37 to 77 MHz for the XCV400. From Fig. 4, it is deduced that significant performance improvements are achieved when critical software parts are mapped on the FPGA. It is noticed that better performance gains are achieved for the ARM7 case than the ARM9-FPGA system. This occurs since the speedup of kernels on the FPGA has greater effect when the FPGA co-exists with a lower-performance processor, as it is the ARM7 relative to the ARM9. Furthermore, the speedup is almost always greater for the MIPSa than the MIPSb processor, since the latter one employs one more integer ALU unit.

For the case of the different Virtex devices, the performance improvements are greater for the XCV400 due to the available number of Control Logic Blocks (CLBs) which permit the implementation of more operations on the FPGA hardware. This leads to better kernels' acceleration through the larger amount of spatial computation due to the increased number of instantiated operations in the configurable logic relative to the smaller FPGA device, the XCV50. The average speedup is 2.1 for the XCV50 and 2.4 for the XCV400.

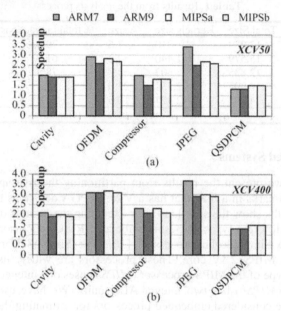

Fig. 4. Speedups from accelerating kernels on (a) XCV50 and (b) XCV400 devices

3.3 Excalibur-Simulated Systems

The results from accelerating the kernels of the five applications of the first set of benchmarks on the Excalibur-simulated SoC [4], are given in this section. In the Excalibur devices, an ARM9 processor is used that it is clocked at 200 MHz, which is also the case in this experimentation. The applications were again optimized for best performance when compiled for the ARM9. The ARM Developer Suite was used for estimating the cycles required for the software execution. Two cases of Altera's APEX FPGAs are utilized for simulating the EPXA1 and the EPXA10 Excalibur devices, where the EPXA10 stands for a larger amount of reconfigurable logic. After the kernels' synthesis with the Synplify Pro, the reported clock frequencies range from 20 to 38 MHz for the EPXA1, and from 22 to 30 MHz for the EPXA10.

The performance gains after partitioning, relative to the execution times of the whole application on the processor ($Cycles_{init}$) are given in Table 2. Greater speedups are achieved for the EPXA10-simulated system, as in the case of the Virtex-based SoCs, where greater performance was achieved for the larger Virtex device. The average speedup is 2.1 for the EPXA1 and 2.3 for the EPXA10. Comparing the speedups of Table 2 with the respective ones for the ARM9-Virtex system, they are approximately the same although the ARM9 is clocked at a lower speed and the clock frequencies after the kernel synthesis on the APEX devices, are smaller than the ones on the Virtex FPGAs.

Table 2. Speedups for the Excalibur-simulated SoCs

Application	Cycles$_{init}$	EPXA1		EPXA10	
		Cycles$_{hw/sw}$	Speedup	Cycles$_{hw/sw}$	Speedup
Cavity	161,441,889	86,301,446	1.9	83,382,000	1.9
OFDM	362,990	143,829	2.5	113,692	3.2
Compressor	20,574,658	11,050,385	1.9	9,631,295	2.1
JPEG	19,951,193	7,413,742	2.7	6,648,191	3.0
QSDPCM	3,895,248,922	3,038,838,531	1.3	3,009,388,039	1.3
Average			2.1		2.3

3.4 FPSLIC-Simulated System

In this section we present the speedups after partitioning the second set of benchmarks on an ATMEL FPSLIC-based system. In FPSLIC devices, an 8-bit AVR core, capable of 30 Millions Instruction Per Second (MIPS), is coupled with a AT40K FPGA. In this experiment, both the AVR and the configurable logic are clocked to 20 MHz. The execution cycles of the non-critical parts of the algorithms on the AVR microcontroller are estimated using the profiling utilities of the Embedded Workbench suite from the IAR Systems Inc. [17]. The algorithms were optimized for best performance when compiled for the AVR. We have used the developed analysis flow for detecting critical basic blocks, although manual kernel detection can be performed since these eight algorithms are relatively small programs. Two basic blocks at maximum were characterized as kernels, which were the case in the FDCT, in the IDCT and in the wavelet algorithm. All the kernels were inner for-loops (without conditional structures) of the main computation bodies in each algorithm.

From Table 3, it is deduced that the speedups are significantly greater than the ones obtained for the five large applications with the 32-bit platforms. This is due to two reasons: (a) the kernel(s) of each algorithm contributes to a larger amount to the total instruction count than the kernels in each of the five real-world applications, (b) the speedups are greater when the reconfigurable logic is coupled with a lower-performance instruction-set processor, as it is the AVR relative to the 32-bit processors.

Table 3. Speedups for the FPSLIC-simulated SoC

Application	Cycles$_{init}$	Cycles$_{hw/sw}$	Speedup
FFT	213,427	66,883	3.2
Matrix Mult.	1,491,462	43,445	34.3
FDCT	308,472	9,528	32.4
IDCT	329,298	9,858	33.4
Convolution	180,385	36,001	5.0
MAD	122,374,066	1,787,826	68.4
FIR complex	259,889	8,289	31.4
Wavelet	193,524	11,403	17.0
Average			28.1

4   Conclusions – Future Work

A partitioning flow for accelerating critical software segments in processor-FPGA SoCs was presented. Significant performance improvements have been achieved. The speedups range from 1.3 to 68.4, where the maximum value corresponds to a low-cost 8-bit system. Future work focuses on the parallel execution of the processor and the FPGA for achieving even greater performance improvements.

References

1. Callahan, T. J., et al.: The Garp Architecture and C Compiler. IEEE Computer, vol. 33, no. 4 (2000) 62-69
2. Hauck, S., et al.: The Chimaera Reconfigurable Functional Unit. IEEE Trans. on VLSI Syst., vol.12, no.2 (2004) 206-217
3. Virtex FPGAs, Xilinx Inc., www.xilinx.com (2005)
4. Excalibur, Altera Inc., www.altera.com (2005)
5. FPSLIC, ATMEL Inc., www.atmel.com (2005)
6. Gajski, D.D., et al.: SpecSyn: An environment supporting the specify-explore-refine paradigm for hardware/software system design. IEEE Trans. on VLSI Syst., vol. 6, no. 1 (1998) 84–100
7. Henkel, J.: A low power hardware/software partitioning approach for core-based embedded systems. Proc. of the 36th ACM/IEEE DAC (1999) 122–127
8. Ye, A., et al.: A C Compiler for a Processor with a Reconfigurable Functional Unit. Proc. of FPGA (2000) 95-100
9. Bazargan, K., et al.: A C to Hardware/Software Compiler. Proc. of FCCM (2000) 331-332
10. Villareal, J., et al.: Improving Software Performance with Configurable Logic. Design Automation for Embedded Systems, Springer, vol. 7 (2002) 325-339
11. Stitt, G. and Vahid, F.: Energy Advantages of Microprocessors Platforms with On-Chip Configurable Logic. IEEE Design & Test of Computers, vol. 19, no. 6 (2002) 36-43
12. Honeywell Inc., http://www.htc.honeywell.com/projects/acsbench (2005)
13. Synplify Pro, Synplicity Inc., www.synplicity.com (2005)
14. SUIF2 compiler infrastucture, http://suif.stanford.edu/suif/suif2/index.html (2005)
15. MachineSUIF, http://www.eecs.harvard.edu/hube/research/machsuif.html (2005)
16. SimpleScalar LLC, www.simplescalar.com (2005)
17. IAR Embedded Workbench, IAR Systems Inc., www.iar.com (2005)

A Thermal Aware Floorplanning Algorithm Supporting Voltage Islands for Low Power SOC Design

Yici Cai, Bin Liu, Qiang Zhou, and Xianlong Hong

EDA Lab, Department of Computer Science and Technology,
Tsinghua University, Beijing, 100084, P.R. China
liubin00@mails.tsinghua.edu.cn

Abstract. This paper presents a new floorplanning algorithm emphasizing power reduction for SOC designs using voltage islands. In this algorithm, the supply voltages and positions of blocks are determined simultaneously for both dynamic and static power reduction. Special notice is taken of the interdependence between power and temperature, and thus a block level power and thermal analyzer is incorporated for thermal aware power estimation. Other goals, including area, wire length, as well as level converters and temperature distribution are taken into account, leading to a multi-objective optimization problem, solved using simulated annealing. Experimental results on a set of modified MCNC benchmarks show that introducing voltage islands can reduce the total power by 15% to 30%, and thermal aware voltage island optimization can further reduce the total power by 4% to 15%, as well as promoting even temperature distribution.

1   Introduction

Power dissipation has become one of the most critical concerns in very deep submicron IC designs for the following two reasons. First, both dynamic and static power increase rapidly because of rising clock frequency, higher leakage current and increasing packing density. Second, more and more devices are powered by batteries with the growth of portable and mobile applications, which urges low power chips. A number of power aware design methodologies and tools have been developed, among which the new design style of voltage island has shown great effectiveness and flexibility for power saving in SOC designs [1].

The motivation of the voltage island approach is that there are usually some modules working unnecessarily fast, and that the supply voltages of such modules can be reduced for power saving without sacrificing performance. In an SOC design utilizing voltage islands, each island refers to a region, usually containing several blocks, that uses a separate supply voltage other than the chip-level supply. By allowing multiple on-chip supplies and flexible placement of voltage islands, the new design style brings forward many opportunities to partially lowering supply voltage for power reduction under performance constraint. Beneficial as it is, some overhead exists: on-chip or off-chip voltage regulators are

V. Paliouras, J. Vounckx, and D. Verkest (Eds.): PATMOS 2005, LNCS 3728, pp. 257–266, 2005.

required to generate suitable supply voltage for each island; level converters are needed if a module is driven by another with lower supply level; power networks may become much complicated with a huge number of islands. All of the above factors are likely to increase the design and manufacturing costs and hence must be carefully managed.

Thermal issue is another concern concomitant with power consumption. Due to increasing packing density and power dissipation, power density is rising rapidly, leading to high temperature across the chip and thus high packing and cooling cost. Moreover, temperatures at different parts of the chip can vary significantly, which challenges reliability, calling for temperature aware design [2]. It should be noticed that while power consumption, especially leakage power, depends heavily on temperature. A 30-degree increase of temperature can cause up to 50% increase of static power in the newest technology generation. In view of the interdependence of power and temperature, integrated power and thermal management is highly desirable in sub-100nm designs.

Many works have been done on the traditional floorplanning problem minimizing a combination of area and wire length. Recently, a few floorplanning algorithms have been proposed considering thermal or power issues. [ ] presents a genetic algorithm for even temperature distribution. [4] develops a thermal driven algorithm for 3D floorplanning. [ ] proposes a method for temperature aware floorplanning with performance considerations. All these works assume that the total power of each module is constant and can be determined before the floorplanning process. [ ] develops an algorithm for low power slicing floorplan design with module selection. This algorithm also considers area, wire length, power density etc.; however, prevalent techniques for low power design are not used and leakage is not explicitly considered. Notably, [ ] defines the problem of architecting voltage island and presents an algorithm including island partition creation, level assignment and physical layout of modules. This is the first published work on floorplanning for voltage island designs to the best of our knowledge; however, many important design aspects are not considered adequately. Wire length and the number of level converters are not measured explicitly, and the area utilization of reported experimental benchmarks are relatively low. Furthermore, like most previous works, this algorithm ignores the temperature dependence of power consumption and every core is assumed to consume a constant power in any circumstance, which can lead to inaccuracy in power and temperature estimation when leakage power is prominent.

The purpose of this paper is to describe a thermal aware floorplanning algorithm for designs with voltage islands, where not only the position, but also the supply voltage of each module is determined. The reason for such combination is that voltage island design is tightly related to the physical implementation due to cost and reliability considerations: it is important to cluster modules with the same supply voltage in order to reduce the number of islands and thus reduce cost; temperature distribution can be determined given the layout and power of every core, which facilitates thermal management, as well as encourages the idea of placing high leakage modules in regions where temperature is relatively low to allay static power. Besides power reduction and hotspot avoidance, the proposed algorithm also tries to optimize the number of level converters, as well as traditional floorplan design objectives including area and wire length. Like many prevalent floorplanning algorithms, the proposed algorithm employs a simulated annealing engine and B* tree is used to represent the floorplan topology[ ].

2   Problem Description

Consider an SOC design with a set of cores[1]. To support the voltage island style, some of the cores in the library are designed to be able to operate at different voltages (possibly with different implementations). We assume that the possible supply voltages of each core are given in a power table, along with the power density at each voltage. Since static power depends exponentially on temperature, we also assume that static power values on a few typical temperatures are known prior to floorplanning. A sample power table is shown in Table 1.

Table 1. A sample power table

name	V_{dd}	dynamic	static(300K)	static(350K)	static (400k)
A	1.0V	800	340	590	940
B	1.0V	200	200	350	550
C	1.0V	1100	380	640	1000
D	0.8V	500	300	420	570
D	0.9V	560	340	490	640
D	1.0V	630	400	570	720
E	0.9V	400	210	300	420
E	1.0V	500	270	390	490

It is necessary to control the number of supply levels since level converters are often needed on interconnects across different voltage islands. Moreover, it is necessary to reduce the number of voltage islands because too many irregularly-arranged islands can make the design much complicated, causing trouble to timing analysis, power routing and clock distribution. Therefore, blocks supplied with the same voltage level is advised to be placed in clusters, each cluster forming an individual island.

The introduction of voltage islands and the consideration of power and thermal issues bring forward more constraints and the goal is usually inconsistent with that of traditional floorplan problems. Fig. 1 illustrates 3 different packing solutions of the 5 blocks whose power table are given in Table 1. The first is an area optimal packing, with an area of 90. Suppose that only two supply levels are allowed, D and E should be supplied with voltage 0.9V and form a voltage island. Then it is favorable to place D and E adjacently and we can get the second packing with area 96. Further, noting that in the second packing, A and

---

[1] Cores, modules and blocks are interchangeable in this paper

C, the two blocks with highest power density, are placed near each other, we are advised to modify the packing and separate A and C to avoid hot spot. Thus, the third packing with more evenly temperature distribution but larger area is preferable. In a larger problem with more blocks, the situation can be much more complex. Different design aspects interact with each other and compromises are usually made between conflicting goals, resulting in a more sophisticated problem. It can be concluded that any algorithm that fails to consider any of the aspects is likely to result in suboptimal solutions.

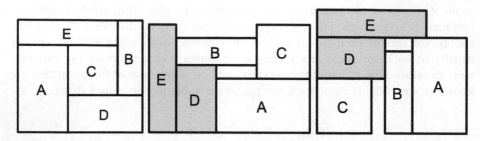

Fig. 1. Three packing solutions for five blocks. The areas are 90, 96, 99 respectively

The problem this paper aims to solve can be described as follows. Given a set of cores $C = \{C_1, C_2, \ldots, C_n\}$, with C_i of width w_i and height h_i, and a power table containing the allowed supply voltages and power density under different temperatures and different voltages. A floorplan with voltage islands is represented with (x_i, y_i, r_i, v_i) for all C_i, where x_i, y_i are the coordinate of the left bottom position of C_i, r_i denotes the rotating or flipping condition of C_i and v_i is the supply voltage for C_i. The goal is to minimize: 1) area; 2) total wire length; 3) total power; 4) number of level converters; 5) maximum on chip temperature; 6) number of voltage islands. Constraints include 1) no overlap between cores; 2) total number of voltage levels does not exceed a predetermined value t.

3   Coupled Power and Thermal Analysis

In the floorplanning process, the selection of supply level and position for every core influences both power and temperature distribution, which are inherently interrelated. When we consider the dependence of power consumption on supply voltage and operating temperature, we get

$$P_i = P_i(v_i, T_i) \tag{1}$$

On the other hand, the steady state temperature distribution satisfies the heat transfer equation.

$$k(x, y, z)\nabla^2 T(x, y, z) + g(x, y, z) = 0 \tag{2}$$

where k denotes thermal conductivity, $T(x, y, z)$ represents the temperature and $g(x, y, z)$ is power density at point (x, y, z). A thermal resistive network can be obtained by modelling each block as a node connected with adjacent nodes by thermal resistors. Then the temperature of each core satisfies the following equation.

$$\begin{pmatrix} T_1 - T_{amb} \\ T_2 - T_{amb} \\ \vdots \\ T_n - T_{amb} \end{pmatrix} = \begin{pmatrix} R_{11} & \cdots & R_{1n} \\ \vdots & \ddots & \vdots \\ R_{n1} & \cdots & R_{nn} \end{pmatrix} \begin{pmatrix} P_1 \\ P_2 \\ \vdots \\ P_n \end{pmatrix} \tag{3}$$

or

$$\mathbf{T} - \mathbf{T}_{amb} = \mathbf{RP} \tag{4}$$

where $\mathbf{T}_{amb}$ is the ambient temperature, and $\mathbf{R}$ is thermal transfer matrix obtained from the positions of modules using *HotSpot* 2.0 [ ].

It is obvious from Eqn. 1 and 3 that the power consumption vector $\mathbf{P}$ and temperature vector $\mathbf{T}$ are interdependent. In [ ] and [ ], an iterative method for coupled power and thermal analysis is used: initial power or temperature estimation are passed to the model and convergence is expected after a few iterations, as shown in Fig 2.

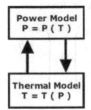

Fig. 2. Flow of coupled thermal and power analysis: iterations between power model and thermal model is needed under given supply voltage and position for every block

The number of iterations needed to achieve certain accuracy is influenced by the initial value of temperature or power. If an initial estimation of temperature distribution that is close to the actual value is used, the iteration is expected to converge quickly. In our algorithm, a linearized power model is used in the coupled power and thermal analysis for quick temperature estimation, and the result are passed to the iterative analyzer as the initial value. The power of each core under given supply voltage is written as

$$P_i = a_i + b_i T_i \tag{5}$$

Such linearization is based on our observation that although subthreshold leakage increases exponentially with temperature, the relationship between total power and temperature is slightly super-linear and it is possible to find a linear approximation within a reasonable range of working temperature. We have carried out a series of experiments using *HSpice* on the error of such approximation. Fig. 3 shows the power consumption of a small circuit under 70nm technology at different temperatures. The maximum error of using a linear model is below 10% between the temperature range from 318K to 383K.

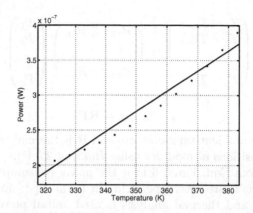

Fig. 3. The trend of total power of a small 70nm circuit working at 1GHz at different temperatures. Results are obtained using HSpice with BSIM4 transistor model

Eqn. 5 can be written in matrix form as

$$
\begin{pmatrix} P_1 \\ P_2 \\ \vdots \\ P_n \end{pmatrix} = \begin{pmatrix} a_1 \\ a_2 \\ \vdots \\ a_n \end{pmatrix} + \begin{pmatrix} b_1 & & & \\ & b_2 & & \\ & & \ddots & \\ & & & b_n \end{pmatrix} \begin{pmatrix} T_1 \\ T_2 \\ \vdots \\ T_n \end{pmatrix} \tag{6}
$$

or

$$
\mathbf{P} = \mathbf{A} + \mathbf{BT} \tag{7}
$$

Combining Eqn. 4 and 7, the estimated temperature vector can be solved as

$$
\mathbf{T} = (\mathbf{I} - \mathbf{RB})^{-1}(\mathbf{RA} + \mathbf{T_{amb}}) \tag{8}
$$

With the initial value given by Eqn. 8, only 1 iteration is performed in our implementation before the distribution of power and temperature is obtained.

4   Algorithm Description

4.1   Solution Representation and Perturbation

B* tree is an elegant floorplan representation scheme proposed in [ ], which is able to represent the topology of admissible floorplans. Block orientation and supply voltage is associated with each node in the B* tree to represent both the position and the supply voltage in our algorithm.

On the neighborhood structure, there are totally 5 kinds of moves operating on an existing solution.

1. Rotate a block clockwise by 90 degree.
2. Flip a block.
3. Swap two blocks.
4. Remove a block from the floorplan and insert it elsewhere.
5. Increase or decrease the supply voltage of a block if allowed.

4.2   Cost Function

We formulate the following cost function to incorporate all the goals of the problem.

$$
\begin{aligned}
cost = \; & c_1 \cdot \exp(power/power_e) + c_2 \cdot lc/N \\
& + c_3 \cdot \exp(area/area_e) \quad + c_4 \cdot wl/wl_e \\
& + c_5 \cdot \max(temp) \quad\quad\quad + c_6 \cdot islands \\
& + c_7 \cdot \mathrm{ramp}(levels - t)
\end{aligned} \tag{9}
$$

Where $power$, lc, $area$, wl, $temp$, $islands$ and $levels$ are the total power, number of level converters needed, chip area, wire length, temperature, number of voltage islands and total number of voltage levels used in the floorplan, respectively. $power_e$ is the total power when all cores works on its lowest allowable voltage under the temperature of 300K; N is the number of cores in the design, $area_e$ is the sum of area of every core; wl_e is estimated total wire length, given by averaging the wire length of 20 randomly generated floorplan. c_1 to c_7 are constants that are 0.3, 0.1, 0.4, 0.2, 0.001, 0.05 and 1.0 respectively in our implementation. Note that c_7 is punitive.

4.3   Two Stage Annealing

To avoid trivial operations of adjusting operating voltages blindly, we divide the algorithm in two stages. In the first stage, several appropriate voltages are selected as candidate supplies for blocks from all possible voltages. Because the selection of voltages influences the cost function prominently, and the layout generated in this stage is not necessarily superior in area or wire length, appropriate voltage levels can be found with relatively smaller search effort. After the first stage, every core is supplied with a low voltage and perturbations on supply voltage can be performed less frequently in the second stage, which focuses on optimization of the complex goal involving all the significant design aspects mentioned above.

The proposed algorithm is outlined in Fig. 4. The two stages differ primarily in the perturbation schemes (referred to as *Perturbation Scheme 1* and *Perturbation Scheme 2* for the two stages respectively). In Perturbation Scheme 1, the supply voltages for all cores are adjusted frequently, so that an optimal voltage configuration can be quickly determined to reduce the search space in the second stage. Perturbation Scheme 2 is mainly adjustment on positions or orientations of blocks. Voltage adjustment is relatively rare (5% in total perturbations in our implementation), and usually associated with spatial changes in order to avoid fragmented voltage island configuration and to save power.

1 Generate an initial solution with random topology and every core working at
its lowest allowed voltage.
2 *temperature* ← *init_temperature*
3 **while** *temperature* > *final_temperature*
4   Perturb the solution with *Perturbation Scheme 1*.
5   Evaluate the new solution and decide whether to accept it.
6   Adjust temperature.
7 **end while**
8 *temperature* ← *init_temperature2*
9 **while** *temperature* > *final_temperature2*
10   Perturb the solution with *Perturbation Scheme 2*.
11   Evaluate the new solution and decide whether to accept it.
12   Adjust temperature.
13 **end while**
14 Output final solution.

Fig. 4. Outline of the algorithm

Another difference between the two stages is in the number of perturbations
performed. The first stage undergoes a fast cooling process since a satisfactory
voltage configuration can be found without very much effort. In the second stage,
however, temperature lowers relatively slow because a large number of solutions
need to be evaluated.

5   Experimental Results

The proposed algorithm has been implemented on a PC with Intel Pentium processor 1.4GHz and 384M primary memory. The area, total wire length, number
of level converters, power consumption, maximum on chip temperature of the
solution generated by the proposed algorithm are measured on a set of MCNC
benchmarks and compared with two other floorplanning algorithms. One is a
traditional floorplanning algorithm optimizing a multiplex goal of area and wire
length under uniform chip-level supply voltage (referred to as Algorithm 1). Another is similar to the proposed algorithm, but without thermal considerations,
assuming the power consumption of a block is constant (referred to as Algorithm 2). In our experiments, the power density when temperature is 343K is
used as the constant power density for every core in Algorithm 2. Since the original MCNC benchmarks do not contain power density information, we generate
the power table based on the predicted value at 70nm technology by ITRS 2003
[ ]. In the following experiments, the maximum number of supply voltages is
limited to 2.

Comparison on area, wire length, power, level converters and running time
between Algorithm 2 and Algorithm 1 are illustrated in Table 2. It can be observed that designs using voltage island can gain a power reduction between 15%
to 30%, and that using voltage islands usually result in larger area or wire length
and some other overheads like level converters. It is interesting to notice that the
xerox benchmark, in which 2 distinct voltage levels are possible, are eventually

Table 2. Results comparison. (Algorithm 1 / Algorithm 2)

name	area (mm^2)	wl (mm)	voltages (V)	power (W)	lc	time (s)
ami33	1.23/1.25	63.0/64.4	1.1/(1.0, 1.1)	1.20/0.95	0/60	25/340
ami49	37.71/38.64	815.2/883.4	1.1/(1.0, 1.1)	42.30/30.12	0/110	51/502
apte	48.45/48.21	440.5/542.0	1.1/(1.0, 1.1)	30.21/25.53	0/90	5/29
hp	9.26/9.57	225.3/226.4	1.1/(1.0, 1.1)	1.30/1.06	0/57	10/75
xerox	20.42/20.49	410.52/505.78	1.1/1.1	12.68/12.68	0/0	12/37

Table 3. Results comparison. (Algorithm 2 / proposed algorithm)

name	area (mm^2)	wl (mm)	voltages (V)	power (W)	$temp_{max}(K)$	time (s)
ami33	1.25/1.28	64.4/64.6	(1.0,1.1)/(1.0, 1.1)	1.02/0.94	412/366	340/710
ami49	38.64/40.22	883.4/834.0	(1.0,1.1)/(1.0, 1.1)	34.12/32.30	452/390	502/1799
apte	48.21/50.04	542.0/550.1	(1.0,1.1)/(1.0, 1.1)	27.02/23.25	408/353	29/79
hp	9.57/10.21	226.4/266.0	(1.0,1.1)/(1.0, 1.1)	1.05/1.01	368/339	75/219

supplied with a single voltage after optimization. This is probably because the
power reduction of dual supply scheme is not very prominent and thus only one
supply is used to avoid overheads.

Algorithm 2 assumes that the power density is constant under different temperatures, which may lead to inaccurate power and temperature estimation. Results of Algorithm 2 are compared with the proposed algorithm with a iterative
power and thermal analyzer. The total power and temperature on the hottest
core are computed with 10 iterations and listed in Table 3. The number of level
converters are equal on all the benchmarks. It is evident that thermal aware
optimization not only avoids hotspot effectively, but also helps in improving the
total power (by 4% to 15% on the above benchmarks).

Fig. 5 shows the floorplan results with voltage islands. The use of rectilinear
voltage islands rather than only allowing rectangular ones places less constraints
on the problem and is probably helpful in improving area utilization and wire
length.

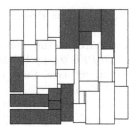 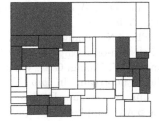

Fig. 5. Voltage islands in ami33 and ami49. The blocks shaded area are supplied with
1.1V, and others 1.0V

6 Conclusion

The introduction of voltage islands provides a flexible way for power reduction. Total power can be reduced effectively at the cost of extra implementation overheads. Voltage islands should be carefully designed with simultaneously consideration of power consumption, temperature, level converters, number of voltage islands, together with area and wire length during the floorplanning process. We have developed a thermal aware floorplanning algorithm supporting voltage islands. Experimental results have demonstrated the effectiveness of our algorithm. The results also indicate that thermal aware power reduction is necessary for sub-100nm designs.

Acknowledgements

This work is supported by the National Natural Science Foundation of China (NSFC) 60476014 and the Hi-tech Research&Development Program (863) of China 2005AA1Z1230.

References

1. D.E. Lackey, P.S. Zuchowski, T.R. Bednar, D.W. Stout, S.W. Gould, J.M. Cohn, "Managing power and performance for system-on-chip designs using voltage islands," in *proc. ICCAD'02*, pp.195-202.
2. W. Huang, M.R. Stan, K. Skadron, K. Sankaranarayanan, S. Ghosh, and S. Velusamy, "Compact thermal modeling for temperature-aware design," in *proc. DAC'04*, pp.878-883.
3. Y. Xie, N. Vijaykrishnan, C. Addo-Quaye, T. Theocharides and M.J. Irwin, "Thermal-aware floorplanning using genetic algorithms," in *proc. ISQED'05*, pp.634-639.
4. J. Cong, J. Wei, Y. Zhang, " A thermal-driven floorplanning algorithm for 3D ICs," in *proc. ICCAD'04*, pp.306-313.
5. . M. Ekpanyapong, M.B. Healy, C.S. Ballapuram, S.K. Lim and H.S. Lee, "Thermal-aware 3D microarchitectural floorplanning," CERCS Technical Reports, GIT-CERCS-04-37, 2004.
6. K.Y. Chao and D.F. Wong, "Floorplanning for low power designs," in *proc. ISCAS'95*, pp.45-48.
7. J. Hu, Y. Shin, N. Dhanwada, R. Marculescu, "Architecting voltage islands in core-based system-on-a-chip designs," in *proc. ISLPED'04*, pp.180-185.
8. Y.C. Chang, Y.W. Chang, G.M. Wu and S.W. Wu, "B* tree: a new representation for non-slicing floorplan," in *proc. DAC'00*, pp.458-463.
9. H. Su, F. Liu, A. Devgan, E. Acar and S. Nassif, "Full chip leakage-estimation considering power supply and temperature variations," in *proc. ISLPED'03*, pp.78-83.
10. International technology roadmap for semiconductors, http://public.itrs.net/Files/2003ITRS/Home2003.htm.

Power Supply Selective Mapping
for Accurate Timing Analysis

Mariagrazia Graziano[1], Cristiano Forzan[2], and Davide Pandini[2]

[1] Dipartimento di Elettronica, Politecnico di Torino, 10129 Torino, Italy
mariagrazia.graziano@polito.it
[2] STMicroelectronics, Central CAD and Design Solutions, 20041 Agrate Brianza, Italy
{cristiano.forzan,davide.pandini}@st.com

Abstract. In Deep Sub-Micron technologies post-layout timing analysis has become the most critical phase in the verification of large System-on-Chip (SoC) designs with several power-hungry blocks. The impact of coupling capacitances has been adequately analyzed, and modern signal integrity analysis tools can effectively consider the crosstalk-induced delay. However, an increasingly important factor that can introduce a severe performance loss is the power supply noise. As technology advances into the nanometer regime, the operating frequencies increase, and clock gating has emerged as an effective technique to limit the power consumption in block-based designs. As a consequence, the amplitude of the supply voltage fluctuations has reached values where techniques to include the effect of power supply noise into timing analysis based on linear models are no longer adequate, and the non-linear dependence of cell delay from supply voltage must be considered. In this work we present a practical methodology that accurately takes into account the power supply noise effects in static timing analysis, which can be seamlessly included into an industrial sign-off design flow. The experimental results obtained from the timing verification of an industrial SoC design have demonstrated the effectiveness of our approach.

1 Introduction

Signal integrity is one of the most important issues as technology rapidly advances in the Deep Sub-Micron (DSM) regime. Of particular importance among the different signal integrity-related problems there is the power supply noise that significantly impairs the performance, functionality, and reliability of VLSI circuits. Power supply noise degrades the driving capability of transistors with smaller feature sizes due to the reduced effective supply voltage seen by the devices. Moreover, a large number of simultaneous switching events in the circuit within a short period of time are the source of current spikes that can cause considerable L·dI/dt simultaneous switching noise (SSN), introducing performance degradation and logic failures, since the noise margins decrease as the supply voltage scales with technology. In the past, several research efforts have been directed towards power supply noise analysis and modeling, and power network optimization. In [1] a design methodology to analyze the on-chip power supply noise for high-performance microprocessors, based on an integrated package-level and chip-level power bus model, and a simulated switching circuit model for each functional block was presented. To reduce the complexity of full-chip analysis, hierarchical power network modeling techniques were proposed [2][3], and multigrid methods were introduced for an efficient simulation of large

V. Paliouras, J. Vounckx, and D. Verkest (Eds.): PATMOS 2005, LNCS 3728, pp. 267–276, 2005.

power grids [4][5]. Design and analysis methodologies for power distribution networks in high-performance microprocessors were presented in [6][7]. The L·dI/dt noise is originated from package and power rail inductance, and following the increasing circuit frequencies and decreasing supply voltages, it is becoming more and more important with each technology generation [8][10][11]. The impact of SSN can be reduced by inserting decoupling capacitors in the proximity of current-hungry blocks [9]. An analysis of all supply voltage fluctuation effects in DSM technologies was presented in [12]. A power supply voltage drop (i.e., IR-drop) can impact the delay of a library cell through one of the following mechanisms:

- A decrease in supply (V_{DD}) and/or increase in ground (GND) voltages diminish the local power supply of the cell, thus reducing the driving strength and increasing the delay;
- A relative V_{DD}/GND voltage shift between the driver and receiver cells on a circuit path will introduce a voltage offset impacting on the delay.

The traditional approach to consider the effect of power supply noise on circuit performances is twofold. Decoupling capacitances are used to limit the impact of SSN, while for the static IR-drop all operating voltages of library cells are uniformly scaled by the expected voltage reduction during characterization. Today, with existing Static Timing Analysis (STA) tools, all V_{DD} (GND) values are set to their minimum (maximum) value. Although this approach is very conservative, sometimes it can be optimistic since it ignores the voltage shift between the driver and receiver cells. Alternatively, the cell delay obtained from STA performed with power supply nominal value is linearly scaled based on the IR-drop values determined from a power network analysis step. However, both these approaches fail to consider the non-linear relation between cell delay and supply voltage, which is rapidly becoming a critical factor of performance degradation for designs of increasing complexity and reduced power supply values with large supply voltage variations. In this work we show that standard approaches cannot be used in present, and, as a consequence, in future technology nodes, and propose a practical methodology for including the power noise effects in STA based on the combination of delay derating factors and a selective library cell timing characterization and remapping for specific supply voltage values. Our approach can be easily inserted into industrial post-layout timing verification flows, and yields a more accurate timing analysis with respect to existing methods.

This paper is organized as follows: in Section 2 approaches that consider the supply variations on delay are overviewed, while in Section 3 the impact of power supply noise on circuit performances is analyzed in detail. Section 4 presents our approach to accurately evaluate the effect of supply voltage fluctuations in post-layout timing analysis, and experimental results showing the effectiveness of our methodology in the timing verification of an industrial SoC design are reported in Section 5. Finally, Section 6 summarizes some conclusive remarks.

2 Previous Work

The problem of considering power supply noise in timing verification was previously addressed. A statistical model that can potentially analyze both static IR-drop and dynamic SSN originated from the package inductance was proposed in [13]. The model is integrated into a statistical timing analysis framework to estimate the performance degradation induced by the power/ground (P/G) noise. Since the power supply voltage fluctuations are pattern-dependent, in [15] a statistical method based on genetic algorithms for path selection and pattern generation for worst-case P/G noise was proposed. In [14] the IR-drop effects on the delay, skew, and slew rates of the clock signals were analyzed. Although a relationship based on the sensitivity of delay to power supply predicted that an $N\%$ change in supply voltage yields approximately an $N/2\%$-$N\%$ change in clock timing, the experimental results demonstrated up to 30% skew variations due to 10% IR-drop value. An iterative approach was proposed, based on linear simulations of the power grid with tap currents, and non-linear simulations of the clock distribution network to model the current sources. In [16] linear closed-form formulas based on the short-channel transistor model [17] were obtained for incremental delay and slope variations of buffers. The increased sensitivity of buffers to P/G noise is dictated by the power supply levels scaling trends, and by the improved device transconductance. The formulas can be included into STA tools to evaluate the P/G noise-induced delay of buffers/repeaters driving heavily loaded nets, but cannot be extended for other library cells. Moreover, the linearity assumption holds when power supply and ground voltage variations are within 10% of their nominal value.

In the work presented in [18] the supply voltages of the cells along a signal path are viewed as independent variables, in order to determine the combination giving the worst-case path delay. The worst-case supply voltage configuration for the cell delay is determined by means of sensitivity analysis. The cell delay and output slope dependence on supply voltages is assumed to be near-linear in a narrow range. Hence, delay and output transition time are represented by a polynomial, whose coefficients for each point in the [input slope, load] tables are determined with a regression method. However, in [19] it was demonstrated that by applying the worst-case supply voltage simultaneously to all cells along a path is quite conservative, thus overestimating the overall path delay by a significant amount. An approach for computing the maximum path delay under power supply fluctuations was proposed in [19], where both the local supply voltage variations and the driver/receiver voltage shifts are considered. The models for the delay and output transition time variations are expressed as linear functions of the supply voltage deviations from their nominal values, and hold when the voltage fluctuations are within a small range of the nominal value. The problem of maximizing the path delay is formulated as a linear constrained optimization problem, where the constraints for the total chip power consumption and for the logic block currents are obtained with gate-level logic simulations. The approaches described in [18] and [19] address the problem of extending timing analysis in order to consider the effect of supply voltage variations on performances. However, when the fluctuations exceed a narrow interval around the power supply nominal value, they do not account for the non-linear dependence of cell delay (and output transition time), which is becoming more and more important in nanometer technologies.

3 Timing Analysis with Power/Ground Noise

Although the works overviewed in Section 2 proposed different approaches to include the P/G noise in timing analysis, today STA in industrial design flows does not correctly evaluate the impact of supply voltage fluctuations on performances, and a conservative methodology that assigns a single worst-case voltage drop/ground bounce value to all cells in the design is typically used. Alternatively, the IR-drop impact on delay is computed by means of the following procedure:

1. Perform STA with a static timing analyzer and obtain the SDF [20] file with the library cell propagation delays related to nominal power supply values;
2. Perform a power network analysis to generate the IR-drop map and evaluate the voltage drop value for each cell;
3. Update the propagation delay for each cell timing arc based on the power supply values obtained at Step 2, and generate an updated SDF' file;
4. Perform STA with the new SDF' file obtained at Step 3, and recompute the path delays.

With the updated SDF' file, STA can be performed considering the impact of power supply voltage fluctuations on circuit performances. Updating the cell propagation delays at Step 3 is carried out by derating the delay evaluated at power supply corner values. Derating based on K-factors is linear:

$$D_{NEW} = D_{NOMINAL} \cdot (1 + K \cdot \Delta V), \tag{1}$$

where $\Delta V = V_{DD\text{-}NEW} - V_{DD}$. However, the relation between delay and power supply is non-linear, as shown in Figure 1, where the supply voltage variations impact on propagation delay for a three-basic inverter chain at different technology nodes (i.e., 180nm, 130nm, 90nm) is illustrated. The power supply voltage drop also impacts on the transition time at the output of the library cells, thus increasing the path delays of the downstream logic cone. As illustrated in Figure 2 technology scaling aggravates the output slope degradation due to supply voltage fluctuations. The voltage supply is varied around the nominal value for each technology. It is clear that as technology scales down (130nm and 90nm) the delay variations increase along with the curve slope, meaning that also the cell delay sensitivity to supply voltage increases. Moreover, the assumption that delay variations originated from supply voltage changes around the nominal value is linear only holds for a narrow voltage range. As a consequence, both the procedure described above based on derating factors (1), and the methods presented in [18] and [19] based on a linear dependence of the delay (and transition time) variations from power supply drop, can be highly inaccurate when the variation exceeds a narrow range around the nominal supply value, which in [18] and [19] is assumed ±10%[1] with respect to the nominal value. The results illustrated in Figure 1 and Figure 2 show that such range is bound to shrink further as technology moves deeper into the nanometer regime. In Section 5 such restricted range where derating factors yield an accurate delay evaluation is demonstrated on an industrial design. Furthermore, supply voltage drop and ground bounce introduce also the following second-order effects, which cannot be adequately captured by linear derating factors:

• Non-homogeneous cell sensitivity to P/G noise at different characterization points;

---

[1] Although this range was obtained empirically and in general it might not be accurate, it is commonly used for derating cell delay

- Sensitivity variation to different P/G-noise states[2];
- Non-linear P/G-noise sensitivity (especially for simultaneous V_{DD} and GND variations).

It is important to notice that in STA the overall path delay is not simply obtained by adding the cell and interconnect propagation delays along the path, but the augmented output transition time due to supply voltage drop further slows down the downstream cells, as the transition time degradation is progressively amplified along the path by the supply voltage drop. In order to develop a robust methodology for post-layout timing verification that includes the impact of P/G noise, the non-linear relation between the cell delay and the supply voltage fluctuations must be taken into account.

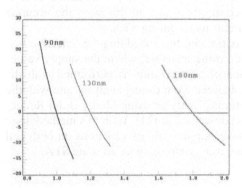   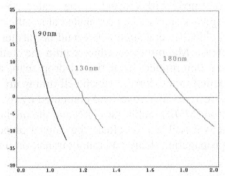

Fig. 1. Propagation delay (%) variations vs. supply voltage variations

Fig. 2. Transition time (%) variations vs. supply voltage variations

4 Selective Power/Ground Noise Mapping

While derating factors or delay-variation linear models can be used to include the P/G noise effects in timing analysis for small supply variations, for large fluctuations the non-linear dependence of delay from supply level dominates, and these models are not adequate.

In this work we propose a practical approach called *Selective P/G Noise Mapping*, which considers the non-linear relation between cell delay and P/G noise with large supply variations. Our approach is based on a specific library cell characterization, based on the power supply values obtained from IR-drop analysis, which is carried out *on-the-fly* during STA. It can be outlined as follows:

1. Perform a power network analysis in order to generate the IR-drop map and evaluate the voltage variation for each cell;
2. Analyze the power supply variations ΔV_{DD} (and ΔGND) for each cell;
3. Only for the cells where the P/G variations exceed the linear range ($\pm10\%$) around the supply nominal value perform a specific timing recharacterization with the supply values obtained from the IR-drop analysis of Step 1;

---

[2] Separate V_{DD} or GND fluctuations, or a combination of the twos

4. Generate a companion P/G-noise library with the cells recharacterized at Step 3;
5. Selectively remap the original gate-level netlist replacing the original library cells with the corresponding cells in the P/G-noise library;
6. Perform STA using the hybrid gate-level netlist obtained at Step 5.

In our approach we are not recharacterizing a complete cell library under all possible P/G combinations, which would be impractical giving the size of modern cell libraries and the P/G configuration space. In contrast, we are carrying out a selective characterization only for those cells whose power supply variation *exceeds* the limits where the delay dependence on supply voltage can be considered as linear (and consequently derating factors can be used), and only for the specific P/G noise configuration obtained from the power network analysis performed at Step 1. It is important to observe that not only the propagation delay, but also the output transition time is measured with respect to the scaled power supply value. In this way, the accurate signal slope can be propagated along the circuit paths during STA.

Ideally, our approach would be further optimized by embedding the Selective P/G Noise Mapping procedure within the static timing analyzer, where the supply values are determined from the IR-drop map, and characterization is performed with the actual input slope for each cell timing arc obtained from timing analysis, and with the effective capacitive load computed by the timing analyzer using Model Order Reduction (MOR) techniques [22] and the method described in [21]. With this methodology it is possible to evaluate the impact of power supply voltage variations on both cell propagation delay and output transition time, thus performing an accurate STA.

5   Experimental Results

We have applied Selective P/G Noise Mapping for the sign-off timing verification of an industrial SoC design in 0.13μm technology. The circuit implements a printer application; it consists of more than 340K library cells, and includes several memory blocks, a microprocessor core, and different clock domains. The maximum IR-drop/ground bounce values for different toggle rates[3] are reported in Table 1, where only a small switching activity causes limited P/G supply variations (around the ±10% range around the nominal supply value). In contrast, when the switching activity (toggle rate) increases, the supply variations soon exceed the linear range boundaries[4].

Table 1. Worst-case P/G supply variations

Switching Activity	5%	10%	30%
ΔIR-drop	106mV	148mV	318mV
ΔGround bounce	119mV	168mV	362mV

[3] The toggle rate is defined as the number of signal transitions per clock cycle. For example, a clock buffer performs two different transitions each cycle, and it has a toggle rate of 2. In contrast, the output of a D Flip-Flop typically changes value after clocking only 50% of the times; hence, it has a toggle rate of 0.5.

[4] It is important to notice that experiments with high values of switching activity were carried out with the purpose of evaluating the power supply voltage behavior

Table 2 shows that for increasing switching activity timing analysis[5] based on derating factors introduces larger errors with respect to the real impact of voltage supply changes on delay, and it can be optimistic thus failing to detect critical delay violations during the sign-off verification phase. When the toggle rate increases there are a larger number of cells whose actual power supply surpasses the narrow range around the nominal value where derating factors yield a sufficiently accurate timing analysis. Therefore, the cell delay (and transition time) non-linearity with respect to large supply voltage variations cannot be neglected.

Table 2. Delay derating accuracy degradation with increasing switching activity

Toggle Rate	0% (Nominal P/G)	5%	10%	30%
PrimeTime™	1.375ns	1.428ns	1.499ns	1.530ns
ELDO™	1.378ns	1.452ns	1.488ns	1.685ns
Error%	0.2%	1.6%	2.6%	9.2%

We have applied our Selective P/G Noise Mapping procedure to the post-layout timing verification, performing STA with PrimeTime™ [24]using derating factors and the method described in Section 3, and the remapped gate-level netlist obtained with our approach presented in Section 4.

We computed the path delay percent variation obtained from larger power supply fluctuations caused by an increased switching activity, with respect to the path delay computed with the supply nominal value. Table 3 and Table 4[6] report the path delay variations caused by P/G noise for four critical paths extracted from a clock domain working at 250MHz in the SoC design, and demonstrate that delay changes derived with our approach are in good accordance with the variations obtained with ELDO™. In contrast, the derating factors present a large error for increasing supply variations (larger switching activity). The experimental results obtained from the analysis of an industrial SoC design in 0.13μm technology, demonstrate that approaches aimed at considering the P/G noise impact on timing based on derating factors or linear models yield accurate results only for small values of supply variations, and the non-linear relation between the cell delay and supply voltage must be accurately evaluated. We implemented our methodology by means of scripts and applied it to the post-layout STA, thus detecting several delay violations that were unnoticed with the standard approach based on derating factors.

When analyzing the experimental results presented in Table 3 and Table 4, it is worth noticing that even a voltage drop range within 10% of the nominal power supply may exceed the limits where derating factors yield and accurate cell delay evaluation. Such range does not have an absolute value, but it was derived empirically and it has been commonly used both in academia and industry. In this work we have considered the same range for the sake of clarity. Therefore, it may be necessary to re-

---

[5] Timing analysis was performed with the static timing analyzer PrimeTime™ [24] and circuit simulations with ELDO™ [23]

[6] In Table 3 and Table 4 we reported the delay percent variation instead of the path delay value, since the purpose of our experiments was not to compare PrimeTime™ against ELDO™, but to demonstrate that our approach can accurately capture the impact of large power supply fluctuations on path delay

duce this empirical range, without impairing the effectiveness of the Selective P/G Noise Mapping methodology.

Table 3. Derating vs. Selective P/G Noise Mapping delay percent (%) variation (Path_A and Path_B)

	Path_A		Path_B	
Switching Activity	10%	30%	10%	30%
PT^7 *Derating Factors*	4.0%	12.7%	3.7%	12.4%
PT *Selective P/G Noise Mapping*	6.7%	23.9%	5.3%	26.7%
ELDO™	8.0%	26.5%	6.8%	25.1%

Table 4. Derating vs. Selective P/G Noise Mapping delay percent (%) variation (Path_C and Path_D)

	Path_C		Path_D	
Switching Activity	10%	30%	10%	30%
PT *Derating Factors*	2.7%	11.0%	2.4%	8.6%
PT *Selective P/G Noise Mapping*	6.4%	28.0%	5.6%	25.9%
ELDO™	7.2%	28.7%	5.4%	27.7%

6 Conclusions and Future Work

As technology moves deeper into the nanometer regime, the power supply-induced noise is a critical source of functional failures, reliability and performance degradation in large SoC designs, which must be accurately evaluated in post-layout verification. Current approaches to consider P/G noise in static timing analysis are showing limits, since they are either very pessimistic, or can only be used when the supply fluctuations are within a narrow range around the nominal value.

In this paper we have presented a practical approach to include the P/G noise impact in sign-off timing verification, based on the combination of delay variation linear modeling, and selective library cell characterization for IR-drop and remapping, in order to accurately consider the non-linear dependence of cell delay on large supply voltage fluctuations. Our approach has been validated in the sign-off timing verification of an industrial SoC design in 0.13μm technology, demonstrating the potential inadequacies of current methods and tools.

Future work will focus on embedding this methodology within a static timing analyzer. We believe that in order to accurately consider the non-linear relation between the cell delay and supply voltage fluctuations, a hybrid methodology including derating factors and *ad-hoc* selective P/G noise cell characterization and remapping must be further explored and included into the post-layout verification phase.

References

1. H. H. Chen and D. D. Ling, "Power Supply Noise Analysis Methodology for Deep-Submicron VLSI Chip Design," in *Proc. of Design Automation Conf.*, Jun. 1997, pp. 638-647.

---

^{7} PrimeTime™

2. H. H. Chen and J. S. Nealy, "Interconnect and Circuit Modeling Techniques for Full-Chip Power Supply Noise Analysis," *IEEE Trans. on Components, Packaging, and Manufacturing Technology-Part B*, vol. 21, pp. 209-215, Aug. 1998.

3. M. Zhao, R. V. Panda, S. S. Sapatnekar, and D. Blaauw, "Hierarchical Analysis of Power Distribution Networks," *IEEE Trans. on Computer-Aided Design*, vol. 21, pp. 159-168, Feb. 2002.

4. S. R. Nassif and J. N. Kozhaya, "Fast Power Grid Simulation," in *Proc. of Design Automation Conf.*, Jun. 2000, pp. 156-161.

5. J. N. Kozhaya, S. R. Nassif, and F. N. Najm, "Multigrid-like Technique for Power Analysis," in *Proc. of Intl. Conf. on Computer-Aided Design*, Nov. 2001, pp. 480-487.

6. M. K. Gowan, L. L. Biro, and D. B. Jackson, "Power Considerations in the Design of the Alpha 21264 Microprocessor," in *Proc. of Design Automation Conf.*, Jun. 1998, pp. 726-731.

7. A. Dharchoudhury, R. Panda, D. Blaauw, and R. Vaidyanathan, "Design and Analysis of Power Distribution Networks in PowerPC™ Microprocessors," in *Proc. of Design Automation Conf.*, Jun. 1998, pp. 738-743.

8. P. Larsson, "Resonance and Damping in CMOS Circuits with On-Chip Decoupling Capacitance," *IEEE Trans. on CAS-I*, vol. 45, pp. 849-858, Aug. 1998.

9. S. Bobba, T. Thorp, K. Aingaran, and D. Liu, "IC Power Distribution Challenges," in *Proc. of Intl. Conf. on Computer-Aided Design*, Nov. 2001, pp. 643-650.

10. H.-R. Cha and O.-K. Kwon, "An Analytical Model of Simultaneous Switching Noise in CMOS Systems," *IEEE Trans. on Advanced Packaging*, vol. 23, pp. 62-68, Feb. 2000.

11. K. T. Tang and E. G. Friedman, "Simultaneous Switching Noise in On-Chip CMOS Power Distribution Networks," *IEEE Trans. on VLSI Systems*, vol. 10, pp. 487-493, Aug. 2002.

12. A.H. Ajami, K. Banerjee, A. Mehrotra, and M. Pedram, "Analysis of IR-Drop Scaling with Implications for Deep Submicron P/G Network Design," in *Proc. of ISQED*, Mar. 2003, pp. 35-40.

13. Y.-M. Jiang and K.-T. Cheng, "Analysis of Performance Impact Caused by Power Supply Noise in Deep Submicron Devices," in *Proc. of Design Automation Conf.*, Jun. 1999, pp. 760-765.

14. R. Saleh, S. Z. Hussain, S. Rochel, and D. Overhauser, "Clock Skew Verification in the Presence of IR-drop in the Power Distribution Network," *IEEE Trans. on Computer-Aided Design*, vol. 19, pp. 635-744, Jun. 2000.

15. J.-J. Liou, A. Krstic, Y.-M. Jiang, and K.-T. Cheng, "Path Selection and Pattern Generation for Dynamic Timing Analysis Considering Power Supply Noise Effects," in *Proc. of Intl. Conf. on Computer-Aided Design*, Nov. 2000, pp. 493-496.

16. L. H. Chen, M. Marek-Sadowska, and F. Brewer, "Buffer Delay Change in the Presence of Power and Ground Noise," *IEEE Trans. of VLSI Systems*, vol. 11, pp. 461-473, Jun. 2003.

17. T. Sakurai and A. R. Newton, "Alpha-Power Law MOSFET Model and its Applications to CMOS Inverter Delay and Other Formulas," *IEEE Trans. on Computer-Aided Design*, vol. 25, pp. 584-594, Apr. 1990.

18. R. Ahmadi and F. N. Najm, "Timing Analysis in Presence of Power Supply and Ground Voltage Variations," in *Proc. of Intl. Conf. on Computer-Aided Design*, Nov. 2003, pp. 176-183.

19. S. Pant, D. Blaauw, V. Zolotov, S. Sundareswaran, and R. Panda, "Vectorless Analysis of Supply Noise Induced Delay Variation," in *Proc. of Intl. Conf. on Computer-Aided Design*, Nov. 2003, pp. 184-191.

20. Standard Delay Format Specification, Open Verilog International, Version 3.0, May 1995.

21. F. Dartu, N. Menezes, and L. T. Pileggi, "Performance Computation of Precharacterized CMOS Gates with RC Loads," *IEEE Trans. on Computer-Aided Design*, vol. 15, pp. 544-553, May 1996.
22. A. Odabasiouglu, M. Celik, and L. T. Pileggi, "PRIMA: Passive Reduced-order Interconnect Macromodeling Algorithm," in *Proc. Intl. Conf. on Computer-Aided Design*, Nov. 1997, pp. 58-65.
23. ELDO™ User Guide, Mentor Graphics, Inc., 2001.
24. PrimeTime™ User Guide, Synopsys, Inc., 2003.

Switching Sensitive Driver Circuit to Combat Dynamic Delay in On-Chip Buses

Roshan Weerasekera[1], Li-Rong Zheng[1],
Dinesh Pamunuwa[2], and Hannu Tenhunen[1]

[1] Laboratory of Electronics and Computer Systems (LECS),
KTH Microelectronics and Information Technology,
ELECTRUM 229, 164 40 Kista, Sweden
{roshan,lirong,hannu}@imit.kth.se
[2] Centre for Microsystems Engineering, Faculty of Applied Sciences,
Lancaster University, Lancaster LA1 4YR, UK
d.pamunuwa@lancaster.ac.uk

Abstract. In this paper, we propose a novel Interconnect Driver circuit scheme for on-chip bus structures, which changes it's drive strength based on the switching pattern of the neighbouring interconnect. The circuit is quite simple compared to driver circuits proposed in the literature. The results show that for the cost of a few transistors, the proposed driver circuit has a wider eye opening (upto a 100% improvement) and reduced jitter (up to a 32% reduction) than a traditional driver for typical DSM technologies.

1 Introduction

Advances in interconnect technologies such as an increase in the number of metal layers, stacked vias, and reduced routing pitch, have played a vital role for continuous improvement of integrated circuit density and operating speed. However, several parasitic effects jeopardize the benefits of scaling.

The interconnects in deep submicron technologies are typically very lossy so that the RC delay dominates. In order to keep the resistance to a minimum, the aspect ratio (height/width) of wires is increased, which gives rise to increased interwire coupling capacitance. This coupling capacitance results in crosstalk which has an effect on the delay, depending on how the aggressor lines switch. To consider two extreme cases, the effective coupling capacitance between two nodes is zero if there is a transition at both nodes at the same time and in the same direction, while it is twice the coupling capacitance if the transition is in the opposite direction (i.e. the Miller effect). In general, the effective coupling capacitance of a signal trace is determined by the signal arrival times and slew times of coupled lines. In [ ], it is estimated that the worst-case effective capacitance is 3 times the coupling capacitance (C_c) for linear ramp voltages, and may be even worse for exponential waveforms. Hence there is a huge variation in charging time for the best- and worst-case situations.

V. Paliouras, J. Vounckx, and D. Verkest (Eds.): PATMOS 2005, LNCS 3728, pp. 277–285, 2005.
© Springer-Verlag Berlin Heidelberg 2005

Reducing the coupling length by inserting buffers is a widely used methodology to reduce not only the RC dependent delay of long interconnects, but also crosstalk noise [ ]. Inserting buffers or repeaters into a design is complicated from a design perspective -especially for semi-global level or global level wires- because they require vias which take up space and complicate the layout process. Also the overall performance is very sensitive to their placement, and the repeaters consume more power. Wire shielding and increasing the inter-wire spacing are other well known options to reduce the effect of crosstalk [ ]. Active shielding methodology employs two shield wires on both sides of the target wire, and maintains the same transition direction as the target wire for fast propagation [ ]. However wire shielding methods use up valuable wiring resources.

As in advanced CMOS technologies, transistors are less expensive than wires, it is more desirable to deal with crosstalk effects by using special circuit techniques. The transition sensitive accelerator [ ] is an alternative to buffer insertion, which works by sensing the transitions on the line and accelerating them by connecting the output node to the relevant logic level. It uses highly skewed inverters, which have a lower noise margin, to directly drive the output. The booster technique proposed in [ ] injects or sinks more current depending on the transition, and a clear timing analysis for interconnects is not shown.

The driver technique proposed in [ ] drives the interconnect in a different manner so that the jitter, due to the effective capacitance variation, is reduced. The repeater circuit proposed in [ ] improves the bus speed by reducing/increasing the threshold of repeaters detecting the switching pattern which may impact the noise margin. Also for this circuit a separate bias circuit is necessary for well and substrate of the pull-up and pull-down transistors.

In this paper we propose an adaptive interconnect driver circuit called Switching Sensitive Interconnect Driver (SSID), which senses the switching of its neighbours to increase/decrease drive capability with an assistant driver. Our driver circuit is quite simple, and the cost for this technique is to use a second driver and couple of minimum sized logic gates.

2   Switching Sensitive Interconnect Driver Circuit

The Switching Sensitive Interconnect Driver (SSID) Circuit proposed in this paper (Figure 1) consists of two drivers: the Main driver, which is similar to a traditional driver (inverter), and an Assistant driver, which can be turned on or off depending on the switching patterns of the aggressor and the victim by means of a logic circuit called *selector*. When the input signals for the victim and the aggressor(s) are in opposite directions, then the assistant would join with the main driver to drive the interconnect, because the effective capacitance is higher, as is mentioned in the introduction. But, when the input signals are in the same direction, the assistant would remain quiet, so the response of the SSID is similar to that of a traditional driver.

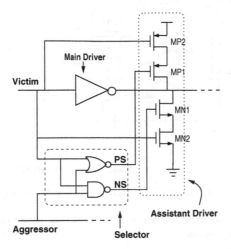

Fig. 1. Circuit Schematic of Switching Sensitive Interconnect Driver for two parallel wires

Table 1. Output of the Selection logic at Various Switching Patterns

Victim State	Aggressor State	Selector Outputs	
		PS	NS
0	0	1	1
0	1	0	1
1	0	0	1
1	1	0	0

2.1   Operation of the SSID

For the sake of simplicity, two parallel wires are considered to explain the operation of the SSID (Figure 1). The selection logic circuitry to control the assistant driver is implemented as:

$$PS = \overline{V_{in} + Agg_{in}}$$
$$NS = \overline{V_{in} \cdot Agg_{in}}$$

where V_{in} means input to the victim line (or the line we are interested in) and Agg_{in} is the input to the aggressor (or the neighbour).

When the inputs are opposite in phase and the victim is in High-to-Low (Low-to-High) transition, PS is at logic zero and NS is at logic one, switching on both MP1 and MN1. Hence the output makes a Low-to-High (High-to-Low) transition through MP2 (MN2). With in phase switching of aggressor and victim, if the victim makes a High-to-Low (Low-to-High) transition, both PS and NS are at logic one (zero), disconnecting MP2 (MN2) from the interconnect. The transistors MN1 and MP1 are mainly used as switches.

The inter-resource communication link for a Network-on-Chip will most likely consists of a larger number of parallel wires, with uniform coupling [ ]. For such

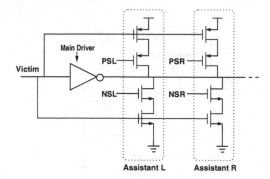

Fig. 2. Circuit Schematic of SSID for victim with two neighbours

Table 2. Effective capacitance variation and operating mode of the Assistant Driver at Various Switching Patterns

Case	Switching Pattern	SF(k)	Status	
			AssistantL	AssistantR
1	$(\downarrow\uparrow\downarrow)$, $(\uparrow\downarrow\uparrow)$	4	ON	ON
2a	$(-\uparrow\downarrow), (-\downarrow\uparrow)$	3	XX	ON
2b	$(\uparrow\downarrow\ -), (\downarrow\uparrow\ -)$		ON	XX
3a	$(-\uparrow\ -), (-\downarrow\ -)$	2	XX	XX
3b	$(\uparrow\uparrow\downarrow), (\downarrow\downarrow\uparrow)$		ON	OFF
3c	$(\uparrow\downarrow\downarrow), (\downarrow\uparrow\uparrow)$		OFF	ON
4a	$(-\uparrow\uparrow), (-\downarrow\downarrow)$	1	XX	OFF
4b	$(\uparrow\uparrow\ -), (\downarrow\downarrow\ -)$		OFF	XX
5	$(\uparrow\uparrow\uparrow), (\downarrow\downarrow\downarrow)$	0	OFF	OFF

a regular wire fabric or a bus configuration, there are two neighbours for middle wires, and for this case, two assistant drivers have to be utilized (see Figure 2). The *AssistantL* is driven from the control signals, PSL and NSL, generated by selection logic between victim and the left aggressor, while *AssistantR* is controlled by the selection logic between victim and the right aggressor. Also, for such a structure the effective capacitance is $C_{ground} + kC_c$ where k may vary between 0 and 4 for typical inputs. Table 2 describes the possible switching patterns and the status of each assistant driver. Here, $\uparrow, \downarrow$ and $-$ are used to represent transitions from zero to one, one to zero, and no transition, respectively. For example the pattern $(\uparrow\downarrow\downarrow)$ describes the transitions of the left aggressor, victim, and right aggressor respectively. The status XX of Assistant Driver(s) depends on the logic value of the quiet aggressor.

3 Driver Comparison and Results

3.1 Simulation Setup

Figure 3 shows the basic simulation setup used for comparing the Switching Sensitive Interconnect Driver (SSID) with a traditional driver. For the simula-

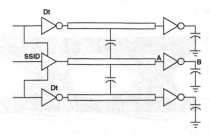

Fig. 3. Simulation setup

tions, we used CADENCE Spectre and UMC 0.18-μm-CMOS process with 1.8 V supply voltage. A 40X inverter has been used as the traditional driver (where minimum or 1X inverter is $W_{pmos} = 2.7 * W_{nmos}$ and $W_{nmos} = 240\ nm$). The main driver of the SSID is also a 40X inverter, and the transistors MN2 and MP2, which forms Assistant Driver of SSID, is selected as same as the NMOS and PMOS transistors of Main Driver respectively. A π5 distributed RC model was used to model the interconnect, with $C_{ground} = 331\ fF/mm$, $C_c = 186\ fF/mm$ and $R = 107\ \Omega/mm$ considering the values for a semi-global wire in 0.18 μm technology [ ]. The driven interconnect is terminated with an inverter at the far-end. To make a fair comparison, the neighbour is driven with a 50X inverter. To get the response of the traditional driver, the same set up was used without the additional driver and the selector.

3.2   Results and Discussion

The effect of distortion on the transmitted signal can be demonstrated by an eye diagram at the output of the victim net, built up over hundreds of cycles, with different pseudo-random bit streams (PRBS) being fed to the three lines. The eye opening indicates the amount of voltage and timing margin available to sample the signal. To easily distinguish between zeros and ones, we should get a clear eye opening in the middle. Figure 4 shows the eye diagram at the output of the victim net for an interconnect driven by a traditional driver, and Figure 5 shows that for an interconnect driven by the SSID. For the shown case in Figure 5, the maximum eye height of the proposed driver is 1.375 times wider than that for the traditional, and the reduction of jitter is 28%.

The variation of maximum eye height and jitter with C_c/C_{ground} for the far-end waveform of the interconnect driven by the SSID is plotted in Figure 6 (a). It is well known that with increasing C_c/C_{ground}, maximum eye height should decrease and the jitter should increase. In terms of physical design parameters the closer the wires to each other, the higher the jitter and the lower the eye height.

From the graphs in Figure 6, the adaptive nature of the SSID is evident; as the coupling quantified by the C_c/C_{ground} ratio changes, the assistant and main drivers share the load appropriately, so that the eye opening is constantly wider than for a traditional driver. The fact that the assistant drivers can switch

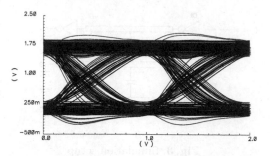

Fig. 4. Eye Diagram at the far-end (Point A) with a traditional driver

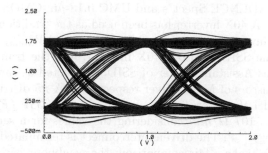

Fig. 5. Eye Diagram at the far-end (Point A) with the Switching Sensitive Interconnect Driver

on selectively allows a much higher drive strength when needed. With a single driver, the drive strength has to be chosen as a medium for the best- and worst-cases, so that the drive strength for the heavily coupled, worst-case switching patterns are generally insufficient, while it is too high for the best-case switching patterns. These graphs show the effectiveness of the SSID for wires with heavy coupling; in other words, the SSID allows wires to be placed closer together with fewer performance penalties. The graph in Figure 7 depicts the eye opening and jitter for the SSID normalized to those for the traditional driver. That is for example for jitter, the values shown represent the jitter for the SSID divided by the jitter for the traditional driver at different coupling strengths. The arrows indicate the appropriate axis.

Figure 8 explores the variation of eye opening and jitter with line length for the technology described in Section 3.1, for both drivers. It can be seen that the eye opening completely closes due to attenuation and signal degradation after 1.5 mm for the traditional driver, and at 2 mm for the SSID.

4 Conclusion

A concept for an interconnect driver methodology, which senses the switching of aggressors is proposed and tested mainly for inter-resource communication links in Network-on-Chip type of architectures for nanometer technologies. We have demonstrated the ability of the SSID to achieve a higher data rate. For example,

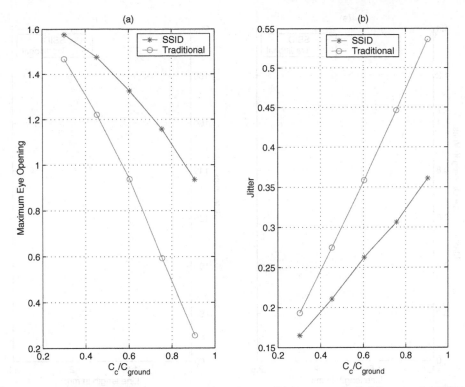

Fig. 6. Variation of (a) Maximum Eye Height and (b) Jitter with C_c/C_{ground}, for SSID and Traditional Driver

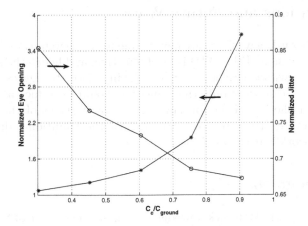

Fig. 7. Maximum Eye opening and Jitter normalized to that of the traditional driver

for a value of 0.75 for C_c/C_{ground} which is typical for semi-global interconnections in DSM circuits, there is an improvement of 100% in the maximum eye height and 32% in the jitter over a traditional driver.

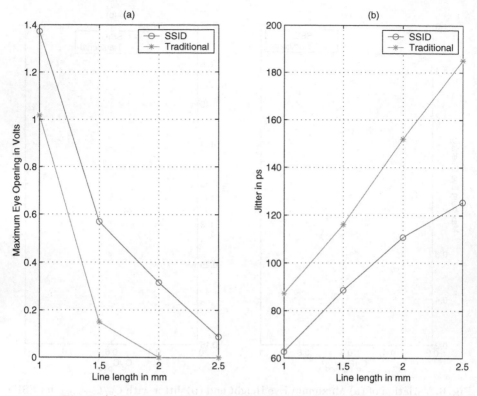

Fig. 8. (a) Maximum Eye height (b) Jitter for various line lengths

5   Limitations and Future Work

The Crosstalk-on-Delay effects are mainly due to the switching of the neighbours. However this version of the SSID is level sensitive, and hence the selector circuit turns on the Assistant Driver even when the input for the aggressors are steady. A circuit which can sense only the transition would introduce a reasonable improvement for the overall performance. Also comparisons of the bandwidth of the bus structure driven by the SSID against the traditional driver would be useful. This work is earmarked for future work.

Acknowledgment

The authors would like to acknowledge the financial support of SIDA under the auspices of the research capacity building project at the Department of Electrical and Electronic Engineering, University of Peradeniya, Sri Lanka.

References

1. Khang, A.B., Muddu, S., Sarto, E.: On switch factor based analysis of coupled rc interconnects. Proceedings of IEEE Int Design Automation Conference (2003) 79–84
2. Backoglu, H.B.: Circuits, Interconnections and Packaging for VLSI. Addison-Wesley (1990)
3. Pamunuwa, D., Zheng, L.R., Tenhunen, H.: Maximizing throughput over parallel wire structures in the deep submicrometer regime. IEEE transactions on Very Large Scale Integration (VLSI) Systems 11 (2003) 224–243
4. Kaul, H., Sylvester, D., Blaauw, D.: Active shields: A new approach to shielding global wires. Proceedings of the 12th ACM great Lakes Symposium on VLSI (2002) 112–117
5. Iima, T., Mizuno, M., Horiuchi, T., Yamashina, M.: Capacitance coupling immune, transient sensitive accelerator for resistive interconnect signals of subquarter micron ulsi. IEEE Journal of Solid-State Circuits 31 (1996) 531–536
6. Nalamalpu, A., Sirinivasan, S., Burleson, W.P.: Boosters for driving long onchip interconnects-design issues, interconnect synthesis, and comparison with repeaters. IEEE Transactions on Computer-Aided Design of Integrated Circuits and Systems 21 (2002) 50–62
7. Weerasekera, R., Zheng, L.R., Pamunuwa, D., Tenhunen, H.: Crosstalk immune interconnect driver design. Proceedings of International Symposium on System-on-Chip (2004) 139–142
8. Katoch, A., Jain, S., Meijer, M.: Aggressor aware repeater circuits for improving on-chip bus performance and robustness. European Solid-State Circuits, 2003. ESSCIRC '03. Conference on (2003) 261–264
9. Liu, J., Zheng, L.R., Tenhunen, H.: Interconnect intellectual property for network-on-chip(noc). Journal of Systems Architecture 50 (2004) 65–79
10. Tenhunen, H., Zheng, L.R.: Introduction to electrical issues in soc/sop. Workshop Notes (2002)

PSK Signalling on NoC Buses

Crescenzo D'Alessandro, Delong Shang, Alex Bystrov, and Alex Yakovlev

Microelectronics System Design Group
School of EECE, University of Newcastle upon Tyne, NE1 7RU, UK
{crescenzo.d'alessandro,delong.shang,a.bystrov,alex.yakovlev}@ncl.ac.uk

Abstract. A novel self-timed communication protocol is based upon phase-modulation of a reference signal. The reference and the data are sent on the same transmission lines and the data can be recovered observing the sequence of events on the same lines. The sender block consists of a reference generator and variable-delay elements, while the receiver includes a delay-locked loop for synchronization and a mutual exclusion element with additional logic (validity bit and FIFO) for data recovery. This protocol exhibits high robustness with respect to transient errors caused by narrow pulse interference, usually associated with crosstalk and radiation.

1 Introduction

The issue of fast and reliable communication fabric is crucial for the successful design of systems-on-chip. An approach to design of such communication fabric is the Network-on-Chip (NoC) [ ]. The synchronization of blocks is a non-trivial aspect of design, and research has delivered interfaces which are *self-timed* and *speed-independent* to address this problem. An example of self-timed interface can be found in [ ], where the communication between two separate clock domains is investigated. Transient errors due to cross-talk, cross-coupling, ground bounce or environment interference become more prominent as integration increases, and this effect can be observed interconnect wires, as underlined by the work of Dupont and Nicolaidis [,]. Quoting Nicolaidis: "it is predicted that single event upsets induced by alpha particles and cosmic radiation will become a cause of unacceptable error rates in future very deep submicron and nanometer technologies. This problem, concerning in the past more often parts used in space, will affect future ICs at sea level" [ ]. This motivates the fault tolerance approach to design. Unfortunately, the implementation of fault tolerance leads to hardware overheads, which in its turn reduces the reliability. It may happen that large systems of the future will be spending most time recovering from transient faults. In order to alleviate this problem, a simple fault masking approach was introduced in [ ]. It was applied to request-acknowledgement handshake protocols, exploiting the redundancy created by the feedback acknowledgement signal. In this paper we also exploit protocol redundancy, but feedback signals are not used. Both approaches leave a small percentage of errors unmasked. Although they significantly reduce error rate, the need in fault-tolerant design remains.

V. Paliouras, J. Vounckx, and D. Verkest (Eds.): PATMOS 2005, LNCS 3728, pp. 286–296, 2005.

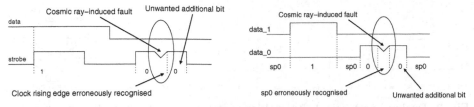

Fig. 1. Protocol errors induced by pulse interference. Left: synchronous single-rail system; right: dual-rail return-to-zero

Multiple-rail encoding is based on the transmission of data on two or more lines; one or more lines go high to indicate one of the possible combinations of data. Typically, these schemes employ one state as no-data (a *spacer*, usually zero). Usual single-rail and dual-rail transmission suffers from the effects of single event upsets (SEU), as the receiver in the channel could lose data, latch wrong data or lose synchronization. Consider, for instance, Figure 1 (left), where data is selected by a strobe. If a pulse is induced on the data line, the only way it could generate an error is if the pulse overlaps with the rising edge of a clock pulse. The occurrence of such an event has a lower probability than that of a narrow pulse upsetting the data line "far" from the strobe pulse. However, consider the case where the strobe signal is corrupted by the same narrow pulse: if the pulse is strong enough to be recognised as a valid transition, the receiver could latch additional unwanted data, possibly not only corrupting the data being sent at the time of the upset, but forcing the system to lose synchronization. In the case of dual-rail using a single spacer protocol, the upset could appear on either lines and still cause unwanted data to enter the communication channel, as in Figure 1 (right).

From the brief description given it is clear that more robust approaches are needed for these asynchronous communication channels. Dual-rail encoding using an alternating spacer protocol offers better resilience to errors; nevertheless, the general encoding is based upon the recovery of data from the value of the lines at a point in time. Our solution extends the concept of alternating spacer protocol to improve robustness on transmission lines. Our focus is on the reliability of on-chip communication channels particularly for NoCs and we propose a methodology to improve the resistance to the special case of SEUs affecting the communication fabric.

2   Approach

2.1   Dual-Rail Encoding

Dual-rail codes are designed so, that data is sent across two lines rather than a single one. The data is encoded by switching high or low one of the lines; the difference in level represents an item of data. The traditional dual-rail protocol employs a single spacer, whereby after each transmission of a valid bit of data the bus returns to the zero state. In [ ] the use of alternating spacers is introduced

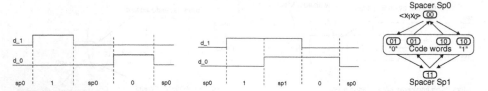

Fig. 2. Dual-rail protocols. Left: single-spacer; center, right: alternating-spacer waveforms and block diagram

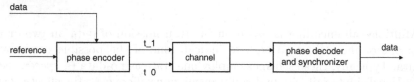

Fig. 3. Block diagram of the overall system

and several implications with respect to security issues (power signatures) are analysed. The paper concludes that the alternating spacer protocol (ASP) is good for security, the circuits implementing it are easy to synthesise in standard gates, and the whole approach can be integrated in the standard design flow. In this paper we use this protocol, as it allows the receiver to distinguish between SEUs and valid transitions.

2.2   PSK Approach

The proposed approach builds upon the ASP. The sender and the receiver, as in Figure 3, *can* be synchronous systems, albeit within completely uncorrelated clock domains; alternatively they can be fully-asynchronous blocks, or combinations of the two. The reference signal is used for sampling the data by its rising and falling transitions. The data being sent modulate the phase of the clock differently on each transmission line by controlling the variable delay elements (VDE). The receiver recovers the data by comparing the signals to each other. Data values are encoded as the sign of the phase difference of the clock signal on the transmission lines. Rather than measuring the phase difference, the receiver decodes the data by observing the sequence of events on the transmission lines. The receiver records the data upon the arrival of the first transition, but the bit validity is recorded *only when the next spacer settles on the transmission lines*. Therefore, the measure of interest is the *differential delay* introduced by the VDEs on the transmission lines, which we indicate with δ.

This differential delay introduce an *event window*, where an imbalance on the lines is present. The size of this event window is determined by δ, which indicates the "nominal value" for this window, and the jitter introduced by the channel on each line γ. Provided that the system is able to reject transient faults appearing *outside the event window*, one such fault will generate an error only if it happens within the window. Effectively, we reduce the event window to a minimum in order to minimise the effect of transient faults, while still recognising data.

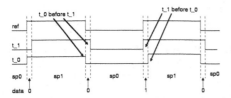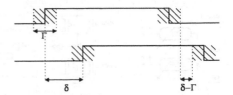

Fig. 4. Left: example of waveforms for Figure 3; right: effect of jitter on choice of δ

In order for the data to be correctly recovered, δ must be recognised at the receiver (but not measured); however if the channel introduces a *systematic differential delay* between the lines, transmission can be impaired. Careful design could minimize this problem, but in order to cancel out the effect of this systematic differential delay some synchronization between the two transmission lines at the receiver is needed. Ginosar and Kol [ ] describe the problem of *adaptive synchronization* and propose an adaptation protocol which employs a *training session*, where the circuit stops operating and the sender transmits dummy data to the receiver; other adaptation protocols are described. In any cases, to ensure reliability, the jitter γ introduced by the channel must be taken into account so that $\delta \gg \Gamma$ where $\Gamma = max(|\gamma|)$ in order to guarantee deterministic behaviour of the system. However, if T is the clock period, $\Gamma \ll T$ and therefore δ can be chosen so that $T > \delta \gg \Gamma$. The value of Γ can be estimated using various techniques [ ].

The fact that valid data is recorded only when the next spacer is generated (and therefore when both transmission lines have changed status) has an important property: a hazard on one of the transmission lines (generated by cross-talk, EM interference, cosmic radiation) will not be recorded as data. This is particularly important if several single-event upsets (SEU) can be generated by the environment; provided the events do not affect both lines *at the same instant*, the system will ignore the error.

An important property of the alternating-spacer protocol described in [ ], from which the following definitions are reported, is the minimisation of the *exposure time* (the variation in energy consumption when processing data values). This is the time where the energy imbalance is exhibited and it is shown in the same paper that the alternating-spacer protocol has a smaller exposure time than the single-spacer protocol, in particular, the lower bound is one gate delay and the upper bound is one clock cycle. The new approach proposed in this paper, however, minimizes the exposure time of the bus, so that the bounds depend on δ and Γ and the following inequality holds: $\delta + \Gamma > exposure\ time > \delta - \Gamma$. Note that using a single-spacer protocol, one could minimize the exposure time by minimizing the width of the pulses representing data; however, this has the drawback of increasing spectrum occupation by the data signal. In terms of robustness and predictability of behaviour, the system is such that a fault will become an error, if it appears through the edge of the second-arriving signal, and when a fault involves both lines at the same time. As a solution to the latter case, the designer could route the two lines so that they are physically apart if

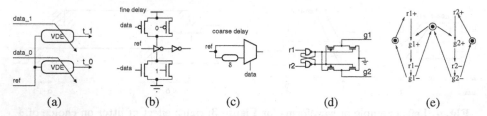

(a)   (b)   (c)   (d)   (e)

Fig. 5. Sender and receiver blocks and possible implementations. (a) dual-rail-based sender; (b) fine and (c) coarse VDE; (d) ME schematic and (e) STG

the synchronizer employed at the receiver has enough capture range. The first case has no solution; however, the probability of such an event occurring is proportional to the ratio of δ and the clock period. Therefore, the approach to avoid such errors would be to employ some high-level protocol to detect and/or correct errors.

3 Operation Principles

3.1 Sender

Figure 5 (a) shows a possible design of the sender having dual-rail input. The VDE implementations are shown in Figure 5 (b) and (c), where the coarse VDE can be used in our sender, and the fine VDE can be used to calibrate a mesochronous system [ ], which is one of application areas for our approach. For the latter, the two data lines can be set to 11 or 00.

3.2 Receiver

The receiver block has the task of recovering the data sent over the communication channel; using the approach described, this task is performed by a Phase Detector (PD). Much literature is devoted to the design and analysis of a number of different PDs. In our case, the quantity of interest is the timing relationship between two related events happening on the two transmission lines, and in particular only the sequence of events rather than their absolute distance in time. Therefore, the PD can be *binary quantized* or *lead/lag*, indicating that one event on one line leads or lags the other corresponding event on the other line.

A Mutual Exclusion element (ME), shown in Figure 5 (d) and (e) [,] is essentially an S-R flip-flop followed by a *metastability filter*, which allows the two inputs to be very close to each other in terms of arrival time. If the requirement $\delta \gg \Gamma$ is met, then the determinism of the system is guaranteed, and this will prevent the S-R flip-flop from entering metastability. However, as the jitter is a random variable, the time $\delta - \gamma$ could be below the expected value, presenting a metastability hazard, hence the use of an ME.

The ME will only recognize events on the rising edge of the inputs; in order to follow the protocol correctly, the falling edge of the input must also encode data.

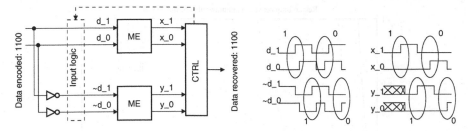

Fig. 6. Double ME receiver

Hence the PD is built around two MEs, one of which has the inputs inverted, as shown in Figure 6. To complete the PD, additional logic is required which perform event detection and indicates the presence of a valid bit at the output. This logic is a combinational function of the input of the PD and the outputs of the MEs. It uses memory elements to ensure stable inputs to the ME and to avoid using more complex circuitry only at the outputs of the MEs.

As in any circuit decoding Phase Modulation, the receiver will require some phase alignment system, in order to make sure that the two incoming modulated clock signals are synchronized. In order to perform phase alignment, a DLL can be used and the PD can be shared between the data recovery system and the DLL, which will resemble a lead/lag-type PLL described in [ ]. For this type of PLLs a class of filters called *sequential filters* is described, which has the attractive property of being governed by statistical equations and, importantly, by a set of observed values rather than a linear combination of a set of inputs. Previous work by the author [ ] illustrates a DLL based on such filters that can be employed in this case.

3.3   Repeaters and Bridges

The system relies on the signals seen from the receiver's PD being aligned with respect to their relative phase. This can deteriorate in long wires. As a counter-measure, the use of repeaters or bridges can be considered. These will recover the data voltage and resend the same data with the correct delay. A bridge can be implemented by putting a receiver and a sender in a back-to-back configuration. A different approach would employ analogue or digital devices to regenerate the appropriate time delay. The bridges could be used to perform additional functions: if used in NoCs, these devices can have several outputs (a combination of a receiver and several senders), thus forming a network switch. A disadvantage of bridges/switches is their latency. The latency is defined by the data validation process, which finishes only after the second wire have switched.

3.4   Bandwidth and Reliability

The International Technology Roadmap for Semiconductors (ITRS) [ ] indicates that future interconnects will be based around high-speed serial links,

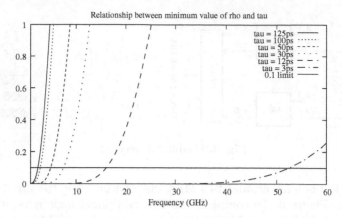

Fig. 7. Minimum value of ρ against frequency of the link in GHz

with bandwidths of several Gigabit/sec. In this context, we evaluate the use of the proposed approach in future technology nodes. The choice of δ and the bandwidth of the circuit are inextricably co-related and depend on the physical characteristics of the receiver and the sender.

From [ ] we can describe the difference in output voltages for a bistable in metastability as $\Delta_V \simeq \theta \cdot \delta \cdot e^{\frac{t}{\tau}}$, where τ is the time constant of the device, θ is the conversion factor from time to initial voltage at the metastable node and Δ_V is the voltage difference at the output which indicates the end of the resolution. This formula is correct for $|\delta| \in \left(0, \frac{\Delta_V}{\theta}\right)$ and is an acceptable approximation for $|\delta| \ll \frac{V_{dd}}{\theta}$. Rearranging the above equation we can write $t \simeq \tau \cdot ln\left(\frac{\Delta_V}{\theta\delta}\right)$. If $T = \frac{1}{f}$ where f is the maximum frequency of the link and t_{res} the resolution time, we can impose the condition $T > t_{res}$ for a reliable link. Note that this inequality holds for one ME as this device only works with one edge of the input. Using the protocol being described, we can relate T to δ by writing $\delta = \rho T$ where $1 > \rho > 0$; substituting in the previous equation we obtain $t_{res} \simeq \tau \cdot ln\left(\frac{\Delta_V}{\theta\rho T}\right)$. Using the inequality indicated and rearranging using basic algebra we obtain $\rho > \frac{\Delta_V}{T\theta}e^{-\frac{T}{\tau}}$. This last equation uses the frequency of the link and the physical characteristics of the technology for the ME to indicate the minimum δ for reliable operation. The values of τ and θ are dependent on the technology, while Δ_V can be chosen in a more or less conservative way. In particular τ is the speed of one NAND gate and θ is related to the switching of the transistors in the gate. From [ ] we find that τ is destined to drop while θ will increase: for example, the speed of a NAND gate is destined to reach 12.13ps in 2009 and 2.81ps in 2018.

Figure 7 shows plots of ρ against the frequency of the link in GHz for various technology nodes indicated by different values of τ. The value of Δ_V is 3.3V and represents an extreme, as the output voltage different is recognised as valid well before the voltage difference reaches 3.3V. θ has been chosen to be 3mV/ps from [ ] as an example and kept constant to simplify the graph. Frequencies

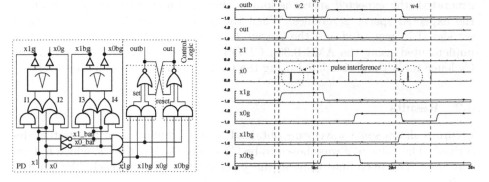

Fig. 8. Implemented receiver and its waveforms

where ρ exceeds 1 are intrinsically unusable, as the resolution time is larger than the period. We can impose a limit on ρ for meaningful safe operation of the proposed protocol of 0.1 (δ one tenth of T) and drive the link at frequencies below the corresponding frequency. Once the maximum frequency has been determined the effective value of ρ can be chosen to be above 0.1, so that any jitter will be ignored, as this will affect δ and therefore ρ. Note, however, that at very high frequencies additional effects become significant; the characterization of such behaviours is under study.

4   Implementation Example

4.1   Description

An implementation example has been designed and simulated as proof-of-concept. The sender is an asynchronous block assembled as shown in Figure 5(a), the coarse VDEs are built as in Figure 5(b); the delay of 500ps was chosen. The receiver design is illustrated in Figure 8. The two ME elements and the input logic (I1-I4) implement the PD in the receiver. It also includes control logic for identifying code-words. The receiver observes the dual-rail pair of signals (x1 and x0), and then based on the order of their switching, records the input code-word. In order to sample 1-to-0 switching, the inverted signals x1_bar and x0_bar are introduced. Under normal operation, if x1 wins, then x1g is generated, alternatively if x0 wins, then x0g signal is generated. The other ME component operates similarly for sampling the falling edges. The corresponding grant signals are x1bg and x0bg.

In order to prevent pulse interference during spacer states, the input logic (gates I1-I4) is used. For example, assuming x1 wins the grant, during the spacer state, if the x1 is changed due to interference, then gate I1 will hold the grant signal until next spacer state coming (x0 and x1g are all high). The control logic converts dual-rail code-words to normal data. Two AN221 gates implement this mechanism. As the alternative spacer protocol is used, after a grant signal is generated, the expected spacer is known. For example, after x1g is generated, the expected spacer is all-ones. When the logic receives the all-one spacer, a set signal is generated and the output (out=1) is produced. The system was implemented using the AMS 0.35u CMOS technology (TECH-CSI) under the Cadence toolkit and simulated using the Cadence analogue simulation tool.

4.2 Results

Figure 8 shows some waveforms obtained during simulation of the system described. The value of δ was chosen to be 5% of the clock period T, which in turn was 10ns. The maximum value of the jitter could therefore reach 500ps in the limit.

We also made an experiment using an unrealistic estimation of jitter to less than 3ps on each transmission line in order to illustrate the behaviour at "extreme" conditions. The value of δ was 15ps, five times bigger then the delay. Even at these short times, the PD does not enter metastability and exhibits correct behaviour. In Figure 8 the windows w1 and w3 indicate the distance δ while w2 and w4 are the windows where interferences are ignored. Note in fact the rejection of two faults happening at times 7ps (w2) and 22ps (w3). In the first case, the fault appear during a transmission clock cycle, but the erroneous data is not latched; in the second case, the fault appears between two clock cycles, but it still does not affect the behaviour of the transmission line. Please note that two transitions on "out" happen close to the faults, but are completely unrelated to them: in fact, they are the result of the decoding of the information appearing at times (respectively) 5ns and 20ns (the transitions on "out" start before the faulty transitions). The time between the reception of a valid data and the production of that data at the output of the receiver block is 1.68ns. This time results from using ME, which are relatively "slow" devices. If δ was chosen to be larger, different techniques could have been used for the design of receiver to recover data, leading to faster response at the expense of a wider event window.

The receiver is not Speed Independent and works under the Fundamental Mode without completion detection logic. Therefore, some timing assumptions are used in the design, for example the inverters used between the two MEs must have identical dynamic behaviour. Apart from this simple one, some others hold:

1. The cycle time of the sender must be greater than 1.68ns (time to generate an output at the PD);
2. $\tau_{\langle x1,x0\rangle = 00/11} \geq \tau_{inv} + \tau_{and2} + 2\tau_{an221}$ after the relevant grant is generated;
3. $\delta < \tau_{grantGenerate}$, where $\tau_{grantGenerate}$ is the time between an edge arriving and the generation of the corresponding grant signal.

The first assumption is expressed in 3.2. the second is necessary to guarantee that the length of the valid spacer is long enough for the signal to propagate through the flip-flop at the right-hand side of Figure 8 and the feedback to reach the flip-flop to keep the value.

5   Conclusions

A novel interconnection approach for SoCs has been presented together with some examples of implementation. The results show high robustness to transient faults of the type described (narrow-pulses) and relative simplicity of implementation. An important feature of the system described is the adaptability to a variety of environments (GALS, NoCs), achieved without the need for sophisticated circuitry. In fact, the system can almost be "plugged in" and work, as long as the synchronization protocol and the buffer stages are designed correctly.

The simulation results show that the circuit works as expected and has the ability to filter out interference. More accurate evaluation of jitter and identification of minimal event windows (possibly on-line) is under consideration. Together with the jitter introduced on the transmission lines, additional sources of jitter are in fact the delay elements themselves, particularly if a delay line is employed. A more analytical description of the design requirements is therefore being carried out, together with a more accurate definition of the effects of a fault appearing through the window w1.

Future work aims to implement more complex protocols employing a larger number of transmission lines in order to increase throughput of the channel and increase reliability. In fact, consider the case where n lines are used for the communication channel. If VDEs are employed on each line the sender could introduce different combinations of delay on the lines and the receiver would, by recognising the sequence of edges of the transmission clock on each line, recover the data being encoded. The work of John Bainbridge, summarised in his thesis [ ], describes the use of a single-rail bus to avoid the overhead imposed by the use of multiple-rail bus implementations; however, he also proposes the use of 1-of-4 encoding [ ] as a possible improvement.

References

1. W.J. Dally and B. Towles. Route packets, not wires: On-chip interconnection networks. In *Procceding. Design Automation Conference. DAC 2001*, pages 684–689, June 2001.
2. A. Chakraborty and M.R. Greenstreet. Efficient self-timed interfaces for crossing clock domains. In *Proceedings. 9th International Symposium on Asynchronous Circuits and Systems*, May 2003.
3. E. Dupont, M. Nicolaidis, and P. Rohr. Embedded robustness IPs. In *Proceedings of the 2002 Design, Automation and Test in Europe Conference and Exhibition (DATE'02)*, 2002.
4. M. Nicolaidis. Time redundancy based soft-error tolerance to rescue nanometer technologies. In *17th IEEE VLSI Test Symposium*, April 1999.
5. A. Yakovlev. Structural technique for fault-masking in asynchronous interfaces. *IEE Proceedings E (Computers and Digital Techniques)*, 140:81–91, March 1993.
6. D. Sokolov, J. Murphy, A. Bystrov, and A. Yakovlev. Design and analysis of dual-rail circuits for security applications. *IEEE Transactions on Computers*, 54(4):449–460, April 2005.

7. R. Ginosar and R. Kol. Adaptive synchronization. In *Proceedings. AINT'2000*, pages 93–101, July 2000.
8. M. A. Kossel and M. L. Schmatz. Jitter measurements of high-speed serial links. *IEEE Design and Test of Computers*, 21:536–543, November-December 2004.
9. C. Molnar and I. Jones. Simple circuits that work for complicated reasons. In *Proceedings. Sixth International Symposium on Asynchronous Circuits and Systems*, volume 1, pages 138–149. IEEE CS, April 2000.
10. J. Cortadella, A. Yakovlev, L. Lavagano, and P. Vanbekbergen. Designing asynchronous circuits from behavioral specifications with internal conflicts. In *Proceedings. ASYNC'94*, pages 106–115. IEEE CS Press, November 1994.
11. J.R. Cessna and D.M. Levy. Phase noise and transient times for a binary quantized digital phase-locked loop in white gaussian noise. *IEEE Transaction on Communications*, COM-20(2):94–104, April 1972.
12. C. D'Alessandro, K.T. Gardiner, D.J. Kinniment, A. Bystrov, and A. Yakovlev. On-chip sub-picosecond phase alignment. Technical report, University of Newcastle upon Tyne, 2005.
13. International Technology Roadmap for Semiconductors (ITRS-2003), 2003.
14. A.M. Abas, A. Bystrov, D.J. Kinniment, O.V. Maevsky, G. Russell, and A.V. Yakovlev. Time difference amplifier. *Electronics Letters*, 38:1437–1438, November 2002.
15. W.J. Bainbridge. *Asynhcronous System-on-Chip Interconnect*. PhD thesis, University of Manchester, UK, March 2000.
16. J. Bainbridge and S. Furber. Delay insensitive system-on-chip interconnect using 1-of-4 data encoding. In *Proceedings. Seventh International Symposium on Asynchronous Circuits and Systems. ASYNC 2001*, pages 118–126, 2001.

Exploiting Cross-Channel Correlation
for Energy-Efficient LCD Bus Encoding

Ashutosh Chakraborty, Enrico Macii, and Massimo Poncino

DAUIN, Politecnico di Torino, Torino, Italy
{ashutosh.chakraborty,enrico.macii,massimo.poncino}@polito.it
http://www2.polito.it/ricerca/eda

Abstract. LCD displays consume a significant portion of the system energy which is a tight constraint for various battery operated portable devices supporting streaming video. With the advent of new LCD technology like Organic-LEDs, and shrinking feature size of I/O controllers, the LCD bus is going to be the major contributor to this power budget for future devices. The data traveling over LCD buses (the images) have high spectral and temporal correlation which can be exploited for reducing the switching activity on the bus. However, there is lack of enough work for exploiting the cross-channel correlation (*CCC*) for low energy encodings. In this work, we prove the effectiveness of *CCC* and propose three new algorithms and associated encoding schemes to save as much as 23% switching activity (average 15.5%) and increase the *Covering Potential* on average by 20% for the LCD bus.

1 Introduction

Historically, battery-operated products have been the most demanding applications for low-power microelectronics. With the tremendous market penetration of devices such as cell phones, PDAs, and other hand-held devices supporting real-time multimedia applications, power management of these devices has become pervasive in the semiconductor industry. Reducing the power demand of such complex devices can be done by applying various power optimization techniques on the different components of the system: computational units (i.e., processors and logic), memories, buses, and peripherals. One of the major components in the energy budget of such devices is the LCD subsystem, which comprises display technology and LCD bus for data transfer. LCD subsystem often consuming as much as 25% to 50% of total system power [1].

For existing LCD subsystems, the energy requirements of the LCD bus is around 25% of the display subsystem power, or roughly 7-8% of the system power. Current LCD technologies are based on back-lit TFT panels, which are intrinsically power-hungry components. Emerging technologies such as organic display (OLED) systems do not require a back-lit to function, thus drawing much less power than conventional TFT displays [11]. With decreasing feature sizes of peripherals such as graphics cards and display controllers, the power consumption in these components is expected to go down. In future systems, we may

V. Paliouras, J. Vounckx, and D. Verkest (Eds.): PATMOS 2005, LNCS 3728, pp. 297–307, 2005.
© Springer-Verlag Berlin Heidelberg 2005

then expect that the power impact of the other components (e.g., frame buffer and LCD buses) will become the actual bottleneck. The LCD bus is typically implemented as a flat cable with capacitance values orders of magnitude more (tens of pF/m) than on-chip buses or PCB buses, and does not show any signs of respite in the near future. The contribution of LCD bus to the total system power will shoot up in near future due to above mentioned reasons. Hence, it can be concluded that the most ubiquitous peripherals which need immediate attention for applications in portable devices are indeed LCD buses.

Besides high capacitances of the wires, another aspect that makes LCD buses a good target for power reduction is related to the fact that most LCDs are based on digital interfaces; this allows to simplify the problem of power reduction on LCD buses from a purely electrical issue (as it would be for analog interfaces) to a simpler "logical" approach based on the *reduction of the number of transitions*. This is a widely studied problem in the domain of low-power EDA; however, most techniques are devoted to parallel buses, whereas LCD buses are based on *serial protocols*. Previous works on energy-efficient data transmission on digital LCD interfaces have been explored in the literature ([4,5]), by exploiting in different ways the well-known correlation existing between adjacent pixel values.

The rest of the paper is organized as follows. We summarize the previous work for low energy LCD bus encoding in Section 2. We list the notation used in the sequel in Section 3 and illustrate the importance of correlation between data traveling over different channels on LCD bus in Section 4. We propose three new algorithms in Section 5 to take advantage of this correlation for low energy transmission schemes. In Section 6, we introduce our encoding scheme, which represents the implementation specification for transfer of data using our Algorithms. We report our results in Section 7 and end the paper with conclusions and directions for carrying out future work in this domain.

2   Previous Work and Background

LCD protocols do not explicitly provide options for energy-efficient transmission. In the literature, most solutions for reducing the energy consumption of LCD sub-systems have focused on circuit-level solutions either in the design of individual components of the display systems [7,8], to specific (hardware or software) control mechanisms inside the LCD controller [1,9], or to the proper design of the frame buffer memory [2].

For LCD bus, which is typically implemented as high capacitance flat cable, the transition count, which is defined as the number of transitions between high and low logical state on LCD buses, is a measure of energy requirement. Recent works ([4,5]) have tried to reduce the bus transition count using energy-efficient data encodings on the digital interface. In order to increase correlation of transmitted data, both schemes transmit pixel differences rather than pixel values, exploiting the well-known Intra-Channel Correlation (ICC) of images. For example, for 8 bit color values, if the consecutive pixels on a particular channel of the data are 130 & 136, instead of sending these values, we send 130 and $6(= 136 - 130)$

In [5], the authors have introduced the concept of Limited Intra Word Transition (*LIWT*) codewords for encoding most commonly occurring differences between two consecutive pixel values. This scheme is possible thanks to Gaussian distribution of the differences between adjacent pixels which is centered at 0 [4] and has a very small variance. The *LIWT* scheme is characterized by a window of pixel differences for which the low transition codewords are transmitted. For any difference lying outside this range, *LIWT* sends out the original pixel value instead of sending the difference. The range in which a pixel difference should lie in order to be sent as a low transition codewords is called *LIWT Window*. A typical *LIWT Window* could be [-10,10]. In the above example, sending 6(00001010) would cause four transitions on the bus. Using *LIWT* we can assign the codeword '11111110' to this difference and thus reduce the transition count to one for sending this difference. However, the complexity of the encoder and decoder at both ends of the LCD bus can become huge for high values of the *LIWT Window*. This is because the decoder is typically implemented as a table and a *LIWT Window* of size N requires storing N codewords in the table of both encoder and decoder. Clearly, the energy savings achieved in this manner is proportional to the number of pixel difference which fall in the LIWT Window. Furthermore, it is immediately derivable that the above scheme does not take into account the correlation that exists between data traveling on different channels of the LCD bus. The first attempt to use Cross-Channel-Correlation (*CCC*) was made in [4], however, the authors concluded that the *CCC* is not interesting property for energy efficient encodings.

Through this work, we show that indeed *CCC* is an important property to be exploited for low power encoding and is at least as good as using *ICC* alone. We use *CCC* to increase the percentage of pixel differences falling in *LIWT* window. Also, to keep the complexity of the encoder and decoder very simple, we focus our attention to very small decode/encode units. For this, we chose a *LIWT* scheme for which the *LIWT Window* is [-4, 4]. This requires storing only 9 codewords for the most frequently occurring difference values at the decoder and encoder. Using both *CCC* and *ICC* leads to considerable reduction in the transition count without any increase in the number of codewords. We propose new algorithms and corresponding encoding schemes to exploit *CCC* and *ICC* together in this paper. We estimate the hardware overhead using our schemes to be much less than the overhead required for implementing a wide range *LIWT Window* using tables at encoder and decoder to accomodate more pixel differences for encoding as differences.

3   Notation

Table 1 lists in brief the notation we have used throughout this paper. Broadly, we have reserved the symbol δ for Intra-Channel properties and the Δ for Cross-Channel properties. As an example, below we show the values travelling on various channels and how various Δ and δ are connected to this data set.

Table 1. Notations used in the sequel

Symbol	Notation
$P_{c,i}$	Pixel value on channel c for i^{th} pixel. $c \in (Red, Green, Blue)$
$\delta_{c,i}$	Difference between i^{th} pixel and i-1th pixel of channel $c(= P_{c,i} - P_{c,i-1})$
$\delta_{c-d,i}$	Difference between i^{th} pixel of channel c and channel d ($= P_{c,i} - P_{d,i}$)
δ	Used for generic description for $\delta_{m,n}$, $m \neq n$
$\Delta_{c-d,i}$	Difference between $\delta_{c,i}$ and $\delta_{d,i}$ ($=$ Cross-Channel Difference)
Δ	Used for generic description for $\Delta_{m,n}$, $m \neq n$

$$R \to 130 \ 140 \ 155 \ 50 \ 65 \qquad \delta_{r,2} = 140 - 130 = 10$$
$$G \to 70 \ 85 \ 120 \ 180 \ 155 \quad \to \quad \Delta_{r-g,3} = \delta_{r,3} - \delta_{g,3} = (155 - 140) - (120 - 85)$$
$$B \to 20 \ 40 \ 140 \ 220 \ 190 \qquad \Delta_{b-g,4} = \delta_{b,4} - \delta_{g,4} = (220 - 140) - (180 - 120)$$

4   Inter-channel Correlation

It is well known fact that there exists considerable amount of correlation amongst the various channels of an image. This correlation has been widely used in Image Processing algorithms for Image compression. Figure 1 shows the original image (*4.2.04* from [10]) along with its decomposition into the three independent channels using *GIMP* [13]. It can be immediately observed that the most of

Original Image   Red Channel

Green Channel   Blue Channel

Fig. 1. Comparison of R,G,B channels for an example Image from [10]

the spatial features, such as edges, gradients are very regular similar on all three channels. Thus, any change occurring in one of the channel, has changes of similar nature occurring on the other channels. Sending out the common portion of this *information* on all three channels is redundant, which can be dealt with by *subtracting* the common information and sending this information on only one channel, and the residues in the rest of the channels. This is typically how an data compression algorithm would deal with such a situation. This observation is basis for our work.

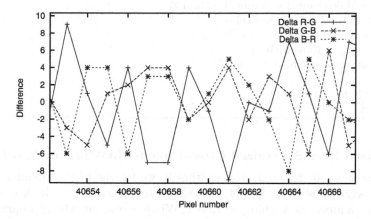

Fig. 2. $\Delta_{r-g,i}$, $\Delta_{g-b,i}$ and $\Delta_{b-r,i}$ for 4.2.04 [10]

To study how this 2-Dimensional spatial correlation amongst various channels manifests itself in case of essentially 1-dimensional (sequential, raster scanned) data traveling over LCD buses, we plotted $\Delta_{r-g,i}$, $\Delta_{g-b,i}$ and $\Delta_{b-r,i}$ traveling over three channels at a given time in Figure 2. It can be observed that the three curves are concentrated near the value 0 for most of the time, suggesting that indeed the *subtraction* operation has left little residues, hence the data must be correlated which we call as Cross-Channel Correlation(*CCC*).

5   Proposed Algorithms

To utilize the *CCC* discussed in Section 4 for reduced transition count, we propose three algorithms which utilize the observation that the most values are close to zero in Figure 2. The basic aim of these algorithms is to bring more pixel difference values within the range of *LIWT Window*.

5.1   Variable Base Cross-Channel LIWT (*VBCC-LIWT*)

In this scheme, as depicted in Algorithm 1 we find the channel which has the middle value for the set of differences $\delta_{r,i}$, $\delta_{g,i}$ and $\delta_{b,i}$. Thus a middle channel as *red* would mean $\delta_{g,i} \leq \delta_{r,i} \leq \delta_{b,i}$ or $\delta_{g,i} \geq \delta_{r,i} \geq \delta_{b,i}$. We assign this channel as the base for the current pixel values. For base channel we apply pure LIWT, whereas for the rest of the two channels, we find $\Delta_{c-base,i}$. In case these $\Delta_{c-base,i}$ lie within the *LIWT Window* we transmit the related code. This scheme does not depend on the values of δ for the non-base channels while considering inclusion into the range of *LIWT Window*. As will be evident in Section 7, the outcome of this algorithm proves that indeed CCC is a strong correlation property.

Algorithm 1 VBCC-LIWT   Task: Encode for Channel c using only Δ

1: $\delta_{c,i} \leftarrow P_{c,i} - P_{c,i-1}$ {$\forall$ c $\in$ [R,G,B]}
2: Find middle channel m out of R, G and B
3: Apply pure LIWT on channel m
4: Calculate $\Delta_{c-m,i}$ {$\forall$ c $\in$ [R,G,B] - m}
5: **if** ($\Delta_{c-m,i}$ in Range) **then**
6:     Send (EncodeCrossChannel(Δ))
7: **else**
8:     Send R_i
9: **end if**

5.2   Fixed Base Synergized Cross-Channel LIWT *(FBSCC-LIWT)*

In this scheme, as depicted in Algorithm 2, we fix one of the channels as base channel, say *red*, without loss of generality. For encoding, we check if $\delta_{c,i}$ or Δ for these channels lie within the LIWT Window and encode appropriately. In case, neither of δ and Δ lie within the range, we transmit the actual pixel value using *EncodePlain*. The benefit of this algorithm stems from the fact that the probability that either δ or Δ will lie inside the *LIWT Window* is very high, leading to low transition codewords to be transmitted for most of the pixels.

Algorithm 2 FBCCS-LIWT   Task: Encode for Channel c using δ or Δ

1: $\delta_{c,i} \leftarrow P_{c,i} - P_{c,i-1}$; $\delta_{r,i} \leftarrow P_{r,i} - P_{r,i-1}$; $\Delta_{c-r,i} \leftarrow \delta_{c,i} - \delta_{r,i}$
2: **if** ($\delta_{c,i}$ in Range) OR (c = Red) **then**
3:     Send EncodeIntraChannel($\delta_{c,i}$)
4: **else if** ($\Delta_{c-r,i}$ in Range) **then**
5:     Send EncodeCrossChannel($\Delta_{c-r,i}$)
6: **else**
7:     Send EncodePlain($P_{c,i}$)
8: **end if**

5.3   Variable Base Synergized Cross-Channel LIWT *(VBCCS-LIWT)*

We present an even stronger algorithm, which combines the above two algorithms to pack extra pixel differences within the *LIWT Window*. This scheme is analogous to shifting origin for the various δ and including Δ for consideration in *LIWT Window*. We present this scheme in Algorithm 3. Choosing the middle channel, m as the base value causes the value of Δ to be small and hence they become more prone to fall inside the *LIWT Window*.

Algorithm 3 VBCCS-LIWT

1: Calculate $\delta_{c,i}$ $\{\forall$ c $\in$ [R,G,B]$\}$
2: Find middle channel m
3: Calculate $\Delta_{c-m,i}$ $\{\forall$ c $\in$ [R,G,B] - m$\}$
4: **if** $(\delta_{c,i}$ in Range) **then**
5:     Send (EncodeIntraChannel(δ))
6: **else if** $(\Delta_{c,i-j}$ in Range) **then**
7:     Send (EncodeCrossChannel(Δ))
8: **else**
9:     Send R_i
10: **end if**

6   Encoding Scheme

For taking full advantage of schemes proposed in Section 5, it is mandatory to use suitable encoding techniques so that the pixels are decodable at the receiver end. We use the last two redundancy bits for this purpose. There are at most five units of information that we need to transmit

1. Whether the pixel is encoded as $\delta_{c,i}$?
2. Whether the pixel is encoded as Δ_{c_i,c_j}?
3. Whether the pixel is plain/un-encoded?
4. Whether the channel in consideration is a base channel?
5. Whether the channel in consideration is not a base channel?

It appears that we need to use at least 3 bits for signaling the above 5 informations. However, it turns out that out of the above questions,

- Algorithm VBCC-LIWT requires only 2, 3, 4 and 5. (Don't use δ)
- Algorithm FBCCS-LIWT requires only 1, 2 and 3. (Red is base)
- Algorithm VBCCS-LIWT requires only 1, 2, 3, 4 and 5. But it can be easily observed that a base channel cannot be encoded as Δ, since it itself contains the base information.

Table 2. Codewords distribution for δ and Δ and plain pixel values

Difference	Used by $\delta = MAP()$	Used by Δ
-4	1000000000	0111111111
-3	1110000000	0001111111
-2	1111100000	0000011111
-1	1111111000	0000000111
0	0000000000	1111111111
1	1111111100	0000000011
2	1111110000	0000001111
3	1111000000	0000111111
4	1100000000	0011111111

Plain Pixel

XXXXXXXA 01   (if A = 0)
XXXXXXXA 10   (if A = 1)

Hence, we are able to pack the required four pieces of information in the last 2 redundant bits used in $LIWT$ scheme. We encode this information in the last two redundancy bits using the scheme shown in Table 2. We reserve codewords ending with "11" for sending cross-channel differences (Δ), "00" for sending inter channel differences (δ). We reserve codewords ending with "01" and "10" for plain pixel transmission depending upon the last bit of the binary equivalent of plain pixel value. The last category of pixels can be quickly recognized by taking XOR of redundancy bits at the decoder side. One salient feature of encoding scheme of Table 2 is that the corresponding codewords for $\delta_{c,i}$ and Δ_{c_i,c_j} for any value are bitwise invert of each other. This information can be used effectively to simplify the Algorithm 2 and Algorithm 3 in Section 5 as

$$EncodeIntraChannel(K) = \text{MAP(k)} \quad /*Defined\ in\ Table\ 2\ */$$

$$EncodeCrossChannel(K) = \text{Invert Bits}\ (EncodeIntraChannel(K))$$

$$EncodePlain(K) = \text{Binary Value}(K) + "01" \ || \ "10"$$

7 Experimental Results

We implemented the algorithms presented in Section 5 using the encoding scheme proposed in Section 6 in C++. We calculated the transition count on the LCD bus and percentage of pixel differences covered by a LIWT Window of range $[-4, 4]$ for all the images.

The percentage of pixel differences covered by $LIWT$ and various schemes proposed in Section 5 for the same size of $LIWT$ $Window$, represents the theoretical bounds on the maximum advantage that can be derived using these schemes. In other words, the percentage of pixel differences lying inside the $LIWT$ $Windows$ (defined now on as Covering Potential, or simply CP) is the maximum benefit that can be achieved using a particular scheme, assuming that other factors such as decodability issues are assumed to be absent. This analysis only takes into account the Algorithms proposed without considering the encoding scheme used for transmission. Undoubtedly, the actual savings will always be less than the this limit because we need to sacrifice some transitions to ensure decodability at the receiver end. We measured the CP for various benchmark images using our schemes and $LIWT$, and are reported in Figure 7.

We observe that using our algorithms, the CP of the encoding schemes has increased substantially. We gain on an average 12% for VBCC-LIWT, 14% for FBSCC-LIWT and 21% for VBCCS-LIWT algorithms.

In Table 3, we present the transition count and savings achieved on top of $LIWT$ scheme (which itself saves 55% transitions) for the three algorithms proposed in Section 5. For the Algorithm 2, we calculate two numbers for savings obtains (Column 4 and 5 in Table 3). These numbers denote the savings for the non-base channels alone, and all three channels taken together. We make the following observation out of this table.

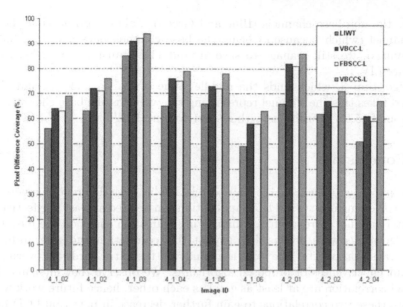

Fig. 3. Comparison of R,G,B channels

Table 3. Transition counts and Savings (in %) for our Schemes Vs *LIWT*

Name	LIWT	FBSCC-L		VBCC-L			VBCCS-L	
		Transition	Saving	Transition	Saving	Saving	Transition	Saving
4.1.01	561148	530379	5.483	510220	9.0756	13.981	482933	13.938
4.1.02	482850	438935	9.094	433364	10.248	15.695	406196	15.875
4.1.03	391362	313138	19.98	325127	16.924	24.754	299309	23.521
4.1.04	509808	445025	12.70	434832	14.706	21.326	415748	18.450
4.1.05	517568	483743	6.535	464281	10.295	15.399	437318	15.505
4.1.06	625735	593608	5.134	572335	8.5339	13.287	549140	12.24
4.1.07	317493	300000	5.509	285326	10.131	15.074	276109	13.034
4.1.08	364651	345764	5.179	328783	9.8362	14.547	319897	12.27
4.2.01	1685211	1564033	7.190	1554661	7.7468	11.072	1422997	15.559
4.2.04	2418165	2317500	4.162	2223288	8.0588	11.472	2073988	14.232
4.2.05	1954931	1715658	12.23	1649956	15.600	23.526	1605643	17.867
			8.747		11.010	16.375		15.647
			Average		Average	Average		Average

1. Algorithm 1, which relies solely on values of Δ (and not δ), achieves a small improvement ($\tilde{9}$%), over the LIWT scheme. This result is important in the sense that it proves that using Δ is at least as important/effective, if not more, as using δ for optimally exploiting correlation between the different channels of an image.

2. Algorithm 2, which transmits codes for values of Δ or δ within range, with the limitation that Red channel is always taken as the base channel(thus $\Delta_{g-r,i}$ and $\Delta_{b-r,i}$ are computed for each pixel) achieves a saving for 16%

for the non-base channels (Blue and Green). Taking into account the Red channel (which because of being the base channel, does not contribute to savings), we still manage to save around 11% transitions with respect to pure LIWT.

3. Algorithm 3, which finds the middle value of $\Delta_{r-g,i}$, $\Delta_{g-b,i}$ and $\Delta_{b-r,i}$ and takes this the channel representing this value as the base channel, saves around 15.5% transitions with respect to pure LIWT.

8   Conclusions

In this paper, we demonstrated that correlation between different channels of an image traversing in a typical LCD application can be used at least as effectively as the inter channel correlation, for minimizing transition count on the bus. Three new algorithms were proposed to exploit this correlation and give up-to 23% lesser transition counts. The work also analyzed the extra hardware overhead to find out effectiveness of proposed algorithms. We observe that inter and intra channel correlation are at least as good as each other, hence future work will be to align these two correlations to gain further decrease in power on LCD buses.

References

1. I. Choi, H. Shim, N. Chang, "Low-Power Color TFT LCD Display for Handheld Embedded Systems," *ISLPED'02: ACM/IEEE International Symposium on Low-Power Electronics and Design*, August 2002, pp. 112–117.
2. H. Shim, N. Chang; M. Pedram, "A Compressed Frame Buffer to Reduce Display Power Consumption in Mobile Systems," *ASPDAC'04: Asia and South Pacific Design Automation Conference*, Jan. 2004, pp. 819–824.
3. W. Kowalsky et al., "OLED Matrix Displays: Technology and Fundamentals," *International Conference on Polymers and Adhesives in Microelectronics and Photonics*, pp. 20-28, 2001.
4. W. C. Cheng, M. Pedram, "Chromatic Encoding: a Low Power Encoding Technique for Digital Visual Interface," *IEEE Transactions on Consumer Electronics*, Vol. 50, No.1, pp. 320-328, January 2004.
5. A. Bocca, S. Salerno, E. Macii, M. Poncino, "Limited Intra-Word Transition Codes: An Energy-Efficient Bus Encoding for LCD Display Interfaces," *ISPLED'04: ACM/IEEE International Symposium on Low-Power Electronics and Design*, Newport Beach, CA, Aug. 2004, pp. 206–211.
6. Digital Display Working Group, Digital Visual Interface, V1.0, www.ddwg.org.
7. K. Meinstein. et al., "A Low-Power High-Voltage Column/Dot Inversion Drive System," *Society for Information Display International Symposium Digest of Technology*, Vol. 28. May 1997.
8. B.-D. Choi, O.-K. Kwon, "Stepwise Data Driving Method and Circuits for Low-Power TFT-LCDs," *IEEE Transactions on Consumer Electronics*, Vol. 46, No. 11, pp. 1155–1160, Nov. 2000.
9. F. Gatti, A. Acquaviva, L. Benini, B. Riccò, "Low Power Control Techniques For TFT LCD Displays," *CASES'02: International Conference on Compilers, Architectures and Synthesis for Embedded Systems*, pp, 218–224, Dec. 2002.

10. A. G. Weber, "USC-SIPI Image Database Version 5," USC-SIPI Report #315, Oct. 1997. sipi.usc.edu/services/database/Database.html.
11. W. Kowalsky et al., "OLED Matrix Displays: Technology and Fundamentals," *International Conference on Polymers and Adhesives in Microelectronics and Photonics*, pp. 20-28, 2001.
12. T. N. Ruckmongathan et al., "Binary Addressing Technique With Duty Cycle Control for LCDs," *IEEE Transactions on Electron Devices*, Vol. 52, NO. 3, March 2005
13. GIMP: "The Gnu Image Manipulation Program", www.gimp.org .

Closed-Form Bounds for Interconnect-Aware Minimum-Delay Gate Sizing

Giorgos Dimitrakopoulos* and Dimitris Nikolos

Technology and Computer Architecture Laboratory
Computer Engineering and Informatics Dept.,
University of Patras, 26500 Patras, Greece
dimitrak@ceid.upatras.gr, nikolosd@cti.gr

Abstract. Early circuit performance estimation and easy-to-apply methods for minimum-delay gate sizing are needed, in order to enhance circuit's performance and to increase designers' productivity. In this paper, we present a practical method to perform gate sizing, taking also into account the contribution of fixed wiring loads. Closed-form bounds are derived and a simple recursive procedure is developed that directly calculate the gate sizes required to achieve minimum delay. The designer, using the proposed method, can easily compare different implementations of the same circuit and explore the energy-delay design space, including in the analysis the effect of interconnect.

1 Introduction

The design of efficient digital circuits requires several decisions that need to be made during the design cycle. One of the major tasks of the designer is to choose the appropriate circuit topology and logic style, determine the sizes of the resulting gates and add extra buffer stages when necessary. Due to the increasing complexity of modern designs and the need for fast and low-power operation, practical and easy-to-apply methods are needed to guide the designer to the best implementation [ ]. In this context, the method of Logical effort [ ] has been presented that allows the formulation of the gate sizing problem in a simple and comprehensive way.

The problem of sizing simple paths of gates is well understood [ ]. However, in almost all cases, simple sizing rules do not suffice, since the effect of intermediate wires should also be taken into account. Handling interconnect effects, when sizing for minimum delay can be performed using sophisticated algorithms or general optimization methods [ ]. Such approaches, although solving the sizing problem exactly, do not give the designer the intuition of how the optimal solution was chosen and how performance will change, when choosing another topology or changing the initial placement of the gates in layout. In order the designer to be able to quickly identify which solution matches better the design's constraints, analytical results are needed that would approximate the optimal

---

* This work has been supported by D. Maritsas Graduate Scholarship of R.A.C.T.I.

V. Paliouras, J. Vounckx, and D. Verkest (Eds.): PATMOS 2005, LNCS 3728, pp. 308–317, 2005.

gate sizes including also the effect of interconnect. In high performance systems, a combined approach is followed. The circuit topology, the number of stages and the logic style are chosen with custom design decisions, while after final placement, a fine tuning step follows that globally optimizes the design for a combined set of constraints [ ]–[ ]. We limit our discussion to small and medium-sized wires, i.e., wires that scale [ ]. Driving longer on-chip wires requires, either the use of repeaters, or the use special signaling techniques [ ]. In these cases, the corresponding driver and receiver circuits are designed separately from the modules they connect, and thus, they are not included in the gate sizing procedure.

The problem of analytically approaching the optimal gate sizes, when considering intermediate interconnect capacitances, is considered hard to solve and only a few solutions exist [ ]–[ ]. However, existing solutions have been derived after making several simplifying assumptions and do not solve the problem in the general case. In this paper, instead of trying to identify an exact solution, we derive closed-form bounds of the gate sizes needed to achieve minimum delay. The proposed bounds are tight enough that allow the approximation of the exact optimal values with almost no loss of accuracy. Also, a simple recursive procedure is developed that solves the problem for paths with multiple levels of intermediate interconnect. The application of the proposed method is straightforward and can help in quickly identifying the optimal number of stages and energy-delay efficient solutions.

The rest of the paper is organized as follows. Section 2 briefly describes the logical effort delay model. Section 3 presents the basic formulation of the proposed approach. In Section 4 the application of the proposed method in selecting the optimal number of stages is described, while in Section 5 a recursive procedure is given that handles gate sizing with multiple levels of interconnect. Finally, conclusions are drawn in Section 6.

2   Gate Delay Model and Simple Path Sizing

Following the logical effort method [ ], the delay of a gate is characterized by its output load, its driving capability and its internal parasitic capacitance. These parameters are modelled using the gate's electrical effort $h = C_{\text{out}}/C_{\text{in}}$, its logical effort g, and parasitic delay p, which are combined as $d = \tau(g \cdot h + p)$. Constant τ is a technology specific parameter that is roughly equal to the delay of unit-sized unloaded inverter. The product of logical and electrical effort $g \cdot h$ is called the stage effort of the gate and it models the delay caused by the gate current charging or discharging the load capacitance. The parasitic delay models the delay needed to charge or discharge the gate's internal parasitic capacitance. The delay model is a first-order approximation of gate's delay yet it is reasonably accurate [ ]. In this paper a standard $0.18\mu m$ technology is used for which $\tau \sim 19ps$ and 1 FO4 inverter delay is roughly equal to 94ps, under typical conditions and supply voltage of 1.8V.

According to the method of logical effort, a path of N gates achieves minimum delay, when all gates of the path have equal stage effort $f = (G \cdot B \cdot H)^{1/N}$.

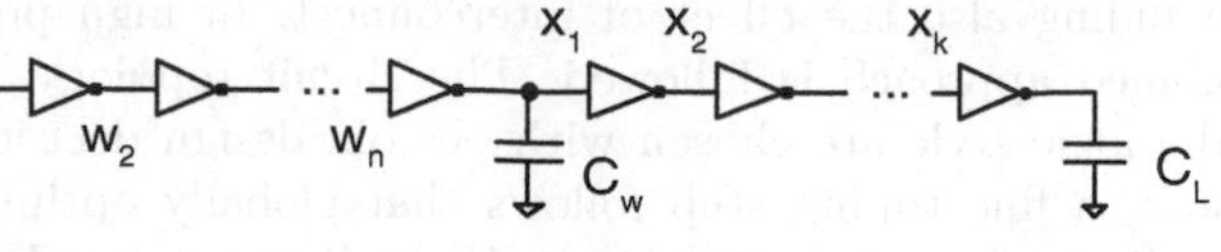

Fig. 1. Two inverter chains separated by fixed wire capacitance C_w

Variables G and B are the product of the logical effort and the branching effort of the gates belonging to the path, and H is the ratio of the final stage loading capacitance to the input capacitance of the first gate of the path. Branching effort b of each gate represents the ratio of all off-path capacitances driven by the gate, to the input capacitance of the following gate of the path.

When fixed off-path wiring capacitances exist in the path, the branching effort of the gate driving the wire cannot be directly estimated [ ]. For small wires, the additional off-path loading capacitances can be safely ignored. However, in reality, the floorplan of the circuit imposes the connection of gates that are placed several hundreds of μm apart. Exact analytical gate sizing under this circumstances is efficiently solved by the proposed method.

3   Gate Sizing with Fixed Wiring Capacitance

At first, the proposed methodology will be presented for the case of two inverter chains connected with a wire with capacitance C_w. In the following, the method will be generalized to handle paths of arbitrary logic gates and multiple levels of intermediate interconnect. The two chains of Fig. 1 consist of n inverters with input capacitance w_i, and k inverters with input capacitances x_i, respectively. The load capacitance C_L and the maximum allowed input capacitance w_1 of the first inverter are considered constant. In many practical cases, the value of the loading capacitance and the input capacitance of the path are not known in advance. Therefore the designer should come up with some reasonable numbers. The input capacitance of the first gate is best described by its maximum allowed value. In this case, the inputs of circuit under consideration can be safely driven without slowing down preceding paths.

Following the logical-effort delay model ($g_{inv} = 1$, $p_{inv} = 1.08$), the path delay of the two inverter chains (normalized by τ) is equal to:

$$D = \frac{w_2}{w_1} + \frac{w_3}{w_2} + \cdots + \frac{C_w + x_1}{w_n} + \frac{x_2}{x_1} + \cdots + \frac{C_L}{x_k} + (n+k)p_{inv}$$

Taking the partial derivatives of D with respect to the input capacitances w_i and x_i, and setting them equal to zero, it is derived that the delay is minimized when equations (1) and (2) are satisfied.

$$f_1 = \frac{w_2}{w_1} = \frac{w_3}{w_2} = \cdots = \frac{C_w + x_1}{w_n} \quad \Rightarrow \quad f_1 = \left(\frac{C_w + x_1}{w_1}\right)^{1/n} \quad (1)$$

$$f_2 = \frac{x_1}{w_n} = \frac{x_2}{x_1} = \cdots = \frac{C_L}{x_k} \quad \Rightarrow \quad f_2 = \left(\frac{C_L}{x_1}\right)^{1/k} \quad (2)$$

Variables f_1 and f_2 represent the stage efforts of the first and the second chain of gates, respectively. Due to the fixed wire capacitance, the gates of the two chains when sized for minimum delay need to have unequal stage efforts. The goal is to find f_1 and f_2 that minimize the total path delay, since $D_{min} = nf_1 + kf_2 + (n + k)p_{inv}$. From (1) and (2) two new equations are derived.

$$f_1^{n-1} (f_1 - f_2) = C_w/w_1 \tag{3}$$

$$f_1^{n-1} f_2^{k+1} = C_L/w_1 \tag{4}$$

The solution of the two non-linear equations cannot be performed analytically. However we will provide tight bounds on the optimal values of f_1 and f_2 that allow the approximation of the exact values very accurately. Then, the optimal input capacitances could be directly computed from the optimal stage efforts f_1 and f_2 given that $w_i = w_1 \cdot f_1^{i-1}$ and $x_i = C_L/f_2^{k-i+1}$.

3.1   Bounds of Stage Efforts

When the path of Fig. 1 is sized for minimum delay, then according to (3) the optimal value of x_1 is given by the solution of $x_1^{n(k+1)} = w_1^k C_L^n (C_w + x_1)^{k(n-1)}$. The appropriate value of x_1 strongly depends on the value of C_w. Increasing C_w, also increases the resulting value of x_1. This makes sense since the delay optimization procedure tries to increase the value of x_1 in order to make the effect of the wire capacitance a small fraction of the total. Input capacitance x_1 assumes its minimum nominal value when C_w is equal to zero. Then, both inverter chains form a single path and the problem is treated exactly the same way as in logical effort, briefly described in Section 2. In this case the optimal size of x_1 is $x_{1,min} = w_1^{k/n+k} \cdot C_L^{n/n+k}$. Based on the definition of stage efforts f_1 (Eq. (1)), for every other non-zero value of the wire capacitance C_w, the value of f_1 should be increased. In this way the last gate of the first chain can better drive both C_w and x_1. Therefore, the value of f_1 should always be greater than $((C_w + x_{1,min})/w_1)^{1/n}$. Defining X_L and X_W as,

$$X_L = (C_L/w_1)^{n/n+k} \qquad X_W = C_w/w_1 \tag{5}$$

and by replacing the minimum value of x_1, it follows that $f_1 > (X_W + X_L)^{1/n}$. According to (4), in order the delay to be minimized, every increase in f_1 should be followed by a proportional decrease of the stage effort of the second path. Hence stage effort f_2 is always less than

$$f_2 < X_L^{1/n} (1 + X_W/X_L)^{(1-n)/n(k+1)} \tag{6}$$

We are interested in identifying more tight bounds for the stage efforts f_1 and f_2. Bounding f_1 suffices since the corresponding upper and the lower bounds of f_2 can be easily derived via (4). We will use equations (7) and (8), derived from (3), (4), which express f_1 as a function of f_2 in two different ways.

$$f_1 = \left(C_w/w_1 + C_L/(w_1 \cdot f_2^k) \right)^{1/n} \tag{7}$$

$$f_1 = f_2 + (C_w/C_L) f_2^{k+1} \tag{8}$$

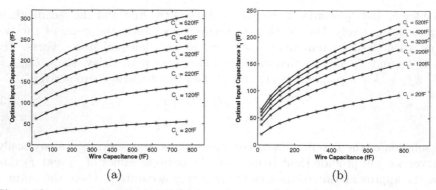

(a)   (b)

Fig. 2. Optimal and estimated input capacitances of the second chain for various final loads and wire capacitances. The path input capacitance is assumed to be 5fF. (a) The case with three inverters in the first chain and one in the second. (b) The corresponding case with three inverters before and after the wire

Replacing the maximum value of f_2 (Eq. (6)) to (7) and after some algebraic manipulations a tighter lower bound of f_1 can be computed as

$$f_1 > (X_W + X_L \cdot \delta)^{1/n}, \quad \text{where} \quad \delta = (1 + X_W / X_L)^{k(n-1)/n(k+1)} \quad (9)$$

It can be observed that the lower bound of f_1 has the same format as in the case where the wire capacitance is ignored. The only difference is the multiplicative term δ that increases the first lower bound closer to the optimal value. The maximum value of f_1 can be calculated via (9), (4) and (8), and is given by

$$f_1 < X_W (X_W + X_L \cdot \delta)^{(1-n)/n} + X_L^{1/n} (\delta + X_W/X_L)^{(1-n)/n(k+1)} \quad (10)$$

The equivalent tight bounds of f_2 are derived by replacing in (4) the minimum and the maximum value of f_1.

$$\left(C_L/(w_1 f_{1,\max}^{n-1}) \right)^{1/(k+1)} < f_2 < \left(C_L/(w_1 f_{1,\max}^{n-1}) \right)^{1/(k+1)} \quad (11)$$

The proposed bounds of the stage efforts are very tight and the expected values of f_1 and f_2 can be derived by computing the geometric mean of their minimum and maximum value, i.e, the square root of the product of the two extreme points, $f_1^* = \sqrt{f_{1,\min} f_{1,\max}}$. Consider for example the case that $n = 2$, $k = 1$ and $w_1 = 10\text{fF}$, $C_w = 50\text{fF}$, and $C_L = 100\text{fF}$. Solving (3) and (4) numerically it is derived that the minimum value of D is $(8.28 + 3p_{inv})\tau$ and it is achieved when $x_1 = 57.2\text{fF}$ and the stage efforts f_1 and f_2 are equal to 3.27 and 1.74, respectively. Following the proposed approach, at first we compute $X_L = 10^{2/3} = 4.642$, $X_W = 5$, and $\delta = 1.2$. Then from (9)–(11) the stage efforts f_1 and f_2 are easily bounded to $3.252 < f_1 < 3.291$ and $1.743 < f_2 < 1.754$. Therefore, the expected values of f_1 and f_2 are equal to 3.271 and 1.748, respectively. It is evident that the derived stage efforts and the precomputed values match exactly.

The proposed solution is accurate irrespective of the values of the design parameters. Figure 2 shows the value of the optimal input capacitance x_1 for

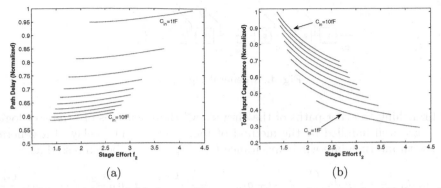

(a)   (b)

Fig. 3. (a) The delay and (b) the total input capacitance of an example path for various values of stage effort f_2. The path consists of three inverters in the first chain and one in the second. The wire and the loading capacitance are 50 and 100fF, respectively

several configurations, derived both exactly (asterisk points) using MATLAB, and using the proposed bounds (solid line). For both cases the maximum absolute error is less than 0.8fF, which is negligible. Sizing for minimum delay in the presence of intermediate wire capacitances leads to increased gate sizes. Several energy efficient solutions can be derived with a very little loss in delay. The proposed formulation can serve as an upper bound to the designer so as to calculate how much he can stress the design gaining a few more picoseconds.

Fig. 3(a) illustrates the delay of an example path, when stage effort f_2 is independently increased up to 2× its minimal value. For each case, stage effort f_1 is recalculated since the input capacitance of the second chain is gradually reduced. Each line of Fig. 3(a) refers to an input capacitance between 1fF and the maximum allowed, that is 10fF. In all cases when varying f_2 the delay loss is negligible (almost flat lines). The form of the delay plots, proves that a wide variety of stage efforts give delays close to the minimum. This is the reason why simplified solutions to the problem of gate sizing with intermediate interconnect [9]–[10] work well in most practical cases. However, even small differences of the input capacitance used by the path negatively affect the delay. In Fig. 3(b) the normalized sum of the input capacitances of all the gates of the circuit is shown. It is evident that the gain in area and energy can be of more than 20% when increasing f_2 1.5× its minimum value. The corresponding delay loss is less than 4%. From both figures we conclude that significant gains can be achieved when varying the stage effort of the rear paths with almost no delay cost. In contrast it is advantageous to use the maximum allowed input capacitance of the path, since for smaller values the delay rapidly increases without offering significant energy savings.

3.2   Extension to Arbitrary Logic Paths

Real circuits contain several types of logic gates, characterized by different values of logical effort and parasitic delay. Also the gates' output are additionally loaded

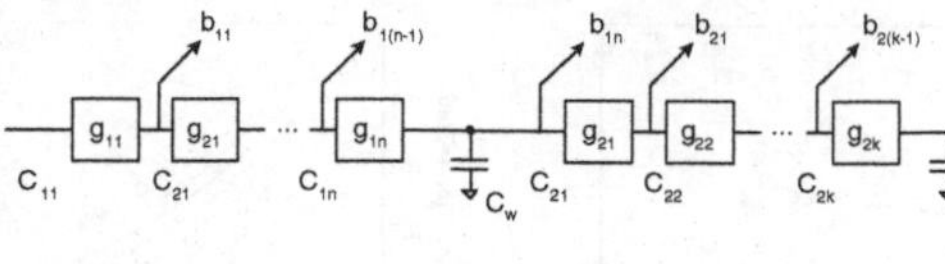

Fig. 4. A general logic path

via branching to other paths of the circuit. All these features of arbitrary logic paths are well handled by the method of logical effort. The delay of a general path with intermediate wire capacitance (see Fig. 4), can be expressed as

$$D = g_{11}b_{11}\frac{C_{12}}{C_{11}} + g_{12}b_{12}\frac{C_{13}}{C_{12}} + \cdots + g_{1n}\frac{C_w + b_{1n}C_{21}}{C_{1n}} + g_{21}b_{21}\frac{C_{22}}{C_{21}} + \cdots + g_{2k}\frac{C_L}{C_{2k}}$$

In order the delay of the path to be minimized, the stage efforts of the first and the second chain of gates should satisfy the following equations.

$$f_1^{n-1}(f_1-f_2) = G_A{\cdot}B_A{\cdot}\frac{C_W}{C_{11}} \quad \text{and} \quad f_1^{n-1}f_2^{k+1} = (G_A{\cdot}G_B){\cdot}(B_A{\cdot}B_B){\cdot}\frac{C_L}{C_{11}} \quad (12)$$

G_A is the product of the logical efforts of the gates of first chain (up to the intermediate wire) and G_B the corresponding logical-effort product of the gates of the second chain (after the wire capacitance). The branching efforts of the two paths B_A and B_B are defined in a similar way. The new equations have the same format as in the case of the two inverter chains (Eq. (3) and (4)). However, the new terms on the right side of both equations contain also the products of the logical and branching effort of each path, which resembles the definition of path effort [ ]. The main difference is that when wire capacitance is considered, two separate definitions of the path effort are needed.

The bounds (9)–(11) can also provide the stage efforts f_1 and f_2 in the case of a general logic path by substituting C_L and C_w with their effective equivalents. The effective wire and load capacitances $C_{w,\text{eff}}$ and $C_{L,\text{eff}}$ are defined as

$$C_{w,\text{eff}} = C_w \prod_{i=1}^{n-1}(g_{1i}b_{1i}) \quad \text{and} \quad C_{L,\text{eff}} = C_L \prod_{i=1}^{n}(g_{1i}b_{1i}) \prod_{i=1}^{k}(g_{2i}b_{2i}) \quad (13)$$

It should be noted that $C_{L,\text{eff}}$ contains the product of the logical and the branching efforts of all the gates from the source up to the end of the path, while $C_{w,\text{eff}}$ uses only the gates before the wiring capacitance, excluding the last one. In the general case, in order to compute the input capacitances of the gates, besides the computed stage efforts, the logical and branching effort of each gate should be also taken into account [ ].

4   Optimal Number of Stages

Having an accurate and easy-to-use method for deriving the optimal stage efforts, the designer can easily and quickly decide the number of stages n and k that

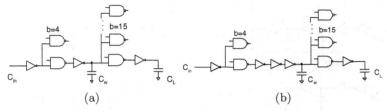

Fig. 5. Selecting the optimal number of stages

are better suited to drive both the wiring and the output load capacitance. It is well known that the best number of stages for the simple chain of gates with no intermediate interconnect capacitance is roughly equal to $\log_4(G \cdot B \cdot C_L/C_{in})$.

When interconnect capacitances are included, it is difficult to derive a simple metric that gives the optimal number of stages. Hence, we propose a simple approach that uses only the derived bounds. Assume for example that $C_{in} = 5$fF, $C_w = 300$fF and $C_L = 100$fF. We would like to simply evaluate in terms of delay the two alternatives shown in Fig. 5. For the first case the effective loading and wiring capacitances are equal to $C_{w,\text{eff}} = 300 \times g_{\text{nand}} \times 4$, $C_{L,\text{eff}} = 50 \times g_{\text{nand}}^2 \times (4 \times 15)$, where in our technology $g_{\text{nand}} = 1.18$ and $p_{\text{nand}} = 1.71$. Therefore, using (9)–(11) the expected stage efforts are equal to $f_1 = 7.75$ and $f_2 = 3.03$ and thus the expected minimum delay, including parasitic delays, is 36τ. Following the same procedure for the second case the expected minimum delay is 31.3τ ($f_1 = 3.66$ and $f_2 = 2.1$). Therefore with just a few value substitutions we conclude that the second alternative, i.e., five stages in the first chain, is advantageous and gives a 13% faster design. When dealing with large intermediate wire loads it is better to add more stages before the wire load so as to minimize its effect.

5   Handling Multiple Levels of Interconnect

In cases that more than two logic paths are separated by fixed interconnect capacitances, the optimal stage effort of each path can be computed using a simple repetitive procedure. In practice, no more than three or four stages of considerable intermediate wire capacitances exist before some form of latching occurs to the outputs of the combinational logic. The proposed method aims in closely approximating the required stage effort of the first chain of gates and then recursively compute the efforts of the remaining gates so as delay to be minimized. Consider for example the path shown in Fig 6. Differentiating the delay of the path with respect to the input capacitances of the gates we get that

$$f_1(f_1 - f_2) = \frac{C_A}{C_{in}}, \quad f_1 f_2(f_2 - f_3) = \frac{C_B}{C_{in}}, \text{ and } f_1 f_2 f_3^2 = \frac{C_L}{C_{in}}. \tag{14}$$

Solving exactly the three non-linear equations gives that the minimum delay equals to $(12.45 + 3p_{inv})\tau$ and the optimal stage efforts f_1, f_2, and f_3 are equal to 4.63, 2.47, and 0.72, respectively.

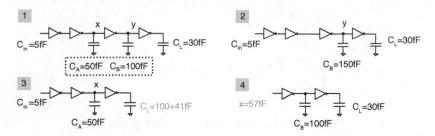

Fig. 6. Steps of the recursive procedure that handles multiple levels of intermediate interconnect

At first, we make the simplifying assumption that $f_1 = f_2$. The assumption that the gates of the first and the second chain have the same effort can be satisfied in two ways. The first one is to ignore the first wiring stage, i.e, C_A is zero [10]. As shown in Section 3, this approach clearly underestimates the input capacitance of the gates closer to the end of the path. In the proposed method we choose an alternative approach. Adding the two first equations of (14) and setting stage effort f_1 equal to f_2 it is derived that $f_1^2(f_1 - f_3) = (C_A + C_B)/w_1$ and $f_1^2 f_3^2 = C_L/w_1$. The new equations can be translated to the equivalent circuit shown in step 2 of Fig. 6, where the effect of the first wire is added to the next level. The resulting path can be easily sized following the procedure described in Section 3. Computing the bounds of f_1 and f_3 it is derived that for minimum delay the input capacitance of the last stage should be $y = 41\text{fF}$.

Since we have one first approximation of the input capacitance of the last stage we continue with step 3, where the loading capacitance is replaced by the combined effect of C_B and the value derived in the first step. Substituting the new parameters to (9)–(11) we get that the input capacitance of the second chain x should be equal to 57fF and the stage effort of the first chain is $f_1 = 4.628$.

At this point we assume that the stage effort of the first chain has been correctly approximated. In fact this is true in our example. Although this seems a rough estimate, it is a valid assumption. According to the form of equations (14) when increasing the stage effort of a chain closer to the input, the following chains immediately require a smaller stage effort. In this way, when one chain gets a near optimal value all the other stage efforts are simultaneously bounded closer to their minimum values too. Continuing with our example, after fixing the input capacitance of the second chain to 57fF, the remaining path needs to be sized (step 4). The resulting stage efforts f_2 and f_3 are equal to 2.48 and 0.725 while the new input capacitance of the last chain equals $y = 41.3\text{fF}$. Since the initial approximation of $y \sim 41\text{fF}$ is roughly equal to the final value, the procedure stops since all stage efforts have converged closely to their minimum values. It can be verified that the computed stage efforts are almost exact.

We have experimentally verified for 1000 randomly generated paths that only one iteration suffices to predict the minimum delay of the circuit with no more than 2% error. If the initial and the final approximation differ significantly one more backward iteration is required to get the exact value. The new iteration should start with step 3, where the output load is increased by the new estimated capacitance. In cases of more intermediate interconnect levels the recursive procedure works exactly the same way. In each step all preceding wiring capacitances are added to the wiring capacitance closer to the output and new estimates are computed until the stage effort of the first path converges to its optimal value.

6 Conclusions

A simple and accurate method for performing gate sizing with intermediate fixed wire capacitances has been presented in this paper. Wiring characteristics are predominant in very-deep submicron technologies, and their effects need to be analyzed and solved early in the design cycle. Therefore, the designer can truly benefit by the adoption of the proposed approach. The development of a practical framework that would treat in a unified manner both wire resistance and capacitance during gate sizing is a subject of ongoing research.

References

1. D. H. Allen *et al.*, "Custom Circuit Design as a Driver of Microprocessor Performance", IBM J. Res. Develop., vol. 44, No. 6, pp. 799 - 822, Nov. 2003.
2. I. Sunderland, B. Sproul, D. Harris, "Logical Effort: Designing Fast CMOS Circuits", Morgan Kaufmann, 1999.
3. N. Hedenstierna, K. O. Jeppson, "CMOS Circuit Speed and Buffer Optimization", IEEE Trans. on CAD, no. 2, pp. 270 - 281 ,March 1987.
4. V. Sundararajan, S. Sapatnekar, K. Parhi, "Fast and Exact Transistor Sizing Based on Iterative Relaxation", IEEE Trans. on CAD, no. 5, pp. 568 - 581, May 2002.
5. K. H. Shepard, *et al.*, "Design Methodology for the S/390 Parallel Enteprise Server G4 Microprocessors", IBM J. Res. Develop., no. 4/5, pp. 515 - 547, Jul./Sep. 1997.
6. Magma Design Automation, "Gain-Based Synthesis: Speeding RTL to Silicon".
7. R. Ho, K. W. Mai, M. A. Horowitz, "The Future of Wires", Proc. of the IEEE, no. 4, pp. 490 - 504, April 2001.
8. W. Dally, J. Poulton, "Digital Systems Engineering", Cambridge Univ. Press, 1998.
9. B. Amrutur, M. A. Horowitz, "Fast Low-Power Decoders for RAMS", IEEE J. Solid-State Circuits, no. 10, pp. 1506 - 1514, Oct. 2001.
10. M. Horowitz, "Logical Effort Revisited", http://eeclass.stanford.edu/ee371
11. S. K. Karandikar, S. Sapatnekar, "Fast Comparisons of Circuit Implementations", Proc. of Design Automation and Test in Europe, pp. 910-915, 2004.

Efficient Simulation of Power/Ground Networks
with Package and Vias*

Jin Shi[1], Yici Cai[1], Xianlong Hong[1], and Shelton X.D. Tan[2]

[1] Department of Computer Science and Technology, Tsinghua University,
Beijing 100084, P.R. China
shi-j03@mails.tsinghua.edu.cn, {caiyc,hxl-dcs}@tsinghua.edu.cn
[2] Department of Electrical Engineering, University of California at Riverside, CA 92521, USA
stan@ee.ucr.edu

Abstract. As the number of metal layers and the frequency of VLSI continue to increase, the voltage droop on both the package and vias is becoming more pronounced. This paper analyzes the numerical problem encountered in simulation of power/ground networks together with C4 package and via model. A new preconditioned iterative method and associated acceleration technique are introduced to overcome shortages of previous methods. The new method not only is effective to speedup simulation without loosing any accuracy, but also extends the usage of preconditioned method to general circuit simulation using only a few additional memories due to its MNA formulation.

1 Introduction

As the operating frequency of modern CPU and ASIC chips is rising rapidly according to Moore's law, the voltage droop on power/ground (p/g) networks must be considered for performance estimations. In addition, recent research has shown that if the package models are ignored in the design of p/g networks, the operation of the chip may suffer from special current surge conditions [1]. In order to maintain a reliable voltage supply, sometimes more than 30% chip area is used for the p/g network. As a result, under present manufacture process, usually the p/g network is assigned to more than 4 metal layers. The excessive use of metal layers leads to larger number of vias, which is hundred times more than before. Also, in many cases provided by industrial, the resistance of the vias can be as hundred times larger than the sheet resistance at the same metal layer. Therefore, today's analysis tool should take via resistance into account. Although modeling and simulation of the p/g network are time consuming computational processes, several efficient techniques have been proposed to tackle the problem [2] [3] [4] [5]. Among these methods, the *preconditioned Krylov-subspace iteration* method is one of the most efficient and stable strategies and it is widely used in today's p/g network simulation. However, existing methods using preconditioner-based scheme requires that the circuit matrix is symmetric positive definite, which is satisfied by *Nodal Analysis* (NA) formulation of the circuit matrix. But NA formulation requires admittance forms for all the circuit elements and is not as general as

---

* This work is supported by National Natural Science Foundation of China (NSFC) 60476014 and National Hi-tech R&D Program of China 2005AA1Z1230

V. Paliouras, J. Vounckx, and D. Verkest (Eds.): PATMOS 2005, LNCS 3728, pp. 318–328, 2005.

Modified Nodal Analysis (MNA) form. If we use a MNA formulation and include the package model and vias, we can find that serious numerical problems are created in simulation.

In our paper, we study these new numerical characters and make some improvements over the classical *Krylov-subspace iteration* based method. The new preconditioned method extends the application of *incomplete choleskey decomposition* from NA form to MNA form. It also speeds up the iteration process by exploring the new numerical characters. Experiment results show that the new method is very efficient and stable for performing demanding simulation of p/g networks in the presence of package and a large number of vias. This paper is organized like this: Section 2 introduces the new p/g network model with package and vias and then analyzes its new numerical characters of the resulting networks; Section 3 briefly reviews the previous NA based methods and their shortages on solving the new problem; Section 4 introduces our new preconditioned method which can be used in MNA form with more efficiency; Section 5 describes a speed up technique based on the new character of the matrix; Finally, Section 6 summarizes the whole paper.

2 New Model of P/G Network and Its Numerical Characters

2.1 Model of C4 Package

For the complexity of VLSI grows rapidly, general wire-bond based package such as CPGA, PPGA is becoming difficult to afford the large quantities of both I/O and power/ground pins due to large power dissipation, it will introduce obvious package resistance and inductance. On the other hand, flip-chip package is becoming more attractive because of its good electrical performance [6]. In our paper, we use a popular C4 package model provided by our industry partners to simulate the voltage droop at each die bump. Figure 1 shows a model for two layer metal trace C4 package. Here each bump on the die connects to a power/ground pin on package through two metal layers and one via. $L_{top}, R_{top}, L_{bot}$ and R_{bot} represent the inductance and resistance

of package metal trace while L_{via} and R_{via} represent the inductance and resistance of package via. There also exist package decoupling capacitance, so $C_{pkg}, R_{cpkg}, L_{cpkg}$ are used to represent its parameters.

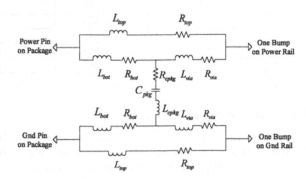

Fig. 1. Two metal layer C4 package model

2.2 Model of Power Rails and Vias

There are many models to describe the power rail, i.e., R or RLC lumped models, transmission line models. Usually, the length of power rail between two adjacent vias is not long enough to cause transmission-line effect in the operating frequency. But this is not always true [7]. This implies that in typical p/g network, all the metal rails in the die are modeled as resistors. Capacitance and inductance are only considered when they affect the power supply significantly, e.g., the decoupling capacitor and the inductance and capacitance of the package. However, most previous modeling of p/g network neglect the vias between two metal layers, this is partially because there is not so many metal layers before, or at that time the via resistance can be neglected when compared to sheet resistance. Table 1 gives typical sheet and via resistance according to an actual CMOS process. As we can see, via resistance exceeds the sheet resistance by more than 10x in layers close to M1. Other typical electrical parameters in different CMOS process can be found in [8]. Figure 2 illustrates our new model considering vias comparing with previous one. We can see in section 2.3 and section 5 that this different topology could introduce both good and bad characters in simualation. Also, this model is easy to extent to a RLC model.

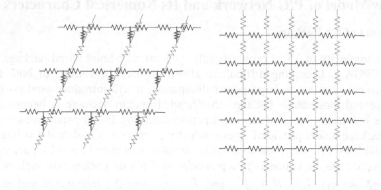

Fig. 2. P/G network models with and without vias

Table 1. Comparison of sheet resistance and via resistance

Metal Layer	Sheet Resistance mΩ	Metal Layer	Via Resistance mΩ
M7	22	M7-M6	300
M6	37	M6-M5	200
M5	55	M5-M4	800
M4	71	M4-M3	1250
M3	90	M3-M2	1300
M2	90		

2.3 Matrix Shape and New Numerical Characters

In this paper, we use MNA method to generate the circuit matrix. It is almost the same as the NA method except for dealing with independent voltage source and extra independent current variable. We'll introduce the matrix form in section 4. Here we just focus on the topology, the stamp strategy and their effect on the matrix shape.

According to MNA methods and trapezoidal discretization [9], any element in the p/g network would stamp a 'quadrangle fill' in the matrix, as shown in Figure 3. In our model, via is modeled as a resistor, so the connectivity of a node in the network is changed. Previously, each node is surrounded by four resistors in a mesh, two in the upper layer and two in the nether layer. But now, each node just connects to two resistors adjacent in the same layer as well as a via, as shown in Figure 2. Therefore, if we make the node sequence number increasing along with each rail, either horizontally or vertically in a certain layer, the matrix shape would be quite different as before which is shown in figure 4.

$$
\begin{bmatrix}
\ddots & \cdots & \cdots & \cdots & \cdots \\
\cdots & g & \cdots & -g & \cdots \\
\cdots & \cdots & \ddots & \cdots & \cdots \\
\cdots & -g & \cdots & g & \cdots \\
\cdots & \cdots & \cdots & \cdots & \ddots
\end{bmatrix}
$$

Fig. 3. 'Quadrilateral fill in' during matrix generation

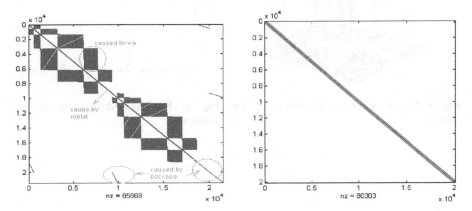

Fig. 4. Comparison between two typical simulation matrixes with and without vias and package

From Figure 4, first of all, we observe that the new matrix is wider in band obvious than previous one, which is usually a poor character in numerical simulation due to more fill-ins and slow convergence. Secondly, the distribution of the non-zero elements is quite different. In previous matrix, because of the mesh topology, non-zero elements caused by metal slices in power rails distribute around the main diagonal, not only stamp the main diagonal but also stamp other sub diagonals nearby. But now, the mesh topology is separated by vias, so the non-zero elements which represent metal slices only stamp in the main diagonal and two sub diagonals adjacent, while other non-zero elements caused by both package and vias are left outside the triple diagonals. This is an attractive numerical character which can be exploited (see Section 5). Thirdly, because both the resistance value of vias and package parameters after discretization are magnitude away from the value of metal slices in power/ground rails, the distribution of large non-zero elements is extremely unbalanced. Figure 5 shows noticeable valleys near and far away from the main diagonal, both of which can overwhelm the other elements in quantity. Also, the metal width usually decreases fast from the M7 to M2, which can cause the conductance varying dramatically in the diagonal of the matrix. This situation could become even worse as the number of layers increases. Figure 6 shows a typical numerical distribution in main diagonal. We can see that the maximum element is hundred times larger than the minimal one. All these unbalanced numerical distributions affect the eigenvalue of the matrix a lot according to *Gershgorin Circle Theorem*. Figure 7 compares the eigen-value of two kinds of matrixes. It is obvious that the cluster effect becomes even worse and the max to min ratio of eigenvalue goes larger, both of which can directly lead to slow convergence rate of the iteration methods. Finally, because of the huge number of vias, the matrix is not as sparse as before.

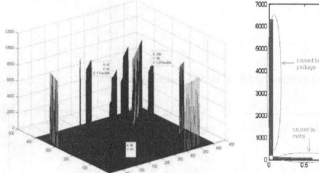

Fig. 5. Typical numerical distribution of new matrix

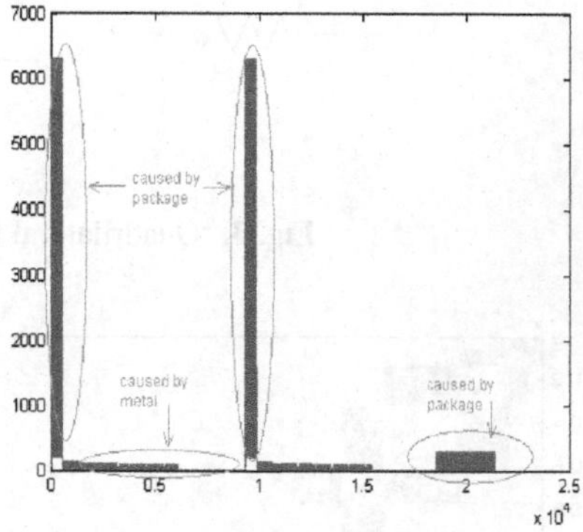

Fig. 6. Numerical distribution in main diagonal

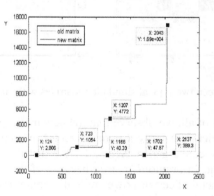

Fig. 7. Comparison of two kinds of matrix on eigenvalue

3   Shortages of Previous Method

Many methods are constructed to perform efficient simulation of p/g network in the past. Among them the *Krylov-subspace iteration* method [2] is proved efficient, accurate and stable by many industrial cases, while the random walk method [4] is criticized for its accuracy and the ADI method [3] is criticized for its stability. Without considering the package parameters and vias, we do not see such a stiff eigenvalue problem. So the system is relative easy to solve using iteration method. However, because method in [2] highly relies on its preconditioner, so changes in matrix shape and numerical characters affect its speed a lot. Usually, two kinds of preconditioners are widely used in this method, one is *incomplete LU decomposition* (ILU), and the other one is *incomplete choleskey decomposition* (ICD). Table 2 lists the iteration times of two kinds of matrixes. Here matrix I is a 12Kx12K matrix with package and vias while matrix II is a 12Kx12K matrix without package and vias. 'Zero fill in' *Preconditioned Conjugate Gradient* (PCG) method is used as the basic iteration method and both two preconditioner are tested. We can see that the inclusion of package and vias increases the iteration times a lot. Another shortage of previous iteration method is that CG method preconditioned by ILU and ICD can not be used to indefinite matrix directly, which limits it to circuits formulated in NA form [2]. This is a fatal weakness of it, which is widely criticized because general circuit simulation requires MNA formulation.

Table 2. Comparison of the iteration times

ILU	Iteration Times		ICD	Iteration Times	
residual	Matrix I	Matrix II	residual	Matrix I	Matrix II
1e-3	165	5	1e-3	166	5
1e-6	258	8	1e-6	258	8
1e-10	377	11	1e-10	375	11

4 New Preconditioned Method Fitting MNA

If the matrix is constructed in MNA form, its shape is given in (1). Here matrix G is a $(n+m) \times (n+m)$ matrix. A is a *sparse positive definite* (s.p.d) $n \times n$ M matrix while B is an $m \times n$ sparse matrix with rank equaling to m

$$G = \begin{bmatrix} A & B^T \\ B & 0 \end{bmatrix} \qquad G' = \begin{bmatrix} A & B^T \\ B & kI \end{bmatrix} \tag{1}$$

Here matrix B is caused by the independent voltage source (pads) and extra independent variable such as the current of inductor. Usually, a lot of bumps exist in C4 package, which can cause m easily to be more than one thousand according to our package model. Also, it causes m zero entries in the main diagnoal. So now matrix G is a indefinite matrix which has m negative eigenvalue. Although we can use another matrix G' to appoximate G and get preconditioner using ILU and ICD, the effectiveness of these preconditioner is degraded becuase m is realtive large. We begin our improvement with ICD method as it uses much less memory than ILU method while shares the same convergence speed.

Theorem 1: $L = \begin{bmatrix} l_{11} & 0 \\ l_{21} & l_{22} \end{bmatrix}$ and $U = \begin{bmatrix} l_{11}^T & l_{21}^T \\ 0 & -l_{22}^T \end{bmatrix}$ are perfect preconditioners of G

matrix, where l_{11} is the choleskey decomposition of matrix A, $l_{21} = B \cdot l_{11}^{-T}$, l_{22} is the choleskey decomposition of matrix $l_{21} \cdot l_{21}^T$

Proof: Because l_{11} is the choleskey decomposition of matrix A, $l_{11} \cdot l_{11}^T = A$ holds.

Also $l_{22} \cdot l_{22}^T = l_{21} \cdot l_{21}^T$ holds (here l_{21} is an mxn matrix while l_{22} is an mxm matrix).

Choose two identity matrix I (nxn) and I' (mxm), considering $l_{21} \cdot l_{11}^T = B$, then we can get (2)-(3) below. Because L and U are triangular matrixes full in rank, so (2) can be rewritten as (3). This means that using L as left pre-multiplier and U as right pre-multiplier, the condition number of iteration matrix can be reduced to 1, so L and U are perfect preconditioners in theory. Also, the replacement of l_{11} and l_{11}^T by LU factors causes no affection to this theorem.

$$L\begin{bmatrix} I & 0 \\ 0 & -I' \end{bmatrix} U = \begin{bmatrix} l_{11} & 0 \\ l_{21} & l_{22} \end{bmatrix} \cdot \begin{bmatrix} I & 0 \\ 0 & -I' \end{bmatrix} \cdot \begin{bmatrix} l_{11}^T & l_{21}^T \\ 0 & -l_{22}^T \end{bmatrix} = \begin{bmatrix} l_{11} & 0 \\ l_{21} & -l_{22} \end{bmatrix} \cdot \begin{bmatrix} l_{11}^T & l_{21}^T \\ 0 & l_{22}^T \end{bmatrix} = \begin{bmatrix} A & B^T \\ B & 0 \end{bmatrix} = G \quad (2)$$

$$L^{-1} \cdot G \cdot U^{-1} = \begin{bmatrix} I & \\ & -I' \end{bmatrix} \quad (3)$$

Theorem 2: Both the time and memory cost to generate L and U is low.

Proof

a) In practical, for A is an M matrix, the *incomplete choleskey decomposition* is guaranteed to exist [10]. So we can use it as approximate decomposition. Here both zero fill in strategy or threshold guided fill in strategy can be used to control the memory usage [10]. $l_{21} \cdot l_{21}^T$ is an s.p.d matrix further because of the connectivity of the pads and other independent variables, it is also a diagonal dominant matrix, so the incomplete choleskey decomposition exists too.

b) Usually, l_{11}^{-T} is not a sparse matrix, if we have to generate the full matrix, the memory usage is prohibited when n is very large. Luckily, we only need to compute $l_{21} = B \cdot l_{11}^{-T}$. Because the B matrix is very sparse (no more than two elements per row), and elements in l_{11}^{-T} can be known in advance without iteration as it is a triangular matrix, just little extra memory is needed when generating l_{21}.

c) Computation cost of l_{21} is low

$$p(i,i) = \frac{1}{l_{11}(i,i)} \qquad l_{11}^{-1}(i,j) = \begin{cases} i=j & p(i,i) \\ i<j & 0 \\ i>j & -p(i,i) \cdot \sum_{k=j}^{i-1} l_{11}(i,k) \cdot p(k,j) \end{cases} \quad (4)$$

$$l_{21}(i,j) = \sum_{b(i,k)\neq 0} b(i,k) \cdot l_{11}^{-T}(k,j) = \sum_{b(i,k)\neq 0} b(i,k) \cdot l_{11}^{-1}(j,k) \qquad b(i,k) \in B \quad (5)$$

Theorem 3: *Symmetrical Lanczos* (symmlq) method [11] is a good method to solve our indefinite system using approximate preconditioner L and L^T

Proof: Because G is a symmetrical indefinite matrix, *Ritz Minimization principle* do not hold any more, which means the *cg* method can not be used to get the solution. Instead, *symmlq method* can solve the project system without minimize anything [11]. Further, *symmlq method* is more stable for matrixes that have extremely eigenvalue. However, *symmlq method* requires positive definite matrix as its preconditioner, direct use of matrix U as a preconditioner will cause problem, but if we use L^T, this problem is overcome. Although *Quasi-Minimal Residual method* (QMR) can take indefinite matrix as its preconditioner, the computation cost is much higher than *symmlq method* according to our experiments. Later, we can see that, by using a speed up technique, the approximate precondioner almost loses little in efficiency.

5 Balance Technique to Speed Up

Usually, threshold guided fill in strategy is used to speed up the iteration. It is a trade off between speed and memory usage. The more fill in we allowed, the faster convergence speed it will reached. Although *Reverse Cuthill-McKee* (RCM) *ordering* [12] and *Minimum Degree Ordering* [13] techniques can be used to partially reduce the fill in and compress the matrix in the band form, they almost can not improve the convergence speed under the same threshold in our application as shown in Table 3. The original reason to use ordering is that under the same memory usage constraint, smaller threshold can be reached. However, in our new method below, we can see that this shortage is overcome. Figure 8 shows the shape of matrix in Figure 4 after RCM ordering, and Table 3 gives brief statistic information about the memory usage and the iteration times. The fill in ratio there is the memory usage of preconditioner divided by the half memory usage of the original matrix.

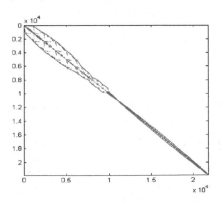

Fig. 8. Matrix after RCM ordering

In section 2, we have analyzed the matrix numerical character and find that the poor eigenvalue is partially caused by the unbalanced distribution of large non-zero elements. Because the new topology has made the metal slice just to stamp in the triple diagonals only, we can utilized this character to reduce the condition number of this matrix which can lead to speed up directly. Our strategy is based on two steps. In the fist step, we reduce the value of non-zero elements outside the triple diagonals using balance transform B. In the second step, we raise the value of non-zero elements in the triple diagonals using balance transform A, as shown in Figure 9. If we design the transform carefully, we can guarantee that the original resistor is equivalent to the transformed two ports network using circuit theory. So no accuracy is sacrificed except introducing some virtual nodes. Because all non-zero elements in admittance matrix is caused by stamp of conductors (caused by metal, via, inductance and capacitance), an ideal situation is that if there is no memory usage constraint, the balance transform will cause the non-zero elements in the triple diagonals to become dramatically larger than that outside the triple diagonals. Therefore the admittance matrix can be treated as a triple diagonal matrix and become very easy to solve. However, this character does not exist in traditional mesh structure p/g network for elements do not just stamp in the triple diagonals of the matrix, and the balance transform even makes the condition number of the old matrix to become worse. In practical usage, sophisticated adaptive algorithm is generated to speed up the solving process considering both the package and via parameters and the memory usage.

Algorithm Name: adaptive balancer

Input: topology of the p/g network, electrical parameters, simulation time step, assistant memory N (T<N<kT). T is original node number, k is a constant.

Output: balanced matrix

1) compute typical conductance of metal slice and via g_m, g_v in all layers

2) discrete in time domain, get typical g_l, g_c

3) sort all typical conductance and get the maximal typical conductance g_{max}

4) using balance transform B to reduce elements' value out of triple diagonals

5) for(i=1;i<=layer number;i++)

{add p_i virtual nodes between two vias using balance transform A

n_i is the number of actual nodes in layer i

$$p_i = \min\left\{\left\lceil \frac{N}{T} \cdot e^{-\frac{n_i}{T}} \right\rceil, \left\lceil \frac{g_{max}}{g_{mi}} \right\rceil\right\} \qquad T = \sum_1^k n_i \}$$

6) output the matrix

Table 3. Comparison between odering and no odering

No odering residual=1e-6		
threshold	Fill in ratio	Avg Iter times
1e-1	1.5	221
1e-2	2.79	93
1e-3	10.96	44
1e-4	48.41	17
Odering residual=1e-6		
threshold	Fill in ratio	Avg Iter times
1e-1	1.1	220
1e-2	2.32	95
1e-3	6.93	46
1e-4	18.55	17

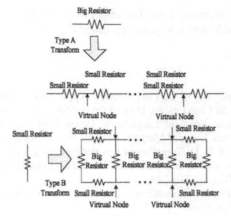

Fig. 9. Balance Transform A & B

In this algorithm, if the assistant memory N we allocated is not enough to perform an ideal balance, layers which have less metal slices will get the higher priority to be balanced. This algorithm is implemented using C++, and is tested on a Linux Work-station with 1 GHz CPU and 512M memory. Table 4 gives out the performance of this algorithm. In left part of the table we use MNA formulation and do iteration in symmlq form using our new preconditioner together with RCM ordering and balance techniques while in right part of the table ordinary pcg method in NA formulation and ICD precondition is used. From the table we can see that new algorithm is one magni-tude faster than general method while the memory usage just rises slightly.

Table 4. Performance of the speed up technique

Node Num	Avg Iter times	CPU time (sec.)	Mem Usage (MB)	Speed Up Ratio	Node Num	Avg Iter times	CPU time (sec.)	Mem usage (MB)
5.5K	15	0.53	65	3.18	5.5K	208	1.69	59
23K	19	1.13	78	9.20	22K	293	10.4	70
130K	30	5.37	148	13.01	100K	456	70.2	136
450K	40	21.2	366	19.79	450K	710	419.6	349

Also, the small average iteration times proved that our new precondioner could fit MNA formulation very well. Further, if the scale of power/ground network and the dense of the bumps increase, or RLC model is used, we can get more benefit by using this strategy as both of those models will cause worse condition number.

6 Conclusion

In this paper, we have analyzed numerical characters of circuit matrices in the pres-ence of package and massive vias in high-performance p/g networks. We then pro-posed a new preconditioned method, which can extent the preconditioned based itera-tive method to MNA form. Also by utilizing special characters, a novel balance technique was generated which can improve iteration convergence rate under a certain threshold without any accuracy loss. The experiment results show that this method is really efficient and stable while its memory usage is reasonable.

References

1. Howard H. Chen and David D. Ling: Power Supply Noise Analysis Methodology for Deep-Submicron VLSI Chip Design, DAC97 Proceedings, 638-643
2. T.Chen and C.C.Chen: Efficient Large-Scale Power Grid Analysis Based on Preconditioned Krylov-Subspace Iterative Methods, DAC01 Proceedings, 559-562
3. Weikun Guo, Sheldon X. D. Tan: Circuit Level Alternationg-Direction-Implicit Approach to Transient Analysis of Power Distribution Networks, International Conference on ASIC Proceedings, 2003, Beijing, 246-249
4. Haifeng Qian, Sani R. Nassif, Sachin S. Sapatnekar: Random Walks in a Supply Network, DAC2003 Proceedings, 93-98
5. J.N.Kozhaya, Sani R.Nassif and Farid N.Najm: A Multigrid-Like Technique for Power Grid Analysis, IEEE Trans. Computer-Aided Design, vol.21, no.10, Oct. 2002, 1148-1160
6. Performance Characteristics of IC Packages, Intel Corp., 2000
   http://www.intel.com/design/packtech/ch_04.pdf
7. Howard Johnson, Martin Graham and etc.: High-Speed Digital Design a Handbook of Black Magic, Pearson Education Press. 1997
8. Michael John, Sebastian Smith: Application-Specific Integrated Circuits, Pearson Education Press, 1997
9. L. W. Nagel: SPICE2: A Computer Program to Simulate Semiconductor Circuits, ERL Memo. No. UCB/ERL Vol M75/520 (1975)
10. Saad, Yousef, Iterative Methods for Sparse Linear Systems, PWS Publishing Company, 1996.
11. Barrett, R., M. Berry, T. F. Chan, et al., Templates for the Solution of Linear Systems: Building Blocks for Iterative Methods, SIAM, Philadelphia, 1994
12. George, Alan and Joseph Liu, Computer Solution of Large Sparse Positive Definite Systems, Prentice-Hall, 1981.
13. Amestoy, P. R., Davis, T. A., and Duff, I. S.: An approximate minimum degree ordering algorithm. SIAM J. Matrix Anal. Applic. 17, 4, 886,905.

Output Resistance Scaling Model for Deep-Submicron Cmos Buffers for Timing Performance Optimisation

Gregorio Cappuccino, Andrea Pugliese, and Giuseppe Cocorullo

Electronics, Computer Science and Systems Department,
DEIS-University of Calabria, 87036 Rende, Italy
{cappuccino,a.pugliese,cocorullo}@deis.unical.it

Abstract. CMOS buffers driving high capacitive loads play an important role in determining the overall performances of present-day complex integrated circuits. Ad hoc optimisation techniques require effective models simple enough to enable the optimisation of multi-million gate circuits.

However, due to size and supply voltage reduction, the behaviour of deep-sub micron devices may differ significantly from the conventional one and the classical models and techniques need to be improved or radically modified. In the paper the Authors show how the well-exploited assumption of a linear relationship between the channel width and the current meanly conducted during the switching transient may be incorrect for deep-submicron buffer transistors. As direct consequence, it follows that the common practice of widening the channel width of the MOSFET transistors may not lead to the expected results in terms of buffer output resistance reduction.

On these bases, a novel expression to estimate the actual effect of a channel widening on the output resistance of CMOS buffer that agrees with HSPICE circuit simulations within 3% error is presented.

1 Introduction

Recent advances in CMOS technology have allowed continued improvements in performances, costs and miniaturisation of integrated circuit.

Obviously the design and optimisation of complex circuits require adequate device models [1,2,3], but solutions providing high accuracy, though suitable for small designs, become inapplicable to a present-day chip that contains several millions of devices [2,3,4].

A typical example is the improvement of speed performances, for which delay minimisation requires accurate but efficient model to capture the transient behaviour of the driver/load systems.

One of the most useful (and used) model for a MOSFET buffer is a linear resistor representing the equivalent resistance [2,4] of the PMOS or NMOS transistor during a low-to-high or high-to-low transition, respectively.

Even though it represents a first order approximation of a complex, strongly non linear relationship between buffer transistor current and output voltage, this model is currently used not only in manual analysis of digital circuits, but it is also widely exploited for developing delay optimisation techniques, such as repeater insertion, buffer tapering, etc [3,5,6].

In accord with this straightforward model, the driver is seen just as a resistor, connected either to the supply voltage V_{DD} or to ground, depending on the logical value of the output.

V. Paliouras, J. Vounckx, and D. Verkest (Eds.): PATMOS 2005, LNCS 3728, pp. 329–336, 2005.

This resistance being strongly time varying, non linear and depending on the operation point of the buffer transistor [2,4], the main aim of the previous works about this topic has been to find an appropriate constant value for it.

However the common conclusion that can be found in literature is that the equivalent resistance is inversely proportional to the W/L (width to length) ratios of the driver transistors, a conclusion that directly comes out from the relationship between channel dimensions and static drain current of a MOSFET.

This latter assumption is at the basis of all conventional buffer sizing criterions [1,2,4,6] and delay models [7,8,9] presented in the past, in which a proportional decreasing of the propagation delay is expected when the channel width of the transistors of the buffer is increased.

The work highlights how this common assumption may be incorrect for present-day technologies. Moreover, on the basis of experimental results, a developed expression to predict the actual effect of transistor widening on the output resistance of the deep-submicron buffer is presented.

The paper is organised as follows: in section 2 a rapid presentation of the classic output resistance model for a CMOS buffer is reported, together with experimental results carried out for different deep-submicron technologies.

In part 3 the proposed expression is presented. In the same section a comparison between the results carried out by proposed model and the conventional one is reported. Finally, some conclusions are drawn in section 4.

2   Channel Width and Output Resistance

One of the most effective way to capture the I/V characteristic of a real MOSFET transistor is to express the drain current by means of three different equations, corresponding to the cut-off, triode and saturation region respectively [2].

The most common are those reported in the following, derived by the Shichman-Hodges model [7], but several examples of more advanced extensions have been presented in literature, as in [8].

The equations refer to an NMOS, but it being understood that all expressions are applicable to the P-Channel MOS, once it is recognized that all the voltage and current polarities in a PMOS transistor are opposite to the corresponding quantities in a NMOS device.

$$
\begin{cases}
I_D = 0 \\
\qquad \text{for} \quad V_{GT} \equiv V_{GS} - V_T \leq 0 \\
\\
I_D = k \dfrac{W}{L}\left(V_{GT}V_{DS} - \dfrac{V_{DS}^2}{2}\right) \\
\qquad \text{for} \quad V_{GT} \geq 0 \quad and \quad V_{DS} \leq V_{min} \equiv \min(V_{GT}, V_{DSAT}) \\
\\
I_D = k \dfrac{W}{L}\left(V_{GT}V_{min} - \dfrac{V_{min}^2}{2}\right)(1 + \lambda(V_{DS} - V_{min})) \\
\\
\qquad \text{for} \quad V_{GT} \geq 0 \quad and \quad V_{DS} \geq V_{min}
\end{cases}
\tag{1}
$$

In (1), V_T is the threshold voltage, k represents the process transconductance parameter and λ takes in to account the channel length modulation effect.

The term W in equation (1) takes into account the direct relationship between the drain current in static condition and the channel width, a relationship confirmed by device measurements also for deep-submicron technologies. The same relationship is exploited for a well-known buffer model widely used for hand calculation but also at the basis of various delay optimisation strategies and techniques [2,3,5,6], namely the effective output resistance.

The output resistance model has been developed on the assumption that in digital circuits the transistors are nothing more than switches with an internal resistance ranging from infinite when the switch is off, to a finite value of resistance when it is completely closed. However the need of a simple model leads to substitute this time varying, non-linear and operating-point-dependent resistance with a constant and linear resistance R_{eq} [2].

The model is as realistic as one is able to fix correctly the value of R_{eq} so that the final results are similar to what would be obtained with original transistor.

A valid approach can be found in [2], where the value of the resistor is carried out as the mean of the value assumed by the drain-to-source/drain-current ratio during an output transition occurring throughout the time interval $\Delta t = t_1 - t_0$:

$$R_{eq} = \frac{1}{\Delta t} \int_{t_0}^{t_1} \frac{V_{DS}(t)}{I_{DS}(t)} dt \tag{2}$$

From (1) and (2), under the hypothesis that the devices stay in saturation for the entire duration of the switching transient, after some algebraic manipulations, the following can be obtained:

$$R_{eq} \approx \frac{3}{4} \frac{V_{DD}}{I_{MAX}} \left(1 - \frac{7}{9} \lambda V_{DD} \right) \tag{3}$$

where I_{MAX} is the value of the drain current given by (1) when $V_{GS} = V_{DS} = V_{DD}$.

The relationship between the drain current and the aspect ratio of the channel (W/L), (a basic assumption made also in all other different expressions found in literature for R_{eq}) implies that the value of the equivalent resistance should be in inverse proportion with the channel width of the devices. In other words, the equivalent resistance R_{eq} of a transistor S times wider than the minimum-sized one should present, during the switching transient, a resistance S times smaller. R_{min}, being the resistance of the minimum sized buffer this can be written as

$$R_{eq} = \frac{R_{min}}{S} . \tag{4}$$

From a practical point of view, it follows that the knowledge of the actual current (or delay) of a reference transistor should suffice to predict the current (or delay) variations obtainable by reducing or increasing the channel width for a specified load.

This latter assumption does not take in to account the effect that a device re-sizing plays on the output voltage waveform: drivers with wider transistors will be characterised by faster output signal transition. Although this influence can be neglected in the case of older, mature technologies, for deep-submicron ones due to reduced margin between the supply voltage and transistor threshold voltage, this effect may alter significantly the device operation.

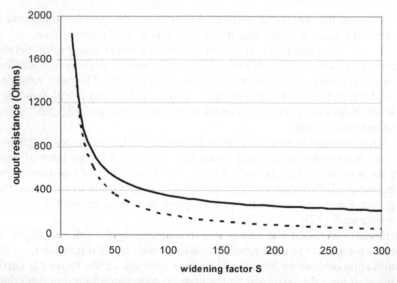

Fig. 1. Output resistance of a 100nm CMOS buffer as function of the channel widening S (R_{min}=1840Ω). In the plots the values carried out by HSPICE simulation (solid line) and those calculated in accord with equation (4), i.e. assuming a linear relationship between channel width and buffer current (dotted line) are compared

To demonstrate this unconventional behaviour, HSPICE analyses have been performed for a wide range of CMOS technologies.

The simulations have been carried out using both real CMOS technology models and the Berkeley Predictive Technology Models [9]. Plots of figure 1 present the trend of the equivalent resistance of a 100nm CMOS buffer driving a capacitive load C_L equal to 100fF.

The minimum-sized buffer (W_P/L_P=240nm/100nm and W_N/L_N=100nm/100nm) is characterised by a R_{min}=1.84kΩ.

The value of the equivalent resistance has been carried out by measuring the propagation delay $t_{p50\%}$ as suggested in [2]:

$$R_{eq} = \frac{t_{p50\%}}{0.69C_L} \quad , \tag{5}$$

where

$$t_{p50\%} = \frac{t_{pLH50\%} + t_{pHL50\%}}{2} \quad , \tag{6}$$

$t_{pLH50\%}$ and $t_{pHL50\%}$ being the 50% delay occurring during a 0-1 and a 1-0 output transition, respectively.

As shown in figure 2, the error between the measured values and those carried out by means of conventional expression (2) becomes higher for a more advanced 70nm micron technology (C_L=80fF). For this latter technology the minimum sized inverter (W_P/L_P=170nm/70nm and W_N/L_N=70nm/70nm) is characterised by a resistance of 2.2kΩ.

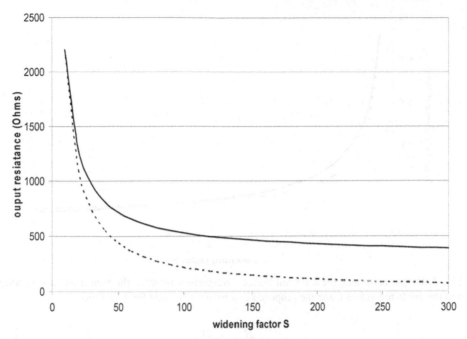

Fig. 2. Output resistance of a 70nm CMOS buffer as function of the channel widening S (R_{min}=2200Ω). In the plots the values carried out by HSPICE simulation (solid line) and those calculated assuming a linear relationship between channel width and buffer current (dotted line) are compared

As clearly highlighted by the plots, the actual value of the output resistance of wider buffer is higher than the expected in accord with (2), demonstrating the linear relationship between channel-width and static-current does not apply to the current effectively conducted by the MOSFET during switching.

In fact, the actual value of the output resistance of the buffer may be significantly higher than the expected one (the mean error is about 50% and 60% for the 100nm and 70nm technology, respectively).

3 The Proposed Expression

The comparison with the ideal trend in figures 1 and 2 gives a useful idea about how the theoretical behaviour differs from the actual one.

As shown, wider the transistor with respect the minimum sized one, higher is the difference between the expected and the actual value of the driver resistance. To take in to account the real influence of a device size increment a modified expression of the equivalent resistance has been developed.

The expression has been developed by interpolating the simulation data carried out for a wide set of load, technology and transistor parameters and taking into account both current and delay variations. As result, for a given technology, the output resistance of a buffer S times greater than the minimum-sized inverter for that technology has be found to be

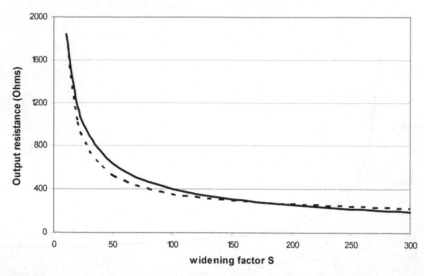

Fig. 3. Output resistance for a 100nm buffer: comparison between the measured (dotted line) and the predicted values using the proposed expression (7) (solid line). α=0.66

$$R_{eq} = \frac{R_{min}}{S^{\alpha}} ,$$ (7)

where the coefficient α, (with α<1), represents the slope of the plot of $log(R_{eq})$ versus $log(S)$ and R_{min} is the output resistance of the minimum-sized inverter.

Fig. 3 compares the actual resistance of the 100nm buffer and the values predicted by the proposed expression. For this technology the α parameter obtained by HSPICE simulation has be found to be equal to 0.66.

A similar comparison for the 70nm technology is presented in fig 4. For this latter technology the value for α has be found to be α=0.57. As shown, simulation results confirm again the correctness of the study and the validity of the proposed expression for the resistance scaling.

In fact, the estimation errors is about 3%, extremely low if compared to the 50-60% error made when the classical expression is used.

4 Conclusions

Delay minimisation in nowadays circuits requires adequate device models to capture the behaviour of the system driver/load efficiently.

All delay minimisation strategies applied to transistor sizing are based on the assumption of an inverse proportion between the final delay of the buffer and the channel width of the MOSFETS.

In the work it has been shown how this assumption may be incorrect for deep-submicron technologies, and how the relationship between transistor channel widening and delay reduction becomes non-linear for these technologies. A developed expression for the actual effect of a rising of channel width on the output resistance is then presented.

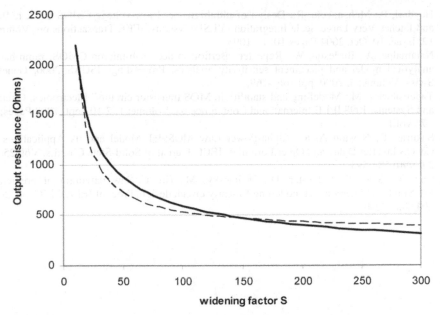

Fig. 4. Output resistance for a 70nm buffer: comparison between the measured (dotted line) and the predicted output resistance using equation (7) (solid line). α=0.57

The expression agrees with HSPICE circuit simulations within 3% error and it allows the actual performance variations obtainable by modifying the transistor size to be carried out. Therefore it can be directly adopted in traditional well-established formulae used for optimisation techniques such as buffer insertion and buffer tapering to improve the effectiveness of these strategies for deep-submicron technologies.

Acknowledgments

The Authors want to acknowledge Dr.Eng. F. Corapi for his precious help. The Authors also want to acknowledge the device group at UC Berkeley (www.device.eecs.berkeley.edu/~ptm) which provided the Predictive Technology Models used through the paper.

References

1. Restle, P. J., Jenkins, K. A., Deutsch, A., Cook P. W.: Measurament and Modelling of On-Chip Transmission Line Effects in a 400 MHz Microprocessor. IEEE Journal of Solid-State Circuits, Vol. 33, n. 4 (1998) 662-665
2. Rabaey, J.M., A.Chandrakasan, and B.Nikolic: 'Digital integrated circuits: a design perspective', 2nd edition, Prentice-Hall International, New Jersey, 2003;
3. Ismail, Y.I., Friedman, E.G.: Optimum repeater insertion based on a CMOS delay model for on-chip RLC interconnect, Proceedings Eleventh Annual IEEE International ASIC Conference, 1998, pp: 369 –373;
4. Cappuccino, G., Cocorullo, G. Corsonello, P.: Cmos buffer sizing for long on-chip interconnects, IEE Electronics Letters, Vol. 34, n.20, Oct. 1998 pp. 1937-1938.

5. Heydari, P.; Mohanavelu, R.: Design of ultrahigh-speed low-voltage CMOS CML buffers and latches Very Large Scale Integration (VLSI) Systems, IEEE Transactions on, Volume: 12, Issue: 10, Oct. 2004 Pages:1081 - 1093

6. Nalamalpu, A., Burleson, W.: Repeater insertion in deep sub-micron CMOS: ramp-based analytical model and placement sensitivity analysis, Proceedings ISCAS 2000, Geneva, Swiss, Volume: 3, 2000, pp: 766 –769;

7. Tadeusiewicz, M.; Modelling and stability in MOS transistor circuits" Electronics, Circuits and Systems, 1998 IEEE International Conference on, Volume: 1, 7-10 Sept. 1998 Pages:71 - 74 vol.1

8. Sakurai, T., Newton A. R.: Alpha-Power Law MOSFET Model and its Applications to CMOS Inverter Delay and Other Formulas. IEEE Journal of Solid-State Circuits, Vol. 25, n. 2 (1990) 584-593

9. Cao, Y., Sato, T., Sylvester, D., Orshansky, M., Hu, C.:New paradigm of predictive MOSFET and interconnect modelling for early circuit design, Proc. of IEEE CICC, pp. 201-204, Jun. 2000

Application of Internode Model to Global Power Consumption Estimation in SCMOS Gates[*]

Alejandro Millán Calderón[1,2], Manuel Jesús Bellido Díaz[1,2],
Jorge Juan-Chico[1,2], Paulino Ruiz de Clavijo[1,2],
David Guerrero Martos[1,2], E. Ostúa[1,2], and J. Viejo[1,2]

[1] Instituto de Microelectronica de Sevilla, Centro Nacional de Microelectronica
Av. Reina Mercedes, s/n (Edificio CICA), 41012 Sevilla, Spain
{amillan,bellido,jjchico,paulino,guerre,ostua,julian}@imse.cnm.es
http://www.imse.cnm.es
Tel.: +34 955056666, Fax: +34 955056686
[2] Departamento de Tecnologia Electronica, Universidad de Sevilla
Av. Reina Mercedes, s/n (ETS Ing. Informatica), 41012 Sevilla, Spain
http://www.dte.us.es
Tel.: +34 954556161, Fax: +34 954552764

Abstract. In this paper, we present a model, Internode, that unifies the gate functional behavior and the dynamic one. It is based on a FSM that represents the internal state of the gate depending on the electrical load of its internal nodes allowing to consider aspects like input collisions and internal power consumption. Also, we explain the importance of internal power consumption (such effect occurs when an input transition does not affect the output) in three different technologies (AMS 0.6 μm, AMS 0.35 μm, and UMC 130 nm). This consumption becomes more remarkable as technology advances yielding to underestimating up to 9.4% of global power consumption in the UMC 130 nm case. Finally, we show how to optimize power estimation in the SCMOS NOR-2 gate by applying Internode to modeling its consumption accurately.

1 Introduction

In the field of verification of digital VLSI systems, it is necessary not only to verify the functional behavior but also the dynamic one, in order to guarantee that the design fulfills frequency and power consumption specifications. The logic level is the best one to carry out this process since, on the one hand, verification at the lower level (transistor level) has a very high computational cost what limits its application to very small systems and, on the other hand, verification at the higher level (RTL, register transfer level) does not obtain the sufficient precision for checking the system dynamic behavior. However, in the logic simulation area, the technology is advancing constantly. This advance influences remarkably in the circuits dynamic behavior causing that: (a) new effects appear that have been obviated due to their low importance in previous technologies, (b) changes

---

[*] This work has been partially supported by the MEC META project TEC 2004-00840/MIC of the Spanish Government

V. Paliouras, J. Vounckx, and D. Verkest (Eds.): PATMOS 2005, LNCS 3728, pp. 337–347, 2005.

appears in the behavior of the effects already considered, and (c) simulation precision get worse because the same absolute errors involve bigger relative errors due to the frequency increase. Thus, in order to maintain/increase the precision of logic simulators, it is necessary to adapt the modeling techniques to this advance taking into account each new aspect (e.g. low voltage [ ], very large scale integration [ ], transition waveforms [ ], and power consumption [ ]). Also, we must denote that these changes in the models behavior imply significant modifications on the simulation algorithms in most cases. Our work has focused on this field by trying to deal with an essential problem that prevent logic simulation from reaching optimal results. The processing that current simulators perform on a gate separates the functional behavior from the dynamic one. This is not a suitable point of view since each behavior type influences in the other one being essential to unify both behaviors into a single model. This can be achieved by considering that a gate can be in different states and that its dynamic behavior depends not only on the input transitions (as usual) but also on the gate state.

The unification of the gate functional behavior and the dynamic one into a single model makes an important headway in logic simulation. In an intuitive way, this approach already starts to be applied when, for example, a different temporal behavior is considered based on what input causes the gate output to change. Nevertheless, it is necessary to develop a methodology that allows to reflect this aspect in a comprehensive and methodical way. In this way, we present a new model (called Internode, internal node logic computational model [ ]) which is based on a finite state machine (FSM) that represents the internal state of the gate depending on the electrical load of its internal nodes. Such model allows to consider aspects unachievable from traditional models like input collisions and internal power consumption, among others. Internal power consumption refers to the consumption caused by any input transition. In traditional models only power consumption when an output change exist is considered. Nevertheless, an input transition causes power consumption always and, although this consumption has been traditionally neglected, this effect becomes more important as the integration scale increases. Thus, Internode is a meta-model that allows modeling the gate behavior in a comprehensive and detailed way but, at the same time, maintaining the simulation at the logic level.

In our first approach to the application of the Internode model, we have studied its usefulness in the estimation of the internal power consumption mentioned. So, the organization of the paper is as follows: Sect. 2 shows the Internode model basis; in Sect. 3 we analyze the need to consider internal power consumption in logic simulation; Sect. 4 presents how Internode can be applied to the estimation of this consumption as well as the usually considered one; and finally we will finish with the main conclusions of this work.

2   Internode Model

The Internode model considers a FSM for the behavior of the gate. The specific FSM depends on the gate structure and the number of states of the internal and the output nodes. The model is based on the notation used in the Moore automata [ ]. Also, the Internode model of a 1-input SCMOS gate is the same as the functional one because such gates does not have internal nodes. So, in this section, we are going to present the model for 2-input SCMOS gates.

In 2-input SCMOS gates, we always have two or more transistors in each MOS-tree with one or more internal nodes. So, the corresponding Internode model must consider the state of these internal node (charged or discharged) as well as the output value. Let us consider, for example, the case of the NOR-2 gate (Fig. 1a). This gate has two transistors in serial mode in the PMOS-tree (P_1 and P_2) and one internal node. On the one hand, considering the output value (Q_2) and the internal node state (Q_1), we have four possible cases producing four states in our Internode model. On the other hand, in order to consider the behavior of the gate, it is necessary to establish transitions between states due to input changes. In Fig. 1b we show the Internode model for a 2-input NOR gate (NOR-2).

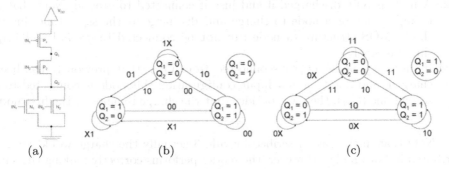

(a)   (b)   (c)

Fig. 1. SCMOS structure for a NOR-2 gate (a), SCMOS NOR-2 Internode model (b), and SCMOS NAND-2 Internode model (c)

Due to the gate structure, the state $Q_1Q_2 = 01$ is impossible to reach because, if Q_2 is charged ($Q_2 = 1$), then it must be connected to V_{DD} and it would imply Q_1 to be connected to V_{DD} too (producing $Q_1 = 1$ instead of $Q_1 = 0$). Also, the state $Q_1Q_2 = 11$ is reached only when the gate receives $IN_1IN_2 = 00$. The input case $IN_1IN_2 = 10$ leads to state $Q_1Q_2 = 00$ always, and the input case $IN_1IN_2 = 01$ leads to state $Q_1Q_2 = 10$. With this operation method we can consider that each state has a characteristic input value. However, it is possible to stay in states $Q_1Q_2 = 00$ or $Q_1Q_2 = 10$ receiving other input value (for example, $IN_1IN_2 = 11$). In a similar way it is possible to get the Internode model for the SCMOS NAND-2 gate (Fig. 1c).

When we compare the Internode model for 2-input SCMOS gates (Fig. 1b and Fig. 1c) with the functional model for the same gates, we can observe that the Internode model deals with the internal state of the gate much better than the functional model. If we intend to apply a dynamic behavioral model, we can reach a higher accuracy using the Internode model because it allows to consider different situations that are considered to be the same in the functional model.

Let us consider, for example, a delay model for the case of a NOR-2 gate. In the functional model we have only one situation in which output raises: from state $Q_2 = 0$ to state $Q_2 = 1$. However, this raise can be reached from two different real states: internal node charged (state $Q_1Q_2 = 10$) or internal node discharged (state $Q_1Q_2 = 00$). That is, in the Internode model we can consider different delay models for these different situations, while in the functional model we have to use the same delay model for them both.

2.1   Extension to N-Input SCMOS Gates

The presented model can be easily extended to SCMOS gates with more than two inputs. In this section, we will extend it to N-input SCMOS gates. In order to do this, in the easiest way, we are going to establish several behavioral rules that we can apply to the implementation of the model. The rules are the next:

1. A node is only charged if and just if connected to V_{DD}.
2. A node is only discharged if and just if connected to ground. Note that it is impossible for a node to charge and discharge at the same time due to the SCMOS structure (a node can not be connected to ground and V_{DD} simultaneously).
3. If there is a node (Q_A) disconnected from V_{DD}, that previously has been charged, and due to a new input configuration this node is connected to a second one (Q_B), then we consider that the charge remains at the first node (Q_A).

Note that in the case described by rule 3 actually the charge would be distributed in both nodes. However, the model performs correctly making the supposition of rule 3, and this rule is necessary in order to maintain the model in a logic level. The two first rules imply that a new input in a gate (that is, to model a gate with $(N + 1)$-input instead of an N-input one) only means a new state in the FSM of the Internode model respect of the N-input case.

Keeping these rules in mind, we are going to study the N-input NOR and NAND gates (NOR-N and NAND-N). As we are going to see, it is possible to build an algorithm for the Internode model in order to establish the specific model for each NOR/NAND gate. The algorithm is a more suitable representation of the model than the state diagram we have used for the 1-input and 2-input gates because, for the N-input case, the $(N + 1)$-state diagram is more difficult to manipulate and understand. Also, the algorithm shows a possible implementation of the Internode model in a logic-level tool. For the NOR-N case, we can establish the next algorithm in order to estimate the new state for a given gate (Fig. 2a):

1. Consider $Q = (Q_1, Q_2, ..., Q_N)$ as the present state of the gate having Q_1, Q_2, etc. as the internal nodes charge indicators and Q_N as the output charge indicator (Q_1 is the internal node nearest to V_{DD}).
2. Consider $IN = (IN_1, IN_2, ..., IN_N)$ as the present input configuration (IN_1 is the input nearest to V_{DD}).

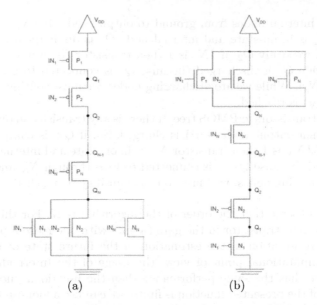

Fig. 2. Structures of a SCMOS NOR-N gate (a) and a SCMOS NAND-N gate (b)

3. Consider $QP = (QP_1, QP_2, ..., QP_N)$ as the future state of the gate.
4. Initially, assume $QP = Q$.
5. Analyze internal nodes from V_{DD} to output node. If IN_1 is 0 then transistor P_1 is in on-state and internal node Q_1 is charged. So, if P_1 is in on-state, we study Q_2: if IN_2 is 0 then transistor P_2 is in on-state and internal node Q_2 is charged (because Q_2 is connected to V_{DD} through P_1 and P_2). While we are charging nodes, we must continue until output node (Q_N) is reached.
6. Analyze transistors in NMOS tree. If there is any transistor in on-state (name it N_J) then output node (Q_N) is discharged. So, if Q_N is discharged, we study Q_{N-1}: if IN_N is 0 then transistor P_N is in on-state and internal node Q_{N-1} is discharged (because Q_{N-1} is connected to ground through P_N and N_J). While we are discharging nodes, we must continue until Q_1 is reached.

For the NAND-N case, we can establish a similar algorithm in order to estimate the new state for a given gate (Fig. 2b):

1. Consider $Q = (Q_1, Q_2, ..., Q_N)$ as the present state of the gate having Q_1, Q_2, etc. as the internal nodes charge indicators and Q_N as the output charge indicator (Q_1 is the internal node nearest to ground).
2. Consider $IN = (IN_1, IN_2, ..., IN_N)$ as the present input configuration (IN_1 is the input nearest to ground).
3. Consider $QP = (QP_1, QP_2, ..., QP_N)$ as the future state of the gate.
4. Initially, assume $QP = Q$.

5. Analyze internal nodes from ground to output node. If IN_1 is 1 then transistor N_1 is in on-state and internal node Q_1 is discharged. So, if N_1 is in on-state, we study Q_2: if IN_2 is 1 then transistor N_2 is in on-state and internal node Q_2 is discharged (because Q_2 is connected to ground through N_1 and N_2). While we are discharging nodes, we must continue until output node (Q_N) is reached.

6. Analyze transistors in PMOS tree. If there is any transistor in on-state (name it P_J) then output node (Q_N) is charged. So, if Q_N is charged, we study Q_{N-1}: if IN_N is 1 then transistor N_N is in on-state and internal node Q_{N-1} is charged (because Q_{N-1} is connected to V_{DD} through N_N and P_J). While we are charging nodes, we must continue until Q_1 is reached.

We must observe that the order of the algorithm is N. For this reason, the inclusion of a new transistor in the gate (a gate with one more input) produces only one more iteration in the estimation of the future state of the gate. So, from the computational point of view, the usage of the Internode model in a logic-level tool has the same performance than the functional model. Also, as the domain of the presented function is finite, we can use a look-up table to store the precalculated future states for all the situations the gate can reach. In this case, the order will be reduced to a unit at simulation time.

Internode can be applied to those processes whose results can be improved by considering the internal state of the gate, such as internal power consumption estimation. On next sections, we show the need to include this consumption into the global estimation process in a SCMOS gate and how to employ Internode to achieve this.

3   Importance of Internal Power Consumption

In the process of power consumption estimation, it is usual to consider only those cases in which output changes. However, as technology advances, the power consumption produced by any input change (although it not affects the output) becomes more relevant. On the next, we will use three different terms referring to the different types of consumption: (1) internal power consumption (produced by an input change that does not affects the output), (2) external power consumption (the traditionally considered one, produced when the output changes its value), and (3) global power consumption (the sum of them both).

In order to approach this effect in a theoretical way, let us consider the SCMOS structure for a NOR-2 gate (Fig. 1a). For our explanation, it is necessary to take into account a parasitic capacitance, C_Y, at node Y (between the two transistors in the PMOS-tree). Assuming this, we are going to analyze the gate behavior under a specific input sequence. Firstly, IN_1 is at high-level and IN_2 is at low-level having transistor P_1 in off-state and transistor P_2 in on-state (Fig. 3). This situation allows C_Y to discharge through N_1. Now, if we raise IN_2 input we will cause P_2 to enter in off-state (step 1), and then, we put IN_1 at low-level causing P_1 to enter in on-state (step 2). As P_1 is in on-state, the parasitic capacitance C_Y is charged; generating a power consumption. Next, we return IN_1 to high-level entering P_1 in off-state (step 3). And finally, we put IN_2 at low-level causing P_2 to enter in on-state (step 4). In this situation, as N_1 transistor is in on-state, the accumulated charge in C_Y is evacuated towards ground through P_2 and N_1 and we lose this energy.

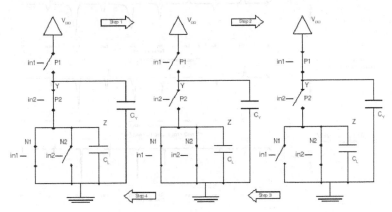

Fig. 3. Firstly, IN_1 is at high-level and IN_2 is at low-level having transistor P_1 in off-state and transistor P_2 in on-state

Although this can appear as an unlikely case, simulations show that, due to such effects, internal power consumption becomes a significant aspect. In Fig. 4, we show power consumption in a NOR-2 SCMOS gate for three different technologies (AMS 0.6 μm, AMS 0.35 μm, and UMC 130 nm). Actually, the input signals pass through a pair of gates in order to drive the gate under study with realistic curves. Simulations are grouped in two cases. In the first one (case A), we have performed a simulation that covers the half of all the possible input transitions in a 2-input gate. In the second one (case B), we cover the other half by interchanging the input curves. As we can observe, power consumption in such conditions reaches important peaks, up to 1.4 mW, 350 μW, and 36 μW (depending on the technology).

In order to study the extension of this aspect, we have carried out this analysis for NOR-2 and NAND-2 gates in the three technologies mentioned. The results obtained are shown in Table 1 and Table 2. Simulations show that, if we neglect the internal consumption, we are underestimating about 4.8%-9.4% of global power consumption. So, integrating this consumption estimation in the gate simulation necessarily leads to an important improvement in results precision.

4 Application of Internode to Internal Power Consumption Estimation

Internode can be applied to global power consumption estimation in a very easy way, including both internal and external consumption. This can be done

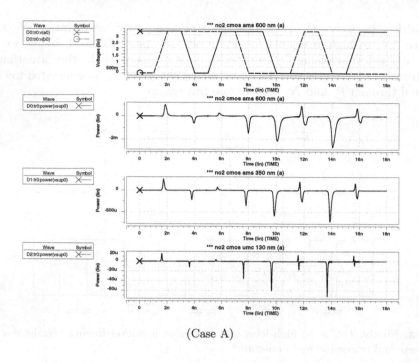

(Case A)

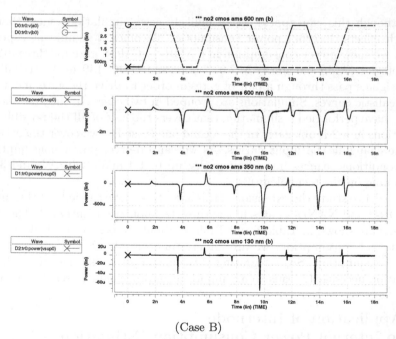

(Case B)

Fig. 4. Power consumption in a SCMOS NOR-2 gate for technologies (AMS 0.6 μm, AMS 0.35 μm, and UMC 130 nm)

Table 1. Power consumption in SCMOS NOR-2 and NAND-2 gates in different technologies distinguishing between internal and external consumption. Results for both cases of study (A and B) are presented

Technology	Consumption	NOR-2(A)	NOR-2(B)	NAND-2(A)	NAND-2(B)
AMS 0.6 μm	Int./Ext. (μJ)	14.3/203.2	15.4/203.5	2.4/46.6	2.3/46.7
AMS 0.35 μm	Int./Ext. (μJ)	2.5/31.7	2.5/31.8	0.7/13.9	0.7/13.9
UMC 130 nm	Int./Ext. (nJ)	101.4/1039.0	113.7/1037.0	35.9/504.4	31.7/505.8

Table 2. Percentage covered by internal power consumption compared to global power consumption for each gate and technology. Results are calculated by adding both cases (A and B) data

Gate	AMS 0.6 μm	AMS 0.35 μm	UMC 130 nm
NOR-2	6.8%	7.3%	9.4%
NAND-2	4.8%	4.8%	6.3%

because Internode considers the internal state of the gate and, so on, includes all the possible input transition cases.

Thus, in order to apply Internode to global power estimation, it is only necessary to choose a suited power model and characterize it for all the cases. Authors have developed a lot of accurate models that suit this task [7–10] but, however, for this explanation it is better to consider the simplest one in order to avoid unnecessary complications because our main interest here is to show how to use Internode for these tasks independently of the model used. The model we are going to use is very simple: energy consumption for each transition is a fixed amount equal to the one obtained by simulation.

Let us consider, for example, the NOR-2 case in UMC 130 nm technology. In order to apply Internode to this task, we need to obtain energy consumption in the gate for each transition of its Internode model (Fig. 1b). However, simulations already presented are not sufficient because they cover all cases of input transitions but they do not distinguish the gate state in what these transitions occur. Thus, we need to perform a specific simulation that produces all Internode transitions measuring energy consumption for each one. In Table 3, the Internode look-up table corresponding to a SCMOS NOR-2 gate is presented by indicating the final state that the gate reaches from a given initial state and input values. In Table 4, we show the data obtained. Each cell of this table presents the energy consumption measured for a specific gate state and input values.

It is important to denote that the inclusion of such model in logic simulators improves remarkably their precision for two reasons. On the one hand, traditional models do not consider a important amount of transition cases that correspond to internal consumption effects (marked as i in Table 3). On the other hand, these models consider only the input behavior and they do not take into account the initial gate state. Thus, such models deal with some transitions as being the same increasing the estimation error: all cases in which output raises (marked as

Table 3. Internode look-up table for the SCMOS NOR-2 gate. Final state is shown depending on the initial gate state and the input values. Cases are marked following these criteria: internal consumption cases (i), raising output cases (r), and falling output cases (f). (State $q_2q_1 = 10$ is impossible to reach)

	in1 in2			
q2 q1	0 0	0 1	1 1	1 0
0 0	1 1_r	0 1_i	0 0_i	0 0_i
0 1	1 1_r	0 1_i	0 1_i	0 0_i
1 1	1 1_i	0 1_f	0 1_f	0 0_f
				Q2 Q1

Table 4. Energy consumption measured in a SCMOS NOR-2 gate (UMC 130 nm technology) for all possible transitions of its Internode model. (State $q_2q_1 = 10$ is impossible to reach)

	in1 in2			
q2 q1	0 0	0 1	1 1	1 0
0 0	-3.954 fJ	-0.6856 fJ	1.029 fJ	-0.001 fJ
0 1	-5.154 fJ	-0.0128 fJ	0.259 fJ	-1.530 fJ
1 1	-0.271 fJ	-0.2653 fJ	0.830 fJ	0.199 fJ

r in Table 3) are modeled in the same way by traditional models and all cases in which output falls (marked as f in Table 3) are considered to be the same too.

Also, we are sure that the importance of internal power consumption will increase in gates with more than two inputs, because these gates contains more internal nodes than the presented ones. So, the inclusion of Internode in logic simulators becomes a very suited technique in order to maintain/improve their precision while technology advances.

5   Conclusions

The unification of the gate functional behavior and the dynamic one into a single model makes an important headway in logic simulation. In this way, we have presented a model (Internode) based on a FSM that represents the internal state of the gate depending on the electrical load of its internal nodes. Such model allows to consider aspects unachievable from traditional approaches like input collisions and internal power consumption, among others.

We have detailed the corresponding Internode model for SCMOS NOR-N and NAND-N gates showing that, from the computational point of view, its usage in a logic-level tool has the same performance than a functional model. Even, by using a look-up table (e.g. Table 3), its order can be reduced to a unit at simulation time.

We have explained the impact of not considering internal power consumption in logic simulators by measuring it in three different technologies (AMS 0.6 μm, AMS 0.35 μm, and UMC 130 nm). This effect becomes more remarkable as technology advances yielding to underestimating up to 9.4% of global power consumption in the UMC 130 nm case. Also, we are sure that internal power consumption will become more important in bigger gates (more than two inputs) because they contain more internal nodes.

Finally, we have show how to apply Internode to power estimation in the SCMOS NOR-2 gate by detailing the corresponding look-up tables for both transition and power consumption behaviors. Also, we have explained the main advantages of our approach because it considers all possible cases and does not confuse them into groups improving logic simulation results remarkably.

References

1. T. Kuroda, "Low-voltage technologies and circuits," in *Low-power CMOS design* (A. Chandrakasan and B. R., eds.), pp. 61–65, New York, NY, USA: Wiley-IEEE Press, 1998.
2. J. M. Daga and D. Auvergne, "A comprehensive delay macromodeling for sub-micrometer CMOS logics," *IEEE Journal of Solid-State Circuits*, vol. 34, no. 1, pp. 42–55, 1999.
3. A. I. Kayssi, K. A. Sakallah, and T. N. Mudge, "The impact of signal transition time on path delay computation," *IEEE Transactions on Circuits and Systems-II: Analog and Digital Signal Processing*, vol. 40, no. 5, pp. 302–309, 1993.
4. A. Chandrakasan, S. Sheng, and B. R., "Low-power CMOS digital design," *IEEE Journal of Solid-State Circuits*, vol. 27, no. 4, pp. 473–484, 1992.
5. A. Millan, M. J. Bellido, J. Juan, D. Guerrero, P. Ruiz-de Clavijo, and E. Os-tua, "Internode: Internal node logic computational model," in *Proc. 36th Annual Simulation Symposium (part of the Advanced Simulation Technologies Conference, ASTC)*, (Orlando, Florida, USA), pp. 241–248, IEEE Computer Society, March 2003.
6. E. F. Moore, "Gedanken experiments on sequential machines," in *Automata Studies*, pp. 129–153, Princeton, New Jersey, USA: Princeton University Press, 1956.
7. L. Bisdounis and O. Koufopavlou, "Analytical modeling of short-circuit energy dissipation in submicron CMOS structures," in *Proc. 6th IEEE International Conference on Electronics, Circuits and Systems (ICECS)*, pp. 1667–1670, September 1999.
8. M. P., N. Azemard, and A. D., "Structure independent representation of output transition time for CMOS library," in *Proc. 12th International Workshop on Power and Timing Modeling, Optimization and Simulation (PATMOS)*, pp. 247–257, Seville, Spain: Springer, September 2002.
9. A. Nabavi-Lishi and N. C. Rumin, "Inverter models of CMOS gates for supply current and delay evaluation," *IEEE Transactions on Computer-Aided Design of Integrated Circuits and Systems*, vol. 13, pp. 1271–1279, October 1994.
10. F. N. Najm, "A survey of power estimation techniques in vlsi circuits," *IEEE Transactions on VLSI Systems*, vol. 2, no. 4, pp. 446–455, 1994.

Compact Static Power Model of Complex CMOS Gates

Jose L. Rosselló, Sebastià Bota, and Jaume Segura

Electronic Technology Group, Universitat de les Illes Balears
Campus UIB, 07122, Palma de Mallorca, Spain
{j.rossello,sebastia.bota,jaume.segura}@uib.es
http://omaha.uib.es

Abstract. We present a compact model to estimate quickly and accurately the leakage power in CMOS nanometer Integrated Circuits (ICs). The model has similar accuracy than SPICE and represents an important improvement with respect to previous works. It has been developed to be used for fast and accurate estimation and optimization of the standby power dissipated by large circuits.

1 Introduction

The static power dissipation has been traditionally neglected when compared to the dynamic component, but as technology scales down the static power is becoming a significant fraction of the total power due to the V_{TH} scaling. For this reason, the development of accurate estimation techniques for this contribution is a must.

Moreover, the estimation of static power in high-density ICs using simple analytical models provides faster estimations, thus minimizing the impact on the total design cycle.

The estimation of the leakage component in complex CMOS gates implies the determination of the I_{OFF} current through an indeterminate number of serially-connected MOS transistors. Recently, an analytical model for leakage power prediction was presented in [1]. The model is only valid for gates with no more than two serially connected transistors. Gu et. al. developed in [2] a model that can only be applied to gates with up to three serially connected transistors. Finally, a more general model for the estimation of the standby leakage for an indeterminate number of serially connected transistors in the stack was presented in [3].

In this work, we develop an accurate analytical model for the estimation of leakage in CMOS gates that is very close to SPICE simulations in terms of accuracy.

2 Transistor Collapsing Technique of CMOS Gates

The leakage current of a MOS device depends exponentially on the gate-source voltage and the transistor threshold voltage [4].

$$I_{OFF} = \frac{W}{L} I_0 \left(\frac{T}{T_{ref}} \right)^2 e^{\frac{V_{GS}-V_{TH}}{nV_T}} \left(1 - e^{\frac{V_{DS}}{V_T}} \right)$$ (1)

where W and L are the channel width and length respectively, T is the temperature of operation, T_{ref} is a reference temperature, I_0 is a process-dependent parameter, while V_T, V_{GS}, V_{DS}, and V_{TH} are the thermal, gate-source, drain-source and threshold voltage respectively. The threshold voltage is expressed as:

V. Paliouras, J. Vounckx, and D. Verkest (Eds.): PATMOS 2005, LNCS 3728, pp. 348–354, 2005.
© Springer-Verlag Berlin Heidelberg 2005

$$V_{TH} = V_{T0} + \gamma' V_{SB} + K_T (T - T_{ref}) - \sigma(V_{DS} - V_{DD}) \qquad (2)$$

where V_{T0} is the zero bias threshold voltage, γ' is related to the body effect, K_T is the sensibility of the threshold voltage with temperature, while σ accounts for the Drain Induced Barrier Lowering (DIBL).

Static current estimation through complex CMOS gates requires a current computation through each branch of transistors connecting the supply and ground nodes. We define an OFF branch as a chain of serially connected transistors (N or P) with at least one transistor being in the OFF state. An ON branch is defined as a chain of transistors with all devices operating in the ON state. For each OFF branch we can find an equivalent transistor with an effective width such that its OFF current (given by (1)) is equal to the current through the OFF branch.

If an OFF branch is in parallel with an ON branch then it is ignored in the static current estimation. Each OFF branch is collapsed to a single equivalent transistor with an effective width W_{eff} while the channel length is supposed to be equal for all the transistors in the chain (and also for the equivalent transistor). Finally, all the parallel-connected OFF branches are collapsed into a single equivalent transistor with an effective width equal to the sum of the effective widths of each OFF branch (see Fig. 1). For the estimation of the effective width of an OFF branch we apply a collapsing technique as described in (Fig.2).

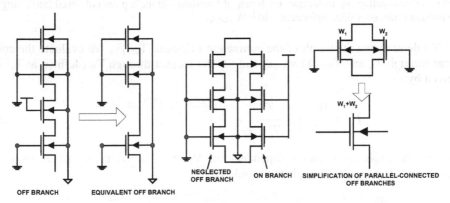

Fig. 1. Simplification process of OFF branches

In Fig. 2 we show an OFF branch having 'N' nMOS transistors (for an OFF branch of pMOS transistors the analysis is equivalent). In the analysis we only consider the OFF transistors, while the ON transistors are neglected (they are considered to be part of the internal nodes of the branch). The closer to ground transistor and the upper transistor are labelled as T_1 and T_N respectively, while the internal nodes are labelled from V_1 to V_{N-1}. The upper transistor is connected to a voltage equal to the supply voltage V_{DD} while the substrate is assumed to have a voltage V_B.

The transistor collapsing method is applied as follows: the pair of transistors at the top of the chain are collapsed into a single equivalent transistor $T_{<N-1,N>}$ with equivalent width $W_{<N-1,N>}$ leading to a chain with N-1 OFF transistors. The process is repeated until we obtain a single transistor $T_{<1,N>}$ with an equivalent width $W_{<1,N>}$ such that its standby current is equal to the OFF current of the original chain.

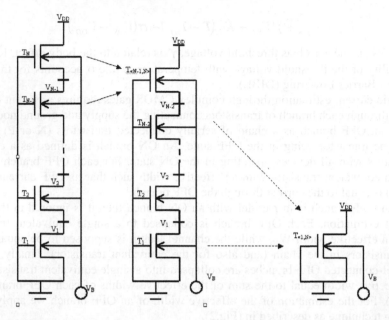

Fig. 2. Chain collapsing technique. Each pair of transistors at the top are collapsed into a single equivalent transistor with equivalent width $W_{<N-1,N>}$

To determinate the width of the equivalent transistor $T_{<N-1,N>}$ we evaluate the current through T_N and T_{N-1}. From (1) and (2) the current through T_N (defined as I_N) is given by:

$$I_N = \frac{W_N}{L} I_0 \left(\frac{T}{T_{ref}}\right)^2 e^{\frac{-V_{T0}+\gamma'V_B-(1+\sigma+\gamma')V_{N-1}}{nV_T}} \qquad (3)$$

where the exponential factor dependent on V_{DS}/V_T can be neglected as long as $V_{DD} \gg V_T$. The current through T_{N-1} would be:

$$I_{N-1} = \frac{W_{N-1}}{L} I_0 \left(\frac{T}{T_{ref}}\right)^2 e^{\frac{-V_{T0}+\gamma'V_B-(1+\sigma+\gamma')V_{N-2}+\sigma(V_{N-1}-V_{DD})}{nV_T}} \left(1-e^{-\frac{V_{N-1}-V_{N-2}}{V_T}}\right) \qquad (4)$$

The equivalent transistor $T_{<N-1,N>}$ has a mathematical expression for the OFF current similar to (3) since its drain-source voltage would be much more larger than V_T.

$$I_{\langle N-1,N\rangle} = \frac{W_{\langle N-1,N\rangle}}{L} I_0 \left(\frac{T}{T_{ref}}\right)^2 e^{\frac{-V_{T0}+\gamma'V_B-(1+\sigma+\gamma')V_{N-2}}{nV_T}} \qquad (5)$$

From (3) and (5) we obtain $W_{<N-1,N>}$ as:

$$W_{\langle N-1,N\rangle} = W_N e^{\frac{-(1+\sigma+\gamma')(V_{N-1}-V_{N-2})}{nV_T}} \qquad (6)$$

Therefore, as stated by (6), the effective width of the two transistors depends exponentially on V_{N-1}-V_{N-2}. For the estimation of V_{N-1}-V_{N-2} we equate (3) and (4). Unfortunately this equation has no analytical solution, although for some special cases we can obtain an approximate solution:

a) V_{N-1} -$V_{N-2} >> V_T$. For this case we obtain the next solution:

$$V_{N-1} - V_{N-2} \cong \alpha V_T f\left(W_N, W_{N-1}\right) \equiv V_A \qquad (7)$$

b) V_{N-1}-$V_{N-2} < V_T$. In this second case we have:

$$V_{N-1} - V_{N-2} \cong V_T e^{f\left(W_N, W_{N-1}\right)} \equiv V_B \qquad (8)$$

where $f(W_N, W_{N-1})$ and α take the form:

$$f\left(W_N, W_{N-1}\right) = Ln\left(\frac{W_N}{W_{N-1}}\right) + \frac{\sigma V_{DD}}{nV_T} \qquad (9)$$

$$\alpha = \frac{n}{1 + \gamma' + 2\sigma}$$

An empirical solution that includes both cases is given by:

$$V_{DS}^{T_{N-1}} \equiv V_{N-1} - V_{N-2} = V_T\left\{1 + \frac{(\alpha - 1)e^f}{\alpha - 1 + e^f}\right\} Ln\left(1 + e^f\right) \qquad (10)$$

In Fig. 2 we show the fitness of the proposed expression with respect to the exact solution of a stack of two transistors for a 0.12μm technology.

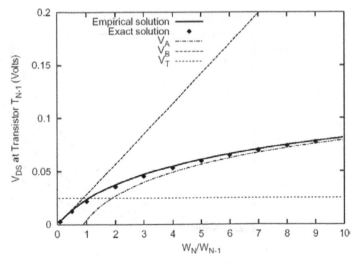

Fig. 3. Variation of the drain-source voltage at transistor T_{N-1}. Expression (10) is found to be a good approximation to V_{N-1}-V_{N-2}

The drain-source voltage of any other transistor T_i is obtained similarly as in (10). Then, the effective channel width of the equivalent transistor of the OFF chain can be obtained from (3) as:

$$W_{\langle 1,N \rangle} = W_N e^{\frac{-(1+\sigma+\gamma')V_{N-1}}{nV_T}} \tag{11}$$

Where V_{N-1} is obtained as:

$$V_{N-1} = \sum_{i=1}^{N-1} V_{DS}^{T_i} \tag{12}$$

For an specific input vector to the gate (say vector 'i') an effective width W_{ieff} may be obtained using the previously described collapsing technique. Finally, the I_{OFF} current of the gate is given by:

$$I_{OFF} = \frac{W_{eff}^i}{L} I_0 \left(\frac{T}{T_{ref}} \right)^2 e^{\frac{-V_{T0}-K_T(T-T_{ref})+\gamma'V_B}{nV_T}} \tag{13}$$

3   Results

We estimated the static current driven by a stack of 'N' nMOS transistors comparing SPICE simulations, the model presented in [3] and the proposed model for a 0.12µm CMOS technology in Fig. 4. Results demonstrate that the model provides an excellent agreement with respect to SPICE, and provides better results than the model in [3]. For small-size circuits, the analytical model shows since three orders of magnitude in terms of speed improvement with respect to SPICE simulations while for large circuits the time improvement is expected to be larger.

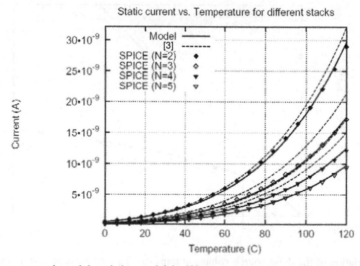

Fig. 4. The proposed model and the model in [3] are compared with SPICE simulations for four stacks of nMOS transistors

The proposed model can be used for fast circuit optimization. In Fig. 5 we show a stack of two transistors such that the two transistor sizes are proportional with an scaling factor 's'. To maintain gate capacitances fixed, the transistor sizing is selected such that the sum $W_1 + sW_1$ is constant and equal to W_{TOTAL}. The proposed formulation can be applied to quickly obtain the I_{OFF} current of the stack as a function of the scaling factor 's'. In Fig.6 we show the I_{OFF} current vs. the scaling factor showing that the maximum current is obtained for a scaling factor between 0 and 1. This result can be combined with the expected propagation delay (that is analytically extracted from [5]) of the gate to obtain the optimum Power-Delay-Product. As can be appreciated in Fig.7, for the technology used (a 120nm technology) the optimum PDP is not obtained at s=1 but at a higher value (s=3.5).

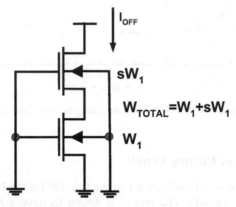

Fig. 5. Stack of two nMOS transistors. The total area of the circuit is constant while a scaling factor 's' is used between the transistor widths

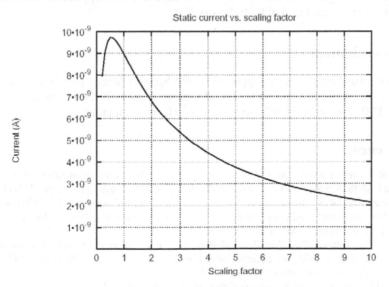

Fig. 6. I_{OFF} current of the stack. It can be appreciated that the maximum static power arises when s~0.5

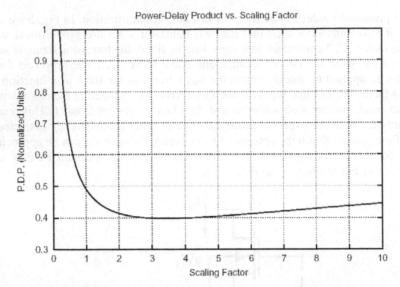

Fig. 7. Power-Delay-Product of the NAND2 gate. The gate is optimized for an scaling factor of s=3.5

4   Conclusion and Future Work

In this paper we propose and validate a compact model for the leakage power estimation in CMOS logic circuits. The model is found to have a similar accuracy than SPICE simulations (within the 1% of error) and represent an important improvement over previously published static models. The model will be applied to a fast and accurate power estimation tool of high-density ICs.

Aknowledgment

This work has been partially supported by the Spanish Ministry of Science and Technology, the Regional European Development Funds (FEDER) from EU project TIC2002-01238, and an Intel Laboratories-CRL research grant.

References

1. Gu R. X., and Elmasry, M. I.: "Power dissipation analysis and optimization of deep submicron CMOS digital circuits", *IEEE J. Solid-State Circuits,* 1996, 31, (5), pp. 707-713
2. Narendra, S, De,V, Borkar, S., Antoniadis, D, and Chandrakasan, A.: "Full-Chip subthreshold leakage power prediction and reduction techniques for sub-0.18mm CMOS", *IEEE J. of Solid-State Circuits*, 2004, 39, (2), pp.501-510
3. Chen, Z., Johnson, M., Wei, L., and Roy, K.: 'Estimation of standby leakage power in CMOS circuits considering accurate modeling of transistor stacks', Proc. ISLPED'98, Monterey, CA, USA, November 1998, pp. 40-41
4. BSIM3. http://www-device.eecs.berkeley.edu/~bsim3/get.html
5. Cite omitted for a blind review.

Energy Consumption in RC Tree Circuits with Exponential Inputs: An Analytical Model

Massimo Alioto[1], Gaetano Palumbo[2], and Massimo Poli[2]

[1] DII – Dip. di Ingegneria dell'Informazione, Università di Siena,
v. Roma n. 56, I-53100 – Siena, Italy
malioto@dii.unisi.it
[2] DEES – Dip. di Ingegneria Elettrica Elettronica e dei Sistemi, Universita' di Catania
viale Andrea Doria 6, I-95125 – Catania, Italy
{gpalumbo,mpoli}@diees.unict.it

Abstract. In this communication, RC tree networks are analyzed in terms of the energy dissipated during an input transition. A closed-form analytical model of the energy consumption is derived for arbitrary values of the input rise time by introducing a suitable first-order equivalent RC circuit, which avoids the explicit pole-zero evaluation. The proposed expression of the energy consumption has an evident meaning, thereby affording a deeper understanding of the network dissipation. Moreover, the energy model is sufficiently simple to be used in pencil-and-paper calculations. Extensive SPICE simulations confirm that the model has an adequate accuracy, as its error is typically within 5%.

1 Introduction

In current Systems-on-Chip (SoCs), the energy consumption is one of the most important design parameters. Indeed, the number of complex blocks integrated on the same chip is limited by the overall energy consumption both in portable systems (to extend the battery lifetime) and in high-speed circuits (to reduce the amount of generated heat). For this reason, it is essential to develop simple energy models of the most common digital blocks to consciously manage the energy-delay trade-offs involved in the design of digital Integrated Circuits (ICs) [1]-[2].

Until now, several models have been proposed to estimate the energy consumption [1]-[3]. However, most of the models proposed in the literature were specifically developed for static and dynamic CMOS gates, by starting from the analysis of the charge of a capacitance [1]-[3]. These models are not valid for many other blocks which are better modeled by an RC tree networks, such as adiabatic, transmission-gate and pass-transistor logic gates, as well as long interconnections [4], [5].

This communication proposes a simple energy model of RC tree networks, and extends the results previously obtained in the particular case of RC ladder circuits [4]-[5]. Analysis is carried out by considering the case of an exponential input waveform, which is well known to be a realistic input waveform in current VLSI circuits [3]-[4]. Since, in RC tree circuits with a typical complexity, a closed-form pole-zero evaluation (and thus of their energy consumption) is not possible, an exact analysis is first restricted to the case of very fast and very slow input signals, since the explicit evaluation of poles and zeroes is not required. Then, a suitable first-order equivalent circuit is introduced, and a general model valid for arbitrary values of the input rise

V. Paliouras, J. Vounckx, and D. Verkest (Eds.): PATMOS 2005, LNCS 3728, pp. 355–363, 2005.

time is derived. Comparison of the predicted energy consumption with SPICE simulations confirms that the model is sufficiently accurate for modeling purposes.

2   General Considerations on the Energy Consumption Evaluation

Let us consider an n-th order RC tree, which is made up of n grounded capacitances C_i with $i=1...n$ (whose non-grounded terminal is connected to node i) and n resistances R_i with $i=1...n$ (whose farthest from input terminal is connected to node i), as shown in Fig. 1. A voltage source $v_{in}(t)$ drives the network input node 0. For the sake of simplicity, it will be assumed that only one resistance R_1 is connected to the input node without loss of generality[1].

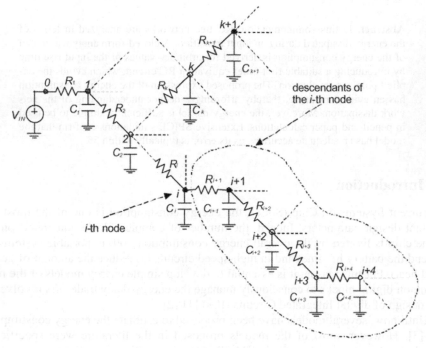

Fig. 1. n-th order RC tree network

To simplify the notation, in the following all nodes belonging to the subtree starting from the generic node i (excepting node i itself) will be referred to as *descendant nodes* of node i (as indicated in Fig. 1). Accordingly, the sum of capacitances C_i associated with a given node i and those of its descendant nodes will be referred to as $C_{DESC,i}$. Let us also define the the total network capacitance C_T as $\sum_{j=1}^{n} C_j$.

---

[1] If two (or more) resistances were connected to the input node, the network could be decomposed into two (or more) different RC trees with only one resistance at this node. Accordingly, the total energy would be equal to the sum of those of these simpler networks

During a transition of the voltage source $v_{in}(t)$ from 0 to the maximum value V_{DD} (usually equal to the supply voltage), the energy dissipated by resistors is equal to the difference of the overall input energy E_{in} and the energy E_C stored in the capacitances

$$E_R = E_{in} - E_C = \int_0^{+\infty} v_{in}(t) \cdot i_{in}(t)\,dt - \frac{1}{2}V_{DD}^2 C_T \qquad (1)$$

where E_{in} was expressed as the integral of the product of the input voltage v_{in} and the input current i_{in}. The input current waveform $i_{in}(t)$ can be evaluated via the network input admittance $Y_{in}(s)$. In RC trees, the poles and zeroes of $Y_{in}(s)$ are real, negative, and alternately placed on the frequency axis, with the first zero being placed at the origin [6], as depicted in Fig. 2. Moreover, $Y_{in}(s)$ can be analytically written as [6]

$$Y_{in}(s) = sC_T \frac{\displaystyle\prod_{i=1}^{n-1}(s\tau_{z_i} + 1)}{\displaystyle\prod_{i=1}^{n}(s\tau_{p_i} + 1)} \qquad (2)$$

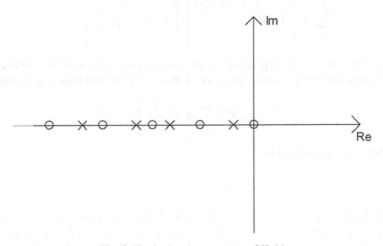

Fig. 2. Typical pole-zero map of $Y_{in}(s)$

being τ_{pi} and τ_{zi} the time constants associated with the poles and the zeroes of the circuit admittance (i.e., their values are obtained as the negative of their reciprocal) ordered such that $\tau_{p1} > \tau_{p2} > ... > \tau_{pn}$ and $\tau_{z1} > \tau_{z2} > ... > \tau_{zn-1}$. The rational form of $Y_{in}(s)$ in (2) can be rearranged through a partial fraction expansion

$$Y_{in}(s) = sC_T \sum_{i=1}^{n} \frac{A_i}{s\tau_{p_i} + 1} \qquad (3)$$

where coefficients A_i (i.e. the residues of $Y_{in}(s)$) are easily found by equating (2) to (3). In particular, by equating the constant term in their numerators, the following property is easily demonstrated [4]-[5]

$$\sum_{i=1}^{n} A_i = 1. \qquad (4)$$

Now, let us assume an exponential input with time constant τ and final value V_{DD}

$$v_{in,\exp}(t)=V_{DD}\left(1-e^{-\frac{t}{\tau}}\right)=V_{DD}\left(1-e^{-2.2\frac{t}{T}}\right) \tag{5}$$

where it was considered that the rise time T of an exponential waveform is equal to $2.2\cdot\tau$ [7]. As is well known, the exponential input is a realistic assumption in current VLSI circuits, as in the practical cases when the input voltage is generated by a buffer driving a long interconnect or a static logic gate driving a transmission-gate circuit [3]. By expressing the Laplace transform of the input current as $Y_{in}(s)\cdot V_{in}(s)$ (i.e. the product of the input admittance and the transform $V_{in}(s)$ of the input voltage (5)), the input current $i_{in}(t)$ is easily evaluated as the inverse Laplace transform of $Y_{in}(s)\cdot V_{in}(s)$, i.e.

$$i_{in,\exp}(t)=2.2\frac{V_{DD}}{T}C_T\cdot u(t)\cdot e^{-2.2\frac{t}{T}}$$

$$\cdot\sum_{i=1}^{n}A_i\left\{\left[1-e^{-\frac{t}{\tau_{Pi}}\left(1-2.2\frac{\tau_{Pi}}{T}\right)}\right]\cdot\left(1-2.2\frac{\tau_{Pi}}{T}\right)^{-1}\right\} \tag{6}$$

being $u(t)$ the Heaviside step function. Relationships (5)-(6) can be substituted into (1) to express the energy consumption as a function of the rise time T according to (7)

$$E_R=C_TV_{DD}^2\sum_{i=1}^{n}A_if\left(\frac{T}{\tau_{Pi}}\right) \tag{7}$$

where function f is defined as

$$f(x)=\frac{1.1}{2.2+x} \tag{8}$$

and is plotted in Fig. 3. It is worth noting that (7) requires the evaluation of poles and zeroes (or equivalently τ_{pi} and A_i), which in general is precluded in practical circuits with $n>2$. However, an exact analytical evaluation is allowed in the particular case of very slow and very fast input waveforms, as discussed in the following section.

3   Energy Consumption for a Very High or Very Low Input Rise Time

When the input rise time T tends to zero (i.e. the input becomes a step waveform), the argument of function f in (7) tends to zero (and f tends to ½), which from (7) leads to the following energy expression

$$E_R(T\rightarrow 0)=\frac{1}{2}C_TV_{DD}^2 \tag{9}$$

As expected, (9) is equal to that obtained in the particular case of a ladder network with a ramp input [4]-[5].

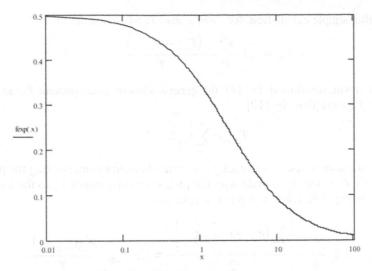

Fig. 3. Plot of function f in (8) versus x

On the other hand, for high values of T the energy (7) can be evaluated by using the Elmore delay approximation, which substantially simplifies the network analysis [8]. Indeed, it is well known that it leads to an exact evaluation of voltage waveforms for very high values of the input rise time [9]. According to the Elmore approximation, for high values of T the transfer function of the voltage v_j of node j is well approximated by a single-pole function with a proper time constant $T_{D,j}$

$$\frac{v_j(s)}{v_{in}(s)} \approx \frac{1}{1 + sT_{D,j}} \qquad (10)$$

from which the Laplace transform $V_j(s)$ of the voltage waveform $v_j(t)$ results to

$$V_j(s) \approx V_{in}(s) \cdot \frac{1}{1 + sT_{D,j}} = \frac{V_{DD}}{s \cdot (1 + sT/2.2)} \cdot \frac{1}{1 + sT_{D,j}} \qquad (11)$$

where the Laplace transform of the input waveform in (5) was substituted. By applying the inverse transform to (11), the voltage waveform $v_j(t)$ results to

$$v_j(t) = V_{DD}\left(1 - \frac{T_{D,j}}{T_{D,j} - T/2.2} e^{-\frac{t}{T_{D,j}}} + \frac{T/2.2}{T_{D,j} - T/2.2} e^{-2.2\frac{t}{T}}\right). \qquad (12)$$

Therefore, since the voltage across the generic resistance R_j between nodes $(j-1)$ and j (see Fig. 1) is equal to $v_{j-1}(t) - v_j(t)$, from (12) the energy E_{Rj} dissipated by R_j results to

$$E_{Rj} = \int_0^{+\infty} \frac{[v_{j-1}(t) - v_j(t)]^2}{R_j} dt$$

$$= \frac{V_{DD}^2}{R_j} \cdot \int_0^{+\infty} \left[-\frac{e^{-\frac{t}{T_{D,j-1}}}}{1 - \frac{T}{2.2 \cdot T_{D,j-1}}} + \frac{e^{-\frac{t}{T_{D,j}}}}{1 - \frac{T}{2.2 \cdot T_{D,j}}} + \frac{T(T_{D,j} - T_{D,j-1})e^{-2.2\frac{t}{T}}}{2.2(T_{D,j} - T/2.2)(T_{D,j-1} - T/2.2)} \right]^2 dt \qquad (13)$$

which, after simple calculation, for $T \gg T_{D,j}$ and $T \gg T_{D,j-1}$ results to

$$E_{Rj} = 1.1 \cdot \frac{V_{DD}^2}{R_j} \cdot \frac{(T_{D,j} - T_{D,j-1})^2}{T}. \tag{14}$$

where (12) was substituted. In (14), the generic Elmore time constant $T_{D,j}$ at the node j (with $j=1 \ldots n$) is given by [10]

$$T_{D,j} = \sum_{k=1}^{n} r_{j,k} C_k \tag{15}$$

being $r_{j,k}$ the sum of resistances along the route obtained by intersecting the path from capacitance C_j to the input node with the path from capacitance C_k to the input node. By substituting (15), relationship (14) results to

$$E_{Rj} = 1.1 \cdot \frac{V_{DD}^2}{R_j} \cdot \frac{\left(\sum_{k=1}^{n} (r_{j,k} - r_{(j-1),k}) C_k\right)^2}{T} = 1.1 \cdot V_{DD}^2 \cdot \frac{R_j (C_{DESC,j})^2}{T} \tag{16}$$

where it was observed that, by definition, term $r_{j,k}$ is equal to the sum of $r_{(j-1),k}$ and the subsequent resistance R_j (i.e. $r_{j,k} - r_{(j-1),k} = R_j$) if node k is a descendant of node $(j-1)$, whereas it is equal to zero otherwise (thus only capacitances connected to node j or its descendants are considered). Thus, the overall energy dissipated by an RC tree network for $T \rightarrow \infty$ is the sum of contributions associated with resistances $R_1 \ldots R_n$

$$E_R(T \rightarrow \infty) \approx 1.1 \cdot \frac{V_{DD}^2}{T} \sum_{j=1}^{n} R_j (C_{DESC,j})^2 \tag{17}$$

4 An Energy Model Valid for Arbitrary Values of the Rise Time

In this section, a first-order equivalent circuit is introduced to evaluate the energy consumption of an RC tree. To this aim, let us first derive an equivalent first-order RC circuit which matches the energy consumption of an RC tree for $T \rightarrow 0$ and $T \rightarrow \infty$. In particular, the energy for $T \rightarrow 0$ in (9) is achieved by an equivalent first-order RC circuit having its capacitance C_{eq} equal to C_T (from comparison of (9) and (7) with $n=1$). Moreover, for $T \rightarrow \infty$ the same energy dissipation as in (17) is achieved by a first-order RC circuit having an equivalent time constant $\tau_{eq} = R_{eq} C_{eq}$ which can be evaluated by equating its energy (given by (17) with $n=1$, resistance R_{eq} and capacitance $C_{eq} = C_T$) to that in (17), thereby leading to the following equivalent time constant

$$\tau_{eq} = R_{eq} C_{eq} = \frac{\sum_{i=1}^{n} R_i (C_{DESC,i})^2}{C_T} \tag{18}$$

This first-order equivalent RC circuit has been derived for very low and very high values of the input rise time, but is also expected to hold for intermediate values of T. Indeed, from the pole-zero map of $Y_{in}(s)$ in Fig. 2, the single-pole behavior certainly holds for very distant poles, but also when two poles are close, since the zero which is in-between tends to null the effect of one of the two poles. Therefore, their overall effect on the input admittance (and hence the energy dissipated) can be represented by a single pole even when poles are not distant from each other.

According to the above observations, the equivalent first-order RC circuit matching the energy consumption of the original RC tree network for $T{\to}0$ and $T{\to}\infty$ can be used for arbitrary values of T. Analytically, the energy consumption of such a first-order equivalent RC circuit is obtained by setting $n=1$ into (7) and substituting the equivalent time constant in (18), thereby yielding

$$E_R = C_T V_{DD}^2 \cdot f\left(\frac{T}{\tau_{eq}}\right) \tag{19}$$

which explicitly depends on the circuit resistances and capacitances, and does not require the closed-form pole-zero evaluation. Note that the model is simple enough for pencil-and-paper calculations through a simple inspection of the RC tree network. By inspection of (19) or (9), it is evident that for low values of T (i.e. $T{<<}\tau_{eq}$) the energy consumption is essentially determined by the overall network capacitance, thus the main contributions are associated with the greatest capacitances. Rather, from (19) or (17), for high values of T (i.e. $T{>>}\tau_{eq}$) the energy consumption is essentially determined by the product of each resistance R_j and the corresponding overall capacitance of its descendants nodes $C_{DESC,j}$, thus the main contributions are associated with high resistances having a high value of $C_{DESC,j}$. As an interesting result, the energy consumption is proportional to $1/T$ for very slow inputs.

5 Simulation Results

The accuracy of the model in (19) was tested by means of SPICE simulations on 1,000 RC tree circuits which were randomly generated by varying the order n from 2 to 20, with resistances and capacitance being varied by five orders of magnitude higher and lower with respect to a reference value. The input rise time was varied by five orders of magnitude around the equivalent time constant of the network.

Some results are reported in Fig. 4, where the energy normalized to that for $T{\to}0$ in (9) is plotted versus the input rise time normalized to the equivalent time constant in (19). This figure confirms that the energy consumption dependence on the input rise time is very close to that of a single-pole circuit reported in Fig. 3, as expected, regardless of the values of resistances, capacitances and the order n of the network.

To evaluate the accuracy of the model in (19), the error with respect to the simulated results was evaluated for each of the 1,000 network. The resulting percentage error is plotted in Fig. 5 versus the input rise time normalized to the equivalent time constant. By inspection of Fig. 5, the error of (19) is always lower than 11%, and typical much lower, as the average error was found to be always lower than 5% even for intermediate values of the input rise time. This confirms that the proposed model is accurate enough for modeling purposes, and it provides results very close to exact ones in typical cases.

Finally, as expected, the error for very low and very high values of T (compared to τ_{eq}) tends to zero, since the equivalent first-order RC circuit is built by matching the exact energy consumption for such values of the input rise time.

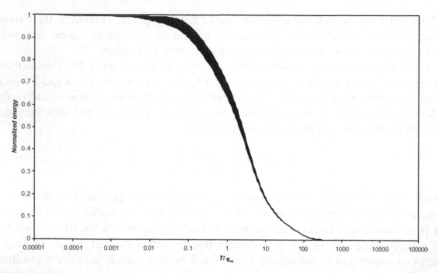

Fig. 4. Simulation results: energy normalized to that with a step input versus the input rise time normalized to the equivalent time constant

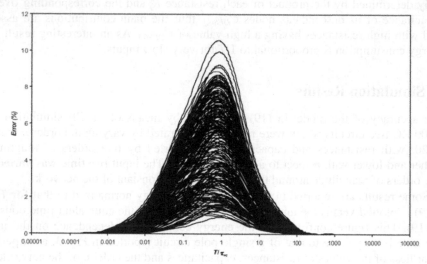

Fig. 5. Error of model (19) with respect to simulated curves in Fig. 4 versus the input rise time normalized to the equivalent time constant

6   Conclusions

In this paper, an analytical model of the energy consumption in RC tree circuits was proposed. The effect of the input rise time on the energy dissipated is analyzed by assuming an input exponential waveform, which is a realistic assumption in current VLSI circuits. To avoid the (unfeasible for $n>2$) pole-zero evaluation, a suitable first-order equivalent RC circuit is derived. To this aim, an exact analysis of energy consumption for asymptotic values of the input rise time is performed, and a first-order circuit matching these expressions is derived. Then, the single-pole approximation is justified even for intermediate values of the input rise time. This strategy leads to a closed-form expression of the energy dissipation which is simple enough to be used for pencil-and-paper evaluation.

Comparison with SPICE simulations shows that the model provides results that are close to simulation results, with a maximum error always lower than 11% and an average error lower than 5%.

References

1. F. Najm, "A survey of power estimation techniques in VLSI circuits," *IEEE Trans. on VLSI*, Vol. 2, N. 4, pp. 446-455, December 1994.
2. R. Gu, M. Elmasry, "Power dissipation analysis and optimization of deep submicron CMOS digital circuits," *IEEE Journal of Solid-State Circuits*, Vol. 31, No. 5, pp. 707-713, May 1996.
3. A. Chandrakasan, W. Bowhill, F. Fox (Eds.), *Design of High-Performance Microprocessor Circuits*, IEEE Press, 2001.
4. M. Alioto, G. Palumbo, M. Poli, "Evaluation of Energy Consumption in RC Ladder Circuits Driven by a Ramp Input," *IEEE Trans. on VLSI Systems*, vol. 12, no. 10, Oct. 2004.
5. M. Alioto, G. Palumbo, M. Poli, "An Approach to Energy Consumption Modeling in RC Ladder Circuits", *Proc. of PATMOS 2002*, Sevilla (Spain), pp. 239-246, Sept. 2002.
6. G. Daryanani, *Principles of Active Network Synthesis and Design*, John Wiley and Sons, 1976.
7. J. Millman, A. Grabel, *Microelectronics (second edition)*, McGraw-Hill, 1987.
8. W. C. Elmore, "The transient response of damped linear networks with particular regard to wideband amplifiers," *J. Appl. Phys.*, vol.19, pp. 55-63, Jan. 1948.
9. R. Gupta, B. Tutuianu, L. Pileggi, "The Elmore Delay as a Bound for RC Trees with Generalized Input Signals," *IEEE Trans. on CAD*, vol. 16, no. 1, Jan. 1997.
10. J. Rubinstein, P. Penfield, Jr., and M. A. Horowitz, "Signal delay in RC tree networks," *IEEE Trans. on Computer-Aided Design*, Vol. CAD-2, pp. 202-211, July 1983.

Statistical Critical Path Analysis
Considering Correlations

Yaping Zhan[1], Andrzej J. Strojwas[1], Mahesh Sharma[2], and David Newmark[2]

[1] Department of ECE, Carnegie Mellon University, Pittsburgh, PA, USA
{yapingz,ajs}@ece.cmu.edu
[2] Advanced Micro Devices Inc., Austin, TX, USA
{mahesh.sharma,david.newmark}@amd.com

Abstract. Critical Path Analysis is always an important task in timing verification. For today's nanometer IC technologies, process variations have a significant impact on circuit performance. The variability can change the criticality of long paths [1]. Therefore, statistical approaches should be incorporated in Critical Path Analysis. In this paper, we present two novel techniques that can efficiently evaluate path criticality under statistical non-linear delay models. They are integrated into a block-based Statistical Timing tool with the capability of handling arbitrary correlations from manufacturing process dependence and also path sharing. Experiments on ISCAS85 benchmarks prove both accuracy and efficiency of these techniques.

1 Introduction

For today's increasing complexity of VLSI designs and tighter timing constraints, timing verification has become a more challenging and important task. Timing information is very useful for both design optimization and yield improvement in manufacturing. Meanwhile, as IC technologies are scaled down to the nanometer regions, circuit delays become highly dependent on manufacturing process variations, especially the intra-die variations, of both gates and interconnects. Due to the correlations among component (gate and interconnect) delays, corner case analysis using traditional Static Timing Analysis (STA) tools is very pessimistic and not capable of finding the circuit delay in variational environments.

As a solution, Statistical Static Timing Analysis algorithms (SSTA) have been proposed recently. Instead of propagating fixed delay values through gates and interconnect, SSTA propagates delay distributions characterized by delay Probability Density Functions (PDFs) [2]-[7].

Path-based SSTA algorithms perform delay analysis path by path [2][3]. They are accurate, good at capturing correlations, and able to report statistical critical paths for circuit optimizations. However, due to the large number of long paths in realistic commercial circuits, we can't afford to analyze all those paths. Hence, the top K longest paths must be selected before a path-based algorithm is applied. The task of selecting the top paths before statistical analyses, when both inter-die and intra-die variations are present, is very challenging. Since non-critical paths in some part of the process space might become critical in the rest part of the process space, no effective approach has been published so far to solve this problem.

V. Paliouras, J. Vounckx, and D. Verkest (Eds.): PATMOS 2005, LNCS 3728, pp. 364–373, 2005.
© Springer-Verlag Berlin Heidelberg 2005

On the other hand, parameterized block-based SSTA approaches [4]-[7] propagate delay statistics level by level with atomic operations *sum* and *max*. All gate/interconnect delays and arrival times are represented as functions of a set of independent process parameters. Therefore, correlations can be preserved during propagations. From these approaches, more accurate final circuit delay distributions are derived and fed back to circuit designers. This helps in predicting the manufacturing yield before the product tape-out.

However, the block-based algorithms often don't provide good estimates on critical paths. From the circuit designers' standpoint, critical paths are more valuable than the final circuit delay distributions, since they can improve circuit performance by optimizing critical path delays. Meanwhile, to ensure correct timing behavior, it is a common practice to include testing of a set of critical paths. Critical paths are very useful for test engineers to generate the test vectors as well. Moreover, the critical paths can be treated as the top K longest paths for path-based SSTA algorithms that enable us to do more specific analyses for better timing accuracy. Therefore, accurate critical path analysis for block-based algorithms will be very useful.

As we know, under statistical delay models, long paths often change their rankings due to process variations. Non-critical paths under nominal delay environment may become critical under certain variations. Therefore, traditional critical path analysis needs to be extended to statistical critical path analysis. Moreover, the correlations resulting from path sharing as well as manufacturing process dependence make this problem even more challenging.

In this paper, we propose two novel approaches, with the capability of handling any delay correlations between any two paths, for path criticality analysis. Both approaches have been integrated into our block-based SSTA algorithm as a post-processing step. The approaches require little CPU overhead and provide very accurate results. The organization of the rest of the paper is as follows: In Section 2, we review previous research on path criticality calculations. We propose our two efficient approaches and discuss our algorithms in Section 3. Experimental results on ISCAS85 circuits are given in Section 4. We conclude this paper in Section 5.

2   Background

Traditionally, critical paths are defined as the longest sensitizable paths under fixed nominal delay models. However, under statistical delay models, each path has a delay distribution. A path may be longer than another path under some process variations, but shorter in other cases. The criticality probability is associated with certain circumstances under which a sensitizable path is critical, i.e., it determines the delay of the circuit. Any path that has a non-zero probability of being critical is defined as one of the statistical critical paths. The sum of the criticality probabilities of all statistical critical paths is 1. The statistical critical paths are ranked by their criticalities. The top statistical critical path with the largest criticality probability is most important and is the one that circuit designers would look at first for performance and yield optimization.

To calculate path criticality in the nanometer process technologies, correlations between paths must be taken into account. There are two major sources of correlations. The first source is from re-convergent fan-outs of the circuit, and the other source comes from the spatial correlations of process parameters, such as inter-die variations and systematic intra-die variations. To consider the correlations, we use the parameterized SSTA technique to represent all the delays and arrival times. Quadratic delay models are used to account for nonlinear delays. Suppose $X=(x_1,x_2,...,x_n)$ is a set of independent process parameters, which can be derived from a Principal Component Analysis (PCA) approach. We can represent any gate/interconnect delay and signal arrival time in the following forms:

$$D_i = X^T A_i X + B_i X + C_i \qquad (1)$$

In a previous research [5], the authors proposed their path criticality evaluation method in a block-based SSTA methodology. In a timing graph, each edge is annotated with an Arrival Tightness Probability (ATP), which is the probability that the edge determines the arrival time of its sink node. The probability of a path being critical is calculated as the product of the ATP's of all edges along the path. However, this method is valid only when the criticality probabilities of any two paths are independent with each other. When complicated correlations are considered, this assumption is not true. To demonstrate this, let's look at a circuit example with re-convergent fan-outs and process spatial correlations, in Figure 1. We assume that the delay of each circuit node consists of two parts, the systematic part due to spatial correlations, and the independent random part. More specifically, we assume the systematic delays of all nodes are equal to d_0 due to their close locations, and the random parts are independent normalized Gaussian variables (d_A through d_E), for example, delay of node A is d_0+d_A. This assumption is not necessary for our analysis, but just to simplify the criticality calculations for the example.

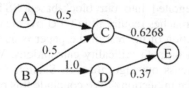

Fig. 1. Circuit example with re-convergent fan-outs

Based on our simplified assumption, the true criticality probabilities of all the paths can be manually obtained from path analyses. For path B-C-E to be longer than path A-C-E, condition $d_B>d_A$ must be satisfied, and for it to be longer than path B-D-E, $d_C>d_D$ should be true. This determines the criticality of path B-C-E to be 0.25. Path A-C-E and B-D-E can be evaluated similarly. The true criticalities of all the paths are listed as the second column of Table 1.

Table 1. Path criticalities from ATP products

Path	True	Independence	Error
A-C-E	0.375	0.5*0.6268=0.3134	16.4%
B-C-E	0.25	0.5*0.6268=0.3134	25.4%
B-D-E	0.375	1.0*0.3736=0.3736	0.4%

The ATP of each edge is marked directly in Figure 1 as obtained from a block-based SSTA algorithm. The last two columns of Table 1 show path criticalities calculated as the products of all ATP's along the paths, and their corresponding errors compared to the true criticalities. From the above example, we find that the method proposed in [5] does not work correctly on the circuits with re-convergent fan-outs.

Let's take path B-C-E as an example to see the cause of the errors. Operator $AT(.)$ denotes the node output arrival time. The probability for path B-C-E to be critical is:

$$p_1 = P\{[\,AT\,(B) > AT\,(A)]\cap[\,AT\,(C) > AT\,(D)]\}$$
$$= P[\,AT\,(B) > AT\,(A)] \bullet P[\,AT\,(C) > AT\,(D)\,|\,AT\,(B) > AT\,(A)] \tag{2}$$

while under the independence assumption, the probability is instead calculated as:

$$p_2 = P[AT(B) > AT(A)] \bullet P[AT(C) > AT(D)] \tag{3}$$

As we know, only when events $AT(C)>AT(D)$ and $AT(B)>AT(A)$ are statistically independent, are (2) and (3) equivalent. Due to the re-convergent fan-out, when $AT(A)>AT(B)$, it is more probable that $AT(C)>AT(D)$, because $AT(C)=d_0+d_C+AT(A)$ and $AT(D)=d_0+d_D+AT(B)$. Therefore, when correlations are present, we have to use conditional probability instead of the product of individual ATP's. On the other hand, for paths like B-D-E, because the tightness at node D and E are independent, the previous method still shows good accuracy.

3 Criticality Calculation

From Section 2, we know that conditional probabilities should be used when correlations exist. Figure 2 shows a general case for path criticality calculation. If a path with n nodes is considered, $AP_i (1\leq i\leq n)$ is the arrival time of the on-path signal of the i-th node. We also assume that at this node, there are k_i side-inputs and $AS_{i,j}$ $(1\leq i\leq n, 1\leq j\leq k_i)$ is the signal arrival time of its j-th side input.

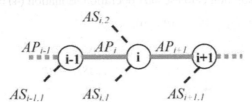

Fig. 2. Path criticality

After a block-based statistical timing analysis is performed, we'll have arrival times of all nodes. Accordingly, the path criticality probability can be expressed as:

$$p = P\{\bigcap_{i=1}^{n}[\bigcap_{j=1}^{k_i}(AP_i > AS_{i,j})]\} \tag{4}$$

In (4), the inner intersection operation defines the local conditions at a particular node, which require the on-path signal to arrive later than all side-input signals. This is consistent with the calculation of the ATP's. The outer intersection operation requires that the local conditions of all nodes along the path to be satisfied at the same time.

The path criticality calculation turns out to be equivalent to the problem of finding the probability of a sub-space formed by all these conditions. Figure 3 illustrates a two-dimensional (with two process parameters) sub-space example with three hypothetical linear conditions. Because all process parameters are assumed as normalized Gaussian random variables, the sub-space is defined by the 3σ bounding box and all the conditions. The shaded area indicates the sub-space for the assumed linear conditions:

$$(X_2 - a_1X_1 - b_1 > 0) \cap (X_2 - a_2X_1 - b_2 > 0) \cap (X_2 - a_3X_1 - b_3 > 0) \tag{5}$$

and its probability is what we want to know.

For general cases with l process parameters and a total of m nonlinear conditions, the sub-space can be similarly represented by a hyper-plane. To derive the sub-space is very complicate and computationally expensive. However, since only the probability of the sub-space is sought, we hereby propose two efficient techniques to solve this problem.

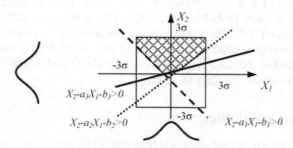

Fig. 3. Subspace example

3.1 *Max* Approach

The first proposed solution is to use *max* operation. Equation (4) can be re-written as:

$$p = P\{\bigcap_{i=1}^{n}[\bigcap_{j=1}^{k_i}(AS_{i,j} - AP_i < 0)]\}$$

$$= P[max(AS_{1,1} - AP_1, AS_{1,2} - AP_1,..., AS_{1,k_1} - AP_1,$$
$$AS_{2,1} - AP_2, AS_{2,2} - AP_2,..., AS_{2,k_2} - AP_2, \tag{6}$$
$$\bullet\bullet\bullet,$$
$$AS_{n,1} - AP_n, AS_{n,2} - AP_n,..., AS_{n,k_n} - AP_n) < 0]$$

Because we've already derived all the m conditions as quadratic functions, we can *max* the conditions pair-wise by moment matching technique [6]. Equations (7)-(10) present a quadratic approximation of the *max* operation, where D_1 and D_2 are quadratic condition functions.

$$D = max(D_1, D_2) = \sum_{i,j} \alpha_{ij}x_ix_j + \sum_i \beta_ix_i + \gamma \tag{7}$$

$$\beta_i = \underset{(x_1,...x_n)}{E}(x_iD) \quad 1 \le i \le n \tag{8}$$

$$\begin{pmatrix} \alpha_{11} \\ \alpha_{22} \\ \vdots \\ \alpha_{nn} \\ \gamma \end{pmatrix} = \begin{pmatrix} 3 & 1 & \cdots & 1 & 1 \\ 1 & 3 & \cdots & 1 & 1 \\ \vdots & \vdots & \ddots & \vdots & \vdots \\ 1 & 1 & \cdots & 3 & 1 \\ 1 & 1 & \cdots & 1 & 1 \end{pmatrix}^{-1} \bullet \begin{pmatrix} E(x_1^2 \cdot D) \\ E(x_2^2 \cdot D) \\ \vdots \\ E(x_n^2 \cdot D) \\ E(D) \end{pmatrix} \tag{9}$$
$$\underbrace{\hphantom{\begin{pmatrix} 3 & 1 & \cdots & 1 & 1 \end{pmatrix}}}_{n+1}$$

$$\alpha_{ij} = \underset{(x_1,...,x_n)}{E} (x_i x_j D) \quad 1 \le i < j \le n \tag{10}$$

The detailed derivation of the moments $E(D), E(x_i D), E(x_i^2 D),$ and $E(x_i x_j D)$ can be found in [6]. Based on this moment match technique, we are able fit the result of a *max* operation with two quadratic inputs back to a new quadratic function of the same parameters. After m-1 *max* operations over all the m conditions, the path criticality can be obtained by numerical integration based on the normalized Gaussian distributed process parameters X:

$$p = P(X^T A X + B X + C < 0) \tag{11}$$

where A, B and C are the fitted quadratic approximation coefficients of the sub-space defined by all the m conditions. As the *max* operation is the core of this approach, both accuracy and efficiency of this solution rely on the *max* operation.

3.2 Monte Carlo Integration

The second proposed solution is to use Monte Carlo integration to get the probability. In (4), we know all the conditions for a given path to become critical after one block-based SSTA run. The path criticality is the probability when all the conditions are satisfied simultaneously. To calculate the criticality, we generate N sets of random process parameters and substitute them into (4). A counter will be used to calculate how many process parameter sets satisfy all the conditions. The path criticality is the final counter value divided by N. The accuracy of Monte Carlo Integration is determined by the number of simulations N due to: [8]

$$\int f \, dV \approx V \cdot E(f) \pm V \cdot \sqrt{\frac{E(f^2) - E^2(f)}{N}} \tag{12}$$

where f is the function that bounds the sub-space defined by all the m conditions, V is the corresponding multidimensional volume, and E(.) is the expectation operator. From (12), the error of Monte Carlo Integration is proportional to $1/\sqrt{N}$. Therefore, a 10,000-point Monte Carlo integration holds the error to the order of 0.01, which usually gives good enough accuracy.

3.3 Algorithm

Both approaches have been integrated into our block-based SSTA tool. The algorithm is illustrated in Figure 4.

```
Statistical_Critical_Path_Analysis (path_set, mode, PT) {
    Block_Based_SSTA();
    if(mode==Monte Carlo) {
        generate_random_vectors();
    }
    While(path_set!=NULL && total_probability<PT) {
        pick_next_path();
        get_all_conditions();
        prob=get_criticality(mode);
        total_probability+=prob;
    }
}
```

Fig. 4. Critical path analysis algorithm

The algorithm takes three input variables: *path_set*, the paths to analyze; *mode*, the working mode; and *PT*, probability threshold. *Path_set* is a set of sorted long paths to be evaluated for criticality. They can be the long paths (both critical and non-critical) under nominal delays from STA tools, in a descending order. The *mode* input indicates whether we use *max* operation or Monte Carlo Integration for criticality calculation. *PT*, the probability threshold can be interpreted as a performance/cost tradeoff criterion. We evaluate the criticality probability path by path. Once the total criticality of those paths that have already been evaluated reaches *PT*, the algorithm exits. If *PT* is too small, some important paths may be missing during the analysis, while if it is set to 1, we will obtain all statistical critical paths including those that only have tiny probabilities of being critical. This will increase the computational cost. For example, if *PT* is set to 0.95, our algorithm will report the top 95% critical paths. Suppose you have a simple circuit with 10 paths. Among them, path A has a criticality probability of 0.70 and path B has a criticality probability of 0.26, while other paths have small criticalities. When *path_set* includes all the 10 paths, the algorithm terminates exactly when both path A and B have been reported.

After the algorithm is launched, a block-based SSTA algorithm is first applied to get arrival times of all nodes of a circuit. Then the first path from *path_set* is selected and the conditions for it to be critical are derived at all nodes. On each node, the conditions are that the on-path signal arrives later than all side-input signals. Next, one of the proposed approaches will be used to calculate the probability when all the conditions are satisfied. After one path is evaluated, the next longest path in *path_set* will be analyzed until either all paths are evaluated or the probability threshold *PT* is reached. In path-based SSTA algorithms, criticality of a certain path is highly dependent to all the other paths being analyzed. Our path criticality analysis is much more robust since a path's criticality is independent of the other paths and totally determined by the analysis on itself. Moreover, due to the probability threshold, not all paths in *path_set* are necessarily evaluated. Therefore, we recommend that *path_set* include non-critical long paths under nominal delays as well.

Table 2 shows how the two proposed approaches work on the example in Figure 1. Column True lists the correct path criticalities, column Max shows the criticalities from proposed *max* approach and column MC is the result from proposed Monte Carlo approach with 10,000-point simulations. Comparing them to the Independence results in Table 1, we can see that the proposed approaches are much more accurate.

Table 2. Algorithm results of the example in Figure 1

Path	True	Max	MC
A-C-E	0.375	0.371	0.379
B-C-E	0.25	0.248	0.246
B-D-E	0.375	0.381	0.375

4   Experiments

The algorithm in Figure 4 has been integrated into our block-based SSTA tool in C++. In this section, extensive experiments are done on the ISCAS85 benchmark circuits under 90nm process technology conditions. The results from different approaches are compared for both accuracy and efficiency. In the experiments, PT is set to 0.95 to get top statistical critical paths up to 95% probability.

Table 3 lists the probabilities of the top statistical critical path of all ISCAS85 benchmark circuits. The Independence experiment assumes path independence and calculates path criticality as the product of the tightness along all the on-path nodes. The Proposed *Max* and the Proposed MC columns show the results from our algorithm in Section 3. To verify the accuracy of these three methodologies, we did a true Monte Carlo experiment as well. In this true Monte Carlo experiment, we randomly select 10,000 variational process parameter sets based on the parameters' distributions, and perform Static Timing Analysis under fixed delay models for every condition. In each STA run, we collect the static critical path of the circuit. The top statistical critical paths are those paths that are reported as critical under most given process conditions. The true Monte Carlo has the best accuracy but is too expensive for large practical circuits. It is hundreds of times slower than the other three statistical approaches. The CPU runtimes of all four experiments to get top 95% statistical critical paths are listed in Table 4.

Compared to the True Monte Carlo approach, the Independence, the proposed *Max*, and the proposed Monte Carlo approach have an average error of 7.16%, 1.49%, and 0.39% respectively over all ISCAS85 benchmark circuits. When we look more closely at the Independence results, we find that for some circuits, it is very accurate, but for some others, the errors are as large as 38.37%. Its accuracy is strongly related to circuit topologies. For those circuits with many re-convergent fan-outs and/or strong spatial correlations, such as C6288, the accuracy deteriorates considerably.

If we compare the two proposed approaches, the proposed Monte Carlo approach excels over the proposed *Max* approach in both accuracy and CPU run times. The reasons can be justified as follows. Based on (12), the accuracy of the proposed Monte Carlo approach is only determined by the number of Monte Carlo Integrations N, and irrelevant to the number of process parameters involved. Moreover, when it verifies the conditions, once a condition is violated, it just skips all the other conditions and goes on with the next Monte Carlo simulation. This makes the algorithm fast. On the contrary, both accuracy and efficiency of the proposed *Max* approach rely on the number of process parameters as well as the number of conditions. Moreover, although we used a non-linear approximation for the *max* operation, errors may still accumulate during the *max* operations of conditions, especially when we have a large number of conditions. Unfortunately, this is often the case in large circuits. The results in Table 5 can justify this issue further. Table 5 shows the criticalities of all statistical critical paths up to 95% probability threshold of circuit C2670. The true Monte Carlo experiment and the proposed Monte Carlo approach both report four statistical critical paths while the proposed *Max* approach reports five. The proposed Monte Carlo approach gets very good results on all top critical paths while the *max* approach doesn't do very well on the third and fourth critical paths. That is because, for shorter paths, the arrival times of on-path signals and side-inputs become closer. This can cause the mean values of the *max* operands, $AS_{i,j} - AP_i$, to be very close, which is the case when *max* operation suffers from largest errors in approximations. Of course, the smaller criticalities on shorter paths also make the relative errors look larger. Nevertheless, we want to point out that the reported statistical critical paths from both proposed approaches match the ones from the true Monte Carlo approach, which gives circuit designers and test engineers correct paths to focus on. Also, if higher order *max* approximation is used, the proposed *max* approach should show better accuracy due to its sound analytical derivation.

Table 3. Probabilities of the top statistical critical path of ISCAS85 benchmark circuits

Circuit	True MC	Independence		Proposed Max		Proposed MC	
	Probability	Probability	Error%	Probability	Error%	Probability	Error%
C17	0.9893	0.9889	-0.04	0.9921	0.28	0.9891	-0.02
C432	0.8682	0.8147	-6.16	0.8696	0.16	0.8686	0.05
C499	0.8603	0.7591	-11.76	0.8664	0.71	0.8611	0.09
C880	0.6966	0.6849	-1.68	0.6844	-1.75	0.6953	-0.19
C1355	0.8108	0.7637	-5.81	0.8241	1.64	0.8148	0.49
C1908	0.8723	0.8810	1.00	0.8874	1.73	0.8739	0.18
C2670	0.3556	0.3790	6.58	0.3504	-1.46	0.3587	0.87
C3540	0.9963	0.9991	0.28	1.0000	0.37	0.9967	0.04
C5315	0.8279	0.8119	-1.93	0.8069	-2.54	0.8253	-0.31
C6288	0.3706	0.2284	-38.37	0.3689	-0.46	0.3720	0.38
C7552	0.3805	0.4000	5.12	0.3603	-5.31	0.3870	1.71

Table 4. CPU run time for top 95% statistical critical paths of ISCAS85 benchmark circuits

Circuit	True MC	Independence	Proposed Max	Proposed MC
C17	34.35	1.29	1.41	1.36
C432	845.58	2.90	6.39	3.01
C499	978.6	1.81	5.07	2.03
C880	1357.01	1.88	3.70	3.65
C1355	2455.24	4.72	4.92	4.84
C1908	3221.11	3.15	3.90	3.28
C2670	4188.79	5.89	11.74	6.73
C3540	6102.67	6.85	7.51	6.78
C5315	9766.77	22.23	25.23	22.77
C6288	11690.84	22.03	28.49	27.80
C7552	12830.86	20.21	19.99	20.19

Table 5. Criticalities of top 95% statistical critical paths of C2670

Path Rank	True MC	Proposed Max		Proposed MC	
	Criticality	Criticality	Error (%)	Criticality	Error (%)
1	0.3556	0.3504	-1.46	0.3587	0.87
2	0.3471	0.3305	-4.78	0.3476	0.14
3	0.2291	0.2056	-10.26	0.2271	-0.87
4	0.0340	0.0457	34.41	0.0326	-4.12
5	N/A	0.0182	N/A	N/A	N/A

5   Conclusions

In this work, we proposed and developed two efficient statistical critical path analysis approaches for block-based SSTA tools. Our algorithm reports good results for statistical critical path analysis in circuits with arbitrary correlations caused by reconvergent fan-outs and process spatial correlations. They enable block-based algorithms to get accurate critical paths with little computing overhead. The algorithm can be added to any block-based SSTA tools as a post-processing step and benefit both circuit designers and test engineers.

Compared to the path-based SSTA algorithms, our approaches are much more robust and efficient in calculating path criticalities. In path-base SSTA, if we don't include all long paths in the analysis, the accuracy of all criticalities will be impaired. However, if we analyze a large number of paths, the computational cost increases considerably, since before all paths are evaluated, we don't know the criticality of any path. In our approaches, even if not all statistical critical paths are considered, the criticalities of all analyzed paths are still accurate. Moreover, a large path set usually doesn't hurt the speed, because most non-critical paths will not be analyzed due to the probability threshold.

References

1. A. Gattiker, S. Nassif, R. Dinakar and C. Long, "Static Timing Analysis Based Circuit-Limited-Yield Estimation", IEEE International Symposium on Circuits and Systems, 2002. Volume 5, 26-29 May 2002
2. M. Orshansky and K. Keutzer, "A General Probabilistic Framework for Worst Case Timing Analysis", Proc. DAC, pp 556-561, June 2002
3. J. A. G. Jess and K. Kalafala et al, "Statistical timing for parametric yield prediction of digital integrated circuits", Proc. DAC, pp. 932-937, June 2003
4. H. Chang, S. S. Sapatnekar, "Statistical timing analysis considering spatial correlations using a single PERT-like traversal", IEEE ICCAD, pp. 621-625 November 2003
5. C. Visweswariah, K. Ravindran, K. Kalafala, S. G. Walker, S. Narayan, "First-Order Incremental Block-Based Statistical Timing Analysis", Proc. 2004 DAC, pp. 331-336, June 2004
6. Y. Zhan, A. J Strojwas, X. Li, L. T. Pileggi, D. Newmark, M. Sharma, "Correlation-Aware Statistical Timing Analysis with Non-Gaussian Delay Distributions", Proc. 2005 DAC, June 2005
7. H. Chang, V. Zolotov, S. Narayan, C. Visweswariah, "Parameterized Block-Based Statistical Timing Analysis with Non-Gaussian Parameters and Nonlinear Delay Functions", TAU workshop, 2005
8. J. M. Hammersley and D. C. Handscomb, Monte Carlo Methods, Methuen's monographs, London, 1964

A CAD Platform for Sensor Interfaces in Low-Power Applications

Didier Van Reeth and Georges Gielen

Department of Electrical Engineering, the Katholieke Universiteit Leuven,
ESAT-MICAS, 3001 Leuven-Heverlee, Belgium
didier.vanreeth@esat.kuleuven.ac.be

Abstract. This paper presents a design automation tool for the generation of sensor interfaces for low-power applications. The tool combines design reuse of known low-power circuits and an efficient optimization algorithm to generate circuits adapted to specific applications. The part of the tool that generates signal filters is described, and the optimization algorithm is compared to other possible optimization methods. Two filters are designed using the tool. Simulation results are compared to existing filters. These results indicate that the filter tool has the desired functionality.

1 Introduction

The increasing role of "ambient intelligence" [1] [2] is creating new challenges in the design of electronic systems. One challenge is the severe limitation of power consumption, imposed by the need for portable, lightweight devices. A large number of these devices will operate at a few hundred micro-watts [3] [4], a constraint that traditionally necessitates full-custom designs. However, design times need to be as short as possible to allow companies to quickly react to changing market demands, which means the design should be automated. Design automation ensures a short design time and correct functionality, eliminating costly redesign and manufacturing reruns, but usually at the expense of the overall performance. This problem can be overcome by using well known, performant circuits and adapting these to new applications.

A tool is presented that will generate the sensor interface in ambient intelligent devices. This tool uses mostly existing circuits, although new circuits can be added, and optimizes these for use in for low-power devices. One part of the tool is a program that generates the optimized circuits for the main subblocks of the sensor interface, such as filters, amplifiers and analog to digital converters (ADCs). This optimization is carried out at two levels, the transistor level and the higher system level. The second part of the tool allows a designer to add new circuits for any type of subblock, along with extra information to improve the optimization process.

In section 2 of this paper, an overview is given of filter module of the interface generation tool. In section 3, the algorithm used in the optimization process is described and compared to other possible algorithms. Design examples and results are presented in section 4, and a general conclusion is in section 5.

V. Paliouras, J. Vounckx, and D. Verkest (Eds.): PATMOS 2005, LNCS 3728, pp. 374–381, 2005.

2   The Interface Tool

The tool for generating the sensor interfaces consists of different modules, one for every building block in a typical sensor interface : an amplifier, a filter and an ADC. Every component will be implemented using known low-power circuits that are optimized for specific applications. The filter module is currently fully developed and will now be described.

2.1   The Filter Module

The filter tool consists of two parts : the first one optimizes individual components out of which the filter is created, the second one builds the actual filter. Typical examples of components that require optimization are transconductors in an active-G_mC topology or the opamps in switched capacitor (SC) filters.

The first part is illustrated in Fig. 1. It uses standard Spice optimization routines for circuit sizing, however Spice is placed in a loop to provide more control over the optimization process. The algorithm used in this loop is described in section 3. First the specifications for the filter are transformed in specifications for the components. In an active-G_mC implementation these new specifications include e.g. transconductor bandwidth and dc gain. These specifications are combined with the predefined (but not optimized) component circuits of the tool and optimized by the Spice optimizer. The result is verified and depending on the outcome of this verification step, the component will either be accepted and used to design the actual filter, or rejected. If it is rejected, which means that not all specifications are met, the starting values of the circuit parameters will be changed. Parameters include transistor sizes, capacitor values and bias currents and voltages. The modifications of the parameter values are made using a combination of straightforward optimization techniques and expert knowledge present in the tool or added by the designer.

A separate module of the filter tool allows the definition of new components. Using this module, all optimization constraints are set; these include minimum

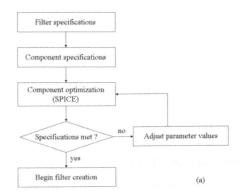

Fig. 1. Spice working in a controlled loop for component optimization

and maximum values for circuit parameters, but also a fixed ratio between variables (e.g. fixed transistor width/length ratios) or the desired operating region of a transistor. It is not necessary to add these constraints, but optimization speed is increased as more circuit knowledge is added by the designer. Along with the main circuit (e.g. a transconductor circuit in an active-G_mC implementation), auxiliary circuits (e.g. a frequency tuning circuit or a common-mode feedback circuit (CMFB)) can be added. The filter tool optimizes both main circuit and auxiliary circuits and integrates all circuits in the complete filter system.

The second part of the filter tool performs the generation of the filter itself (see Fig. 2). First a filter structure is chosen : this is a combination of a specific approximation (Butterworth, Chebyshev or Bessel) and a certain implementation (biquad cascade or active RLC simulation). Then the filter is generated (order and component values are determined), followed by a simulation and evaluation of the results. Depending on whether the filter meets the required specifications, it is either accepted or adapted and resimulated to make it meet the specifications. In the next step, all information concerning the filter's performance is collected and reported, so the designer can compare different structures. When all structures have been tested, the filters that meet the required specifications are sorted according to power consumption. By default the one with the lowest power consumption will be selected as the best solution, but the designer can override this automatic choice.

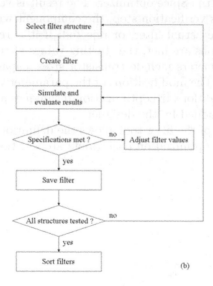

Fig. 2. Generation of the complete filter

2.2   Implementation of the Filter

The filters are implemented as active-G_mC filters. Advantages of this implementation are the signal frequency range, the low power consumption and typically lower noise compared to SC filters. A disadvantage is the need for a tuning circuit to achieve good accuracy. Filters are build as cascaded biquads or as active simulations of RLC ladder filters.

To maximize flexibility, different transconductors are provided, each one performing better in a certain range of frequencies. The transconductor in Fig. 3 [ ] is based on the pseudodifferential $V_{GS}V_{DS}$-type four-quadrant multiplier [ ]. This transconductor can be used when very low frequencies are required, as in biomedical applications.

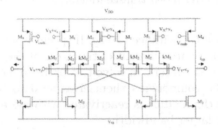

Fig. 3. Four-quadrant multiplier based transconductor, for very low frequencies

Figure 4 shows another transconductor [ ], which has a higher G_m value and bandwidth, making it useful for filters operating at frequencies in the kHz range. It uses source degeneration to set the transconductance value. Both this and the previous transconductor consume very little power, in the order of a few μW.

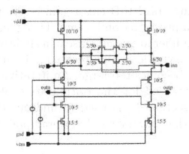

Fig. 4. Linearized transconductor using source degeneration, for low frequencies

The last transconductor, Fig. 5 [ ], has the highest bandwidth, due to the absence of internal nodes. The transconductor is based on the CMOS inverter and is ideal for filters with a higher center frequency (MHz range).

The filter tool supports established filter prototypes (Butterworth, Chebyshev and Bessel). It calculates the filter order for a given specification and generates the circuit netlist. The circuit is simulated and parameters are automatically extracted to determine if the filter meets the specifications. These parameters include the center or cutoff frequency, the stopband attenuation, the harmonic

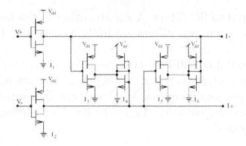

Fig. 5. Inverter based transconductor, for high frequencies

distortion and the total capacitance (area). If specifications are met, the filter is accepted, otherwise there are different options, e.g. if the stopband attenuation is too small, a higher order filter can solve this problem; however, if the problem is a too low SNR, the components themselves need to be redesigned. In this case, the first part of the module is reactivated with more severe specifications, e.g. higher dc gain or larger bandwidth.

2.3 Integration of the Different Modules

On a higher level, research is being done to optimize the integration of the different subblocks of the system. During the optimization process of the individual components, information is recorded to better adapt these subblocks to each other. One example of this process is the gain of the amplifier : this gain can actually be spread over the amplifier and the filter. E.g. in a ladder filter implemented with transconductors, 10 dB extra gain is easily attained by increasing the number of parallel input transconductors. This gain can be used to ease the requirements on the amplifier at the expense of a slightly higher power consumption in the filter. The downside of this technique is the increased distortion in the filter when the gain becomes too large. An iterative optimization algorithm, much like the algorithm used for the optimization of individual components (which is described in section 3), is used to find out the configuration for the individual components that minimizes the overall power consumption.

3   The Optimization Algorithm

There is a large variety of optimization methods and algorithms available. The first choice that must be made when choosing an appropriate optimization method, is whether an equation-based or a simulation-based approach is used. These methods will now be compared.

Equation-based optimization. This is the fastest method since it doesn't need Spice-like simulations that demand much processor power. Instead, it relies on a set of equations that describe the circuit that is optimized. These equations are usually derived manually by the designer. This approach benefits from the knowledge of the designer who provides the equations. Furthermore, these equations are usually well suited for use in an advanced optimization algorithm, making it possible to find an optimal circuit in a relatively short time. The disadvantage of this approach is a lack of flexibility and the necessity to have a full set of equations to describe the circuit that needs optimizing. This lack of flexibility hinders design reuse. A circuit that was developed ten years ago may not be described accurately by the same equations now as it was then, due to the differences in technology. This is certainly a concern when deep-submicron technologies are considered. A designer therefore has to provide equations for every new circuit and also for most existing circuits whenever a new technology is used. Deriving these equations is not a trivial task and may be too time-consuming in many applications. The accuracy of these equations is also a concern : the final optimized circuit is only as good as the equations used in the optimization process.

Simulation-based optimization. This approach relies on direct simulation of the circuit that is optimized. The simulator used is usually some variant of Spice that is placed in a loop. This approach is naturally much slower than equation-based optimization, but has few other disadvantages. It is as accurate as the simulator used and it is very flexible, requiring little or no knowledge at all of the circuit. Therefore it is very easy to set up the optimization process for any given circuit. Furthermore, the long simulation times are not a real problem anymore with the current computers's processing power. However, even with the increased processing power, not every optimization algorithm is suited to the simulation-based optimization. This is discussed next.

The standard optimization routines in Spice variants like Eldo or HSpice are in most cases not powerful enough to optimize a somewhat larger circuit. Therefore, these optimizations are placed in a loop and between successive simulations the circuit parameters are adapted, which means that the next simulation is started with initial conditions that are sufficiently different from the previous initial conditions. These adaptions depend on the outcome of the simulations, as well as on the algorithm used to make the adaptions. Different types of algorithms exist : there are blind optimization methods and heuristically informed methods, and algorithms that provide the best possible solution or simply any acceptable solution.

Blind methods will simply adapt parameter values until a solution has been found or all possible combinations of parameter values have been tried by the simulator optimization routines. Eventually this method can be used to find the best possible solution. In that case, all combinations have to be tried even after a solution has been found. For the problem at hand, blind methods are clearly useless. Even in a very small circuit, there are too many parameters that can have too many values to explore all possible combinations.

Heuristically informed methods use a certain plan to speed up the optimization. The method used in the tool presented in this paper is parameter-oriented hill climbing, a method in which a parameter is changed in different ways, after which the effect is observed. When the parameter has been adapted in all possible ways, the change that improved the the circuit's performance the most is retained. E.g. a transistor's length could first be decreased and then increased, after which simulation results corresponding to these changes are compared with each other and with the original simulation (before the changes were made). Depending on the outcome, it can then e.g. be decided that this parameter needs to be increased for best results. This process is repeated for every parameter until a set of 'best evolutions' for all parameters has been extracted. Once this is done, the real optimization process is started, in which all parameters are continually adapted according to this set. This algorithm will stop when an acceptable solution is found, not necessarily the best solution. Other heuristic methods do find the best solution, but they are not suited to this type of optimization problem. Even with the current available processing power, the simulation-based optimization takes too much time if too many combinations of parameter values need to be examined. The proposed method achieves good results in a limited time.

4 Results

For demonstration purposes, two filters have been created with the tool : a lowpass filter with a cutoff frequency of 4.4 kHz which may be used in audio communication applications and a bandpass filter with a center frequency of 3 MHz and a bandwidth of 0.5 MHz. The aim was to keep the total distortion under 1% (-40 dB). Both filters have been simulated in a 0.35 μm technology. The filters are compared to those published in [ ] and [ ]. Table 1 shows the characteristics of these filters; the results indicate that the filter tool generates filters that closely match the given specifications. For both filters, a CMFB circuit was automatically generated; the lowpass filter also contains a separate biasing circuit.

5 Conclusions

A tool has been presented to automate the design of sensor interfaces for low-power applications. The main focus was on the module that generates filters for these interfaces. Instead of designing a filter from scratch, known low-power

Table 1. Performance of the simulated filters

Parameter	Value LPF	Value [ ]	Value BPF	Value [ ]
V_{DD}	2.5 V	3 V	2.5 V	2.5 V
Power dissipation	21.5 μW	54 μW	3.8 mW	4 mW
Cutoff frequency	4.48 kHz	—	—	22, MHz
Center frequency	—	5.57 kHz	3.06, MHz	—
3 dB Passband	—	1.52 kHz	0.48 MHz	—
Order	4th	6th	6th	3th
THD ($V_{PP} = 0.5$ V)	-41 dB	—	-42 dB	—
Optimization time	2.5 hrs	—	0.5 hrs	—

circuits are adapted to the new application. The designer only needs to give the functional specifications, whereafter an optimized filter is generated. This tool allows comparison of multiple filter structures and technologies, is fully customizable and guarantees a low-power solution that meets given specifications. After optimization of the individual components, a higher level optimization routine attempts to find the ideal combination of low level components, to further decrease power consumption.

To demonstrate the working of the tool, two filters have been designed. These filters incorporate a CMFB circuit. One filter has a cutoff frequency of 4.5 kHz and consumes 21.5 µW, the other has a center frequency of 3.1 MHz while consuming 3.8 mW. Both filters meet the required specifications.

References

1. K. Ducarel, M. Bogdanowicz, F. Scapolo, J. Leijten, J-C. Burgelman, "Scenarios for ambient intelligence in 2010. Final report", *IPTS-Seville, ISTAG*, February 2001
2. Michael Friedewald, Olivier Da Costa, "Science and Technology Roadmapping: Ambient Intelligence in Everyday Life (AmI@Life)", *European Science and Technology Observatory (ESTO)*, August 2003
3. Emile Aarts, Raf Roovers, "IC design challenges for ambient intelligence", *Proc.DATE'03*, pp. 2 - 7, 2003
4. Christian C. Enz, Amre El-Hoiydi, Jean-Dominique Decotignie, Vincent Peiris, "WiseNET : an ultralow-power wireless sensor network solution", *Computer*, vol. 37, pp. 62 - 70, August 2004
5. Anand Veeravalli, Edgar Sànchez-Sinencio, José Silva-Martinez, "A CMOS transconductance amplifier architecture with wide tuning range for very low frequency applications", *IEEE J. Solid State Circuits*, vol. 37, pp. 776 - 781, June 2002
6. Gunhee Han, Edgar Sànchez-Sinencio, "CMOS transconductance multipliers - a tutorial", *IEEE Trans. Circuits Syst. II*, vol. 45, pp. 1550-1562, December 1998
7. Elvi Räisänen-Ruotsalainen, Kimmo Lasanen, Markus Siljander, Juha Kostamovaara, "A low-power 5.4 kHz CMOS gm-C bandpass filter with on-chip center frequency tuning", *Proceedings of the 2002 IEEE International Symposium on Circuits and Systems*, Phoenix, Arizona, U.S.A., vol. 4, pp. 651-654, May 2002
8. Bram Nauta, "A CMOS transconductance-C filter technique for very high frequencies", *IEEE J. Solid State Circuits*, vol. 27, pp. 142 - 153, February 1992
9. Esther Rodríguez-Villegas, A.J. Payne, C. Toumazou, "A 290nW, weak inversion, Gm-C biquad", *IEEE International Symposium on Circuits and Systems (ISCAS02)*, pp. 221-224

An Integrated Environment
for Embedded Hard Real-Time Systems
Scheduling with Timing and Energy Constraints

Eduardo Tavares[1], Raimundo Barreto[1], Paulo Maciel[1],
Meuse Oliveira Jr.[1], Adilson Arcoverde[1], Gabriel Alves Jr.[1],
Ricardo Lima[2], Leonardo Barros[1], and Arthur Bessa[1]

[1] Centro de Informática, Universidade Federal de Pernambuco
Recife, PE, Brazil
{eagt,rsb,prmm,mnoj,aoaj,gaaj,lab2}@cin.ufpe.br, arthur.bessa@fucapi.br
[2] Departamento de Sistemas Computacionais, Universidade de Pernambuco
Recife, PE, Brazil
ricardo@upe.poli.br

Abstract. Embedded hard real-time systems have stringent timing constraints that must be satisfied for the correct functioning of the system. Additionally, there are systems where energy is another constraint that must also be satisfied. In order to satisfy such requirements, a pre-runtime scheduling is presented to find a feasible schedule satisfying both constraints. The proposed approach uses state space exploration for finding feasible schedules taking into account timing and energy constraints. The main problem with such method is the space size, which can grow exponentially. This paper shows how to minimize this problem, and presents a depth-first search method on a timed labeled transition system derived from the time Petri net model. EZPetri is an Eclipse perspective for Petri nets based on PNML. It provides facilities for integrating Petri net applications and existing Petri net tools. In this paper we demonstrate how the plug-in technology of Eclipse was employed to integrate the pre-runtime scheduling synthesis framework with the EZPetri environment. In order to depict the practical usability of the proposed approach, a pulse-oximeter case study is adopted to show how to find a feasible schedule.

1 Introduction

Embedded hard real-time systems are dedicated computer applications having to satisfy stringent timing constraints. In many cases, energy is another constraint to be considered. Scheduling plays an important role for meeting these requirements. There are two general approaches for scheduling tasks: runtime and pre-runtime scheduling. In *runtime scheduling*, the schedule is computed online as tasks arrive, using a priority-driven approach. However, there are cases where this approach may constrain the possibility of finding a feasible schedule, even if such schedule exists [11, 12]. On the other hand, *pre-runtime schedulers* compute schedules entirely off-line. This paper adopts the *pre-runtime scheduling* approach. Such an approach makes the execution predictable, reduces the

V. Paliouras, J. Vounckx, and D. Verkest (Eds.): PATMOS 2005, LNCS 3728, pp. 382–392, 2005.

context switching, and excludes the need for complex operating systems. Due to the use of arbitrary precedence and exclusion relations, predictability is an important matter in safety-critical systems. In accordance with [ ], pre-runtime scheduling is often the only means of providing predictability in complex systems.

This work adopts *state space exploration*, since it presents a complete automatic strategy for verifying finite-state systems [ ]. In spite of the fact that a scheduling can be found using this strategy, this may be limited by the excessive size of its state space. The proposed approach tackles this problem by applying techniques for state space reduction, and a depth-first search algorithm.

EZPetri [ ] is an Eclipse-based [ ] environment for integrating existing Petri net tools and applications. The integration is made using a standard XML format, namely, PNML [ ]. In order to provide an easy-to-use tool, the pre-runtime scheduler is integrated with the EZPetri environment. The specification of tasks is performed through a graphical interface. The energy consumption and timing diagram are rendered and presented to the user. Details about Petri nets and the other technologies behind the pre-runtime scheduler are transparent to the EZPetri user.

2 Related Work

The issue of energy constraints in real-time systems has been extensively studied and reported in the literature. In [ ], Wang presents an algorithm that finds a schedule with optimized real-time performance and energy consumption. It adopts a runtime scheduling approach and only soft timing constraint is considered. The algorithm does not support preemption and inter-task relations. AlEnawy and Aydin [ ] introduce static (pre-runtime) and dynamic (runtime) scheduling mechanisms for deal with energy-constrained scheduling. The proposed approach does not guarantee that all tasks will be executed, but only selected tasks with high priority. Preemption is supported, but inter-task relations are not taken in account.

Xu and Parnas [ ] present a branch-and-bound algorithm that finds an optimal pre-runtime schedule on a single processor for real-time process segments with release, deadlines, and arbitrary exclusion and precedence relations. Despite the importance of their work, it does not present real-world experimental results and does not consider energy constraint.

Several works (e.g. [ ]) models the scheduling problem, whilst our work models the tasks of a system and searches for a feasible schedule. Thus, these works may have better performance in some situations. Nevertheless, time efficiency is not a critical concern when con- sidering schedules computed off-line. Moreover, our so- lution may generate timely and predictable scheduled code [ ], which is difficult in the mechanism adopted, for instance, in [ ]. Another interesting feature of our approach is that the use of Petri net analysis techniques allows one to check several system properties. Most works (e.g. UPPAAL [ ]) deal with runtime scheduling approaches. However, as presented before, this approach may not be able to find a feasible schedule, even if such schedule exists.

3   Computational Model: Syntax and Semantics

The computational model syntax is given by a time Petri net [ ], which is a Petri net extended with time, and its semantics is given by its time labeled transition system. A time Petri net (TPN) is a bipartite directed graph represented by a tuple $\mathcal{P} = (P, T, F, W, m_0, I)$. P (places), and T (transitions) are two types of nodes. The edges are represented by $F \subseteq (P \times T) \cup (T \times P)$. $W : F \to \mathbb{N}$ represents the weight of the edges. A TPN marking m_i is a vector $m_i \in \mathbb{N}^{|P|}$, and m_0 is the initial marking. $I : T \to \mathbb{N} \times \mathbb{N}$, represents the timing constraints, where $I(t) = (EFT(t), LFT(t)) \; \forall t \in T$. A time Petri net extended with power constraints is represented by $\mathcal{P}_v = (\mathcal{P}, \mathcal{V})$. $\mathcal{P}$ is the underlying TPN, and $\mathcal{V} : T \hookrightarrow \mathbb{R}^+$ is a partial function that assigns transitions with power consumption values. Note that some transitions may not have any associated power consumption value. $ET(m_i)$ is a set of enabled transitions in marking m_i. Let M be the set of all reachable markings of $\mathcal{P}$. $C \in \mathbb{N}^{|ET(M)|}$ is a clock vector, which represents the time elapsed since the respective transition enabling. In order to facilitate the TPN's analysis, it is defined the dynamic firing interval $(I_D(t) = (DLB, DUB)$, where $DLB(t) = max(0, EFT(t) - c(t))$ and $DUB(t) = LFT(t) - c(t)$. $I_D(t)$ is dynamically modified whenever the respective clock variable is incremented, and t does not fire.

The set of states S of $\mathcal{P}$ is given by $S \subseteq (M \times \mathbb{N}^{|ET(M)|} \times \mathbb{N})$, that is, a single state is defined by a tuple (m, c, v), where m is a marking, c is its respective clock vector for $ET(m)$, and v is the accumulated power consumption up to this state. The initial state is $s_0 = (m_0, c_0, v_0)$, where $c_0(t) = 0 \; \forall t \in ET(m_0)$, and $v_0 = 0$. $FT(s, v_{max})$ is the set of firable transitions at state s defined by: $FT(s, v_{max}) = \{t_i \in ET(m) | DLB(t_i) \le min(DUB(t_k)) \wedge v \le v_{max} \; \forall t_k \in ET(m)\}$, where $FT \subseteq ET \subseteq T$, and v_{max} is the power constraint. The firing domain for t at a specific state s, is defined by: $FD_s(t) = [DLB(t), min(DUB(t_k))]$, $\forall t_k \in ET(m)$.

The semantics of a TPN $\mathcal{P}$ is defined by associating a TLTS $\mathcal{L}_{\mathcal{P}} = (S, \Sigma, \to, s_0)$ such that: (i) S is a finite set of discrete states of $\mathcal{P}$; (ii) $\Sigma \subseteq (T \times \mathbb{N})$ is an alphabet of labels representing activities. The labels are (t, θ) corresponding to the firing of a firable transition (t) at a specific time value (θ) in the firing interval $FD(s)$, $\forall s \in S$; (iii) $\to \subseteq S \times \Sigma \times S$ is the transition relation; and (iv) s_0 is the initial state of $\mathcal{P}$.

Let $\mathcal{L}_{\mathcal{P}}$ be a TLTS derived from a time Petri net $\mathcal{P}$, and $s_i = (m_i, c_i, v_i)$ a reachable state. $s_j = \texttt{fire}(s_i, (t, \theta))$ denotes that firing a transition t at time θ from the state s_i, a new state $s_j = (m_j, c_j, v_j)$ is reached, such that:
(i) $\forall p \in P, \; m_j(p) = m_i(p) - W(p, t) + W(t, p)$;

(ii) $v_j = v_i + \mathcal{V}(t)$;

(iii)$\forall t_k \in ET(m_j), \; C_j(t_k) = \begin{cases} 0, & if(t_k = t) \\ 0, & if(t_k \in ET(m_j) - ET(m_i)) \\ C_i(t_k) + \theta, & else \end{cases}$.

The firing of a transition t_i, at a specific time θ_i in the state (s_{i-1}) defines the next state (s_i).

Let $\mathcal{L}_{\mathcal{P}}$ be a TLTS of a TPN $\mathcal{P}$, where s_0 its initial state, $s_n = (m_n, c_n, v_n)$ a final state, and $m_n = M^F$, which is the desired final marking. $s_0 \xrightarrow{(t_1,\theta_1)} s_1 \xrightarrow{(t_2,\theta_2)}$ $s_2 - - \to s_{n-1} \xrightarrow{(t_n,\theta_n)} s_n$ is defined as a *feasible firing schedule*, where $s_i = \texttt{fire}(s_{i-1}, (t_i, \theta_i))$, $i > 0$, if $t_i \in FT(s_{i-1}, v_{max})$, and $\theta_i \in FD_{s_{i-1}}(t_i)$. As it is presented later, the modeling methodology guarantees the final marking M^F is well-known since it is explicitly modeled.

4   Task Timing Specification

Let $\mathcal{T}$ be the set of tasks in a system. Let τ_i be a periodic task defined by $\tau_i = (ph_i, r_i, c_i, d_i, p_i)$, where ph_i is the initial phase (delay associated to the first time request of a task after the system starting); r_i is the release time (interval between the beginning of a period and the earliest time that a task execution can be started); c_i is the worst case computation time; d_i is the deadline (interval between the beginning of a period and the time when the task must be completed); and p_i is the period (time interval in which the task must be executed). Let $\tau_k = (c_k, d_k, min_k)$ be a sporadic task, where c_k is the worst case computation time; d_k is the deadline; and min_k is the minimum period between two activations of task τ_k. A task is classified as sporadic if it can be randomly activated, but the minimum period between two activations is known. As pre-runtime approaches may only schedule periodic tasks, the sporadic tasks have to be translated to an equivalent periodic task [10]. A task τ_i *precedes* task τ_j, if t_j can only start execution after t_i has finished. A task τ_i *excludes* task τ_j, if no execution of t_j cannot start while task t_i is executing. Exclusion relations may prevent simultaneous access to shared resources. Each task $\tau_i \in \mathcal{T}$ consists of a finite sequence of *task time units* $\tau_i^0, \tau_i^1, \cdots, \tau_i^{c_i-1}$, where τ_i^{j-1} always precedes τ_i^j, for $j > 0$. A task time unit is the smallest indivisible granule of a task, during which it cannot be preempted by any other task. A task can also be split into more than one *subtasks*, where each subtask is composed by one or more task time units.

5   Modeling Real-Time Systems

Hard real-time systems are those that besides its functional correctness, timeliness must be satisfied. The modeling phase is very important to attain such constraints.

5.1   Scheduling Period

The proposed method schedules the set of periodic tasks occurring in a period that is equal to the least common multiple (LCM) of the periods of the given set of tasks. The LCM is also called *schedule period* (P_S). Within this new period, there are several *tasks instances* of the same task, where $N(\tau_i) = P_S/p_i$ gives the instances of τ_i. For example, consider the following task model consisting of two

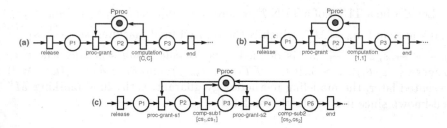

Fig. 1. Modeling Scheduling Methods

tasks: $\tau_1 = (0, 0, 2, 7, 8)$ and $\tau_2 = (0, 2, 3, 6, 6)$. In this particular case, $P_S = 24$, implying that the two periodic tasks are replaced by seven new periodic tasks ($N(\tau_1) = 3$, and $N(\tau_2) = 4$), where the timing constraints of each task instance has to be transformed to consider that new period [ ].

5.2   Petri Net Models for Scheduling Methods

Figure 1 presents three ways for modeling scheduling methods, where $c = cs_1 + cs_2$ is the task computation time (cs_1 and cs_2 are computation times for the first and last subtask, respectively):

a) *all-non-preemptive*: processor is just released after the entire computation be finished. Figure 1(a) shows that computation transition timing interval has bounds equal to the task computation time (i.e., $[c, c]$);

b) *all-preemptive*: tasks are implicitly split into all possible subtasks. This method allows running other *conflicting tasks*, meaning that one task could preempt another task. It is worth observing, the difference between the timing interval for the computation transition and the arc weight in Figures 1(a) and 1(b).

c) *defined subtasks*: tasks are split into more than one explicitly defined subtasks. Figure 1(c) shows two subtasks.

5.3   Tasks Modeling

Figure 2 is also used to show (in dashed boxes) the three main *building blocks* for modeling a real-time task. These blocks are: *(a) Task Arrival*, which models the periodic invocation for all task's instances. Transition t_{ph} models the initial phase, whilst transition t_a models the periodic arrival for the remaining instances; *(b) Deadline Checking* captures deadline misses. Some works (e.g. [ ]) extended the Petri net model for dealing with deadline checking. *(c) Task Structure*, which models time releasing, processor granting, computation, and processor releasing. Figure 2 presents a non-preemptive TPN model for the example described in previous subsection. It does not model the seven task instances. Instead, it models only the two original tasks, and the time period of every task instances.

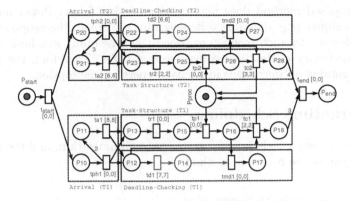

Fig. 2. Petri net model

6   Energy Consumption

Considering the Petri net task model (see Figure 2), system energy consumption is associated with transitions representing context-switching and task computation (t_c). Context-switching takes place if a granting-processor transition (t_p) is fired and the processor has been used by a different task. For the purpose of this work, energy dissipation value in context-switching is constant for a task set model.

Table 1. Energy Values

Description	Values
τ_1	1 nJ
τ_2	2 nJ
Context-Switching	1 nJ
Energy Constraint	18 nJ

The energy consumption value in context-switching and in each task computation must be known beforehand. In this work, the values were measured through a real prototype. The worst-case energy consumption value has been adopted. Table 1 shows the worst-case energy consumption values for each task. The figure also presents the context-switching and the energy constraint values for the schedule period of the model depicted in Figure 2. The sum of energy dissipated in context-switching and in fired computation transitions results the total energy consumed during an execution of a schedule period.

The usage of the pre-runtime scheduler improves the accuracy of timing and energy consumption estimation. On the other hand, runtime approach cannot assure such an accuracy, since unpredictability of tasks arrival leads to more context-switching, increasing the energy consumption substantially.

The proposed method does not substitute other Lower-Power and Power-Aware techniques (e.g. dynamic voltage-scaling). Instead, the proposed method is a complement to such techniques, since the scheduling synthesis algorithm avoids unnecessary context-switching between tasks. Therefore, the generated schedule contains optimizations in terms of energy consumption.

7 Pre-runtime Scheduling

This section shows a technique for state space minimization, and the algorithm that implements the proposed method.

7.1 Minimizing State Space Size

Partial-Order Reduction. If activities can be executed in any order, such that the system always reaches the same state, these activities are *independent*. Partial-order reduction methods exploit the independence of activities [ ]. An independent activity is one that is not in conflict with other activity, that is, when it is executed it does not disable any other activity, such as: arrival, release, precedence, computation, processor releasing, and so on. This reduction method proposes to give for each class of activities a different *choice-priority*. Dependent activities, like processor granting and exclusion relations, have lowest priority. Therefore, when changing from one state to another, it is sufficient to analyze the class with highest choice-priority and pruning the other ones. This reduction is important due to two reasons: (i) it reduces the amount of storage; and (ii) when the system does not have a feasible schedule, it returns more rapidly.

Undesirable States. Section 5 presents how to model undesirable states, for instance, states that represent missed deadlines. The proposed method is interested for schedules that do not reach any of these undesirable states.

7.2 Pre-runtime Scheduling Algorithm

The algorithm proposed (Fig. 3) is a depth-first search method on a TLTS. The *stop criterion* is obtained whenever the desirable final marking M^F is reached. Considering that, (i) the Petri net model is guaranteed to be bounded, and (ii) the timing constraints are bounded and discrete, this implies that the TLTS is finite and thus the proposed algorithm always finishes. When the algorithm reaches the desired final marking (M^F), it implies that a feasible schedule satisfying both timing and energy constraints was found (line 3). The state space generation is modified (line 5) to incorporate the state space pruning. PT is a set of ordered pairs $\langle t, \theta \rangle$ representing for each firable transition (post-pruning) all possible firing time in the firing domain. The *tagging scheme* (lines 4 and 9) ensures that no state is visited more than once. The function `fire` (line 8) returns a new generated state (S') due to the firing of transition t at time θ. The feasible schedule is represented by a TLTS generated by the function

```
1  scheduling-synthesis(S,M^F,TPN, V_max)
2  {
3    if (S.M = M^F) return TRUE;
4    tag(S);
5    PT = remove-undesirable(partial-order(firable(S,V_max)));
6    if (|PT| = 0) return FALSE;
7    for each (⟨t,θ⟩ ∈ PT) {
8      S'= fire(S, t, θ);
9      if (untagged(S') ∧ scheduling-synthesis (S',M^F,TPN,V_max)){
10       add-in-trans-system (S,S',t,θ);
11       return TRUE;
12     }
13   }
14   return FALSE;
15 }
```

Fig. 3. Scheduling Synthesis Algorithm

`add-in-trans-system` (line 10). The whole reduced state space is visited only when the system does not have a feasible schedule.

8   Integrating Pre-runtime Scheduler with EZPetri

The pre-runtime scheduler is integrated with EZPetri for the purpose of simplifying the input of tasks information and facilitating the visualization of the results, which are represented by a timing diagram and a energy chart.

As an example, suppose we have the task timing specification depicted in Table 2, and the task energy specification in Table 1. These specification are entered into the EZPetri framework by using a task/message editor plug-in, that is a graphical editor for specifying timing and energy information of tasks, inter-tasks message passing, as well as, inter-task relations. After that, the time Petri net model (Figure 2) is automatically generated by a Petri net model generator plug-in. Such a model is used for finding a feasible schedule. When successful, the EZPetri framework shows a timing diagram (Figure 4) representing the feasible schedule found by using a schedule rendering plug-in. Also, an energy chart (Figure 5) is exhibited, showing the accumulated energy consumption in each task time unit. Using the EZPetri framework is easier for integrating several tools. In this specific situation, we specify the tasks of a system and have as result the timing diagram that represents a feasible schedule found, and a energy chart, showing the energy consumption in the schedule. Moreover, all formal activities, from the specification up to the final result, are hidden from the final user.

9   Case Study

In order to show the practical usability of the proposed approach integrated with EZPetri, a pulse-oximeter [ ] is used as a case study. This electronic equipment

Fig. 4. Timing Diagram   **Fig. 5.** Energy Chart

is responsible for measuring the blood oxygen saturation through a non-invasive method. A pulse-oximeter may be employed when a patient is sedated during surgical procedures. It checks the oxygen saturation to guarantee an acceptable level of body oxygenation. This equipment is widely used in center care units (CCU). The architecture of this system can be seen in Figure 6.

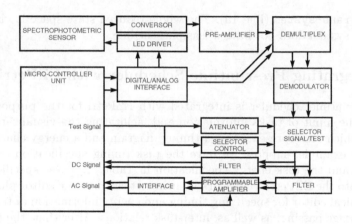

Fig. 6. Pulse-oximeter architeture

In this paper, for sake of simplicity, only two processes represent the oximeter task model: an excitation process (PA), compounded by 4 tasks, and an acquisition-control process (PB), compounded by 10 tasks. Additionally, 11 inter-task relations are also considered. The LCM (Least Commom Multiple) is 80.000, implying 178 task's instances.

The PA is responsible for dispatching pulse streams to the leds in order to generate radiation pulses. The PB captures radiations crossing patient's fin-

Table 2. Simple Task Timing Specification

Task	Release	Computation	Deadline	Period
τ_1	0	2	7	8
τ_2	2	3	6	6

ger, and computes the oxygen saturation level. Both processes are divided into threads. Each thread represents a task. A thread of a process cannot be interrupted by any other thread of the same process. Therefore, context saving and restoring are not performed between tasks of the same processes, but only between threads (tasks) of different processes.

The steps followed by EZPetri to find a feasible schedule for the oximeter specification are described below:

1. the user writes the tasks' specification;
2. a PNML file, representing the Petri net model, is automatically generated;
3. the Petri net model is sent to the pre-runtime scheduler;
4. if a feasible schedule is found, the schedule (TLTS) is returned to EZPetri;
5. based on the TLTS received, EZPetri generates the timing diagram and energy consumption chart;

Other experimental results can be seen on Table 3. These experiments were performed on a Duron 900 Mhz with 256 MB RAM, OS Linux, and compiler GCC 3.3.2.

Table 3. Experimental Results Summary

Example	instances	state-min	found	time (s)
Simple Control Application	28	50	50	0.0002
Robotic Arm	37	150	150	0.014
Xu&Parnas (example 3)	4	171	1566	0.121
Xu&Parnas (figure 9)	5	281	2387	0.222
Pulse Oximeter	**178**	**850**	**850**	**0.256**
Mine Pump Control	782	3130	3255	0.462
Heated-Humidifier	1505	6022	6022	0.486
Unmanned Ground Vehicle	433	4701	14761	2.571

10 Conclusions

This paper has proposed a formal modeling methodology based on time Petri nets, and a framework for pre-runtime scheduling satisfying timing and energy constraints through a reduced state space exploration algorithm. In spite of this analysis technique is not new, to the best of our knowledge, there is no work reported similar to ours that formally models hard real-time systems and finds (whether one exists) the corresponding pre-runtime schedule. The proposed method has been applied into a practical system, namely the pulse oximeter, showing that this is a promising approach for finding feasible pre-runtime schedules in real-world applications. In addition, we demonstrated the pre-runtime scheduling synthesis framework integrated with the EZPetri environment. Such integration, using PNML, facilitated the construction of various features :

– the inputting of tasks' information through a graphical editor;
– the visualization of the Petri net representing the tasks of a system;
– the graphical representation of the results using a timing diagram and an energy chart;

Although the algorithm is quite simple and easy to understand, we plan to perform a proof asserting its correctness. Another extension is to add some flexibility to the pre-runtime scheduling by providing a small runtime scheduler for selecting different operational modes, where each operational mode has a different pre-runtime schedule associated.

References

1. A. Arcoverde Jr, G. Alves Jr, R. Lima, P. Maciel, M. Oliveira Jr, and R. Barreto EZPetri: A Petri net interchange framework for Eclipse based on PNML. First International Symposium on Leveraging Applications of Formal Method (ISoLA'04), Oct 2004.
2. K. Altisen, G. Göbler, A. Pnueli, J. Sifakis, S. Tripakis, and S. Yovine. A framework for scheduler synthesis. IEEE Real-Time System Symposium, pages 154–163, Dec 1999.
3. T. A. AlEnawy and H. Aydin On Energy-Constrained Real-Time Scheduling. Proceedings of the 16th EuroMicro Conference on Real-Time Systems (ECRTS '04),Catania, Italy, June 2004.
4. Blind Review
5. P. Godefroid. Partial Order Methods for the Verification of Concurrent Systems: An Approach to the State-Explosion Problem. PhD Thesis, University of Liege, Nov. 1994.
6. K. Larsen, P. Pettersson, and W. Yi. Uppaal in a Nutshell Int. Journal on Software Tools for Technology Transfer, 1(1–2):134–152, Oct. 1997.
7. Lipcoll, D. and Lawrie, D. and Sameh, A. Eclipse Platform Technical Overview Object Technology International Inc, July, 2001
8. P. Merlin and D. J. Faber. Recoverability of communication protocols: Implicatons of a theoretical study. IEEE Transactions on Communications, 24(9):1036–1043, Sept. 1976.
9. M. N. Oliveira Júnior Desenvolvimento de Um Protótipo para a Medida Não Invasiva da Saturação Arterial de Oxigênio em Humanos - Oxímetro de Pulso. MSc Thesis, UFPE, August, 1998 (in portuguese).
10. A. K. Mok. Fundamental Design Problems of Distributed Systems for the Hard-Real-Time Environment. PhD Thesis, MIT, May 1983.
11. J. Xu and D. Parnas. Scheduling processes with release times, deadlines, precedence, and exclusion relations. IEEE Trans. Soft. Engineering, 16(3):360–369, March 1990.
12. J. Xu and D. Parnas. On satisfying timing constraints in hard real-time systems. IEEE Trans. Soft. Engineering, 1(19):70–84, January 1993.
13. J. Wang, B. Ravindran, and T. Martin. A Power-Aware, Best-Effort Real-Time Task Scheduling Algorithm. IEEE Workshop on Software Technologies for Future Embedded Systems, p. 21, May 2003.
14. Weber M., Kindler, E. The Petri Net Markup Language. Petri net Technology Communication Systems.. Advances in Petri Nets, 2002.

Efficient Post-layout Power-Delay Curve Generation

Miodrag Vujkovic, David Wadkins, and Carl Sechen

Dept. of Electrical Engineering, University of Washington, PO Box 352500,
Seattle, WA 98195, USA
{miodrag,d90mhz,sechen}@ee.washington.edu

Abstract. We have developed a complete design flow from Verilog/VHDL to layout that generates what is effectively a post-layout power versus delay curve for a digital IC block. Post-layout timing convergence is rapid over the entire delay range spanned by a power versus delay tradeoff curve. The points on the gate-sizing generated power-delay curve, when actually laid out, are extremely close in transistor-level simulated power and delay, using full 3D extracted parasitics. The user can therefore confidently obtain any feasible post-layout power-delay tradeoff from the power-delay curve for a logic block. To the best of our knowledge, this is the first report of such a post-layout capability.

1 Introduction

Circuit designers spend a large part of their time executing and optimizing logic-synthesis and physical-implementation (place-and-route) tasks. Timing closure refers to the application of electronic design automation (EDA) tools and design techniques to minimize the number of iterations required between these tasks to meet chip-timing specifications. The advent of deep-submicron (DSM) technologies shifted the design paradigm from a conventional logic-dominated to an interconnect-dominated design process [1]. Today, the timing closure problem is largely caused by the inability to predict interconnect loading with adequate precision prior to physical design [2]. Different methodologies have been proposed to address DSM timing closure challenges [3][4][5][6][7][8][9][10].

We present an automated design flow that generates an optimized power-delay plot for a design. One of the critical concerns is how close the power and delay of an actually laid out block would compare to the power-delay points generated by gate sizing, since the sizing process invalidates the layout from which it began. That is, an appreciable number of cell width changes surely requires a re-routing, and at least a local placement refinement. In this work, we go well beyond that reported in [14] in that the points on our gate-sizing generated power-delay curve, when actually laid out, are extremely close in transistor-level simulated power and delay, using full 3D extracted parasitics. Thereby, the user can confidently obtain any desired post-layout power-delay tradeoff from the power versus delay curve for a logic block, with a small number (usually two) of ECO iterations, and only a single global placement. To the best of our knowledge, this is the first report of such a post-layout capability.

Not only do the ECO iterations converge the timing rapidly, but they also appreciably improve the overall power efficiency of a design. A common metric of power efficiency is the energy (per operation) times delay squared or EDD. Note that EDD is largely independent of the power supply voltage. For a variety of blocks, we show that the average improvement in EDD after the two ECO iterations was 26%.

V. Paliouras, J. Vounckx, and D. Verkest (Eds.): PATMOS 2005, LNCS 3728, pp. 393–403, 2005.

2 Design Flow

The basic steps in our design flow are shown in Figure 1. This flow is foundry independent. However, in this paper some steps are illustrated using the TSMC 0.18um technology, with native design rules, featuring a drawn channel length of 0.18um and a fanout-of-four inverter delay of 84 ps.

1) Iterative design synthesis using Synopsys DC
2) Add wire load model to synthesized netlists
3) Delay/power analysis (Pathmill/NanoSim)
4) "Optimized" curve extraction after synthesis
5) Select point(s) from the "optimized" curve (e.g., target delay, min PD point, etc.)
6) Decompose synthesis cells to the set of basic functions in our optimized library
7) Initial Power/Delay optimization using AMPS
8) "Optimized" P-D curve extraction
9) Calculation of total cell width for each point on P-D plot
10) Choose desired points from one or more delay sub-ranges
11) Extract cell layouts from library
12) High quality (iTools) placement run for each point
13) iTools routing (variable die approach)
14) Import design to Cadence database and run Assura LVS
15) Hierarchical parasitic (RC) extraction using Assura RCX from Cadence (for cells not already in library and top-level)
16) Logical-effort based clock-tree sizing
17) Clock skew measurements
18) Delay/power/area measurements for each generated layout
19) If converged (usually need 2 iterations), stop; else do 20-23
20) Modification of the smallest wire capacitances
21) New AMPS optimizations for one or more layout points
22) "Optimized" curve extraction over all optimizations
23) Re-execute steps 9-18 where step 12 is now an ECO-based placement run (iTools)

Fig. 1. Basic steps in our Power-Delay Optimization design flow

2.1 Design Synthesis

Design Compiler (DC) from Synopsys was used for synthesizing the RTL descriptions into optimized, technology-dependent, gate-level designs). DC was run iteratively to generate optimized designs for various target delays. DC starts from the minimum area point that usually corresponds to the lowest frequency range, and reduces the maximum delay in each subsequent iteration *(step 1)*. Our experiments have shown that 20-30 iterations are needed to obtain a wide power-delay range. We use both DC Ultra and power optimization to produce an optimized netlist.

Library functional composition is critically important to power efficiency. We extended the work in [13] to cover a complete range of digital IC blocks, not just combinational circuits. We found that the best power efficiency was obtained if the library includes only the following 14 combinational functions: INV, NAND2-4, NOR2-3, AOI21, AOI22, OAI21, OAI22, XOR2/XNOR2, MUX2:1, FAsum (sum gate for a full adder), and FAcarry (carry gate for a full adder). Note that power efficient cells, at least with respect to dynamic power, have relatively few transistors in series, and are quite simple functions.

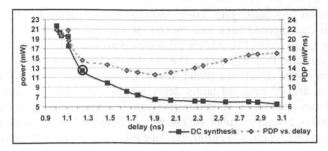

Fig. 2. Iterative design synthesis – each point on the "optimal" curve represents one possible synthesized design. The circled point was chosen since it provides a good trade-off between delay and PD product (minimum EDD)

We found that DC produces better synthesis results if it is also provided with compounds of the 14 base functions. We therefore supply DC with a generic cell library consisting of 160 combinational cells, which actually corresponds to all of the combinational cells in the Artisan library that directly map (decompose) to our 14 base functions. We provide the various library functions in four different drive strengths. Inverters and buffers are available in nine drive strengths. We include a total of 20 flip-flops; five different topologies in four different drive strengths.

We use a basic linear fanout-length relationship for estimating wire loads. Based on past experience, we estimated the average wire lengths to be 17um per fanout for 0.18um technology. Since the capacitance per unit length for wires averages approximately 0.2fF/um for all technology nodes, the wire load estimation for 0.18um technology is 3.4fF per fanout. After design synthesis these wire loads were added to each node *(step 2)*.

2.2 Power-Delay Curve Generation

For each synthesized design, we run static timing analysis using PathMill and dynamic power simulation using NanoSim to determine the worst-case delay and average power consumption *(step 3)*. For very large circuits, we replace dynamic power simulation with static power simulation, based on the switching activity from gate-level simulation.

Among all synthesized points, only the dominant (or Pareto) points are extracted to define the "optimized" power versus delay curve *(step 4)*. An optimized power versus delay curve obtained after the synthesis step for benchmark *des* is shown in Figure 2. After "optimized" curve extraction, we pick *(step 5)* one or more points that satisfy user-specified criteria (*e.g.*, the minimum EDD product, as shown here, or the minimum power-delay-area product, or a specific target delay). The chosen implementations are the starting point for the gate sizing step, which is actually a gate selection step from a very rich library of cell sizes.

2.3 Gate Sizing Optimization

The first step in transistor-level optimization is netlist conversion *(step 6)*. While the inclusion of compounds of the base cells yields better synthesis results, we have found that better and faster gate sizing optimization is obtained if these compound cells are first decomposed back into the set of 14 base cells. We use AMPS from Synopsys for gate sizing [11]. AMPS estimates power and delay based on the PowerMill/Nanosim and PathMill tools. The AMPS optimization process consists of a series of delay-mode (DM) runs (dmr1, dmr2...) and power-mode (PM) runs (pmr1, pmr2...) for a desired range of target delay and power values, as shown in Fig. 3. Each AMPS run consists of 10 iterations *(step 7)*. During the optimization process AMPS generates a large number of points in (power, delay) space, and only the dominant points are retained to define the "optimized" power-delay curve *(step 8)*.

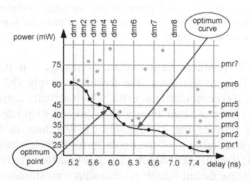

Fig. 3. AMPS optimization process consists of target delay (dmr) and target power (pmr) optimization runs. Each run generates a set of points in power-delay space

The gate sizing is performed such that all channel-connected nMOS devices for a cell are sized as a group, and all channel-connected pMOS devices are sized as a separate group. AMPS is provided with a discrete set of allowable transistor sizes. For larger widths, the sizes correspond to an integral number of half folds (*i.e.* transistor widths are 1.0, 1.5, 2.0, 2.5, *etc.*, times the maximum nonfolded size).

After "optimized" curve extraction, netlists for selected points on the power-delay curve are generated. The area of a design point is estimated by the total cell area *(step 9)*. In next step *(step 10)*, we divide the delay range of interest into one or more subranges and choose one point from each sub-range that satisfies a specified criteria (*e.g.*, minimum PD product, minimum PDA product, or a specific target delay).

2.4 Cell Layout Generation

All of the variants (approximately 1300) of the 14 base functions in our cell library are generated using our automatic, parameterized standard-cell layout generator *Cellgen (step 11)*. *Cellgen* also produces the layouts for the 5 sequential cell types.

Although gate sizing optimization is performed on the 14 base cells, we found it highly beneficial to create compound or merged cells prior to global placement and routing. *Cellgen* extracts two (sized) base cells from the library and merges them into a single place and route instance. "Merged cells" are created from basic cells (*e.g.* a NAND2 and an output inverter make a new AND2 cell, whereas an input inverter and NAND2 cell constitute a NAND2B cell). The cell generator wires up the sub-cells, avoiding the use of metal layers as much as possible. This improves subsequent global placement and routing, since vastly fewer nets need to be routed by the global router. The use of the compound cells reduced the number of nets to be routed by an average of 60% for a set of benchmark circuits.

2.5 Placement and Routing

All attempts at meeting timing should reside with the cell sizer, since the cells can have their sizes optimized for the actual extracted wire loads. A critical advantage is that the placement tool can focus on only a single objective, namely minimizing wire length. Consequently, much lower wire lengths are achieved since the placement algorithm is not distracted by various other concerns such as timing, congestion, *etc.* Our experience is that the iTools placer [12] yields the lowest wire lengths.

Our flow also requires a variable die router; we use the iTools router. As opposed to the fixed die routers, variable die routers are able to create routing space wherever needed in order to complete the routing. Hence, variable die routers inherently complete 100% of the routing. The variable die router has another advantage: each net is routed in near minimum length, every time. Congestion has minimal impact on the variable die router. It simply creates space (*e.g.* increasing row separations or adding feed-throughs) wherever needed to wire up a net using near minimum total wire length. Thus, a minor placement change, such as what might occur after an ECO placement run, will yield similar lengths for each net.

The fixed die router, however, is simply bullied by congestion. If a congested area lies between two pins, the router has no choice but to detour around it. Invariably this net seems to lie on a critical path and now the load for the driving gate is dramatically larger than before, causing timing to diverge. Furthermore, a minor placement change often dramatically changes the congestion landscape, causing nets to become much longer (or shorter) than previously. As a consequence, significant cell sizing must be performed; this process seems to loop forever.

The iTools placer is used to produce a global placement for each design point (*step 12*). Following global placement, design is ready for the routing phase (*step 13*).

Parasitic extraction is done hierarchically with Assura RCX from Cadence (*step 15*). At the end of routing phase, the design is imported to the Cadence database (*step 14*) for final LVS and RCX extraction of top-level parasitics (*step 15*).

In traditional flows, clock tree synthesis is performed after placement. This greatly disturbs cell placement, and consequently wire lengths, making timing closure extremely difficult. We augmented the iTools placer to add a hierarchical H-tree of symmetrical inverters during the actual placement process. Initially, iTools equalizes the transistor capacitive loads on the leaf inverters; however, it cannot assure an equal RC wire load to each flip-flop during placement since routing is yet to be performed.

After global placement and routing, each inverter in the tree is sized using logical effort, including the capacitive loads of the wires. The result is that the fanout capacitance is constant for each leaf inverter, which in turn tends to equalize the rise and fall time for each buffer and reduces the overall skew. However, sizing cannot compensate for the skew between flip-flop loads due to wire resistance effects. The residual clock skews are accurately simulated using Hspice and skew information is incorporated into the next optimization step, where it is used to optimize the timing (*step 17*).

2.6 Post-layout Power/Delay Analysis and Resizing

After the layout and 3D parasitic extraction phase, we perform timing and power measurements using PathMill and NanoSim. We then compare these "actual" delay and power values with the "predicted" values from the "optimized" curve after AMPS optimization. Figure 4 shows a power-delay curve after initial AMPS optimization, a set of chosen points from the curve, and also a set of points obtained after the first layout phase (including parasitic extraction) for the design *complex_multiplier_16x16*. The initial "optimized" curve is produced using the simple wire load model, and therefore the delay/power deviation of the actual points from predicted points is expected to be the largest in this iteration. Delay and power changes (deviations) are computed after the first ECO iteration relative to the global placement and routing, as well as from the second ECO iteration relative to the first ECO iteration, and so forth.

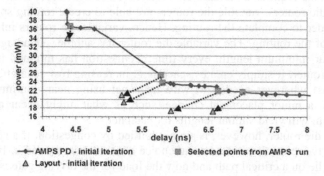

Fig. 4. After initial AMPS optimization, 5 points were chosen from the predicted "optimized" curve for the initial layout phase that produces realistic points in power-delay space (step 10)

Next, we run one or more AMPS optimizations *(step 21)* and extract the "optimized" curve over all optimizations *(step 22)*. "Optimized" curves from individual optimizations contribute to part of the overall curve (in the vicinity of the starting point). Contrary to the initial iteration, the circuit optimizer is provided with full 3D extracted wire parasitics in any subsequent iteration. Steps 9-18 defined in Fig. 1 are repeated if necessary. The iTools placement is executed in ECO mode using wire length constraints from the previous iteration. The iTools ECO mode initializes the placement of all the cells according to the previous iteration, and then it makes very small moves necessary to even up rows and remove the cell overlap created by cells that changed sizes. Because of the wire length constraints and the ECO mode, the resulting wire lengths were very similar to what they were in the previous iteration. Typically, two ECO iterations are needed to produce a converged timing result.

3 Results

We now demonstrate that the points on our gate-sizing generated power-delay curve, when actually laid out, are extremely close in transistor-level simulated power and delay, using full 3D extracted parasitics.

Figure 5 shows the predicted "optimized" power-delay curve, the set of selected points from the curve and the set of actual points after layout and parasitic extraction for two iterations of our design flow for design *complex_multiplier_16x16*

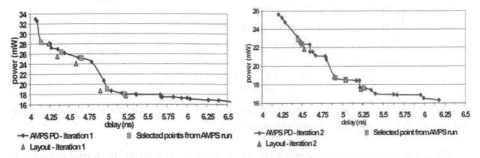

Fig. 5. Predicted "optimized" power-delay curve, set of chosen points based on the minimum EDD in the specified delay sub-range and the set of actual points after the layout phase for iterations 1 and 2 – (*complex_multiplier_16x16*)

Figure 6 shows the predicted "optimized" power-delay curves following the initial and two subsequent ECO iterations. The initial wire loads were overestimated and all layout points were below the predicted curve (Figure 4). The following two refinement iterations lowered the predicted power-delay curve and reduced the deviation between predicted and actual points after layout.

Some of the designs were optimized using three ECO iterations to evaluate the benefit of an additional refinement iteration. We chose to apply this third ECO iteration to the designs that had more significant EDD improvements from iteration 1 to iteration 2. For design *des,* the minimum EDD product improved another 4% from iteration 2 to 3 (compared to 7% from iteration 1 to iteration 2). Average delay and power deviations decreased from 2.2% to 1.3% and from 1.0% to 0.7%, respectively. For design *C6288,* the minimum EDD product improved 5% from iteration 2 to 3 (compared to 16% from iteration 1 to iteration 2). Average delay and power deviations decreased from 1.9% to 0.5% and from 2.4% to 0.6%, respectively. For designs where the improvement from iteration 1 to iteration 2 was less significant, another refinement iteration was evidently not needed.

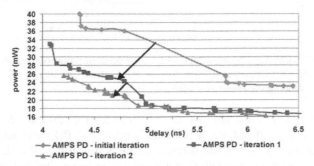

Fig. 6. Predicted "optimized" power-delay curves for each iteration (design *complex_multiplier_16x16*). In the initial iteration, the wire load is overestimated. Points on the final "predicted" PD curve have the lowest power of all iterations

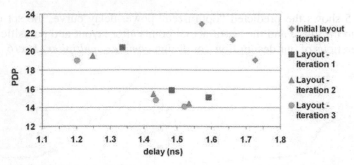

Fig. 7. PDP vs. delay points measured after the layout phase. Our optimization flow effectively reduces the PD product from iteration to iteration

We compared power-delay products for a given design from the initial iteration to the last iteration. Figure 7 shows a typical progression of our design flow and the performance improvement in power-delay product (PDP) from iteration to iteration. Note that our optimization flow is effective at reducing the power-delay products for the same delay target. Typically the largest reduction in PDP is from the initial layout iteration to the first ECO iteration.

Figure 8 overlays the predicted "optimized" power-delay for the initial and three subsequent ECO iterations for design *des*. As opposed to the design shown in Figure 6, wire loads were underestimated in the initial iteration and all initial layout points were above the predicted curve. The predicted PD curve for iteration 1 is above the initially predicted curve as expected, but the two refinement iterations lowered the PD curve and further reduced the deviation between predicted and actual layout points.

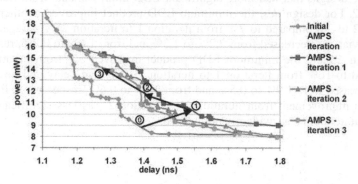

Fig. 8. Predicted "optimized" power-delay curves for each iteration (design *des*). In the initial iteration, the wire load is underestimated. Points on the final "predicted" PD curve have the lowest power (excluding the initial iteration)

Figure 9 shows how the iterative process reduces the absolute delay and power deviation between predicted points from the "optimized" PD curve and actual points measured after the layout and full parasitic extraction. Initially, this deviation can be relatively large (positive or negative), but it quickly reduces after iteration 1. We have invariably found that after only two iterations the deviations are suitably small (below 2%), and under 1% after 3 ECO iterations.

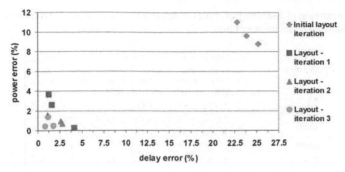

Fig. 9. Delay and power deviation between predicted points from the "optimized" PD curve and actual points measured after layout (design *des*)

We also compared the energy times delay squared (EDD) for a given design from the initial iteration to the last iteration. Note that EDD is largely independent of the power supply voltage. Results for eight designs are given in Table 1. The average improvement in EDD after two refinement iterations was 26%.

Our key goal was to allow the user to choose any desired power-delay tradeoff point from the final "optimized" PD curve, and to obtain post-layout power and delay characteristics very close to those on the "optimized" PD curve. In other words, the goal was to minimize the deviation between the predicted points on the transistor-level "optimized" PD curve and the actual points measured after layout and full parasitic extraction. Table 2 shows the average delay/power deviations after each iteration for eight designs. The average delay/power error after the final (second) iteration was 1.7% and 1.5%, respectively.

Table 1. Comparison of minimum energy*delay^2 for different designs from iteration to iteration

Benchmark	Min EDD (initial iter.)	Min EDD (iter. 1)	Min EDD improv.	Min EDD (iter. 2)	Min EDD improv.
K2	1.7	1.6	6%	1.4	21%
C5315	17.8	16.6	8%	14.5	23%
C7552	29.7	25.5	17%	23.2	28%
C6288	547.0	428.7	28%	369.2	48%
Des	32.9	23.5	40%	22.0	49%
mult_16x16	38.3	34.9	10%	34.3	12%
Complexmult_16x16	526.6	451.7	17%	448.6	17%
32-bit FIR	1179.2	1089.89	8%	1071.734	10%
Avg improvement:			16.6%		26.1%

4 Conclusion

We have developed a complete design flow from Verilog/VHDL to layout that generates what is effectively a post-layout power versus delay curve for a digital IC block. Post-layout timing convergence is rapid over the entire delay range spanned by a power versus delay tradeoff curve.

Table 2. Comparison of average power/delay deviations between predicted points from the PD curve and points after the layout phase

Benchmark	Initial iteration		Iteration 1		Iteration 2	
	Aver. delay error	Aver. power error	Aver. delay error	Aver. power error	Aver. delay error	Aver. power error
K2	13.5%	3.0%	2.2%	0.6%	1.7%	1.9%
C5315	23.5%	19.4%	12.1%	3.6%	3.0%	2.0%
C7552	13.9%	3.9%	0.9%	0.7%	1.7%	3.3%
C6288	9.9%	11.8%	4.4%	4.8%	1.9%	2.4%
Des	23.9%	9.8%	2.3%	2.2%	2.2%	1.0%
mult_16x16	20.8%	22.8%	1.3%	0.4%	1.0%	0.2%
Complexmult_16x16	8.5%	21.1%	1.4%	2.4%	0.4%	0.9%
32-bit FIR	5.1%	12.1%	3.6%	0.5%	2.1%	0.3%
Avg error:	14.9%	13.0%	3.5%	1.9%	1.7%	1.5%

The points on our gate-sizing generated power-delay curve, when actually laid out, are extremely close in transistor-level simulated power and delay, using full 3D extracted parasitics. The average changes in both power and delay are under 1.7% between the first and second ECO iterations. The user can therefore confidently obtain any feasible post-layout power-delay tradeoff from the power-delay curve for a logic block. To the best of our knowledge, this is the first report of such a post-layout capability. Finally, we have demonstrated an average improvement of 26% in EDD as a result of the ECO iterations.

References

1. Cheng, C.K.: Timing Closure Using Layout Based Design Process. (www.techonline.com/community/tech_topic/timing_closure/116)
2. Bryant, R., et al.: Limitations and Challenges of Computer-Aided Design Technology for CMOS VLSI. Proc. IEEE, Vol. 89, No.3, (2001)
3. Coudert, O.: Timing and Design Closure in Physical Design Flows. Proc. of the International Symposium on Quality Electronic Design (ISQED) (2002)
4. Hojat, S., Villarrubia, P.: An Integrated Placement and Synthesis Approach for Timing Closure of PowerPC TM Microprocessors. Proc. IEEE Int. Conference on Computer Design (ICCD) (1997) 206-210
5. Halpin, B., Sehgal, N., Chen, C.Y. R.: Detailed Placement with Net Length Constraints., Proc. of the 3rd IEEE Int. Work. on SOC for Real-Time Applications (2003)
6. Posluszny, S., et al.: Timing Closure by Design, A High Frequency Microprocessor Design Methodology. Proc. of Design Automation Conf. (DAC) (2000) 712-717
7. Northrop, G., Lu, P.F.: A Semi-Custom Design Flow in High-Performance Microprocessor Design. Proc. of Design Automation Conference (DAC) (2001) 426-431
8. Yoneno, E., Hurat, P.: Power and Performance Optimization of Cell-Based Designs with Intelligent Transistor Sizing and Cell Creation. IEEE/DATC Electronic Design Processes Workshop, Monterey (2001)
9. Hashimoto, M., Onodera, H.: Post-Layout Transistor Sizing for Power Reduction in Cell-Base Design. IEICE Trans. Fund., Vol.E84-A. (2001) 2769-2777
10. Gosti, W., Khatri, S.R., Sangiovanni-Vincentelli, A.L.: Addressing the timing closure problem by integrating logic optimization and placement. IEEE/ACM International Conference on Computer Aided Design (ICCAD) (2001) 224 - 231

11. AMPS User Guide Version W-2004.12, Synopsys (2004)
12. iTools, InternetCAD.com Inc. (2004)
13. Vujkovic, M., Sechen, C.: Optimized Power-Delay Curve Generation for Standard Cell Ics. Proc. Int. Conf. Comp. Aided Design (ICCAD) San Jose (2002)
14. Vujkovic, M., Wadkins, D., Swartz, B., Sechen, C.: Efficient Timing Closure Without Timing Driven Placement and Routing. Proc. Design Automation Conference (DAC) San Diego (2004)

Power – Performance Optimization
for Custom Digital Circuits

Radu Zlatanovici and Borivoje Nikolić

University of California, Berkeley, CA 94720
{zradu,bora}@eecs.berkeley.edu

Abstract. This paper presents a modular optimization framework for custom digital circuits in the power – performance space. The method uses a static timer and a nonlinear optimizer to maximize the performance of digital circuits within a limited power budget by tuning various variables such as gate sizes, supply, and threshold voltages. It can employ different models to characterize the components. Analytical models usually lead to convex optimization problems where the optimality of the results is guaranteed. Tabulated models or an arbitrary timing signoff tool can be used if better accuracy is desired and although the optimality of the results cannot be guaranteed, it can be verified against a near-optimality boundary. The optimization examples are presented on 64-bit carry-lookahead adders. By achieving the power optimality of the underlying circuit fabric, this framework can be used by logic designers and system architects to make optimal decisions at the microarchitecture level.

1 Introduction

Integrated circuit design has seamlessly entered the power-limited scaling regime, where the traditional goal of achieving the highest performance has been displaced by optimization for both performance and power. Solving this optimization problem is a challenging task due to a combination of discrete and continuous constraints and the difficulty in incorporating costs for both energy and delay in the objective functions.

System designers typically bypass this problem by forming a hybrid metric, such as MIPS/mW, for evaluating candidate microarchitectures. Similarly, designs at the circuit level are evaluated based on metrics that combine energy and delay, such as the energy-delay product (EDP). A circuit designed to have the minimum EDP, however, may not be achieving the desired performance or could be exceeding the given energy budget. As a consequence, a number of alternate optimization metrics have been used that generally attempt to minimize the $E^m D^n$ product [1]. By choosing parameters n and m a desired tradeoff between energy and delay can be achieved, but the result is difficult to propagate to higher layers of design abstraction.

In contrast, a more systematic and general solution to this problem minimizes the delay for a given energy constraint [2]. Note that a dual problem to this one, minimization of the energy subject to a delay constraint yields the same solution.

Custom datapaths are an example of power-constrained designs where the designers traditionally iterate in sizing between schematics and layouts. The initial design is sized using the wireload estimates and is iterated through the layout phase until a set delay goal is achieved. The sizing is refined manually using the updated wireload estimates. Finally, after minimizing the delay of critical paths, the non-critical paths

V. Paliouras, J. Vounckx, and D. Verkest (Eds.): PATMOS 2005, LNCS 3728, pp. 404–414, 2005.

are balanced to attempt to save some power, or in case of domino logic to adjust the timing of fast paths. This is a tedious and often lengthy process that relies on the designer's experience and has no proof of achieving optimality. Furthermore, the optimal sizing depends on the chosen supply and transistor thresholds. An optimal design would be able to minimize the delay under power constraints by choosing supply and threshold voltages, gate sizes or individual transistor sizes, logic style (static, domino, pass-gate), block topology, degree of parallelism, pipeline depth, layout style, wire widths, etc.

Custom circuit optimization under constraints has been automated in the past. IBM's EinsTuner [3] uses a static timing formulation and tunes transistor sizes for minimal delay under total transistor width constraints. It uses simulation instead of modeling for best accuracy, but it only guarantees local optimality. TILOS [4] solves a convex optimization problem that results from the use of Elmore's formula for gate delays. While the models are rather inaccurate due to their simplicity, the result is guaranteed to be globally optimal.

This paper builds on similar ideas and presents a modular design optimization framework for custom digital circuits in the power – performance space that:

- Formulates the design as a mathematical optimization problem;
- Uses a static timer to perform all circuit-related computations, thus relieving the designer from the burden of providing input patterns;
- Uses a mathematical optimizer to solve the optimization problem numerically;
- Adjusts various design variables at different levels of abstraction;
- Can employ different models in the timer in order to balance accuracy and convergence speed;
- Handles various logic families (static, dynamic, pass-gate) due to the flexibility of the modeling step;
- Guarantees the global optimality of the solution for certain families of analytical models that result in the optimization problem being convex;
- Verifies a near-optimality condition if optimality cannot be guaranteed.

2 Design Optimization Framework

The framework is built around a versatile optimization core consisting of a static timer in the loop of a mathematical optimizer, as shown in Fig. 1.

The optimizer passes a set of specified design variables to the timer and gets the resulting cycle time (as a measure of performance) and power of the circuit, as well as other quantities of interest such as signal slopes, capacitive loads and, if needed, design variable gradients. The process is repeated until it converges to the optimal values of the design parameters, as specified by the desired optimization goal. The circuit is defined using a SPICE-like netlist and the static timer employs user-specified models in order to compute delays, cycle times, power, signal slopes etc.

Since the static timer is in the main speed-critical optimization loop, it is implemented in C++ to accelerate computation. It is based on the conventional longest path algorithm. The current timer does not account for false paths or simultaneous arrivals, but it can be easily substituted with a more sophisticated one because of the modularity of the optimization framework.

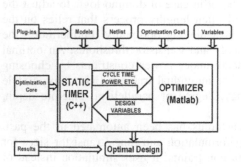

Fig. 1. Design optimization framework

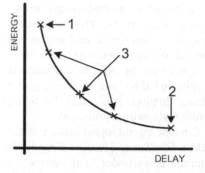

Fig. 2. Typical optimal energy – delay tradeoff curve for a combinational circuit

Adjust GATE SIZES in order to *Minimize DELAY* subject to:

Maximum ENERGY PER TRANSITION, Maximum internal slopes, Maximum output slopes, Maximum input capacitances, Minimum gate sizes

Additional constraints on signal slopes and minimum gate sizes are inserted in order to ensure manufacturability and correct circuit operation. By solving this optimization problem for different values of the energy constraint, the optimal energy-delay tradeoff curve for that circuit is obtained, as shown in Fig. 2.

The optimal tradeoff curve has two well defined end-points: point 1 represents the fastest circuit that can be designed; point 2 represents the lowest power circuit, mainly limited by minimum gate sizes and signal slope constraints. The points in-between the two extremes (marked "3" on the graph) correspond to minimizing various $E^m D^n$ design goals (such as the well known energy – delay product, EDP).

3 Models

The choice of models in the static timer greatly influences the convergence speed and robustness of the optimizer. Analytical or tabulated models can be used in the optimization framework, depending on the desired accuracy and speed targets. Table 1 shows a comparison between the two main choices of models.

3.1 Analytical Models

In our initial optimizations we use a simple, yet fairly accurate analytical model. This model allows for a convex formulation of the resulting optimization problem, where the gate sizes are the optimization variables. The model has three components: a delay equation (1), a signal slope equation (2), and an energy equation (3):

$$t_D = p + g\frac{C_{load}}{C_{in}} + \eta \cdot t_{slope\_in} \tag{1}$$

$$t_{slope\_out} = \lambda + \mu\frac{C_{load}}{C_{in}} + \nu \cdot t_{slope\_in} \tag{2}$$

$$E = \sum_{all\_nodes} \alpha_i C_i V_{DD}^2 + T_{cycle} \sum_{all\_gates} W_j P_{leak,j} \tag{3}$$

Equation (1) is an extension of the simple linear model used in the method of logical effort [5], or the level-1 model with limited accuracy in commercial logic synthesis tools[6]. Equations (1) and (2) are a straightforward first order extension to these models that accounts for signal slopes.

The capacitance of a node is computed using (4):

$$C_{node} = \sum_{gate\_inputs\_at\_node} k_i W_i + C_{wire} \tag{4}$$

where W_i are the corresponding gate sizes.

Each input of each gate is characterized for each transition by a set of seven parameters: p, g, η for the delay, λ, μ, ν for the slope and k for the capacitance. Each gate is also characterized by an average leakage power P_{leak} measured when its relative size is $W=1$. Each node of the circuit has an activity factor α, which is computed through logic simulation for a set of representative input patterns.

All the above equations can be written as posynomials in the gate sizes, W_i:

$$t_D = p + g \frac{\sum_{driven\_gate\_inputs} k_i W_i + C_{wire}}{kW_{current}} + \eta \cdot t_{slope\_in} \tag{5}$$

$$t_{slope\_out} = \lambda + \mu \frac{\sum_{driven\_gate\_inputs} k_i W_i + C_{wire}}{kW_{current}} + \nu \cdot t_{slope\_in} \tag{6}$$

If t_{slope_in} is a posynomial, then t_D and t_{slope_out} are also posynomials in W_i. By specifying fixed signal slopes at the primary inputs of the circuit, the resulting slopes and arrival times at all the nodes will also be posynomials in W_i. The maximum delay across all paths in the circuit will be the maximum of several posynomials, hence a generalized posynomial. A function f is a generalized posynomial if it can be formed using addition, multiplication, positive power, and maximum selection starting from posynomials [7].

The energy equation is also a generalized posynomial: the first term is just a linear combination of the gate sizes while the second term is another linear combination of the gate sizes multiplied by the cycle time, that in turn is related to the delay through the critical path, hence also a generalized posynomial.

The optimization problem described in Sect. 2 using the above models has generalized posynomial objective and constraint functions:

Adjust W_i in order to Minimize $max(t_{arrival, primary\_outputs})$ subject to:

$E \leq E_{max}$, $t_{slope, primary\ outputs} \leq t_{slope\_out,max}$, $t_{slope, internal\ nodes} \leq t_{slope\ internal, max}$, $C_{primary\ inputs} \leq C_{in,max}$, $W_i \geq 1$.

Table 1. Comparison between analytical and tabulated models

ANALYTICAL MODELS	TABULATED MODELS
- limited accuracy	+ very accurate
+ fast parameter extraction	- slow to generate
+ provide circuit operation insight	- no insight in the operation of the circuit
+ can exploit mathematical properties to formulate a convex optimization problem	- can't guarantee convexity; optimization is "blind"

Such an optimization problem with generalized posynomials is called a *generalized geometric program* (GGP) [7]. It can be converted to a convex optimization problem using a simple change of variables:

$$W_i = \exp(z_i) \tag{7}$$

With this change of variables the problem is tractable and can be easily and reliably solved by generic commercial optimizers. Moreover, since in convex optimization any local minimum is also global, the optimality of the result is *guaranteed*.

This delay model applies to any logic family where a gate can be represented through channel-connected components [8], as in the case of complementary CMOS or domino logic. The limitation of this approach is that it uses linear approximations for the delay, signal slopes, and capacitances. Fig. 3 shows a comparison of the actual and predicted delay for the rising transition of a gate for a fixed input slope and variable fanout. Since the actual delay is slightly concave in the fanout, the linear model is pessimistic at low and high fanouts and optimistic in the mid-range.

3.2 Tabulated Models

If the accuracy of such linear, analytical models is not satisfactory tabulated models can be used instead. For instance, (1), (2) and their respective parameters can be replaced with the look-up table shown in Table 2.

The table can have more or less entries, depending on the desired accuracy and density of the grid. Actual delays and slopes used in the optimization procedure are obtained through linear interpolation between the points in the table. The grid is non-uniform, with more points in the mid-range fanouts and slopes, where most designs are likely to operate. Additional columns can be added to the tables for different logic families – for instance relative keeper size for dynamic gates.

The resulting optimization problem, even when using the change of variables from (7), cannot be proven to be convex. However, since the analytical models describe the correct behavior of the circuits (although not absolutely accurate), the resulting optimization problem is *nearly-convex* and can still be solved with very good accuracy and reliability by the same optimizers as before [9]. The result of the nearly-convex problem can be checked against a *near-optimality boundary*. The example in Fig. 4 shows a comparison of the analytical and tabulated models and the corresponding near-optimality boundary.

The figure shows the energy-delay tradeoff curves for an example 64-bit Kogge-Stone carry tree in static CMOS using a 130nm process. The same circuit is optimized using each of the two model choices discussed in this section.

Table 2. Example of a tabulated delay and slope model (NOR2 gate, input A, rising transition)

C_{load}/C_{in}	$t_{slope\ in}$	t_D	$t_{slope\ out}$
1	20 ps	19.3 ps	18.3 ps
...	...	...	...
10	200 ps	229.6 ps	339.8 ps

Both models show that the fastest static 64-bit carry tree can achieve the delay of approx. 560ps, while the lowest achievable energy is 19pJ per transition. The analytical models are slightly optimistic because the optimal designs exhibit mid-range gate fanouts where the analytical models tend to underestimate the delays (Fig 3.). However, the models indeed exhibit the correct behavior without being grossly inaccurate.

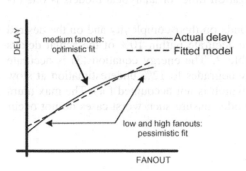

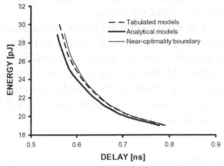

Fig. 3. Accuracy of fitted models

Fig. 4. Analytical vs. tabulated models and near-optimality boundary

The near optimality boundary is obtained by using tabulated models to compute the delay and energy of the designs resulted from the optimization with analytical models. This curve represents a set of designs optimized using analytical models, but evaluated with tabulated models. Since those designs are guaranteed to be optimal for analytical models, the boundary is within those models' error of the actual global optimum. However, if an optimization using the correct models (tabulated) converges to the correct solution, it will always yield a better result than a re-evaluation of the results of a different optimization using the same models. Therefore, if the optimization with tabulated models is to converge correctly the result must be within the near-optimality boundary (e.g. smaller delay for the same energy).

If a solution obtained using tabulated models is within the near-optimality boundary it will deemed "near-optimal" and hence acceptable.

In a more general interpretation, optimizing using tabulated models is equivalent to optimizing using a trusted timing signoff tool whose main feature is very good accuracy. The result of such an optimization is not guaranteed to be globally optimal. The near-optimality boundary is obtained by running the timing signoff tool on a design obtained from an optimization that can guarantee the global optimality of the solution. The comparison is fair because the power and performance figures on both curves are evaluated using the same (trusted and accurate) timing signoff tool.

3.3 Model Generation and Accuracy

Tabulated models are generated through simulation. The gate to be modeled is placed in a simple test circuit and the fanout and input slope are adjusted using perl scripts. The simulator is invoked iteratively for all the points in the table and the relevant output data (delay, output slope) is stored. This can be lengthy (although parallelizable) if the grid is very fine and the number of points large. This characterization is similar to the one performed for the standard-cell libraries, and yields satisfactory accuracy.

For the analytical models data points are obtained through simulation in the same manner as for tabulated models. Least squares fitting (in Matlab) is used to obtain the parameters of the models. The number of points required for a good fit (50 – 100, depending on the model) is less than the number of points needed for tabulated models (at least 1000) and thus the characterization time for analytical models is one order of magnitude shorter.

The error of the analytical models depends on their complexity and on the desired data range. The models in (1) and (2) are accurate within 10% of the actual delays and slopes for the range specified in Table 2. The energy equation (3) is accurate within 5% for fast slopes but its accuracy degrades to 12% underestimation at slow input sloped due to the crowbar current (which is not accounted for). The maximum slope constraints for output and internal nodes ensure such worst cases do not occur in usual designs.

4 Results

We use the presented optimization framework to optimize a 64-bit adder, which is a very common component of custom datapaths. The critical path of the adder consists of the carry computation tree and the sum select [10]. Tradeoffs between the performance and power can be performed through the selection of circuit style, logic design of carry equations, selection of a tree that calculates the carries, as well as through sizing and choices of supply voltages and transistor thresholds.

Carry-lookahead adders are frequently used in high-performance microprocessor datapaths. Although adder design is a well-documented research area [11,12,13,14], fundamental understanding of their energy-delay performance at the circuit level is still largely invisible to the microarchitects. The optimization framework presented in this paper provides a means of finding the energy budget breakpoint where the architects should change the underlying circuit design.

Datapath adders are good example for the optimization because their layout is often bit-sliced. Therefore, the critical wire lengths can be estimated pre-design and are a weak function of gate sizing. The optimization is performed on two examples:

1. A 64-bit Kogge-Stone adder carry tree implemented in standard static CMOS, using analytical models to tune gate sizes, supply and threshold voltages;
2. 64-bit carry lookahead adders implemented in domino and static CMOS, using tabulated models.

4.1 Tuning Sizes, Supply and Threshold Using Analytical Models

In order to tune supply and threshold voltages, the models must include their dependencies. A gate equivalent resistance can be computed from analytical saturation current models (a reduced form of the BSIM3v3 [15,16]):

$$R_{EQ} = \frac{1}{V_{DD}/2} \int_{V_{DD}/2}^{V_{DD}} \frac{V_{DS}dV_{DS}}{I_{DSAT}} = \frac{3}{4} \frac{V_{DD}(\beta_1 V_{DD} + \beta_0 + V_{DD} - V_{TH})}{W \cdot K(V_{DD} - V_{TH})^2}(1 - \frac{7V_{DD}}{9V_A}) \qquad (8)$$

Using (8), supply and threshold dependencies can be included in the delay model. For instance (1) becomes (9), with (2) having a very similar expression:

$$t_D = c_2 R_{EQ} + c_1 R_{EQ} \frac{C_{load}}{C_{in}} + (\eta_0 + \eta_1 V_{DD}) \cdot t_{slope\_in} \tag{9}$$

The model is accurate within 8% of the actual (simulated) delays and slopes around nominal supply and threshold, over a reasonable yet limited range of fanouts (2.5 – 6). For a +/- 30% range in supply and threshold voltages the accuracy is 15%.

Fig. 5 shows the optimal energy-delay tradeoff curves of a 64-bit Kogge-Stone carry tree implemented in static CMOS in three cases:

1. Only gate sizes are optimized for various fixed supplies and the nominal threshold;
2. Gate sizes and supply are optimized for nominal threshold;
3. Gate sizes, supply and threshold voltage are optimized jointly.

Fig. 6 shows the corresponding optimal supply voltage for case 2 and Fig. 7 shows the corresponding optimal threshold for case 3 normalized to the nominal threshold voltage of the technology.

A few interesting conclusions can be drawn from the above figures:

- The nominal supply voltage is optimal in exactly one point, where the $V_{DD} = 1.2V$ curve is tangent to the optimal V_{DD} curve. In that point, the sensitivities of the design to both supply and sizing are equal [2];
- Power can be reduced by increasing V_{DD} and downsizing if the V_{DD} sensitivity is less than the sizing sensitivity;
- The last picosecond is very expensive to achieve because of the large sizing sensitivity (curves are very steep at low delays);
- The optimal threshold is well below the nominal threshold. For such a high activity circuit, the power lost through increased leakage is recuperated by the downsizing afforded by the faster transistors with lower threshold. Markovic et al, [2], came to a similar conclusion using a slightly different approach.

4.2   Tuning Sizes in 64-Bit CLA Adders Using Tabulated Models

Using tabulated models as described in Sect. 3, various adder topologies implemented in different logic families are optimized in the energy–delay space under the typical loading for a microprocessor datapath. Details about the logic structure of the adders can be found in [17]. Fig. 8 shows the energy – delay tradeoff curves for a few representative adder configurations. Radix-2 (R2) adders merge 2 carries at each node of the carry tree. For 64 bits, the tree has 6 stages of relatively simple gates. Radix-4 (R4) adders merge 4 carries at each stage, and therefore a 64-bit tree has only 3 stages but the gates are more complex. In the notation used in Fig. 8 classical domino adders use only (skewed) inverters after a dynamic gate, whereas compound domino use more complex static gates, performing actual radix-2 carry-merge operations [18].

Based on these tradeoff curves, microarchitects can clearly determine that under these loading conditions radix-4 domino adders are always preferred to radix-2 domino adders. For delays longer than 12.5 FO4 inverter delays, a static adder is the preferred choice because of its lower energy.

Static adders are generally low power but slow, while domino logic is the choice for short cycle time. The fastest adder implements Ling's pseudo-carry equations in a domino radix-4 tree with a sparseness factor of 2 [17].

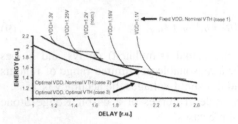

Fig. 5. Energy - delay tradeoff curves for different sets of optimization variables

Fig. 6. Optimal supply voltage for designs sized with nominal threshold voltage

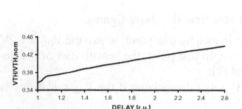

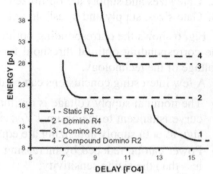

Fig. 7. Optimal threshold voltage when all optimizations are performed simultaneously

Fig. 8. Energy – delay tradeoff curves for selected 64-bit CLA adders

4.3 Runtime Analysis

The complexity and runtime of the framework depend on the size of the circuit. Small circuits are optimized almost instantaneously. A 64-bit domino adder with 1344 gates (a fairly large combinational block) is optimized on a 900MHz P3 notebook computer with 256MB of RAM in 30 seconds to 1 minute if the constraints are rather lax. When the constraints are particularly tight and the optimizer struggles to keep the optimization problem feasible, the time increases to about 3 minutes. A full power – performance tradeoff curve with 100 points can be obtained in about 90 minutes on such a machine. For grossly infeasible problems the optimizer provides a "certificate of infeasibility" in a matter of seconds.

For large designs the framework allows gate grouping. By keeping the same relative aspect ratio for certain groups of gates, the number of variables can be reduced and the runtime kept reasonable. Gate grouping is a natural solution for circuits with regular structure. All the adders optimized in Sect. 4.1 and 4.2 use gate grouping for identical gates in the same stage.

5 Conclusions

This paper presents a design optimization framework that tunes custom digital circuits based on a static timing formulation. The framework can use a wide variety of models and tune different design variables. The problem solved is generally an energy-constrained delay minimization. Due to the flexibility in choosing models, the framework can easily handle various logic families.

If analytical models are used the optimization is convex, can be easily and reliably solved, and its results are guaranteed to be optimal. The accuracy of the modeling can be improved by using look-up tables, at the cost of the optimality guarantee as well as increased characterization time and complexity. More generally, the optimization can be run on any trusted and accurate timing signoff tool, with the same tradeoffs and limitations as for tabulated models. Results obtained using tabulated models (or with the said "trusted and accurate timing signoff tool") can be verified against a near-optimality boundary computed from results guaranteed optimal in their class. If the results fall within that boundary they are considered near-optimal and therefore acceptable.

The framework was demonstrated on 64-bit carry-lookahead adders in 130nm CMOS. A static Kogge-Stone tree was tuned using analytical models by adjusting gate sizes, supply voltage, and threshold voltage. Complete domino and static 64-bit adders were also tuned in a typical high performance microprocessor environment using tabulated models by adjusting gate sizes.

We are extending this framework to optimize the latch positions in pipelined datapaths. By building on the combinational circuit optimization, this tool would allow microarchitects a larger freedom in trading off cycle time for latency.

Acknowledgement

This work was supported in part by NSF grant ECS-0238572.

References

1. Penzes, P. I; Martin, A. J.: Energy – Delay Efficiency of VLSI Computations, Proc. Great Lakes Symposium on VLSI, 2002,104-111
2. Markovic, D. et al.: Methods for True Energy – Performance Optimization. IEEE Journal of Solid State Circuits vol. 39 Issue 8 (Aug. 2004) 1282 – 1293
3. Conn, A. R.,et al.: Gradient – Based Optimization of Custom Circuits Using a Static Timing Formulation. Proceedings of Design Automation Conference DAC'99, 452 – 459
4. Fishburn, J. P.; Dunlop, A. E.:TILOS: A Posynomial Programming Approach to Transistor Sizing. IEEE International Conference on Computer – Aided Design ICCAD'85, 326-328
5. Sutherland I., Sproul R., Harris D.: Logical Effort, Morgan-Kaufmann, 1999
6. Synopsys® Design Compiler User's Manual Version 2004.12
7. Boyd, S.; Vandenberghe, L: Convex Optimization, Cambridge University Press, 2003
8. Zlatanovici, R; Master thesis, UC Berkeley, 2002
9. Mathworks, Matlab® Optimization Toolbox User's Guide Version 3
10. Rabaey, J. M.; Chandrakasn, A.; Nikolic, B: Digital Integrated Circuits: A Design Perspective, 2nd edition, Prentice-Hall 2003
11. Kogge, P. M; Stone, H. S.: A Parallel Algorithm for Efficient Solution of a General Class of Recursive Equations, IEEE Transactions on Computer,s August 1973, 786-793
12. Park, J.; Ngo, H. C.; Silberman, J. A.; Dhong, S. H.: 470ps 64bit Parallel Binary Adder, 2000 Symposium on VLSI Circuits,192-193
13. Han T.; Carlson, D. A.: Fast Area Efficient VLSI Adders, 8th Symposium on Computer Arithmetic 1987, 49-56

14. Naffziger, S.: A Sub-nanosecond 0.5μm 64b Adder Design, International Solid-State Circuits Conference, 1996, 210-211
15. Toh, K. Y.; Ko, P. K.; Meyer R. G.: An Engineering Model for Short-channel CMOS Devices. IEEE Journal of Solid State Circuits vol. 23 Issue 4 (Aug. 1998) 950 – 958
16. Garrett, J; Master thesis, UC Berkeley, 2004
17. Zlatanovici, R.; Nikolic, B.: Power – Performance Optimal 64-bit Carry-lookahead Adders. European Solid State Circuit Conference ESSCIRC 2003, 321 – 324
18. Dao, H. Q; Zeydel, B. R.; Oklobdzija, V. G.: Energy Minimization Method for Optimal Energy – Delay Extraction. European Solid State Circuit Conference ESSCIRC 2003, 177-180

Switching-Activity Directed Clustering Algorithm for Low Net-Power Implementation of FPGAs

Siobhán Launders[1], Colin Doyle[1], and Wesley Cooper[2]

[1] Department of Electronic and Electrical Engineering,
University of Dublin, Trinity College, Dublin, Ireland
launders@tcd.ie
[2] Department of Computer Science,
University of Dublin, Trinity College, Dublin, Ireland

Abstract. In this paper we present an innovative clustering technique which is combined with a simple tool configuration search aimed at power minimisation in LUT (look-up table)-based FPGAs. The goal of our technique is to reduce the capacitance on high power consuming nets by including as many of these nets as possible inside clusters wherein they can be routed on low capacitance lines. We introduce two new metrics for identifying power critical nets based on the switching activity and the number of net-segments that can be totally absorbed by a cluster. The results of our method show an average reduction of 32.8% with a maximum reduction of 48.9% in the net power over that achieved by Xilinx's ISE 5.3i tools.

1 Introduction

Power consumption has become a limiting factor in the mass acceptance of FPGAs and the ever increasing range of FPGA applications, especially in battery-powered devices. FPGAs are popular due to their short design cycle, reprogrammability, growing capacities and decreasing costs [1, 2, 3, 4]. However, limitations such as battery life and device temperature are becoming critical and power consumption is therefore emerging as a major concern for FPGA vendors [1, 6].

One of the main sources of power dissipation in an FPGA is dynamic switching power with *interconnect power* being responsible for 60% or more of this [6, 7] and any attempt to reduce FPGA power must focus on how to minimise interconnect power [7]. Switching power is caused by the charging and discharging of parasitic capacitances during logic transitions with supply voltage, switching frequency, and net capacitance being the main contributors. A reduction in any of these factors will lead to a reduction in interconnect power and hence overall circuit power.

There exists a wide range of clustering or mapping algorithms aimed at FPGA power minimisation. Clustering LUTs connected by low fanout nets can reduce the number of nets routed on the inter-CLB (configurable logic block) lines in the design [5, 8] and consequently, the net power consumption. The main limitation is that only circuits whose high switching nets are also low fanout nets will benefit from a significant power reduction. The SACA (switching activity directed clustering algorithm) technique by contrast offers a net power-conscious solution which targets the high-switching / high-fanout nets in the circuit and aims to reduce the power consumption of these nets. Similar to the techniques described in [5, 8] we exploit the presence of

V. Paliouras, J. Vounckx, and D. Verkest (Eds.): PATMOS 2005, LNCS 3728, pp. 415–424, 2005.

low capacitance lines within the CLBs to achieve power optimisation. However, our clustering goal, methodology and the core optimisation algorithm are distinct.

Reprogramming and rewiring LUTs connected in cascade can reduce the switching activities of the inter-LUT nets for that function [2]. Concurrent reprogramming of multiple LUTs after place and route can reduce the switching activities of the LUT output nets [3, 4]. Re-ordering the input signals to the LUTs can reduce the internal switching activity of the LUTs [9]. In each of these works, the reduction in switching activity resulted in a corresponding reduction in power consumption. Similar to these methods, we use switching activity to guide our process towards a low power solution. Our technique is orthogonal to these methods [2, 3, 4, 9] and can be used in conjunction with such methods to achieve further reductions in power. While [2, 3, 4] decreased the power by reducing the switching activities on individual nets in the circuit, the SACA technique concentrates on both the switching activity and the other main contributor to net power consumption: net capacitance.

In LUT-based FPGAs there exist dedicated local lines connecting the LUTs within the CLB cluster. These lines are short and have a comparatively small capacitance associated with them making them very power efficient [5] and an appropriate target for use when routing high power consuming nets. The SACA approach aims to reduce the net capacitance on power critical nets by routing these nets on the low-power local lines available. Increasing the number of high switching net-segments internal to the CLBs will have a two-fold effect on lowering the net power. Firstly, this will reduce the power consumption of those nets routed on the local lines. Secondly, this will reduce the number of high switching net-segments routed on the inter-CLB lines which will reduce the switching activity and hence power consumption on these higher capacitance lines. Both of these factors lead to a reduction in net power and overall power. This is achieved by careful choice of which LUTs are clustered together in each CLB. The SACA technique clusters those LUTs strongly connected by power critical nets and places them in a minimum number of CLBs. The SACA algorithm is applied at the beginning of the floorplanning process and works in conjunction with existing layout tools. It combines a power-aware algorithm with technology specific data to direct the layout process towards an optimal low-power solution.

The next section introduces some basic concepts and definitions. Section 3 describes the SACA technique and how it is applied to produce a low-power solution. Section 4 shows the experimental results and Section 5 concludes the paper and gives direction for future work.

2   Basic Concepts and Definitions

The fundamental building block for most FPGAs is a k-input LUT which acts as a function generator that can implement any combinational logic function of up to k variables. A CLB cluster consists of a fixed number of LUTs placed in close proximity. Cluster-based FPGAs consist of an array of such CLBs surrounded by a ring of IOBs (input / output blocks) connected together by the general routing matrix.

Interconnect power is the power consumed by the nets on this general routing matrix with switching frequency, fanout, net length, and routing resources being the main contributors. Because the SACA algorithm clusters the LUTs before final placement and routing are carried out, the actual length of the nets, or resources used to route them is not known and so is not considered any further. Switching frequency and fanout are both used to guide the clustering process as explained below.

In order to effect a power conscious algorithm we require a knowledge of which nets in the circuit will produce the highest power savings if targeted. This requires a measure of the relative power consumed by each net as well as an indication of how much this power can be reduced by targeting that net. In CLB-based FPGAs only a portion of each net can be absorbed by the CLBs (i.e. routed on the local lines) and this needs to be factored into our decision process.

Number of Absorbable Net-Segments, N_{abs}. This gives a measure of the relative power savings achievable by targeting each net. The number of LUTs that can be placed in a CLB is bounded above by the maximum cluster size, N_{max}. This restricts N_{abs} for any net as follows:

Consider the simple case where a net is wholly terminated by LUTs (i.e. where all terminals on the net connect to LUTs). The number of LUTs attached to the net, $N_{lut} =$ number of sink LUTs + the source LUT = fanout + 1. Extending this to the situation where the source or some of the sinks on the net involve input / output blocks then:

$$N_{lut} = (\text{fanout} + 1) - N_{IOB} . \tag{1}$$

where $N_{IOB} = $ the number of input or output blocks attached to the net.

For a net with $N_{lut} <= N_{max}$ it is possible to route the inter-LUT net-segments entirely within a single CLB and $N_{abs} = $ fanout $- N_{IOB} = N_{lut} - 1$ (from equation 1). However, for nets with $N_{lut} > N_{max}$ some of the net-segments will have to be routed on the general inter-CLB lines as more than one CLB will be needed to accommodate the LUTs. N_{abs} will be reduced by 1 for each additional CLB used. The number of additional CLBs used $ = $ quotient$[(N_{lut} - 1), N_{max}]$, thus the number of absorbable nets can be calculated as:

$$N_{abs} = (N_{lut} - 1) - \text{quotient}[(N_{lut} - 1), N_{max}] . \tag{2}$$

$N_{abs\text{-}local}$ is a modified version of N_{abs} relating to the particular sub-network of the circuit currently being analysed and is defined by:

$$N_{abs\text{-}local} = (N_{lut\text{-}local} - 1) - \text{quotient}[(N_{lut\text{-}local} - 1), N_{max}] . \tag{3}$$

where $N_{lut\text{-}local}$ is the number of LUTs this net-segment connects to that exist within the sub-network being analysed.

'PS' Power-Sensitive Rating. Once N_{abs} for each net has been calculated we can identify the power critical nets in the circuit. To accomplish this we are introducing two new metrics, PS and PS_{local}. PS is a global metric that rates a net according to the product of its switching frequency and N_{abs} to indicate which nets in the circuit consume the most power and will benefit most (i.e. attract the most power savings) from clustering:

$$PS = switching\text{-}frequency_{net} * N_{abs} . \tag{4}$$

PS_{local} is a modified version of PS relating to the particular sub-network of the circuit currently being analysed. It rates a net-segment according to the product of its switching frequency and the number of absorbable net-segments that exist in the local sub-network, $N_{abs\text{-}local}$ and is defined by:

$$PS_{local} = switching\text{-}frequency_{net} * N_{abs\text{-}local} . \tag{5}$$

The PS rating will indicate which nets in any circuit or sub-section of a circuit are likely to consume the most power, i.e. are the most 'power-sensitive'. It does not indicate an absolute value of power consumption.

3   SACA Technique

The aim of the switching-activity directed clustering algorithm is to seek out the most power-sensitive nets in the circuit and cluster the LUTs attached to them together in the minimum number of CLBs possible. The net parameters used to guide this process are: switching frequency, fanout, N_{lut}, N_{abs}, and PS. Our clustering technique is based on a maximum cluster size, N_{max}, equal to four, but may be adapted for other cluster sizes. Clustering is achieved in three stages: netlist ranking process, cluster ranking process, and cluster formation as illustrated in Figure 1 below.

3.1   Netlist Ranking Process

In the first stage, the circuit is simulated to provide an accurate estimation of the switching activity of each individual net. The entire circuit netlist is parsed and this information is combined with the switching activity data to generate a net-array containing the switching frequency, fanout, N_{lut}, N_{abs} and PS, for each net in the circuit. The goal of this stage is to rate every net in the circuit in terms of how much power it is likely to consume, and then to rank the nets in order of their power sensitivity.

The power consumption potential for each net is calculated in the form of the PS metric described above (equation 4) and the array is then sorted in descending order of PS values. For nets with equal PS values, they are further sorted in descending order of N_{lut} as a measure of which nets will produce greater power saving results if prioritized.

The output of this stage is a fully ranked net-power sensitive netlist. The nets ranked highest are those estimated to use the most power, and which will benefit most from clustering the LUTs attached to them. For clustering purposes they are considered to be the most critical nets. Nets are then chosen sequentially in the order they appear in the array to direct the actual clustering process in stages two and three in Figure 1.

3.2   Cluster Ranking Process

In the second stage, we focus on the network of LUTs and nets that are associated with the highest ranked, i.e. most critical, net from above. We are concerned now only with (i) the most critical 'power-sensitive' net in the circuit, (ii) all unclustered LUTs attached to this net, and (iii) all net-segments connecting to or from those LUTs.

The goal of this stage is to determine which other power-sensitive net-segments can be absorbed at the same time when the LUT clustering for this net is achieved in stage three of our process. Stage two accounts for the secondary and even tertiary power reduction benefits available by careful choice of the LUT groups.

There are three main sections of this process, each of which work on the local sub-network only. They involve the forming of a new local LUT array, ranking of the local nets involved, and generation and ranking of the ideal LUT clusters.

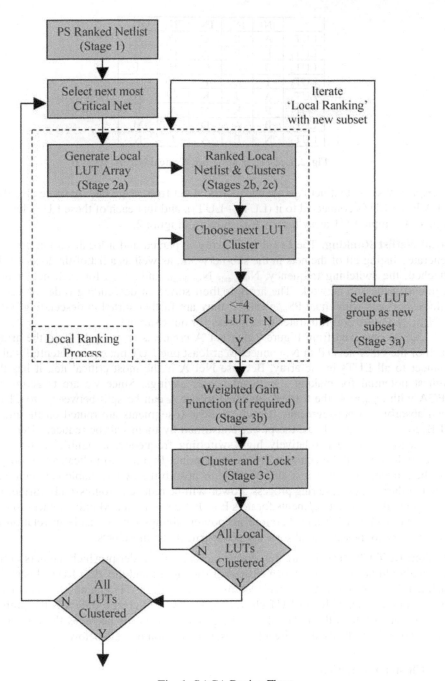

Fig. 1. SACA Design Flow

Local LUT Array. This array lists each of the LUTs within the sub-network and the nets connected to them, as illustrated by the following example.

	IN_1	IN_2	IN_3	IN_4	OUT
LUT_0	A	B	C	D	E
LUT_1	L	E	A	C	G
LUT_2	B	A	G	H	I
LUT_3	I	C	E	A	J
LUT_4	A	J	K	G	H
LUT_5	F	K	H	A	B
LUT_6	B	A	D	J	M
LUT_7	N	E	A	J	N

Fig. 2. Local LUT array for 'Critical Net' A

Example: Assume that net A is the most critical net from stage one. Assume also that net A has 8 LUTs connected to it ($LUT_0 - LUT_7$), and that each of these LUTs have 4 inputs. A sample LUT array for net A is shown in Figure 2.

Local Netlist Ranking. The Local LUT array is parsed and a localised net array is generated listing all of the nets in the sub-network, as well as a list of the local LUTs attached, the switching frequency, $N_{lut\text{-}local}$, $N_{abs\text{-}local}$, and PS_{local} for each net or net-segment in the sub-network. The array is then sorted in descending order of PS_{local} values. For nets with equal PS_{local} values, they are further sorted in descending order of $N_{lut\text{-}local}$. Figure 3(a) illustrates this process for our example.

It can be observed from Figure 2 that net A connects to every LUT in the array. Each of the other nets (B to N) connect to at least one LUT but could potentially also connect to all LUTs in the array. Because Net A is the most critical net, it has the highest potential for making significant power savings. Since we are targeting an FPGA with $N_{max} = 4$, the 8 LUTs attached to net A can be split between two CLBs each absorbing 3 net-segments. If each of these 6 segments are routed on the intra-CLB, low-capacitance lines, the power consumed by them will be reduced. Because these segments have a relatively high switching frequency, a significant level of power reduction will be achieved by this clustering. Figure 3(a) indicates that power will be reduced further if it is also possible to absorb the two available net-segments of net G during the clustering process. Power will be reduced progressively further by absorbing all of the net-segments for nets E to I. Nets F, L, and M only connect to this sub-network once each and therefore no power savings can be made in relation to them as it is not possible to absorb any net-segments on those nets.

Ranked LUT Clusters. Figure 3(a) above indicates the order in which the nets in the circuit should be considered when clustering the circuit and the ideal LUT cluster for each net. Clustering is actually achieved by grouping the LUTs and not the nets them-selves, and the ranked list of LUT clusters in Figure 3(b) clearly indicates the preferential order in which the LUTs should be grouped. This list becomes the output of stage two and will be used to direct the cluster formation process below.

3.3 Cluster Formation

The goal of the third stage is to cluster the LUTs to produce an optimal low net-power layout. This is an iterative process which cycles through the LUTs from the 'ranked LUT clusters' list from stage two above and places them into CLBs in the order they appear in the list until all viable clusters in the sub-network have been formed.

Net	$N_{abs\text{-}local}$	SF	PS_{local}	$N_{lut\text{-}local}$	LUTs attached	LUT Clusters
A	6	50	300	8	0 – 7	
G	2	50	100	3	1,2,4	1,2,4
E	3	30	90	4	0,1,3,7	0,1,3,7
B	3	25	75	4	0,2,5,6	0,2,5,6
C	2	35	70	3	0,1,3	0,1,3
H	2	30	60	3	2,4,5	2,4,5
D	1	60	60	2	0,6	0,6
K	1	40	40	2	4,5	4,5
N	1	35	35	1	7	7
J	3	10	30	4	3,4,6,7	3,4,6,7
I	1	25	25	2	2,3	2,3
F	0	50	0	1	5	5
L	0	30	0	1	1	1
M	0	10	0	1	6	6

Fig. 3. (a) Ranked local netlist for net A, (b) Ranked LUT Clusters for net A

Cluster-Size Limitations. The cluster size is bounded above by N_{max}, the maximum number of LUTs per CLB, which depends on the type of FPGA being targeted. LUT clusters > N_{max} are subdivided so that they can fit into a CLB. This is achieved by iteratively repeating stage two of the process outlined above until the 'ranked LUT clusters' list contains no more than N_{max} elements in any row. The cluster size is bounded below by a user defined variable, N_{min}, the minimum number of LUTs needed to form a viable cluster. The value of N_{min} is defined during the clustering process and clusters are formed in the range $N_{min} - N_{max}$.

Cluster and Lock Process. The array is read on a row-by-row basis. Starting from the top row, the LUTs specified there are clustered together and 'locked' to prevent accidental re-clustering by a later iteration of the process. Any remaining, unclustered LUTs on the 2nd row are then grouped together and 'locked'. This process is repeated for all remaining rows until all LUTs within the array have been clustered or there are not enough LUTs available to form a viable cluster.

Weighted Gain Function. A complication occurs when the unclustered LUTs listed on any particular row cannot be clustered together due to space limitations. For the example shown in Figure 3(b), the first two rows of the table indicate that LUT_1, LUT_2 and LUT_4 will form the 1st cluster for net A, while LUT_0, LUT_3 and LUT_7 will form the 2nd cluster (LUT_1 is already clustered and so is 'unavailable' for inclusion in the 2nd cluster). The 3rd row indicates that LUT_5 and LUT_6 should be clustered together but net A only needs 2 clusters and these are already partially specified. As a result, neither cluster has room left to fit both LUT_5 and LUT_6. These two LUTs must be split up with each one becoming part of an existing cluster. It is quite easy to see intuitively from Figure 3(b) that LUT_5 should be clustered along with LUT_1, LUT_2 and LUT_4 and LUT_6 clustered with LUT_0, LUT_3 and LUT_7 as net-segments from nets H, D, K and J can all be additionally absorbed as a result. This LUT-Cluster matching is achieved algorithmically by the application of a weighted gain function. This function matches each listed LUT to its best existing cluster based on the additional power reductions that can be gained as a result. When this weighted gain function is required, it needs to be carried out before the 'cluster and lock' step for the particular row involved described above. If not needed, it is simply skipped by the process.

When all the LUTs in the sub-network are clustered or the number of remaining unclustered LUTs is less than N_{min}, the next most power sensitive net from stage 1 is then selected to form the nucleus for a new sub-network and stages 2 and 3 are repeated for the unclustered LUTs in this section of the circuit. This process is iterated until no more viable clusters can be formed in any sub-network. The final output of this clustering process is a *user constraint file* (.ucf) which is then used to direct the PAR tools regarding which LUTs to place together in each CLB.

4 Experimental Results

We have implemented our new clustering technique using Xilinx ISE5.3i and Model-Sim SE5.7d software, targeting the XCV300-pq240-4 Virtex FPGA, using three of the ITC'99 benchmarks circuits [10] as a testbench.

Each of the VHDL circuits was first synthesised, translated into 4-input LUTs, mapped to the Virtex FPGA and placed and routed with ISE tools using default ISE settings. To find the optimal low-power tool configuration settings, a range of synthesis and PAR options was applied to each circuit and the circuit was re-synthesised and re-implemented for each of 196 combinations of these options. XPower was run in each case, the power consumption recorded and the 30 combinations which produced the lowest net-power and total-power measurements were selected for clustering. The SACA clustering technique outlined in Section 3 was used to cluster each of these with two clustering options applied in each case ($N_{min} = 3$ and $N_{min} = 4$). The circuit was re-implemented for each of the 60 resulting combinations and again XPower was run and the power consumption recorded.

Table 1 shows the number of clusters (CLBs), the worst individual net-delay, and the total net-power measurements, for each of the three testbench circuits, with respect to both the original circuit, N, using default ISE configuration settings, and the lowest net-power clustered circuit, N_C, achieved as a result of the extensive synthesis / PAR / clustering configuration options search. The configuration options which resulted in the lowest net power are also shown, where columns A - E list, respectively, the synthesis optimisation goal, the synthesis optimisation effort level, the FSM encoding algorithm applied, the PAR effort level, and the cluster-size lower bound, N_{min}. The average reduction is also shown for CLB, delay and power measurements.

The average reduction in the number of clusters of 18.6% was due to the use of optimised tool configuration settings combined with clustering the LUTs into fewer CLBs. While the B08 circuit actually experienced an increase in the individual net delays, an average reduction of 23.9% was achieved across the three circuits. The operational frequency for all power measurements was 12.5MHz and our approach shows an average reduction of 32.8% in the net power when compared to the ISE5.3i default settings with a maximum reduction of 48.9%.

5 Conclusions and Future Work

In this paper we have presented a fundamentally novel clustering technique which provides a solution for reducing power consumption in an FPGA. Our algorithm involves routing high switching nets on the dedicated, low-capacitance lines that exist within CLB clusters thus reducing the power consumed on these nets and also reducing the switching activity of the nets routed on the general routing matrix leading to a reduction in interconnect power and overall power. This was achieved by careful selection of which LUTs to place in each CLB cluster.

Table 1. SACA technique results

Circuit	# CLBs		Net Delay (ns)		Net Power (mW)		Synthesis / PAR / Clustering Options				
	N	N_C	N	N_C	N	N_C	A	B	C	D	E
B03	25	20	5.640	3.184	2.378	1.727	Speed	1	None	2	3
B08	21	19	3.896	4.716	1.222	0.952	Area	1	Sequential	3	4
B09	19	14	6.312	3.214	1.328	0.678	Area	1	Compact	2	3
Avg. Red	18.6 %		23.9 %		32.8 %						

Two new metrics were introduced to rate the power sensitivity of each net in the circuit based on the switching activity and number of absorbable net-segments of that net. LUTs were clustered according to the number and rating of power sensitive net-segments that connected them together. Using this approach we have successfully demonstrated a reduction in the net power, which has a direct impact on overall power consumption, across a range of circuits with an average reduction of 32.8% and a maximum reduction of 48.9%.

Work has begun on a new version of the clustering algorithm based on a more accurate model for the net power estimations available from the Xilinx XPower tool instead of the switching activity estimations used above. This will enable a more accurate calculation of the power-sensitive rating for each net in the circuit. It is expected that this higher level of accuracy will result in a further reduction of net power.

References

1. F. G. Wolff, M. J. Knieser, D. J. Weyer, C. A. Papachristou, "High-Level Low Power FPGA Design Methodology," *NAECON 2000: IEEE National Aerospace and Electronics Conference,* pp. 554-559, Dayton, OH, October 2000.
2. C-S. Chen, T-T. Hwang, C. L. Liu, "Low Power FPGA Design – A Re-engineering Approach," *DAC-34: ACM/IEEE Design Automation Conference,* pp. 656-661, Anaheim, CA, June 1997.
3. J-M. Hwang, F-Y. Chiang, T-T. Hwang, "A Re-engineering Approach to Low Power FPGA Design Using SPFD," *DAC-35: ACM/IEEE Design Automation Conference,* pp. 722-725, San Francisco, CA, June 1998
4. B. Kumthekar, L. Benini, E. Macii, F. Somenzi, "Power Optimisation of FPGA-based designs without rewiring," *IEE Proceedings on Computers and Digital Techniques,* Vol. 147, No. 3, pp. 167-174, May 2000.
5. C. Doyle, S. Launders, "A Circuit Clustering Technique Aimed at Reducing the Total Amount of Interconnect Resource used in an FPGA," *Proceedings of IEEE Design and Diagnostics of Electronic Circuits and Systems Workshop*, pp 211-214, Stará Lesná, Slovakia, April 18-21, 2004.
6. L. Shang, A. S. Kaviani, K. Bathala, "Dynamic Power Consumption in Virtex-II FPGA Family," *FPGA '02: ACM/SIGDA 10th Int. Symposium on FPGAs,* pp. 157-164, Monterey, CA, February 2002.

7. E. Kusse, J. Rabaey, "Low-Energy Embedded FPGA Structures," *ISLPED 98: IEEE International Symposium on Low Power Electronics and Design*, pp. 155-160, Monterey, CA, August 1998.
8. A. Singh, M. Marek-Sadowska, "Efficient Circuit Clustering for Area and Power Reduction in FPGAs," *FPGA '02: ACM/SIGDA 10th Int. Symposium on FPGAs*, pp. 59-66, Monterey, CA, February 2002.
9. M. J. Alexander, "Power Optimization for FPGA Look-Up Tables," *ISPD-97: ACM/IEEE International Symposium on Physical Design,* pp. 156-162, Napa Valley, CA, April 1997.
10. "A Set of RT- and Gate-level Benchmarks" by *The CAD Group at Politecnico di Torino:* download http://www.cad.polito.it/tools/#bench

Logic-Level Fast Current Simulation for Digital CMOS Circuits[*]

Paulino Ruiz de Clavijo[1,2], Jorge Juan-Chico[1,2],
Manuel Jesús Bellido Díaz[1,2], Alejandro Millán Calderón[1,2],
David Guerrero Martos[1,2], E. Ostúa[1,2], and J. Viejo[1,2]

[1] Instituto de Microelectronica de Sevilla, Centro Nacional de Microelectronica
Av. Reina Mercedes, s/n (Edificio CICA), 41012 Sevilla, Spain
paulino@imse.cnm.es
http://www.imse.cnm.es
Tel.: +34 955056666, Fax: +34 955056686
[2] Departamento de Tecnologia Electronica, Universidad de Sevilla
Av. Reina Mercedes, s/n (ETS Ingenieria Informatica), 41012 Sevilla, Spain
http://www.dte.us.es
Tel.: +34 954556161, Fax: +34 954552764

Abstract. Nowadays, verification of digital integrated circuit has been focused more and more from the timing and area field to current and power estimations. The main problem with this kind of verification is on the lack of precision of current estimations when working at higher levels (logic, RT, architectural levels). To solve this problem it is not only necessary to use good current models for switching activity but, also, it is necessary to calculate this switching activity with high accuracy. In this paper we present an alternative to estimate current consumption using logic-level simulation. To do that, we use a simple but accurate enough current model to calculate the current consumption for each signal transition, and a delay model that obtains high accuracy when it is used to measure the switching activity (the Degradation Delay Model -DDM-). In the paper we present the current model for CMOS inverter, the characterization process and the model implementation in the logic simulator HALOTIS that includes the DDM. Results show a high accuracy in the estimation of current curves when compared to HSPICE, and a potentially large improvement over conventional approaches.

1 Introduction

Verification of digital integrated circuits is a key point during the design process. Verification tasks take place at the different design level: layout, logic-level, architectural and system levels. Due to the increasing scale of integration developed during the last two decades, more and more interest is devoted to verification at higher levels, including system and software levels [[–]].

---

[*] This work has been partially supported by the MEC META project TEC 2004-00840/MIC of the Spanish Government

V. Paliouras, J. Vounckx, and D. Verkest (Eds.): PATMOS 2005, LNCS 3728, pp. 425–435, 2005.

On the other hand, verification interest has been moving more and more from the timing and area field to current and power estimations, the main motivations being switching noise and energy consumption calculations. The first one is an important factor limiting the accuracy of the analog part in mixed-signal circuits [ ], while the second is becoming a major design condition due to the increasing difficulties to dissipate the power generated by high performance processors and the need to operate low-power devices on batteries [ ].

In this scenario, electrical simulation and verification is mainly limited to the optimization of basic building blocks. Filling the gap between electrical and system verification is logic-level simulation. While whole system simulation is not feasible at the logic level any more, major system parts, up to several millions of transistors, are still manageable at the logic-level. That makes logic-level simulation the more accurate alternative available in many practical cases when the rough system-level estimations are not useful.

Facing the new challenges cited above, EDA (Electronic Design Automation) vendors have been incorporating current and power estimation facilities to their logic-level simulation tools (PRIMEPOWER [ ], XPOWER [ ], POWERTOOL [ ]). Current/power estimations at the logic level are generally obtained in a two-phase procedure: first, logic-level simulation takes place and switching activity at every node is computed. Then, power consumed at every node is computed by using charge-based power models. On the other hand, current curves may be generated during logic-level simulation by using current models in a event-driven basis [,].

Since these approaches are basically correct, switching activity estimations at the logic level has been traditionally largely overestimated [,], mainly due to the fact that conventional behavioural models are not able to accurately handle the propagation of glitches and input collisions [,]. This way, well implanted commercial tools may easily overestimate switching activity (then current and power) from 30% to 100% [,].

On the contrary, it has been shown in [ ] that it is possible to develop behavioural models that catch the dynamic nature of the switching process at the logic level (e.g. Degradation Delay Model -DDM- [ ]). The immediate consequence of this is a natural ability to accurately handling the generation, propagation and filtering of glitches.

The objective of this paper is to show that using this kind of dynamic behavioural modelling, the accuracy of logic-level estimations of the supplied current is largely improved. To do that, current estimation capabilities have been incorporated to a previously developed logic-level simulation tool named HALO-TIS [ ]. Since the tool already incorporates the DDM, switching activity is accurately handled. Current estimation is computed during the simulation process by applying a current model to every signal transition in the internal nodes of the circuit, generating a set of partial current curves. These current curves are summed up after simulation to generate the global current profile of the circuit.

Although this work is still in its early stages, results are very promising if we consider that, in our opinion, there is still plenty room to improve the process, specially the current modelling. In the next section, we introduce the current model that is incorporated in the simulator in order to obtain current curves. Model characterization and validation is done in section 3. Implementation issues are discussed in section 4, while preliminary results are presented in section 5. Finally, we will summarize the main conclusions.

2   Current Model

Global current calculation is obtained by summing up the current driven by each circuit cell individually from the power supply. For a given cell we assume that the current only circulates during the logic switching of the cell's output. Fig. 1 shows the currents involved in a CMOS inverter structure when a raising output transition takes place. The main currents taken from V_{cc} during the inverter's switching process are I_{cl} and I_{SC}. I_{cl} is the current that charges the inverter output node, and I_{SC} is the short-circuit current which goes directly from V_{cc} to GND while the NMOS part is still conducting. When we deal with raising output transitions, the total current driven from the power supply is the sum of I_{cl} and I_{SC}. For falling output transitions, the capacity C_L is discharged, and the only component taken from the power supply is I_{SC}.

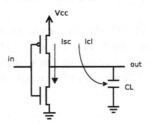

Fig. 1. CMOS Inverter currents during logic switching

The actual current curve in the gate, as obtained with HSPICE [ ] electrical simulator, is shown in Fig. 2-b. Our model proposes to fit the actual current curve by a triangle-shaped, two-pieces linear curve. The triangle approximation can be seen in Fig. 2-a.

This approach is similar to that proposed by other authors [,], but using a simpler characterization method. The triangular shape is defined by three points: the triangle starting point instant (T_b), the current maximum value and the instant when it takes place (T_{max}, I_{max}), and the instant time where the triangle ends (T_e).

These points can be approximated as follows: T_b is the instant when the input transition starts and the point T_e is the instant when the output transition ends, which are both known. To calculate I_{max} and T_{max} we use the model proposed in [ ]. In that work, the authors obtain the following equations:

$$I_{max} = \sqrt{\frac{K_p \times W_p \times V_{DD}^2 (C_L + C_{SC})}{\tau_{in}}} \qquad (1)$$

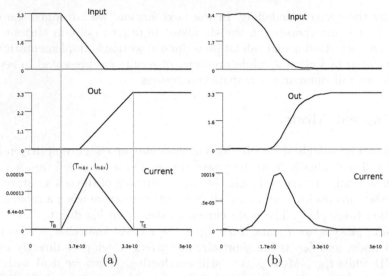

Fig. 2. Inverter switching curves. a) Triangle current model. b) HSPICE simulation

$$T_{max} = \frac{(C_L + C_{SC})V_{DD}}{I_{max}} \tag{2}$$

where K_p and W_p are respectively the transconductance factor and width of the PMOS transistor, V_{DD} is the supply voltage, C_{SC} is the short-circuit capacitance as defined in [20] and τ_{in} is the transition time of the input.

On the other hand, for falling output transitions we only have to consider the short-circuit current I_{SC}, so that:

$$I_{max} = \sqrt{\frac{K_n \times W_n \times V_{DD}^2 \times C_{SC}}{\tau_{in}}} \tag{3}$$

In this case, T_{max} is the time when input transition crosses the inverter input threshold.

The proposed model allows the calculation of an approximated, triangle-shaped current curve for every output transition of a cell, only based on cell parameters and timing data provided during logic simulation.

3   Current Model Validation and Characterization

To validate the model proposed in the previous section, we have simulated the behavior of CMOS inverter built using a $0.35\mu m$ technology. This simulation has been carried out with HSPICE. The objective is to check the I_{max} dependence with respect to τ_{in} and C_L. For this reason, we will express the equation1 as follow:

$$I_{max}^2 = \frac{V_{DD}^2}{\tau_{in}}(QC_L + R) \tag{4}$$

In this equation we can see the linear dependence of I^2_{max} with C_L and $1/\tau_{in}$, where Q and R are gate-level model parameters that hides the internals of the cell. This dependence has been checked by plotting I^2_{max} values versus C_L for different τ_{in}, and also, I^2_{max} values versus τ_{in} for different C_L. Fig. 3 shows the circuit configuration used to obtain simulation data. The first and second inverter in the chain are used to achieve realistic input transition waveforms at the input of $inv3$, which is the cell under study. The interest of working with realistic transitions in $inv3$ is in the high variation of I_{max} values observed when linear (artificial) input transitions are used. On realistic transitions, τ_{in} is measured taking the 30% and 70% reference points with respect to the full supply rail.

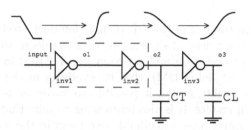

Fig. 3. Inverter characterization circuit

The simulations have been carried out for a range of typical values of τ_{in} and C_L. τ_{in} is altered in a realistic way by modifying C_T in the circuit. The design space explored includes fast and slow input transitions and light and heavy loads. To get this, C_T and C_L values take the inverter input capacity (C_{in}) as a reference, where C_{in} is:

$$C_{in} = (W_n + W_p) \times L \times C_{ox} \tag{5}$$

where W_n and W_p are the NMOS and PMOS widths respectively, L is the MOS length and C_{ox} is the oxide capacitance by unit area.

In Fig. 4-a we show the curves representing I^2_{max} versus C_L/C_{in} for different τ_{in} and its linear regressions. The Fig. 4-b shows I^2_{max} the curves I^2_{max} versus τ_{in} for different C_L/C_{in} and its linear regressions.

Based on these linear regressions it is possible to extract values for Q and R parameters of equation 4. Q and R values are mostly independent of loading and input transition time conditions. Moderate input ramps and typical loading conditions are used to compute Q and R values. As an example, inverter parameters for a raising output under conditions given by $C_T = 2 \times C_{in}$ and $C_L = 2 \times C_{in}$ gives:

$$Q = (4.67 \pm 0.28) \times 10^{-5} A/V^2, R = (5.94 \pm 0.46) \times 10^{-19} A^2 s/V^2 \tag{6}$$

4   Model Implementation in HALOTIS

HALOTIS [ ] is a logic-level simulator that is being developed in our research group. It's main feature is the inclusion of the Degradation Delay Model (DDM).

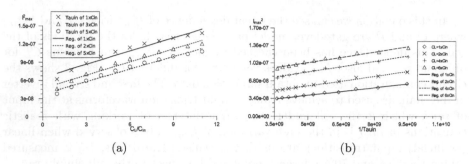

Fig. 4. (a) I_{max}^2 vs. C_L for different τ_{in}. (b) I_{max}^2 vs. τ_{in} for different C_L

HALOTIS, through the use of the DDM, has demonstrated a drastic accuracy improvement over conventional simulation tools thanks to the novel and precise treatment of glitch generation and propagation. HALOTIS also implements a Conventional Delay Model (CDM), that it is a delay model without the degradation effect. This facility makes it possible to easily evaluate how the DDM improves simulation results. It is also interesting to note that accounting for the degradation effect introduces a negligible overhead in the logic simulation process. Currently, HALOTIS is able to read digital circuit netlists in VERILOG [21] format and read patterns in MACHTA [22] format. It is designed in a modular way to enable for new functionality to be easily added as research on models and algorithm improves.

The current model described in section 2 has been included in the simulation engine through a new source code module that implements equations (2) and (4) corresponding to T_{max} and I_{max} respectively. The rest of the necessary parameters to compute the current waveform associated to each transition (T_e and T_b in Fig. 2-a) are already available during the simulation process. Additionally, the set of parameters for each gate now also includes Q and R parameters measured using the characterization process described in section 3.

When a new output transition is generated during simulation process, the new module processes this transition and calculates the points of the triangle that defines the current waveform associated with that output transition. All triangles are stored in a data type which is processed after simulation to generate the general current profile of the circuits, i.e. the current generated by the power supply as a function of time. The data available can also be used to locate hot spots or noisy nodes, although these applications have not been explored yet.

5   Simulation Results

The effect of summing up individual current components for every signal transition is better appreciated in multi-level digital circuits, where transitions and current waveforms overlap for a period of time. Also, glitch conditions, evolution and eventual filtering will more likely take place as logic depth increases. Thus, a simple but adequate circuit to demonstrate the possibilities of the proposed model and its combination with the degradation effect is a chain of gates like that depicted in Fig. 5. The quality of the results is measured by comparing to circuit-level simulation using HSPICE.

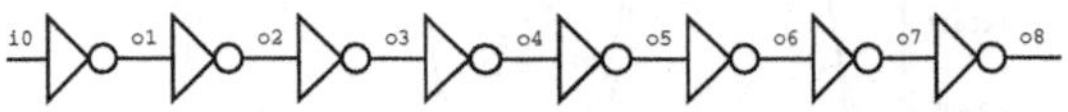

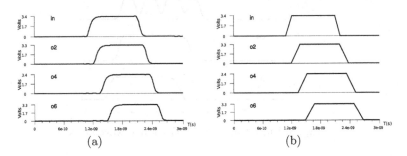

Fig. 5. Inverter chain

Two types of input patterns are simulated. The first example is intended to show the current model accuracy in the absence of degradation effect. The second pattern will make degradation effect to take place and will rise up the benefits of using a dynamic behavioural model like the DDM.

The first case is a wide pulse applied to the first inverter in the chain. This pulse is fully propagated through the inverter chain and no degradation effect takes place since the input pulse is wide enough to avoid that. The voltage waveforms obtained at selected internal nodes of the chain with HSPICE and HALOTIS are shown in Fig. 6-a and Fig. 6-b respectively. Logic-level simulation with HALOTIS matches HSPICE results quite well, and it can be seen how HALOTIS takes input transition times into account and reflects that in the plot. We note here that logic simulation results using the Degradation Delay Model (DDM) or a Conventional Delay Model (CDM) without degradation are the same since degradation effect does not take place for these stimuli.

Fig. 6. Single input pulse voltage waveforms. (a) HSPICE (b) HALOTIS

As it has already be explained, current components for every transition is computed during simulation. The complete current waveform obtained by summing up these components is plotted in Fig. 7, besides the current curve generated by HSPICE. It can be easily observed how the current curve generated by HALOTIS quite well matches HSPICE results. Specially, current peaks are quite accurately reproduced.

The second example simulates a pulse train applied at the input of the chain. In this case, the pulses are narrow enough to degrade from logic level to logic. Some of the pulses will be filtered at some point in the chain. Simulated voltage

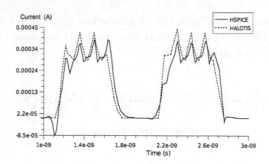

Fig. 7. Single input pulse current waveform

waveforms at even nodes using HSPICE are depicted in Fig. 8, while logic simulation considering the degradation effect (DDM) is in Fig. 9. It can be seen that
the initial pulse train is degraded through the chain and some pulses are filtered
until a single wide pulse is propagated to the output of the chain. This effect is
reproduced both by HSPICE and HALOTIS when using the DDM. However, a
conventional delay model (CDM) would propagate the full pulse train unaltered

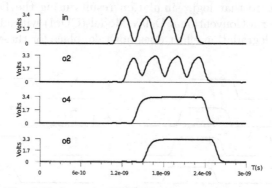

Fig. 8. Pulse train HSPICE voltage waveforms

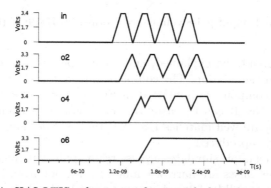

Fig. 9. Pulse train HALOTIS voltage waveforms with degradation effect (DDM)

(Fig. 10). This behaviour and its implication on the switching activity has been reported in detail in [16, 18]. This time, the current waveforms calculated using the DDM and the CDM differ since the CDM will consider additional transitions and current componets that are actually filtered. Fig 11 compares HSPICE and HALOTIS results using both DDM and CDM models. Like in the previous example, simulation using the DDM gives results that are similar to HSPICE results. In particular, current peaks are again very accurately determined.

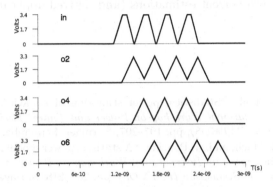

Fig. 10. Pulse train HALOTIS voltage waveforms without degradation effect (CDM)

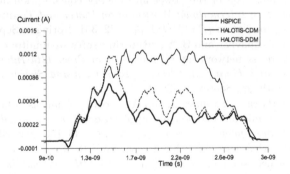

Fig. 11. Pulse train current waveforms. HSPICE, DDM and CDM comparison

Simulation without degradation effect (CDM) gives much worse results, even though the core behavioural model is the same that the one used by the DDM.

6   Conclusions

In this contribution a simple current model for logic transitions that can be used in logic-level simulators to obtain current estimations has been presented. The proposed model has been implemented in a experimental logic-level simulation tool that combines the current equations with the so-called Degradation Delay Model to obtain accurate current waveforms at the logic level when compared to electrical simulators like HSPICE. It has been demonstrated that, despite its simplicity, the proposed model provides with accurate results while keeping the speed up of logic-level over electrical simulation (2 to 3 orders of magnitude). It has also been shown that including the degradation effect in logic simulation is a key point in order to achieve some level of accuracy, since conventional delay models will largely overestimate the switching activity, thus current, yielding to useless results despite the accuracy of the internal current model.

The combination of the DDM and logic-level current models appears as a good alternative to current estimations (and derived applications) at the logic level.

References

1. B. Arts and et al., "Stadistical power stimation of behavioral descriptions," in *Proc. 13th International Workshop on Power and Timing Modeling, Optimization and Simulation (PATMOS)*, pp. 197–207, Springer, September 2003.
2. M. Bruno, A. Macii, and M. Poncino, "A statistic power model for non.synthetic rtl operators," in *Proc. 13th International Workshop on Power and Timing Modeling, Optimization and Simulation (PATMOS)*, pp. 208–218, Springer, September 2003.
3. E. Seen, J. Laurent, N. Julien, and M. E., "Softexplorer: Estimation, characterización and optimization of the power and energy consumption at the algoritmit level," in *Proc. 14th International Workshop on Power and Timing Modeling, Optimization and Simulation (PATMOS)*, pp. 342–351, Springer, September 2004.
4. E. Lattanzi, A. Acquaviva, and B. A., "Run-time software minitor of the power consumption of wireless network interface cards," in *Proc. 14th International Workshop on Power and Timing Modeling, Optimization and Simulation (PATMOS)*, pp. 352–361, Springer, September 2004.
5. R. Jimenez, P. Parra, P. Sanmartin, and A. Acosta, "Analysis of high-performance flips-flops for submicron mixed-signal applications," *Int. Journal of Analg Integrated Circuits ans Signal Processing*, vol. 33, pp. 145–156, Nov 2002.
6. A. Abbo, R. Kleihorst, V. Choudhary, and S. L., "Power consuption of performance-scaled simd processors," in *Proc. 14th International Workshop on Power and Timing Modeling, Optimization and Simulation (PATMOS)*, pp. 532–540, Springer, September 2004.
7. "Primepower reference manual.," in *http://www.synopsys.com, Synopsys, Inc.*
8. "Xpower reference manual.," in *http://www.xilinx.com, Xilinx, Inc.*
9. "Powertool reference manual.," in *http://www.veritools.com, Veritools Inc.*
10. A. Bogliolo, L. Benini, G. De Micheli, and B. Riccò, "Gate-level power and current simulation of cmos integrated circuits," *IEEE Trans. on very large scale of integration (VLSI) systems*, vol. 5, pp. 473–488, December 1997.
11. S. Nikolaidis and A. Chatzigeorgiou, "Analytical estimation of propagation delay and short-circuit power dissipation in cmos gates.," *International journal of circuit theory and applications*, vol. 27, pp. 375–392, 1999.
12. C. Baena, J. Juan, M. J. Bellido, P. Ruiz-de Clavijo, C. J. Jimenez, and M. Valencia, "Simulation-driven switching activity evaluation of CMOS digital circuits," in *Proc. 16th Conference on Design of Circuits and Integrated Systems (DCIS)*, November 2001.

13. A. Ghosh, S. Devadas, K. Keutzer, and J. White, "Estimation of average switching activity in combinational and sequiential circuits," in *Proc. 29th Design Automation Conference*, pp. 253–259, June 1992.
14. C. Metra, B. Favalli, and B. Ricco, "Glitch power dissipation model," in *Proc. 5th International Workshop on Power and Timing Modeling, Optimization and Simulation (PATMOS)*, pp. 175–189, September 1995.
15. M. Eisele and J. Berthold, "Dynamic gate delay modeling for accurate estimation of glitch power at logic level," in *Proc. 5th International Workshop on Power and Timing Modeling, Optimization and Simulation (PATMOS)*, pp. 190–201, September 1995.
16. J. Juan-Chico, M. J. Bellido, P. Ruiz-de Clavijo, C. Baena, and M. Valencia, "Switching activity evaluation of cmos digital circuits using logic timing simulation," *IEE Electronics Letters*, vol. 37, pp. 555–557, April 2001.
17. M. J. Bellido-Díaz, J. Juan-Chico, A. J. Acosta, M. Valencia, and J. L. Huertas, "Logical modelling of delay degradation effect in static CMOS gates," *IEE Proc. Circuits Devices and Systems*, vol. 147, pp. 107–117, April 2000.
18. P. Ruiz-de Clavijo, J. Juan, M. J. Bellido, A. J. Acosta, and M. Valencia, "Halotis: High Accuracy LOgic TIming Simulator with Inertial and Degradation Delay Model," in *Proc. Design, Automation and Test in Europe (DATE) Conference and Exhibition*, March 2001.
19. "Cadence home page.," in *http://www.cadence.com*.
20. P. Maurine, R. Poirier, N. Azémard, and D. Auvergne, "Switching current modeling in cmos inverter for speed and power estimation," in *Proc. 16th Conference on Design of Circuits and Integrated Systems (DCIS)*, pp. 618–622, November 2001.
21. "Verilog resources page.," in *http://www.verilog.com*.
22. "Mentor graphics home page.," in *http://www.mentor.com*.

Design of Variable Input Delay Gates
for Low Dynamic Power Circuits

Tezaswi Raja[1], Vishwani D. Agrawal[2], and Michael Bushnell[3]

[1] Transmeta Corp., Santa Clara, CA
traja@transmeta.com
[2] Auburn University, Aubrun, AL
vagrawal@eng.auburn.edu
[3] Rutgers University, Piscataway, NJ
bushnell@caip.rutgers.edu

Abstract. The time taken for a CMOS logic gate output to change after one or more inputs have changed is called the *output delay* of the gate. A conventional multi-input CMOS gate is designed to have the same input to output delay irrespective of which input caused the output to change. A gate which can offer different delays for different input-output paths through it, is known as a variable input delay(VID) gate and the maximum difference in delay between any two paths through the same gate is known as "u_b". These gates can be used for minimizing the active power of a digital CMOS circuit using a previosuly described technique called variable input delay(VID) logic. This previous publication proposed three different designs for implementating the VID gate.
In this paper, we describe a technique for transistor sizing of these three flavors of the VID gate for a given delay requirement. We also describe techniques for calculating the u_b of each flavor. We outline an algorithm for quick determination of the transistor sizes for a gate for a given load capacitance.

1 Introduction

We first describe the prior work and motivation for this work in this section. We then describe the sizing procedures and algorithms in the following sections followed by contributions and conclusion.

1.1 Prior Work

Dynamic power consumed in the normal operation of a circuit consists of essential power and also *glitch power*. Glitches are spurious transitions caused by imbalance in arrival times of signals at the input of a gate. Techniques such as *delay balancing, hazard filtering, transistor sizing, gate sizing* and *linear programming* have been proposed for eliminating glitches [-]. For further reference the reader is directed to recent books and articles [-]. Our focus in this paper is a recent technique known as *variable input delay logic* [,]. Raja

V. Paliouras, J. Vounckx, and D. Verkest (Eds.): PATMOS 2005, LNCS 3728, pp. 436–445, 2005.
© Springer-Verlag Berlin Heidelberg 2005

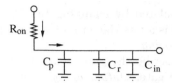

Fig. 1. The RC components along the charging path

et al. described a technique for reducing glitches using special gates are known as *variable input delay(VID) gates* where the delay through any input-output path can be manipulated without affecting the delays of the other paths upto a certain limit. This limit is known as the *differential delay upper bound* or u_b. This u_b is determined by the technology in which the circuit is implemented and is needed for finding the optimal solution to the linear program.

1.2   Motivation

Raja *et al.* describe three new ways of implementing the VID gate *viz. Capacitance manipulation, nMOS transistor insertion* and *CMOS transistor insertion* [12, 13]. Each of these gate designs can be used for efficient manipulation of input delay without altering the output delay of the gate. However, the paper has the following shortcomings.

– How are the transistor sizes determined from the delay assignment?
– How is the u_b calculated for every gate type?
– What is the algorithm for finding the right sizes and what are the trade-offs?

These are the questions we try to answer in this paper.

1.3   Components of RC Gate Delay

Gate Delay is the time taken for the output signal at the output of the gate to reach 50% of *Vdd* after the signal at the input of the gate has reached 50% of *Vdd* [25, 26].

Consider the path shown in Figure 1. The delay of a gate is a function of the on resistance R_{on}(ignoring saturation effects) and the load capacitance C_L. The load capacitance is given by:

$$C_L = C_p + C_r + C_{in} \tag{1}$$

where C_p is the parasitic capacitance due to the on transistor, C_r is the routing capacitance of the path and C_{in} is the input capacitance of the fanout transistors. C_{in} is the major component of C_L. C_r and C_p are non-controllable and hence, we ignore them in the current discussion. The delay of the path during a signal transition is given by:

$$Delay = R_{on} \times C_L \tag{2}$$

The delay can be manipulated by changing the C_L or the R_{on} by sizing the transistor accordingly. This alters the gate delay along all paths equally. This is called *conventional gate sizing*.

For VID logic, we describe the gate delay as the sum of *input delay* and *output delay* through the gate. *Output delay* is the common delay component of the gate no matter which input has caused the transition. *Input delay* is the delay component on input 1 that is present only on the input-output path through input 1 of the gate. Both input and output delays should be independent. Clearly, conventional gate sizing cannot be used for designing a VID gate. In this paper, we describe the *variable input delay gate sizing* for the VID gates proposed by Raja *et al.* [12, 13].

2   Gate Design by Input Capacitance Manipulation

The overall gate delay is given by Equation 2. In the new gate design we need to manipulate the input delay of the gate without affecting the output delay too much. Substituting Equation 1 into Equation 2, we get:

$$Delay = R_{on} \times (C_p + C_r) + R_{on} \times C_{in} \qquad (3)$$
$$= Output\ Delay + Input\ Delay \qquad (4)$$

From the above analysis we separate the input and output delays of the gate. The output delay depends on C_p and C_r, which are unalterable. The input delay is a function of R_{on} and C_{in} of the transistor pair. Thus, *input delay* of an input X can be changed by increasing the C_{in} offered by the transistor pair connected to X. Note that this does not alter the input delays of the other inputs of the gate(this is not always true as shown in Sec 2.2).

2.1   Calculation of u_b

The delay of the transistor pair can be calculated by using Equation 3. The input capacitance of a transistor pair is given by:

$$C_{in} = W \times L \times C_{ox} \qquad (5)$$

where W is the transistor width, L is the transistor length and C_{ox} is the oxide capacitance per unit area, which is technology and process dependant. The range of manipulation for C_{in} is limited by the range of W and L of the transistors allowed. The range of dimensions for digital design, in any technology, is governed by second-order effects, such as *channel length modulation, threshold voltage variation, standard cell height* etc [25, 26]. We have chosen the limit of the transistor length for 0.25μ technology as $3\mu m$, which is determined by the standard cell height. The minimum gate length in the same technology is $0.3\mu m$. Hence, the maximum difference in input capacitance is $2.7 \times C_{ox}$. The maximum differential delay d_{dif} and the minimum differential delay d_{min} obtainable in the technology can thus be:

$$Maximum\ Differential\ Delay\ d_{dif} = R_{on} \times 2.7 \times C_{ox}$$
$$Minimum\ Gate\ Delay\ d_{min} = R_{on} \times 0.3 \times C_{ox}$$

Thus, the gate differential delay upper bound u_b is given by:

$$u_b = \frac{d_{dif}}{d_{min}} = \frac{R_{on} \times 2.7 \times C_{ox}}{R_{on} \times 0.3 \times C_{ox}} = 9$$

Thus, the u_b of the technology can be calculated by using the bounds on the dimensions of the transistors in the particular technology. There are several design issues in this gate design as described below.

2.2 Design Issues

The gate design proposed in the previous section has several drawbacks.

- In this gate design output and input delays are not independent for both falling and rising transitions. For example, the NAND gate consists of two pMOS transistors in parallel and two nMOS transistors is series. The gate has different rising delays along both inputs if pMOS transistors are sized differently. But the same is not true for a falling delay. Altering the size of one of the nMOS transistors affects the R_{on} of the output discharging path and, thus, the output delay. This dependancy makes the sizing, for a given delay, a non-linear problem which can be difficult to converge.
- The parasitic capacitance C_p is assumed to be constant and independent of the transistor sizes. But in reality, C_p is a function of the transistor sizes. Altering the sizes of one transistor can affect C_p and the output gate delay.
- When the transistors are connected in series to one other, some of them are ON and some are OFF. This causes the threshold voltages of the transistors to change drastically due to *body effect* [25, 26]. This makes the output delay of the gate, input pattern dependant. This is a problem as the LP gives a single delay for every gate output [10, 11].

3   Gate Design with nMOS Pass Transistors

In the design proposed in Sec. 2, the main problem was the inter-dependence of output and input delays. In this second design, we propose to leave the input capacitance unaltered, and increase the resistance of the path.

3.1   Effects of Increasing Resistance and Input Slope

Consider the charging path shown in Figure 1. Energy is drawn from the supply to charge the C_L through R_{on}. The energy consumed by a signal transition is

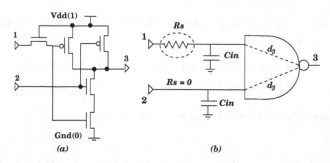

Fig. 2. The proposed single added nMOSFET VID NAND gate. (a) Transistor Level showing the nMOS transistor added and (b) charging path for transitions along the different paths through the gate

given by $0.5C_L V_{dd}^2$, where C_L is the load capacitance and V_{dd} is the supply voltage. Note that the energy expression does not include resistance R_{on} in it. The resistance governs the switching time but the overall energy per transition remains the same. Hence, *increasing the resistance of the path does not alter the energy consumed per transition*. Increasing resistance of the slope however, degrades the slew of the input waveform. This increase in input slope affects gate delay and needs to be acounted for.

$$Gate\ Delay = t_{step} + t_{slew}$$

where t_{step} is the gate delay when the input is a step waveform and t_{slew} is the gate delay due to the input slope or *slew*. Thus, by increasing R_{on} we manipulate t_{slew} part of the gate delay. But increasing the input slew decreases the *robustness* and noise immunity of the circuit [ ]. A large input slope means that the circuit is *in transition* for a longer period of time and is more susceptible to noise and short-circuit power. The input slope is *restored* or improved by using *regenerative gates*. The CMOS logic gates are regenerative as they improve the slope of the waveform while passing the signal transition from the input to the output. In our new VID gate design by inserting resistance, we use this regenerative property of the CMOS gates in the output for restoring the slope. However, the slope restoration also has limits and hence, there is a practical limit to degrading the input slope. This is one of the major factors that influence the practical value of u_b of a given technology.

3.2   Proposed Gate Design

We insert a single nMOS transistor that is always ON, with resistance R_s, in the series charging path. A modified NAND gate is shown in Figure 2. The delays of the gate along both I/O paths are given by:

$$d_{2\to3} = R_{on} \times C_L \tag{6}$$
$$d_{1\to3} = R_{on} \times C_L + R_s \times C_L \tag{7}$$
$$= Ouput\ Delay + Input\ Delay \tag{8}$$

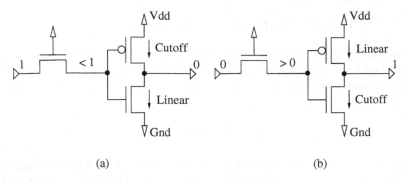

(a)   (b)

Fig. 3. The logic degradation of the single nMOS transistor addition (a) When logic 1 is passed through and (b) When logic 0 is passed through the gate

Thus, the input and output delays are separated completely from each other. The output delay can be controlled by sizing the gate transistors and the input delay can be controlled through R_s. $d_{2\rightarrow3}$ is not affected by altering $d_{1\rightarrow3}$. This concept can be extended to a n-input gate. The differential delay of path x with respect to the other $n - 1$ paths, can be controlled by inserting $n - 1$ transistors in series with the inputs. These paths can be independently controlled by sizing the $n - 1$ transistors. Thus, we have a VID gate design that is extendible to all multi-input gate types.

3.3   Calculation of u_b

As seen from Equation 8, the input delay can be controlled independently by altering the size of the nMOS transistor. The nMOS transistor passes logic 0 effectively but degrades the signal when passing logic 1. Let us assume that there is a degradation of voltage λ when a logic 1 is passed through the transistor [,
 ]. When the transistor is acting as a resistor, there is an IR voltage drop also across the capacitor. The drop can be significant for two reasons:

- If the drop is too large, then the transistors in the fanout will not switch OFF completely. This increases short circuit dissipation of the fanout gate.
- The leakage power of the transistors is a function of the *gate to source voltage* (V_{gs}). Hence, larger drop would increase leakage current of fanout gate.

The circuit in Figure 3(a) shows a single transistor pair at the output of the nMOS. The operating regions for the transistors are as shown. The critical condition in this configuration is the pMOS transistor remaining in cutoff. If this condition is not met, the pMOS transistor is also ON and, hence, there is a direct path from the supply to the ground. This increases the short circuit dissipation. To meet the condition, we need to make sure that $V_g > V_{dd} - V_{tp}$, where V_{tp} is the threshold voltage of the pMOS transistor. There are two factors that control the input voltage V_g is this case, $(1)I_{ds}R_s$, where I_{ds} is the drain to source stand-by current through the series transistor and (2)the signal degradation λ [ ].

$$V_{dd} - \lambda - I_{ds}R_s > V_{dd} - V_{tp} \text{ or } R_s < \frac{V_{tp} - \lambda}{I_{ds}} \tag{9}$$

Consider the input configuration in Figure 3(b). The nMOS transistor passes a logic 0 without any degradation($\lambda = 0$). The critical condition here is the nMOS transistor in cutoff. By using a similar analysis as above, the condition is given by:

$$I_{ds}R_s < V_{tn} \text{ or } R_s < \frac{V_{tn}}{I_{ds}} \tag{10}$$

Equations 9 and 10 give the upper bound on R_s. This limits the amount of resistance that can be added to the charging path. Thus, the amount of input delay that can be added is also limited by this condition.

$$u_b = \frac{d_{diff}}{d_{min}} = \frac{R_{max} \times C_L}{R_{on} \times C_L} = \frac{R_{max}}{R_{on}} \tag{11}$$

where R_{max} is the maximum resistance that can be added and C_L is the load capacitance of the gate. This is the *theoretical limit* of u_b but the practical limit is governed by signal integrity issues as explained in Sec. 3.1.

3.4   Design Issues

This new VID gate design, although an improvement over the design in Sec. 2 has the following issues.

- Theoretical u_b can be further reduced by dimension limits on the series nMOS transistors.
- The short circuit dissipation is a function of the ratio of the input and output waveform slopes [25]. By inserting resistance we are increasing the input waveform slope thereby increasing the short circuit dissipation.
- The leakage power is a function of the gate to source voltage (V_{gs}). Since $\lambda > 0$ when passing a 1, the leakage power of the fanout transistors increases. This drawback is alleviated in the design discussed in the next section.
- This design has an area overhead due to extra transistors added.

4   Gate Design with CMOS Pass Transistors

In the gate design described in Sec. 3, the single nMOS transistor degrades logic 1, thereby increasing leakage power. This disadvantage can be alleviated by adding a CMOS pass transistor instead. The CMOS pass transistor consists of an nMOS and a pMOS transistor connected in parallel. Both transistors are kept always ON and $\lambda = 0$ while passing either logic 1 or logic 0.

4.1   Calculation of u_b

The u_b calculation is similar to the single nMOS added design but with $\lambda = 0$. Note that the resistance R_s is the effective parallel resistance of both the transistors together.

4.2   Design Issues

The design issues involved in this gate design are:

- R_s is the effective series/parallel resistance of both the nMOS and the pMOS transistors. Hence, effective resistance per unit length reduces and the transistors have to be longer to achieve the same resistance as a single nMOS transistor.
- Larger area overhead than the design in Sec. 3.

5   Technology Mapping

The process of designing gates that implement a given delay by altering the dimensions of the transistors is called *technology mapping* or *transistor sizing*. In this section we describe the transistor sizing of VID gates. From Eqn. 2, gate delay is dependant on C_L of the gate, which is dependant on the dimensions of the fanout gate size. Hence, to obtain a valid transistor sizing for delay at a gate G, the sizes of the gates in the fanout of G have to be decided. Therefore, to design an entire circuit, we use a *reverse breadth first search* methodology and first design the gates connected to the primary outputs and work towards the inputs of the circuit.

The objective is to design a gate with a load capacitance C_L in a particular instance, in order to have a required delay d_{req}. The procedure involves searching for the appropriate sizes for all of the transistors in the gate. The dimensions for the search space of an n-input gate are load capacitance, $2n$ transistor widths and $2n$ lengths for a total of $4n+1$ dimensions. This can be a time consuming process to do for large circuits. So we propose to do this in two stages. The first stage is to generate a look-up table of sizes by simulation, for different d_{req} and C_L. For every gate type, we simulated the gate with the smallest sizes to find rising delay d_{rise} and falling delay d_{fall}. The objective function is to minimize $\epsilon = \frac{|d_{req}-d_{rise}| + |d_{req}-d_{fall}|}{d_{req}}$. The d_{rise} and d_{fall} can be increased by increasing the length of the transistors and decreased by increasing the width. Thus, by an iterative process an implementation for the given d_{req} and C_L can be achieved(to within acceptable values of error ϵ) and noted in the look-up table. Thus, the look-up table has size assignments for all different gate types and some values of C_L. This look-up table can be used for all circuits.

When a particular circuit is being optimized, the look-up table may not have the exact C_L. In such cases, we go to the second stage of fine tuning the sizes. We start with the closest entry in the look-up table. Each dimension is perturbed by one unit(since dimensions are discrete in a technology) and the sensitivty is calculated where:

$$Sensitivity = \frac{d_{current}}{|d_{req} - d_{rise}| + |d_{req} - d_{fall}|}$$

where $d_{current}$ is the present measured gate delay, and d_{rise} and d_{fall} are the rise and fall delays after a perturbation in the dimension. There can be 8 perturbations, two for each of the dimensions. The perturbation with the highest sensitivity is incorporated and the gate is simulated again. The objective function is to minimize ϵ given earlier. This procedure is called the *steepest descent* method as the objective function is minimized by driving the dimensions based on sensitivities. The complexity is greatly reduced by using the lookup table as the search is limited to the neighborhood of the solution. Hence, local minima will not be a problem. The procedure can also be tuned for including the area of the cell in the objective function.

6   Summary

In this paper, we explained why conventional CMOS gates cannot be used as VID gates. We presented three new implementations of the VID gate. We presented an analysis of each of the gates and listed their shortcomings. Then we proposed a two-step approach for fixing the transistor sizes of every instance in the circuit. The main idea of this paper is to present the transistor level implementation details of the *variable input delay logic*. The advantages of the technique, its power reduction results and comparisons with other techniques are the same as presented in earlier publications and are not duplicated here [11–13].

References

1. Agrawal, V.D.: Low Power Design by Hazard Filtering. In: Proc. of the International Conference on VLSI Design. (1997) 193–197
2. Agrawal, V.D., Bushnell, M.L., Parthasarathy, G., Ramadoss, R.: Digital Circuit Design for Minimum Transient Energy and Linear Programming Method. In: Proc. of the International Conference on VLSI Design. (1999) 434–439
3. Berkelaar, M., Jacobs, E.: Using Gate Sizing to Reduce Glitch Power. In: Proc. of the ProRISC Workshop on Circuits, Systems and Signal Processing, Mierlo, The Netherlands (1996) 183–188
4. Berkelaar, M., Buurman, P., Jess, J.: Computing Entire Area/Power Consumption Versus Delay Trade-off Curve for Gate Sizing Using a Piecewise Linear Simulator. IEEE Transactions on Circuits and Systems **15** (1996) 1424–1434
5. Berkelaar, M., Jess, J.A.G.: Transistor Sizing in MOS Digital Circuits with Linear Programming. In: Proc. of the European Design Automation Conference, Mierlo, The Netherlands (1990) 217–221
6. Berkelaar, M., Jacobs, E.T.A.F.: Gate Sizing Using a Statistical Delay Model. In: Proc. of the Design Automation and Test in Europe Conference, Paris, France (2000) 283–290
7. Sathyamurthy, H., Sapatnekar, S.S., Fishburn, J.P.: Speeding up pipelined circuits through a combination of gate sizing and clock skew optimization. In: Proc. of the International Conference on Computer-Aided Design. (1995) 467–470
8. Benini, L., DeMicheli, G., Macii, A., Macii, E., Poncino, M., Scarsi, R.: Glitch power minimization by gate freezing. In: Proc. of the Design Automation and Test in Europe Conference. (1999) 36
9. Kim, S., Kim, J., Hwang, S.Y.: New Path Balancing Algorithm for Glitch Power Reduction. IEE Proceedings: Circuits, Devices and Systems **148** (2001) 151–156

10. Raja, T., Agrawal, V.D., Bushnell, M.L.: Minimum Dynamic Power CMOS Circuit Design by a Reduced Constraint Set Linear Program. In: Proc. of the International Conference on VLSI Design. (2003) 527–532
11. Raja, T., Agrawal, V.D., Bushnell, M.L.: CMOS Circuit design for Minimum Dynamic Power and Highest Speed. In: Proc. of the International Conference on VLSI Design. (2004) 1035–1040
12. Raja, T., Agrawal, V.D., Bushnell, M.L.: Variable Input Delay Logic and Its Application to Low Power Design. In: Proc. of the International Conference on VLSI Design. (2005) 598–605
13. Raja, T.: Minimum Dynamic Power Design with Variable Input Delay Logic. PhD thesis, Rutgers University, Dept. of ECE, Piscataway, New Jersey (2004)
14. Datta, S., Nag, S., Roy, K.: ASAP: A Transistor Sizing Tool for Area, Delay and Power Optimization of CMOS Circuits. In: Proc. of the IEEE International Symposium on Circuits and Systems. (1994) 61–64
15. Musoll, E., Cortadella, J.: Optimizing cmos circuits for low power using transistor reordering. In: Proc. of the European Design Automation Conference. (1995) 219–223
16. Hashimoto, M., Onodera, H., Tamaru, K.: A Practical Gate Resizing Technique Considering Glitch Reduction for Low Power Design. In: Proc. of the Design Automation Conference. (1999) 446–451
17. Chandrakasan, A.P., Brodersen, R.W.: Low Power Digital CMOS Design. Kluwer Academic Publishers, Boston (1995)
18. Chandrakasan, A.P., Sheng, S., Brodersen, R.W.: Low Power CMOS Digital Design. IEEE Journal of Solid-State Circuits 27 (1992) 473–484
19. Chandrakasan, A., Brodersen, R., eds.: Low-Power CMOS Design. IEEE Press, New York (1998)
20. Nebel, W., Mermet, J.: Low Power Design in Deep Submicron Electronics. Kluwer Academic Publishers, Boston (1997)
21. Rabaey, J.M., Pedram, M., eds.: Low Power Design Methodologies. Kluwer Academic Publishers, Boston (1996)
22. Rabaey, J.M., Pedram, M.: Low Power Design Methodologies. Kluwer Academic Publishers, Boston (1995)
23. Roy, K., Prasad, S.C.: Low-Power CMOS VLSI Circuit Design. Wiley Interscience Publications, New York (2000)
24. Yeap, G.: Practical Low Power Digital VLSI Design. Kluwer Academic Publishers, Boston (1998)
25. Rabaey, J., Chandrakasan, A., Nikolic, B.: Digital Integrated Circuits: A Design Perspective. Prentice Hall, Upper Saddle River, NJ (2003)
26. Weste, N., Eshraghian, K.: Principles of CMOS VLSI Design: A Systems Approach. Addison Wesley Publications, Reading, MA (1985)

Two-Phase Clocking and a New Latch Design for Low-Power Portable Applications

Flavio Carbognani, Felix Bürgin, Norbert Felber,
Hubert Kaeslin, and Wolfgang Fichtner

ETH, Zurich, Switzerland
carbo@iis.ee.ethz.ch
http://www.iis.ee.ethz.ch

Abstract. The energy efficiency of a $0.25\,\mu$m general-purpose FIR filter design, based on two-phase clocking, versus a functionally equivalent benchmark, based on one-phase clocking, is demonstrated by means of measurements and transistor level simulations. Architectural improvements enable already a 20% energy savings of the two-phase clocking implementation. Yet, for the first time, the limitations imposed by the supply voltage ($< 2.1\,$V) and the operating frequency ($< 10\,$MHz) on the actual energy efficiency of this low-power strategy are investigated. Transistor level re-design is undertaken: a new slew-insensitive latch is presented and replaced inside the two-phase implementation. Spectre simulations point out the final 30% savings.

1 Introduction

Energy efficiency is of primary importance in CMOS circuits intended for the portable market. The clocking strategy does have a great influence on it, as it directly affects the power consumption of the clock network(s) and all the sequential cells, which typically represent a large fraction of the whole chip. The one-phase single-edge-triggered clocking strategy has imposed as a standard over the years, because it guarantees easiness of design, simple testability and it is fully compatible with all tools for VLSI implementation. Yet, some publications [1, 2] have already underlined that the intrinsic skew-insensitivity of the two-phase clocking, while saving energy-consuming clock buffers, may often translate into more energy efficient implementations. On the other hand, other works [3, 4] have concentrated on the re-design of low-power latch circuits. However, very few papers [5] take into consideration both aspects together, the two-phase clocking strategy and the latch design. Pursuing this task on a well-established low-power architecture [6], the present work confirms the energy efficiency of the two-phase clocking over the one-phase counterpart, but points out, for the first time, the intrinsic supply voltage and frequency limitations. A new latch design is hereby introduced, which circumvent these tight restrictions, by adding to the traditional skew- a substantial slew-insensitivity. The results of this work address specifically to a large variety of low- and middle-frequency (up to some tens of MHz) portable applications, ranging from hearing aids to bio-medical devices and audio streaming (MP3 players, etc. . .).

V. Paliouras, J. Vounckx, and D. Verkest (Eds.): PATMOS 2005, LNCS 3728, pp. 446–455, 2005.

The rest of the paper is organized as follows: in sec. 2 the different design implementations are described; in sec. 3 the results of the measurements are presented. Sec. 4 discusses the interesting points of the measurements, through a three-step analysis: at architectural level, at gate level and, finally, at transistor level. In sec. 5 the operating frequency limitations are addressed. Sec. 6 presents a novel low-power latch circuit, and sec. 7 shows the final results out of simulations. The paper is completed by the conclusions in sec. 8.

2   Chip Design

Three versions (**A**, **B** and **C**) of the same 100-tap 16-bit general-purpose FIR filter, a typical candidate for adaptive audio processing in the time-domain [ ], together with latch-based data and coefficient memories, have been manufactured in a 0.25 μm process. In fig. 1 the block diagram of the implemented FIR filter is presented. The same basic structure has been repeated for all three designs. The main architectural differences among them are the number representation, the presence of clock gating, the clocking strategy and the depth of the clock distribution network, as summarized in tab. 1. Designs **B** and **C** use a hybrid number representation (sign-magnitude in the multiplier and two's complement in the adder), which has been reported to be more efficient in MAC units for audio data processing [ ]. The lack of clock gating makes design **A** a very inefficient reference. As a matter of fact, the un-gated clock net capacitance of **A**, which is charged/discharged during each period, is very large: therefore, the clock-tree generator is forced to instantiate many energy consuming clock buffers (see tab. 1). This fact together with the increased FF internal power consumption cause an overall dissipation of **A** 4 to 5 times higher than the others:

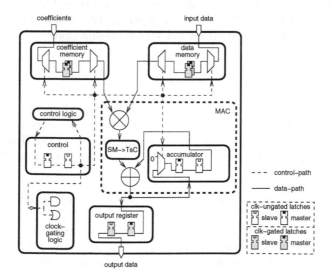

Fig. 1. Block diagram of the implemented designs

Table 1. Characteristics of the designs (simulated entries are in italics)

Design Characteristic	A	B	C	*D*
Number repr.	TwosC	Hyb	Hyb	*Hyb*
Clock-gating	No	Yes	Yes	*Yes*
Clocking Strategy	SET-FF	SET-FF	2-ph	*2-ph*
Clock Tree Cells	9 buf&10 del.	1 buf&13 del.	None	*None*
Core Area (mm^2)	0.71	0.55	0.56	*0.34*
Standard Cell Density	82%	82%	82%	*n.a.*
Energy (μW/MHz) at 2.5 V/1.25 V	1030/-	220/51	280/41	*186.5/36*
Clock Ramp Time (ns) at 2.5 V	*n.a.*	*0.8*	*8.5 Master*	*7.1 Master*

it will not be considered any longer. **B** should be regarded as a robust state-of-the-art design; **C** is the first proposed improvement, whose benefits, however, are exploitable only at low supply voltages, as shown later. The cores of **A**, **B** and **C** have been designed using the VHDL and then synthesized using Synopsys, targeting a 0.25 μm standard cell library. The timing constraints during the synthesis have been kept very relaxed, namely a clock period of 500 ns has been chosen, minimum requirement for the elaboration of data sampled at 20 kHz (the 100-tap FIR filters combine full time sharing with iterative decomposition). Clock tree generation, placement and routing have been performed using Silicon Ensemble with the same timing constraints as during the synthesis. Latch-based clock gating has been implemented as in [,]. All three designs have been fabricated on a common die (fig. 2): **B**, **C** and **A** appear as three distinct blocks with independent power rings, located from left to right respectively. Beside these three designs actually implemented on silicon, a fourth one **D**, not integrated yet but simulated in Cadence environment, is proposed as a further improvement (last column in tab. 1). It shares the same architecture as **C**, but enhances the intrinsic energy efficiency of the latter, by replacing the latches with functionally equivalent sub-circuits optimized for low energy dissipation.

3   Measurements

To emphasize the new results of this work, the measured energy consumptions (in μW/MHz) of **B** and **C** are reported in fig. 3 (upper curves) as functions of the supply voltage, for a sequence of typical speech signal. The two curves present a quite unexpected behavior, which can be summarized in the following three statements:

1. The difference in the energy consumption of the two designs depends strongly on the supply voltage.
2. The curve related to **C** rises more steeply than the other one.
3. The two curves present an unexpected cross-point at around 2.1 V: for larger voltages **B** dissipates less, for lower voltages **C** is more efficient (20% energy saving at 1.25 V).

Fig. 2. DIE photo, 2.435 mm x 2.435 mm (including pads and seal-ring)

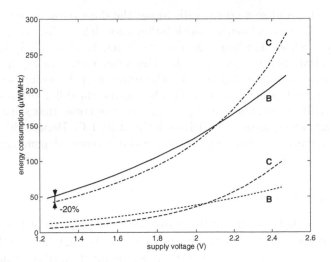

Fig. 3. Measured energy dissipation as function of the supply voltage (upper curves); simulated sequential energy (lower curves)

4   Discussion

The following discussion explains the unexpected cross-point in fig. 3 by comparing the results of architectural-level, gate-level and transistor-level simulations with the measurements.

4.1   Architectural-Level Analysis

The architectural differences (sketched in fig. 4) of **B** compared with **C** have been investigated first. **B** includes a huge central clock buffer, a clock tree with

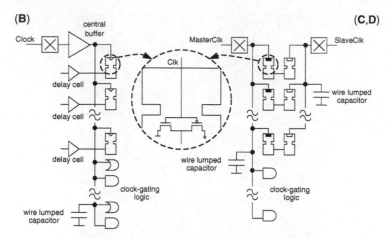

Fig. 4. Architectural differences between **B** and **C, D**

FFs as leaf cells, and some delay cells along the data paths to meet hold time requirements; **C** presents neither clock buffers nor delay cells, and replaces each FF with two latches. Simulating the circuits in fig. 4, whose dissipations will be called sequential energy later on, yields the lower curves in fig. 3. With good approximation, the separation between the upper and the lower curves in fig. 3 is constant, regardless of the design. This is because the difference represents the dissipation of all the combinational gates and the memories that are implemented basing on the same pieces of VHDL code in **B** and **C**. Hence, the "sequential energy" is actually the only responsible for the different dependency of **B** and **C** on the supply voltage.

4.2 Gate-Level Analysis

Simulating a single FF of **B** and the corresponding latch pair of **C** revealed that up to 80% of the total dissipation takes place in the first inverter (zoom-in of fig. 4) at the clock input (energy consumption as function of the supply voltage in fig. 5).

Therefore, the difference between **B** and **C** is mainly caused by the FFs/ latches with no clock gating in the two designs. In **C**, the lack of a central buffer makes the clock ramp times exceedingly slow (around 8.5 ns the master clock at 2.5 V, more than ten times larger than the clock in **B**). While the non-overlap phases provide ample margins against malfunctioning, this causes much energy to be dissipated by cross-over currents in the first clock inverter inside each latch.

4.3 Transistor-Level Analysis

To confirm this, calculations have been carried out using an α-power-law transistor model [ ]. This model has been calibrated on our NMOS and PMOS current/voltage characteristics. The cross-over energy E_{CO} spent in a CMOS

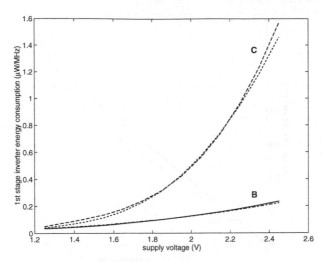

Fig. 5. Simulation: first stage clock inverter inside the FF in **B** (solid) and α-power-law approximation [ ] (dash-dotted); first stage inverters in the corresponding latch pair in **C** (dashed) and α-power-law approximation [ ] (dotted)

inverter that discharges a small output capacitance (small enough to consider the output transition time negligible compared with the input ramp time) can be expressed as [ ]:

$$E_{co} = V_{dd} \cdot t_t \cdot I_{d0p}(V_{dd}) \cdot \frac{1}{\alpha_p + 1} \cdot \frac{1}{2^{\alpha_p}} \cdot \frac{(1 - v_{tn} - v_{tp})^{\alpha_p + 1}}{(1 - v_{tp})^{\alpha_p}}. \tag{1}$$

where t_t is the input transition time, I_{d0p} is the saturated drain current of the PMOS depending on V_{dd}, α is the velocity saturation index of the PMOS, v_{tp} is the normalized threshold voltage of the PMOS, and v_{tn} is the normalized threshold voltage of the NMOS. The dotted and dash-dotted curves in fig. 5 show the good accuracy obtained from simulations with this model. Eq. 1 suggests two conclusions: E_{CO} increases linearly with the input ramp time and the dependency of E_{CO} on the supply voltage is much more than quadratic since I_{d0p} increases more than linearly with V_{dd}. Slow clock ramps (as in **C**) in conjunction with high supply voltages thus greatly inflate the contribution of cross-over currents to the overall energy dissipation. Using eq. 1 to estimate the ratio between cross-over and switching energies dissipated by the inverter inside the FF of **B** and inside the corresponding latch of **C** yields the following surprising result: at 2.5 V, this ratio passed from 0.7 (in **B**) to more than 9 (in **C**)!

5   Frequency Limitations

As a direct consequence, we address for the first time the problem of extending efficiently the two-phase clocking to large-throughput applications. There is typically a frequency bound, above which two-phase clocking dissipates more

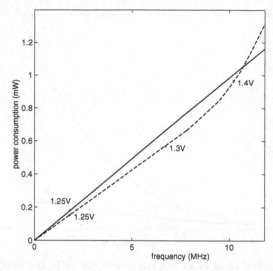

Fig. 6. Minimum power consumption to operate **B** (solid) and **C** (dashed) at a given frequency, with the corresponding minimum supply voltage

than the one-phase equivalent. Fig. 6 shows the minimum measured power that is necessary to operate designs **B** (solid) and **C** (dashed) at a given frequency. The points along the curves specify the supply voltages that are needed to reach the corresponding frequency value on the x-axis. According to fig. 6, **C** is more energy efficient only up to 10.5 MHz; for higher frequencies, it is necessary to increase the supply voltage (1.4 V or more in **C** against 1.25 V in **B**) and the design is no more competitive.

6   New Low-Power Latch Circuit

In order to mitigate this problem, a new D-latch was needed. The idea was to substitute simple pass-transistors for transmission-gates in the feedback and in the D paths in order to avoid local clock inversion or, alternatively, the need for distributing a pair of complementary clocks (fig. 7). This avoided large cross-over contributions and partially traded speed for improved energy efficiency. Yet, in order to keep wide margins of reliability, we rejected:

1. All dynamic and semi-static solutions [ ].
2. All ratioed schemes, in which the functionality itself depends on transistor sizing.

The proposed circuit (fig. 7 right) is basically a 6-transistor latch [ ] with the addition of: an output inverter to provide isolation from glitches that might back-propagate from pin Q (2 transistors) to node B; an asynchronous reset to keep the compatibility with our primitive latch (2 transistors). Hence, the final circuit counts 10 transistors against the 21 transistors of the old latch

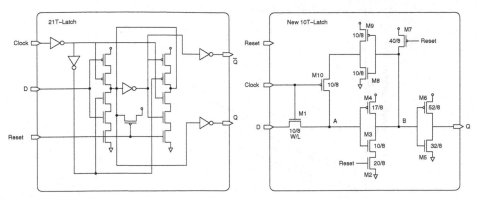

Fig. 7. Old latch circuit out of the technology library (left); new low-power latch circuit (right)

circuit out of the technology library (fig. 7 left), occupying roughly one half of the area. No inverted output was included because, in two-phase asymmetric clocking strategy, at least half of the latches, namely the master-clocked ones, make use only of the Q output. An external low-power inverter was added only when necessary. It should also be mentioned that the transition, for which the novel latch is more attractive in terms of energy dissipation, is when the clock toggles without any change in the input signal D, a very frequent transition in our design. While the new 10-transistor latch would dissipate nearly no energy in this case, the original latch would be highly inefficient due to the inverter at the clock input.

7 Final Simulation Results

By substituting the new slew-insensitive latch circuit for the original one in design **C**, a more energy efficient design **D** has been derived. Spectre simulations confirm the functionality and the energy savings of the new design: each latch dissipates from three to six times less, enabling an overall energy saving of around 30% at 1.25 V of **D** when compared to **B** (fig. 8). Moreover, the new slew-insensitive latch circuits avoid any cross-point in the curves of fig. 8, relaxing the frequency limitations discussed in sec. 5.

8 Conclusions

The conclusions that can be drawn from this work are the following:

1. Level-sensitive two-phase clocking is more energy efficient than SET one-phase clocking (up to 20% in the reported circuits), due to its weaker timing constraints (mainly in skew time and insertion delay), making clock buffers and delay cells unnecessary.

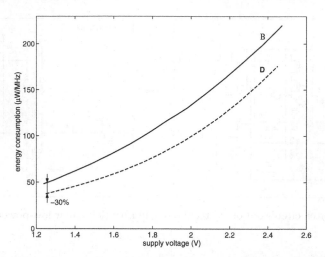

Fig. 8. Simulated energy consumption of **D** (dashed) compared to the energy consumption of **B** (solid)

2. Yet, due to the slower clock ramps, a side effect of the lack of clock buffers, the latches tend to dissipate more energy at higher supply voltages, compromising the savings.
3. If traditional latch circuits are used, low supply voltage and large clock ramp time restrict the energy savings to circuits with low clock frequency (< 10.5 MHz in the reported circuits).
4. Designing chips for nominal supply voltage and then simply scaling it down leads to non-optimal solutions energy-wise. As a matter of fact, when limiting the analysis at 2.5 V in fig. 3, better choice would have been to go for architecture **B**. Yet, at 1.25 V, which appears a natural option for the supply voltage as both designs are still meeting the requested throughput, **C** dissipates 20% less energy.
5. In two-phase low-power designs, the latches tend to become the critical energy sinks. Clock-slew-rate insensitive structures strongly favor low energy dissipation (30% less than the SET one-phase clocking reference of this report).
6. Clock-slew-rate insensitive latches enable extending two-phase intrinsic energy efficiency to designs that work at higher operating frequency. Fig. 8 shows no cross-over point between the two curves; hence, no frequency bound in the energy efficiency is expected for **D**, as it was for **C** (see fig. 6).

References

1. Mosch, P., et al.: A 660-μW 50-Mops 1-V DSP for a Hearing Aid Chip Set. IEEE Journal of Solid-State Circuits **35** (2000) 1705-1712
2. Arm, C., Masgonty, J., Piguet, C.: Double-Latch Clocking Scheme for Low-Power I.P. Cores. PATMOS (2000) 217-224

3. Zyuban, V.: Optimization of Scannable Latches for Low Energy. IEEE Trans. on VLSI **11** (2003) 778-788
4. Ching, L., Ling, O.: Low-power and low-voltage D-latch. IEE Electronics Letters **34** (1998) 641-642
5. Zyuban, V., Meltzer, D.: Clocking Strategies and Scannable Latches for Low Power Applications. ISLPED (2001) 346-351
6. Erdogan, A.T., Zwyssig, E., Arslan, T.: Architectural trade-offs in the design of low power FIR filtering cores. IEE Proc. Circ. Dev. Syst. **151** (2004) 10-17
7. Wassner, J., et al.: Waveform Coding for Low-Power Digital Filtering of Speech Data. IEEE Trans. on Signal Processing **51** (2003) 1656-1661
8. Nose, K., Sakurai, T.: Analysis and Future Trend of Short-Circuit Power. IEEE Trans. on CAD of IC and Systems **19** (2000) 1023-1030

Power Dissipation Reduction During Synthesis of Two-Level Logic Based on Probability of Input Vectors Changes

Ireneusz Brzozowski and Andrzej Kos

AGH University of Science and Technology, Mickiewicza 30, 30-059 Kraków, Poland
{brzoza,kos}@agh.edu.pl
http://scalak.elektro.agh.edu.pl/

Abstract. Detailed analysis of CMOS gate behaviour shows that a gate load and input capacitance depends on number of their activated inputs and kind of applied signals i.e. steady state, edge. And in consequence it has an influence on power dissipation. So a reason of the gate switching can be called as *the gate driving way*. Based on the probability of the gate driving way and associated portion of dissipated energy more accurate model of power dissipation for CMOS gates and circuits is presented. Example of circuits synthesis show power reduction possibilities during design of two-level logic circuits based on information about primary input vector changes – a new circuit activity measure.

1 Introduction

Power consumption of ordinary CMOS gates is mainly connected with their switching although static power dissipation takes significant values especially in deep-submicrometer technology [1]. Dynamic power dissipation is usually divided into two kinds. The first one is connected with capacitance existing in the CMOS gate and circuit and the second one, quasi-short power dissipation is connected with current path occurring through the gate during switching. Considering dynamic, capacitive power dissipation one can find that many authors in recent years usually use capacitive power dissipation model for CMOS gates consists of load, input capacitance and switching activity information [2], [3]. Power Prime Implicant are introduced in [3] and appropriate methods for low power two-level synthesis are developed. However these methods base on static probability of input signals and did not take into account temporal and spatial correlations between input signals. Some authors extend this model using transition probability and propose power reduction techniques based on this information [4], [5]. Taking into account transition probability beside static probability of input signals its temporal correlations are considered. In consequence it leads to better low power solution of designed circuits.

Careful analysis of CMOS gates behaviour during switching and consumed power measurements in real circuits, done by authors, show that dynamic power dissipation of the CMOS gate depends on reason of switching. Moreover energy is consumed even when the gate output does not change its logical state under the influence of inputs changes – the gate is driven but switching does not occur. In order to take into

V. Paliouras, J. Vounckx, and D. Verkest (Eds.): PATMOS 2005, LNCS 3728, pp. 456–465, 2005.
© Springer-Verlag Berlin Heidelberg 2005

consideration these dependencies authors propose new model of power dissipation for CMOS gates and circuits. The model is capacitive type and consists of equivalent capacitance and probability of changes of the gate input vectors [6]. Introduced measure of circuit activity called *driving way probability* includes static probability and transition probability of gate input signals. Moreover, because changes of vectors are considered, so some spatial correlations between input signals are considered.

In section two motivation and results of real circuit measurements are presented. Section three includes description of introduced model and section four describes method for assessment of CMOS gates energetic parameters for the model. In section five authors discuss possibilities of using proposed model and driving way probability for two-level synthesis of enhanced energetic circuits. Moreover the model of power dissipation presented in the paper gives more accurate power dissipation estimation then traditional one [7].

2 Motivation

Let us consider two inputs CMOS NAND gate driven in such a way as shown in Fig. 1. In the first case the gate output remains in steady state 1 because of 0 at input B. In the second case the NAND gate is switching because of steady state 1 being at input B. There is a question, is the same input capacitance of input A (C_{InA}) in both cases?

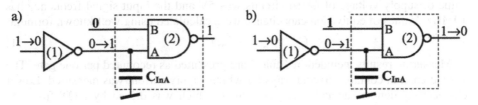

Fig. 1. Two kinds of 2-inputs NAND gate driving: a) (0↑), b) (1↑)

In order to answer the question authors decided to make measurements of supply current of driving NOT gate (1) in both cases. Authors used fabricated chip with some static CMOS gates [8]. The chip was designed in Microelectronics Systems Design Group at the Institute of Electronics AGH-UST for investigations of CMOS gates power dissipation using AMIS CMOS 0.7μm C07M-D technology. The integrated module consists of four kinds of CMOS gates: NOT, 2- and 3-inputs NAND, and 3-inputs NOR. Dimensions of transistors of designed gates were chosen in such a way that gates have maximum noise margin (Table 1). Additionally it ensures nearly equal rising and falling times of gates output voltages. Circuits for tests were designed as blocks of one hundred in parallel-connected gates for increasing measured current.

Table 1. Dimensions of transistors in tested gates in fabricated chip [μm]

Transistor type	NOT	2-inputs NAND	3-inputs NAND	3-inputs NOR
P-chan	2.6/0.7	1.5/0.7	1.1/0.7	7.9/0.7
N-chan	1.0/0.7	1.0/0.7	1.0/0.7	1.0/0.7

The 2- and 3-inputs NAND, and 3-inputs NOR gates in the chip were put to the test. Measuring circuit is presented in Fig. 2. Supply current of the first chip was measured while gates from the second one were driven with a few driving ways. One of inputs of tested gate was driven by the NOT but for remaining ones logic 0 or 1 were applied.

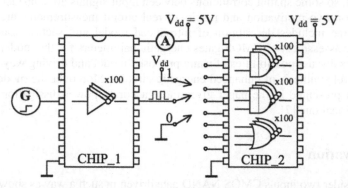

Fig. 2. Measuring circuit for comparison of gate driving way influence on input capacitance

Table 2 contains results of measured current and differences of calculated equivalent capacitance for considered driving ways for 3-inputs NAND and NOR gates. The value of supply voltage of tested circuits was 5V and the input signal frequency has 10MHz. Values of equivalent capacitance were calculated using well-known formula:

$$C_{equ} = I_{dd}/(f \cdot V_{dd}).$$ (1)

Measuring results included in Table 2 are presented as recounted per one gate. The second column contains driving ways for which supply current was measured. Lower character 'a' denotes actually activated input, which was driven by NOT from the CHIP_1. The sequence of inputs 'ABC' means that input A is a gate of NMOS (PMOS) transistor, which is the first connected to ground (supply V_{dd}) of serially connected transistors in NAND (NOR) gate. Next column it is supply current of the NOT gate. So it is sum of consumed current by the NOT and input of driven gate from CHIP_2. Supply current of the NOT without load was about 6.5μA it corresponds to equivalent capacitance equal 130fF and it was constant for all measuring cases. This value is greater than simulation result. But in real chip are some parasitic capacitance not taken into account during simulations i.e. capacitance of internal lines, bonding pads in the chip, pins, external interconnections in testing circuit etc. So the right side of Tab. 2 includes differences of measured current and calculated equivalent capacitance for all measuring cases. When input A is driven, which is most distant to the gate output, small interaction between inputs can be observed. For remains inputs (B or C) interactions are greater. When the gate is switching then input capacitance becomes greater due to output load influence on inputs because of coupling capacitance. This phenomenon is described by Miller effect. Even though good accuracy multimeter (KEITHLEY 2010, 7½ digits) was used for measurements obtained results differ from simulations and rather should be treated as showing of general trend in CMOS gates and start point for further investigation. Differences can be explained by relatively big load occurred in real circuit, which was transferred at inputs due to Miller effect.

Table 2. Measuring results of 3-inputs NAND and NOR gates

Nand3	ABC	I_{dd} [μA]						
input A	1) a00	15.2185		1-2	1-3	1-4	2-3	2-4	3-4
	2) a10	15.2181	ΔI_{dd} [μA]	0.0004	0.0003	1.0321	0.0001	1.0325	1.0324
	3) a01	15.2182	ΔC_{equ} [fF]	**0.01**	**0.01**	**20.64**	**0.00**	**20.65**	**20.65**
	4) a11	16.2506							
input B	1) 0a0	15.1647		1-2	1-3	1-4	2-3	2-4	3-4
	2) 1a0	15.2003	ΔI_{dd} [μA]	0.0356	0.016	0.1233	0.0516	0.0877	0.1393
	3) 0a1	15.1487	ΔC_{equ} [fF]	**0.71**	**0.32**	**2.47**	**1.03**	**1.75**	**2.79**
	4) 1a1	15.2880							
input C	1) 00a	15.3847		1-2	1-3	1-4	2-3	2-4	3-4
	2) 10a	15.3872	ΔI_{dd} [μA]	0.0025	0.0001	0.1179	0.0024	0.1154	0.3322
	3) 01a	15.3848	ΔC_{equ} [fF]	**0.05**	**0.00**	**2.36**	**0.05**	**2.31**	**6.64**
	4) 11a	15.5026							
Nor3	ABC	I_{dd} [μA]						
input A	1) a00	15.6410		1-2	1-3	1-4	2-3	2-4	3-4
	2) a10	15.5490	ΔI_{dd} [μA]	0.092	0.089	0.084	0.003	0.008	0.005
	3) a01	15.5520	ΔC_{equ} [fF]	**1.84**	**1.78**	**1.68**	**0.06**	**0.16**	**0.10**
	4) a11	15.5570							
input B	1) 0a0	15.4777		1-2	1-3	1-4	2-3	2-4	3-4
	2) 1a0	14.9922	ΔI_{dd} [μA]	0.4855	0.1156	0.4848	0.3699	0.0007	0.3692
	3) 0a1	15.3621	ΔC_{equ} [fF]	**9.71**	**2.31**	**9.70**	**7.40**	**0.01**	**7.38**
	4) 1a1	14.9929							
input C	1) 00a	15.8743		1-2	1-3	1-4	2-3	2-4	3-4
	2) 10a	14.9758	ΔI_{dd} [μA]	0.8985	0.897	0.8975	0.0015	0.001	0.0005
	3) 01a	14.9773	ΔC_{equ} [fF]	**17.97**	**17.94**	**17.95**	**0.03**	**0.02**	**0.01**
	4) 11a	14.9768							

3 Model of Dynamic Power Dissipation

Classical models of dynamic power dissipation in CMOS gates assume constant values of load and input capacitance. Though, input capacitance can take different values for different inputs. But these values are still constant and gate driving way independent. Nevertheless when we carefully make deeper insight into CMOS gate structure and coupling capacitance of MOS transistors to be taken into account, one can imagine that changing voltage at input can affect other inputs. The coupling consists of gate-to-drain and gate-to-source capacitance of transistors and occurs between each gate input and the output, directly or through turned on transistors. So due to Miller effect output voltage affects inputs, and an extent of this influence will depend on action of the output. It either changes or remains in steady state. Therefore for accurate estimation of power dissipation constant values of load and input capacitance cannot be taken. The capacitances should be treated as a function of gate driving way. The gate driving way is defined as a change between two input vectors. That change will cause switching of the gate or not, but evaluation of particular power consumption is possible. Also equivalent capacitance regarding dissipated power can be calculated considering all possible changes of input vectors – all gate driving ways.

3.1   The CMOS Gate Model

The proposed power dissipation model (Fig. 3) is similar to classical capacitive one but, as mentioned above, values of capacitance included in the model are function of the gate driving way probability. Capacitors of the model presents current consumed from supply or from inputs so they are called *equivalent*. The capacitor C_{Lint} at the gate output represents current, which flows from supply and is consumed by total internal load capacitance of the gate. It is the internal load. The capacitors appeared at the inputs also represent current consumed by this gate but flowing from previous ones. These are loads of previous, driving gates.

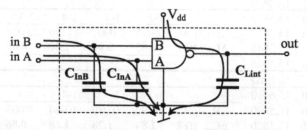

Fig. 3. Proposed model of dynamic power dissipation for 2-inputs NAND gate

Values of these equivalent capacitors will be calculated based on measurements (or simulations) of average current flowing as a result of the gate input signals change. These values depend on the gate driving way, i.e. number of switching inputs and a kind of input slope, rise or fall. Let us consider two inputs gate and all possible changes of its input vectors (Fig. 4). It will be convenient write down a table with all possible input vectors and their changes. In this way we obtain all possible driving ways for two input gate.

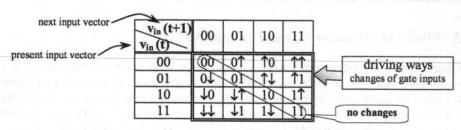

Fig. 4. All possible input vector changes for 2-input gate – *the gate driving ways*

Arrows denote rising and falling edges at the gate input respectively. The diagonal of this table represents no changes of the input vector and all these cases can be treated as one driving way regarding vectors changes at primary circuit inputs. So the total number of driving ways N_{DW} for n-inputs gate can be expressed as follows:

$$N_{DW} = 2^{2n} - 2^n + 1. \tag{2}$$

Let us define *the gate driving way probability* as probability of the occurrence of the particular gate driving way. It can be calculated as the number of particular gate driving way occurrences divided by number of all vector changes at the circuit primary inputs. In this definition a specific distribution of input vectors changes can be taken into account. Additionally this definition includes no change cases and therefore sum of probabilities for all driving ways is one. Driving ways caused by glitches can be counted too. So from this definition driving way probability $p(dw_j)$ of the circuit node j can be expressed by following formula:

$$p(dw_j) = \frac{1}{k} \lim_{k \to \infty} \sum_{i=1}^{k} dw_j(i). \tag{3}$$

And is calculated as a sum of all particular driving ways dw_j, including glitches transition, divided by number of the primary input changes k when its value goes to infinity.

Above defined gate driving way probability can be treated as coefficients, which describe participation of driving way dependent equivalent capacitance in total gate power dissipation. So the total equivalent capacitance of the gate can be calculated as sum of product of particular equivalent capacitance and probability of appropriate driving way considering all possible driving ways of the gate.

3.2 The CMOS Circuit Model of Power Dissipation

Above gate model can be easily applied to logic circuit modelling. Let us consider a node j of a circuit presented in Fig. 5. The node is driven by 2-inputs NAND gate and is loaded by 3-inputs NAND and NOT gates. Interconnections are represented by capacitance $C_{CON}(j)$. Values of capacitors should be calculated as sum of products of a gate particular capacitance and probability of appropriate gate driving way considering all driving ways according to proposed gate model. But input and output capacitance for a gate depends on the same driving ways so we can sum all gate capacitance obtaining one parameter for the gate C_{Lall}. For instance for $g1$ it is sum of $C_{InA}(g1)$, $C_{InB}(g1)$, and $C_{Lint}(g1)$. It can be treated as moving the input capacitors at the gate output. This is in force regarding total power dissipation because capacitors represent energy consumed from supply. So nodal equivalent capacitance $C_{equ}(j)$ can be now calculated as sum of product of particular capacitance c_{Lall} by appropriate driving way of gate which drives the node considering all possible driving ways dw_j:

$$C_{equ}(j) = \sum_{dw_j} c_{Lall}(dw_j) \cdot p(dw_j). \tag{4}$$

Total power dissipation of a circuit is represented by equivalent capacitance obtained by sum of all nodal equivalent capacitance over the circuit.

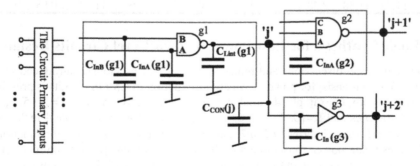

Fig. 5. Dynamic power dissipation model of an example circuit node j

4 Assessment of Energetic Parameters for CMOS Gates

In this paper only dynamic capacitive power dissipation is considered so we need a method to determine all values of particular equivalent capacitance c_{Lall} for a gate. Only capacitive component is needed to measure without quasi-short. It is known that if CMOS gate is driven by input voltages with enough short rising and falling edges the quasi-short cannot occurs. Because of coupling capacitance an overshoot can occurs at the gate output for fast input rising edge. During the overshoot the output voltage is greater than the supply. If the input edge is so fast that PMOS transistor turns off after its linear region without enters saturation before the overshoot finishes than shot-circuit current no occurs. So enough fast input signals are needed and measurements should be done for rising and falling edges separately for all possible input vectors changes. The requirements are much easier to carry out using simulation technique than measuring one. Using measuring circuits similar to these presented in Fig. 1 and Fig. 2 appropriate simulations were done for NOT, 2-, 3-, and 4- inputs NAND and NOR gates. All tested gates were designed in AMIS CMOS 0.7µm C07M-D technology. Netlists for simulations were generated taking into account all parasitic elements coming from layouts. Simulations were done using PSPICE software. Firstly, using Matlab software appropriate input function was generated as PWL – piecewise linear waveform. All possible vector changes were included. So during one simulation cycle all possible driving ways for one gate can be considered. Obtained data were transferred to Matlab and using specially created scripts values of equivalent capacitance were calculated. For instance Table 3 include obtained results for the 2-inputs NAND gate. Negative values in the table are acceptable because equivalent capacitance represents current so it means that current flows to the supply. It is possible due to discharging of coupling capacitance and it occurs when a gate is driven with rising edge at inputs and does not switching.

Table 3. Results of 2-input NAND assessment – values of equivalent internal load and input

	C_{Lint} **[fF]**					C_{InA} **[fF]**					C_{InB} **[fF]**		
AB	00	01	10	11	AB	00	01	10	11	AB	00	01	10	11
00	0	-2.80	-2.47	1.95	00	0	0	6.95	6.56	00	0	6.63	0	7.58
01	2.80	0	0.38	3.69	01	0	0	7.98	7.58	01	0	0	0	0.85
10	2.47	2.72	0	3.72	10	0	0	0	-0.39	10	0	7.23	0	8.06
11	11.20	11.88	8.58	0	11	0	0	0.39	0	11	0	-0.85	0	0

5 Logic Synthesis for Low Power in Two-Level Circuits Design

In this paragraph discussions and remarks on usage of previously introduced model, during logic synthesis, for power dissipation reduction in CMOS circuits is presented. Let us start with an example, 4-variable logic function given by sets: F={1, 5, 7, 8, 9, 10, 15} for ones and R={0, 2, 3, 4, 6, 11, 12, 13, 14} for zeros. Table 4 includes all implicants ($i1 \div i6$), minterms ($m1 \div m7$) and five realizations ($y1 \div y5$) of the function. In order to estimate power consumption of designed circuit and to find minimum power realization, information about input vector changes is needed. Fig. 6 shows an example distribution of 4-input vector changes probability, called *dist16*.

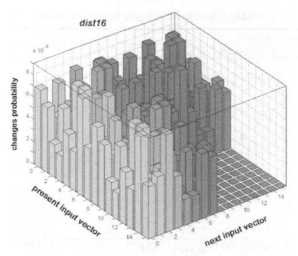

Fig. 6. Example distribution probability of 4-input vector changes

At this point we should note that no each theoretically generated distribution is realizable. At the beginning we consider table as shown in Fig. 6 but including numbers of input vector changes instead of probability. So we can say that given distribution of input vector changes is practically realizable if for any possible input vector the total number of changes from other vectors to this one is equal the total number of changes from the considered vector to others. The $L_{IN}(v_i)$ denotes number of changes from any other vectors to the vector v_i and $L_{OUT}(v_i)$ denotes number of changes from the vector v_i to others. Considering all possible N-inputs vectors $(V=2^N)$ following equality is required for realization of given distribution:

$$\forall_{v_i \in V} L_{IN}(v_i) = L_{OUT}(v_i).$$ (5)

Using probability similar statement can be writing down:

$$\forall_{v_i \in V} p_{IN}(v_i) = p_{OUT}(v_i).$$ (6)

In practical it means that for both number of vector changes and probability sum of i-th column must be equal sum of i-th row in distribution table. The *dist16* satisfies this realizability requirement.

For all minterms and implicants of example function equivalent capacitance was calculated for distribution *dist16* and obtained values were written down to Table 4. The first three realizations ($y1 \div y3$) are minimum area solutions but the others are an effect of disjoint cover. All circuits are built with NAND gates. For each realization equivalent capacitance of the first-level gates was calculated except possible inverters (it was assumed that negated input variables are available). Next gate driving way probability and equivalent capacitance of output gate were calculated for all five realizations. And equivalent capacitances for whole circuits were calculated finally.

The first possibility of circuit optimization comes from analysis of equivalent capacitance values and relies on inputs reordering of gates. This optimization step was done firstly for gates realizing all implicants and minterms.

Table 4. Results of example function synthesis

minterm, implicant	covered vectors 1	5	7	8	9	10	15	C_{equ} [fF]	min area y1	y2	y3	disjoint cover y4	y5
m1	x							7.832					
m2		x						7.743				x	
m3			x					7.761					
m4				x				6.761					
m5					x			6.821					x
m6						x		6.761					
m7							x	6.763					
i1	x			x				8.099	x	x		x	
i2	x	x						9.086		x	x		x
i3			x	x				9.456	x				
i4		x					x	7.991	x	x	x	x	x
i5					x	x		8.125	x	x	x	x	x
i6				x	x			7.993			x		
equ.cap of 1-level gates $C_{equ\_1\text{-}lev}$:									33.671	33.301	33.195	31.957	32.024
profit:									0.00%	1.11%	1.43%	5.36%	5.14%
equ.cap of output gate C_{equ_out}:									7.9126	7.9082	7.8183	7.8123	7.5947
profit:									0.00%	0.06%	1.21%	1.28%	4.19%
total equ.cap of circuit C_{equ_tot}:									**41.583**	**41.209**	**41.013**	**39.770**	**39.618**
profit:									**0.00%**	**0.91%**	**1.39%**	**4.56%**	**4.96%**

Realizations of the 1-level can be divided into two groups, minimum area where each circuit consists of four 3-inputs gates and the second group with three 3-inputs and one 4-inputs gates. It can be observed that if equivalent capacitance of 1-level gates is the lowest in the group that the output gate equivalent capacitance is the lowest too. The equivalent capacitance is sum of product of particular driving way probability and appropriate values of gates equivalent capacitance. So if total equivalent capacitance is lower for the same realization it means that in the sum of products some values of probability are lower than in other cases. Values of particular equivalent capacitance are the same because each function realization in the group has the same gates. Lower driving way probability of a gate causes lower activity of the gate output, in consequence lowering driving way probability of the second level gate – function output. The second observation is that disjoint cover is better because simultaneous and compatible changes at two or more inputs of the output gate are avoided.

Based on above presented conclusions and authors experiments following procedure can be proposed for design of two-level circuits with reduced power dissipation:

1) all implicant of the function generation,
2) calculation of equivalent capacitance for all implicants and minterms for given probability distribution of input vector changes,
3) optimization of each gate using gates inputs reordering,
4) generating of all disjoint covers of the function,
5) covers sorting by realization cost (number and kind of used gates),
6) evaluation of equivalent capacitance of 1-level gates for each cover,
7) choosing realization with minimum equivalent capacitance from each group,
8) calculation of gate driving way and equivalent capacitance of the output gate for each above chosen realization,
9) selection of the best realization.

6   Conclusions

New model of dynamic power dissipation in CMOS circuits and its usefulness for power dissipation reduction is presented. Some spatial correlations between inputs variables are taken into account. Presented new idea should be treated as a start point for further investigations. However, obtained results are promising and show possibility of a few percent of power dissipation reduction. Proposed procedure needs further improvements to avoid generation of all implicants of a function, which is quite time consumed especially for bigger functions.

References

1. Roy, K., Mukhopadhyay, S., Mahmoodi-Meimand, H.: Leakage Current Mechanisms and Leakage Reduction Techniques in Deep-Submicrometer CMOS Circuits. Proceedings of the IEEE, Vol. 91, No. 2. (2003) 305-327
2. Wang. Q., Vrudhula. S. B. K., Yeap. G., Gangulu. S.: Power Reduction and Power-Delay Trade-Offs Using Logic Transformations. ACM Trans. On Design Automation of Electronic Systems, Vol. 4. No. 1. (1999) 97-121
3. Iman, S., Pedram, M.: Logic Synthesis for Low Power VLSI Designs. Kluwer, Boston (1998)
4. Tseng, J-M., Jou, J-Y.: Two-Level Logic Minimization for Low Power. ACM Trans. On Design Automation of Electronic Systems, Vol. 4. No. 1. (1999) 52-69
5. Theodoridis, G., Theoharis, S., Soudris, D., Goutis, C.: A New Method for Low Power Design of Two-Level Logic Circuits. VLSI Design Journal of Custom-Chip Design, Simulation, and Testing, Vol. 9. No. 2. (1999) 147-158
6. Brzozowski, I., Kos, A.: Modelling of Dynamic Power Dissipation for static CMOS Gates and Logic Networks. Proc. 10th MIXDES Conf., Łódź Poland (2003) 336-341
7. Brzozowski, I., Kos, A.: Simulation Methods for Estimation of Dynamic Power Dissipation in Digital CMOS Circuits. Proc. 11th MIXDES Conf., Szczecin Poland (2004) 435-440
8. Brzozowski, I., Kos, A.: New Idea of Objective Assessment of Energy Consumed by Real VLSI Gates. Proc. the 9th MIXDES Conf., Wrocław Poland (2002) 579-584

Energy-Efficient Value-Based Selective Refresh for Embedded DRAMs

K. Patel[1], L. Benini[2], Enrico Macii[1], and Massimo Poncino[1]

[1] Politecnico di Torino, 10129-Torino, Italy
[2] Università di Bologna, 40136 Bologna, Italy

Abstract. DRAM idle power consumption consists for a large part of the power required for the refresh operation. This is exacerbated by (i) increasing amount of memory devoted to cache, that filter out many accesses to DRAM, and (ii) increased temperature of the chips, which increase leakage and thus data retention times. The well-known structured distribution of zeros in a memory, combined with the observation that cells containing zeros in a DRAM do not require to be refreshed, can be constructively used together to reduce the unnecessary number of required refresh operations. We propose a value-based selective refresh scheme in which both horizontal and vertical clusters of zeros are identified and used to selectively deactivated refresh of such clusters. As a result, our technique significantly achieves a net reduction of the number of refresh operations on average of 31%, evaluated on a set of typical embedded applications.

1 Introduction

Embedded DRAM (EDRAM) is viewed as viable design option for applications with significant memory requirements, tight performance constraints and limited power budgets. Embedded DRAM has lower density, requires a more expensive mask set and fabrication process, but it offers a drastically improved energy-per-access [ ]. The energy efficiency of EDRAMs advantage may be reduced, or even worse, compromised, if adequate countermeasures are not taken to address its idle power consumption, caused mainly by the periodic refresh operation. Refresh is a more serious problem for EDRAMs than for standard DRAMs for two main reasons. First, technology options to reduce cell leakage cannot be as aggressively pursued in EDRAM as in standard DRAMs, because of cost reasons, and fundamentally as a consequence of the tradeoff between efficient logic and efficient memory. Second, the presence of fast switching logic on the same die causes higher-temperature operation, which increases leakage, thereby implying an increase in refresh rate.

From the architectural viewpoint, faster refresh rates imply larger idle power for EDRAM. The importence of idle power is magnified by the fact that DRAMs are often used with a low duty cycle of busy periods, since DRAM accesses are filtered by the higher levels of the memory hierarchy (i.e, fast and small SRAM based caches). For these reasons, several researchers have proposed techniques

V. Paliouras, J. Vounckx, and D. Verkest (Eds.): PATMOS 2005, LNCS 3728, pp. 466–476, 2005.

for idle power minimization in EDRAM memories [ ], which are also applicable to generic DRAMs [ ]. Most of these techniques aim at providing very low-power *shutdown* states, either with loss of information or with significant access time penalty. Alternative approaches aim at reducing power less aggressively than with shutdown, while at the same time minimizing access time overhead [,].

In this paper, we propose a low-overhead refresh power reduction technique based on the concept of *selective refresh*. We exploit the well-known dominance of zero bits in the data stored in DRAM, and by adding a limited amount of redundant storage and logic (which is the overhead in our technique) to index the memory blocks that contain a zero value, so as to eliminate refresh to these blocks. As a result, our technique significantly reduces the number of refresh operations, decreasing idle power. One important design exploration parameter is the granularity at which we apply zero value detection and tagging. In this work, we propose two alternative mechanisms, namely horizontal and vertical zero clustering and we analyze the granularity at which they can be applied. Our results, demonstrate that an average reduction of 31% of the refresh operations, measured for different granularities.

2   Previous Work

The non-uniform distribution of values in memories have been exploited in several ways, although this property has been mainly used in the context of caches, with the objective of reducing the average access time or the total energy by reducing the cost of memory reads using the "common-case" principle.

The frequent value (FV) cache [ ] is based on the analysis of application data usage, that allows to identify few frequently accessed data values; these are stored in a small buffer, resulting in frequent access of the small buffer. The experimentally observed dominance of zeros in a cache has been exploited by the value-conscious cache [ ], where a sort of hierarchical bitline scheme is used to avoid discharging of the bitline whenever a zero value is stored. A similar approach is used in the dynamic zero compression (DZC) architecture [ ], where, zero bytes are encoded using one bit, achieving energy reduction while accessing zero bytes in cache.

Concerning architectural techniques that aim at reducing the energy impact of refresh, Ohsawa et al. [ ] propose two refresh architectures. The first one, called *selective refresh*, is based on the observation that data need to be retained (and thus refreshed) only for the duration of their *lifetimes*. The difficulty in the implementation of this architecture lies in that it is not immediate to extract this information, which may require some support by the compiler. The second architecture, called *variable refresh period*, uses multiple refresh periods for different regions of the array, based on the fact that the data retention time of the cells is not constant. This property was first exploited at the circuit level [ ], and has been refined in [ ] by employing a refresh counter for each row, plus a register that stores the refresh period of a given row. This idea has been elaborated into a more sophisticated scheme in [ ], where the idea variable refresh period is applied on a granularity (a "block") smaller than a row.

3   Value-Based Selective Refresh

3.1   Motivation

The technique proposed in this paper is motivated by the observation of two properties of DRAM memory systems. The first one is a consequence of the fact that, since most of the memory references are filtered out by caches, only few accesses go to main memory (normally a DRAM). This causes the contribution of idle power consumption to become dominant, since refresh is a mandatory operation. The plot in Figure 1 endorses this fact: it shows, for a set of benchmarks, the split between refresh and access energy as a function of the refresh period, referred to a system with 16KB of L1 Cache and no L2 cache. We notice that the relative importance of refresh becomes dominant for refresh periods around few million cycles; assuming a 200MHz frequency typical of a SoC, this figure is equivalent to a few tens of ms, comparable to the refresh periods of common EDRAM macros [2, 3]. Notice that the addition of a L2 cache would make refresh energy even more dominant.

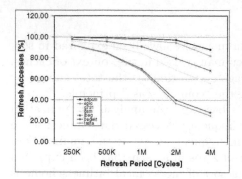 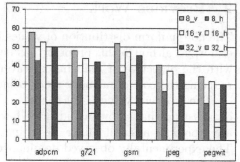

Fig. 1. (a) Refresh vs Access Energy Split (b) Distribution of Zero Clusters

The second property is that memories, in general, tend to exhibit structured distribution of 0's and 1's. This also holds for DRAMs, that besides the classical high *horizontal* occurrence of 0's (e.g., a large number of zero bytes), exhibit an even more relevant *vertical* occurrence of 0's. We will use the term *clusters* to denote these subsets of rows or columns.

Figure 1-(b) shows the occurrence frequency of 0's in a DRAM, relative to a set of benchmarks. Values are relative to different granularities (8, 16, and 32) of 0-valued clusters, either vertically or horizontally. The plot shows that, while the number of horizontal zero clusters decreases quite rapidly as the cluster size increases, vertical clusters are more frequent and do not decrease much for larger cluster sizes: on average, 38% of the vertical 32-bit clusters contain all zeros.

Our idea is to use the latter property to reduce the number of refresh operations by observing that *cells containing a zero do not need to be refreshed*. Since they can be grouped into clusters, it is possible to avoid the refresh of an entire cluster. In other words, we transform the operation of refresh, that is is done independent of the value contained in the cell, into a *value-dependent* operation.

From the architectural standpoint, our *value-based selective refresh* consists of grouping 0's in clusters. The information regarding the value of each cluster is stored in an extra DRAM cell. This cell stores a one if all bits in the cluster are zero, and zero otherwise. From the observation of the data of Figure 1-(b), clustering of zeros can be either horizontal or vertical, and hence, in this paper, we investigate two different approaches to clusters zero cells.

3.2   Horizontal Zero Clustering(HZC)

Our approach to cluster zeros in the horizontal direction is similar to one proposed by Villa *et. al.* [ ], in which clustering of zeros is exploit to reduce dynamic power dissipation in caches (SRAM). Figure 2-(a) shows the conceptual architecture of HZC. Each word line is divided in number of clusters each one having its *Zero Indicator Bit (ZIB)*. The ZIBs are placed in extra columns in the cell array, depicted as vertical gray stripes in Figure 2-(a). As shown in Figure 3-(a), depending on the content of ZIB cell, the local word line of the corresponding cluster can be disconnected from the global word line. The operations of the memory with HZC architecture can be summarized as follows.

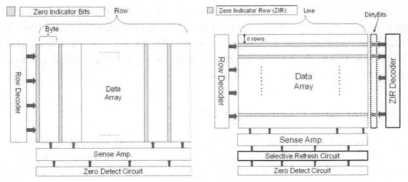

Fig. 2. Architecture of (a) Horizontal and (b) Vertical Zero Clustering

- **Refresh:** During the refresh operation, the local wordline of the cluster is connected or disconnected from the global wordline based on the content of ZIB cell. Figure 3-(a) shows that refresh is not performed if the ZIB is one (M2 is turned off, and M1 is turned on, grounding the local word line).
- **Read:** Read operation similar to refresh. During read, the ZIB is read and depending on its value, the bits in the cluster will be read or not. If ZIB stores a one, then the bits in the cluster are not read (we know that they are all zero), whereas if it stores a zero they will be read out.
- **Write:** During write operation the ZIB bit is updated when its cluster is written. The *Zero Detect Circuit* (ZDC) will detect if all cluster bits are zero; if so, a '1' is written into the ZIB. The design of ZDC is very similar to the to the one found in [ ], and thus it is not reported here.

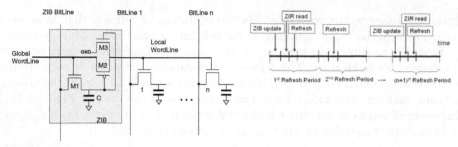

Fig. 3. (a) ZIB circuit [5] (b) Refresh Period Operations : VZC Architecture

Notice that during the read and refresh operations the ZIB is read; since the read operation is destructive, the proposed scheme imposes a small read time overhead. When the ZIB is '0', the cluster bits have to refreshed along with ZIB, during the refresh operation. Referring to Figure 3-(a), we notice that, during refresh, the bitline is pre-charged to $V_{dd}/2$ thus partially charging the capacitance C of the ZIB and hence possibly turning transistor M2 off and hence cutting off the local wordline of rest of the n bits. If we wait to restore the value of ZIB, the local wordline will again be active and connecting the cluster bits to their bitlines.

When the ZIB is '1', this indicates a zero cluster. This will cut off the local wordline of the cluster during read. Thus, the sense amplifiers of the cluster's columns will remain in a meta-stable state. To avoid this problem, the sense amplifier design has to be modified, as done in [ ] (the modified circuit is not shown here for space reasons).

3.3   Vertical Zero Clustering(VZC)

Vertical Zero Clustering(VZC) aims at detecting and exploiting the presence of clusters of zeros in columns of the DRAM array. Figure 2-(b) shows the conceptual architecture for VZC. Depending on the granularity of clustering every n rows will have one *Zero Indicator Row* (ZIR). Each ZIR contains one ZIB for each column in the DRAM array.

Since we add one ZIR for n rows, we need a separate decoder for ZIRs, which will use higher-order address bits to access it, depending on the granularity. We also add an additional column to the array, containing a set of *dirty bit* indicators, one for each ZIR. These bits will be used to track writes and to ensure that the ZIB bits are correctly updated (as discussed in detail later). Similarly to HZC, VZC also requires support circuitry for each column of the array. These circuits are shown in Figure 4-(a) and Figure 4-(b), that are used in different moments by the different operations on the DRAM. Memory operations in the VZC can be summarized as follows.

- **Read:** Read is performed as a normal read, the only difference being the presence of the circuit depicted in Figure 4-(a).

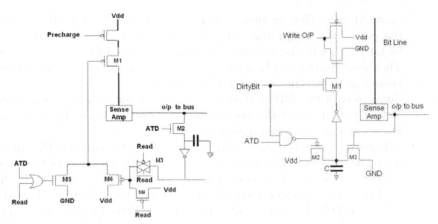

Fig. 4. (a) Selective Refresh Circuit: VZC (b) Write Circuit of ZIB

During read operation, the *Read* signal is held high, ensuring that the transistor M1 is turned on. Then, a normal pre-charge takes place, followed by the read of the desired cell. Notice that the ATD signal in the figure is used only during refresh to selectively disable the pre-charge of bitlines.

– **Write:** Write is also done in the usual way. Notice that as long as data is just read from memory, the status of the clusters will not change. Conversely, when a datum is written into the memory, it may modify the status of the clusters; therefore, the status of the ZIB must be changed accordingly. This may be a problem, since, in order to change the status of the ZIB, all rows of the cluster must be read. Rather, we avoid this overhead by postponing the ZIB update operation until the next refresh operation by zeroing the ZIB of the cluster corresponding to the current write address. This operation also sets the dirty bit of that ZIR by writing '1' to it. Based on the value of the dirty bit, the next refresh operation of this cluster will determine the latest status of ZIBs.

– **Refresh:** Refresh operation has two modes. One is the *normal* refresh operation where zero clusters are not refreshed and the other one is the *ZIB update* mode.

  • *Normal Refresh:* Before starting the refresh of the first row of the cluster, corresponding ZIR is read. Reading of ZIR is triggered by the *Address Transition Detection (ATD)* signal. The ATD signal goes to one every n rows that are refreshed, that is, when we cross the boundary of a ZIR. The ATD signal triggers the ZIR read operation using the ZIR decoder, shown in Figure 2-(b). As shown in Figure 4-(a), ATD will turn on transistor M2 and depending on the value of ZIB the capacitor will be charged or discharged. At the end of the read operation on ZIR, ATD will go low.
  If the ZIB is '1', a '0' will be produced at the output of inverter of Figure 4-(a). The read signal will be held low during the refresh operation, so that the transmission gate M3 will be turned on, putting also tran-

sistor M6 on. This, in turn, will turn the transistor M1 off cutting of the pre-charge from bitline. Hence if the ZIB is '1' the pre-charge will be for that particular column, will remain disabled for the next n rows. Therefore, during row-based refresh operation, the bits of this column belonging to the cluster will not be refreshed.

Conversely, if the ZIB is '0' this results in discharging the capacitance, forcing the output of the inverter to '1'. This will turn M6 off and and M1 on, so that a normal refresh will occur.

- *ZIB update mode:* During the ZIB update mode, the status of the clusters has to be determined to update the value of ZIB. This part of the circuit is shown in Figure 4-(b). As explained above, before starting to refresh of the first row of a cluster, the content of the ZIR for that cluster is read by raising ATD high. Reading ZIR will read the dirty bit corresponding to that ZIR. If the dirty bit is is set (i.e., the status of the ZIR is unknown) this will turn transistor M2 on using the NAND gate. This will result in charging the capacitance C. This will, in turn, result in putting '1' at the Write Output, using transistor M1. All these operations occur when the ZIR is read. Assuming ZIR is reset and its dirty bit is set, the regular row based refresh will follow the ZIR read operation.

  Notice that the ATD signal will go low before the row-based refresh starts. During refresh if any of the bits of this given column is '1', then it will put transistor M3 on and it will ground all the charged capacitances, setting output of the inverter to '1'. This will make Write O/P to go to '0'. For those ZIRs which have the dirty bit set, the value of Write O/P will be written back to the corresponding ZIR at the end of the refresh of all the rows of the cluster. The end is actually detected, again, by ATD, since after the refresh of the last row of the cluster it will move to the first row of the next cluster.

The sequence of operations occurring during refresh of the VZC DRAM is summarized in Figure 3-(b). When refreshing the first row of the cluster, refresh operation will be performed after ZIB update and ZIR read. The following $n-1$ row refresh operations will be normal refresh operations. Notice that since the content of the ZIR is built based on the content of DRAM cells, it has to be refreshed as well. This is done when *reading* the ZIR, and does not require an explicit refresh.

3.4   Write Impact on Refresh

In the VZC architecture, whenever a write comes, it resets the ZIR of the cluster to which this write belongs and sets the dirty bit. Hence, on the next refresh, this cluster will have to be refreshed. If during the same refresh period another write comes to the same cluster then it will not change the status of ZIR since this cluster has already been declared as "dirty". Instead, if the write goes to another cluster it results in destroying another cluster by resetting its ZIR. Hence, on the next refresh this cluster will have to be refreshed as well. If many writes are distributed over different clusters this will jeopardize the opportunities to save refresh to these clusters. This is also strictly related to the cluster size. Experiments show that as we move towards coarser granularity, the percentage of dirty clusters increases.

This is due to the fact that, even though distribution of writes to different clusters is reduced, the total number of clusters is reduced. In general, however, this percentage remains quite small and hence, dirty writes do not reduce significantly the impact of VZC.

4   Overhead Analysis

Both HZC and VZC architectures have some area and timing overhead. Here we briefly discuss it with approximate quantification.

Concerning area, Table 1 summarizes different components contributing to area overhead for HZC and VZC architectures for different cluster granularities n. The percentage overhead is with respect to the area of that component in a regular, non-clustered DRAM.

Table 1. Relative Percentage Area Overhead

	HZC			VZC		
Components	$n=8$	$n=16$	$n=32$	$n=8$	$n=16$	$n=32$
Data Matrix	37.5	18.7	9.3	12.5	6.25	3.12
Bitline	12.5	6.25	3.12	100		
Wordline	100			12.5	6.25	3.12
Sel. Refresh	negligible constant overhead					
Row Decoder	No Overhead			9–10	3–4	1–2

- *Data Array:* For the data array in HZC architecture every n bits we have three extra transistors (2 n-MOS and 1 p-MOS, Figure 3-(a)), i.e., an overhead of $3/n$. In the VZC architecture we have an extra additional row for every n rows, hence an overhead of $1/n$. Notice that in HZC architecture the p-MOS transistor drives the local wordline, and depending on the granularity of the cluster, it has to be wide enough to drive it without introducing an extra delay in reading.
- *Bitlines:* In the HZC architecture, we have an extra bitline for every n bitlines, while in the VZC architecture we have an extra wire running parallel to every bitline (Figure 4-(a)). Though this wire is not an extra bitline, for the sake of the overhead calculation we considered it as a bitline overhead.
- *Wordlines:* Due to divided wordline type of architecture (Figure 3-(a)) of HZC, we have extra local wordlines, which has total length per row approximately equal to the length of the global wordline. In the VZC architecture we have an extra row for every n rows.

- *Row Decoders:* While the HZC architecture does not need an extra row decoder, the VZC architecture has an extra row decoder for decoding ZIRs. Though the complexity of this extra row decoder is significantly smaller than the main row decoder, its cost depends on n. As shown in the table, with respect to the regular row decoder this extra row decoder has marginal overhead. And its contribution to overall area overhead is very small, since the dynamic logic based decoders themselves have complexity of 5 to 8 % with respect to data matrix (considering transistor counts).

Delay, on the contrary, is slightly more critical for read operations. For the HZC architecture read operation has a slight increase in delay since the ZIB has to be read out to determine the value of the rest of the n bits. Whereas, in case of VZC architecture, read operation is performed in the normal way, and hence there is no visible increase in delay. Concerning the write operation, in the HZC scheme during each write the zero detect circuit has to determine if there are zero clusters or not, before committing the write to the lines. Hence there is an increase in write time, determined by the delay of the zero detect circuit. Conversely, in the VZC architecture, write operation is carried as normally, followed by the resetting ZIR and the setting of dirty bit. Hence, there is no sizable delay increase during the write operation as well. Overall, the overhead of the VZC architecture is smaller, and, even more important, it does not impact normal read and write operations. This fact, coupled with the statistics of Figure 1-(a) seems to make VZC a more competitive architecture than HZC.

5   Experimental Results

For our experiments we have used a modified version of the SimpleScalar-3.0/PISA [ ]. All our experiments are run using `sim-outorder`. We configured simulator to have separate L1 Instruction (direct mapped) and Data cache (4-way set associative), both of size 16KB with 32-byte block size. L2 cache has been disabled, since the relatively limited execution time of the applications did not allow to see sufficient traffic towards the DRAM.

During all our experiments we have kept data retention time of a DRAM cell to be one million CPU cycles. Assuming a 200MHz frequency of a typical SoC, this is equivalent to 5 milliseconds. We have used the MediaBench suite [ ], which includes various multimedia, networking and security related applications. Most of the benchmarks have separate encoding and decoding applications.

Figure 5 plots the percentage of refreshes avoided by two the HZC and VZC architectures, for different granularities of cluster. The plots correspond to encoding and decoding applications of the MediaBench benchmarks, Notice that the reported values already account for the refresh overheads brought by HZC and VZC, and are in fact equivalent to reductions of refresh energy.

In the plots, x_v represent the relative savings brought by VZC architecture, where x is the granularity of horizontal (h) or vertical (v) clustering. As can be seen from plots, at the byte granularity both VZC and HZC bring almost the same percentage of savings, but as we move towards granularity of 16 and

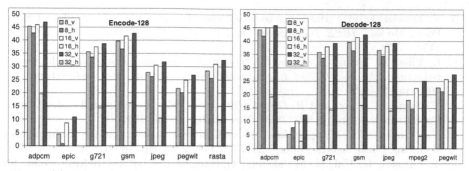

Fig. 5. (a) Relative Refresh Energy Savings in Encoding (b) Decoding applications

32 bits, the dominance of VZC architecture becomes visible. As the plots show, savings with VZC architecture for granularities of 8,16 and 32 bits are not too different, whereas in case of HZC architecture the difference is large.

The average savings for the best architectures are 26.5% for HZC (cluster size = 8) and 33% for VZC (cluster size = 32). Notice that VZC with cluster size of 32 provides the best results, due to much smaller overhead.

6   Conclusions

In this paper, we have proposed two value-conscious refresh architectures suitable for embedded DRAMs. Based on the observation that zeros do not need to be refreshed, we group bits into clusters to avoid refresh of entire cluster. We have explored clustering in both horizontal and vertical directions, and various cluster sizes. Our experiments show that as we move towards higher granularity vertical clustering become more effective than horizontal one. Due to smaller overhead, for higher granularity vertical clustering offers substantial advantage. Experimental results show that the best overall architecture, that is, vertical clustering with cluster size of 32 provides a 33% reduction of refresh energy, evaluated on a set of embedded multimedia applications.

References

1. D. Keitel-Schulz, N. Wehn, "Embedded DRAM Development: Technology, Physical Design, and Application Issues," *IEEE Design and Test*, Vol. 18, No. 3, pp. 7–15, May 2001.
2. C.-W. Yoon et al., "A 80/20MHz 160mW Multimedia Processor integrated with Embedded DRAM MPEG-4 Accelerator and 3D Rendering Engine for Mobile Applications," *ISSCC'04*, pp. .202- 522, Feb. 2004.
3. R. Woo, et al. "A Low-Power Graphics LSI integrating 29Mb Embedded DRAM for Mobile Multimedia Applications," *ASPDAC'04*,pp. 1758–1767, Feb. 2004.
4. F Morishita, et al., "A 312MHz 16Mb Random-Cycle Embedded DRAM Macro with 73/spl mu/W Power-Down Mode for Mobile Applications," *ISSCC'04*,pp. .202- 522, Feb. 2004.

5. V. Delaluz, et al., "Hardware and Software Techniques for Controlling DRAM Power Modes," *IEEE Transactions on Computers*, Vol. 50, No. 11, Nov. 2001, pp. 1154 - 1173.

6. L. Villa, M. Zhang, K. Asanoivc, "Dynamic zero compression for cache energy reduction," *Micro-33: 33rd International Symposium on Microarchitecture*, Dec. 2000, pp. 214–220.

7. Y. Zhang, J. Yang, and R. Gupta, "Frequent Value Locality and Value-Centric Data Cache Design," *ASPLOS'00*, Nov. 2000, pp. 150–159.

8. Y.J. Chang, C.L. Yang, F Lai, "Value-Conscious Cache: Simple Technique for Reducing Cache Access Power," *DATE04*,Feb. 2004. pp. 16–21.

9. T. Ohsawa, K. Kai, K. Murakami, "Optimizing the DRAM Refresh Count for Merged DRAM/Logic LSIs," *ISLPED'98*,Aug. 1998, pp. 82–87.

10. Y. Idei, et al., "Dual-Period Self-Refresh Scheme for Low-Power DRAMs with On-Chip PROM Mode Register," *IEEE Journal on Solid-State Circuits*, Vol. 33, No. 2, Feb. 1998, pp. 253–259.

11. J. Kim, M.C. Papaefthymiou, "Block-Based Multiperiod Dynamic Memory Design for Low Data-Retention Power," *IEEE Transactions on VLSI Systems*, Vol. 11, No. 6, Dec. 2003, pp. 1006–1018.

12. SimpleScalar home page, http://www.simplescalar.com/

13. C. Lee, M. Potkonjak, W. Mangione-Smith, "MediaBench: A Tool for Evaluating and Synthesizing Multimedia and Communications Systems", *International Symposium on Microarchitecture*, Dec. 1997, pp. 330–335.

Design and Implementation
of a Memory Generator for Low-Energy
Application-Specific Block-Enabled SRAMs

Prassanna Sithambaram, Alberto Macii, and Enrico Macii

Politecnico di Torino, Torino, Italy 10129

Abstract. Memory partitioning has proved to be a promising solution
to reduce energy consumption in complex SoCs. Memory partitioning
comes in different flavors, depending on the specific domain of usage and
design constraints to be met. In this paper, we consider a technique that
allows us to customize the architecture of physically partitioned SRAM
macros according to the given application to be executed. We present
design solutions for the various components of the partitioned memory
architecture, and develop a memory generator for automatically generating layouts and schematics of the optimized memory macros. Experimental results, collected for two different case studies, demonstrate the
efficiency of the architecture and the usability of the prototype memory
generator. In fact, the achieved energy savings w.r.t. implementations
featuring monolithic architectures, are around 43% for a memory macro
of 1KByte, and around 45% for a memory macro of 8KByte.

1 Introduction

In current systems-on-chips (SoCs), memory takes, on average, more than half
of the silicon real estate. Since avoiding expensive external memory accesses is
a key objective for high-performance system-level design, these on-chip memories bear the brunt of memory accesses by the processor cores. With memories
accounting for the largest share of power consumption in SoCs, an emphasis has
been placed on the design of low power memories. Since process compatibility
and fast operation make SRAMs the favored choice for small-sized on-chip memories, this work focuses on a new low-power architecture for embedded SRAMs.
In particular, it describes the design and implementation of an automatic layout
generator for the new architecture.

One of the most successful techniques for memory energy optimization is
partitioning. The rationale of this approach is to sub-divide a large memory into
many smaller memories that can be accessed independently. Energy-per-access
is reduced because on every access only a single small memory consumes power,
while all the others remain quiescent. The partitioning approaches aim at finding
the ideal balance between the energy savings (accruing due to localized access
of small memory banks) and the energy, timing and area overhead (due to the
control logic and wiring needed to support the multiple partitions).

Partitioning of the memory can be at two levels: *Logical* and *physical*. The
former involves instantiating several smaller memory macros instead of the original, monolithic one, and then synthesizing the control logic to activate one

V. Paliouras, J. Vounckx, and D. Verkest (Eds.): PATMOS 2005, LNCS 3728, pp. 477–487, 2005.

macro at a time [-]. In the latter approach, the internal circuitry of the memory macro is modified to enable mutually exclusive activation of sub-blocks in the cell array [,]. In physical partitioning, the block select logic is merged with the word-line decoder, unlike logical partitioning. It also avoids duplication of precharge and read-write circuitry, as long as the blocks are in the same *segment* [ ] or partition, and it is thus proposed as a successful approach for implementing energy-efficient memories. These memories are targeted for general purpose applications and are composed of optimal *uniform-sized* partitions. Logical partitioning approaches, however, have also attempted to use the predictability and non-uniformity of memory access traces in embedded software to tailor the number and size of partitions to a particular application [ ], and achieved good results. The authors of [ ] call this the *MemArt* approach.

Our proposal combines the benefits of physical partitioning with application-specific tuning of the partitions. The work of [ ] introduces energy models for such an architecture. Based on these models, the authors formulate and solve the problem of optimally choosing the granularity of the partitions and the boundaries for a given address trace, where the optimization objective is to minimize energy under some given time constraints. Theoretical results show how the application-specific physically-partitioned memories (referred in [ ] and in the sequel as *Application-Specific Block-Enabled* (ASBE) memories) outperform their general-purpose counterparts, and how this solution is superior to application-specific logical partitioning, thanks to its reduced overhead.

Here, we propose the design and implementation of a prototype memory generator that creates ASBE memories, taking care to keep power consumption of the peripheral circuitry to a minimum. In its preliminary implementation, the memory generator design is equipped to handle memory sizes upto 8KByte (256 rows by 256 columns) and word sizes from 8 to 32 bits, although its extension to handle larger memory macros does not present major difficulties and it is currently underway in an industrial setting. It uses a modified *MemArt* algorithm to generate the partitions and provides complete flexibility in the number of partitions to be included in the memory macro.

We demonstrate the feasibility of the proposed ASBE memory architecture by using the memory generator to realize two different memory macros for the STMicroelectronics 130nm HCMOS9 process; the first is a 1KByte (256 x 32) SRAM, customized for an application running on an ARM core, the second is a 8KByte (256 x 256) SRAM for a program executed by a LX-ST200 VLIW processor. The average energy savings w.r.t to monolithic implementations are around 43% and 45% for the 1KByte and 8Kbyte SRAM macro's respectively.

2   Previous Work

Constructing a memory using a memory module generator usually involves creation of a "basic" memory block, which is further combined architecturally to form low-power memory configurations [ ].

Memories designed for general purpose systems, supporting an assorted mix of applications, need to be partitioned into uniform, optimally sized partitions. For memories in embedded SoCs, however, the access patterns can typically be profiled at design time. Since these access patterns are non-uniform, a partitioned architecture with the most frequently accessed addresses concentrated in the smaller banks can be developed with great energy-delay savings [ ]. In order to reduce the number of memory blocks, and to improve the percentage accesses in the smaller blocks, application-specific partitioning may require a *re-ordering* of the memory addresses [,]. On-the-fly reordering will originate a clustered access trace that will yield a smaller number of partitions with improved access frequency for the smaller blocks and predominantly dormant larger blocks. Because of its considerable advantages over the uniform partitioning approach, we will focus on application-specific partitioning, as done in [ ].

Logical partitioning is based on the precept that smaller memories have smaller access cost than larger ones. This scheme requires control circuitry to decode the address inputs and select the appropriate block, and some wiring and area overhead in placing the blocks. So the partitioning should strike an optimal balance between the energy savings from having smaller blocks and the overhead involved in supporting them. In [ ], starting from the execution profile of an embedded application running on a given processor core, the authors synthesize a multi-banked SRAM architecture optimally fitted to the profile. Results on a set of benchmark applications run on an ARM processor showed that the application-specific, logical partitioning approach provides significant power reductions (34% on an average). However, there are a few crucial avenues of improvement which we seek to address.

- In case we need to perform address clustering, reordering of addresses will require additional logic to do the necessary address swapping. This additional logic adds power, delay, area and wiring cost to the existing overhead. If the reordering is significant, it can seriously offset the achievable energy benefits.
- Every memory bank is a complete memory block in itself, possessing its own decoders, precharge, sensing and column multiplexing circuitry. This peripheral circuitry accounts for almost half the power consumed by SRAMs. Having multiple banks will cause redundancy of this circuitry, and greater consumption of power (with smaller blocks having higher overhead-per-block).

Both the above factors limit severely the extent of partitioning, and provide a tight upper bound on the number of partitions. Physical partitioning, on the other hand, involves creating partitions using bit-line and word-line division.

- A hierarchical divided bit-line (DBL) is presented in [ ] by Karandikar and Parhi. Multiple rows of 6T SRAM cells are grouped together hierarchically to form blocks. Each block's bit-lines are connected to the Global Bit-line through pass-transistors controlled by the Block Select signal. By thus reducing the capacitance on the bit-lines, active power and access time are reduced by about 50% and 20% respectively, at a penalty of about 5% extra transistors. This technique partitions the array horizontally.

- In [ ], Yoshimoto *et al.* describe a divided word-line (DWL) structure to divide the Word-Line (WL) and select it hierarchically with minimal area penalty. This division results in only the required bit-lines being selected and precharged, thus saving considerable column current. It results in decrease in word-selection delay. This technique partitions the memory array vertically.

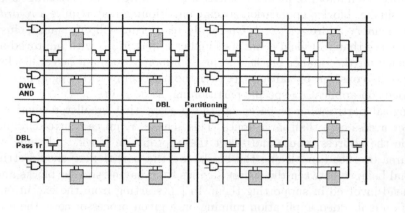

Fig. 1. Physical Partitioning using DBL and DWL

Using a combination of the above two techniques, it is possible to create a memory array partitioned horizontally and vertically as shown in Figure 1. Together, they provide the framework within which physically partitioned memories optimally designed for specific applications are implemented.

3   ASBE Memory Macros

The two shortcomings of logical partitioning were duplication of peripheral circuitry and cost of control logic for address reordering. To minimize these, and for memory partitioning to yield maximum benefits, it is necessary that the internal circuitry of the SRAM be tailored to the application. This would avoid unnecessary decoding and, as it will be shown later, enable us to perform address clustering at no cost for some addresses, and at a little cost for others.

The DBL and DWL physical partitioning techniques are general purpose methods used to generate the "basic" memory block in [ ]. As such, they are applied to originate uniform partitions of the memory cell array. The distinguishing feature of the ASBE approach of [ ] is that the DBL+DWL-based physical architecture of the memory is driven by the knowledge of the memory access profile in such a way that the memory partitions fit around the addresses that are mostly required, following the original *MemArt* principle described in [ ] and properly modified in [ ] to account for the different costs of the various components of the memory architecture. More specifically, the modified *MemArt* algorithm, which is in charge of determining the optimal partitions of a ASBE memory macro based on the clustered access profile, requires accurate models for the evaluation of access time and energy consumption of the ASBE memory architecture. Such models were derived in [ ] taking those used in Cacti [ ] as the starting point.

The theoretical results of [ ] (energy savings for the ASBE memory architecture were evaluated using the aforementioned models) showed significant promise in terms of achievable energy reductions. However, making ASBE memories usable in practice calls for the availability of an automatic memory generator, whose purpose would be to take, as inputs, the aspect of the memory and the partitions generated by the modified *MemArt* algorithm, and generate the memory array and the corresponding peripheral logic, including also the address reorderer as part of the decoder.

The next sections outline a possible structure for the various parts of the ASBE memory architecture, conceived in order to achieve minimum energy consumption, and details on the design and implementation of a prototype tool for automatic layout generation of such an architecture.

4   ASBE Memory Architecture

The ASBE memory architecture consists of the following major components: The cell array, the row decoder (including also the address reorderer and the block selector), the read circuitry, the precharge logic, the column multiplexers and the write circuitry. Details on our implementation of such components are provided in the remainder of this section.

4.1   Cell Array

The cell array consists of the following leaf cells:

- A 6T SRAM cell: This cell has 6 transistors with 2 cross coupled inverter pairs. The noise margins for the cell are large and it can work at low V_{DD} [ ]. Other major advantages are higher speed and lower static current (consequently lower power consumption) [ ]. The layout is itself compact and an advantage that accrues from this is that the length of the cell is more than its breadth, so that the overall layout is finally skewed in favour of longer, less expensive bit-lines. The sizing is such that, even for the largest memory array we can manage at this time (i.e., 256 x 256), the bit-lines are longer than the more expensive word-lines.
- A pass transistor for every block: It connects the local bit-lines of a bank to the global bit-lines, according to the DBL approach. The cell incurs a 15% area penalty over the normal SRAM cell.
- An AND gate: It splits a row into local word-lines. According to the DWL approach, the global word-line and the column select line are AND-ed to decide the word to be accessed. Since it lies in the critical path for access-time, a high speed AND from the technology library is adapted.

482   Prassanna Sithambaram, Alberto Macii, and Enrico Macii

To minimize power and delay due to parasitic capacitances, all signals spanning multiple cells, like the global word-line, the global bit-line, the column select and so on, involve higher metal layers. All cells also share maximum possible contacts to minimize chip area and wiring of the array [ ]. A layout of the cell array is shown in Figure 2.

4.2   Row Decoder

For a normal access, the row decoder consumes upto half the access time and a significant fraction of the total RAM power [ ]. While the logical function of the decoder is simple – it is equivalent to 2^n n-input AND gates – there is a large number of options for how to implement this function.The option we have considered assumes hierarchical decoding.

Since the row decoder is on the critical path, a fast wide-NOR implementation for the decoder is proposed by Caravella in [ ]. Amrutur in [   ] also favors the speed gains and scalability of a wide-NOR implementation over others. However, scalability is not a major issue as the target basic block for an embedded environment does not need to be very big. Also, our focus is on low-power design, at possibly a slight expense of delay. In [ ], Hezavei et al. have compared the performance of six different circuit styles, static, dynamic and skewed implementations of NAND and NOR logic, for row decoders. Based on the results reported in their paper, we opt for a *dynamic, NAND-based pre-decoder*, which consumes the least power and is quite competitive in terms of delay. A layout of the two stage NAND decoder is shown in Figure 3.

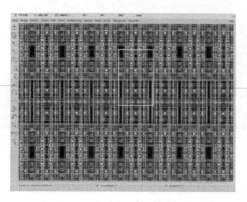

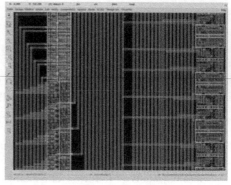

Fig. 2. Memory Cell Array   **Fig. 3.** Two Stage NAND Decoder

The work of [ ] also outlines how dynamic logic enables the implementation of a *pulsed word line* (PWL) scheme with no additional hardware overhead. The dynamic NAND gates provides pulsed outputs, the duration of which can be controlled. This is used to minimize the duration of active input on word lines, by deactivating the Word-Lines (and SRAM cells) before the bit-line voltages make a full swing. This leads to reduced power consumption and greater speed.

The last point to be considered for the decoder architecture regards the address reordering and clustering function. Figure 4 illustrates how address reordering is achieved *at no cost* by swapping the pre-decode lines that are AND-ed at the decoder stage (in the example, addresses of row 0 and row 1 are exchanged.) This step is performed run-time when the application-specific decoder is created.

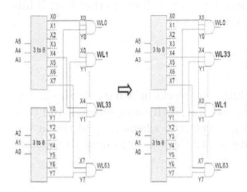

Fig. 4. Address Re-ordering   **Fig. 5.** Layout of Sense Amplifier

4.3 Read Circuitry

It consists of the sense amplifiers and of the isolated bit-line (IBL) logic. A full swing on the bit-line is a slightly slow operation. Hence, sense amplifiers are used to sense the slight voltage difference and amplify it to a correct data value [ ]. Since our primary purpose here is to demonstrate the memory architecture and build a modular generator for it, we have opted for a simple single-stage implementation as in [ ] (Figure 5). Concerning, the IBL logic, in [ ] a pair of isolation transistors are used to disconnect the sense amplifier from the bit-lines by the time the correct data is detected. It helps reducing read power as it prevents complete swing on bit-lines. It also disconnects the sense amplifiers during write operation as they are not needed. The isolation transistors can be turned off after a minimum 10% voltage difference is sensed between the lines.

4.4 Precharge, Column MUX, Write Circuitry

For precharge, we use a pair of pull-up transistors per column [ ], and one more to equalize bit-line and inverted bit-line. These are turned on before each read/write operation, pre-charging and equalizing both bit-line and inverted bit-line to a high voltage, and then will be switched off at the beginning of the operation leaving the bit-lines at the same voltage.

The column MUXes are used to avoid replication of the read/write circuitry and also to selectively precharge only the required words in a row (DWL). When a precharge operation is performed, the column MUX's also disconnect the bit-lines from the read/write circuitry. This minimizes the occurrence of a direct V_{dd}/Gnd path in the column, which is one of the major sources of power dissipation [ ]. A simple tree MUX [ ] is adapted for this work.

The write circuitry involving two pass transistors controlled by the *write* and *inverted write* signals are used to control the write operation [ ]. During the write operation, only one of *write* and *inverted write* is active, and the connected pass transistor grounds the relative line forcing a zero voltage to that line. If bit-line is forced low, a zero is written, and vice-versa.

5   ASBE Memory Generator

The ASBE memory generator for the architecture works in two steps: Simulating the application and using its trace to generate the partitioning schema, and then assembling custom-built basic blocks to generate the schematic and the layout of the memory. The sub-sections below deal with the two steps in greater detail.

5.1   Obtaining the Partitions

The source code of the application is compiled and the executable fed to an instruction-level simulator (ISS) for the processor architecture of choice, together with the memory specification parameters. The ISS generates the dynamic address trace, which is provided as input to the modified *MemArt* partitioning engine to obtain the partitions (or horizontal cuts) of the ASBE memory.

5.2   Generating the ASBE Memory

The ASBE memory macro generator produces the schematic and layout of an SRAM, given its dimensions and partition sizes.

The leaf cells of the SRAM core, decoders, sense amplifiers, precharge and write circuitry were custom laid-out using Cadence Virtuoso. Cadence SKILL code was used to instantiate the leaf cells and to make necessary routings at appropriate coordinates, thus generating the schematic and the layout of the complete SRAM partition. The simulations were performed using the Cadence Analog Artist simulator, and a 130nm HCMOS technology provided by STMicroelectronics was used for the design.

The entire memory is composed of partitions, each complete with its own decoding and sensing circuitry. Every partition consists of numerous DBL "banks" or "blocks", each of which can be independently enabled, one at a time. The shape of the partition, ie the memory core, is also a critical parameter in the final performance of the SRAM [ ].

The width of the partitions is limited by the word-line load. Since the word-line is in the critical path of the decoder, the word-line capacitance and the resulting delay and energy consumption becomes unacceptable if it is too long. The maximum width supported by the current version of the memory generator is 256 cells, with the minimum word size equal to 8. Two factors limit the length of the partition. One is the bit-line load, which grows with the number of banks into which the partition is divided. The second factor is the row-decoding capacity, ie maintaining a low energy consumption, and still meeting the timing constraints. The current maximum size of the partition is 256 x 256 bits, or 8KByte. However, since the bit-lines are in metal 3 and the word-lines in metal 2 (which has higher parasitic capacitance), having bit-lines longer than the word-lines may offer better results.

A few considerations that went into designing the memory generator were:

1. Focus on low power architecture and implementation.
2. A modular, leaf-cell independent design. Any changes in the layout of the leaf cells requires only a change of specified parameters in the memory generator. This enables the generator code to be generic and easily adaptable to different technologies. For eg, we can replace a 6T SRAM cell with a 4T SRAM cell, or the NAND pre-decoders with wide-NOR pre-decoders. The generator design is modular, and hence not concerned with how a particular functionality is achieved.
3. Complete flexibility in the location of the 'cuts', or partition sizes of the memory array.
4. Flexibility in word size and words-per-line in the array.
5. Flexibility in the memory sizes that can be generated.

6   Experimental Results

To validate the applicability of the ASBE memory architecture and the quality of the layouts obtained through the ad-hoc memory generator, we have considered the source code of a typical embedded application (i.e., a video decoder), and we have compiled it for two different platforms: An ARM7TDMI RISC core with a code memory of 1KByte and a LX-ST200 VLIW core with a code memory of 8KByte.

We have run memory profiling by means of the ARMulator for the ARM core and LX-ISS for the LX-ST200 core and captured the memory access traces. Both traces were fed to the modified *MemArt* engine of [ ] to obtain number, sizes and addressing spaces of the memory partitions.

For each of the two case studies, memory size and partition specification have been provided to the ASBE memory generator, in order to obtain the final layout of the complete ASBE memory architecture. Parasitic extraction, as well as timing and power simulations were performed using Cadence Virtuoso ICFB. The reference technology for both logic and cell array is the 130nm HCMOS9 by STMicroelectronics.

Area, delay and energy results for the two ASBE memory macros are reported in Tab. 1. For comparison reasons, the same table includes also data regarding the monolithic SRAM macros, as well as two instances of uniformly partitioned memories (one made of 32 partitions of 8 rows each and one containing 16

Table 1. Results: Energy, Access Time and Area

Platform	Monolithic			Uniform (32 part.)			Uniform (16 part.)			ASBE		
	E [nJ]	D [ns]	A [μm^2]	E [nJ]	D [ns]	A [μm^2]	E [nJ]	D [ns]	A [μm^2]	E [nJ]	D [ns]	A [μm^2]
ARM7TDMI	4.3e06	0.31	144052	2.8e06	0.28	145279	2.7e06	0.28	145485	2.4e06	0.29	144751
LX-ST200	2.1e07	1.36	518393	1.3e07	1.34	520664	1.2e07	1.33	521089	1.1e07	1.34	519809

Table 2. Results: Savings w.r.t. Monolithic Memory

Platform	Uniform (32 part.)			Uniform (16 part.)			ASBE		
	Δ E [%]	Δ D [%]	Δ A [%]	Δ E [%]	Δ D [%]	Δ A [%]	Δ E [%]	Δ D [%]	Δ A [%]
ARM7TDMI	34.83	7.96	-0.85	37.15	9.55	-0.99	43.34	6.05	-0.48
LX-ST200	38.86	1.97	-0.43	42.03	2.63	-0.52	45.34	1.68	-0.27

partitions of 16 rows each). Savings (positive numbers) and overheads (negative numbers) w.r.t. the monolithic memories are collected in Tab. 2.

From the results we can observe, first, that the best energy savings for both case studies come from the ASBE macros, and not from the uniformly-partitioned implementations; this is in line with the theoretical results obtained in [ ], thus confirming the viability and effectiveness of the ASBE memory architecture and its superiority if compared to uniform memory partitioning. From the data we can also evince the quality of the generated layouts; in fact, area occupation is slightly better than in the case of the uniformly partitioned memories, while timing is basically the same (and, obviously, better than for the monolithic architectures).

7   Conclusions

In this paper, we have pursued the objective of combining profile-driven techniques with physical partitioning in order to automatically generate energy-efficient partitioned memory architectures specifically customized for a given software application. We have contributed design solutions for the various components of the ASBE memory architecture, and described the implementation details for a prototype ASBE memory generator.

We have validated the results of automatic memory generation by means of two realistic case studies; results have confirmed the quality of the generated layouts and their efficiency in terms of energy, delay and area.

Concerning future work, we plan to investigate alternative architectures for low-power, low-overhead implementation of the memory decoder. This would allow us to release the constraint on the size of the ASBE macros that the generator can currently handle (256 x 256 bits). In addition, we intend to explore more efficient schemes for address clustering than those presented in [ ], as coming up with an optimal reordering scheme would result in an improved quality of the partitions. On a longer term, we will work towards the integration of the ASBE memory generator algorithms into an industrial framework for SRAM synthesis.

References

1. S. L. Coumeri, D. E. Thomas, "An Environment for Exploring Low-Power Memory Configurations in System-Level Design", *ICCD-99*, pp. 348-353, Oct. 1999.
2. N. Kawabe, K. Usami, "Low-Power Technique for On-Chip Memory using Biased Partitioning and Access Concentration", *CICC-00*, pp. 275-278, May 2002.
3. L. Benini, L. Macchiarulo, A. Macii, M. Poncino, "Layout-Driven Memory Synthesis for Embedded Systems-on-Chip", *IEEE Transactions on VLSI Systems*, Vol. 10, No. 2, pp. 96-105, Apr. 2002.
4. A. Karandikar, K. K. Parhii, "Low Power SRAM Design using Hierarchical Divided Bit-Line Approach", *ICCD-98*, pp. 82-88, Oct. 1998.
5. M. Yoshimoto, K. Anami, H. Shinohara, T. Yoshihara, H. Takagi, S. Nagao, S. Kayano, T. Nakano, "A Divided Word-Line Structure in the Static RAM and Its Application to a 64K Full CMOS RAM", *IEEE Journal of Solid-State Circuits*, Vol. 18, No. 5, pp. 479-485, Oct. 1983.
6. L. Benini, A. Ivaldi, A. Macii, E. Macii, "Block-enabled memory macros: design space exploration and application-specific tuning", DATE 2004, Vol. 1, pp. 698-699, Feb. 2004.
7. A. Macii, E. Macii, M. Poncino, "Improving the Efficiency of Memory Partitioning by Address Clustering", *DATE-03*, pp. 18-23, Mar. 2003.
8. P. Shivakumar, N. P. Jouppi, "CACTI 3.0: An Integrated Cache Timing, Power and Area Model", WRL Research Report 2001/2, Compaq Western Research Labs, Dec. 2001.
9. J. Rabaey, *Digital Integrated Circuits: A Design Perspective*, Prentice Hall, 1996.
10. J. Hezavei, N. Vijayakrishnan, M. Irwin, "A Comparative Study of Power Efficient SRAM Designs", *GLS-VLSI-00*, pp. 117-122, Mar. 2000.
11. E. Sicard, *Microwind and Dsch User's Manual*, http://www.microwind.org.
12. B. Amrutur, M. Horowitz, "Fast Low Power Decoders", *IEEE Journal of Solid-State Circuits*, Vol. 18, No. 5, pp. 479-485, Oct. 1983.
13. J. Caravella, "A Low Voltage SRAM For Embedded Applications", *IEEE Journal of Solid-State Circuits*, Vol. 32, No. 2, pp. 428-432, Mar. 1997
14. M. Jagasivamani, *Development of a Low-Power SRAM Compiler*, Virginia Polytechnic Institute, Blacksburg, VA,

Static Noise Margin Analysis of Sub-threshold SRAM Cells in Deep Sub-micron Technology

Armin Wellig and Julien Zory

STMicroelectronics – Advanced System Technology
Chemin du Champ-des-Filles 39, 1228 Geneva, Switzerland
{armin.wellig,julien.zory}@st.com

Abstract. Reducing leakage current in memories is critical for low-power designs in deep submicron technology. A common architectural technique consists of lowering the supply voltage to operate SRAM cells in sub-threshold (V_{th}). This paper investigates stability aspects of sub-V_{th} SRAM cells, both analytically and by simulation in STMicroelectronics' 90nm CMOS technology. For the first time analytical expressions for the Static Noise Margin in sub-V_{th} as a function of circuit parameters, operating conditions and process variations are derived. The 3G receiver case study illustrates the leakage saving potential of stable sub-V_{th} SRAM designs resulting into energy savings of up to 65%.

1 Introduction

In most of today's battery-operated system on chip (SoC) designs power management is driven by three factors: mobility, process technology and complexity. An emerging SoC design paradigm is to introduce on-chip "power islands" where individual functions are isolated in terms of both voltage and frequency (e.g. [1]) tailored for optimal power, performance and leakage current. Especially in memory-dominated SoC designs such as cellular phones or PDAs, an efficient memory leakage reduction scheme is critical to guarantee a long battery lifetime.

Various techniques have been proposed to reduce the leakage power of 6T SRAM cells (Fig. 1). At the device level, leakage reduction can be achieved by controlling the dimensions (length, oxide thickness, junction depth, etc.) and doping profile in transistors. At the circuit level, leakage reduction is achieved by controlling statically or dynamically the voltage of the different device terminals (i.e., gate, drain, source and substrate). An overview of existing techniques is beyond the scope of this paper but can be found in e.g. [2][3]. From a system architect's perspective those techniques are a design constraint and part of the low-power standard cell library specification. In this paper, we focus on architecture level techniques. One key concept is to lower the supply voltage (V_{DD}) to operate the SRAM cell in the sub-V_{th} region (generally referred to as *drowsy* mode) or gating off the V_{DD} at the expense of destroying the cell state. An exponential reduction in leakage power as a function of V_{DD} can be exploited while ensuring stable operating conditions. Modifications of the memory layout are kept to a minimum which guarantees a low area and integration overhead. These techniques have been widely used in drowsy cache designs with assistance of static or dynamic activity analysis (e.g., [4][5]) and are also the basis of this work.

Knowledge of the minimum V_{DD} preserving the SRAM cell stability allows a designer to exploit the maximum achievable leakage reduction for a given technology under different operating conditions. Little work has been done on the stability analy-

V. Paliouras, J. Vounckx, and D. Verkest (Eds.): PATMOS 2005, LNCS 3728, pp. 488–497, 2005.

sis of SRAM cells operating in sub-V_{th}. In [6], an analytical expression for the minimum data retention voltage (DRV) is derived and verified against simulations and actual test chip measurement results. The theoretical minimum DRV is obtained in the (somewhat unrealistic) case of having no noise margin at all. From there the actual DRV for realistic conditions is estimated by offsetting the theoretical results by a constant value. This paper formalizes for the first time in literature the static noise margin (SNM) of SRAM cells operating in the sub-V_{th} region and explores stability aspects. We derive closed form expressions for both active (i.e., read/write) and standby modes. The SNM expressions are verified by means of SPICE simulations.

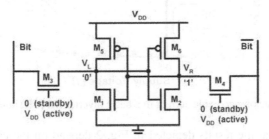

Fig. 1. 6T SRAM cell with 2 inverter pairs M_1/M_5 and M_2/M_6 and access transistors M_3 and M_4

The remaining of this paper is organized as follows: Section 2 introduces the leakage current components relevant for SRAM cells operating in sub-V_{th}. In Section 3, analytical expressions for the SNM in standby and active modes are derived. In Section 4, the stability of sub-V_{th} SRAM cells is analyzed followed by leakage power reduction discussions in Section 5. Finally, we conclude the paper in Section 6.

2 Leakage Current Components

Sub-threshold I_{sub} and gate-oxide I_{ox} currents are the main leakage components in today's SRAM cells [2]. The sub-threshold leakage is a weak inversion (i.e., $V_{gs} < V_{th}$) current across the device. In this paper, we use the model developed in [7] which shows how I_{sub} depends on the threshold and supply voltages:

$$I_{sub} = S \cdot I_0 \cdot \exp\left(\frac{V_{gs} - |V_{th}|}{n \cdot V_T}\right)\left(1 - \exp\left(\frac{-V_{ds}}{V_T}\right)\right) \tag{1}$$

where S is the transistor (W/L) ratio, $V_T = kT/q$ denotes the thermal voltage, I_0 the process-specific current at $V_{gs} = V_{th}$, T the chip temperature and n the sub-V_{th} swing coefficient. Note that the threshold voltage V_{th} is a function of circuit (e.g., reverse body biasing [3]) and device parameters. To simplify the following analysis, short channel effects impacting V_{th} (e.g., drain-induced barrier lowering where $\Delta V_{th} \propto V_{ds}$) are neglected since we assume V_{th} to be constant for circuits operated at $V_{DD} \leq V_{th}$. Nevertheless, the analytical models developed match well the SPICE (BSIM4 models) simulations in a 90nm process. Further note that the thermal voltage V_T increases linearly as temperature raises. If I_{sub} grows enough or other parts of the SoC build up heat, V_T will also start to rise, further increasing I_{sub} causing a vicious cycle. From (1) we conclude that I_{sub} can be reduced by lowering the supply voltage, increasing V_{th} (traded off with speed reductions) and/or cooling the design.

The gate-oxide current I_{ox} is mainly due to tunneling through the thin gate oxide insulation. Due to an exponential decline with V_{DD} and independence on the operating temperature [7], I_{ox} does not significantly increase the leakage current in the sub-V_{th} operating region (Fig. 2); I_{ox} is therefore neglected in the following analysis.

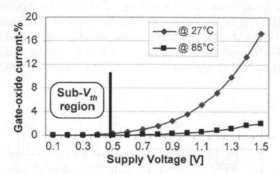

Fig. 2. Percentage of gate-oxide current in a basic inverter (SPICE simulations in 90nm)

It is evident that the results depicted in Fig. 2 depend on the circuit model accuracy, process technology, transistor sizing and operating conditions. Throughout this paper we use the low-power high-V_{th} 90nm CMOS process technology from STMicro-electronics; it offers very good performance-leakage trade-offs (from a device and circuit perspective) in portable SoC applications such as mobile phones [8].

3 Analytical SNM Models for Sub-V_{th} SRAM Cells

A common metric to characterize the noise immunity of SRAM cells is the static noise margin (SNM, [9]) defined as the minimum dc noise voltage necessary to flip the state of a cell. An estimate of the SNM can be obtained by drawing the static voltage transfer characteristic (VTC) as illustrated in Fig. 3 and finding the maximum possible square between them. Its side length L represents the minimum noise voltage present at each of the cell storage nodes necessary to flip the state of the cell. We observe that the eye opening is different for the SRAM cell in *standby mode* (i.e., access transistors are off) and *active mode* (i.e., access transistors are turned on). Needless to say that the cell is most vulnerable to noise during a read access, since the *'0'* storage node rises to a voltage higher than ground due to the bit-line discharge.

In [9], analytical SNM models for 6T SRAM cells were introduced, but limited to long-channel square law device behavior. In [10], the SNM models were refined to study the impact of intrinsic device fluctuations. All the models are limited to cells in *active* mode operating in either the linear or saturation region. In this paper, we extend the analysis to sub-V_{th} operation in standby and active modes. As in [10], analytical expressions for the static VTC functions F_1 and F_2 of the two cell inverters in the neighborhood of the corners Q/R and Q'/R' of the maximum squares (Fig. 3) can be derived by solving Kirchhoff's current law at the cell storage nodes V_R and V_L:

$$\text{Storage Node } V_L: \ I_1 = I_3 + I_5 \tag{2}$$
$$\text{Storage Node } V_R: \ I_2 = I_4 + I_6 \tag{3}$$

Since all six transistors operate in the sub-V_{th} region and considering that the leakage current I_i is dominated by the sub-threshold current, I_i is modeled by equation (1).

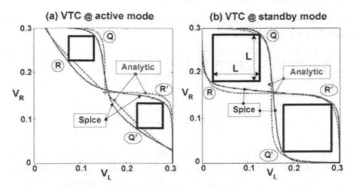

Fig. 3. VTC in (a) active and (b) standby mode (where $SNM = L$)

To derive closed form expressions the set of Kirchhoff current equations must be simplified first. The different sub-V_{th} current components were simulated with SPICE; they are plotted in Fig. 4 for the right inverter circuit M_2, M_4 and M_6 (see Fig. 1) in both active and standby modes. We observe that some currents can be neglected depending on the access mode of the cell.

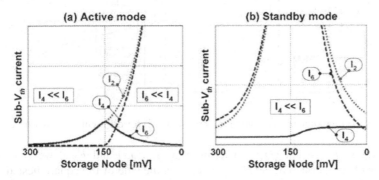

Fig. 4. Sub-V_{th} current components I_2, I_4 and I_6 in (a) active and (b) standby ($V_{DD} = 300$ mV)

3.1 Standby Mode

Without loss of generality we assume that the cell stores a '0' in storage node V_L and a '1' in storage node V_R. Moreover, the bit-lines are set to V_{DD} in standby mode. Both access transistors M_3 and M_4 are switched off, i.e. the gate terminal voltage is set to ground. Since the sub-V_{th} current is dominated by its *diffusion* component, i.e. movement of carriers due to a concentration gradient controlled by V_{gs}, I_3 and I_4 can be neglected irrespective of the storage node voltages (i.e., the *drift* component is negligible). SPICE simulations confirm those observations as illustrated in Fig. 4b. Thus, (2) and (3) can be simplified to $I_1 = I_5$ at node V_L and $I_2 = I_6$ at node V_R. To eliminate the voltages $V_{gs,i}$ and $V_{ds,i}$ of transistor M_i in (1) we can write the Kirchhoff voltage equations as follows

$$V_{gs,1} = V_{ds,2} = V_R \,, \quad V_{gs,2} = V_{ds,1} = V_L \,,$$
$$V_{gs,5} = V_{ds,6} = V_{DD} - V_R \,, \quad V_{ds,5} = V_{gs,6} = V_{DD} - V_L \tag{4}$$

Substituting (1) into (2, 3) and applying (4), we can express the static VTC functions $F_1(V_L = x)$ and F_2 $(V_R = y)$ like this

$$F_1(x) = F_s^{1,5} + \frac{n_1 \cdot n_5 \cdot V_T}{n_1 + n_5} \ln\left(\frac{1 - e^{((x - V_{DD})/V_T)}}{1 - e^{(-x/V_T)}}\right), \, F_2(y) = F_s^{2,6} + \frac{n_2 \cdot n_6 \cdot V_T}{n_2 + n_6} \ln\left(\frac{1 - e^{((y - V_{DD})/V_T)}}{1 - e^{(-y/V_T)}}\right) \tag{5}$$

where

$$F_s^{i,j} = \frac{n_i \left(V_{DD} - \left|V_{th}^j\right|\right) + n_j \cdot \left|V_{th}^i\right|}{n_i + n_j} + \frac{n_i \cdot n_j \cdot V_T}{n_i + n_j} \ln\left(\frac{S_j}{S_i}\right) \tag{6}$$

Note that the optimization problem of finding the maximum square enclosed in the VTC side lobes cannot be reduced to a single dimension since analytical expressions can be derived neither for $F_1^{-1}(y)$ nor $F_2^{-1}(x)$. Thus, the resolution of SNM needs to be formulated as a multivariate optimization problem: Maximize $f(x,y)$ subject to $g(x,y)=0$ over $0 \le x,y \le V_{DD} \in \mathfrak{R}$.

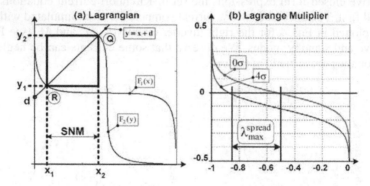

Fig. 5. (a) Lagrange optimization problem – (b) $\chi(\lambda)$ for different process variations σ

To express $f(x,y)$ and $g(x,y)$ in terms of x, y, $F_1(x)$ and $F_2(y)$ we can make the following observation: The SNM is modeled as the side of the square nested between $F_1(x)$ and $F_2(y)$ with the longest diagonal (refer to Fig. 5a), whose equation is given by $y = x + d$. The intersection of this family of diagonals with $F_1(x_1)$ and $F_2(y_2)$ becomes $F_1(x_1) = x_1 + d$ and $y_2 = F_2(y_2) + d$. Thus, the constraint $g(x_1,y_2)$ can be derived by eliminating offset d. The objective function $f(x_1,y_2)$ is simply the length of the nested square enclosed by the VTC side lobe, yielding

$$f(x_1,y_2) = F_2(y_2) - x_1 \text{ subject to } g(x_1,y_2) = y_2 - F_1(x_1) - F_2(y_2) + x_1 = 0 \tag{7}$$

Since $f(x_1,y_2)$ is differentiable within a convex optimization problem delimited by either VTC side lobe, a maximizing value of (x_{max}, y_{max}) exists and can be found with the *Lagrange Multiplier method* which can be expressed as follows

$$\Phi(x,y,\lambda) = F_2(y) - x - \lambda(y - F_1(x) - F_2(y) + x) \text{ with } \vec{\nabla}\Phi(x_{max}, y_{max}, \lambda_{max}) = \vec{0} \tag{8}$$

where λ is referred to as the Lagrange multiplier and $\vec{\nabla}$ denotes the multivariate gradient vector. Because it is sufficient to restrict the analysis to one VTC side lobe (left one as an arbitrary choice, i.e. $x < V_{DD}/2$ and $y > V_{DD}/2$), the transfer functions $F_1(x)$ and $F_2(y)$ can be simplified as follows

$$F_1(x) \cong F_s^{1,5} - \frac{n_1 \cdot n_5 \cdot V_T}{n_1 + n_5} \ln\left(1 - e^{(-x/V_T)}\right), \quad F_2(y) \cong F_s^{2,6} + \frac{n_2 \cdot n_6 \cdot V_T}{n_2 + n_6} \ln\left(1 - e^{((y - V_{DD})/V_T)}\right) \quad (9)$$

Solving $\vec{\nabla}\Phi = \vec{0}$ yields x and y as a function of λ

$$x = \phi(\lambda) = -V_T \cdot \ln\left(-\frac{\lambda + 1}{(r-1)\lambda - 1}\right) \quad \text{with } r = \frac{n_1 \cdot n_5}{n_1 + n_5} \quad (10)$$

$$y = \varphi(\lambda) = V_T \cdot \ln\left(\frac{-\lambda}{(s-1)\lambda + s}\right) + V_{DD} \quad \text{with } s = \frac{n_2 \cdot n_6}{n_2 + n_6} \quad (11)$$

Finally, substituting (10) and (11) into constraint $g(x_{max}, y_{max})$ results in the following condition for λ_{max}

$$\gamma(\lambda_{max}) = \varphi(\lambda_{max}) - F_1(\phi(\lambda_{max})) - F_2(\varphi(\lambda_{max})) + \phi(\lambda_{max}) = 0 \quad (12)$$

The function $\gamma(\lambda)$ is plotted in Fig. 5b. The range of the Lagrange Multiplier λ mainly depends on the process variations σ and can be quite significant. Given the complexity of equation (12) it is not possible to find a good approximation for $\gamma(\lambda)$ within this λ range; thus, $\lambda_{max} \in [-1, 0]$ needs to be solved by numerical methods. Finally, the SNM can be expressed in terms of λ_{max} by substituting (9-11) into (7) as follows

$$SNM = F_s^{2,6} + V_T \cdot \ln\left(\left[\frac{s \cdot \lambda_{max} + s}{(s-1) \cdot \lambda_{max} + s}\right]^s \cdot \frac{\lambda_{max} + 1}{1 + (1-r) \cdot \lambda_{max}}\right) \quad (13)$$

where the parameters $F_s^{2,6}$, r and s were defined in (6), (10) and (11), respectively.

3.2 Active Mode

We assume that the bit-lines are pre-charged to V_{DD} prior to each cell access and that the node V_L stores a '0' and V_R a '1'. From Fig. 4a we conclude that I_5 can be neglected in the left inverter and I_4 in the right one. With these approximations, (2) and (3) can be simplified to $I_1 = I_3$ at node V_L and $I_2 = I_6$ at node V_R. Following the same step-by-step derivations as in the previous section, a closed-form SNM can be derived in terms of λ_{max}

$$SNM = F_s^{2,6} + V_T \cdot \ln\left(\left[\frac{s \cdot \lambda_{max} + s}{(s-1)\lambda_{max} + s}\right]^s \cdot \frac{(q+1)\lambda_{max} + 1}{(q+1-n_1)\lambda_{max} + 1}\right) \quad \text{and } q = \frac{n_1}{n_3} \quad (14)$$

where the parameters $F_s^{2,6}$ and s were defined in (6) and (11), respectively. Due to the limited space the derivation of λ_{max} as in (12) is left to the reader.

3.3 Accuracy

The accuracy of the developed SNM models is illustrated in Fig. 6 by plotting the relative error with respect to the values obtained through SPICE simulations. For a $V_{DD} < 150\ mV$, the model introduces some errors due to approximating the transfer functions F_1 and F_2 (refer to (9)). Indeed lowering V_{DD} down to the data retention value implies an SNM of zero; thus, F_1 and F_2 approach a linear function and the approximations in (9) are not valid anymore. For a $V_{DD} > 400\ mV$, the SRAM cell is at the edge of operating in the sub-V_{th} region. Thus, our current model (defined by (1)) is obviously no more appropriate.

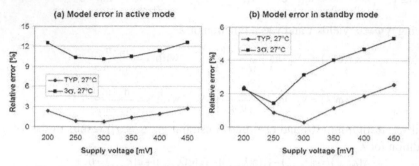

Fig. 6. Sub-V_{th} SNM model accuracy in (a) active and (b) standby mode

Moreover, we observe that the SNM model degradations are more important when introducing process variations. This is mainly due to the Kirchhoff current law simplifications at the storage nodes. For instance, shifting V_{th} by 3σ impacts the relationships (as shown in Fig. 4) among the different sub-V_{th} current components of the SRAM cell. To conclude, the derived SNM models offer a fairly reliable means for the designer to either validate circuit simulations or to forecast noise immunity characteristics of new process technology variants where device and circuit models are not (yet) available.

4   Stability Analysis

The static noise margin is a measure of the cell's ability to retain its data state. As pointed out in Section 3, the worst-case situation is under "read-disturb" conditions as shown in Fig. 7a. For instance, setting the stability criterion to $SNM = 100\ mV$ (application-specific), the SRAM cell cannot be safely operated in the sub-V_{th} region during active mode. In this scenario, without relaxing the stability criterion, more sophisticated devices and circuits would be needed to support stable sub-V_{th} read operations tailored for ultra-low power applications.

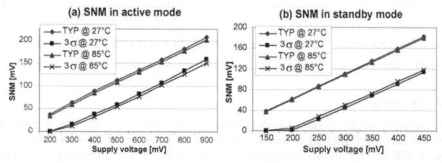

Fig. 7. Stability of an SRAM cell in (a) active and (b) standby modes in 90nm (from SPICE)

In standby mode, an (off-the-shelf) low-power SRAM can be operated in sub-V_{th} even for a process variation of up to 3σ and operating temperature of $85°C$. Thus, incorporating dynamic supply voltage control in memory-dominated SoC designs is of great interest for applications characterized by long idle periods (e.g., mobile phone during standby). Finally, it should be noted that an increase in temperature is considerably less disturbing ($\Delta SNM \cong 1mV/15°C$) to the cell stability than variations in the die ($\Delta SNM \cong 50 - 65mV$ @ 3σ).

5   Leakage Reduction Analysis

So far, we have not yet addressed the leakage power reduction achievable for embedded SRAMs operating in sub-V_{th}. In this section, we discuss the leakage power dissipation in a SRAM cell for different operating conditions in a 90nm process. Moreover, advanced receiver designs implementing the 3G cellular system [11] will serve as case study to demonstrate the potentials for energy savings.

As pointed out in Section 4, the voltage supply of SRAM cells can be reduced below threshold V_{th} without altering their data states during standby. The leakage power (left ordinate) and the resulting reduction with respect to "full-V_{DD}" operation (right ordinate) are shown in Fig. 8 for one SRAM cell.

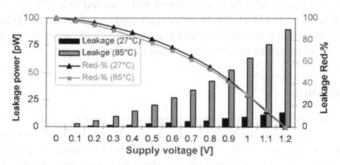

Fig. 8. SRAM cell's leakage power (bars) and relative reduction (lines) in 90nm

A reduction of more than **80%** can already be achieved for a V_{DD} around $400\ mV$ ensuring a SNM of about $100\ mV$ taking into account process variations of up to 3σ at a chip temperature of $85°C$. Thus, completely gating off V_{DD} gives (only) an extra 20% reduction in leakage power with the cost of data state destruction (e.g., cache design). This motivates once more the recent efforts of incorporating embedded multi-rail V_{DD} SRAMs in advanced processor and SoC architectures (e.g. [12]).

5.1   Case Study: Advanced 3G Receiver

The credo of emerging "3G and beyond" cellular systems is to enhance the service flexibility and QoS. One immediate consequence is an increase in processing complexity and memory size. For instance, the introduction of a physical layer retransmission scheme such as the H-ARQ [11] increases the memory up to a factor 6. A receiver supporting this feature is composed of roughly 80% memory.

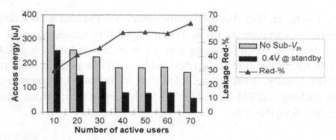

Fig. 9. Energy reduction for advanced 3G baseband receiver

To optimize throughput and dynamic power a multi-bank SRAM architecture is selected for H-ARQ implementation, i.e. 6 banks with a 64-bit data path. Depending on the link configuration, application requirements and system capacity, each SRAM bank can be either in *Read/Write*, *Standby* or *Sleep* mode. The banks in *Sleep* mode are idle over a long period and can be gated off. The memory content during *Standby* mode must be preserved. The energy savings when operating the SRAM banks in the stable sub-V_{th} region is shown in Fig. 9 for an application requirement of *64 kbps*. Note that we assume that the network implements *Proportional Fair* packet scheduling [13] and that all users have the same application requirements.

Depending on the number of (active) users to be served in a cellular sector, implementing a dual-V_{DD} SRAM results in an overall energy reduction of up to **65%**. This demonstrates that on top of reducing leakage at device and circuit level, architectural techniques are necessary to enable energy-efficient SoC designs.

6 Conclusion

A common architectural technique to reduce leakage power consists of lowering the supply voltage to operate the SRAM cells in sub-threshold (V_{th}). This paper investigated stability aspects of sub-V_{th} SRAM cells. For the first time in literature, a closed-form expression was derived to characterize the static noise margin (SNM) in standby (i.e., access transistors are off) and active (i.e., access transistors are on) mode in the sub-V_{th} operating region. The SNM models were verified against SPICE simulations in STMicroelectronics' 90nm CMOS technology. It was shown that a leakage power reduction of up to *80%* per cell can be achieved while preserving the cell state in sub-V_{th}. As a result, operating the memory-dominated processing blocks of advanced 3G receiver in drowsy mode yields energy savings of up to *65%*.

References

1. J. Hu *et al.*: Architecting Voltage Islands in Core-based System-on-a-Chip Designs. In: Int. Symp. on Low Power Electronics and Design, Newport Beach, USA (2004) 180-185
2. Y. Nakagome, M. Horiguchi, T. Kawahara and K. Itoh.: Review and future prospects of low-voltage RAM circuits. In: IBM R&D Journal, Vol. 47, No. 6, (2003) 525-551
3. K. Roy *et al.*: Leakage Current Mechanisms and Leakage Reduction Techniques in Deep-Submicrometer CMOS Circuits. In: Proceedings of the IEEE, Vol. 91, No 2, (2003) 305-327

4. M. Powell *et al.*: Gated-Vdd - A circuit technique to reduce leakage in deep-submicron cache memories. In: Int. Symp. on Low Power Electronics and Design, Rapallo, Italy (2000) 90-95
5. K. Flautner *et al.*: Drowsy caches - simple techniques for reducing leakage power. In: 29th Annual Int. Symposium on Computer Architecture, Anchorage, USA (2002) 25-29
6. H. Qin *et al.*: SRAM Leakage Suppression by Minimizing Standby Supply Voltage. In: 5th Int. Symposium on Quality Electronic Design, San Jose, USA (2004) 55-60
7. A. Chandrakasan, W. Bowhill and F. Fox.: Design of High-Performance Microprocessor Circuits. In: IEEE Press, (2001)
8. STMicroelectronics, http://www.st.com/stonline/prodpres/dedicate/soc/asic/90plat.htm
9. E. Seevinck, F.J. List and J. Lohstroh.: Static Noise Margin Analysis of MOS SRAM Cells. In: IEEE Journal of solid-state circuits, Vol. 22, No. 5, (1987) 748-754
10. A.J. Bhavnagarwala *et al.*: The Impact of Intrinsic Device Fluctuations on CMOS SRAM Cell Stability. In: IEEE Journal of solid-state circuits, Vol. 36, No. 4, (2001) 658-665
11. 3rd Generation partner project, http://www.3gpp.org
12. K. Flautner and D. Flynn.: A Combined Hardware-Software Approach for Low-Power SoCs - Applying Adaptive Voltage Scaling and Intelligent Energy Management Software. In: DesignCon 2003, http://www.arm.com
13. H.J. Kushner and P.A. Whiting.: Convergence of proportional-fair sharing algorithms under general conditions. In: IEEE Trans. on Wireless Communications, Vol. 3, Issue 4, (2004) 1250-1259

An Adaptive Technique for Reducing Leakage and Dynamic Power in Register Files and Reorder Buffers

Shadi T. Khasawneh and Kanad Ghose

Department of Computer Science
State University of New York, Binghamton, NY 13902-6000, USA
{shadi,ghose}@cs.binghamton.edu

Abstract. Contemporary superscalar processors, designed with a one-size-fits-all philosophy, grossly overcommit significant portions of datapath resources that remain unnecessarily activated in the course of program execution. We present a simple scheme for selectively activating regions within the register file and the reorder buffer for reducing leakage as well as dynamic power dissipation. Our techniques result in power savings in excess of 60% in these components, on the average with no performance loss.

1 Introduction

Modern superscalar processors exploit instruction level parallelism by using a variety of techniques to support out-of-order execution, along with the dispatching of multiple instructions per cycle. Register renaming is at the core of all such techniques. In register renaming, a new physical register is allocated for every instruction that produces a result into a register to maintain the true data dependencies.

The allocation of the register takes place as instructions are fetched in program order, decoded and dispatched into an issue queue, regardless of the availability of the source register values. As register values are generated, instructions waiting in the issue queue are notified. An instruction is ready to start execution or issue when all of its source operands are available (or are soon to become available). Such instructions are issued to the execution units and read their source operands from the physical register file and/or from a bypass network.

In high end microprocessors, to support a large instruction window for the sake of performance, the physical register files typically have a large number of registers, often as many as 80 to 128 integer or floating point registers. As performance demands continue, the number of such registers is likely to go up. Furthermore, in a S-way superscalar processor that can issue S instructions per cycle, the integer or floating point register file (RF) needs to accommodate at least 2S read ports and S write ports. Moreover contemporary register file structures are heavily accessed. All of these factors make the register file a significant source of power dissipation and a localized hot spot within the die. The hot spot - a localized high-temperature area within the die - results from the large power dissipation within the relatively small area of the register file. As device sizes continue to shrink, leakage power becomes a significant part, often as much as 50% of the total power dissipation within a CMOS device [7, 3]. Furthermore, as leakage power increases dramatically with temperature, hot spots - such as the physical register file - are likely to become hotter with an increase in the leakage dissipation. This is essentially a positive feedback system and

V. Paliouras, J. Vounckx, and D. Verkest (Eds.): PATMOS 2005, LNCS 3728, pp. 498–507, 2005.

causes a thermal runaway, leading to temperature increases that can cause the devices to malfunction as the junction temperatures exceed safe operating limits [3]. The reorder buffer (ROB), a large, multi-ported register file like structure, also dissipates a significant amount of leakage power. The techniques presented in this paper emphasize a simple implementation and has no impact on the performance.

2 Relevant Circuit/Organizational Techniques

The circuit techniques employed by our scheme for reducing static power dissipation in the bitcells used within the register file (RF) and the reorder buffer (ROB) borrows from the Drowsy cache design of [4]. A drowsy cache design effectively switches the supply voltage to a bitcell in-between a lower ("drowsy") level (that preserves the contents of the bitcell but reduces leakage) and a normal (and higher) operating level. To access the contents of a bitcell, the supply voltage has to be raised to the normal level. The transition time in-between a drowsy state and a normal state can be limited to a single cycle [4]. We extend this mechanism to add a second lower supply level (close to zero) where the bitcell loses its contents but can be switched to a normal mode in two to three cycles. Leakage power is reduced further in this mode (deep-sleep mode) compared to the drowsy state. For the simulated data on the 0.07 micron CMOS process used in our studies, the normal operating voltage was 1.0 volts, the state-preserving drowsy supply voltage was assumed as 0.3 volts and the ultra-low leakage non-state-preserving supply voltage was 0.15 volts. (The technology data was scaled from a 0.13 micron process using the linear scaling approach taken in [5]. We also used a modified version of the e-CACTI tool [5] to compute leakage and dynamic dissipation in the bitcell arrays and other components.) Leakage within other components of the RF and the ROB – such as address decoders, drivers, prechargers, sense amps, control circuitries for reads and writes - is assumed to be minimized using devices with two different threshold voltages. The e-CACTI tool also models these dissipations.

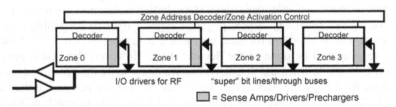

Fig. 1. A register file with four zones

For the purpose of our coarse-grained leakage reduction technique, we re-organize the RF and ROB from their usual monolithic design to a segmented design with multiple zones, as shown in Figure 1 (for a RF). Registers within the monolithic RF structure is broken down into 4 groups or zones in Figure 1, each zone having its own drivers, decoders, sense amps and local control logic. Each zone contains the same number of contiguous registers. A register address is broken down into two parts – a zone address and a zone-local register address. The zone address is decoded using a decoder external to the zones, as shown. This decoder also includes additional logic to control the state of all the zones. A single zone can be in one of the following states: normal or active, content-preserving standby, volatile or deep-sleep mode. This structure permits zones to be kept in different states to minimize the overall leakage power. The zoned structure also reduces dynamic dissipation, as one set of "super" bit lines and through buses are used by each active and accessed zone; inactive zones do not load up these buses. (In a monolithic implementation, bitcells in all registers load up the common bit lines.) Further reduction in dynamic power occurs through the use of smaller prechargers, sense amps and buffers within each zone: these components are activated only within a zone that is accessed. If the number of registers within a zone is small, one can altogether dispense with the sense amp for the zones. The reorder buffer can be segmented into zoned structures in a similar fashion.

3 Reducing Leakage in the Register File

The technique proposed for reducing leakage dissipation in the register file exploits the following observations:

1. A significant number of cycles elapses between the time of register allocation (at instruction dispatch) and the time a result is written into the register. As the register does not hold any valid information during this period, it can be kept in a deep sleep mode to avoid leakage dissipations.
2. After a register has been written and read out by instructions that consume its value, a significant number of cycles elapse before the register is deallocated. However, in this period the register contains a potentially useful value. In this case, we reduce leakage dissipation by keeping the register in a standby mode that preserves its contents but also reduces the leakage.

3.1 Activating RF Zones Dynamically

In this scheme, called the *on-off scheme*, zones within a register file are in either an active state or in the (volatile) deep-sleep state. Initially, all zones are turned off. With the use of register renaming, a new physical register (or two for multiple precision instructions) will be allocated for each instruction being dispatched. The allocation of the new register will be done in the decode stage. However, the first access to the register will be made when the dispatched instructions write the computed result into this register from the write back stage. To reduce RF leakage power, we thus attempt to allocate the destination registers at the time of instruction dispatch within a zone that is already active, to minimize any overhead associated with turning on an inactive zone. If an active zone is not available for any allocation, one (or more) zone(s) in the deep sleep are used for the allocation and these zones are then activated. Once activated, a zone remains in that state till it is completely free, i.e., till it can be deallocated. The two cycle delay in activating the zone has no consequence on the performance, as the dispatched instructions do not produce a result into their destination for at least two cycles following the dispatch (the time needed to reach write-back stage). A 1-bit wide auxiliary register file is maintained, with a single entry for each zone to indicate the status of each zone (as active or in the deep-sleep mode). The logic for looking up a free RF is adapted with very little change to permit us to make allocations preferentially within a targeted zone.

Our studies show that the policy of allocating a new register to a zone has little impact on the overall power savings. We therefore use a policy that is easy to implement in hardware – registers are allocated within the first active (FA) zone that is found in the free register list. If all active zones are full, or none are active, then the first inactive (in deep sleep state) zone is activated.

3.2 Putting RF Zones into the Standby Mode

The main idea in this scheme, called the *standby scheme*, is to put all zones in standby mode, and activate zones on a need-to basis. Registers for destinations are allocated at the time of instruction dispatch within zones that are kept in the standby mode till the first access is needed to that zone – when the first of the dispatched instructions issue and write the result into the zone. To reduce leakage within a zone containing valid data, we keep the zone activated for only a small number of cycles, say M cycles, before we revert it back to the standby mode. (We have used values of M = 2 and M =3 in our studies). This is done by using a small 2-bit counter with each zone; these counters are part of the status array that holds the status of each zone. Performance penalties are avoided in this scheme by making simple modifications to two pipeline stages.

First, the writeback stage needs to be able to identify the zone being written to by each instruction one (or a few) cycles before the actual writeback takes place. Such a requirement is not unusual and is routinely implemented in high-clock rate superscalar machines, where the destination address (wakeup tag) is broadcasted to the issue queue one (or a few) cycles before the result is actually needed. The only change needed in the writeback stage is to have it look up the status of the target zone from a status array (similar to scheme described in section 3.1) and activate that zone before the result is written to it in a later cycle. For the zone sizes used in this study, it takes just one cycle to change the state of a zone from standby to active, thus the transition time can be completely hidden with no impact on performance.

The second simple modification is to the issue logic. The issue logic needs to identify instructions that need to read the register file to access one or more source operands. (These are ready instructions that could not be selected for issue in the cycle following the broadcast of the tag that waked up the instruction). As such instructions are selected for issue, the selection logic reads the status of the zones that contain registers to be read and activates them if they are on the standby mode. If such zones are already active and are to remain active for an additional cycle, no additional steps are needed. If the zone is found to be active for just the current cycle, then the zone's associated counter is reset to M to guarantee that the zone remains active till the cycle where the source operands are read out. Doing this ensures that a zone remains in an active state when back-to-back requests to access the zone happen to occur. Switching glitches caused in the course of switching often between a standby state and an active state are thus avoided when requests to access a zone are clustered over an interval that exceeds M.

The one cycle delay in transitioning a zone from the standby mode to the active mode to allow an issuing instruction to read source register(s) from the zone is effectively hidden by overlapping this transition with the 1 cycle needed to move the instruction to the execution unit. This is possible because of the following reasons. As soon as the selection logic grants the request for a ready instruction to issue, it starts activating the required zone from a standby state to an active state. This is possible as the *zone* address of the source registers are kept in a dedicated RAM bank, adjacent to the issue queue (IQ) entries; the remaining part of the register addresses are within the IQ entries. As the grant signal comes down the selection tree and the selected instruction is read out on the IQ port and moves to the execution unit, the issued instruction presents the source register addresses to the register file and the register address is decoded. In parallel with all of these events, and starting with the propagation of the grant signal down the selection tree, the narrow bank containing the zone addresses is read out and the required zone is activated if needed, requiring an additional cycle (Figure 2). Thus by the time the word line for bit cells in the RF are to be driven, these bitcells are already activated. Consequently, the one cycle needed for activating a zone is effectively hidden and there is no impact on performance. We are assuming a contemporary issue mechanism where wakeup, selection and issue are spread over two clock cycles.

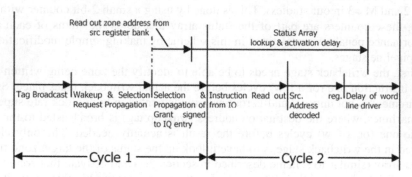

Fig. 2. Timing associated with instruction issue and zone activation

Standby scheme provided more savings than on-off scheme, as shown in section 5. Finally, we discuss a *hybrid scheme* where the on-off and standby schemes are combined by putting any unused zone into the off (deep-sleep) mode.

3.3   Extending the RF Leakage Management Scheme to the ROB

The standby scheme can be also applied to the ROB in a fashion to that deployed for the register file. In a P4-style pipeline, the ROB is accessed in the dispatch, writeback and commit stages. Assuming a 4-way CPU, in the worst case and a ROB with a total of 18 zones (as studied here), 4 zones could be accessed from any of these stages, thus up to 12 zones can be active each cycle, providing a *minimum* of 22.22% reduction in ROB usage and the associated leakage power.

In the dispatch stage, and ROB entry is allocated for each instruction, and since a zone needs 1 cycle to be activated, the allocation is done in fetch/decode where the activation is triggered so that a zone will be ready 1 cycle later to maintain a 0% IPC loss. Similarly, in the writeback stage, the ROB entries corresponding to the instructions in writeback stage will be activated 1-cycle before writeback. At commit, all possible commit entries are activated to simplify the circuitry needed to maintain performance; these entries could span 1 or 2 zones. Each zone is assumed to be active for M cycles (see section 3.2). The first cycle is for the transition from standby into active mode. The read/write access is done in the second cycle. The third cycle is for the transition from active to standby unless the same zone is being accessed by a different instruction, in that case, the zone is assumed (in the simulations) to be active for more 3 cycles. The allocation of ROB entries is done in a circular fashion, and thus there is no room to optimize this policy to gain extra power.

4 Experimental Results

We used a modified version of the well-known SimpleScalar simulator for our studies. We simulated a superscalar CPU with a fetch width of 4, an issue width of 4, a dispatch width of 4 and a commit width of 4. The IQ was set to 64 entries, and a ROB of 144 entries. The RF configuration used had 80 registers in each of the integer and floating point RFs (80INT + 80FP registers). The size of the load/store buffer was set at 128. A large subset of the integer and floating point benchmarks of SPEC2000 was used and executed for 100 million cycles, after skipping the first 400 million cycles.

4.1 Register File On/Off Results

The average number of cycles between register allocation and actual usage is 20.26 cycles, as shown in Figure 3. In figure 4, we show the impact of using alternative allocation policies: FA – First active zone (see section 3), MRU – allocate within the most recently used zone first, BF – allocate within the zone that best matches the allocation size. Figure 4 shows that the number of turned-off zones for the MRU, FA and BF are 4.52 (28.25%), 4.35 (27.19%) and 4.35 (27.19%) respectively. Figure 5, shows the number of turned-off zones for a RF configuration with 16 zones each in the integer and floating point RF. Here, for MRU, FA and BF, the average number of turned off zones are 10.03 (31.34%), 10.08 (31.5%) and 10.07 (31.47%) respectively.

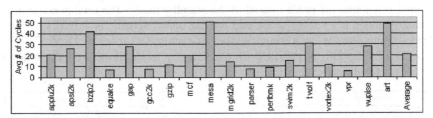

Fig. 3. Average # of Cycles between Register Allocation and Access

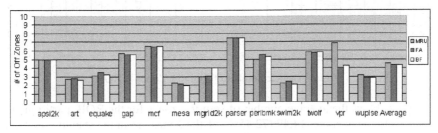

Fig. 4. Average number of zones turned off (8 INT + 8 FP)

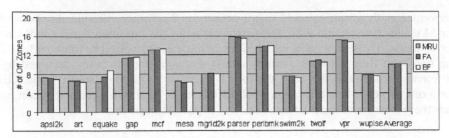

Fig. 5. Average number of zones turned off (16 INT + 16 FP)

4.2 RF and ROB Standby Scheme Results

In this section we will show the results for the standby scheme along with the combined hybrid scheme for the RF and ROB.

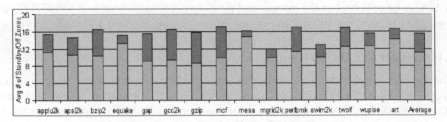

Fig. 6. Average number of standby/off zones for register file

Figure 6 shows the results for the standby mode (the entire bar). It also shows how many of these zones can be turned-off (upper half of each bar). There are 16 zones in each of the integer and floating point RFs. The hybrid scheme provides the same total number of standby/off zones but realizes added power savings by putting the unallocated zones into off mode instead of the standby mode. The total average number of standby zones is 15.56 (77.8%), and for the on/off scheme is 4.39 (21.95%). The hybrid scheme provides an average number of standby/off zones as 15.56 (77.8%), of which 11.17 (55.85%) is provided by the standby mode alone, and the other 4.39 (21.95%) is for the zones that can be turned off.

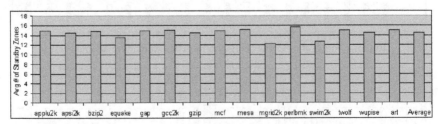

Fig. 7. Average number of Standby Zones in ROB

Figure 7 shows the average number of standby zones for the ROB, partitioned into 18 zones. On the average, 14.49 (72.45%) of the zones are in the standby mode. This percentage is slightly lower than that for the RF.

5 Power Savings

We modified the e-CACTI tool of [5], which is designed for estimating the dynamic and leakage power of caches, to measure the dynamic and leakage power of the RF and the ROB. Assumptions made in this regard were noted in Section 2. All of the reported measurements are for a 0.07 micron CMOS technology.

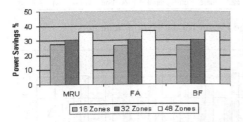
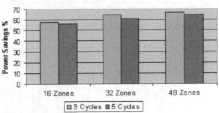

Fig. 8. On/Off RF Results using MRU, FA and BF allocation schemes

Fig. 9. Standby Scheme Power Savings Percentage for Register File (with FA)

Figure 8 shows the RF leakage power savings for different allocation schemes. FA provides the best results and it is also simpler to implement than the other allocation schemes. FA is used in all of the subsequent results for the RF. Figure 9 shows the leakage power savings for the standby scheme and how it varies with the activation period. Extending the activation period of a zone to 5 cycles will decrease the savings to 56.34% (from 57.85%), 61.01% (from 64.09%) and 64.15% (from 66.41%) for 16, 32, 48 zones respectively.

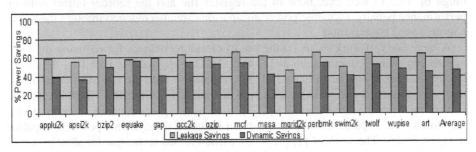

Fig. 10. Register File Leakage and Dynamic Power Savings for Standby Mode

The standby scheme provides more power savings (compared to on-off scheme). The total average leakage power savings is 59.81%, as shown in Figure 10, and Dynamic power savings 45.56%. Turning off the unused zones (the hybrid scheme), as shown in Figure 11 - increases the leakage power savings up to 64.89% (an additional 8.49%) compared to using the standby mode alone, which is expected since turned-off zones do not leak power.

Figure 12 shows a total power savings of 61.99% leakage power and a 43.26% of dynamic power on the average. It is also possible to use the hybrid approach in ROB to increase the savings (as in RF). Furthermore, the commit logic could also be enhanced to activate the commit-zones only if it contains ready-to-commit entries.

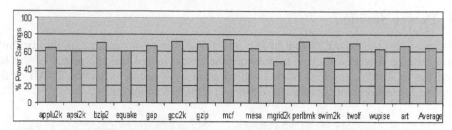

Fig. 11. Register File Leakage Power for the Hybrid Scheme

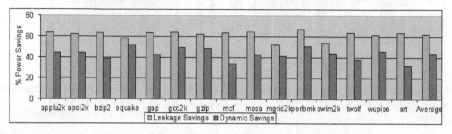

Fig. 12. Leakage and Dynamic Power Savings in ROB

6 Conclusions and Related Work

We proposed a set of simple microarchitectural techniques for reducing leakage power dissipation in the register file and the reorder buffer of contemporary superscalar processors. The techniques proposed achieve a leakage power reduction in the range of 47% to well over 60% in the register file and the reorder buffer with no performance penalty. Dynamic power dissipations are also reduced through the use of a multi-zoned organization.

A large body of work exists on the use of circuit techniques for reducing the leakage energy within bitcells, such as [1, 4, 6, 7]. Our approach is based on the use of circuit techniques similar to that of [4] in conjunction with the use of microarchitectural techniques. Some leakage reduction techniques for register files/bitcells also exploit microarchitectural statistics [1, 3], such as the predominance of zeros within the stored data. A plethora of work exists on reducing the dynamic power dissipation in register files. The work of [2] proposes a fine-grained technique for shutting down unused registers to save leakage power. Once a register is activated, it stays in this mode whether the contents are accessed or not. The work presented here relies on a coarse-grained approach that controls the state of zones within the register file as active, drowsy and deep-sleep and thus saves additional power by putting zones that are not being accessed into the drowsy mode when they contain useful data. The work of [3] proposes a cell design that permits fine-grained activation and deactivation of bitcells to reduce leakage dissipation and shows how energy savings are possible using such bitcells in register file banks and caches. Our approach, in contrast to the work of [3] uses standard bitcells with support for supply voltage management, as used in Drowsy caches [4]. We also achieve dynamic power savings in our techniques because of the use of multi-segmented structures for the register file and the reorder buffer.

References

1. Azizi, N. et al, "Low-leakage Asymmetric-cell SRAM", in Proc, ISLPED 2002, pp. 48-51.
2. Goto, M. and Sato, T., "Leakage Energy Reduction in Register Renaming", in Proc. 1st Int'l Workshop on Embedded Computing Systems (ECS) held in conjunction with 24th ICDCS, pp.890-895, March 2004.
3. Heo, S. et al, "Dynamic Fine-grain Leakage Reduction using leakage-biased bitlines", in Proc. ISCA 2002, pp. 137-147.
4. Kim, N. S. et al, "Drowsy Instruction Caches - Leakage Power Reduction using Dynamic Voltage Scaling and Subbank Prediction", in Proc. MICRO-35, 2002, pp. 219-230.
5. Mamidipaka, M. and Dutt, N., "eCACTI: An Enhanced Power Estimation Model for On-chip Caches", University of California, Irvine, Center for Embedded Computer Systems, TR-04-28, September 2004.
6. Narendra, S. et al, "Scaling of Stack Effect and its Application for Leakage Reduction", in Proc. ISLPED, 2001, pp.195-200.
7. Powell, M. et al, "Gated Vdd - A Circuit Technique to Reduce Leakage in Deep Submicron Cache Memories", in Proc. ISLPED 2000, pp. 90-95.

Parameter Variation Effects
on Timing Characteristics
of High Performance Clocked Registers

William R. Roberts and Dimitrios Velenis

Electrical and Computer Engineering Department
Illinois Institute of Technology
Chicago, IL 60616-3793
robewil1@iit.edu, velenis@ece.iit.edu

Abstract. Violations in the timing constraints of a clocked register can cause a synchronous system to malfunction. The effects of parameter variations on the timing characteristics of registers that determine the timing constraints are investigated in this paper. The sensitivity of the setup time and data propagation delay to parameter variations is demonstrated for four different register designs. The robustness of each register design under variations in power supply voltage, temperature, and gate oxide thickness is determined.

1   Introduction

Scaling of on-chip feature size has been the dominant characteristic in the evolution of integrated circuits. This trend has produced phenomenal improvements in circuit functionality and performance by increasing the number of transistors on-chip and the transistor switching speed. The continuous quest for higher circuit performance has pushed clock frequencies deep into the gigahertz frequencies range, reducing the period of the clock signal well below a nanosecond. This development has resulted in hundreds of thousands of clocked registers that control the flow of data within a circuit under strict timing constraints [ ]. Violation of the timing constraints at a register can cause incorrect data to be latched within a register, resulting in a system malfunctioning.

To satisfy the timing constraints at the clock registers, tight timing control upon the delay of signals arriving at the registers is required. Effects such as variations in process and environmental parameters [,] and interconnect noise [ ] can introduce uncertainty in signal delay. Delay uncertainty can cause a violation of the tight timing constraints in a register, especially at the most critical data paths within a system [ ].

In addition to uncertainty in signal propagation delay, parameter variation effects may also affect the performance of devices *within* a clocked register. Variations in transistor switching speed [ ] can affect the timing characteristics of a register, thereby causing timing constraint violations. Therefore, a system malfunction may occur, not only due to uncertainty in the signal propagation delay, but also due to variations in the timing characteristics within a register.

V. Paliouras, J. Vounckx, and D. Verkest (Eds.): PATMOS 2005, LNCS 3728, pp. 508–517, 2005.
© Springer-Verlag Berlin Heidelberg 2005

In this paper the effects of process and environmental parameter variations on the timing characteristics of clocked registers are investigated. The sensitivity of the timing constraints is evaluated for four different clocked register designs. A brief summary of the timing characteristics of clocked registers is presented in Section 2. The clocked register designs are introduced in Section 3. The effects of variations in power supply voltage, temperature, and gate oxide thickness on the clocked registers are discussed in Section 4. Finally some conclusions are presented in Section 5.

2 Timing Characteristics of Clocked Registers

A common feature among clocked registers is the use of a common synchronizing signal, namely the *clock* signal, to control the timing of the data storage process. Depending upon the temporal relationship between the clock signal and the data being latched, registers are classified into latches and flip-flops [ ]. In latches, input data is latched while the clock signal is maintained at a specific voltage level. Therefore, latches are characterized as *level-sensitive* registers. In flip-flops, the input data signals are latched by a transition edge in the waveform of the clock signal. For that reason flip-flops are also described as *edge-triggered* registers. The four clocked register designs considered in this paper are edge-triggered registers.

Data is successfully stored within a register if the following timing constraints between the input data signal and the clock signal are satisfied:

- *Setup time* (T_{SETUP}), which determines the minimum time that the value of the data signal should be valid before the arrival of a latching clock signal.
- *Hold time* (T_{HOLD}) that specifies the minimum time that the data signal should remain at a constant value after data storage is enabled by the clock signal.

Furthermore, the propagation delay of the output signal of a register is determined in terms of the temporal relationship between the input data and clock signal:

- *Clock propagation* delay (T_{C-Q}) is defined as the delay between a latching event of the clock signal and the time the latched data is available at the output Q of a register.
- *Data propagation* delay (T_{D-Q}), is the delay between the arrival of a data signal at the input D of a register and the time the data is available at the register output Q. Data propagation delay can also be determined as the sum of the setup time T_{SETUP} and the clock propagation delay T_{C-Q}.

In this paper, the effects of parameter variations on the timing characteristics and constraints of clocked registers are considered. In particular the variations in the clock propagation delay (T_{C-Q}) and setup time (T_{SETUP}) are investigated. These two parameters determine the temporal relationship among the input and output signals of a register and the clock signal.

3   Flip-Flop Designs

The four flip flop designs considered in this paper are introduced in this section. The flip flops are simulated for $0.18\mu m$ technology at 1.8 Volts nominal power supply voltage. The timing characteristics of each flip flop are listed in Table 1. The optimal setup time T_{SETUP} is determined in order to achieve the minimum data propagation delay T_{D-Q}. Notice in Table 1 that the minimum data propagation delay T_{D-Q} for each flip flop is the sum of the optimum setup time T_{SETUP} and the corresponding clock propagation delay T_{C-Q}. The values for T_{SETUP}, T_{C-Q}, and T_{D-Q} are determined for "infinite" hold time, meaning that the input data signal is not switched until the end of the clock period. The T_{HOLD} values listed in Table 1 are determined such as the clock propagation delay is maintained within 3% of the nominal T_{C-Q} value.

Table 1. Timing characteristics at nominal parameter values

Flip Flop Design	Optimal $T_{SETUP}(ps)$	$T_{C-Q}(ps)$	Minimum $T_{D-Q}(ps)$	$T_{HOLD}(ps)$
MUX	56	120.7	176.7	20
PPC603	80	191.4	271.0	0
SAFF	-5	194.4	189.4	80
HLFF	-10	165.2	155.2	130

The flip flop designs listed in Table 1 are the Multiplexer Flip Flop, the PowerPC 603 Flip Flop, the Sense Amplifier Flip Flop, and the Hybrid Latch Flip Flop. These designs are presented in the following sections.

3.1   Multiplexer Flip-Flop

The multiplexer master-slave flip flop design (MUX) [ ] has the shortest T_{C-Q} delay among all of the flip-flops listed in Table 1. A schematic diagram of the MUX flip-flop is illustrated in Figure 1. As shown in Figure 1, two transmission gate latches are connected in series and form a master-slave configuration. Both phases of the clock signal (CLK and $\overline{CLK}$) are used as the select signals for the transmission gate multiplexers.

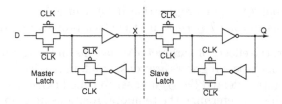

Fig. 1. Multiplexer Flip Flop

3.2   PowerPC 603 Flip Flop

The PPC603 flip flop [ ] illustrated in Figure 2 is originally designed for the PowerPC 603^{TM}microprocessor[1]. As shown in Figure 2, the PPC 603 flip flop is based on a master-slave configuration and both phases of the clock signal are required for latching data. The PPC603 flip flop has the longest T_{D-Q} propagation delay among the considered register designs as listed in Table 1.

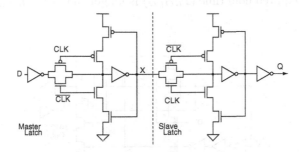

Fig. 2. PPC 603 Flip Flop

3.3   Sense Amplifier Flip Flop

The Sense Amplifier flip flop (SAFF)[ ] is also based upon the master-slave principle, however, a differential sense amplifier is used for the master stage. A diagram of the SAFF is illustrated in Figure 3. It is shown in Figure 3 that a single phase of the clock signal is required, albeit the use of differential input data signals. As listed in Table 1, the optimum setup time for the SAFF is negative. Therefore, in order to achieve minimum T_{D-Q} delay the input data signal should arrive later than the positive edge of the clock signal.

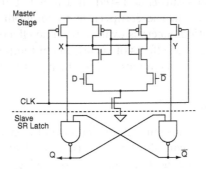

Fig. 3. Sense Amplifier Flip Flop

---

[1] PowerPC 603^{TM}is a registered trademark of International Business Machines Corporation

3.4 Hybrid Latch Flip Flop

The hybrid latch flip flop design (HLFF) [10] is illustrated in Figure 4. HLFF is a pulsed latch register, different from the master-slave designs presented so far. An inverter chain is used to generate a delayed pulse of the clock signal. During that pulse the precharged node X within the flip flop evaluates and the input data is latched. As listed in Table 1, the optimum setup time of the HLFF is negative, and the T_{C-Q} delay is the shortest among the four flip flop designs. However, the required hold time (T_{HOLD}) is longer than the T_{HOLD} of any other design.

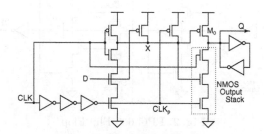

Fig. 4. Hybrid Latch Flip Flop

4 Delay Variations

The effects of parameter variations on the timing characteristics of the flip flops presented in Section 3 are investigated in this section. In particular, the effects of variations in power supply voltage (V_{DD}), temperature ($Temp$), and gate oxide thickness (t_{ox}) on the clock propagation delay (T_{C-Q}) and setup time (T_{SETUP}) of each register are demonstrated. The variation in T_{C-Q} delay due to parameter variations is determined when the setup time of the data signal is equal to the optimal setup time listed in Table 1. Furthermore, for each flip flop design, the increase in the setup time to maintain constant T_{C-Q} delay under parameter variation effects is evaluated. The effects of V_{DD} variations are discussed in section 4.1. The effects of variations in temperature are presented in section 4.2 and the variations in gate oxide thickness is discussed in section 4.3.

4.1 V_{DD} Variations

In high performance synchronous integrated circuits, effects such as the IR voltage drop and $L\frac{di}{dt}$ noise can affect the voltage level at the power supply [11]. Furthermore, power saving mechanisms such as clock gating and system standby that control the switching activity of large circuit blocks within an IC may also affect the power supply voltage level. The effects of V_{DD} variations on the timing characteristics of registers are demonstrated in this section.

The effect of power supply voltage drop on the timing characteristics of the MUX flip-flop are illustrated in Figure 5. The T_{SETUP} and T_{C-Q} delays under nominal V_{DD} value are shown in Figure 5(a). With decreasing V_{DD} the clock propagation delay increases, as shown in Figure 5(b). Further reduction in the power supply voltage causes the flip flop to malfunction as illustrated in Figure 5(c). However, the effects of V_{DD} drop can be compensated by increasing the T_{SETUP} time as shown in Figure 5(d).

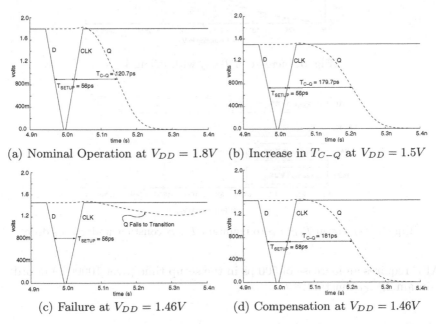

(a) Nominal Operation at $V_{DD} = 1.8V$ (b) Increase in T_{C-Q} at $V_{DD} = 1.5V$

(c) Failure at $V_{DD} = 1.46V$ (d) Compensation at $V_{DD} = 1.46V$

Fig. 5. Effect of V_{DD} drop on clock propagation delay T_{C-Q}

The increase in the T_{C-Q} of each flip flop design is evaluated for a drop on the power supply voltage of 33% below the nominal V_{DD} value and the results are illustrated in Figure 6. Notice in Figure 6 that the operation of the HLFF flip flop fails for power supply drops below the nominal V_{DD} value. Furthermore, the MUX and PPC603 flip flops are operational for up to 18% drop in the power supply. As shown in Figure 6 the SAFF is the most robust design under V_{DD} variations. The SAFF flip flop remains operational for a full drop in power supply voltage of 33% below the nominal V_{DD} value. Furthermore, the percent increase in the T_{C-Q} delay is the smallest compared with the other designs.

In addition, the increase in the setup time of the flip flops in order to maintain a constant T_{C-Q} value under power supply variations is evaluated. The increase in T_{SETUP} for each flip flop, and also the range of V_{DD} voltage drop that the clock propagation delay remains constant are presented in Figure 7. As shown in Figure 7, increasing the setup time for the PPC603, SAFF, and HLFF flip flops can only maintain T_{C-Q} constant for a 6% drop in power supply. Increasing the setup time of the MUX flip flop is much more efficient in keeping the value of T_{C-Q} constant as shown in Figure 7. In addition, notice in Figure 7 that the

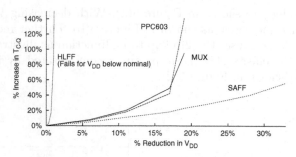

Fig. 6. Increase in T_{C-Q} with falling V_{DD}

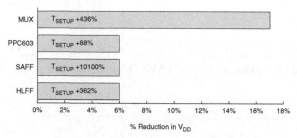

Fig. 7. Increase in T_{SETUP} to maintain T_{C-Q} constant under V_{DD} drop

SAFF requires an increase of 500 ps in the setup time (over 10000%) in order to maintain T_{C-Q} constant.

4.2   Temperature Variations

With increasing die area, circuit density, and on-chip power dissipation the variation of temperature across an integrated circuit becomes significant. The operating temperature of a circuit block depends upon the proximity to a hot spot within a die. Therefore, the effects of temperature on circuit performance are non-uniform across a die. The effects of temperature variations on the register clock propagation delay and setup time constraints are investigated in this section.

Variations in the operating temperature within the range of 27°C to 90°C (an increase of 245% above the nominal value) are considered for the four flip flop designs. The increase in T_{C-Q} with increasing temperature while T_{SETUP} is maintained constant is illustrated in Figure 8. As shown in Figure 8, the operation of the HLFF fails for operating temperatures above the nominal value. Increasing the operating temperature has similar effect to T_{C-Q} of the MUX, PPC603 and SAFF designs. As shown in Figure 8, the T_{C-Q} delay of SAFF is the most robust under operating temperature variations.

In addition, the effect of increasing the setup time to compensate for temperature variations is investigated. The increase in T_{SETUP} and operating temperature for which the clock propagation delay remains constant are presented

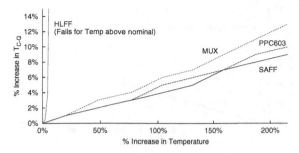

Fig. 8. Increase in T_{C-Q} with increasing temperature

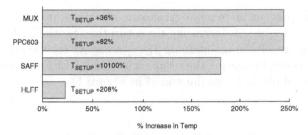

Fig. 9. Increase in T_{SETUP} to maintain T_{C-Q} constant under temperature variations

in Figure 9. It is shown in Figure 9 that increasing the setup time for the MUX and PPC603 flip flops results in maintaining T_{C-Q} constant for the entire range of temperature variation. Furthermore, HLFF is the most sensitive design to temperature variations since increasing the setup time can compensate for an increase in the operating temperature of only 20% above the nominal value.

4.3   Gate Oxide Thickness Variations

As the on-chip feature size is decreased the effects of variations in the manufacturing process are aggravated. One of the device parameters that is susceptible to imperfections of the manufacturing process is the gate oxide thickness (t_{ox}) with a nominal value below 20 angstroms for the current technology nodes [ ]. The slightest variation in the gate oxide deposition process can create a significant variation in t_{ox}. In this section the effects of t_{ox} variation within 10% of the nominal t_{ox} value on the clock propagation delay T_{D-Q} and the setup time T_{SETUP} of the registers are demonstrated.

The increase in T_{C-Q} of the flip flop designs due to variations in t_{ox} while T_{SETUP} is constant is presented in Figure 10. As shown in Figure 10 the operation of HLFF fails for t_{ox} values above nominal. It is also shown in Figure 10 that T_{C-Q} of the MUX flip flop is more sensitive to t_{ox} variations compared with the clock propagation delay of the PPC603 and SAFF designs.

Furthermore, the effect of increasing setup time in order to compensate for variations in t_{ox} is investigated. The increase in T_{SETUP} to achieve constant

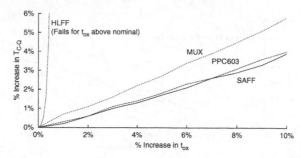

Fig. 10. Increase in T_{C-Q} with increasing gate oxide thickness

T_{C-Q} and the percent increase in gate oxide thickness are illustrated in Figure 11. As shown in Figure 11, increasing the setup time for the MUX, PPC603, and SAFF flip flops results in maintaining T_{C-Q} constant for the entire range of t_{ox} variation. However, increasing the setup time of the HLFF can only compensate for a variation of 6% in t_{ox} as illustrated in Figure 11.

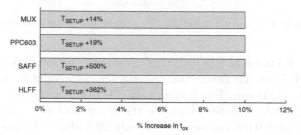

Fig. 11. t_{ox} Increase for which T_{C-Q} is Maintainable

5   Conclusions

The effects of variations in power supply voltage, temperature, and gate oxide thickness on the timing characteristics of high performance clocked registers are demonstrated in this paper. It is shown that the hybrid latch flip flop (HLFF) is the most sensitive design to parameter variations due to violations of internal timing conditions. In addition, it is shown that the clock propagation delay T_{C-Q} of the sense amplifier flip flop (SAFF) is the most robust under parameter variations, when the setup time T_{SETUP} is constant. However, maintaining the T_{C-Q} of SAFF constant required the greater percent increase in the setup time of the input data signal. The multiplexer and PowerPC 603 flip flops demonstrated similar sensitivity to variations in V_{DD}, temperature, and t_{ox}. Maintaining the T_{C-Q} of the MUX flip flop constant required the smallest increase in setup time among the four register designs.

References

1. I. S. Kourtev and E. G. Friedman, *Timing Optimization Through Clock Skew Scheduling.* Norwell Massachusetts: Kluwer Academic Publishers, 2000.
2. S. Sauter, D. Schmitt-Landsiedel, R. Thewes, and W. Weber, "Effect of parameter variations at chip and wafer level on clock skews," *IEEE Transactions on Semiconductor Manufacturing*, vol. 13, no. 4, pp. 395–400, November 2000.
3. S. Natarajan, M. A. Breuer, and S. K. Gupta, "Process variations and their impact on circuit operation," *Proceedings of the IEEE International Symposium on Defect and Fault Tolerance in VLSI Systems*, pp. 73–81, November 1998.
4. K. T. Tang and E. G. Friedman, "Delay and noise estimation of cmos logic gates driving coupled rc interconnections," *Integration, the VLSI Journal*, vol. 29, no. 2, pp. 131–165, September 2000.
5. D. Velenis, M. C. Papaefthymiou, and E. G. Friedman, "Reduced delay uncertainty in high performance clock distribution networks," *Proceedings of the IEEE Design Automation and Test in Europe Conference*, pp. 68–73, March 2003.
6. R. Sitte, S. Dimitrijev, and H. B. Harrison, "Device parameter changes caused by manufacturing fluctuations of deep submicron mosfet's," *IEEE Transactions on Electron Devices*, vol. 41, no. 11, pp. 2210–2215, November 1994.
7. N. H. Weste and D. Harris, *CMOS VLSI Design: A Circuits and Systems Perspective*, 3rd ed. Boston: Addison-Wesley, May 2004.
8. G. Gerosa, S. Gary, C. Dietz, D. Pham, K. Hoover, J. Alverez, H. Sanchez, P. Ippolito, T. Ngo, S. Litch, J. Eno, J. Golab, N. Vanderschaaf, and J. Kahle, "A 2.2 w, 80 mhz superscaler risc microprocessor," *IEEE Journal of Solid-State Circuits*, vol. 29, no. 12, pp. 1440–1454, December 1994.
9. V. Stojanovic and V. G. Oklobzijia, "Comparative analysis of master-slave latches and flip flops for high-performance and low-power systems," *IEEE Journal of Solid-State Circuits*, vol. 34, no. 4, pp. 536–548, April 1999.
10. H. Partovi, R. Burd, U. Salim, F. Weber, L. DiGregorio, and D. Draper, "Flow-through latch and edge-triggered flip-flop hybrid elements," *IEEE International Solid-State Circuits Conference*, pp. 138–139, Febuary 1996.
11. A. V. Mezhiba and E. G. Friedman, *Power Distribution Networks in High Speed Integrated Circuits.* Norwell Massachusetts: Kluwer Academic Publishers, 2004.
12. S. I. A., "The national technology roadmap for semiconductors," Semiconductor Industry Association, Tech. Rep., 2003.

Low-Power Aspects
of Nonlinear Signal Processing

Konstantina Karagianni and Vassilis Paliouras

Electrical and Computer Engineering Department
University of Patras, Greece

Abstract. This paper discusses techniques that reduce the complexity
and power dissipated by nonlinear operators in a digital signal processing
system. In particular, a technique is presented that allows the designer
to trade signal quality measured in SNR for power dissipation reduction.
Signal quality refers here to the severity of the impact of overflow. The
proposed approach is based on the nonlinear operation of saturation,
as a means to control overflow. Furthermore, the organization of digital
circuits for the computation of powers is discussed.

1 Introduction

Nonlinear behavior arises in a system due to its nature or is deliberately introduced by the designer. While the variety of possible nonlinear elements in a
system is infinite, it is possible to distinguish several categories:

1. Computation of value nonlinearities, such as as powers, sinusoids, and expo-
   nentials;
2. Bilinear systems, linear in the state and input separately;
3. Piecewise linear systems; and
4. Nonlinearities with memory.

The particular categorization is implied by both the theoretical techniques employed to analyze the behavior of these systems, as well as implementation-
related issues. This paper focuses on value nonlinearities from the digital system implementation viewpoint. This category comprises nonlinearities that exist
practically in all digital signal processing systems, due to the finite-word length
limitations of data representation.

A common technique to combat overflows in digital signal processing systems
is the use of saturated arithmetic. Saturation is a nonlinear behavior exhibited in
several diverge contexts [ ]. The digital counterpart is commonly used to mitigate
the effect of potential overflows on signal quality [ ]. This paper discusses a
technique to reduce both the complexity of saturation and the underlying power
dissipation.

Other kinds of memoryless nonlinearity, usually found in the implementation
of a digital signal processing system include the computation of powers of the
signal. This paper proposes a design technique to derive new low-complexity
circuits for the computation of the Nth power of a number.

V. Paliouras, J. Vounckx, and D. Verkest (Eds.): PATMOS 2005, LNCS 3728, pp. 518–527, 2005.

The remainder of this paper is organized as follows: Section 2 discusses the basics of saturated arithmetic and proposes a technique to minimize the corresponding power dissipation. Section 3 discusses a design technique to minimize the complexity of the computation of powers. Finally conclusions are discussed in section 4.

2 Saturated Arithmetic

Quantization is a form of nonlinearity inherent in digital signal processing systems, due to the nature of the finite word length representations.

Saturation is a common nonlinearity in signal processing. When a partial result exceeds the available dynamic range offered by the adopted representation, it is replaced by a number of the same sign and of the maximum absolute value that can be represented in a given representation. The hardware unit that implements saturation arithmetic is called *limiter* [ ]. The operation of saturation performed by the limiter, is described by:

$$Q(x) = \begin{cases} x_{max}, & x \geq x_{max} \\ x, & x_{min} \leq x \leq x_{max} \\ x_{min}, & x \leq x_{min} \end{cases}, \tag{1}$$

where x_{min} and x_{max} denote the minimum and maximum representable numbers. For n-bit two's-complement representation, it holds that

$$x_{min} = -2^{n-1} \tag{2}$$

$$x_{max} = 2^{n-1} - 1. \tag{3}$$

In case the word is organized as a fixed-point word of k integral bits and l fractional bits, i.e., $n = k + l$, (2) and (3) become

$$x_{min} = -2^{k-1} \tag{4}$$

$$x_{max} = 2^{k-1} - 2^{-l}. \tag{5}$$

2.1 Impact on Signal Quality

In case no provision is taken to combat overflow, the signal quality degrades. Signal quality degradation is modeled as the superposition of a noise signal on the ideal signal. The noise signal $r[n]$ is computed as

$$r[n] = x[n] - \widehat{x}[n], \tag{6}$$

where $x[n]$ is the desired ideal signal and $\widehat{x}[n] = Q(x[n])$, where $Q(\cdot)$ denotes the application of finite word length, including overflow. The power of an N-point noise sequence is defined to be

$$P_r = \sum_N (r[n])^2, \tag{7}$$

while the signal power is

$$P_x = \sum_N (x[n])^2. \tag{8}$$

A common measure of signal quality is the Signal-to-Noise Ratio (SNR), defined as:

$$SNR = \log_{10} \frac{P_x}{P_r}, \tag{9}$$

when measured in dB.

It is noted that there exist different approaches than the application of saturation to combat overflow. For example proper input signal scaling, or scaling at particular nodes in a signal-flow graph representing a computation can be employed to prevent overflow (cf. [ ]). A more straightforward strategy is to increase the available dynamic range as the computation evolves.

2.2   Saturated Addition

The implementation of saturated addition does not require comparisons. It is well known that an overflow flag (OVF) can be easily produced by most adder organizations. OVF can be exploited in combination with a sign bit to determine the output of a signed saturated addition, as shown in Fig. 1. The area complexity

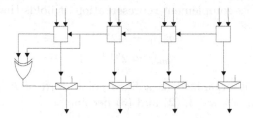

Fig. 1. Organization of a saturated binary adder

A of the two's-complement implementation of saturated addition, assuming that it is based on a ripple-carry adder is

$$A = nA_{\mathrm{FA}} + nA_{\mathrm{mux}} + A_{\mathrm{xor}} + A_{\mathrm{inv}}, \tag{10}$$

where A_{FA}, A_{mux}, A_{xor}, and A_{inv} are the area complexities of a 1-bit full adder, a multiplexer, a xor gate, and an inverter. Eq. (10) is actually a lower bound due to ignoring fan-out limitations.

2.3   Proposed Scheme

It is here shown that a simplification of the saturated adder scheme allows for a reduction of the power dissipated by saturated addition at the cost of a moderate decrease of signal quality.

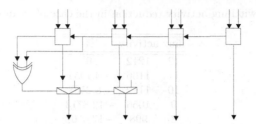

Fig. 2. Proposed simplification of a saturated binary adder

The proposed saturated adder organization is depicted in Fig. 2. The main idea of the proposed simplification is that multiplexers are placed at the m most significant positions of the output of the adder only. The complexity of this scheme is

$$A = nA_{\mathrm{FA}} + mA_{\mathrm{mux}} + A_{\mathrm{xor}} + A_{\mathrm{inv}}, \qquad (11)$$

with $m < n$.

The impact of the proposed scheme on the noise power is quantified in Table 1, for the 12-bit case.

Table 1. Noise power with respect to the number of employed multiplexers, in a 12-bit two's complement saturated adder

mux	noise power	% difference
12	184655477	0
11	184859916	0.1107
10	185180628	0.2844
9	186057492	0.7593
8	187536084	1.5600
7	190721940	3.2853
6	196617748	6.4782
5	210244628	13.8578
4	236002324	27.8068
3	296789012	60.7258
2	438789140	137.6258
1	964113428	422.1147

Data in Table 1 reveal that for the particular adder, the reduction of the number of multiplexers from 12 to 8 approximately increases signal noise power by a moderate 1.5%. The particular simplification reduces the hardware complexity of the saturated adder by reducing the number of the multiplexers, since $m < n$. Naturally the proposed scheme reduces the corresponding switching activity. The reduction of the switching activity is reported in Table 2.

The trade-off between complexity and signal quality is experimentally explored with respect to the data word length. Experimental results show that for

Table 2. Switching activity reduction in the case of saturated addition

mux	activity	%
12	1212	0
11	1166	−3.7954
10	1112	−8.2508
9	1056	−12.8713
8	998	−17.6568
7	948	−21.7822
6	902	−25.5776
5	852	−29.7030
4	802	−33.8284
3	746	−38.4488
2	698	−42.4092
1	652	−46.2046

two's-complement saturated adders of word length 8, the use of 7 multiplexers imposes a noise power increase of less than 2%. However, for two's-complement saturated adders of word length of 12, 16, 24, and 32 bits only 8 multiplexers, placed at the most significant bit positions, imply signal power increase of less than 2%.

Assuming that less that 2% of increase in signal noise power is acceptable, significant switching activity reduction is achieved, as depicted for various adder word lengths in Fig. 3.

2.4   Saturated Multiplication

Assume that x and y are two's complement data words with k integral bits and l fractional bits. Then it holds that

$$- 2^{k-1} \leq x, y \leq 2^{k-1} - 2^{-l}. \tag{12}$$

The product xy is limited as follows

$$- 2^{k-1} \left(2^{k-1} - 2^{-l}\right) \leq xy \leq \left(2^{k-1}\right)^2 \Longleftrightarrow \tag{13}$$
$$-2^{2k-2} + 2^{k-1-l} \leq xy \leq 2^{2k-2}. \tag{14}$$

Since the range described by (14), in which a product assumes values, is wider than the range of (12), which is the range of representable numbers, overflow occurs.

For the common case of $k = 1$, (14) gives

$$- 2^{2 \cdot 1-2} + 2^{1-1-l} \leq xy \leq 2^{2 \cdot 1-2} \Longleftrightarrow -1 + 2^{-l} \leq xy \leq 1. \tag{15}$$

When compared to (12), it is obtained that only a positive overflow can possibly occur.

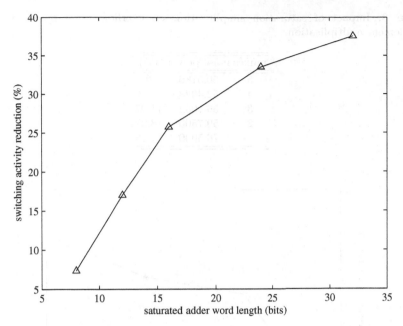

Fig. 3. Percentage reduction of switching activity for various word length values in an adders limiter. Overflow noise power increase is limited to less than 2%, for each case

A positive overflow occurs when

$$2^{2k-2} > 2^{k-1} - 2^{-l} \Longleftrightarrow \tag{16}$$
$$(2^{k-1})^2 - 2^{k-1} > -2^{-l}, \tag{17}$$

which is true for all integers, $k \geq 1$. Therefore positive oveflows can always occur. A negative overflow occurs when

$$-2^{2k-2} + 2^{k-1-l} < -2^{k-1} \quad \Longleftrightarrow \tag{18}$$
$$-(2^{k-1})^2 + 2^{k-1} + 2^{k-1}2^{-l} < 0 \quad \Longleftrightarrow \tag{19}$$
$$2^{k-1} > 1 + 2^{-l}. \tag{20}$$

Hence for all $k > 1$, a negative overflow is also possible. Notice that for $k = 1$, a negative overflow does not occur.

Table 3 presents experimental results for the case of multiplication of two numbers organized into two's-complement words of two integral bits and five fractional bits. Assuming that a 2% increase in overflow signal noise is tolerable, the switching activity reduction in the limiter for various integral word lengths is depicted in Fig. 4.

Table 3. Impact of saturation simplification on overflow noise power in two's-complement multiplication

muxes	noise power	(%)
5	50.5186	0
4	52.4023	3.73
3	56.1094	11.07
2	59.7500	18.27
1	76.5000	51.43

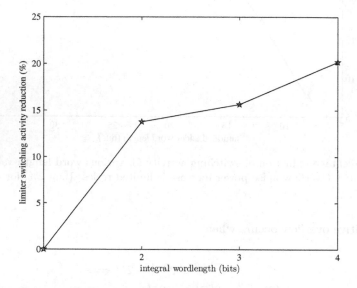

Fig. 4. Percentage reduction of switching activity for various integral word length values in a multiplier's limiter. Overflow noise power increase is less than 2%

3   Computation of Powers

The computation of x^2 is usually performed by exploiting the property of the bit products

$$x_i x_j = x_j x_i, \tag{21}$$

therefore,

$$x_i x_j + x_j x_i = 2 x_i x_j, \tag{22}$$

with $i < j$ and $x_i \in \{0, 1\}$. The computation of higher powers is usually decomposed to cascading computation of multiplication and squarings (computations of x^2) [ ]. Here it is shown that by extending the design concepts usually applied to the design of units for x^2, significant reduction in the complexity of x^3 is achieved.

Assume the unsigned four-bit computation of x^2. Since x is assumed unsigned, it can be written as

$$x = \sum_{i=0}^{n-1} x_i 2^i,$$ (23)

where $x_i \in \{0, 1\}$.

The output of the squarer is given by

$$x^2 = \sum_k c_k 2^k,$$ (24)

where $c_k = \sum_{i+j=k} x_i x_j$. Therefore, in order to derive x^3, the expression can be used:

$$x^3 = x \cdot x^2 = \sum_l x_l 2^l \sum_k c_k 2^k = \sum_l \sum_k x_l c_k 2^{l+k}.$$ (25)

By using the bit product property that

$$x_i x_i = x_i,$$ (26)

and, generally, $x_i^k = x_i$, k, integer, $k > 0$, the expression for x^3 can be significantly simplified. The sought result is expressed as

$$x^3 = \sum_k d_k 2^k,$$ (27)

where d_k denotes a sum of bit products and may assume values larger than one; therefore d_ks are not single-bit quantities. Values d_k are important in the design procedure because they reveal the particular bit products that need to be added at the kth digital position. By defining the bit products f_k, as follows

$$f_1 = x_0 x_1$$ (28)
$$f_2 = x_0 x_2$$ (29)
$$f_3 = x_0 x_3$$ (30)
$$f_4 = x_1 x_2$$ (31)
$$f_5 = x_1 x_3$$ (32)
$$f_6 = x_2 x_3,$$ (33)

and by using (25), it can be obtained that the required d_ks are given as

$$d_0 = x_0$$ (34)
$$d_1 = 3x_0 x_1 = 3f_1$$ (35)
$$d_2 = 3x_0 x_2 + 3x_0 x_1 = 3f_2 + 3f_1$$ (36)
$$d_3 = 3x_0 x_3 + 6x_0 x_1 x_2 + x_1 = 3f_3 + 6x_2 f_1 + x_1$$ (37)
$$d_4 = 6x_0 x_1 x_3 + 3x_0 x_2 + 3x_1 x_2 = 6x_3 f_1 + 3f_2 + 3f_4$$ (38)
$$d_5 = 6x_3 f_2 + 3f_5 + 3f_4$$ (39)

$$d_6 = 3f_3 + 6x_3f_4 + x_2 \tag{40}$$

$$d_7 = 3f_5 + 3f_6 \tag{41}$$

$$d_8 = 3f_6 \tag{42}$$

$$d_9 = x_3. \tag{43}$$

The constant multipliers that scale the various bit products in the above equations, declare that a particular bit product should be allocated in several bit positions. For example, since $d_1 = 3f_1 = f_1 + 2f_1$, bit product f_1 should be allocated to bit positions of significance 1 and 2. No explicit multiplication by the factor 3 is required. The proposed design strategy for the computation of x^3 comprises the following steps:

1. Expression of the multiplicative scaling factors as sums of powers of two;
2. Allocation of bit products to positions starting from the least significant and proceeding to the most significant ones; and
3. Derivation of an adder-based reduction structure.

By iteratively decomposing the constant multiplication factors into sums of powers of two, a significant reduction of bit products is achieved. Bit products multiplied by powers of two are appropriately moved to more significant d_ks. When combined with existing bit products, new scaling factors occur, which are subsequently decomposed into powers of two. The procedure is clarified with an example.

$$
\begin{array}{ll}
d_0 = x_0 & d'_0 = d_0 \\
d_1 = 2f_1 + f_1 & d'_1 = f_1 \\
d_2 = 2f_1 + f_1 + & \Longrightarrow \quad d'_2 = f_1 + f_1 + f_2 \\
\quad\quad 2f_2 + f_2 & d'_3 = d_3 + f_2 + f_1
\end{array}
\tag{44}
$$

In (44), bit product f_1 is moved from d_1 to d_2, as is appears multiplied by 2. Similarly, f_1 and f_2 are moved from d_2 to d_3. The particular bit product re-allocations define d'_1 and d'_2.

$$
\begin{array}{ll}
d'_0 = d_0 & d'_0 = d_0 \\
d'_1 = f_1 & d'_1 = f_1 \\
d'_2 = f_1 + f_1 + f_2 & \Longrightarrow \quad d'_2 = f_2 \\
d'_3 = d_3 + f_2 + f_1 & d'_3 = d_3 + f_2 + f_1 + f_1
\end{array}
\tag{45}
$$

In (45) the new scaling factors are decomposed into sums of powers of two. Therefore the sum $f_1 + f_1$ for d'_2 is reduced to the addition of f_1 in d'_3.

$$
\begin{array}{ll}
d'_0 = d_0 & d'_0 = d_0 \\
d'_1 = f_1 & d'_1 = f_1 \\
d'_2 = f_2 & \Longrightarrow \quad d'_2 = f_2 \\
d'_3 = d_3 + f_2 + f_1 + f_1 & d''_3 = d'_3 + f_2 \\
& d'_4 = d_4 + f_1
\end{array}
\tag{46}
$$

In (46) the sum $f_1 + f_1$ in the computation of d'_3 is reduced to the addition of f_1 to d_4. The procedure continues until all d_k have been processed. Subsequently, an adder structure is employed to add the allocated bit products.

For the $n = 4$ case, the procedure leads to a structure with 15 full adders and 7 half adders, compared to 25 full adders and 8 half adders required by a multiplier-squarer combination.

4   Conclusions

A major obstacle in the wider use of nonlinear techniques in signal processing is the underlying complexity. In this paper techniques have been presented that can lower the complexity of certain simple memoryless nonlinear operations, which are of practical interest even in linear signal processing applications.

In particular, the implementation of the limiter for addition has been studied. It is shown that a trade-off exists between signal noise power and limiter complexity.

The proposed techniques have been shown to reduce complexity without imposing any impact on signal quality, or with minor increase of noise power.

References

1. P. A. Cook, *Nonlinear dynamical systems*. Prentice Hall, 1986.
2. P. Lapsley, J. Bier, A. Shaham, and E. A. Lee, *DSP Processor Fundamentals: Architectures and features*. IEEE Press, 1997.
3. K. Parhi, *VLSI Digital Signal Processing Systems*. Wiley, 1999.
4. B. Parhami, *Computer Arithmetic - Algorithms and Hardware Designs*. New York: Oxford University Press, 2000.

Reducing Energy Consumption of Computer Display by Camera-Based User Monitoring

Vasily G. Moshnyaga and Eiji Morikawa

Dept. of Electronics Engineering and Computer Science, Fukuoka University
8-19-1 Nanakuma, Jonan-ku, Fukuoka 814-0180, Japan
vasily@fukuoka-u.ac.jp

Abstract. This paper proposes an approach for reducing energy consumption of computer display. Unlike existing power management schemes, which link the display operation to a key press or movement of the mouse, we employ a video camera to bind the display power state to the actual user's attention. The proposed method keeps display active only if its user looks at the screen. When the user detracts his or her attention from the screen, the method dims the display down or even switches it off to save energy. Experiments show that the method can reduce the display energy significantly in environments which frequently detract the display viewer.

1 Introduction

1.1 Motivation

Display is one of the major power consumers in modern computer system. The 19" Sony SDM-S93 LCD monitor (1280x1024pixels), for instance, burns in active mode 50W or almost 38% of the total PC system power (130W). Although laptop displays do not use as much power, it is still a relatively big consumer. With new applications like mobile video players, electronic books, etc., LCD makers are being called on to cut power consumption while pro-viding better images. Portability, however, is by no means the sole driving force behind the push for low-power displays. Rapid utilization of multiple displays - each consuming tens of watts - throughout homes and buildings increases cost and environmental impact of energy consumption significantly. Although most PC displays support power management, new robust methods are needed for evolving display usage scenarios.

1.2 Related Research

Several schemes have been proposed to optimize energy consumption of computer displays. Modern PCs, which conform to VESA's Display Power Management Signaling[ ], turn display to low power states (standby, suspend, and off) after a specified period of inactivity on mouse and/or keyboard. The Advanced Configuration and Power Interface Specification [ ], developed by HP, Intel, Microsoft,

V. Paliouras, J. Vounckx, and D. Verkest (Eds.): PATMOS 2005, LNCS 3728, pp. 528–539, 2005.

Table 1. Default APM setting for Sony Vaio PCG-Z1V/P Monitor

Mode	APM off	Word/Text	DVD	Music	Camera
Brightness level	9	7	9	5	9
Inactivity interval	never	10	never	10	10

Phoenix, and Toshiba, links the display brightness and inactivity intervals to application. One problem with this adaptive power management (APM) is that it strongly depends on inactivity intervals, either set by default (see 1) or by the user. If set improperly (e.g. to shut the display down after 1 minute of idle time), the APM can be quite troublesome, switching the display off when it must be on. At the same time, if inactivity interval is long, the APM efficiency is low. Modifying the inactivity interval can be done only through system settings, which many users consider annoying. As result, they set it to several (10, 20 or even 30) minutes and so shrink the energy savings.

Another approach to save the display's energy is to scale down or dim the backlight luminance. This approach is based on observation that transmissive and transflective color TFT LCD panels [ ] do not illuminate itself but filter a backlight, the primary source of display energy dissipation. Because simply dimming the backlight degrades the display visibility, Choi, et al [ ] proposed to maintain brightness or contrast of the LCD panel when the backlight is dimmed down. To reduce the average energy demands of the backlight, Gatti, et al [ ] suggested the backlight auto regulation scheme. Cheng and Pedram[ ] showed that a concurrent brightness and contrast scaling (CBCS) technique further enhances image fidelity with a dim backlight, and thus saves an extra power. Chang, et al [ ] introduced a dynamic luminance scaling or DLS technique that dimmed the backlight while allowing more light to pass through the screen panel to compensate for the loss of brightness in the original image. Shim, et al [ ] combined the DLS technique with dynamic contrast enhancement and applied it for transflective TFT LCD panels. Pasricha, et al[ ] presented an adaptive middleware-based technique to optimize backlight power when playing streaming video. A modification of the LCD panel to permit zoned backlighting has been reported in [ ].

There are also a variety of schemes for automated adjustment of brightness in high dynamic range panoramic images, e.g. [ ]-[ ]. These schemes dynamically brighten or darken image regions depending on the scene content and average local luminance. The view-port center (the center of area with dimensions $(width/2 \times height/2)$ is constantly monitored and the average pixel luminance of the view-port is calculated. This value is then compared with a preset optimum value and brightness is adjusted accordingly.

Despite differences, the proposed brightness and/or contrast adjustment techniques have one feature in common. Namely, they work independently of the viewer attention. The techniques are able to lower the display e nergy in active mode especially when showing images or video. But if a display delivers text or idles for some time, the APM remains the only energy savior. If the relation between the inactivity interval (T) and the time of keyboard (K) and/or

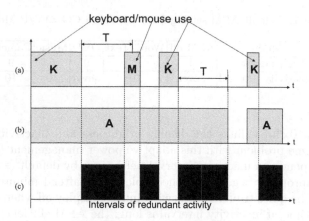

Fig. 1. An illustrative example

mouse (M) typing is as in Fig. 1(a), the display will be active (A) almost all the time (see Fig1.b). Surely, if nobody looks at the screen in between key presses, activating the display is unnecessary for all the intervals shown in Fig.1(c).

We claim that ignorance of the viewer's attention is the main problem of existing display power management schemes. While the display operation must depend on the viewer, none of the schemes, up to our knowledge, takes the viewer's focus into account. Because existing power management depends on a key press but not user's eyes, it can not distinguish whether the user looks at screen or not. As a result, it may either switch the display off inappropriately (i.e. when the user looks at screen without pressing a key) or stay in active mode while idling. We propose a method which can solve this problem.

1.3   Contribution

In this paper, we introduce a new Camera-Driven Display Power Management (CDDPM) method and describe techniques of its implementation in personal computer system. Unlike existing approaches, which only 'senses' a user through keyboard and/or mouse, the CDDPM 'watches' a user through a video camera. It detects the viewer's focus and keeps the display bright only if he or she looks at the screen. When the viewer detracts his or her attention from the screen, the method dims the display down or even switches it off to save energy.

The CDDPM method is based on well-known techniques used for face extraction and eye-gaze detection[13]-[21]. We contribute to these techniques by extending their application to a new field, namely, energy reduction of computer systems. Amid possible solutions, we chose those which provided real-time eye-gaze detection with a single camera and low computational overhead.

This paper is organized as follows. In Section 2 we discuss the proposed approach and outline its implementation scheme. In Section 3 we describe the experimental results. Section 4 summarizes our major findings and outlines the future work.

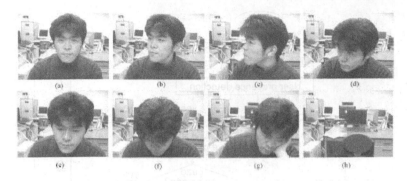

Fig. 2. Example images of the PC user

2   The Proposed Approach

2.1   Main Idea and Overview

The approach we propose is based on the observation that PC users do not view the display quite often. When detracted by phone, colleague, kids, coffee, etc., they usually do not switch the display onto the power saving mode, leaving it active while talking, reading, or even stepping away from PC for a break. Even though screen savers do not save power, many users leave monitors on screen savers for hours. This is especially a case with PCs operated at home, office or school where a user averts his/her attention frequently. Surely, keeping the monitor active when nobody looks at it is unreasonable. The main goal of our CDDPM method is to increase the PC energy efficiency by enabling the computer to "see" its user and lower the display power whenever the user's attention is detracted from the screen. The main idea of CDDPM is simple. When the PC user looks at screen, as illustrated in Fig.2 (a), the display is kept active in 'power-up' mode to provide the best visibility. If the user detracts his/her attention from the screen, as in Fig.2 (b-g), the method dims the backlight luminance to decrease energy consumption. Finally, if the user has not been looking at screen for a long time or disappeared from the camera's range (see Fig.2,h), the display is turned off.

The CDDPM uses the following assumptions:

1. The PC is equipped with a color video camera. Even though some PCs now do not have a camera, the recent advances in CMOS camera development and miniaturization allow us to assume that all PCs will be enabled with a color CMOS video camera in the near future. This can be an image sensor, such as [ ], embedded in display for viewer monitoring, or a general purpose visual communication camera,connected via USB port for video capture, conferencing, etc. In both cases we assume that the viewer monitoring mode is optional to the user. That is, the user can activate the viewer monitoring whenever the camera is not engaged in other applications.

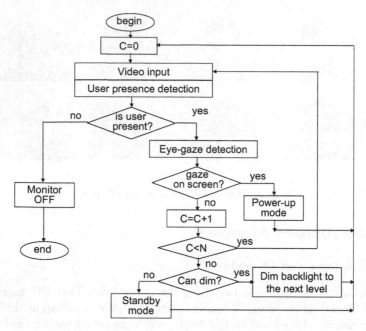

Fig. 3. Flowchart of the proposed approach

2. The display has a number of backlight intensity levels with the highest level corresponding to the largest power consumption and the lowest level to the smallest power, respectively. The highest level of backlight intensity is enabled either initially or whenever the user looks at the screen.

3. If the user has not been looking at the screen for N consecutive frames, the backlight intensity goes down to the next level. If the lowest level has been reached and the user is still not looking at the screen for N consecutive frames, the display is put onto standby mode. Returning back from either standby or OFF modes is done by pushing the ON button.

Figure 3 shows the CDDPM flowchart. Here, C counts the number of consecutive image frames in which the user's gaze is off the screen. If no user is detected in the current video frame, the display is turned off. Otherwise, the method tracks the user's eye-gaze. If the gaze has been off the screen for more than N consecutive frames, the current backlight luminance is dimmed down to the next level. Any on-screen gaze reactivates the initial backlight luminance by moving the display onto power up mode. However, if no on-screen gaze has been detected for more than N frames and the backlight luminance has already been lowest, the display enters the standby mode. Below we describe the method in detail.

2.2   User Presence Detection

The goal of this task is to determine from the camera readings whether or not the user is currently present in front of display. To solve the task, we employ

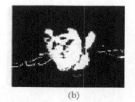

(a)   (b)   (c)

Fig. 4. An illustration of the user detection process

well-known face detection techniques [21]. Several approaches such as shape-based, feature-based, pattern-based, color-based, motion based etc., have been investigated. Due to high efficiency and a comparatively small computation cost of skin-color oriented techniques [18]-[20], we chose color as a primary tool in detecting a face in an image with complex background. To reduce computation load further, we use a look-up table instead of Gaussian model to detect the user presence in the frame.

Figure 4 illustrates the user detection process. After receiving a new color frame (see Fig.4,a) we first transform the RGB color image obtained from video camera into the $\{Y, r, b\}$ representation. Chromatic colors (r, b) known as pure colors in the absence of brightness (Y), are defined as

$Y = 0.299 \times R + 0.587 \times G + 0.144 \times B; \; r = R - Y; \; b = B - Y.$

Next, we digitize the $\{Y, r, b\}$ image. For the segmentation of skin color region, we consider hue and saturation values as discriminating color information because of reducing the lightening effect on images. If the $\{Y, r, b\}$ representation of a pixel satisfies the following thresholds: $10 < Y < 240$, $10 < r < 50$, $-20 < b < 10$, we replace it by 1, else by 0. (The thresholds have been determined empirically).

Figure 4(b) shows the image after binarization. As we observe, the resulted image has a number of isolated spot regions which are caused by light variation and noise. To eliminate these noised and extraneous background pixels that may have been present in the image, we divide the image into blocks of 3x3 pixels in size and then apply area growing/shrinking techniques based on contents of each block. We assume that pixels of the block will become white (black) if more than half of its pixels are white (black). Figure 4 (c) illustrates the results after noise reduction.

If the total number of white pixels in the resulted image exceeds the predefined threshold (W), the user is detected in the frame. Otherwise, the method detects that the user is absent and turns the display onto low-power (standby) mode.

2.3 Eye-Gaze Detection

The goal of this task is twofold: to find eyes in a face region and then based on their number and position determine direction of eye-gaze. The eye gaze-detection is activated if and only if the previous task has detected the user in front of display. The task is solved in the following steps:

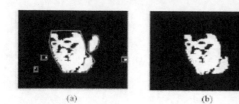

(a)   (b)   (c)

Fig. 5. An illustration of the eye tracing: (a) image labeling; (b) face extraction; (c) eyes localization

1. **Labeling.** All pixels belonging to a separate continuous region are labeled by the same (region's) number. Based on the labeling results we detect a number of image regions (e.g. 5 in Fig.5,a) and compute the size (in number of pixels) for each of them.

2. **Face extraction.** The largest connected region of the white colored pixels in the camera image is selected. Figure 5(b) illustrates the results.

3. **Eyes localization.** To solve this task we search for circle-shaped regions within the extracted face and for each region compute the circumference ratio $E = 4 \times \pi \times S/l2$. Here, S is the area of region, and l is the circumference length of the region. Since the higher E corresponds to a more true circle, we select those regions which have $0.3 < E < 1.0$ and $30 < S < 150$. The centers of the circles define eye positions in the current frame. Figure 5(c) shows the result. Located eye positions are marked by overlay graphics.

4. **Gaze detection.** The distance (D) between the eye centers is computed. If the distance can not be established (e.g. the number of detected eyes is less than two) or it is smaller than a threshold, L, the eye-gaze is detected to be off the screen. In this case, if $C > N$, we dim the backlight luminance down to the next level by decrementing the backlight voltage ($V = V - \Delta V$). Otherwise, the eye-gaze is considered "on the screen". In this case, the backlight luminance is set-up to the highest level by (the backlight voltage returns to the initial value, V_0).

5. **Search window optimization.** Once we have located the positions of eyes on the first frame, we use this knowledge to reduce complexity of processing the successive frames. Since the user face motion is slow in practical circumstances, we limit the search region in the next frame to plus or minus 16 pixels in vertical and horizontal direction of the original pixel locations of the eyes in the current frame.

3   Experimental Results

To evaluate the proposed method, we developed prototype software and run it in the Linux (Fedora project) OS on two computers: a desktop and a notebook. Table 2 specifies computers used in the experiment. Each computer was equipped with a CCD Logitec Quick CAM PRO4000 video camera (160x120 frame size, 30fps) installed above the display.

Table 2. Tested PC systems

Specification	CPU	Memory	Display
Desktop PC	Pentium-4 @2.53GHz	512MB (RAM) 100GB (HDD)	15" TFT Iiyama TXA3822JT 50W(max),0.63W(standby)
Notebook PC Sony Vaio PCG-Z1V/P	Pentium-M @1.6GHz	512MB (RAM) 60GB (HDD)	14.1" SXGA LCD 5W(max),0.63W(standby)

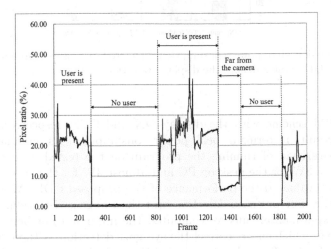

Fig. 6. The user presence monitoring results

First, we tested the user presence detection program (i.e. without running the eye-gaze detection). After extensive profiling in different conditions (user sits/stands/walks away, head moves, hands move, distance change, etc.) we found that $W = 0.06$ allowed correct user presence detection almost at 1.5 meters distance from the user and the display/camera.

Figure 6 demonstrates the results obtained during a user monitoring test of the following scenario: the user was present in front of the display (frames 1-299, 819-1491, 1823-2001); moved a little from the display but still present from the camera perspective (frames 1300 to 1491); stepped away from PC, disappearing from the camera (frames 300-818, 1492-1822). In this figure, the ordinate shows the ratio of pixels representing the user to the total number of pixels in the frame; the abscissa is the frame number. We observe that the program clearly differentiated events when the user was present in front of display from those when he was absent. The abrupt variation of the plot between 819-1491 frames is due to the user movement (head turns up-down, left-right, user is close/distant to the screen, hand moves, etc.) tested in the experiment. Detecting the user presence on the desktop PC took less than 20 ms per frame on a peak and 12.2 ms on average.

We measured the total power consumed by the systems (computer, display and camera) during the program run and compared it to that consumed by

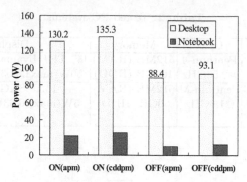

Fig. 7. The peak system power consumption measured during user presence detection. The ON and OFF states denote the display operation modes

existing APM scheme with the display ON and OFF, respectively. Figure 7 shows the results in terms of peak power consumption. As one may conclude, the power overhead of running the program on the tested hardware is quite visible: over 5.3W for the desktop PC and almost 3W for the notebook.

Next, we evaluated the performance of the proposed CDDPM method running both the user presence detection and the eye-gaze detection. Based on a variety of experimental runs we computed the minimum number of pixels between the eyes which ensure correct the eye-gaze detection. Although it depends on the user farness from the camera, lighting, display size, etc., the values of $L = 18$ pixels ensured robust eye-gaze detection at the distance up to 1.5m (ordinary lighting, 15" monitor size). The execution time was 30ms per frame (on peak) and 24ms (on average).

Finally, we report on energy efficiency of the CDDPM method. Due to current inability to vary backlight voltage from the Linux OS (a specific board is needed), we measured the system power at nine different levels of screen brightness and then used them to simulate the energy consumption of the method. Table 3 shows the power levels used in the simulation. The columns labeled by "cddpm" and "apm" reflect the power consumption obtained with and without running the CDDPM software, respectively.

In the simulation we used the following assumptions:
(i) Moving from one power level to another takes 20ms[ ].
(ii) Stepping down from a power level can be done only if the eye-gaze has been off screen for more than 15 frames (i.e. longer than 0.5 sec).

Figures 8-9 illustrate the results obtained for a test in which the user was typing from a book (frames 1-700); reading the book without looking at display (frames 700-1200); and walking away from the room, leaving the PC on. Even though the proposed CDDPM method takes more power than APM and the tested sequence lasts only 66.7sec, the method saves 23.3% and 34.1% of the total energy consumed by the desktop and the notebook, respectively. If the APM inactivity interval is set to 5min, and the above scenario (see Figures 8-9) continues for 5 min, the energy savings could be significant!

Table 3. Power levels (Watt)

Power level	Desktop PC apm	Desktop PC cddpm	Notebook PC apm	Notebook PC cddpm
P8	130.2	138.5	22.44	26.64
P7	129.2	137.7	18.83	22.46
P6	127.5	136.2	15.75	18.66
P5	126.4	134.2	14.09	17.28
P4	124.9	131.6	13.90	15.90
P3	120.8	127.9	12.11	14.80
P2	116.3	124.4	12.18	14.09
P1	114.9	122.8	11.04	13.06
P0	113.1	120.3	10.01	12.08
Monitor off	88.0	95.1	9.8	11.2
PC standby	64	64	4.1	4.1

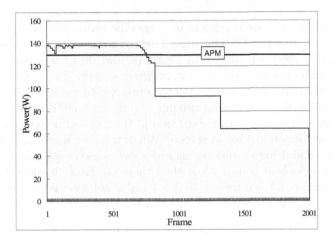

Fig. 8. Power consumption of the desktop system

4   Summary

In this paper we presented a new camera-driven approach to display power management. Although the results showed that a large display energy reduction is possible, the power overhead of the current software implementation was large. To reduce the power overhead, several ways can be explored. The first and the most important one is to reduce computational complexity of the method; namely, the range of image scans during noise reduction and eye localization. A possible solution could be in applying a two step gaze tracking method [ ] that detects a point "between the eyes" first, and then locates the eyes. Because finding the point "between the eyes" is easier, and more stable than detecting the eyes, reduction in computations could be possible. Replacing the CCD camera (0.5W), which we used in the experiment, with a less power-hungry CMOS video sensor (20-40mW) would also reduce the power overhead. Another approach is

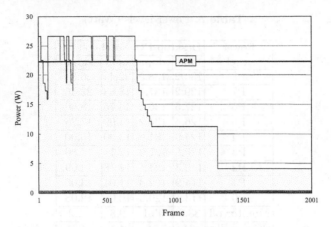

Fig. 9. Power consumption of the notebook system

to implement the proposed scheme in a specific hardware. By embedding the CDDPM hardware and a CMOS camera on the same chip, we will omit energy-expensive CPU operations and data transfers, and thus lower power.

The research presented is a work in progress and the list of things to improve it is long. In the current work we restricted ourselves to a simple case of a singular user monitoring. However, when talking about the CDDPM in general, some critical issues arise. For instance, how should the method behave when handling more than one person looking at screen? When tracking a group, (e.g. a PC user and viewers behind his shoulders) should some person's gaze be ignored? The user might not look at screen while the others do. How should this be solved? Concerning this point, we believe that a feasible solution is to keep the display active while there is someone looking at the screen. We are currently working on this issue.

References

1. Display Power Management Signaling Specification (DPMS), available from VESA, http://vesa.org/vbe3.pdf
2. Advanced Configuration and Power Interface Specification, Rev.3.0, Sept.2004, http://www.acpi.info/spec.htm
3. Display Modes(Transmissive/Reflective/Transflective), Sharp Microelectro- nics of the Americas, 2002, see http://www.sharpsma.com/sma/Products/displays/App RefGuide/DisplayModes.htm
4. L. Choi H.Shim, N.Chang, "Low Power color TFT LCD display for hand-held embedded systems", *Proc. of Int. Symp.on Low-Power Electronics and Design*, Aug.2002, pp.112-117.
5. F.Gatti, A.Acquaviva, L.Benini, B.Ricco, "Low-Power control techniques for TFT LCD displays", *Proc. of Int. Conf. Compilers, Architecture and Synthesis for Embedded Systems*, Oct. 2002, pp.218-224
6. W.-C. Cheng, M.Pedram, "Power minimization in a Backlight TFT-LCD display by concurrent brightness and contrast scaling", *Proc. of Design Automation and Test in Europe*, Feb. 2004, Vol. 1, pp. 16-20.

7. N.Chang, I.Choi, and H.Shim, "DLS: dynamic backlight luminance scaling of liquid crystal display", *IEEE Trans. VLSI Systems*, vol.12, no.8, Aug.2004, pp.837-846.
8. H. Shim, N. Chang, and M. Pedram, "A backlight power management framework for the battery-operated multi-media systems." *IEEE Design and Test Magazine*, Sept/Oct. 2004, pp. 388-396.
9. S.Pasricha, M.Luthra, S.Mohapatra, N.Dutt, and N.Venkatasubramanian, "Dynamic backlight adaptation for low-power handheld devices", *IEEE Design and Test Magazine*, Sept/Oct. 2004, pp. 398-405.
10. J.Flinn and S.Satyanarayanan, "Energy-aware adaptation for mobile applications", *Proc. Symp.Operating Systems Principles*, 1999, pp.48-63.
11. K.S.N. Pattanai, J.E. Tumblin, H.Yee, and D.P. Greenberg, "Time dependent visual adaptation for realistic image display", *SIGGRAPH*, pp.47-54, 2000.
12. J.E. Tumblin, J.K.Hodgins, and B.K.Guenter, "Two methods for display of high contrast images", *ACM Trans. on Graphics*, Vol.18, no.1, pp. 56-94, Jan. 1999.
13. A.Gee and R.Cipolla, "Determining the gaze of faces in images", *Image and Vision Computing*, vol.12, no.18, pp.639-647, 1994.
14. B.Schiele, and A.Waibel, "Gaze tracking based on face color", *Proc. Int. Workshop on Automatic Face- and Gesture- Recognition*, 1998.
15. J.Yang and A.Waibel, "A real time tracker", *Proc. 3rd IEEE Workshop on Application of Comp. Vision*, pp.142-147, 1996.
16. K. Sobottka and I. Pitas, "A novel method for automatic face segmentation, facial feature extraction and tracking", *Signal Proc. Image Com.*, Vol. 12, No. 3, pp. 263-281, June, 1998.
17. R.Stiefelhagen and J.Yang, "Gaze tracking for multimodal human-computer interaction", *Proc. IEEE Int. Conf. Acoustics, Speech and Signal Proc.*, 1997.
18. J.C. Terrilon, M. David, and S. Akamatsu, "Automatic detection of human faces in natural scene images by use of a skin color model and of invariant models", *Proc. IEEE 3rd Int. Conf. on Automatic Face and Gesture Recognition*, pp.88-93, 1998.
19. S.Kawato and J.Ohya, "Two-step approach for real time eye-tracking with a new filtering technique", *IEEE Int. Conf. on Systems, Man & Cybernetics*, pp.1366-1371, 2000.
20. S.Amarnag, R.Kumarran, J.Gowdy, "Real-time eye-tracking for human computer interfaces", *Proc. ICME*, pp. 557-560, 2003.
21. E.Hjelmas, B.K.Low, "Face detection: A survey", *Computer Vision and Image Understanding*, Vol.83, no.3, pp.236-274, 2001.
22. Beyond Logic: CMOS Digital Image Sensors, http://www.beyondlogic.org/imaging/camera.htm

Controlling Peak Power Consumption During Scan Testing: Power-Aware DfT and Test Set Perspectives

Nabil Badereddine, Patrick Girard, Arnaud Virazel,
Serge Pravossoudovitch, and Christian Landrault

Laboratoire d'Informatique, de Robotique et de Microélectronique de Montpellier
LIRMM, Université de Montpellier II / CNRS
161, rue Ada – 34392 Montpellier Cedex 5, France
{Badereddine,Girard,Virazel,Pravossoudovitch,Landrault}@lirmm.fr
http://www.lirmm.fr/~w3mic

Abstract. Scan architectures, though widely used in modern designs for testing purpose, are expensive in power consumption. In this paper, we first discuss the issues of excessive peak power consumption during scan testing. We next show that taking care of high current levels during the test cycle is highly relevant so as to avoid noise phenomena such as IR-drop or Ground Bounce. Next, we discuss a set of possible solutions to minimize peak power during all test cycles of a scan testing process. These solutions cover power-aware design solutions, scan chain stitching techniques and pattern modification heuristics.

1 Introduction

While many techniques have evolved to address power minimization during the functional mode of operation, it is now mandatory to manage power during the test mode. Circuit activity is substantially higher during test than during functional mode, and the resulting excessive power consumption can cause structural damage or severe decrease in reliability of the circuit under test (CUT) [1-4].

The problem of excessive power during test is much more severe during scan testing as each test pattern requires a large number of shift operations that contribute to unnecessarily increase the switching activity [2]. As today's low-power designs adopt the approach of "just-enough" energy to keep the system working to deliver the required functions, the difference in power consumption between test and normal mode may be of several orders of magnitude [3].

In this paper, we first discuss the issues of excessive peak power consumption during scan testing. As explained in the next section, peak power consumption is much more difficult to control than average test power and is therefore the topic of interest in this paper. We present the results of an analysis performed on scan version of benchmark circuits, showing that peak power during the test cycle (i.e. between launch and capture) is in the same order of magnitude than peak power during the load/unload cycles. Considering that i) logic values (i.e. test responses) have to be captured/latched during the *test cycle* (TC) while no value has to be captured/stored during the load/unload cycles, and ii) TC is generally operated at-speed, we highlight the importance of reducing peak power during TC so as to avoid phenomena such as IR-drop or ground bounce that may lead to yield loss during manufacturing test.

Next, we present a set of possible solutions to reduce peak power during TC. We first discuss straightforward power-aware design solutions that consist in over sizing

V. Paliouras, J. Vounckx, and D. Verkest (Eds.): PATMOS 2005, LNCS 3728, pp. 540–549, 2005.
© Springer-Verlag Berlin Heidelberg 2005

power and ground rails. Then, we discuss two other classes of solutions: those based on some modifications of the scan chain and its associated clock scheme (scan chain stitching) and those based on power-aware assignment of don't care bits in patterns of the deterministic test sequence (X filling techniques). All these solutions can achieve significant reductions in peak power consumption during TC. Selecting one of them depends on its impact on some parameters such as additional DfT features, circuit design modifications, design flow adjustments, test application time and pattern volume increase, at-speed testing capabilities, etc.

The rest of the paper is organized as follows. In the next section, we discuss peak power issues during scan testing. In Section 3, we analyze peak power during the test cycles of scan testing and we highlight the importance of reducing this component of the power. In Section 4, we discuss possible solutions to reduce peak power during all test cycles of scan testing. Section 5 concludes this paper.

2 Peak Power Issues

Power consumption must be analyzed from two different perspectives. Average test power consumption is, as the name implies, the average power utilized over a long period of operation or a large number of clock cycles. Instantaneous power or peak power (which is the maximum value of the instantaneous power) is the amount of power required during a small instant of time such as the portion of a clock cycle immediately following the system clock rising or falling edge. In [4], it is reported that test power consumption tends to exceed functional power consumption in both of these measures.

Average power consumption during scan testing can be controlled by reducing the scan clock frequency – a well known solution used in industry. In contrast, peak power consumption during scan testing is independent of the clock frequency and hence is much more difficult to control. Among the power-aware scan testing techniques proposed recently (a survey of these techniques is given in [5] and [6]), only a few of them relates directly to peak power. As reported in recent industrial experiences [3], scan patterns in some designs may consume much more peak power over the normal mode and can result in failures during manufacturing test. For example, if the instantaneous power is really high, the temperature in some part of the die can exceed the limit of thermal capacity and then causes instant damage to the chip. In practice, destruction really occurs when the instantaneous power exceeds the maximum power allowance during several successive clock cycles and not simply during one single clock cycle [3]. Therefore, these temperature-related or heat dissipation problems relate more to elevated average power than peak power. The main problem with excessive peak power concerns yield reduction and is explained in the sequel.

With high speed, excessive peak power during test causes high rates of current (di/dt) in the power and ground rails and hence leads to excessive power and ground noise (V_{DD} or Ground bounce). This can erroneously change the logic state of some circuit nodes and cause some good dies to fail the test, thus leading to unnecessary loss of yield. Similarly, IR-drop and crosstalk effects are phenomena that may show up an error in test mode but not in functional mode. IR-drop refers to the amount of decrease (increase) in the power (ground) rail voltage due to the resistance of the devices between the rail and a node of interest in the CUT. Crosstalk relates to capacitive coupling between neighboring nets within an IC. With high peak current demands during test, the voltages at some gates in the circuit are reduced. This causes these gates to exhibit higher delays, possibly leading to test fails and yield loss [7]. This phenomenon is reported in reports from a variety of companies, in particular when at-speed transition delay testing is done [3]. Typical example of voltage drop and ground bounce sensitive applications is Gigabit switches containing millions of logic gates.

3 Analysis of Peak Power During Scan

During scan testing, each test vector is first scanned into the scan chain(s). After a number of load/unload clock cycles, a last shift in the scan chain launches the test vector. The scan enable (SE) signal is switched to zero, thus allowing the test response to be captured/latched in the scan chain(s) at the next clock pulse (see Figure 1). After that, SE switches to one, and the test response is scanned out as the next test vector is scanned in.

There can be a peak power violation (the peak power exceeding a specified limit) during either the load/unload cycles or during TC. In both cases, a peak power violation can occur because the number of flip-flops that change value in each clock cycle can be really higher than that during functional operation. In [7], it is reported that only 10-20 % of the flip-flops in an ASIC change value during one clock cycle in functional mode, while 35-40 % of these flip-flops commutate during scan testing.

In order to analyze when peak power violation can occur during scan testing, we conducted a set of experiments on benchmark circuits. Considering a single scan chain composed of n scan cells and a deterministic test sequence for each design, we measured the current consumed by the combinational logic during each clock cycle of the scan process. We pointed out the maximum value of current during the n load/unload cycles of the scan process and during TC (which last during a single clock cycle). Note that current during TC is due to transitions generated in the circuit by the launch of the deterministic test vector V_n (see Figure 1).

Identification of peak power violation cannot be done without direct comparison with current (or power) measurement made during functional mode. However, this would require knowledge of functional data for each benchmark circuit. As these data are not available, the highest values of current we pointed out are not necessarily peak power (current) violations. There are simply power (current) values that can lead to peak power (current) violation during scan testing. Reports made from industrial experiences have shown that such violations can really occur during scan testing [3] [4].

The benchmarking process was performed on circuits of the ISCAS'89 and ITC'99 benchmark suites. We report in Table 1 the main features for some of these circuits. We give the number of scan cells, the number of gates, the number of test patterns and the fault coverage (FC). All experiments are based on deterministic testing from the ATPG tool "TetraMAX™" of Synopsys [8]. The missing faults in the FC column are the redundant or aborted faults. Primary inputs and primary outputs were not included in the scan chain, but were assumed to be held constant during scan-in and scan-out operations. Random initial logic values were assumed for the scan flip-flops.

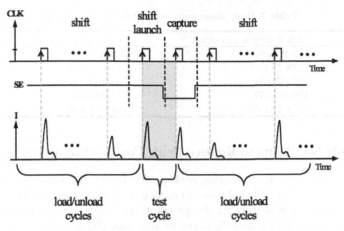

Fig. 1. Scan testing and current waveform

Table 1. Features of experimented circuits

Circuit	# cells	# gates	# patterns	FC (%)
b14s	245	4444	419	99.52
b17s	1415	22645	752	98.99
s9234	228	5597	161	99.76
s13207	669	7951	255	99.99
s38417	1636	22179	145	100

Results concerning peak power consumption are given in Table 2. We have reported the peak power consumed during the load/unload cycles and that consumed during TC. These values are a maximum over the entire test sequence. Power consumption in each circuit was estimated by using PowerMill® of Synopsys [9], assuming a power supply voltage of 2.5 Volts and technology parameters extracted from a 0.25μm digital CMOS standard cell library. These results show that peak power consumption is always higher during the load/unload cycles than during TC. This result was quite predictable as the number of clock cycles during the load/unload phase is much more than one. More importantly, these results show that even if peak power is higher during the load/unload cycles, peak power during TC is in the same order of magnitude. This may lead to problematic noise phenomena during TC whereas these phenomena do not impact the load/unload process. Let us consider again the IR-drop phenomenon. As discussed earlier, it is due to a high peak current demand that reduces the voltages at some gates in the CUT and hence causes these gates to exhibit higher delays. The gate delays do not affect the load/unload process as no value has to be captured/stored during this phase. Conversely, the gate delays can really affect TC because the values of output nodes in the combinational logic have to be captured in the scan flip-flops. As this operation is generally performed at-speed, this phenomenon is therefore likely to occur during this phase and negatively impact test results and thus yield. We can therefore conclude that taking care of peak power during TC and trying to minimize the switching density of the circuit during this phase are really relevant and requires new development of dedicated techniques.

Table 2. Peak power consumption during scan testing

Circuit	Peak during load/unload	Peak during test cycle
b14s	395.55 mW	319.83 mW
b17s	1038.35 mW	1118.68 mW
s9234	358.68 mW	339.88 mW
s13207	499.68 mW	483.30 mW
s38417	1121.80 mW	1074.33 mW

4   Possible Solutions to Reduce Peak Power During the Test Cycle

Considering the fact that minimizing peak power during TC is needed, we present in this section a set of possible solutions to reduce peak power during TC. We first discuss straightforward power-aware design solutions. Then, we discuss two classes of solutions: those based on modifications of the scan chain and its associated clock scheme (scan chain stitching) and those based on power-aware assignment of don't care bits in patterns of the deterministic test sequence (X filling techniques).

4.1   Power-Aware Design Solution

In order to reduce peak power during the test cycles, a straightforward approach would consist in reducing the resistance of the power/ground nets by over sizing power and ground rails. This solution has the advantage to be simple to implement and has limited side effect, i.e. low area overhead. However, this solution requires early in the design flow an estimation of the increase in power consumption during test with respect to power consumption during functional mode. As test data are generally not available at the early phases of the design process, this solution may not be satisfactory in all cases.

4.2   Scan Chain Stitching

A possible solution to reduce peak power during TC consists in partitioning the scan chain into multiple sub scan chains, called "scan segments", and acting on clock signals accordingly. Each scan segment is activated alternately, thus allowing a reduction of the CUT activity during both load/unload cycles and the test cycle. From this basic mode of operation, two types of solutions have already been imagined.

The low power scan architectures proposed in [10] and [11] are based on a gated clock scheme for the scan chain and the clock tree feeding the scan chain. The idea is to reduce the clock rate on the scan cells during shift operations without increasing the test time. For this purpose, a clock whose speed is half of the normal speed is used to activate one half (first scan segment) of the scan cells during one clock cycle of the scan operation. During the next clock cycle, the second half of the scan cells (second scan segment) in the scan chain is activated by another clock whose speed is also half of the normal speed. The two clocks are synchronous with the system clock and have the same but shifted in time (non-overlapping) period during scan operations. During capture operations, the two clocks operate as the system clock so that a unique clock signal is used. Capture operations are operated using a single clock cycle. In case more than two segments are defined during scan chain partitioning, the process is modified accordingly. Experiments performed on benchmark circuits have shown that this technique can reduce peak power during TC by approximately a factor of two, without affecting the test time and the performance of the circuit.

Another technique has been presented in [12] that achieve power reduction during both shift and capture operations with no impact on the performance of the design and with a minimal impact on area and testing time. This technique consists in splitting the scan chain into a given number of length-balanced segments and in enabling only one scan segment during each test clock cycle. The difference with the previous proposed techniques [10,11] is that in this case, each test response is captured using more than one clock cycle, e.g. two clock cycles for a two-segment scan chain. Peak power during TC is reduced by a factor which is roughly equal to the number of scan segments. Such a solution requires each scan segment to be structurally independent from the others so as to insure no capture violation.

These two types of solutions can reduce peak power during TC because in both cases, test launch is operated in more than one clock cycle. However, the main disadvantage is that they do not allow at-speed testing, due to multiple scan shifting and/or capture phases, and hence prevent detection of timing defects.

Another possible solution to reduce peak power during TC is to use scan cell reordering. Scan cell reordering consists in determining the order in which the scan cells of a scan chain have to be connected to minimize the occurrence of transitions during all test cycles. It can be demonstrated that this combinatorial optimization problem is NP-hard - the number of possible solutions is n! where n is the number of scan cells in the scan chain. Due to its exponential nature, this problem cannot be solved by an exact method. Heuristics have therefore to be used.

In [13], we have proposed such a scan cell reordering solution that this time allows at-speed testing. From the set of scan cells and a pre-computed sequence of deterministic test vectors, a heuristic process provides a scan chain order that minimizes the occurrence of transitions and hence the peak power during TC. The problem has been formulated as a global optimization problem (we try to minimize peak power during TC for all vectors of the test sequence) and a polynomial-time approximation solution based on a greedy algorithm has been implemented.

Experiments performed on ISCAS'89 and ITC'99 benchmark circuits have been done to measure the reduction in peak power obtained during TC. A sample of these results is shown in Table 3. For each circuit, we report the peak power during TC obtained first from an ordering provided by an industrial tool and next with the ordering technique proposed in [13]. For the evaluation in both cases, the deterministic test sequences presented in Table 1 were used assuming random initial logic values for the scan flip-flops. The industrial ordering has been performed by using the layout synthesis tool Silicon Ensemble® of Cadence Design System [14]. This synthesis tool allows first to perform scan insertion in the design corresponding to the experimented circuit and next the placement and routing of flip-flops in the design with respect to delay and area constraints. For each circuit, the design and the ordering of the scan chain have been carried out with a random placement of the scan-in and scan-out pins. Peak power is expressed in milliWatts and the values reported for each circuit are a mean of peak power (or instantaneous power) consumed during each test cycle of the scan process. Note that these values differ from those in Table 2 which represent a maximum over the entire test sequence. The last column in Table 3 shows the reduction in peak power dissipation expressed in percentages. Complete results on benchmark circuits have shown that peak power reduction up to 30% can be achieved with the ordering technique proposed in [13]. Note that many other algorithms can be used to perform low power scan cell ordering. We are currently working on a constrained global optimization problem (minimizing peak power during TC for all vectors of the test sequence while maintaining each vector under a "violation" limit) by developing and implementing a heuristic solution based on Simulated Annealing.

Table 3. Peak power savings in the CUT by scan cell reordering

Circuit	Peak – Industrial solution	Peak - Proposed solution	Reduction
b14s	197.17 mW	172.87 mW	12.3 %
b17s	949.47 mW	837.70 mW	11.8 %
s9234	247.32 mW	200.74 mW	18.8 %
s13207	405.56 mW	337.03 mW	16.9 %
s38417	993.22 mW	746.08 mW	24.9 %

Compared with other low power scan techniques, this solution offers numerous advantages. It works for any conventional scan design - no extra DfT logic is required – and both the fault coverage and the overall test time are left unchanged. However, several practical implications of this solution have to be discussed.

First, the heuristic procedure does not explicitly consider constraints such as the placement of scan in and scan out pins or the existence of multiple scan chains with multiple clock domains in the CUT. In this case, the proposed technique has to be modified to allow these constraints to be satisfied. For example, scan chain heads and tails may be predefined and pre-assigned in the case of constraints on scan in and scan out pin position. This kind of pre-assignment may be important to avoid long wires between external scan/out pins and scan chain heads/tails.

In the case of circuits with multiple scan chains and multiple clock domains, which are common in industrial designs, almost no modification of the proposed technique is required. Actually, each scan chain can be considered separately and the heuristic procedure has to be applied successively on each scan chain.

The main drawback of this scan ordering technique is that power-driven chaining of scan cells cannot guarantee short scan connections and prevent congestion problems during scan routing. To avoid this problem, several solutions can be proposed depending on the DfT level at which the peak power problem is considered. First, if scan reordering can be performed before scan synthesis (in this case, flip-flop placement is not already done), the solution is to consider a DfT synthesis tool that can accept a fixed scan cell order (produced by our heuristic) and from which it can optimally place and route the scan resources. Now, if scan reordering cannot be done before scan synthesis (in this case, flip-flop placement is known and fixed), a solution to consider routing is to apply a clustering process as the one developed in [15] that allows to design power-optimized scan chains under a given routing constraint. In this case, the routing constraint is defined as the maximum length accepted for scan connections. Results given in [15] have shown very good tradeoff between test power reduction and impact on scan routing. Note that in all situations, ATPG is done earlier in the design flow.

4.3 X Filling Techniques

In conventional ATPG, don't care bits (Xs) are filled in randomly, and then the resulting completely specified pattern is simulated to confirm detection of all targeted faults and to measure the amount of "fortuitous detection" – faults which where not explicitly targeted during pattern generation but were detected anyway. It is interesting to note that the fraction of don't care bits in a given pattern is nearly always a very large fraction of the total available bits [16]. This observation remains true despite the application of state-of-the-art dynamic and static test pattern compaction techniques. The presence of significant fraction of don't care bits presents an opportunity that can be exploited for power minimization.

In order to avoid congestion problems inherent to scan chain modification techniques and to allow at-speed testing, pattern modification techniques can be used to reduce peak power during TC. Here, the idea is to use a test generation process during which non-random filling is used to assign values to don't care bits (Xs) of each test pattern of the deterministic test sequence. For example, it is possible to apply the following non-random filling heuristics:

- Don't care '0': all don't care bits in a test pattern are set to '0'
- Don't care '1': all don't care bits in a test pattern are set to '1'
- Adjacent filling: all don't care bits in a test pattern are set to the value of the last encountered care bit (working from left to right).When applying adjacent filling, the most recent care bit value is used to replace each `X' value. When a new care bit is encountered, its value is used for the adjacent X's

For example, consider a single test pattern that looks like the following: 0XXX1XX0XX0XX. If we apply each one of the three non-random filling heuristics, the resulting pattern will be:

- 0000100000000 with '0' filling
- 0111111011011 with '1' filling
- 0000111000000 with adjacent filling.

These non-random filling heuristics (among few others) have been evaluated in [7] to measure the reduction in average power consumption during scan shifting (load/unload cycles). Results reported in [7] indicate that the adjacent filling technique does an excellent job of lowering overall switching activity while still maintaining a reasonably increase in pattern volume.

From our side, we have evaluated these heuristics to measure the reduction in peak power consumption during TC with respect to a random filling of don't care bits. Results obtained with the adjacent filling heuristic are reported in Table 4. As for results in Table 3, the values reported for each circuit are a mean of peak power (or instantaneous power) consumed during each test cycle of the scan process. These results show significant reduction in peak power during TC (up to 89%) with an increase of the overall pattern volume ranging from 0 to 15% (3.9% on average for the complete set of experimented circuits).

Pattern modification techniques are therefore promising solutions to reduce peak power during TC. In addition, these techniques require no modification of the basic design of the circuit and no additional DfT features are required to implement these solutions. Finally, at-speed testing is possible so that the defect coverage of the initial test sequence can be maintained. Our future work will consist in investigating much more on this type of solutions.

Table 4. Peak power savings in the CUT with adjacent filling

Circuit	Peak – Random filling	Peak – Adjacent filling	Reduction
b14s	178.00 mW	131.23 mW	26.3 %
b17s	961.86 mW	191.07 mW	80.1 %
s9234	240.02 mW	80.39 mW	66.5 %
s13207	402.62 mW	42.33 mW	89.5 %
s38417	978.10 mW	275.80 mW	71.8 %

5 Conclusion

In this paper, we have shown that excessive peak power consumption during all test cycles of scan testing has to be controlled to avoid noise phenomena such as IR-drop or ground bounce. Without caution, these phenomena may lead to yield loss during manufacturing test as test cycles are generally operated at-speed.

The reduction of peak power during TC can be addressed from different perspectives. First, we can perform circuit design implementation considering test power constraints. Next, modifications applied on the scan chain, the scan clock scheme or the test patterns can also be used.

References

1. Semiconductor Industry Association (SIA): International Technology Roadmap for Semiconductors (ITRS). 2003 Edition
2. Bushnell, M.L., Agrawal, V.D.: Essentials of Electronic Testing. Kluwer Academic Publishers, 2000
3. Shi, C., Kapur, R.: How Power Aware Test Improves Reliability and Yield. IEEDesign.com, September 15, 2004
4. Saxena, J., Butler, K.M., Jayaram, V.B., Kundu, S., Arvind, N.V., Sreeprakash, P., Hachinger, M.: A Case Study of IR-Drop in Structured At-Speed Testing. IEEE Int. Test Conf. (2003) 1098-1104
5. Girard, P.: Survey of Low-Power Testing of VLSI Circuits. IEEE Design & Test of Computers, Vol. 19, N° 3 (2002) 82-92
6. Nicolici, N., Al-Hashimi, B.: Power-Constrained Testing of VLSI Circuits. Springer Publishers (2003)
7. Butler, K.M., Saxena, J., Fryars, T., Hetherington, G., Jain, A., Lewis, J.: Minimizing Power Consumption in Scan Testing: Pattern Generation and DFT Techniques. IEEE Int. Test Conf. (2004) 355-364
8. TetraMAX™. Version 2001.08, Synopsys Inc. (2001)
9. PowerMill®. Version 5.4, Synopsys Inc. (2000)
10. Saxena, J., Butler, K.M., Whetsel, L.: An Analysis of Power Reduction Techniques in Scan Testing. IEEE Int. Test Conf. (2001) 670-677
11. Bonhomme, Y., Girard, P., Guiller, L., Landrault, C., Pravossoudovitch, S.: A Gated Clock Scheme for Low Power Scan Testing of Logic IC's or Embedded Cores. IEEE Asian Test Symp. (2001) 253-258

12. Rosinger, P., Al-Hashimi, B., Nicolici, N.: Scan Architecture with Mutually Exclusive Scan Segment Activation for Shift- and Capture-Power Reduction. IEEE Trans. on CAD, Vol.23, N°7 (2004) 1142-1153
13. Badereddine, N., Girard, P., Pravossoudovitch, S., Landrault, C., Virazel, A.: Peak Power Consumption During Scan Testing: Issue, Analysis and Heuristic Solution. IEEE Design and Diagnostics of Electronic Circuits and Systems Workshop (2005) 151-159
14. Silicon Ensemble®. Cadence Design System (2000)
15. Bonhomme, Y., Girard, P., Guiller, L., Landrault, C., Pravossoudovitch, S.: Efficient Scan Chain Design for Power Minimization During Scan Testing Under Routing Constraint. IEEE Int. Test Conf. (2003) 488-493
16. Wohl, P., Waicukauski, J.A., Patel, S., Amin, M.B.: Efficient Compression and Application of Deterministic Patterns in a Logic BIST Architecture. ACM/IEEE Design Automation Conf. (2003) 566-569

A Design Methodology for Secured ICs Using Dynamic Current Mode Logic

François Macé, François-Xavier Standaert,
Jean-Jacques Quisquater, and Jean-Didier Legat

UCL Crypto Group, Microelectronics Laboratory, Universite Catholique de Louvain

Abstract. This paper presents principles and concepts for the secured design of cryptographic IC's. In order to achieve a secure implementation of those structures, we propose to use a Binary Decision Diagrams (BDDs) approach to design and determine the most secured structures in Dynamic Current Mode Logic. We apply a BDD based prediction to the power consumption of some gates, validate our model using SPICE simulations, and use it to mount efficient power analysis attacks on a component of a cryptographic algorithm. Moreover, relying on our simulation results, we propose a complete methodology based on our BDD model to obtain secured IC's, from the boolean function to the final circuit layout.

Keywords: Differential Pull Down Networks, Binary Decision Diagrams, Differential Power Analysis, Side-channel attack.

1 Introduction

Cryptographic electronic devices such as smart cards, FPGAs or ASIC's are taking an increasing importance to ensure the security of data storage and transmission. However, they are under the threat of attacks taking advantage of side-channel leakages caused by the physical implementation of any given algorithm. Among these leakages, the use of power consumption proved to be very efficient to recover information about the data handled by any given circuit [1].

To prevent the attacker from using efficiently this kind of data, some high level and algorithmic countermeasures were developed. Random process interrupts, dummy instructions were used to avoid the sequential execution of the algorithm but were shown to be inefficient in [2]. Other randomization techniques were also proposed, such as random noise addition but they do not provide any countermeasure as the signal is still present and simple statistical methods can be applied to recover it. Masking methods [3], that consist in masking the data with random boolean values, were proposed at the algorithmic level, but they still leak some information and reduce the implementation efficiency of the algorithm.

Interesting alternatives were presented which propose to tackle the problem directly at the transistor level, by using specific logic styles. Their purpose is to decorrelate the power consumption of a circuit from the data it handles by

V. Paliouras, J. Vounckx, and D. Verkest (Eds.): PATMOS 2005, LNCS 3728, pp. 550–560, 2005.

obtaining steady activity and power consumption. In [ ] and [ ], it was proposed to use particular dynamic and differential logic styles which achieve a very regular power consumption. Even if they were not able to totally suppress the power consumption variations relative to the data handled, and thus do not provide a theoretical countermeasure against power analysis attacks, it remains that they help to make the attack significantly harder and weaken its efficiency.

We try here to propose a particular model for the power consumption of such logic styles, and more specifically, we apply our model and predictions to the Dynamic Current Mode Logic.. Indeed, to obtain high efficiency attacks, the power consumption behavior of particular implementations of algorithms must be predicted with quite a good accuracy. So far, predictions were efficiently made on CMOS implementations, using a model based on the number of switching events occurring within the circuits. But, because of their different power consumption behavior, this model is unapplicable to dynamic and differential logic styles [ ].

In [ ], Binary Decision Diagrams (BDD) were used as a tool to optimize the design of these DPDN. We thus show here how these graphs can also be used to predict the power consumption of such gates and, afterwards, choose the structure that achieves the best resistance against power analysis attacks.

This paper is thus be structured as follows. In section 2 we give a short introduction to Binary Decision Diagrams. Next, in section 3, we present some dynamic and differential logic styles and the interest of Dynamic Current Mode Logic (DyCML). Afterwards, we propose our power consumption behavior model in section 4. The section 5 presents the experiments (validation of the power consumption model, use of this model to mount power analysis attacks) and the achieved results (choice of the most secured implementation) and we conclude in section 6.

2 Binary Decision Diagrams

2.1 Structure and Principles

Binary Decision Diagrams were firstly introduced by Akers in [ ] as a method to define, represent, analyze, test and implement large digital functions. These graphs can be easily developed using the Shannon expansion theorem from which a boolean function can be recursively divided into other functions, each one depending on all input variables but one The graph structure of a function f is thus obtained by applying the Shannon expansion recursively for each input, until the function to expand is the constant function 1 or 0.

Using this representation for boolean functions allows their easy manipulation and easy combinations between them. Among the abundant literature presented on BDDs, Bryant proposed in [ ] to refine their representation and algorithms used to manipulate them. The basical representation of boolean functions by the mean of such graphs was defined as follows.

A function graph is a rooted, directed, acyclic graph with a vertex set V containing two types of vertices. A nonterminal vertex v has for attributes a particular input x_i ($i \epsilon \{0, ..., m - 1\}$) of the implemented boolean function (f :

$\{0,1\}^m \to \{0,1\}$), an argument index $index(v) \in \{1, ..., n\}$ and two children vertices $low(v)$ and $high(v)$. A terminal vertex v has for attributes a particular index value $index(v)$ and a value $value(v) \in \{0,1\}$, representing the function's result. Moreover, if a child vertex of a nonterminal vertex ($low(v)$ or $high(v)$) is nonterminal, its index should be higher than the one of its parent.

The BDD can be reduced from its first structure based on Shannon expansion. As a matter of fact, two conditions allow a vertex to be removed from the graph or replaced by another vertex, leading to a different structure representing the same function.

(1) If the two child vertices of a vertex v are identical ($low(v) = high(v)$), then vertex v should be removed from the graph and its mother should receive the child of v in replacement of it.

(2) The replacement of one or more vertex by a particular one occurs when these two or more vertices define the exact same function as the particular vertex. Then, all the duplicated vertices and all their children should be removed from the graph and replaced by the particular one.

These two principles are shown in Figure 1. In this Figure, we have represented non-terminal vertices by a circle and terminal vertices by a square. Sub-Figures 1-a and -b illustrate the suppression of one vertex having identical children (the vertex with index value 4), while sub-Figures -c and -d show the fusion of two vertices implementing the same function (vertices 7 and 8). The BDD can also

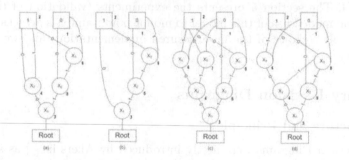

Fig. 1. Reduction of binary decision diagrams

be modified, by changing the data ordering, in order to minimize the number of vertices within the BDD, and thus reduce the mean path length from the root vertex to the terminal ones. A data ordering, or input ordering, corresponds to the sequence in which the inputs are considered to apply the Shannon's expansion.

2.2   Use of BDDs

Binary Decision Diagrams are often used to find minimal representations of boolean functions and to manipulate them very efficiently [ ], using algorithms like the ITE (If Then Else) algorithm to combine different functions (and thus different BDDs). In the same way, those algorithms have been used to build logic optimization systems capable to handle very large circuits with high performances in term of runtime [ ].

3   Dynamic and Differential Logic Styles

3.1   Their Use in Cryptographic Hardware

In [ ], Tiri *et. al* recall that it is necessary to use a dynamic and differential logic style to build logic gates achieving a power consumption independent from the data handled. In first approximation, the power consumption behavior of CMOS leaks information because it consumes power only for output transitions. On the basis of this model, they recommended to use a dynamic and differential logic style to produce a uniform switching activity for and thus ensure a power consumption event for each evaluation cycle, whatever the inputs. They also showed that, in order to really balance the power consumption, we should use a dynamic and differential logic style that achieves a power consumption being the most possible constant. This can be obtained discharging the whole internal capacitance of the gate. It is why they developed Sense Amplifier Based Logic (SABL). The advantage of this logic style is that, by discharging the whole internal capacitance, it produces a power consumption with reduced variations in function of the inputs. Its main drawback is that it yields to a high power consumption.

To overcome this problem, it was proposed in [ ] to use Dynamic Current Mode Logic (DyCML) developed by M. Allam et al. [ ] , to achieve the needed steady power consumption with good performances (in term of power consumption and delay). Thanks to this dynamic current source, DyCML gates produce a low output swing of which value is a function on the total (intrinsic and extrinsic) output load capacitance and on the size of one transistor acting like a virtual ground.

In [ ], it has been shown that, even if the two implementations of circuits present the same high security margins, according to criterions defined in [ ], DyCML was recommended because of its better performances. Moreover, DyCML presents the feature of being directly connectible through a asynchronous scheme, which is usually considered to improve the security of implementations of cryptographic algorithms [ ].For an illustration purpose, we give the structure of 2 inputs XOR gates implemented in both DyCML and SABL logic style in Figure 2. As SABL needs a domino-interconnect structure between gates to counteract the charge sharing effect, we presented the gate with the needed output inverters.

3.2   Gate Design

One important step, while designing dynamic and differential gates, consists in deriving an efficient structure for the part of the gate that effectively computes

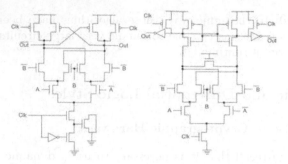

Fig. 2. Structure of XOR gate: a- DyCML, b- SABL

the implemented function. Depending on the type of logic style used, this can be done using several methods like using the Karnaugh Map (K-Map Procedure) or the tabular methods proposed in [ ]. However, these methods do not always lead to a structure using differential pairs (Differential Pull Down Network - DPDN) like it is needed for DyCML or SABL.

A method proposed in [ ] exploits the isomorphism existing between DPDN and the BDD implementing the function. A boolean function implemented with a DPDN is computed by the means of logical switches, which determine which one of the two outputs of the gate will be discharged. Each logical switch is composed of a transistor pair, with connected sources, of which one of the gate is driven by the logical input relative to the vertex and the other one by the complement of this logical input. The drain of the transistors are connected to the sources of particular switches of the structure. The isomorphism is clear to see if you assimilate a switch of the DPDN to a particulart non-terminal vertex of the BDD. To illustrate this, Figure 3 gives the implementation of two basic functions (AND/NAND and XOR/XNOR) with the corresponding BDD.This isomorphism between the structures of the graph and the DPDN can be used to design very efficient DPDN with optimization of their performances in term of area, delay [ ] and power consumption.

4   BDD Based Tool

This isomorphism gave us the idea to use the BDD not only as a tool of optimization of classical performances of the gates, but also as a tool to predict its power consumption and which structure implemented in the gate would be the most secured. Indeed, Can't we use this graph to help to determine the structure having the lower leakage of information?

In order to do so, we proposed to model the power consumption relative to one single input sequence using the following assumptions. Firstly, we consider that variations in the power consumptions are caused by variations of the number of internal capacitances within the DPDN that are charged/discharged at each evaluation cycle, like it was proposed in [ ]. Secondly, we focus our interest on the diffusion capacitances of the transistors. The third hypothesis we make is to

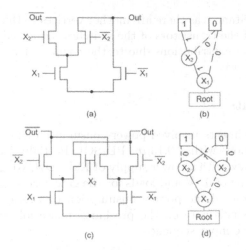

Fig. 3. Isomorphism between DPDN and BDD for functions AND (a-b) and XOR(c-d)

consider drain-gate capacitances and source-drain capacitances equal and being the only ones to play a role. A precise modeling of their value is complex because it evolves along the time, depending on the voltages at the source, drain and gate and on the form of their evolution. It is why to predict the power consumption relative to a particular input sequence, we only counted the number of normalized capacitances that were *activated* (by this, we mean effectively connected to the discharged output of the gate), considering each one equal.

To achieve rapid predictions for all possible implementations (as we show in the experiments, the considered boolean functions are defined as following $f : \{0,1\}^4 \rightarrow \{0,1\}^4$), we designed a program taking an hexadecimal truth table for input and producing several outputs (these outputs are produced for each 1 bit function obtained considering one 4 bit function as the combination of four 1 bit functions $\{0,1\}^4 \rightarrow \{0,1\}$, as a 1 bit function is implemented in a single gate):

- The graph structures for each input ordering
- The Spice netlist relative to the transistor implementation of the graphs
- The predictions of the connected capacitances for each input sequence, for each graph

Indeed, for a function defined on $\{0,1\}^4 \rightarrow \{0,1\}$, there are $!4 = 24$ possible input orderings in such a graph. The applied procedure consisted thus in building the graphs corresponding to the different input orderings, reducing them, determining the discharged output node for each input sequence and finding the total number of connected normalized capacitances for each particular input sequence. The number of normalized capacitances connected to one vertex has tree contributors:

- 2 capacitances for the 2 source-gate capacitances of the transistors forming the switch associated to the vertex.

- As many capacitances as there are mother vertices to this vertex (drain-gate capacitances of the transistors of the switches connected to this vertex).
- All capacitances contributions due to the connected vertices in function of the input sequence.

5   Experiments

The different experiments involved power consumption extractions realized using SPICE simulations with a $0.13\mu m$ PD-SOI (Partially Depleted Silicon-On-Insulator) technology, with a power supply of $1.2V$. DyCML gates were designed with an output swing of 0.8V and loads for the gate corresponding to one active transistor. We extracted the power consumption by averaging the power consumption of each single gate on the precharge phase following the evaluation corresponding to the input sequence.

5.1   Validating the Power Consumption Model

To achieve the validation of our power consumption model, we use the correlation coefficient to mesure how our predictions compare to the actual Spice simulations of the power consumption. We selected the correlation coefficient because it is a usual tool to mount practical power analysis attacks [ ]. However, other statistical tools could be considered.

We firstly validated our model on a simple 2 inputs DyCML AND gate. The predictions, structure and normalized connected capacitances of the gate are summarized in Figure 4 and table 1. Exhibiting a correlation value of 0.9696, the model and predictions match pretty well. We then evaluated the validity of

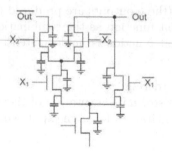

Fig. 4. Parasitic Capacitances of a AND gate

Table 1. Prediction features in function of the input sequence

x_2	x_1	f	Discharged Node	Connected Capacitances
0	0	0	$\overline{OUT}$	8
0	1	0	$\overline{OUT}$	8
1	0	0	$\overline{OUT}$	5
1	1	1	OUT	7

our model on more complex gates. To do so, we used the function P and Q of a Khazad Sbox [ ]. These functions are defined on $GF(2^4)$. We thus determined the 8 one bit functions because each 1 bit function can be implemented in a single gate. For each of these 1 bit functions there are 24 possible implementations (input orderings). We then correlated these predictions and simulated power consumptions to finally obtain an average correlation of 0.8387 and a mean standard deviation of the correlation of 0.1433 between the power consumption and its prediction.

5.2   Simulated Attacks

We then evaluated the utility of such a model in the context of power analysis attacks. To do so, we mounted simulated attacks using the predictions of connected capacitances and the simulations of the power consumptions. We ran these on a single Sbox for which the inputs resulted in a XOR operation between plaintexts and a key.

We computed the total power consumption M_i of the Sbox, for a plaintext i. The second step of this simulated attack consisted in predicting the number of connected capacitances $P_{k,i}$ using guesses of the inputs obtained by realizing a XOR operation between each possible key k and the plaintext i. The last step consisted in calculating the correlation between the simulated power consumption M and the predictions P_k for a key k.

We used the data collected after the SPICE simulations for the measurements and, using this attack methodology, we extracted the correlation after the encryption of a number of plaintexts variating from 1 to 256, and for all the possible keys. We extracted the value of this correlation and also correlation margins for 256 encrypted plaintexts. The correlation margin 1 is defined here as the difference in correlation between the right key guess and the wrong key guess having the highest correlation, while correlation margin 2 is the difference between the correlation for the right key guess and the mean correlation of the wrong ones. This was done for 4 particular implementations of the Sbox:

- the one with the total simulated power of each function P and Q having the largest variance (Implementation 1)
- the one with the total simulated power of each function P and Q having the smallest variance(Implementation 2)
- the one for which the implementation of each 1 bit function is the most correlated with the power consumption prediction (Implementation 3)
- the one for which the implementation of each 1 bit function is the less correlated with the power consumption prediction (Implementation 4).

The simulated attack results are given in table 2. In this table, we give the values of both correlation for the right key guess and correlation margin, for each implementation. A negative value for the correlation margin 1 corresponds to an unsuccessful attack. We also give the mean number of encryptions needed to achieve a discrimination between the right key guess and the wrong ones.

As we can see, the first three implementations were successfully attacked, as the all had a positive correlation margin. We also can see that they all have high correlation and sufficient correlation margin to allow successful attacks, while for the fourth implementation, 2 key guesses (the right one and a wrong one) remaind correlated to the key, leading to a negative value of the correlation margin and preventing us from being able to obtain the right key guess. The difference in correlation between the right key guess and the second most correlated wrong one is 0.0453.

Table 2. Simulated Attack Results

Implementation	Correlation	Margin 1	Margin 2	♯ of texts
1	0.8393	0.4401	0.8429	15
2	0.6186	0.2855	0.6210	40
3	0.9199	0.5305	0.9235	7
4	0.3462	-0.0037	0.3469	N.A.

5.3 Discussion

It is quite obvious that the efficiency of a power analysis is dependant of the obtained correlation for the right keyguess and on the difference in correlation between this right keyguess and the wrong ones. Moreover, as it was presented previously, these values are dependant of the power consumption model used to mount the attack and on the chosen algorithm.

These simulated attack results thus clearly emphasize the possibility of choosing the implementation for which the power consumption is the harder to precisely model, and thus being the most resistant towards a power analysis attack based on this model. Indeed, correlation and correlation margin can be significantly decreased in comparison to the other implementations, while the number of cleartexts needed is highly dependent on the predictability of the chosen implementation.

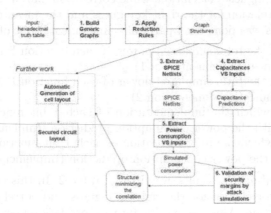

Fig. 5. Proposed methodology for the design of secured implementation of cryptographic functions

With all the tools, we get the possibility to automatically choose the implementation of a circuit that limits at the maximum the side channel information leakage of a circuit. Moreover, as suggested in [ ], the graph structure generated by the tool can help for automated cell layout generation. This was done in [ ], for some non-complementary digital VLSI cells for which transistor placement can be efficiently achieved from a graph structure describing the cell, applying adapted selection algorithms to choose the best structure. The adjunction of all this concepts yields to a design methodology that can be described like in Figure 5.

6 Conclusion

In this paper, we investigated a model for the power consumption of Dynamic and Differential Logic styles. These circuits have previously been proven to offer a good alternative to CMOS in terms of security against side-channel attacks. However, their security against such attacks highly depends on the impossibility to efficiently predict their power consumption: a previously uninvestigated problem.

As a main result, we demonstrated that for one specific logic style, namely the DyCML, it is possible to derive a simple and efficient leakage model and to build practical attacks based on it. Then, we showed that it is possible to search exhaustively among the possible representations of a logic function, in order to find the hardest to predict one. In other words, we looked for the structures offerering the best resistance against power analysis. Confirmed by simulated experiments, we finally illustrated that the best resulting implementations allow an improved security, as the original power analysis attacks could not be applied anymore. Remark that, event if it is not discussed in the paper, such circuits also exhibit lower power consumption variances and are consequently harder to measure.

An open question is to know if and how the methodology presented can be generalized to other DDLs. The improvement of the power consumption models is another concern. Finally, we plan to perform practical experiments in order to evaluate how our simulation-based conclusions relate to practice.

References

1. P. Kocher, J. Jaffe, B. Jun, Differential Power Analysis, In *Advances in Cryptology - CRYPTO'99*, LNCS, vol 1666, pp 388-397, Springer, 1999
2. C. Karlof, D. Wagner, Hidden Markov Model Cryptanalysis, In   *CHES'2003*, LNCS, vol 2779, pp 17-30, Springer, 2003
3. T.S. Messerges, Using second-Order Power Analysis to Attack DPA Resistant Software, In   *CHES'1999*, LNCS, vol 1965, pp 71-77, Springer, 1999
4. S.B. Akers, Binary Decision Diagrams, *IEEE Transactions on Computers*, vol. C-27, No. 6, pp.509-516, June 1978
5. R.E. Bryant, Graph Based Algorithms for Boolean Function Manipulation, *IEEE Transactions on Computers*, vol. C-35, No. 8, pp. 677-691, August 1986
6. C. Yang, M. Ciesielski, BDS: A BDD-Based Logic Optimization System, In *ACM/IEEE Design Automation Conference - DAC'2000*, pp. 92-97, June 2000
7. K. Tiri, M. Akmal, I. Verbauwhede, A Dynamic and Differential CMOS Logic with Signal Independent Power Consumption to Withstand Differential Power Analysis on Smart Cards, In *The Proceedings of ESSCIRC 2002*, pp.403-406
8. F. Mace, F.-X. Standaert, I. Hassoune, J.-D. Legat, J.-J. Quisquater, A Dynamic Current Mode Logic to Counteract Power Analysis Attacks, In *The Proceedings of DCIS 2004*, pp.186-191

560   François Macé et al.

9. M.W. Allam, M.I. Elmasry, Dynamic Current Mode Logic (DyCML): A New Low-Power High-Performances Logic Styles, *IEEE Journal of Solid-State Circuits*, vol. 36, No. 3, pp. 550-558, March 2001

10. J. J.A. Fournier, S. Moore, H. Li, R. Mullins, G. Taylor, Security Evaluation of Asynchronous Circuits, In   *CHES'2003*, LNCS, vol 2779, pp 137-151, Springer, 2003

11. K.M. Chu, D.I. Pulfrey,Design Procedures for Differential Cascode Voltage Switch Circuits, *IEEE Journal of Solid State Circuits*, vol. SC-21, no. 6, pp. 1082-1087, December 1986

12. J. Cortadella, Mapping BDDs into DCVSL gates, UPC/DAC Technical Report No. RR 95/04, February 1995

13. E. Brier, C. Clavier and F. Olivier, Optimal Statistical Power Analysis, *IACR e-print archive 2003/152*, http://eprint.iacr.org, 2003

14. P. Barreto, R. Rijmen, The KHAZAD Legacy-Level Block Cypher, NESSIE Project Home Page, https://www.cosic.esat.kuleuven.ac.be/nessie, 2001

15. M. A. Riepe, K. A. Sakallah, Transistor Placement for Noncomplementary Digital VLSI Cell Synthesis, *ACM Transactions on Design Automation of Electronic Systems*, vol. 8, no. 1,pp. 81-107, January 2003.

Power Consumption Characterisation of the Texas Instruments TMS320VC5510 DSP

Miguel Casas-Sanchez, Jose Rizo-Morente, and Chris J. Bleakley

Department of Computer Science
University College Dublin, Dublin 4, Ireland
{miguel.casassanchez,jose.rizo-morente,chris.bleakley}@ucd.ie

Abstract. The increasing popularity of mobile computers and embedded computing applications drives the need for analysing and optimising power in every component of a system. This paper presents a measurement based instruction-level power consumption model of the popular Texas Instruments TMS320VC5510 Digital Signal Processor. The results provide an accurate and practical way of quantifying the power impact of software. The measurement and characterisation methodology are presented along with the results extracted. Finally some power saving techniques are introduced together with an assesment of their impact.

1 Introduction

Power consumption has become an important issue in the design of embedded applications, especially in recent years with the popularisation of handheld devices such as cameras, mobile phones and PDAs. The application software executing on these devices has a substantial impact on their power consumption [1]. Traditionally, time consuming logic-level simulations of processor models were used to evaluate the power dissipation of a processor when executing benchmark programs. While utilised by processor designers, such models are typically not available to software designers. Hence, power consumption due to software execution can only be measured a posteriori. This leads to last minute software tweaking, longer design cycles and a lack of predictability in the design process.

In this article, a power consumption model for the popular TMS320VC5510 Digital Signal Processor from Texas Instruments (TI) is developed. The C5510 is targetted at low power, medium performance DSP applications. Succesful systems based on this DSP include modems (3Com), cell phone handsets (Nokia, Ericsson), portable MP3 players (Sanyo), digital still cameras (Sony) and digital video recorders (JVC). The power consumption model introduced herein is based on experimental measurements of assembly instructions executing on an actual processor. To the best of our knowledge this is the first time that such a model has been presented for this processor.

The paper is structured as follows. Firstly, the measurement methodology is presented including the theoretical background of instruction costs and the actual physical measurement scheme. Secondly, the target architecture is described

V. Paliouras, J. Vounckx, and D. Verkest (Eds.): PATMOS 2005, LNCS 3728, pp. 561–570, 2005.

detailing its relevant features. Thirdly, the experiments and results are presented together with an analysis. The current consumption of different instruction components is detailed along with the inter-instruction cost, the impact of parallelism and other processor-specific effects. In the fourth section, this information is used as the basis of proposing some power saving techniques. Finally, the conclusions are presented and future work is outlined.

2   Methodology

The goal of the research is to measure the energy and power consumed by the processor as it executes a program. The average power consumed by a program is given by $P = I \times V$ where I is the average current and V is the core voltage. The energy consumed by the program is given by $E = P \times T = (I \times V) \times (N \times \tau)$ where T is the total duration of code execution. This in turn can be calculated as $N \times \tau$ where N is the number of cycles of duration τ. In this context V and τ are fixed, unless otherwise noted. Hence given N for a program, there is only need to measure the average current I in order to calculate the energy consumed.

It has been shown [,] that the total energy cost of a program can be modelled as a base cost for each instruction plus an inter-instruction effect. The inter-instruction effect is an overhead incurred when specific pairs of instructions are executed consecutively. This overhead arises from changes in the circuit state. The base instruction cost is measured by putting several instances of the target instruction in an infinite loop and measuring the current drawn. If a pair of instructions is put into an infinite loop, the current is larger than the average of the individual base instruction costs. This increase is measured as the inter-instruction cost. Special instruction sequences must be executed in order to isolate other effects, such as special addressing modes, various memory access types and operational modes. Taking these costs into account, the energy consumption of a program can be expressed in the form of (1).

$$E_p = \sum_i (B_i \times N_i) + \sum_{i,j} (O_{i,j} \times N_{i,j}) + \sum_k E_k \tag{1}$$

where B_i represents the base cost of instruction i, N_i is the number of occurrences of instruction i, $O_{i,j}$ stands for the circuit state overhead for the instruction pair i,j and E_k are the energy costs of other effects. This formula has been applied to several processors giving accurate results for Motorola DSP56K [ ], ARM7 [ ], M3DSP [ ], Hitachi SH-4 [ ] and i960 [ ].

The physical measurement methodology was applied to the 5510 DSK Development Software Kit [ ], that provides $1.1 - 1.6V$ core voltage and 24MHz frequency reference connected to a PC running the TI Code Composer Studio (CCS). The tool was used to download and run the test programs, as illustrated in Fig. 1. External software routines were used to trigger the measurements using the digital storage scope. The current drawn was measured with a non intrusive, 0.1mA resolution current probe. The probe bandwith is around 50MHz providing enough resolution for the present purposes. The measurements were taken at $1.6V, 24MHz$ unless otherwise noted. For each measurement the mean

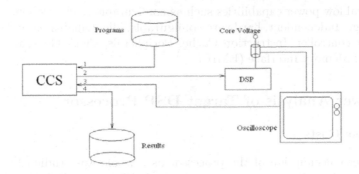

Fig. 1. Instruction Energy Consumption Measuring Scheme

of between ten and twenty results was taken. The measurements are completely repeatable.

3 Target Architecture

The target embedded DSP used for the study is a Texas Instruments TMS320 VC5510, with variable core voltage and frequency up to 200MHz. It implements several special architectural features relevant to the work at hand. These are as follows:

- Four functional units, namely Instruction, Program, Address and Data units, which form a four stage prefetch pipeline in the Instruction unit and a four stage pipeline, expanded up to eight stages in the case of memory operands. All instructions are executed in one clock cycle, except branches, and the CPU is able to execute two instructions simultaneously under some restrictions.
- Four 40 bit accumulator registers, eight 24 bit addressing registers and four 16 bit temporal registers, along with registers for circular addressing, hardware loops, stack and data page pointers, status registers and others.
- Two independent 40 bit D-unit MAC units, one 40 bit D-unit ALU, one 16 bit A-unit ALU, one A-unit and one D-unit swap, and one D-unit barrel shifter. One A-unit address generation unit, featuring one extra ALU for indirect addressing modes.
- Twelve independent buses to memory, divided into two program buses (data and address), six data read buses and four data write buses, along with internal buses for constant passing between units.
- Instruction Buffer Queue (IBQ), that fetches four program bytes a cycle, up to 64, passes six bytes at a time to the instruction decoder, and is flushed each time the Program Counter jumps. One configurable 24Kbyte Instruction Cache, 64 Kbytes of dual RAM (DARAM) able to read and write in the same cycle, and 256 Kbytes of static RAM (SARAM) divided into 32 blocks that can be accessed independently. System ROM of 32 Kbytes and almost 16 Mbytes of external addressing. One DMA controller.

– Several low power capabilities such as configurable clock frequency and core voltage. Independent, hardware-configurable, idle domains off and on: CPU, DMA controller, Instruction Cache, Peripherals, Clock Generator and External Memory Interface (EMIF).

4   Power Analysis of Target DSP Processor

4.1   Base Costs

The power consumption of the processor as a whole was studied by means of the Idle modes. When running an infinite NOP loop at 20MHz the current consumption is 12.3mA. The functional units can be assumed inactive and the fetch and decode units consume a minimum amount of power. By means of the Idle domains this NOP consumption figure can be separated into the components due to Clock, CPU, Memories, DMA, EMIF and Peripherals sub-systems. The results are provided in Fig. 2 and Table 1. It is clear that the power consumption of the Clock sub-system dominates.

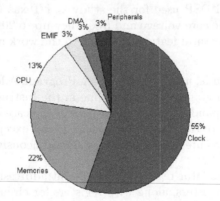

Fig. 2. Background Consumption Proportions

Table 1. Current Saving with Idle Domains (mA)

Unit	DMA	CPU	Peripherals	EMIF	Memories	CLKGEN
Current Saving	0.4(3.25%)	1.6(13.00%)	0.4(3.25%)	0.4(3.25%)	2.7(21.95%)	6.8(55.28%)

An extensive series of measurements were made to determine the base costs of the VC5510 instructions. Measurements were made for the entire instruction set, including all combinations of operations and addressing modes. In total around 500 instructions were measured. The results were grouped according to the taxonomy in the Texas Instruments manuals [ ], namely Arithmetic, Bit Manipulation, Logical, and finally Extended Registers and Move Operations. However, given the large current consumption difference between internal and external instructions to the CPU, that is, operations without or with memory access [ ], a further separation between RR (register and register) and MR (memory and register) instructions is made. The aggregate current consumptions for the groups can be seen in Table 2.

Table 2. Basic Cost for Instruction Groups (mA)

	Arithmetical Ops.	Bit Ops.	Logical Ops.	Move Ops.
RR	13.4-15.5 (15.67%)	14.2-16.1 (13.38%)	13.4-15.7 (17.16%)	13.5-15.7 (17.16%)
MR	15.5-21.1 (36.13%)	16.4-18.1 (10.37%)	16.9-19.7 (16.57%)	15.1-20.0 (33.56%)

Varying the registers used in a single instruction within the same register bank or between banks, has no significant impact on the power consumption. Results for the Logical, Arithmetical and Move operations vary by less than 3.6%. Different inmediate values or operands lead to different bus switching activities. However variation in power due to data dependency is only significant for the D-unit when processing 40 bit arithmetic. Results for Logical operations show a 5.69% variation for 16 bit arithmetic and 13.69% for 40 bit arithmetic. It is also worth noting that experiments changing data and registers at the same time showed no further relation between them. Surprisingly, operations involving 16 bit immediate data consume less power than those involving 8 bit inmediates. Experiments for Logical instructions show 8.10% less consumption in the long case, and 5.63% less in the long case for MAC operations. This appears to be due to the different instruction lengths - 3 bytes for 8 bit and 4 bytes for 16 bit. The latter fitting more easily in the IBQ.

Use of indirect, register based, addressing modes is significant, since they imply costly memory accesses. A further division can be made based on the use of an extra Address Unit ALU, between the modes that modify the base register and those that do not. The former is $0.5 - 0.8$mA greater for Logical, Bit and Move operations, and $2.0 - 3.0$mA for Arithmetic operations. The power consumption of DARAM and SARAM read only accesses are the same. Simultaneous DARAM read and write operations consume more power than their SARAM equivalents by 2.98% and 15.6%, respectively. Finally, the instruction bytewidth, ignoring data dependency, is not significant, as are the bit shifts included in some instructions.

The energy consumption significance of different instruction components was calculated by means of multilinear regression within the Arithmetical operations and can be found in Fig. 3. The selected set of values were the instruction or pipeline base cost, the instruction bytewidth (2,4,5), the inmediate data bytewidth, the addressing modes of the operands (being 0 if inmediate, 1 if register, 2 if register indirect and 3 if postincremented register based indirect), the bit shift option, the extra memory access cycle option, the extra dual access and the extra ALU used for parallel instructions. The mean square error was 0.74 and the R^2 parameter 99.78%, with 129 samples composing the regression. It

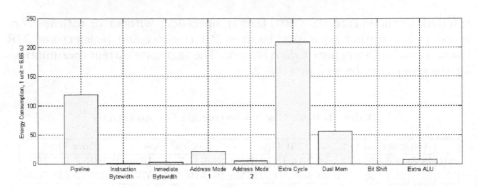

Fig. 3. Instruction Component Energy Consumption

is worth noting the similarity between the pipeline energy consumption and the previously measured NOP consumption (11.9 and 12.3mA respectively). As is evident from the results, the extra memory cycle cost dominates.

4.2   Inter-instruction Costs

Given the large number of measurements necessary for testing all combinations of instructions, only inter-instruction effects between groups of instructions were considered. The inter-instruction cost between instructions belonging to different classes are shown in Table 3. The entry in row i and column j gives the overhead cost when an instruction belonging to class j occurs after an instruction belonging to class i. This table is symmetric, since the results show no perceptible variation. Each result is once again divided between RR and MR operations. It is worth noting that the in-group overhead values are constant (diagonal values).

Table 3. Overhead Cost for Instruction Groups RR/MR (mA)

	Arithmetical Ops	Bit Ops.	Logical Ops.	Move Ops.
Arithmetical Ops	1.1/ 3.0-3.6	1.0-2.0/ 3.0-5.0	1.0-2.0/ 3.0-4.0	1.5-2.5/ 3.5-4.0
Bit Ops		1.0-1.1/ 3.8-4.2	1.0-2.0/ 3.0-5.0	1.5-2.3/ 3.5-5.0
Logical Ops			1.0-1.1/ 3.0-4.0	1.0-2.5/ 3.5-4.0
Move Ops				1.0/ 3.4

4.3   Parallel Instructions Costs

One special feature of the DSP is the possibility to execute two instructions simultaneously provided some conditions are met. The assembler supports some single instructions in which the parallelism is Implied. Explicitly parallel instructions may also be used provided that the same resource is not re-used, the total instruction code is not longer than six bytes and at least one of the instructions is parallelisable.

Experiments were conducted to measure the overhead due to parallelism relative to the cost of the original, non-parallel, instruction with maximum base cost.
To reduce the number of measurements to a reasonable number, it was decided to
measure the overhead for combinations of representative instructions from each
group. The results are provided in Fig. 4. Combinations for RR instructions
show lower overheads than combinations of MR-RR instructions. Implied parallel instructions tend to have current consumption very close to the maximum of
the original instructions.

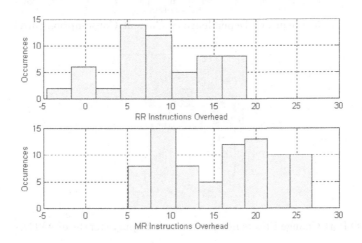

Fig. 4. Parallel Instruction Overhead Distribution (tenths of mA)

4.4   Other Effects Costs

Other effects to be considered for the present DSP include the current consumption frequency effects, penalty of using circular addressing, Instruction Buffer
Queue flush/fill, and I/O port read/write. The behaviour of power consumption
versus frequency for the NOP instruction can be found in Fig. 5. The expected
linear relationship between power and frequency is evident. Circular addressing
modes are based on modulo N arithmetic and a special set of registers. After
several measurements, no observable current consumption change was measured
relative to the linear addressing modes. Current consumption is also unaffected
by Instruction Buffer Queue flushes or unconditional branches. These are simply implemented by changes in the value of the Program Counter and an extra
pipeline flush that inserts five or six idle cycles. Thus the energy consumed by
this instructions is five times higher, although the current consumption remains
constant. The change in consumption due to I/O port accesses is significant,
even though the lines are themselves driven from a separate 3.3V supply. A
glitch of around $0.12\mu s$ duration, as in Fig. 6, is observed when the I/O port
value toggles, for both reading and writing.

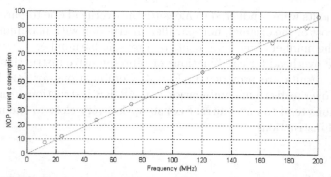

Fig. 5. Frequency Dependence of Current Consumption (mA)

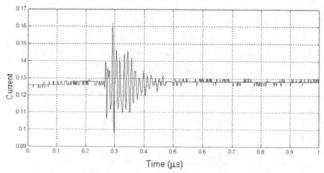

Fig. 6. I/O Port Change Effects in Core Current Supply (tenths of A) ($1\mu s = 20$cycles)

5   Power Saving Techniques

An understanding of these power measurements may be used to derive energy saving assembly-level code transformations.

Adjacent instructions should be parallelised wherever possible. For instance two voice channels should be filtered with the same set of coefficients. Two contiguous MAC instructions with post modification of pointer registers consume 19.1mA each, while the parallelised version consumes 22.8mA but only lasts one cycle, thus saving 40.3% energy.

Another easy but powerful technique is using separate memories and several SARAM banks so that wait cycles do not have to be inserted. One example is applying an XOR masking operation directly to data in memory. That is, taking the data, applying the mask and saving the result back in the same position, while at the same time increasing the pointer. Using SARAM this operation takes two cycles, consuming 18mA. Meanwhile DARAM consumes 20.9mA saving a cycle. Thus an energy saving of 43.1% is achieved. Alternatively, different SARAM banks can be accesed at the same time. For instance in a dual MAC operation, putting both data and coefficients in the same bank consumes 22.1mA and lasts two cycles. When the coefficients are in different banks the operation lasts one cycle and consumes 22.9mA, thus saving 48.2% of the energy.

Switching off Idle domains also leads to power savings, for instance during DMA transfer the CPU and Peripherals can be switched off saving 16% of power. During system wait periods both CPU and DMA can be switched off. This in turn switches off the memories, thus consuming 38% less power. The complete core can be switched off, waking up by means of a hardware interrupt, saving 100% of the power consumption.

Frequency and voltage scaling can be used to save power, by means of the formula $P_{consumed} \propto f \times V_{cc}^2$. When running at 24MHz, a 1.1V core voltage could be used instead of 1.6V, saving 52.7% of the power consumption. Note that reducing the frequency alone save power but not energy, since the program would last longer at lower frequency.

6 Conclusions and Future Work

In this contribution, an instruction-level power and energy consumption model and experimental measurements for the Texas Instruments TMS320VC5510 DSP were presented. Attention was paid to the behaviour of the various instruction groups and to the effects of consecutive instructions and instruction parallelism. As has been showed in the last section, the data presented is applicable to the problem of power aware software design.

Future work includes the application of results to the development of an automated energy consumption estimator for the current processor. It is planned that this estimator will be used as part of a more extensive power aware software design methodology.

Acknowledgements

This work is sponsored by Enterprise Ireland (EI) under agreement number PC/2004/311. The authors would like to thank the Electrical and Electronic Engineering Department at University College Dublin, Ireland, for their support.

References

1. Roy, K., Johnston, M.: 6.3. ASI Series. In: Software design for low power. Kluwer - NATO Advanced Study Inst on Low Power Design in Deep Submicron Electronics (1996)
2. Tiwari, V., Malik, S., Wolfe., A.: Power analysis of embedded software: A first step towards software power minimization. IEEE Transactions on VLSI Systems (1994)
3. Tiwari, V., Lee, M., Malik, S., Fujita, M.: Power analysis and low-power scheduling techniques for embedded dsp software. In: International Symposium on System Synthesis. (1995)
4. Klass, B., Thomas, D., Schmit, H., Nagle, D.: Modeling inter-instruction energy effects in a digital signal processor. In: Power Driven Microarchitecture Workshop and XXV International Symposium on Computer Architecture. (1998)

5. Sinevriotis, G., Stouraitis, T.: Power analysis of the arm 7 embedded microprocessor. In: 9th International Workshop on Power And Timing Modeling, Optimization and Simulation (PATMOS). (1999)
6. Lorenz, M., Wehmeyer, L., Drager, T., Leupers, R.: Energy aware compilation for dsps with simd instructions. In: Proceedings of LCTES/SCOPES. (2002)
7. Sinha, A., Chandrakasan, A.: Jouletrack - a web based tool for software energy profiling. In: Design Automation Conference. (2001) 220–225
8. Russell, J., Jacome, M.: Software power estimation and optimization for high performance, 32-bit embedded processors. In: Proceedings of ICCD. (1998)
9. Spectrum Digital Inc.: TMS320VC5510 DSK Technical Reference. (2002)
10. Texas Instruments Inc.: TMS320C55x DSP Mnemonic Instruction Set Reference Guide. (2002)
11. Texas Instruments Inc.: TMS320C55x DSP CPU Reference Guide. (2004)

A Method to Design Compact DUAL-RAIL Asynchronous Primitives

Alin Razafindraibe[1], Michel Robert[1], Marc Renaudin[2], and Philippe Maurine[1]

[1] LIRMM, UMR CNRS/Université de Montpellier II,
(C5506), 161 rue Ada, 34392 Montpellier, France
{razafind,robert,pmaurine}@lirmm.fr
[2] TIMA Laboratory 46, avenue Félix Viallet, 38031, France
marc.renaudin@imag.fr

Abstract. This paper aims at introducing a method to quickly design compact dual-rail asynchronous primitives. If the proposed cells are dedicated to the design of dual-rail asynchronous circuits, it is also possible to use such primitives to design dual-rail synchronous circuits. The method detailed herein has been applied to develop the schematics of various basic primitives. The performances of the 130nm obtained cells have been simulated and compared with more traditional implementations.

I Introduction

If asynchronous circuits can outperform synchronous ICs in many application domains such as security, the design of integrated circuits still remains essentially limited to the realization of synchronous chips. One reason can explain this fact: no CAD suite has been proposed by the EDA industry to provide a useful and tractable design framework. However, some academic tools have been or are under development [1], [2], [3], [6].

Among them TAST [3] is dedicated to the design of micropipeline (µP) and Quasi Delay Insensitive (**QDI**) circuits. Its main characteristic is to target a standard cell approach. Unfortunately, it is uncommon to find in typical libraries (dedicated to synchronous circuit design) basic asynchronous primitives such as Rendezvous cells also called C-elements. Consequently, the designer of dual-rail asynchronous IC, adopting a standard cell approach, must implement the required boolean functions with AO222 or Majority gates [1], [4], [5]. It results in sub optimal physical implementations. Within this context, we developed a method allowing the quick design of the main combinatorial functionalities in dual-rail CMOS style.

Our goal is here to introduce the proposed design method and to compare the performances of the resulting cells with those of more traditional implementation styles. The remainder of the paper is organized as follows. Section II is dedicated to the introduction of the method itself. Before concluding, section III is devoted to the comparisons of performances (propagation delays, power consumption and realization cost) for the various boolean functions implemented with our design technique and with more usual implementation styles.

V. Paliouras, J. Vounckx, and D. Verkest (Eds.): PATMOS 2005, LNCS 3728, pp. 571–580, 2005.
© Springer-Verlag Berlin Heidelberg 2005

II Dual-Rail Cell Design

In this paragraph, we introduce the dual-rail cell design technique which can be either used to design new functionalities or to quickly translate in dual-rail the content of any single rail library.

II.a Targeted Cell Topology

Before to detail the design technique, let us remind the main specificities of the primitives required to design dual-rail logic blocks exchanging data one with another according to a four phase handshake protocol, and to the dual-rail encoding given in Fig.1.

For such circuits, the data transfer through a channel starts by the emission of a request signal encoded into the data, and finishes by the emission of an acknowledge signal. During this time interval, which is a priori unknown, the incoming data must be hold in order to guarantee the quasi-delay-insensitivity property. This implies the intensive use of logical gate including a state holding element usually latch or feedback loops.

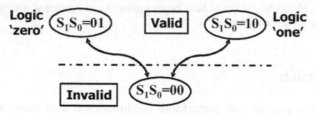

Fig. 1. Dual-rail encoding

As we target a CMOS implementation, it results from the preceding consideration that any dual-rail primitive may be considered as the juxtaposition of two complex gates, each one driving a latch as illustrated in Fig.2.

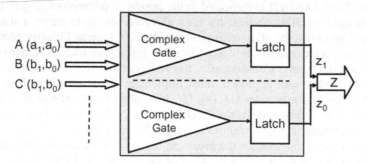

Fig. 2. Targeted cell topology

II.b Dual-Rail Cell Design Method

Adopting the dual-rail cell topology of Fig.2, the design of dual-rail asynchronous primitives amount to the identification of the two complex cells controlling the latches. This can be realized in six successive steps that constitute the proposed design method.

Step n°1
The first step consists in identifying the three boolean expressions characterizing the functionality to be realized.

Two of those three expressions are related to the settling conditions of the output rails z_1 and z_0 to V_{DD}, indicating respectively that the output has a boolean value of '1' or '0'.

The third expression is related to the settling conditions z_1 and z_0 to Gnd which correspond to the invalid state of the dual-rail encoding. So, in order to illustrate this identification step, let us realize a 3-input OR gate: $A(a_1,a_0)+B(b_1, b_0)+C(c_1,c_0)$. The three expressions characterizing this functionality in dual-rail logic are:

$$Z = A + B + C \tag{1}$$

$$\overline{Z} = \overline{A} \cdot \overline{B} \cdot \overline{C} \tag{2}$$

$$Z^I = A^I \cdot B^I \cdot C^I \tag{3}$$

If expressions (1) and (2), that are related to the settling conditions of z_1 and z_0 to V_{DD} are well known, expression (3) is more specific to the dual-rail encoding of Fig.1 since it defines the settling conditions to Gnd of both rails. In this expression, the character 'I' is used to stipulate that the logical bits Z, A, B and C are all in the invalid state. In the same way, the character 'V' will be used to stipulate that a logical bit is in a valid state (see fig. 1).

Table 1. Truth table of a 3-input OR gate

A	B	C	Z
0	0	0	0
0	0	1	1
0	1	0	1
0	1	1	1
1	0	0	1
1	0	1	1
1	1	0	1
1	1	1	1

Step n°2
After the identification stage of the three characteristic expressions, the second step of the method consists in rewriting these three expressions under a canonical form. This can easily be done starting from the truth table of the functionality to be realized (see Table 1). For the considered 3-input OR gate, this step leads to the following expressions:

$$Z = A\cdot B\cdot C + A\cdot B\cdot \overline{C} + A\cdot \overline{B}\cdot C + A\cdot \overline{B}\cdot \overline{C} + \overline{A}\cdot B\cdot C + \overline{A}\cdot B\cdot \overline{C} + \overline{A}\cdot \overline{B}\cdot C \tag{4}$$

$$\overline{Z} = \overline{A} \cdot \overline{B} \cdot \overline{C} \tag{5}$$

$$Z^I = A^I \cdot B^I \cdot C^I \tag{6}$$

Step n°3
In the third step, the canonical expressions obtained in the second step are reformulated so that the dual-rail encoding appears explicitly. This translation, from a natural encoding to the dual-rail encoding of Fig.1, is done using the conversion table (see Table 2) that defines the equivalences between the traditional simple rail logic and the dual-rail logic. Note that in table 2, 'bit' and 'bit!' mean that the 'bit' logical variable correspond respectively to '1' logical value and '0' one. So, for example, if we consider the following expression: $\overline{A} \cdot \overline{B} \cdot \overline{C}$, the dual-rail equivalent will be: $A_0 B_0 C_0$. As a result, for the considered 3-input OR gate, this step leads to the following dual-rail expressions:

$$z_1 = a_1 \cdot b_1 \cdot c_1 + a_1 \cdot b_1 \cdot c_0 + a_1 \cdot b_0 \cdot c_1 + a_1 \cdot b_0 \cdot c_0 + a_0 \cdot b_1 \cdot c_1 + a_0 \cdot b_1 \cdot c_0 + a_0 \cdot b_0 \cdot c_1 \quad (7)$$

$$z_0 = a_0 \cdot b_0 \cdot c_0 \quad (8)$$

$$\overline{z}_1 \cdot \overline{z}_0 = \overline{a}_1 \cdot \overline{a}_0 \cdot \overline{b}_1 \cdot \overline{b}_0 \cdot \overline{c}_1 \cdot \overline{c}_0 \quad (9)$$

Expressions (7-9) are sufficient to derive in a traditional way the topologies of the complex cells. However, for the considered 3-input OR gate, expression (9) leads to serially stack an excessive number of P transistors. In order to overcome this problem, it is necessary to introduce some additional variables. This is done in a fourth step.

Table 2. Single rail to dual-rail conversion table

	Type	bit	bit!	Bit^V	Bit^I
A(a1,a0)	Input	a_1	a_0	$a_1 \oplus a_0 \equiv a_1 + a_0$	$a_1 \,!\cdot a_0 \,!$
B(b1,b0)	Input	b_1	b_0	$b_1 \oplus b_0 \equiv b_1 + b_0$	$b_1 \,!\cdot b_0 \,!$
C(c1,c0)	Input	c_1	c_0	$c_1 \oplus c_0 \equiv c_1 + c_0$	$c_1 \,!\cdot c_0 \,!$
Z(z1,z0)	Output	z_1	z_0	$z_1 \oplus z_0 \equiv z_1 + z_0$	$z_1 \,!\cdot z_0 \,!$

Step n°4
If it is possible to insert any additional variables to implement the P transistor arrays, it is also possible to take advantage of the dual-rail encoding specificities. Among those main specificities of the dual-rail encoding, two are particularly interesting. The existence of forbidden state (1,1) constitutes the first one, while the second is related to the mutually exclusive behaviour of the rails conveying the boolean value of a logical bit (see Table 2). These two specificities confer some interesting properties to the 'exclusive-or' operation.

Table 3. Truth table of $a_1 \oplus a_0$

a_1	a_1	A	$a_1 \oplus a_0$	$a_1 + a_0$
0	0	Invalid	0	0
0	1	Valid, '0'	1	1
1	0	Valid,'1'	1	1
1	1	forbidden	Impossible	Impossible

As shown in Table 3, the existence of the prohibited state allows defining a bijection between the state of validity of the logical bit A, and the value of $a_1 \oplus a_0$ which is strictly equal to $a_1 + a_0$.

This bijection, that justifies the single-rail to dual-rail conversion table (see Table 2), can be used to redefine the settling conditions to the invalid state of the output bit. For the considered 3-input OR gate, this implies the definition of three additional variables:

$$U = a_1 \oplus a_0 = a_1 + a_0$$
$$V = b_1 \oplus b_0 = b_1 + b_0 \tag{10}$$
$$W = c_1 \oplus c_0 = c_1 + c_0$$

These three intermediate values being defined, it is afterward possible to simplify the expressions (7), (8), (9) to obtain:

$$z_1 = a_1 \cdot V \cdot W + a_0 \cdot b_1 \cdot W + a_0 \cdot b_0 \cdot c_1 \tag{11}$$

$$z_0 = a_0 \cdot b_0 \cdot c_0 \tag{12}$$

$$\overline{z}_1 = \overline{z}_0 = U \cdot V \cdot W \tag{13}$$

Expressions (11), (12), (13) enable implementing the N and P transistor arrays without transgressing the rules defining the maximum number of transistor that can be serially stacked.

Step n°5
The fifth step consist in designing the schematic of the two complex cells starting from expressions (11) (12) (13) and considering that the additional variables are delivered by the neighbourhood of the cell. This can be performed in traditional way. In the case of the 3-input OR gate, this leads to the schematic of the Fig.3.

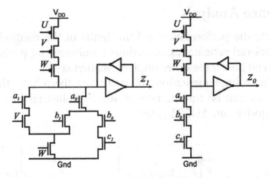

Fig. 3. Proposed dual-rail 3-input OR gate schematic

Step n°6
The final step aims at obtaining the complete schematic of the dual-rail primitive. This is achieved by implementing the calculus of the additional variables, i.e. by integrating the exclusive-or of all the pairs of rails carrying a single binary value.

These 2-input XOR gates, being equivalent to classical 2-input OR gates, as demonstrated earlier, they can be integrated at a low cost using usual single-rail 2-input OR gate. Consequently, it is very easy to obtain the complete schematic (see Fig. 4) of the dual-rail 3-input OR gate which is constituted of 42 transistors. This schematic clearly highlights that the incoming signals will have to cross in the worst case: two layers of transistors (elementary layer) allowing the evaluation of the intermediate values U, V and W, a layer related to the complex gates and finally a layer relating to the output latches.

Consequently, if the transistors are properly sized, the forward latency, i.e. the propagation delay of the 3-input OR gate under consideration will be in the worst case of four transistor layers what is very little with respect to more usual implementations, as we will see in section III.

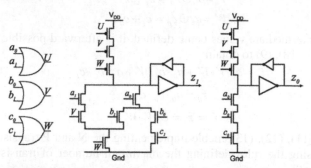

Fig. 4. Pseudo-static implementation of the 3-input OR gate.

It should be noted that this pseudo static implementation of the 3-input OR gate can easily be transformed into a fully static one (see Fig.5). However, if this can reduce significantly the propagation delays of the considered dual-rail 3-input OR gate, this increases its realization cost.

III Performance Analysis

In order to evaluate the performance and the limits of the method suggested herein, we derived the dual-rail schematics of various combinatorial primitives adopting the proposed method and also three other implementation styles.

Among these three implementation styles, two of them have the characteristic of using only cells that can be traditionally found in industrial libraries such as INV, NAND, NOR, Majority and AO222 gates.

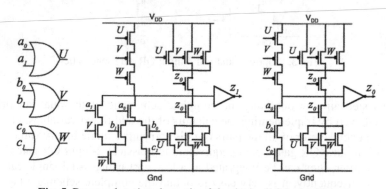

Fig. 5. Proposed static schematic of the dual-rail 3-input OR gate

Those two styles are afterwards denoted by AO222 and Majority gate based styles since they do use (Fig.6) respectively AO222 gates and Majority gates to implement the electrical rendezvous (C-element) of two signals. The third implementation style considered in the remainder of this paper is the one proposed in [5].

III.a   Implementation Cost

The implementation of those various boolean functions has allowed evaluating the integration cost of the four implementation styles considered. Table 4 that gives the obtained results highlights the benefits offered by the suggested design technique. As shown, following this technique we are able to design the most widely used functionalities with up to 65% less transistors.

This reduction of the realization cost, ranging from 43% to 73%, is due to the fact that several electrical paths share the same state holding element. This pooling of the state holding element appears clearly while comparing the structure of Fig.4 to that of Fig.6. Indeed, one can note that only one state holding element by rail is required for the 3-input OR gate we proposed, whereas eight state holding elements (8 of the C-element) are necessary to guarantee the correct behaviour of the structures represented in fig. 6.

If the pooling of the state holding elements explains partially this reduction of the integration cost, this one is mainly due to the fact that the suggested method enables the implementation of 3 and 4-input gates without necessarily cascading several 2 input gates. This constitutes a decisive advantage in term of cost, but also in term of speed, since the number of elementary layers crossed by the signals is also reduced. This point is illustrated by Table 5.

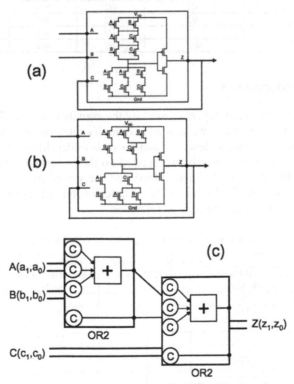

Fig. 6. Wiring (a) an AO222 and (b) a majority cell to obtained a C-element, (c) synoptic scheme of a dual-rail OR3 realized with AO222 or Majority cells; © is the symbol of the C-element

Table 4. Realization cost

Integration cost (transistor count)					
Dual-rail Cell	**Proposed Pseudo-static style**	**Proposed Static Style**	**AO222 based style**	**Majority based style**	**[5]**
Or2 / And2 / Nor2 /Nand2	30	38	64	56	42
Or3 / And3 / Nor3 /Nand3	42	54	128	112	84
Or4 / And4 / Nor4 /Nand4	56	72	192	168	126
Xor2 / Xnor2	32	40	68	60	44
Xor3 / Xnor3	52	64	136	120	88
Xor4 / Xnor4	72	88	204	180	132
AO21/AOI21	43	55	128	112	84
AO22/AOI22	58	74	192	168	126

Table 5. Min and Max numbers of elementary layers crossed by the signals

Min and Max numbers of elementary layers crossed by the signals					
Dual-rail Cell	**Proposed Pseudo-static style**	**Proposed Static Style**	**AO222 based style**	**Majority based style**	**[5]**
Or2 / And2 / Nor2/Nand2	4/2	4/2	4/2	4/2	2
Or3 / And3 / Nor3 /Nand3	4/2	4/2	8/4	8/4	4
Or4 / And4 / Nor4 /Nand4	4/2	4/2	8/4	8/4	4
Xor2 / Xnor2	4/2	4/2	4/2	4/2	2
Xor3 / Xnor3	4/2	4/2	8/4	8/4	4
Xor4 / Xnor4	4/2	4/2	8/4	8/4	4
AO21/AOI21	4/2	4/2	8/4	8/4	4
AO22/AOI22	4/2	4/2	8/4	8/4	4

III.b   Speed Performances

If the number of elementary layers crossed by the signals is a good first order indicator of the speed performances, we wanted to quantify them more precisely. Therefore we simulated the propagation delays of various cells under different loading and controlling conditions. More precisely, we applied on theirs inputs linear ramps of various durations (10, 50, 100, 200 and 500ps) and we varied the values of the output loading capacitance (5, 10, 20, 50, 100fF). Table 6 summarizes the results obtained taking the AO222 based design style as reference.

Table 6. Average propagation delay reduction

Average propagation delay reduction				
Cell	**Proposed Pseudo Static Style**	**AO222 based style**	**Majority based style**	**[5]**
Or2	5.2 %	0	22.6 %	18.1 %
Or3	38.7 %	0	38.1 %	26.6 %
Or4	35.8 %	0	38.9 %	25.3 %
Xor2	7.3 %	0	27.2 %	19.0 %
Xor3	39.0 %	0	38.9 %	28.1 %
Xor4	30.7 %	0	37.7 %	27.4 %
AO21	36.8 %	0	34.3 %	28.0 %
AO22	28.5 %	0	31.8 %	26.9 %

As expected from table 5, for 2-input cells the propagation delays of our cells are equivalent to those obtained for the AO222 and Majority based cells while they are smaller for 3 and 4 input gates.

III.c Energy Consumption

If the integration cost, evaluated as the number of transistors necessary to realize of boolean function, is a first order metric of the energy consumption, we characterized it more precisely. Therefore, we simulated the energy consumption for various cells controlled by voltage ramps of different durations (10, 50, 100, 200 and 500ps) and loaded by different capacitances (5, 10, 20, 50, 100fF). Table 7 gives the average energy consumption reduction of the various implementation styles considered herein taking the AO222 based design style as reference.

Table 7. Average energy consumption reduction

Cell	Average energy consumption reduction			
	Proposed Pseudo Static Style	AO222 based style	Majority based style	[5]
Or2	-30.3 %	0 %	17.0 %	14.7 %
Or3	7.2 %	0 %	26.6 %	31.9 %
Or4	11.3 %	0 %	29.2 %	34.6 %
Xor2	-36.1 %	0 %	17.6 %	19.1 %
Xor3	2.2 %	0 %	29.1 %	32.5 %
Xor4	6.1 %	0 %	30.7 %	33.5 %
AO21	5.5 %	0 %	29.8 %	27.7 %
AO22	7.5 %	0 %	30.9 %	27.0 %

As expected, the 3 and 4-input gates that we proposed have smaller energy consumption than the AO222 based cells have. However, this is not the case for our 2-input cells that consumes significantly more than the other implementation styles. This is mainly explained by the energy dissipated in the 2-input OR gates generating the internal signals. They do have, indeed a high toggling rate that penalizes the energy consumption.

IV Conclusion

We have introduced a method to quickly design dual-rail CMOS asynchronous primitives. The primitives obtained thanks to this design technique have better or similar performances than those of more traditional implementation styles such as the AO222 based style. Moreover, this technique allows designing 3 and 4 input CMOS dual-rail cells. To our knowledge, this was not possible up to now. As a result, the cells obtained with the design techniques introduced in this paper are good candidates for the design of dual-rail asynchronous and synchronous circuits.

References

1. T. Chelcea, A. Bardsley, D. A. Edwards, S. M. Nowick" A Burst-Mode Oriented Back-End for the Balsa Synthesis System." Proceedings of DATE '02, pp. 330-337 Paris, March 2002

2. J. Cortadella, M. Kishinevsky, A. Kondratyev, L. Lavagno and A. Yakovlev "Logic synthesis of asynchronous controllers and interfaces ", Springer-Verlag, ISBN: 3-540-43152-7 (2002)
3. M. Renaudin et al, "TAST", Tutorial given at the 8th international Symposium on Advanced Research in Asynchronous Circuits and Systems, Manchester, UK, Apr. 8-11, 2002.
4. C. Piguet, J. Zhand "Electrical Design of Dynamic and Static Speed Independent CMOS Circuits from Signal Transistion Graphs" PATMOS '98, pp. 357-366, 1998.
5. P. Maurine, J. B. Rigaud, F. Bouesse, G. Sicard, M. Renaudin "Static Implementation of QDI asynchronousprimitives", PATMOS 2003
6. J. Sparso, S. Furber "Principles of Asynchronous Circuit Design – A Systems Perspective", Kluwer Academic Publishers (2001)

Enhanced GALS Techniques for Datapath Applications

Eckhard Grass[1], Frank Winkler[2], Miloš Krstić[1],
Alexandra Julius[2], Christian Stahl[2], and Maxim Piz[1]

[1] IHP, Im Technologiepark 25, 15236 Frankfurt (Oder), Germany
{grass,krstic,piz}@ihp-microelectronics.com
[2] Humboldt-Universität zu Berlin, Institut für Informatik,
Unter den Linden 6, 10099 Berlin, Germany
{fwinkler,julius,stahl}@informatik.hu-berlin.de

Abstract. Based on a previously reported request driven technique for Glob-ally-Asynchronous Locally-Synchronous (GALS) circuits this paper presents two significant enhancements. Firstly, the previously required local ring oscilla-tors are avoided. Instead, an external clock with arbitrary phase for each GALS block is used. Details of the required wrapper circuitry, the proposed design flow and performance are provided. Secondly, to reduce supply noise, a novel approach applying individual clock jitter for GALS blocks is proposed. A simu-lation using the jitter technique shows that for a typical GALS system, the power spectrum of the supply current can be reduced by about 15 dB.

1 Introduction

New communication systems are being defined, developed and standardized at a rate that was not conceivable in the past. The rapidly changing markets put ever increas-ing pressure on the system designers. Reducing the time-to-market to its absolute limit is in many cases a necessity for successful commercial application.

This leads to an increasing demand for Intellectual Property (IP) blocks in the area of signal processing and communications. The use of verified and tested IP blocks can significantly reduce the design time of a system. However, achieving timing clo-sure for a complex system is a nontrivial task. This problem is further amplified when several clock domains and clock gating are used.

The main aim of this paper is to alleviate the problem of system integration for complex designs in the area of communications and signal processing. Another issue which becomes more and more important for designers of complex systems is the reduction of Electromagnetic Interference (EMI). In particular for mixed-signal de-signs, the noise which is introduced into the system via power supply lines, substrate as well as by wire capacitance and inductance, can severely degrade the analog per-formance.

A possible solution to the problems above is the deployment of Globally-Asynchronous Locally-Synchronous (GALS) circuits. GALS can contribute to sim-plify and speed-up the system integration and reduce EMI as well as power dissipa-tion. On the basis of GALS, rapid development of complex systems using existing IP blocks is facilitated [1, 2, 3].

In the past we have introduced a request driven GALS technique which is well suited for datapath applications [4, 5]. It benefits from low latency, and low power

V. Paliouras, J. Vounckx, and D. Verkest (Eds.): PATMOS 2005, LNCS 3728, pp. 581–590, 2005.

dissipation. This is achieved since no redundant clock pulses are supplied to the locally synchronous module (LSM). Each GALS block is driven with the request signal of it's predecessor. For emptying internal pipeline stages each GALS block is fitted with a local ring oscillator This system was successfully verified and tested in a GALS implementation of an IEEE 802.11a compliant baseband processor [5].

In this paper we propose two enhancements of this GALS technique. Firstly, the deployment of local ring oscillators turned out to be awkward and time consuming, both during design and test, since each ring oscillator needs individual tuning. Furthermore, the area and power penalty of the ring oscillators is deteriorating system performance. Therefore, the new versions of wrappers presented here will use an *external clock source* rather than ring oscillators.

Secondly, we propose the introduction of *jitter* for de-synchronizing the operation of different GALS blocks. This will reduce EMI and hence facilitate the integration of complex mixed-signal chips. Such introduction of blockwise jitter cannot easily be achieved with synchronous circuits.

The paper is organized as follows: In Section 2, after a brief review of existing solutions, the new generalized GALS wrapper is introduced. Section 3 describes possible solutions for locally introducing clock jitter to GALS blocks. In Section 4, we give some information on the design flow and verification. In Section 5, simulation results are reported and a comparison with previous implementations is given. Finally, conclusions are drawn in Section 6.

2   Concepts for Wrapper Implementation

2.1   Internally Clocked Wrapper

Our previous GALS designs were based on asynchronous wrappers, which had an integrated clock generator [5]. Those clock generators were implemented as ring oscillators. The advantage of this approach is that each GALS block can run on its own speed and the Electromagnetic Radiation (EMR) frequency spectrum of the complete chip is spread out due to different oscillator frequencies and phases. However, it is very tedious to implement the ring oscillators and to get their timing right. Furthermore, they may require re-tuning during operation. From a design automation point of view, it would be beneficial to use external oscillators instead. This would also reduce the power dissipation and silicon area of the wrapper.

2.2   New Externally Clocked Wrapper

Externally clocked GALS wrappers deliver the same advantages as internally clocked wrappers since it is not necessary to deliver the global clock to all GALS blocks at the same clock phase. Hence, there is neither a global clock tree nor tight skew control needed. The request driven operation is not affected, whereas the operation for emptying pipeline stages in the LSM (time-out mode) is driven by an external clock rather than a ring oscillator. During time-out mode, neighbouring GALS blocks will operate in a *mesochronous* way, i.e. at the same frequency but at different phases. The principal architecture for the new externally clocked GALS wrapper is shown in Fig. 1. Instead of the ring oscillator, an arbiter circuit is needed.

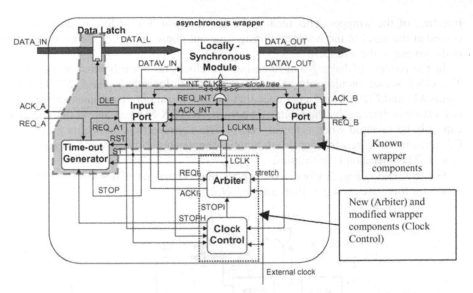

Fig. 1. Proposed new GALS wrapper with external clock source

In the following, we briefly discuss the issues related to the mesochronous clocking scheme.

Clock Synchronisation: Unlike for a ring oscillator, the external clock cannot be stopped. The provision of the clock to the locally synchronous block has to be gated with an appropriate signal shown as *STOPI* in Fig. 1.

Since the external clock is not synchronized with the gating signal *STOPI*, their logic combination, potentially can result in a glitch. This glitch in turn could lead to a malfunction of the synchronous block. Hence, signal *STOPI* must change its value only during the phase when the clock signal is low. For this purpose some arbitration of the clock signal with *STOPI* is required. This is done in the new arbiter block shown in Fig. 2. The operation is explained in the following sections.

Stretching the Clock Signal: Another issue related to controlling the external clock, arises at the output of a LSM. While the GALS wrapper is outputting data vectors, halting the clock signal at level zero is required. This is needed since a downstream GALS block may still be busy processing previous data and cannot acknowledge the current token.

Therefore, we have to arbitrate between incoming request signals and the external clock signal. If a request arrives and the clock signal line is high, no acknowledge signal will be granted. Alternatively, if request is high, the clock signal will be stretched until the request line is released.

In general, to avoid metastability during data transfer, the wrapper cannot receive any new data while the stretch signal is active. Compared to the wrapper based on a ring oscillator, we face new problems with the external clock arbitration. Halting and releasing the clock signal has to be realised in such a way that no hazards can occur.

Operation of New Wrapper with Arbiter: In order to prevent metastability, the arbiter is responsible for stretching the clock signal while there is handshake activity at the interface of the wrapper. This means, the clock signal is kept low when data is received at the input of the wrapper, during time out phase and when data is transferred at the output of the wrapper.

In the process of halting and releasing the clock, no hazards must be generated. The clock signal should only be stopped during its logic low phase. MUTEX elements M1 and M2 in Fig. 2 guarantee that the clock can only be stopped when *external_clk=0* is detected by the arbiter. The third MUTEX M3 guarantees that the clock can only be released when *external_clk=0*. The asymmetric C-elements C1, C2 and C3 perform the glitch-free logic combination between various internal signals of the arbiter. The operation is as follows:

For example in Fig. 2, when *Stretch* goes high, signal *ste* will go to low. The condition for this state change is that *external_clock=0*.

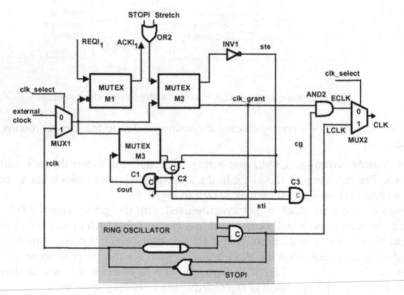

Fig. 2. Arbiter block for clock synchronisation

After that, the following scenario happens: *C2- -> sti- -> cg-* and hence, the clock *ECLK* is disabled via AND2. When *stretch* goes low than *ste+ -> C1+ -> C2+* (when *external_clock* is low) *-> sti+ -> cg+* and clock *ECLK* is enabled again. In this scheme a possible race can occur between signals *cg* and *clk_grant* during arbitration. To avoid a hazard on *ECLK,* it is required that the path *external_clock -> M2 -> clk_grant* is faster than path *stretch -> M2 -> ste -> sti -> cg.* However, this one-sided timing constraint will normally be fulfilled anyway or can easily be achieved.

2.3 Dual-Mode Wrapper with External and Internal Clock Source

A special version of the wrapper which can either be driven by a local ring oscillator or an external clock is termed 'dual-mode wrapper'. Here, the advantages of the external clock source and the internal clock source are combined by fitting the ring-oscillator into the arbiter unit. Two multiplexers, MUX1 and MUX2, shown in Fig. 2, allow the selection of the appropriate clock source. In our implementation, the clock source can only be changed by a *clk_select* signal when the circuit is in idle state i.e. when all internal switching activity has ceased. The dual clock system allows during normal operation, depending on the environment, to use either the external or the internal clock source. This way, the circuit can be optimised in terms of data through-put, latency, power dissipation and EMI for different applications. If, for instance, reduction of EMI is of utmost priority, the internal oscillators can be used and the external clock can be completely switched off. If, on the other hand, reduction of power dissipation is of highest importance, the ring oscillators can be switched off and the external clock can be used. It can also support testing of GALS systems. Fur-thermore, this dual clock wrapper can be used for fault tolerant systems. If, for in-stance, an external unit detects that a quartz oscillator is broken, the local ring oscilla-tors can be activated to keep the system function intact. The cost of the dual-mode wrapper is slightly higher power dissipation and circuit area.

3   Introduction of Clock Jitter in a GALS System

GALS systems using an external clock source do offer a lot of advantages. They are easy to design, consume low power and have small silicon area. However, compared to a system with local ring oscillators they partly re-introduce one disadvantage: EMI is increased since large parts of the circuit work exactly at the same frequency but with different phase. With local ring oscillators this is not the case. In some synchro-nous designs global *clock jitter* is used to reduce Electromagnetic Radiation (EMR) [6], [7]. However, this global clock jitter does not reduce peak power. In GALS sys-tems we can introduce a blockwise clock jitter since the wrappers will take care of the data transfer between blocks. This blockwise approach will have the advantage of reducing both, the spectral energy of the clock frequency as well as peak power. In this case, we will combine clock jittering and clock phasing in one approach. Two techniques are conceivable:

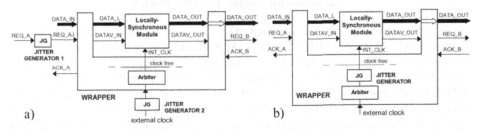

Fig. 3. Possible positions of jitter generator

a) It is possible to put one jitter generator (JG) in the signal path of the incoming REQ line of each block. Additionally, for operation in time-out mode, another JG should be inserted into the signal path of the external clock signal (see Fig. 3a).

b) It is possible to place one jitter generator (JG) in front of the clock input of the LSM (Fig. 3b). Both versions were tested and show similar results. However, the latter version requires less silicon area and has lower power dissipation.

Jitter Generator: In our GALS system, a jitter generator creates a random signal delay of a clock or request input signal. It consists of a True Random Number Generator (TRNG) or a Pseudo Noise Generator (PNG) with a sufficiently long period, and a programmable Delay Element (DE), shown in Fig. 4. The traditional jitter generation method is the Phase Modulation (PM). A voltage-controlled delay line with buffers was presented in [7]. For digital programmable delay elements, inverter chains are used. A multiplexer randomly selects a delay line output. Each signal event starts the next Pseudo Noise (PN) generation. For high clock frequencies simple Linear Feedback Shift Registers (LFSR) are preferred. In a given technology, they work up to nearly the maximum flip-flop toggle frequency. The effect of clock jitter was described in [6]. Using clock phase modulation, the spectral energy peaks were significantly reduced (up to 16 db). A higher modulation index gives lower spectral peaks, but reduces the performance slightly. Best results are achieved with random-noise phase-modulation. A high cycle-to cycle clock jitter can cause setup time violations in synchronous systems. Therefore, often triangular waveforms are used instead of PN sequences. They have lower clock-to-clock jitter, but less spectral improvements. Since the asynchronous wrappers do guarantee correct communication between GALS blocks, such systems can operate for a wide range of timing and jitter parameters.

Jitter Simulation Results: In order to estimate the effect of the clock jitter, we have created a MATLAB model of the supply current of a typical GALS system and compared it with the equivalent synchronous system. For the synchronous circuit, we assumed a triangular shape of the current waveform within one clock cycle with a 5 ns rise time and 10 ns fall time.

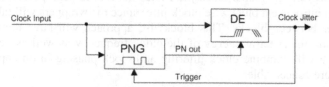

Fig. 4. Principal structure of a jitter generator

The peak supply current was assumed to be $I_{DD(peak)} = 1$ A and the clock period was assumed to be 20 ns (50 MHz). For the GALS circuit, we assumed that the synchronous circuit is split into 10 blocks. The 'size' of those blocks was chosen to be such that their respective peak supply current is distributed as follows: Two blocks of 200 mA each, four blocks of 100 mA each and four blocks of 50 mA each. Rise time and fall time were not altered. Each of the 10 current waveforms was given an equally distributed random phase in the range of ±10 ns. Afterwards, for each of the waveforms an equally distributed random jitter in the range of ±1 ns was introduced. The ten waveforms, each representing the supply current of a GALS block, were added to compute the total supply current. Subsequently, the power spectrum of the total supply current was calculated using a Fourier Transform. The results of this Fourier Transform given in Fig. 5 show, that for a wide range of frequencies, the spectral components in the GALS system with jitter are reduced by about 15 dB when compared to the synchronous circuit. Furthermore, in the time domain, the supply current peaks are reduced to about 30% of the synchronous system.

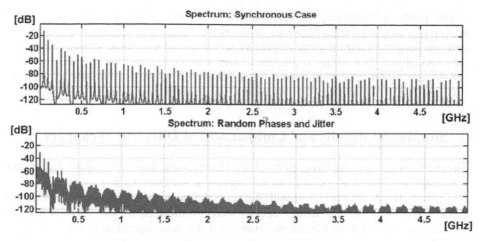

Fig. 5. Power spectrum of supply current for synchronous circuit (top) and its equivalent GALS implementation (bottom)

4 Design Flow and Verification

The design flow used is a combination of the standard synchronous design flow in conjunction with the application of some asynchronous tools. For the synthesis of asynchronous controllers, we have used the 3D tool [8] for translation of burst mode specifications into Boolean equations. Asynchronous wrappers are modelled in structural VHDL code that can be mapped into gates, using synchronous synthesis tools. Using specific attributes and directives, the hazard free behaviour can be preserved.

Finally, for the layout, standard CAD tools were used. After wrapper synthesis, a formal verification was performed using the model checker LoLA.

Wrapper Verification Using LoLA: In order to validate the correctness of our proposed concept, and to investigate possible hazards in the system, we decided to perform a formal analysis of the asynchronous wrapper. Our modelling approach presented in [10] is a combination of event-based and level-based modelling (see [9]). For each gate type of the given wrapper, a Petri net pattern is built which describes edges and levels of all signals in the pattern. This way, a pattern preserves all information needed to detect hazards. The wrapper is a connected composition of several instances of the corresponding patterns. The question whether a hazard can occur in a gate is reduced to a model checking problem: the reachability of a particular marking in the Petri net.

To verify the wrapper's model we employed LoLA [11], an explicit model checker that features powerful state space reduction techniques. We also used the known reduction technique of *abstraction* to alleviate the problem of state space explosion (see [10]). As a result we detected a number of potential hazards, signal races, e.g. if in the arbiter *cout* is low and a falling edge at *lclk* occurs then there is a race between *clk_grant-* and *cg+* (see Fig. 2), and a deadlock which is caused by a nondeterministic choice in the input AFSM. In most cases the calculation whether a specific hazard marking is reachable takes LoLA less than one second.

Designing and analyzing the wrapper was an iterative procedure. For example, the arbiter was designed and afterwards analyzed. With the help of our analysis procedure we detected a hazard in the respective subcircuit which would strongly influence the clock signal. Due to this result, the arbiter was re-designed and the wrapper analyzed once again. Our hierarchical, pattern-based modelling approach allowed an easy and fast update of the Petri net model after design modifications.

5   Simulation Results and Comparison of Different Wrappers

The main parameters we have analyzed are silicon area, data rate, latency and power consumption. A comparison of those parameters is shown in Table 1.

Three different wrappers, one with the ring oscillator, the second using an external clock signal, and the dual mode version, were synthesized for our 0.25 m CMOS technology. For all three circuits we simulated the data dependent power consumption of the wrappers with realistic gate-level timing and a switching activity based on 50 Msps data rate. The tool PrimePower was used for this purpose.

The largest silicon area is required for the circuit combining internal and external clock generation since it requires both the ring oscillator and additional logic to switch modes. The external clock generation requires by far the smallest silicon area since no ring oscillator is required. The throughput of the GALS system after synthesis was determined by simulation.

Table 1. Performance comparison of different GALS wrappers

Parametar \ Component	Int. wrapper (Ring oszi.)	Ext. wrapper (External clk)	Int/Ext wrapper (Dual-mode)
Total silicon area (μm^2):	79929	29879	84682
Number of gates:	1332	498	1411
Max. throughput - request mode (Msps)	119	133.9	148 / 148
Max. throughput - local mode (Msps)	86.9	79,4	96.2 / 82.6
Latency (ps)	7590	5150	6320 / 6290
Power (mW)	1.12	1.01	1.4 / 1.26

Results of this evaluation are given in Table 1. In our technology, the wrappers are operational up to about 150 Msps in request-driven mode. However, this number is due to some manual gate resizing, done after initial gate mapping. In local clock generation mode, the maximal throughput is in the range of 85 to 100 Msps. With external clocking, we achieve slightly lower speed of about 80 to 85 Msps. We have also generated latency figures for the different wrapper structures. The latency is defined as the time that data needs to pass from the last register stage of one GALS block to the first register stage of the subsequent GALS block in the data path. As we can see from Table 1, this time is in the range of 5.1 to 7.6 ns, whereby the best result is achieved with the externally clocked wrapper.

The power consumption figures presented in Table 1 are extracted from the simulation of different wrapper configurations using a realistic scenario of receiving, processing and transferring one data burst with 50 Msps datarate. In general, one wrapper consumes around 1 to 1.5 mW. This means, the wrapper does not cause any significant power overhead in our GALS system.

6 Conclusions

This paper proposes two significant enhancements to a previously reported request-driven technique for Globally-Asynchronous Locally Synchronous (GALS) circuits. To empty internal pipeline stages within locally synchronous modules our previously reported designs have used a ring oscillator for each GALS block. This technique causes much effort to design the oscillators and tune them to the correct frequency. Therefore, in this work, we have deployed an external clock instead. This clock can have an arbitrary phase for each GALS block. The attendant modifications to the GALS wrapper lead to reduced silicon area and lower power consumption when compared to the version with ring oscillators.

Secondly, to reduce power supply noise, electromagnetic radiation (EMR) and substrate noise, we propose the application of clock jitter to each individual GALS block. This reduces the spectral energy of the supply current variation as well as the peak supply current of a circuit. The blockwise application of clock jitter cannot easily be achieved for synchronous systems since the data transfer between blocks would be problematic. To our knowledge, this is the first paper where a blockwise application of clock jitter in conjunction with GALS systems is proposed. Using a MATLAB simulation model we have shown that for a typical GALS circuit the spectral power of the supply current can be reduced by about 15 dB. In the time domain, supply current variations are reduced to about 30 % when compared to an equivalent synchronous circuit.

Further work will focus on implementation, fabrication and test of a real chip using the proposed techniques.

References

1. J. Muttersbach, *Globally-Asynchronous Locally-Synchronous Architectures for VLSI Systems*, Doctor of Technical Sciences Dissertation, ETH Zurich, Switzerland, 2001.
2. Robert Mullins, George Taylor, Peter Robinson, Simon Moore, „*Point to Point GALS Interconnect*", IEEE Proc. ASYNC'2002, pp. 69-75, Manchester (UK), April 2002.
3. D. S. Bormann, P. Y. K. Cheoung, *Asynchronous Wrapper for Heterogeneous Systems*, In Proc. International Conf. Computer Design (ICCD), October 1997.
4. M. Krstić, E. Grass, *New GALS Technique for Datapath Architectures*, In Proc. International Workshop PATMOS, pp. 161-170, September 2003.
5. M. Krstić, E. Grass, C. Stahl, *Request-driven GALS Technique for Wireless Communication System*, Proceedings 11th IEEE International Symposium on Asynchronous Circuits and Systems (ASYNC'2005), New York City, pp. 76-85, March, 2005.
6. Mustafa Badaroglu, Piet Wambacq, Geert Van der Plas, Stéphane Donnay, Georges Gielen, and Hugo De Man. *Digital Ground Bounce Reduction by Phase Modulation of the Clock*. Proceedings of the Design, Automation and Test in Europe Conference and Exhibition (DATE'04) 2004.

7. Tian Xia Peilin Song, Keith A.Jenkins Jien-Chung Lo. *Delay Chain Based Programmable Jitter Generator.* Proceedings of the Ninth IEEE European Test Symposium (ETS'04) 2004.
8. Kenneth Yun, David Dill: *Automatic synthesis of extended burst-mode circuits: Part I and II.* IEEE Transactions on Computer-Aided Design, 18(2), pp. 101-132, Feb. 1999.
9. A. Yakovlev and A. M. Koelmans. *Petri Nets and Digital Hardware Design. LNCS: Lectures on Petri Nets II: Applications,* 1492, 1998.
10. Ch. Stahl, W. Reisig, M. Krstić. *Hazard Detection in a GALS Wrapper: a Case Study.* Accepted for ACSD 2005.
11. K. Schmidt. *Lola – a low level analyser.* In Nielsen, M. and Simpson, D., editors, International Conference on Application and Theory of Petri Nets, LNCS 1825, page 465 ff. Springer-Verlag, 2000.

Optimizing SHA-1 Hash Function for High Throughput with a Partial Unrolling Study*

H.E. Michail, A.P. Kakarountas, George N. Selimis, and Costas E. Goutis

Electrical & Computer Engineering Department,
University of Patras, 25600 Patras, Greece
{michail,kakaruda,gselimis,goutis}@ee.upatras.gr

Abstract. Hash functions are widely used in applications that call for data integrity and signature authentication at electronic transactions. A hash function is utilized in the security layer of every communication protocol. As time passes more sophisticated applications arise that address to more users-clients and thus demand for higher throughput. Furthermore, due to the tendency of the market to minimize devices' size and increase their autonomy to make them portable, power issues have also to be considered. The existing SHA-1 Hash Function implementations (SHA-1 is common in many protocols e.g. IPSec) limit throughput to a maximum of 2 Gbps. In this paper, a new implementation comes to exceed this limit improving the throughput by 53%. Furthermore,power dissipation is kept low compared to previous works, in such way that the proposed implementation can be characterized as low-power.

1 Introduction

Due to the essential need for security in networks and mobile services, as specified in various standards, such as the WTLS security level of WAP in [1], IPsec and the 802.16 standard for Local and Metropolitan Area Networks [2], an efficient and small-sized HMAC [3] implementation, to authenticate both the source of a message and its integrity, is very important. Moreover year-in year-out Internet becomes more and more a major economical parameter of world's financial and thus whole new applications are being created that presuppose authentication services.

One recent example is the Public Key Infrastructure (PKI) that incorporate authenticating services providing digital certificates to clients,servers,etc. PKI increases citizen's trust to public networks and thus empowers applications such as on-line banking,B2B applications,electronic payments,stock trading etc. The PKI that is considered as a must-have mechanism for the burst of e-commerce worldwide involves the use of the SHA-1 hash function.However the implementations that will be used in the PKI should have a much higher throughput

---

* We thank European Social Fund (ESF), Operational Program for Educational and Vocational Training II (EPEAEK II) and particularly the program PYTHAGORAS, for funding the above work

V. Paliouras, J. Vounckx, and D. Verkest (Eds.): PATMOS 2005, LNCS 3728, pp. 591–600, 2005.

comparing to the present implementations in order to be able to correspond to all requests for digital certificates.

On the other hand applications like SET (Secure Electronic Transactions) have started to consecrate on mobile and portable devices. SET is a standard for secure electronic transactions via public networks that has been deployed by VISA,MASTERCARD and many other leading companies in financial services. SET presupposes that an authenticating module that includes SHA-1 hash function is embedded in any mobile or portable device.This means that the implemented authentication core must be low-power.

This paper is mainly focused on SHA-1 [4] due to its major use in standards, although other hash functions, like MD5 [5], can also be considered. Various techniques have been proposed to minimize the SHA-1 implementation size. The most common techniques are operation rolling loop and/or re-configuration. On the other hand, alternative design approaches have been proposed to increase throughput for a variety of hash function families. The most common techniques are pipeline and parallelism. Design approaches that meet both constraints of high-performance and small-size were presented in [7] and [8], where SHA-1 was implemented applying simultaneously the re-use and pipeline techniques. A novel design approach is proposed to increase SHA-1 throughput, which exceeds by 53% the throughput of the implementation presented in [8]. Conservative estimations also show that a 30% of power saving can also be achieved.

This paper is organized as follows: In section 2 previous implementations of the SHA-1 are presented. In section 3 the proposed design approach is detailed and in section 4 the choice of unrolling two operations is justified. In section 5 power issues concerning the SHA-1 are presented.Throughput and area results of the proposed SHA-1 are offered in section 6 and it is compared to the other implementations. Finally, conclusions are offered in section 7.

2   Existing Implementations of SHA-1

The Secure Hash Standard [4] describes in detail the SHA-1 hash function. It requires 4 rounds of 20 operations each, resulting in a total of 80 operations, to generate the Message Digest. In Fig.1, the interconnection of two consecutive operations is illustrated. Each one of the a_t, b_t, c_t, d_t, e_t, is 32-bit wide resulting in a 160-bit hash value. K_t and W_t are a constant value for iteration t and the t_{th} w-bit word of the message schedule, respectively. The architecture of a SHA-1 core is formed as illustrated in Fig. 2. In the MS RAM, all message schedules W_t of the t_{th} w-bit word of the padded message are stored. The Constants Array is a hardwired array that provides the constant values K_t and the constant initialization values H_0 - H_4. Additionally, it includes the W_t generators. Throughput is kept low due to the large number of the required operations. An approach to increase significantly throughput is the application of pipeline. However, applying or not pipeline, the required area is prohibitive for mobile and portable applications. Thus, various techniques have been proposed to introduce to the market high-speed and small-sized SHA-1 implementations

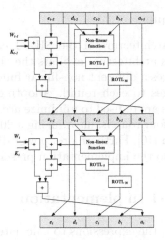

Fig. 1. 2 consecutive SHA-1 operations

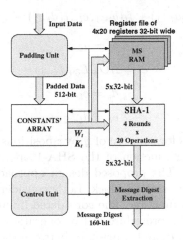

Fig. 2. Typical SHA-1 core

2.1   Rolling Loop Technique

In [9] the rolling loop technique was used in order to reduce area requirements. The proposed architecture of [9] requires only 4 operation blocks, one for each round. Using a temporal register and a counter, each operation block is re-used for 20 iterations. After 20 clock cycles the value of the first round is ready and propagated to the next round. This approach is considerably area-efficient, throughput is kept low due to the requirement of 81 clock cycles to generate the Message Digest. In [10] a re-use technique was applied to the non-linear functions, exploiting the similarity of the operation block. Modifying the operation block to include the four non-linear functions, the non-linear function that corresponds to the time instance t is selected through a multiplexer.

2.2   Pipeline Technique

In [7], [8] and [11] the architecture of the SHA-1 core is based on the use of four pipeline stages. This architecture exploits the characteristics of the SHA-1 hash function that requires a different non-linear function every 20 clock cycles, assigning a pipeline stages to each round. Adopting design elements from [9] and [10], operation blocks are re-used to minimize area requirements. This allows parallel operation of the four rounds, introducing a 20 cycle latency to quadruple the throughput of that in [10]. Furthermore, power dissipation and area penalty are kept low, compared to the implementation presented in [9].

3   Proposed SHA-1 Implementation

From [4] and from Fig. 1, the expressions to calculate a_t, b_t, c_t, d_t, e_t, are given in Eq. 1-5

$$a_t = ROTL_5(a_{t-1}) + f_t(b_{t-1}, c_{t-1}, d_{t-1}) + e_{t-1} + W_t + K_t \qquad (1)$$

$$b_t = a_{t-1} \qquad (2)$$

$$c_t = ROTL_{30}(b_{t-1}) \qquad (3)$$

$$d_t = c_{t-1} \qquad (4)$$

$$e_t = d_{t-1} \qquad (5)$$

where $ROTL_x(\text{y})$ stands for rotation of y by x positions to the left, and $f_t(\text{z,q,r})$ represents the non-linear function of the SHA-1 operation block which is applicable on operation t. The proposed design approach is based on a special property of the SHA-1 operation block. Let's consider two consecutive operations of the SHA-1 hash function. The considered inputs a_{t-2}, b_{t-2}, c_{t-2}, d_{t-2}, e_{t-2} go through a specific procedure in two operations and after that the considered outputs a_t, b_t, c_t, d_t and e_t arise. In between the signals a_{t-1}, b_{t-1}, c_{t-1}, d_{t-1}, e_{t-1} that are outputs from the first operation and inputs for the second operation have been computed. Except of the signal a_{t-1}, the rest of the signals b_{t-1}, c_{t-1}, d_{t-1}, e_{t-1} are derived directly from the inputs a_{t-2}, b_{t-2}, c_{t-2}, d_{t-2} respectively. This means consequently that also c_t, d_t and e_t can be derived directly from a_{t-2}, b_{t-2}, c_{t-2} respectively. Furthermore,the fact that a_t and b_t calculations require the d_{t-2} and e_{t-2} inputs respectively, which are stored in temporal registers is observed. The output a_t requires only d_{t-2} whereas b_t requires only e_{t-2}. It is clear enough that these these two calculations can be performed concurrently. In Fig. 3, the consecutive SHA-1 operation blocks of Fig. 1, have been modified so that a_t and b_t are calculated concurrently.

The gray marked areas on Fig. 3 indicate the parts of the proposed SHA-1 operation block that operate in parallel. Examining the execution process it is noticed that only a single addition level has been introduced to the critical path. This is necessary because during the computation of a_t the b_t value has to be known.So in three addition levels the b_t value is known and in parallel the two

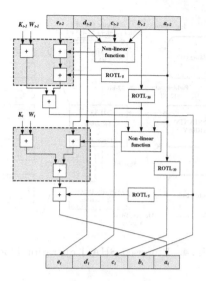

Fig. 3. Proposed SHA-1 operation blocks

addition levels of a_t computation have already been performed. An extra addition level is required in order for the a_t value to be fully computed. Considering the above facts it is obvious that the critical path in the proposed implementation consists of four addition levels instead of the three addition levels consisting the critical path of a non-concurrent implementation. Although, this fact reduces the maximum operation frequency in the proposed implementation, the throughput is increased significantly as it will be shown. In Eq. 6, the expression of throughput is given.

$$Throughput = \frac{\#bits * f_{operation}}{\#operations} \tag{6}$$

For the above equation the theoretical expected operating frequency is about 25% lower since the critical path has been exceeded from three to four addition levels comparing to non-concurrent implementations. However the hash value in the proposed implementation is computed in only 40 clock cycles instead of 80 in the non-concurrent implementations. This computations lead to the result that theoretically the throughput of the proposed implementation increases by 50%.

In Fig. 4, the modified structure of the hash core is illustrated, as it is proposed in [7] and [8], where there are four pipeline stages and the proposed operation block for each round.

The partially unrolled expressions that give a_t, b_t, c_t, d_t and e_t, are now described from Eq.7-11.

$$a_t = ROTL_5(a_{t-1}) + f_t(a_{t-2}, c_{t-2}, ROTL_5(b_{t-2})) + d_{t-2} + W_t + K_t \tag{7}$$

$$b_t = ROTL_5(a_{t-2}) + f_t(b_{t-2}, c_{t-2}, d_{t-2}) + e_{t-2} + W_{t-1} + K_{t-1} \tag{8}$$

$$c_t = ROTL_{30}(a_{t-2}) \tag{9}$$

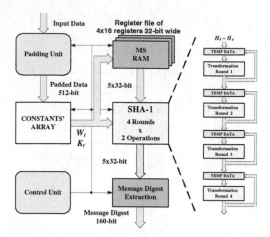

Fig. 4. Proposed SHA-1 operation blocks

$$d_t = ROTL_{30}(b_{t-2}) \tag{10}$$

$$e_t = c_{t-2} \tag{11}$$

From Eq. 7-11, it can be assumed that the area requirements are increased. Thus, the small-sized constraint is violated. However, the hardware to implement the operation blocks of the SHA-1 rounds is only a small percentage of the SHA-1 core. Moreover considering the fact that the SHA-1 hash core is a component and not entire the authenticating scheme obviously the proposed implementation satisfies the design constraint for small-sized, and high-performing operation. Besides that in the next section it will be shown that the proposed implementation also meets the design constraints for the characterization as low-power.

4   Number of Operations to Be Partially Unrolled

The selection to unroll two operations and not more can he answered by the analysis of the resulted critical path and the corresponding throughput compared to the area that is required for the implementation of the SHA-1 hash core. The results of this analysis are presented in Fig.5 where the ratio Throughput per Area for several unrolled operations is illustrated.It is useful to point out that the number of operations to he unrolled when dividing 20 must give an integer result for implementation's shake.From Fig.5 obviously the best ratio of unrolled operations per achieved frequency was given for two operations unrolling. Partial unrolling does not always present the same best fitting point. i.e. two operations unrolled,but there are many parameters that determine the best solution.One parameter is the algorithm since it specifies the ratio of throughput gain per area penalty. Another parameter is the rest security scheme that is incorporated along with the hash function.If there is another module that operates on a quite

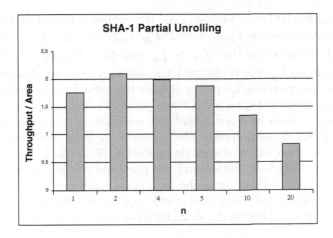

Fig. 5. SHA-1 Unrolling Operation Analysis

low frequency and requires a significant amount of area for its implementation then it is better to unroll more operations since that results to a lower frequency hut also to a higher throughput for the hash core. Moreover the introduced area penalty as percentage is smaller.In our implementation we tried to implement a hash core that has a higher throughput than presented implementations and thus is small-sized and low-power addressing in this way to a great variety of applications.

5 Power Issues

The proposed SHA-1 operation block not only results to a higher throughput for the whole SHA-1 core but it also leads to a more efficient implementation as long as the power dissipation is concerned. The reduction of the power dissipation is achieved due to a number of reasons.

First of all the decrease of the operating frequency of the SHA-1 core results to lower dynamic power dissipation for the whole SHA-1 core. This can easily be seen regarding the relevant power equations. Moreover the adopted methodology for the implementation of each SHA-1 operation block combines the execution of two logical SHA-1 operations in only one single clock cycle. This means that the final message digest is computed in only 40 clock cycles and thus calls for only 40 write operations in the temporal register that save all the the intermediate results until the final message digest has been fully derived.

It is possible to estimate the total power savings considering that the initial power dissipation was calculated as $P_{init}=80P_{op}(f_{op}) + 80P_{WR}(f_{op})$, where $P_{op}(f_{op})$ is the dynamic power dissipation of a single operation (depends from the operation frequency f_{op}) and $P_{WR}(f_{op})$ is the power dissipated during write/read operation of the registers (also depends from f_{op}). Both $P_{op}(f_{op})$ and $P_{WR}(f_{op})$'s values are proportional to the operating frequency f_{op}. This means for a decreased f_{op} both $P_{op}(f_{op})$ and $P_{WR}(f_{op})$ result to decreased values.

According to the latter assumptions, the proposed operation block's power dissipation is estimated as $P_{prop} = 40(2*P_{op}(f'_{op})) + 40P_{WR}(f'_{op}) = 80P_{op}(f'_{op}) + 40P_{WR}(f'_{op})$. Considering that $f_{op} > f'_{op}$ and thus $P_{op}(f_{op}) > P_{op}(f'_{op})$ and $P_{WR}(f_{op}) > P_{WR}(f'_{op})$ (according to what was previously mentioned), it can be derived that the operating frequency defines the overall power savings and that the proposed implementation has a lower power dissipation.

The above calculations are considered as conservatives since the proposed operation block's dynamic power dissipation is for sure less than the twofold dynamic power dissipation of a single operation. This can be easily realized if the conventional single-operation and the proposed double-operation block are examined thoroughly. However in the theoretical analysis the factor 2 was used in order to cover the worst case that could happen including any power leakages that could be revealed due to the extra hardware used in the proposed operation block.

If a similar implementation is intended to be used in a device that does not exploits the extra throughput (i.e some portable or mobile devices for certain use) then this fact can lead to an even more low-power device. This can be achieved if a certain targeted technology(in ASIC) is used where the operating frequency can be slowed down and at the same time the supplying voltage Vdd can be also decreased. Obviously this leads to a significant reduction of the total power consumption whereas the throughput of the device fluctuates to the desirable limits of conventional high-throughput implementations.

In the proposed implementation a 53% higher throughput is achieved comparing to competitive implementations. s a result of this the operating frequency can be reduced about 53% and have the same throughput with the other competitive implementations. The reduction of the operating frequency also leads to reduction of the supplying voltage Vdd (in ASIC designs)at about 40% taking in consideration conservative aspects. On the other hand a significant increase in the effective capacitance of the circuit occurs by a factor of two, that has to be taken in consideration. Considering that the power dissipation in a circuit is proportional to the effective capacitance,to the operating frequency and to the square of the supplying voltage,it can be assumed that in this way an extra 60% power saving can be achieved meeting this way the constraint for extended autonomy.

6   Experimental Results and Comparisons

In order to evaluate the proposed SHA-1 design approach, the XILINX FPGA technology was used. The core was integrated to a v150bg352 FPGA device. The design was fully verified using a large set of test vectors. The maximum achieved operating frequency is equal to 55 MHz, an expected decrease of 25% compared to [8] that correspond to the extra addition level introduced to the critical path. Although the operating frequency of the proposed implementation is lower than that of [7], [8] and [10], the achieved throughput exceeds 2,8 Gbps. In Table 1, the proposed implementation is compared to the implementations of [7], [8],

[9], [10], [11] and [12]. From the experimental results, there is a range of 53% - 2266% increase of the throughput compared to the previous implementations. It has to be noticed that the implementation of [10] was re-designed for the specific technology, for fair comparison. In [10], the reported operating frequency was 82 MHz and the throughput was 518 Mbps.

Table 1. Operating Frequencies and Throughput

Implementations	Operating Frequency(Mhz)	Throughput(Mbps)
[7]	71	1731
[8]	72	1843
[9]	43	119
[10]	72(82)	460(518)
[11]	55	1339
[12]	38.6	900
Prop. Arch.	55	2816

Furthermore, regarding the overall power dissipation to process a message, the proposed implementation presents significant decrease, approximately by 30% compared to the nearest performing implementation [8]. Power dissipation was calculated using Synopsys Synthesize Flow for the targeted technology. The activity of the netlist was estimated for a wide range of messages so that the gathered values of the netlist activity can be considered as realistic. Then, from the characteristics of the technology, an average wire capacitance was assumed and the power compiler gave rough estimations. The results were also verified on test boards, measuring the overall power consumed for a given set of messages, for each implementation. Power dissipation is decreased primarily due to the lower operating frequency, without compromising performance. Also, power dissipation decrease is achieved due to the reduction by 50% of the write processes to the temporal registers.

In the case of the introduced area, the implementation of a SHA-1 core, using the proposed operation block, presented a 20% overall area penalty, compared to the implementation of [8]. The introduced area is considered to satisfy the requirements of the small-sized SHA-1 implementations, meeting in parallel the high-performance and low-power constraints.

7   Conclusions and Future Work

A high-speed and low power implementation of the SHA-1 hash function was proposed in this paper. It is the first known small-sized implementation that exceeds the 2 Gbps throughput limit (for the XILINX FPGA technology - v150bg352 device). From the experimental results, it was proved that it is performing more than 50% better than any previously known implementation. The introduced area penalty was approximately 20% compared to the nearest performing implementation. This makes it suitable for every new wireless and mobile communication application [1], [2] that urges for high-performance and small-sized solutions. However, the major design advantage of the proposed design approach is the low power dissipation that is required to calculate the hash value of any given message. Compared to other high-performing implementations, approximately 30% less power per message is required. The proposed design approach will be used to form a generic methodology to design low-power and high-speed implementations for various families of hash functions.

References

1. WAP Forum, Wireless Application Protocol, Wireless Transport Layer Security, Architecture Specifications, 2003.
2. IEEE Std. 801.16–2001, IEEE Standard for Local and Metropolitan Area Networks, part 16, "Air Interface for Fixed Broadband Wireless Access Systems," IEEE Press, 2001.
3. HMAC Standard, The Keyed-Hash Message Authentication Code, National Institute of Standards and Technology (NIST), 2003.
4. FIPS PUB 180-1, Secure Hash Standard (SHA-1), National Institute of Standards and Technology (NIST), 1995.
5. R. L., Rivest, The MD5 Message digest Algorithm, IETF Network Working Group , RFC 1321, April 1992.
6. H., Dobbertin, The Status of MD5 After a Recent Attack, RSALabs' CryptoBytes, Vol.2, No.2, Summer 1996.
7. N., Sklavos, P., Kitsos, E., Alexopoulos, and O., Koufopavlou, "Open Mobile Alliance (OMA) Security Layer: Architecture, Implementation and Performance Evaluation of the Integrity Unit," New Generation Computing: Computing Paradigms and Computational Intelligence, Springer-Verlag, 2004, in press.
8. N., Sklavos, E., Alexopoulos, and O., Koufopavlou, "Networking Data Integrity: High Speed Architectures and Hardware Implementations," IAJIT Journal, vol. 1, no. 0, pp. 54–59, 2003.
9. S., Dominikus, "A Hardware Implementation of MD-4 Family Hash Algorithms," in Proc. of ICECS, pp. 1143–1146, 2002.
10. G., Selimis, N., Sklavos, and O., Koufopavlou, "VLSI Implementation of the Keyed-Hash Message Authentication Code for the Wireless Application Protocol," in Proc. of ICECS, pp. 24–27, 2003.
11. N., Sklavos, G., Dimitroulakos, and O., Koufopavlou, "An Ultra High Speed Architecture for VLSI Implementation of Hash Functions," in Proc. of ICECS, pp. 990–993, 2003.
12. J.M., Diez, S., Bojanic, C., Carreras, and O., Nieto-Taladriz, "Hash Algorithms for Cryptographic Protocols: FPGA Implementations," in Proc. of TELEFOR, 2002.

Area-Aware Pipeline Gating for Embedded Processors

Babak Salamat and Amirali Baniasadi

Electrical and Computer Engineering, University of Victoria
{salamat,amirali}@ece.uvic.ca

Abstract. Modern embedded processors use small and simple branch predictors to improve performance. Using complex and accurate branch predictors, while desirable, is not possible as such predictors impose high power and area overhead which is not affordable in an embedded processor. As a result, for some applications, misprediction rate can be high. Such mispredictions result in energy wasted down the mispredicted path. We introduce area-aware and low-complexity pipeline gating mechanisms to reduce energy lost to possible branch mispredictions in embedded processors. We show that by using a simple gating mechanism which comes with 33-bit area overhead, on average, we can reduce the number of executed instructions by 17% (max: 30%) while paying a negligible performance cost (average 1.1%).

1 Introduction

Modern embedded processors aim at achieving high-performance while maintaining die area and power consumption at a low level. While technology advances continue to provide embedded processors with more resources, we are still far away from affording advanced complex techniques. It is due to such restrictions that embedded processors do not exploit many techniques frequently used by high-performance desktop machines or servers. One way to narrow this gap is to revisit solutions introduced for high performance processors to develop affordable implementations.

Branch prediction is essential as it provides steady instruction flow at the fetch stage which in turn results in shorter program runtime. However, unfortunately, predictors are not perfect and make mispredictions. Such branch mispredictions result in longer program runtimes and energy wasted down the mispredicted instruction path. Compared to a high-performance processor, the energy lost to mispredictions appears to be more costly in embedded processors. This is due to the fact that in an embedded processor power resources are limited to batteries, making efficient power management even a more critical task.

In addition, embedded processors are bound to exploit simple and possibly less accurate branch predictors compared to high performance processors. Using more simple and less accurate branch predictors by embedded processors compared to high performance processors is exemplified by the XScale processor: Intel's XScale which was introduced in 2001 [2] uses a 256-bit bimodal predictor while the Alpha EV6 [3] that was released in 1997 used 36Kbits.

More importantly, it is expected that as future embedded processors start exploiting deeper pipelines, branch misprediction cost will increase even further.

To address these concerns finding techniques to reduce energy lost to speculation while maintaining performance is essential.

V. Paliouras, J. Vounckx, and D. Verkest (Eds.): PATMOS 2005, LNCS 3728, pp. 601–608, 2005.

A previous study has suggested using pipeline gating to reduce energy lost to branch misspeculation [1]. Unfortunately, previously suggested pipeline gating mechanisms are either too complex to be used in an embedded processor or depend on complex branch predictors which are not affordable in an embedded processor.

The goal of this work is to introduce a set of power-efficient and area-aware pipeline gating methods to reduce misprediction cost while maintaining performance. In particular, we introduce three very low-overhead pipeline gating mechanisms to reduce energy lost to branch misprediction while maintaining performance.

Pipeline gating relies on accurate branch confidence estimation [6]. We also introduce low-overhead branch confidence estimation techniques to be used in embedded processors.

Our most aggressive technique reduces the number of mistakenly fetched instructions by 17% (max: 30%) with an average performance loss of 1.1%.

The rest of the paper is organized as follows. In section 2 we discuss our motivation. In section 3 we explain our techniques in more details. In section 4 we present methodology and results. In section 5 we review related work. Finally, in section 6 we offer concluding remarks.

2 Motivation

As explained earlier, modern processors lose energy due to possible branch mispredictions. We refer to the mistakenly fetched instructions as *wasted activity* (WA). Note that mispredicted instructions do not commit and are flushed as soon as the mispredicted branch is resolved. As such, we define WA as:

$$WA = \frac{fetched - commited}{fetched}$$

Where *fetched* and *committed* are the numbers of instructions fetched and committed during execution of a program, respectively.

In figure 1 we report the percentage of instructions fetched down the mispredicted path. We report for a subset of MiBench benchmarks [5] and for a processor similar to Intel's XScale. We include benchmarks with both high and low WA in figure 1 and through this study. As presented, WA is more than 20% for three of the applications studied here. To understand why different applications come with different WAs it is important to take into account other parameters including branch misprediction rate.

To explain this further, in figure 2 we report misprediction rate of the bimodal predictor used in XScale for the selected benchmarks. As reported in figures 1 and 2, the four applications with higher WAs also show higher misprediction rates.

While both high-performance processors and embedded processors lose energy to branch mispredictions, it is important to take a different approach in embedded processors. Our study shows that front-end gating should be done more carefully in an embedded processor where front-end buffers are usually small in size. In a high-performance processor, exploiting large size buffers makes a steady instruction flow possible in the event of a fetch stall (*e.g.*, stalls caused by a cache miss). However, in embedded processors, where less resource is available to the processor front-end, stalling the front-end for correctly predicted instructions can impact performance dramatically (*more on this later*).

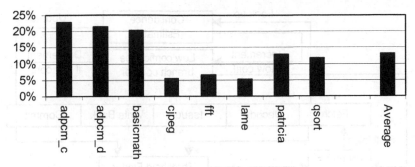

Fig. 1. Wasted activity in fetch stage

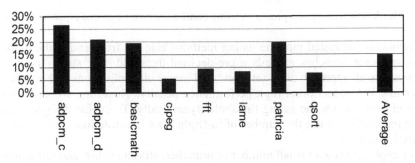

Fig. 2. Misprediction rate for XScale's bimodal predictor

3   Area-Aware Confidence Estimation

We present the schematic of a processor using pipeline gating in figure 3.

Our goal is to stall instruction fetch when there is a high chance that the fetched instructions will be flushed. To do this we gate the pipeline front-end when there is low-confidence in the executed instructions. In order to gate the pipeline while maintaining performance we need a mechanism to identify low confidence branches.

Several studies have introduced accurate confidence estimators. However, previously suggested estimators rely on exploiting complex structures which may not be affordable in an embedded processor. To apply pipeline gating to an embedded processor, we introduce three accurate confidence estimation techniques which impose very little overhead.

3.1   History-Based Confidence Estimation

In this method we assume that recently mispredicted branches are more likely to be mispredicted in the future. As such we keep track of recently fetched branch instructions' confidence using a very small 16-bit structure. This structure is a PC-indexed 8-entry table where there is a 2-bit counter associated with each entry. The 2-bit counter is incremented for accurately predicted branches. We reset the associated counter if the branch is mispredicted. We look up this structure at fetch and in parallel to probing the branch predictor. If the 2-bit counter is not saturated we consider the branch as low confidence.

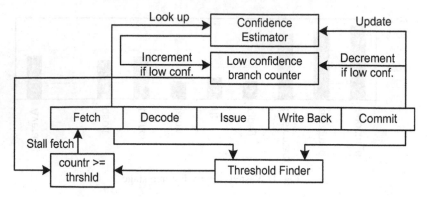

Fig. 3. A schematic of a processor using confidence estimation and pipeline gating

Previously suggested pipeline gating methods gate the front-end if the number of low-confidence branches exceeds a pre-decided threshold. Our study shows that a one-size-fits-all approach does not work well across all applications and may result in either low WA reduction or high performance penalty. Therefore, we add a level of adaptivity and decide the gating threshold dynamically. To decide the gating threshold dynamically, we use the number of in-flight branch instructions and average misprediction rate.

For applications with small number of branches, aliasing is low and our confidence estimator is more likely to do a more effective job. This is particularly true if average misprediction rate for an application is low. As such, for applications with lower number of branches or low misprediction rate, we gate the pipeline if the number of low-confidence branches exceeds one. For applications with higher number of branches and higher misprediction rates, we gate the pipeline if the number of low confidence branches exceeds two. Intuitively, for application with high number of branches and high misprediction rates, we need to see at least two low-confidence branch instructions before losing confidence in the following instructions.

Accordingly, history-based confidence estimation requires measuring the number of instructions per branch (IPB) and the number of mispredictions occurring during regular intervals. We measure IPB every 256 instructions and set the threshold to two if IPB drops below 4 (indicating a high number of branch instructions) and if the misprediction rate is above 12.5%. Misprediction rate is measured by shifting the number of branches 3 bits to right (divide by 8) and comparing the result with the number of mispredicted branches.

3.2 Predictor-Based Confidence Estimation

In the second method we assume that the saturating counters which are already being used by the branch predictor indicate branch instruction confidence. By using the already available structures we minimize the hardware overhead associated with pipeline gating.

At fetch, and while probing the branch predictor to speculate the branch outcome, we mark a branch as low confidence if its corresponding branch predictor counter is not saturated. Similar to the history-based method we gate the pipeline if the number of low-confidence branches exceeds a dynamically decided threshold. We increase the gating threshold from 1 to 2 if IPB drops below 4.

3.3 Combined Confidence Estimation

Our study shows that, often, each of the two methods discussed above captures a different group of low-confidence branch instructions. To identify a larger number of low-confidence branches, in this method we use a combination of the two techniques. A branch is considered low-confidence if either the history-based or predictor-based confidence estimator marks it as low-confidence. By using this technique we are able to achieve higher WA reduction while maintaining performance. Similar to the methods discussed above, we also maintain area overhead at a very low-level.

3.4 Area Overhead

In the history-based technique we use an 8-entry confidence estimator which contains 8 2-bit counters. Besides, we need an 8-bit counter to count the instruction intervals, a 6-bit saturating counter with shift capability to count the number of branches in each interval and a 3-bit saturating counter to keep track of mispredictions. The total area requirement is equivalent to 33 bits which is very small.

The area overhead is even lower for the predictor-based method. For this method we need only an 8-bit counter and a 6-bit saturating counter to keep track of instruction intervals and the number of mispredicted branches respectively. Thus, the total required area is only 14 bits.

For the combined method, we use the same structures as we used in the history-based technique. We also look up branch predictor counters which already exist in the processor. Thus, the area requirement is the same (*i.e., 33-bits*) as the history-based technique.

4 Methodology and Results

In this section we present simulation results and analysis for the three proposed methods. We report WA reduction is section 4.1. We report the impact of pipeline gating on performance in section 4.2.

To evaluate our techniques, we used a subset of MiBench benchmark suite compiled for MIPS instruction set. We picked benchmarks with both high and low WA. The results were obtained for the first 100 million instructions of the benchmarks. We performed all simulations on a modified version of the SimpleScalar v3.0 tool set [4]. We used a configuration similar to that of intel's XScale processor for our processor model. Table 1 shows the configuration.

4.1 Wasted Activity Reduction

As explained earlier a considerable number of instructions fetched in an embedded processor are flushed due to branch mispredictions. As expected, WA is higher for benchmarks with higher misprediction rates and low IPBs. An immediate consequence of WA is higher power dissipation. Thus, reducing extra work will ultimately result in lower energy consumption as long as performance is maintained.

Table 1. Configuration of the processor model

Issue Width	In-Order:2
Functional Units	1 I-ALU, 1 F-ALU, 1 I-MUL/DIV, 1 F-MUL/DIV
BTB	128 entries
Branch Predictor	Bimodal, 128 entries
Main Memory	Infinite, 32 cycles
Inst/Data TLB	32 entries, fully associative
L1 - Instruction/Data Caches	32K, 32-way SA, 32-byte blocks, 1 cycle
L2 Cache	None
Load/Store queue	8 entries
Register Update Unit	8 entries

In figure 4 we report WA reduction for the three proposed confidence estimation techniques. As it can be seen, the highest WA reduction is achieved using the combined confidence estimation technique. Maximum WA reduction is 30% and achieved by the combined method for *fft*.

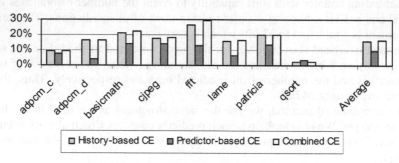

Fig. 4. WA reduction. *Higher is better*

4.2 Performance

In figure 5 we report performance. Bars from left to right report performance for history-based, predictor-based and the combined method compared to a processor that does not use pipeline gating. Reportedly, the predictor-based technique has the lowest amount of performance loss among the three techniques. Average performance loss is only 0.2% for this technique. Our study shows that, for *patricia*, applying pipeline gating results in an increase in L1 cache hit rate which explains why we witness about 0.4% performance improvement for this benchmark. Average performance loss for history-based and combined techniques is 1.1% and 1.2% respectively.

The predictor-based method maintains performance cost below 1% for all applications. History-based and combined maintain performance cost below 1% for 6 of the 8 applications. Both techniques result in a performance cost of about 3% for *basicmath*. This is the result of frequent changes of behavior for branch instructions in *basicmath*. This makes capturing branch instruction confidence very challenging by using simple confidence estimators designed to perform under resource and area constraints.

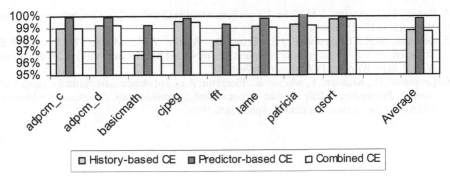

Fig. 5. Performance compared to a conventional processor. *Higher is better*

5 Related Work

Several studies proposed power efficient architectures for embedded processors. Our work focuses on reducing wasted activity by using pipeline gating.

S. Manne, A. Klauser and D. Grunwald [1] have investigated effects of pipeline gating on reducing wasted activity in high-performance processors. They proposed techniques for branch confidence estimation and used them for pipeline gating in high performance processors. Our study is different from theirs as we propose low overhead techniques for embedded processors. We also propose a dynamic method to change the gating threshold.

6 Conclusion

We proposed three low-overhead pipeline gating techniques for embedded processors. To evaluate our techniques, we used a representative subset of MiBench benchmarks and simulated a processor similar to Intel's XScale processor.

We showed that by using simple confidence estimation techniques, it is possible to reduce the number of mispredicted instructions fetched by up to a maximum of 30%. All proposed techniques maintain average performance cost below 1.2%.

Acknowledgements

This work was supported by the Natural Sciences and Engineering Research Council of Canada, Discovery Grants Program, Canada Foundation of Innovation, New Opportunities Fund and the University of Victoria Fellowship.

References

1. Manne, S., Klauser, A., Grunwald, D.: Pipeline Gating: Speculation Control for Energy Reduction. *Proc. 25th Ann. Int'l Symp. Computer Architecture,* pp. 132-141, June 1998.
2. Intel, Intel XScale Microarchitecture. 2001.
3. Digital Semiconductor: DECchip 21064/21064A Alpha AXP Microprocessors: Hardware Reference Manual, June 1994.

4. Burger, D.C., Austin, T.M.: The SimpleScalar tool set,version 2.0. *Computer Architecture News*, 25(3):13–25, June 1997.
5. Guthaus, M.R., Ringenberg, J.S., Ernst, D., Austin, T.M., Mudge, T., Brown, R.B.: MiBench: A free, commercially representative embedded benchmark suite. *IEEE 4th Annual Workshop on Workload Characterization.* 2001.
6. Grunwald, D., Klauser, A., Manne, S., Pleszkun, A.: Confidence estimation for speculation control. *Proceedings 25th Annual International Symposium on Computer Architecture, SIGARCH Newsletter, Barcelona, Spain, Jun. 1998.*

Fast Low-Power 64-Bit Modular Hybrid Adder

Stefania Perri, Pasquale Corsonello, and Giuseppe Cocorullo

Department of Electronics, Computer Science and Systems
University of Calabria, Arcavacata di Rende – 87036 – Rende (CS), Italy
perri@deis.unical.it, {p.corsonello,g.cocorullo}@unical.it

Abstract. This paper presents the design of a new dynamic addition circuit based on a hybrid ripple-carry/carry-look-ahead/carry-bypass approach. In order to reduce power, the usage of duplicated carry-select stages is avoided. High computational speed is reached thanks to the implemented two-phase running. The latter makes the proposed adder able to exploit the time usually wasted for precharging dynamic circuits to accelerate the actual computation. Limited power dissipation and low area occupancy are guaranteed by optimizations done at both architecture and transistor levels.

When realized using the UMC 0.18μm 1.8V CMOS technology, the new 64-bit adder exhibits a power-delay product of only 30.8pJ*ns and requires less than 3400 transistors.

1 Introduction

Addition is a fundamental arithmetic operation in almost any kind of processor and it usually has the largest impact on the overall performance and power dissipation. For this reason, improving the addition efficiency continues to be an attractive research topic. As it is well known, for obtaining a good trade-off between high-speed and low power, the logic design (i.e. static or dynamic logic) and the adder style (i.e. ripple-carry, carry-skip, carry-look-ahead, carry-select, carry-increment, etc.) play a crucial role.

To get fast adding results, one of the widely used methods is the Carry-Look-Ahead addition (CLA) technique, which ensures high computational speed but at the expense of high power dissipation and layout area. In the last few decades, many variations of this approach have been proposed to improve the speed and/or to reduce the power consumption [1, 2].

Another way to reduce the addition time is to avoid long carry propagation paths by employing either a *carry-select* (CSEL) or a *carry bypass* (CBA) technique, which lead with power dissipation lower than the CLA but also with lower speed.

Alternative widely used addition architectures are based on hybrid carry-look-ahead carry-select schemes (HCLA-CSELs) [3, 4, 5, 6]. These structures use a carry-tree unit that quickly computes carries into appropriate bit positions. The carries produced by the carry-tree are then inputted to sum-generation units typically organized as carry-select blocks, which produce the final output. The usage of duplicated carry-select logic causes consistent power consumption.

To reach the highest speed-performance, in realizing addition circuits dynamic logic design styles are typically preferred. The running of a dynamic circuit consists of alternating precharge and evaluation phases that take place during each clock cycle. This implies that, when a typical clock signal is used with a duty cycle of about 50%,

V. Paliouras, J. Vounckx, and D. Verkest (Eds.): PATMOS 2005, LNCS 3728, pp. 609–617, 2005.

just a half period is actually available for computation. In other words, a half clock period is wasted for precharging dynamic modules.

The above considerations suggested that the exploitation of a two-stage dynamic adder able to completely use the clock cycle for each operation can lead with higher speed performance. Moreover, in order to reduce the power consumption, the usage of pure carry-select blocks should be avoided.

The circuit proposed here is based on a dynamic structure customized for efficiently exploiting both carry-tree and reduced-area carry-select [7] based approaches. One important feature also provided by the new addition architecture is the modularity. In fact, wide adders can be easily realized using a cascade of 16-bit basic modules. This approach has been chosen for simplifying the carry-tree architecture, thus reducing the power dissipation and the area requirement. Moreover, it requires less effort for realizing custom layouts. The case analyzed here refers to a 64-bit adder realized using four replicas of identical 16-bit blocks. Obviously, carry propagations occur between contiguous 16-bit blocks. Thus, the 64-bit adder here examined is expected to be slower than the equivalent non-modular implementation. To minimize the power consumption ensuring high-speed, optimizations at both architecture and transistor levels have been performed. They allowed obtaining an average energy consumption of just 39pJ and a 1.25GHz running frequency, with occupying only 0.045mm^2 of silicon area. Comparison with several known adders demonstrated the efficiency of the proposed two-stage addition architecture.

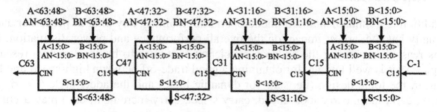

Fig. 1. The top-level architecture of the new adder

2 The New Adder

As depicted in Fig.1, the 64-bit adder here presented receives the operands $A<63:0>$ and $B<63:0>$, their inverted version $AN<63:0>$ and $BN<63:0>$, and the carry-in C-1 as input and generates the sum word $S<63:0>$ and the carry-out $C63$. It can be seen that the designed circuit consists of four cascaded 16-bit addition modules, each organized as shown in Fig.2.

The two main modules M1 and M2 run in opposite manner: when the clock signal is high, M1 performs its evaluation phase, whereas M2 is in its precharging phase; on the contrary, when the clock signal is low, M2 evaluates, whereas M1 precharges. In order to ensure the adder runs correctly, dynamic latches have been used on signals generated by M1 and then inputted to M2. In Fig.2, the latched signals are labelled through the _L subscript.

M1 receives the operands, $A<15:0>$ and $B<15:0>$, and their inverted versions, $AN<15:0>$ and $BN<15:0>$. M2 receives the carry-in CIN, the latched signals coming from M1 and generates the final sum $S<15:0>$ and the carry-out $C15$.

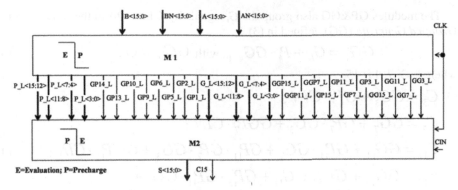

Fig. 2. The architecture of the basic 16-bit module

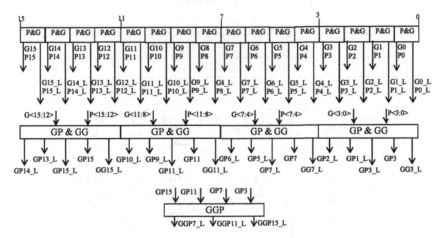

Fig. 3. The module M1

Observing Fig.3, it can be seen that M1 computes the propagate (P_i) and generate (G_i) terms defined in (1) through the modules P&G for each bit position i, with $i=0,...,15$.

$$P_i = a_i \oplus b_i ; \; G_i = a_i \cdot b_i \qquad (1)$$

The P_i signals are then grouped through the modules GP&GG by 2-bit, 3-bit and 4-bit to form the signals *Grouped Propagate* (GP) defined in (2), where $j=1,...,15$, and with $GP_0 = P_0$, $GP_4 = P_4$, $GP_8 = P_8$, $GP_{12} = P_{12}$. This implies that the generic label GP_j indicates $P_{\left\lfloor \frac{j}{4} \right\rfloor \cdot 4} \cdot ... \cdot P_j$ and not $P_0 \cdot ... \cdot P_j$ as in a conventional radix-4 prefix circuit. For example, GP_6 means $P_4 \cdot P_5 \cdot P_6$.

$$GP_j = P_j \cdot GP_{j-1} \qquad (2)$$

The GP_j signals are then grouped again by 4-bits through the GGP module to form the signals GGP_h, with $h=7, 11, 15$.

The modules GP&GG also group the G_i signals by 4-bit to form the signals named *Grouped Generate* (GG) defined in (3)

$$GG_k = G_k + P_k \cdot GG_{k-1}, \text{ with } GG_0 = G_0 \qquad (3)$$

where k=3, 7, 11, 15.

$$C_3 = GG_3 + GP_3 \cdot CIN \; ;$$

$$C_7 = GG_7 + GP_7 \cdot GG_3 + GGP_7 \cdot CIN \; ;$$

$$C_{11} = GG_{11} + GP_{11} \cdot GG_7 + GP_{11} \cdot GP_7 \cdot GG_3 + GGP_{11} \cdot CIN \; ; \qquad (4)$$

$$C_{15} = GG_{15} + GP_{15} \cdot GG_{11} + GP_{15} \cdot GP_{11} \cdot GG_7 +$$

$$+ GP_{15} \cdot GP_{11} \cdot GP_7 \cdot GG_3 + GGP_{15} \cdot CIN$$

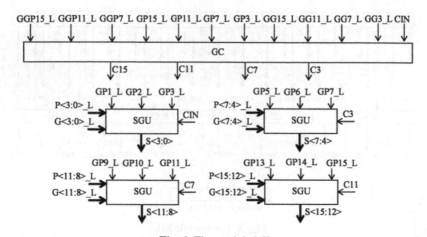

Fig. 4. The module M2

In Fig.4, the organization of the M2 module is depicted. It consists of a carry-tree unit GC and four 4-bit sum generation units (SGUs). The GC module receives from M1 the latched versions of appropriate *grouped propagate* and *grouped generate* signals and produces the carries *C3*, *C7*, *C11* and *C15* as shown in (4). The latched versions of *propagate*, *generate* and *grouped propagate* signals produced by M1, and the carry signals coming from GC are inputted to the SGUs. Each of them generates four bits of the final sum.

Figs.5 and 6 illustrate the transistor level schematics designed for realizing the proposed adder. In order to reduce the power dissipation reaching sufficiently high speed with low area occupancy, new architectures have been designed for both the GC and the SGU modules used in M2. Structures similar to that used for the new adder typically exploit carry-select schemes to quickly form the output signals. Unfortunately, this approach requires several logic modules to be duplicated thus causing a non-negligible increase in energy consumption and silicon area occupancy. To save both energy and area without compromising speed performance, the GC and SGU modules have been structured as reduced-area carry-select circuits [7, 8].

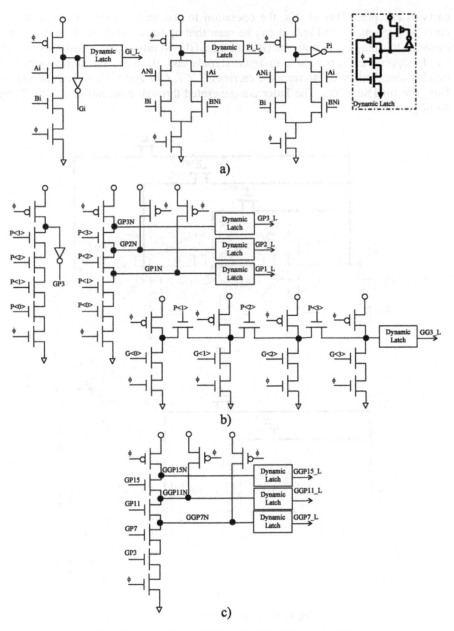

Fig. 5. The modules a) P&G, b) GP&GG and c) GGP

Note that, the time needed for computing the signals *propagate, generate, grouped propagate* and *grouped generate* are different (i.e. *propagate* and *generate* signals are computed earlier). However, all these signals are computed by M1 when the clock is high and during this phase M2 is till precharging. When the clock becomes low, M2 starts its evaluation phase and, due to the latch action, all its inputs are stable, the carry-in excepted. Due to this, the operation to accelerate in the M2 module is the carry absorption. From Fig.6, it can be seen that by-pass transistors have been introduced to this purpose in both the GC and the SGU circuits. These transistors are gated by the appropriate *grouped propagate* signals and allow an incoming carry to be quickly absorbed by GC to form the carries *C3*, *C7*, *C11* and *C15*, and by the SGUs to form the final sum bits. The latter are generated through conventional static 2-input XOR gates.

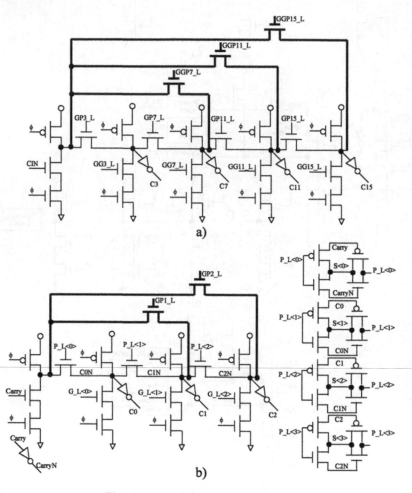

Fig. 6. The modules a) GC and b) SGU

3   VLSI Implementation and Comparison Results

The new proposed 64-bit adder has been realized with the UMC 0.18um 1.8V CMOS technology and using just 3340 transistors (including transistors needed for buffering the clock signal). The realized custom layout depicted in Fig.7 occupies about 0.045mm².

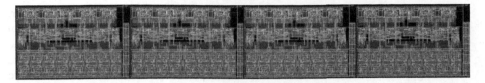

Fig. 7. The realized layout

NanoSim Synopsys tools have been used to measure the worst-case delay and the average energy dissipation. Estimated wire capacitances have been taken into account and each output signal has been loaded by a D-type flip flop that, for the used technology, corresponds to a load capacitance of about 7fF. Obtained results demonstrated that, as expected, the worst-case delay occurs when a carry generated into the least significant bit position of the adder has to be propagated through all the subsequent bit positions to form the most significant sum bit S_{63} (i.e. $G_0=1$, $P_0=0$, $P_1,...,P_{62}=1$, $P_{63}=0$). The new addition circuit performs its computation in the following two phases. When the clock signal is high, all the M1 modules inside the adder evaluate, while all the M2 modules precharge. BSIM3V3 HSPICE post-layout simulations have shown that this first step requires at most 350ps. When the clock signal is low, the M1 modules precharge, while the M2 modules evaluate. In this phase, about 560ps are required to form the carry signal $C47$ and further 230ps are needed to generate the valid sum word. It is worth noting that in the worst case analyzed above, the carry-out $C63$ is quickly computed. In fact, $P63=0$, thus it does not depend on carries propagating from previous bit positions. It can be concluded that the new adder exhibits a worst-case delay of about 790ps. Nanosim Synopsys tools estimated an average energy dissipation of just 39pJ.

The new adder has been compared to several addition circuits known in literature [5, 6, 9, 10, 11]. In [5], a fast dynamic adder based on an efficient quaternary carry-tree is described. In [6], a 64-bit static adder is demonstrated that exploits hybrid quaternary carry-look-ahead carry-skip tree and sum units organized as carry-increment blocks. In [9], a hybrid carry-lookahead carry-select scheme is proposed. In [10], a static carry-select adder (SCSA), a dynamic carry-lookahead adder (DCLA), a dynamic Kogge-Stone adder (DKSA), and a dynamic Ling/conditional sum adder (DLCNSA) are characterized. In [11], a branch-based logic carry-select adder (BBL), a complementary pass-logic carry-select adder (CPL), and a complex BBL carry-select adder (CBBL) are described.

All the above cited adders are referenced in Table 1 that summarizes comparison results. Computational times, energy consumption, number of transistors, silicon area, and logic design style have been evaluated. For the sake of completeness, several standard-cells based adders fully synthesized by means of the most recent Synopsys tools have been also characterized. Data reported in Table 1 clearly demonstrate the efficiency of the new addition architecture.

It is important to note that the compared adders are realized using different bulk CMOS and SOI technologies. Therefore, for comparison reliability, it has to be taken into account that characteristics of the adders are significantly influenced by the used technology. In fact, as shown in [12], SOI technologies compared to bulk CMOS counterparts allow 35% and 33% reduction in computational time and power dissipation, respectively. Additional effects on both energy consumption and computational delay are due to the supply voltage [13]. For example, let's examine the CPL adder, very recently presented in [11], for estimating its delay and energy characteristics on the basis of the percentage improvements given in [12]. Estimation shows that, if the 0.18μm 1.8V bulk CMOS technology is used, the CPL adder would have delay time and energy dissipation 26% and 89% higher, respectively, than the 64-bit adder here proposed. Moreover, from Table 1, it can be seen that the CPL circuit requires 139% more transistors. Obviously, similar considerations are valid for all the circuits described in [9] and [11].

Table 1. Comparison with known 64-bit adders

	Tech	Supply voltage [V]	Delay [ns]	Energy [pJ]	ExD [nsxpJ]	Logic	#transis- tors	Area [um^2]
[5]	0.25	2.5V	0.67	200	134	dyn	5460	160000
[6]	0.18	1.8V	0.98	54	52.9	static	N.A.	15000
[9]	0.1 SOI	1.2V	0.35	N.A.	-	dyn	N.A.	N.A.
SCSA [10]	0.18	1.8V	1.06	N.A.	-	static	N.A.	N.A.
DCLA [10]	0.18	1.8V	1.1	N.A.	-	dyn	N.A.	N.A.
DKSA [10]	0.18	1.8V	0.58	N.A.	-	dyn	N.A.	N.A.
DLCNS A [10]	0.18	1.8V	0.74	N.A.	-	dyn	N.A.	N.A.
BBL [11]	0.18 SOI	1.5V	0.72	96	69.1	static	18219	205800
CPL [11]	0.18 SOI	1.5V	0.7	50	35	static	7991	N.A.
CCBBL [11]	0.18 SOI	1.5V	0.54	29	15.7	static	6026	N.A.
Paral- lelpre- fix*	0.18	1.8V	1.6	83	132.8	static	N.A.	26683
CSEL*	0.18	1.8V	3.06	39	119.3	static	N.A.	13036
CLA*	0.18	1.8V	2.1	25	52.5	static	N.A.	9814
HCLA- CSEL*	0.18	1.8V	1.5	84	126	static	N.A.	26911
New	**0.18**	**1.8V**	**0.79**	**39**	**30.8**	**dyn**	**3340**	**45000**

* Fully synthesized static CMOS standard-cells based adder

4   Conclusions

A new 64-bit sub-nanosecond low-power adder has been presented. The proposed circuit exploits a hybrid ripple-carry/carry-look-ahead/carry-bypass architecture.

Comparisons with several known static and dynamic addition circuits have demonstrated that the new adder allows the best area-time-power trade-off to be achieved. In fact, when realized using the UMC 0.18um CMOS technology, it exhibits a worst-case delay of only 0.79ns, dissipates just 39pJ on average and occupies ~0.045mm^2 of silicon area.

References

1. Kogge, P.M., Stone,H.S.: "A parallel algorithm for efficient solution of a general class of recurrence equations", IEEE Trans. on Computers, 1973, C22 (8)
2. Brent, R.P., Kung, H.T.:"A regular layout for parallel adders", IEEE Trans. on Computers, 1982, C31 (2)
3. T. Lynch, E.E. Swartzlander, "A spanning-tree carry-look-ahead adder", *IEEE Trans. on Comp.*, Vol. 41, n°8, 1992.
4. V. Kantabutra, "A recursive carry-look-ahead/carry-select hybrid adder", *IEEE Trans. on Comp.*, Vol. 42, n°12, 1993.
5. Woo, R., Lee S., Yoo,H.:"A 670ps, 64-bit dynamic low-power adder design", Proc. of IEEE ISCAS Conference, May 2000, Geneva, pp.28-31
6. S.Perri, P.Corsonello, G.Staino, "A low-power sub-nanosecond standard-cells based adder", Proceedings of the IEEE ICECS Conference, December 2003, Sharjah, Arab Emirates, pp.296-299.
7. Tyagi A., "A reduced-area scheme for carry-select adders" *IEEE Trans. on Comp.*, Vol. 42, n°10, 1993, pp.1163-1170.
8. S.Perri, P.Corsonello, F.Pezzimenti, V.Kantabutra, "Fast and Energy-Efficient Manchester Carry-ByPass Adders", *IEE Proceedings on Circuits, Devices and Systems*, Vol.151, n°6, 2004, pp.497-502.
9. J.J. Kim, R. Joshi, C.T. Chuang, K. Roy, "SOI-Optimized 64-bit High-Speed CMOS Adder Design", *Proceedings of Symposium on VLSI Circuits Digest of Technical Papers*, June 2002, pp.122-125.
10. H.Dao, V.G.Oklobdzija, "Application of Logical Effort Technique for Speed Optimization and Analysis of Representative Adders", *Proceedings of the 35th Annual Asilomar Conference on Signals, Systems and Computers*, California (USA), Nov. 2001, Vol.2, pp.1666-1669.
11. A.Neve, H.Schettler, T.Ludwig, D.Flandre, "Power-Delay Product Minimizations in High-Performance 64-bit Carry-Select Adders", *IEEE Trans. On VLSI Systems*, Vol. 12, N° 3, 2004, pp.235-244.
12. P.Simonen, A.Heinonen, M.Kuulusa, J.Nurmi, "Comparison of bulk and SOI CMOS technologies in a DSP Processor Circuit Implementation", *Proceedings of the 13th International Conference on Microelectronics*, Rabat, Marocco, Oct. 2001.
13. J.M.Rabaey, A. Chandrakasan, B.Nikolic, "*Digital Integrated Circuits – A Design Perspective*", Second Edition, Prentice Hall, New Jersey, 2003.

Speed Indicators for Circuit Optimization

Alexandre Verle, A. Landrault, Philippe Maurine, and Nadine Azémard

LIRMM, UMR CNRS/Université de Montpellier II, (C5506),
161 rue Ada, 34392 Montpellier, France
{pmaurine,azemard}@lirmm.fr

Abstract. Due to the development of high performance portable applications associated to the high-density integration allowed by deep submicron processes, circuit optimization under delay constraints has emerged as a critical issue for VLSI designers. The objective of this work is to avoid the use of random mathematical methods (very CPU time expensive), by defining simple, fast and deterministic indicators allowing easy and fast implementation of circuits at the required speed. We propose to extend the method of equal sensitivity, previously developed for combinatorial paths [1], to circuit sizing in order to solve the circuit convergence branch problem. We propose a coefficient based approach to solve the divergence branch problem. Validation is given by comparing with an industrial tool the performance of different benchmarks implemented in a standard 180nm CMOS process.

1 Introduction

Delay bound determination and sizing under constraints for a complete circuit is one of the most difficult task to be achieved. A lot of solutions has been proposed for paths, but, because a circuit can be assimilated to overlapping paths, the resulting interdependence between delays and input capacitances of the different circuit paths imposes the solution of a NP complete problem [2].

The only solution to optimize a circuit in an optimal or quasi-optimal way, consists in using mathematical approaches [3-6]. Unfortunately, these approaches are very CPU time expensive and are quickly limited by the circuit size. An another approach consists in evaluating and then in sizing the slowest path of the circuit [7]. This procedure is repeated until constraint satisfaction on all the paths. However, the path overlap requires many iterations. In this case, the convergence of the algorithm is not guaranteed and the risk of infinite loops is very high [8].

To effectively reduce the number of loops, we extend, in this paper, the equal sensitivity method defined on a path [1] to the sizing of all the circuit paths. To solve the path sizing problem two particular structures must be considered, the divergence and re-convergence branches. For that, we study these two structures in order to propose indicators allowing to define an accurate and deterministic circuit sizing methodology for reducing the number of optimization loops.

In section 2, we define the delay bounds of a circuit. For that, we study the problem of re-convergences and determine, from an initial sizing, the most probable real critical path [9] to get an abstraction of the circuit delay performance. In section 3, we treat the transistor sizing problem under delay constraint. We study the problems induced by the divergence branches. Finally we apply and validate the proposed approach to full circuits before concluding in section 4.

V. Paliouras, J. Vounckx, and D. Verkest (Eds.): PATMOS 2005, LNCS 3728, pp. 618–628, 2005.
© Springer-Verlag Berlin Heidelberg 2005

2   Critical Path Evaluation: Convergence Problem

Before sizing a circuit under delay constraint, it is necessary to be able to estimate the feasibility of its constraint. For that we define bounds from physical indicators.

- For the maximum delay (T_{MAX}), we set all transistors of the circuit at minimum size. This technique was already used to obtain a pseudo-maximum delay of a path.
- For the minimum delay (T_{MIN}), the problem is different, because this minimum delay depends on divergences and path overlaps: it is a NP complete problem.

The problem is the determination of the critical path of the circuit (i.e. the path with the longest delay for a given sizing) and the value of its minimum delay. The lower delay bound of the circuit (T_{MIN}) and the feasibility of circuit delay constraint are determined from the first path identified as a critical path. Thus, if this path is badly defined at the beginning, iterations are necessary. The most frequently used technique consists in determining the critical path starting from a minimum sizing but it presents an obvious lack of effectiveness. Indeed, this approach does not reflect the path ability to be sized.

2.1   Critical Path Problem Illustration

To illustrate this problem, let us consider three independent paths (path1, path2 and path3 described in Fig.1). For these paths, gate type and parasitic capacitances (C_P units) of each node are given, inv, nrx and ndx for inverters, nor and nand x inputs, respectively. First we define the delay bounds.

For the upper bound, a realistic approach consists in setting all transistors at the minimum size.

For the lower bound, we apply the method of equal sensitivity previously developed for combinatorial paths [1]. So, for a path, the inferior delay bound is easily obtained by canceling the derivatives of the path delay with respect to the input capacitance of its gates. This results in a set of linked equations where the size of a gate depends on the sizes of its previous and next gates. Instead of solving this linked equations, we use an iterative approach starting from any local solution (equal to the local minimum on Fig.1) in order to save CPU time. With this approach, we reach quickly the minimum delay. The evolution of these iterations is illustrated in Fig.1.

We can compare the minimum and maximum delays of these paths. The curves of Fig.1 represent for each path, the delay versus area with an implementation at minimum area (T_{MAX}) and an implementation at global minimum (T_{MIN}). In this figure, each point represents an iteration, starting from the 1st iteration for the local solution, to the last iteration for the delay min.

In Fig.1, we note that for the implementation at minimum area, the critical path of the circuit (the slowest path) is path1. This path must represent the circuit performance when it is implemented at minimum size, i.e. the value of the minimum delay, T_{MIN1}, for the circuit.

However if we compare the delay evolution with minimum sizing of the other paths, we can note that path2 limits the delay performance of the circuit because parth1 is faster.

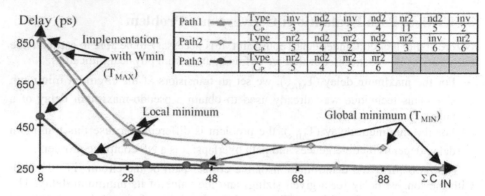

Fig. 1. Design space exploration for three paths of a circuit

Thus, to extract the critical path of a circuit, the implementation at minimum area is not sufficient, because a defined critical path (path1) can have better delay performances than a defined sub-critical path. Only the global implementation allows to exactly evaluate the circuit critical path: this is not easy to realize because an exhaustive research is CPU time expensive.

To avoid these problems, we propose to use a circuit implementation at local minimum. As we can verify in Fig.1, an evaluation at local minimum also allows to have a good idea of the path delay performances.

Tab.1 gives the delay values obtained for path1, path2 and path3 with implementations at minimum area, at local minimum and finally at global minimum (the exact technique). For each implementation, the critical path is represented by a gray box. The implementation at local minimum limits the risk of error in the determination of the critical path and gives a satisfactory solution.

Table 1. Critical path identification for different implementations

	T_{MAX} (ps) for Minimum area	T (ps) for Local minimum	T_{MIN} (ps) for Global minimum
Path 1	855	369	274
Path2	794	440	352
Path 3	494	299	264

In the following, we define an approach, allowing to evaluate the minimum delay bound of a circuit, by using a path classification obtained with an implementation at local minimum.

2.2 Local Minimum Definition

The definition of a local minimum for a circuit requires to define sizing rules taking into account re-convergences.

For a simplified delay model [10], based on the transition time of a gate, the local minimum of an array of two gates (Fig.2) is obtained by sizing the input capacitance of each gate, following:

$$C_{INk} = \sqrt{\frac{S_k}{S_{k-1}} \cdot C_L \cdot C_{REF}} \tag{1}$$

where Sk represents the current possibilities of the gate K, C_L the load of this gate and C_{REF} a reference capacitance value, defined from the minimum value available in the process.

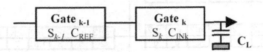

Fig. 2. A cell with two gates

To generalize this approach to a circuit, it is necessary to manage the re-convergence sizing, in order to obtain a correct local minimum evaluation.

To formalize the input capacitance of a re-convergent gate, we consider in Fig.3, a reconvergence of N gates.

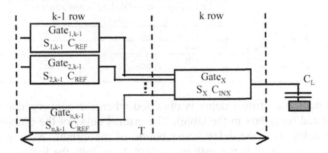

Fig. 3. A generic re-convergence

Considering (1) it is obvious that the sizing of the re-convergent gate must be defined with respect to the slowest preceding gate:

$$C_{INx} = \sqrt{\frac{S_X}{\max\{S_{1,k-1}; S_{2,k-1}; ..., S_{n,k-1}\}} \cdot C_L \cdot C_{REF}} \tag{2}$$

This C_{INx} value allows to determine the maximum value of the input capacitance of the re-convergent gate, with no slowing down of the delay of the gates of the k-1 row.

2.3 Validation

We validate this critical path search approach by implementing at minimum global, minimum local and minimum sizing the circuit represented in Fig.4.

For this circuit, gate type and parasitic capacitances (C_p units) of each node are given, inv, nrx and ndx for inverters, nor and nand x inputs, respectively.

In order to determine the exact critical path for an implementation at global minimum, we apply an exhaustive research by sizing each path of the circuit (in$_i$ → OUT) at global minimum. For each case, the re-convergence branches not belonging to the path under study will be also sized at delay global minimum. Tab.2 compares the delays obtained according to the selected path (in$_i$ → OUT sized at global minimum) and according to the sensitized input of convergence branch (in$_i$). For each path, we also give the sum of circuit transistor sizes ($\sum$W) in μm.

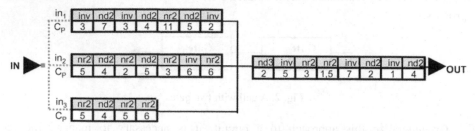

Fig. 4. Combinatorial circuit with re-convergence branches

Table 2. Exhaustive research of critical path for an implementation at global minimum

		Path delay (in$_i$ →OUT) / sensitized input (ps)			$\sum$W (μm)
		in$_1$	in$_2$	in$_3$	
Path sized at	in$_1$→OUT	448	541	459	370
global	in$_2$→OUT	448	540	459	374
minimum	in$_3$→OUT	448	541	459	370

For each path, the critical delay is obtained when the convergence branch input, in2, is sensitized (gray box in the table). The critical delay of the circuit is the lowest path critical delay: it is the delay lower bound of the circuit. As a result, the exact critical path of the circuit is the path in$_2$ → OUT. It exhibits the lowest sizing sensitivity.

Tab.3 compares the approach with an implementation at minimum sizing and the approach with an implementation at local minimum, together with the delay and area ($\sum$W). It gives the critical delay and path (gray box) of these implementations.

Table 3. Delay and area comparison for different implementations

		Circuit delay / sensitive input (ps)			$\sum$W (μm)
		in$_1$→OUT	in$_2$→OUT	in$_3$→OUT	
Implementation	Minimum sizing	1556	1487	1190	37
at	Local minimum	693	762	625	102

The critical path obtained for an implementation at global minimum (Tab.2), shows that the implementation at minimum sizing does not give the good critical path. On the other hand, the implementation at local minimum gives the same critical path as the approach with an implementation at global minimum.

After determination of the critical path, the next step consists in sizing the circuit to respect a delay constraint. For that, it is essential to take into account and to manage divergence branches.

3 Sizing Under Constraint: Divergence Problem

The sizing under constraint of a path with divergence branches can involve loop problems in the sizing algorithm. Indeed, the re-sizing of a gate with a size modified at a preceding iteration can strongly increase the CPU time and put the algorithm in failure. So, to solve this problem, for the path under optimization, it is essential to fix the input capacitance size of each divergence branch. The proposed approach consists in evaluating these input capacitance sizes by using an analytically calculated coefficient.

3.1 Description of the Approach with Coefficient

The proposed technique consists in determining a γ coefficient for each input of divergence branch of the critical path. This coefficient allows to calculate and to fix the input capacitance of each divergence branch. Let us detail the principle of this approach with the circuit of Fig.5.

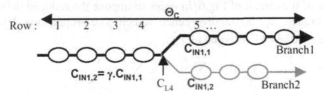

Fig. 5. Illustration of the approach with divergence coefficient

In Fig.5, the critical path is represented in bold and it has a divergence branch (Branch2). To fix the input capacitance of the divergence branch, we use a γ coefficient defined by the ratio of the input capacitance of the divergence branch (Branch2) to the input capacitance of the critical path gate of Branch1: $\gamma = C_{IN1,2}/C_{IN1,1}$. Then path sizing is processed backward from the path output to the path input. As an example, the load seen by the gate located at row 4 (C_{L4}) is then $C_{L4}=C_{IN1,1}\cdot(1+\gamma)$. This approach allows to size easily the circuit. Let us define, and evaluate now, the coefficients of divergence branches.

3.2 Coefficient Definition and Evaluation

Circuit sizing problems are due to the circuit complexity and the significant number of unknown parameters. To calculate divergence coefficient, instead of using slow and blind mathematical approaches, we propose to define a simple indicator γ. To determine the value of the γ coefficient, let us consider the circuit of Fig.6.

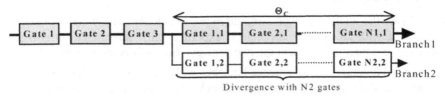

Fig. 6. Circuit with a critical path and its divergence branch

The objective is to extract a metric to obtain a ratio between the input capacitances of gate1,2 and gate1,1. At this level, any sizing information exists. So, some assumptions are made.

- Initially, on the critical path, each gate is sized at a reference value, C_{REF}, equal to the minimum size allowed by the process or the library.
- To reduce the CPU time, to obtain a fast evaluation of the path delays and to propose a simple heuristic, we use a first order delay model [10], based on the gate transition time, as

$$T_{HL,LH} = \tau_{ST} \cdot S_{HL,LH} \cdot \frac{C_L}{C_{IN}} \tag{3}$$

where, τ_{ST} is a time unit that characterizes the process, $S_{HL,LH}$, $S_{HL,LH}$ represent the symmetry factor of the falling, rising edges. C_L, and C_{IN} represent, respectively, the output load including the cell parasitic capacitance and the gate input capacitance.

The principle of the proposed approach is to equalize the propagation delay of the two branches of the circuit of Fig.6, in order to impose the reduced delay value of the critical path branch, Θ_C, on the divergence branch. The propagation delay of the critical path branch (Branch1) is equal to

$$\Theta_C = \frac{T_C}{\tau_{ST}} = \sum_{i=1}^{i=N_1} \Theta_{i,1} = \sum_{i=1}^{i=N_1} \frac{C_{Li,1}}{C_{INi,1}} = \sum_{i=1}^{i=N_1} S_{i,1} \cdot \frac{C_{Pi,1} + C_{INi+1,1}}{C_{INi,1}} \tag{4}$$

where $C_{Pi,1} + C_{INi+1,1} = C_{Li,1}$ is the output capacitance of the Gatei,1 of Branch1.

Now we apply this delay constraint, Θ_C, on the divergence branch (Branch2). The goal is to size the gates of Branch2 to obtain the value of the input capacitance of the divergence branch ($C_{IN1,2}$), and then the value of the γ coefficient. Consequently, we cannot apply the constant sensitivity method, because with this method the input capacitance of the branch is fixed [1] in order to have a convex delay function. So, to obtain an approximate but fast sizing, we use the heuristic approach of [11].

We apply the equal repartition of delays on Branch2 and we obtain the analytical expression of the γ coefficient ($\gamma = C_{IN1,2}/C_{IN1,1}$) to apply the input gate of each divergence branch.

$$\Theta_C = \sum_{i=1}^{i=N_2} \Theta_{i,2} = N_2 . \Theta_{1,2} = N_2 . \frac{S_{1,2} . C_{L1,2}}{C_{IN1,2}} \Rightarrow \gamma = \frac{S_{1,2} . C_{L1,2} . N_2}{C_{IN1,1} \sum_{i=1}^{i=N_1} S_{i,1} . \frac{C_{Li,1}}{C_{INi,1}}} \tag{5}$$

Now we compare this analytical coefficient to an experimental optimal coefficient.

3.3 Coefficient Validation

We validate the analytical expression allowing to calculate the γ coefficient (5), by comparison with an optimal coefficient, experimentally determined on two simple circuits. These circuits represent the two configurations of divergence branches:

- balanced divergences: divergence branches with same topology (circuit test1).
- unbalanced divergences: divergence branches with different characteristics (circuit test 2).

Fig.7 presents the characteristics of the two test circuits. L_{depth} represents the path logical depth, i.e. the number of gates in each path. *Gate1* is the input gate type of path under study and θ_{MAX} is the propagation delay for an implementation at minimum area (W_{MIN}).

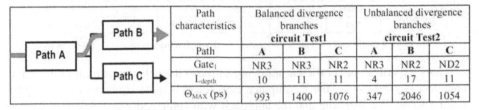

	Path characteristics	Balanced divergence branches circuit Test1			Unbalanced divergence branches circuit Test2		
	Path	**A**	**B**	**C**	**A**	**B**	**C**
	Gate$_1$	NR3	NR3	NR2	NR3	NR2	ND2
	L$_{depth}$	10	11	11	4	17	11
	Θ_{MAX} (ps)	993	1400	1076	347	2046	1054

Fig. 7. Principal characteristics of the two test circuits

First we validate the approach on circuit Test1 (balanced divergence branches) and then on circuit Test2 (unbalanced divergence branches).

a) Validation on Balanced Divergence Branches (Circuit Test1)
Circuit test1, presented on Fig.7, allows to check that the computed value of the γ coefficient (5) is not under evaluated and is equivalent to its experimental value.

For that, Tab.4 shows the area evolution (ΣC_{IN}) of circuit Test1 for different delay constraints (Θ_C) and for different values of the γ coefficient. The γ experimental coefficient allowing to reach the delay constraint at minimum area cost is $\gamma=0,43$.

Table 4. Experimental results for different γ coefficients on the circuit Test1

Θ_C(ps)	ΣC_{IN} for different γ Coefficients						
	$\gamma=0,65$	$\gamma=0,55$	$\gamma=0,48$	$\gamma=0,43$	$\gamma=0,38$	$\gamma=0,3$	$\gamma=0,2$
2393	70	70	70	70	70	70	70
2000	79,6	78,81	78,81	78,81	78,81	78,81	78,81
1500	118,7	117,93	117,81	117,81	117,9	117,9	117,9
800	592,54	582,14	579,96	579,14	579,29	583,06	605,86

Computed coefficient : $\gamma=0,51$ ⟸   ⟸ Experimental coefficient : $\gamma=0,43$

We can note that the computed γ value is equivalent to the experimental value. Now let us validate this approach on a circuit unbalanced divergence branches.

b) Validation on Unbalanced Divergence Branches (Circuit Test2)
We apply the same procedure on the circuit Test2 presented in Fig.7. We want to check that the computed value of γ coefficient (5) is not overestimated and is equivalent to its experimental value.

In Tab.5, we can see the area evolution (ΣC_{IN}) of circuit Test2 for different delay constraints (Θ_C) and for different values of the γ coefficient.

Table 5. Experimental results for different γ coefficients on circuit Test2

Θ_C(ps)	ΣC_{IN} for different γ Coefficients				
	γ=0,35	γ=0,3	γ=0,2	γ=0,15	γ=0,09
2393	71	71	71	71	71
1100	188,17	187,54	186,94	186,94	186,94
1000	236	235	233,71	233,63	233,6
750	786,01	776,5	759,92	752,5	748,95

Computed coefficient : γ=0,298   Experimental coefficient   γ=0,09

The γ experimental coefficient allowing to reach the delay constraint at minimum area cost is γ =0,09. So, the γ computed coefficient is quite equivalent to its experimental value. We also note that the approach with computed γ coefficient request an area (ΣC_{IN}) slightly higher than the experimental approach.

After validation of the γ coefficient calculated by an analytical expression, we now compare this approach to that of an industrial tool (AMPS).

3.4 Comparison with AMPS

In order to check the effectiveness of the coefficient based approach, we compare this approach to an implementation given by an industrial physical synthesis tool (AMPS tool from Synopsys company). We have just seen on the two previous circuits (circuit Test1 and circuit Test2) that the approach with coefficient gives satisfying results. These two circuits, with a critical path and a divergence branch, have allowed to apply the sizing protocol on simple examples. Thus, comparison with AMPS will be applied on a circuit with several divergence branches (balanced and unbalanced divergence branches); we used the circuit Test3, illustrated in Fig.8. This circuit is constituted of a critical path (in bold) with the a, b, c and d sub-circuits and three divergence branches (E, F and G). Each sub-circuit is represented by a box with its gate number. g_{in} indicates the input gate type of sub-circuit.

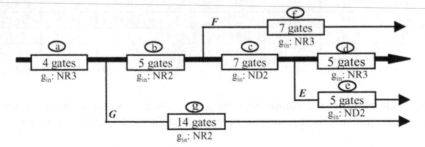

Fig. 8. Circuit Test 3

So, circuit Test3 is a mixture of the two previous test circuits (balanced and unbalanced divergences). This circuit is used to evaluate the effectiveness of the approach with coefficient on an example with multi-divergence branches. Contrary to the two preceding circuits (Fig.7), it is very difficult to obtain an experimental coefficient on a general circuit, due to problems of CPU time. Thus circuit Test3 may give evidence of the interest in using the coefficient based approach. For illustration, we compare in Tab.6 the γ coefficients calculated using the approach with coefficient (5), for each divergence branches (E, F and G) of the circuit Test3.

Table 6. γ coefficients for divergence branches of circuit Test3

γ coefficients		
Divergence branch G	Divergence branch F	Divergence branch E
0.32	0.42	0.48

The curves of Fig.9. represent the delay evolution versus the area (ΣWi), for the circuit test3, using the approach with coefficient and the tool AMPS. These results are obtained with Hspice for a standard 0,18μm process.

On Fig.9, we can note that the implementation obtained by the approach with coefficient, gives better results than AMPS, at once for the evaluation of the minimum delay and for area saving under delay constraint. We noticed that the proposed approach enable to reach the best results.

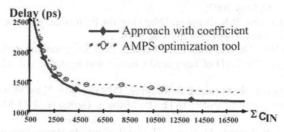

Fig. 9. Space design exploration for the circuit test 3 with AMPS

4 Conclusion

In this paper, we have presented an indicator based solution for circuit sizing under delay constraint. We have given solutions to the sizing of re-convergence (determination of delay bounds) and divergence branches (sizing under delay constraint).

We have defined a methodology in which we firstly propose to determine the circuit delay bounds from critical path search (the path with the worst sizing sensitivity). Then we have demonstrated that an implementation at minimum area does not reflect the real performances of a circuit, and proposed an implementation at local minimum to obtain a better idea of the circuit performance and a robust path classification. To determine this local minimum, that defines the circuit performance, we have proposed and validated an approach allowing to manage the re-convergence branches.

In the second time, for an achievable constraint, we have proposed a complete circuit sizing solution. Highlighting the divergence branch problem, this solution allows to fix the size of the input capacitances of divergence branches in a deterministic way, allowing to avoid iterations.

This circuit sizing protocol is under implementation for validation on ISCAS benchmarks.

References

1. A.Verle, X. Michel, N. Azémard, P. Maurine, D. Auvergne, "Low Power Oriented CMOS Circuit *Optimization Protocol*", DATE'05, Munich, Germany, 7-11 March 2005, pp 640-645.
2. R. Murgai, "On the Global Fanout Optimization Problem", In IWLS, Granlibakken, USA, 1999.
3. D. Marple, "*Transistor Size Optimization in Tailor Layout System*", Proc of the 26th IEEE/ACM Design Automation Conference, pp43-48, June 1989.
4. V. Sundarajan, S. S. Sapatnekar and, K. K. Parhi, "*Fast and Exact Transistor Sizing Based on Iterative Relaxation*," IEEE Transactions on Computer Aided Design of Integrated Circuits and Systems, Vol. 21, $N°$ 5, pp568-581, May 2002.
5. A. R. Conn and, *al*, "JiffyTune: Circuit Optimization Using Time-Domain Sensitivities," IEEE Transactions on Computer Aided Design of Integrated Circuits an Systems, Vol. 17, NO. 12, pp1292-1308, December 1998.
6. H. Tennakoon and, C. Sechen, "Gate Sizing Using Lagrangian Relaxation Combined with a Fast Gradient-Based Pre-Processing Step," Proc of the IEEE/ACM International Conference on Computer-Aided Design, ICCAD'02, pp395-402, November 2002.
7. Y. Y. Yu, V. Oklobdzija and, W. W. Walker," An Efficient Transistor Optimizer for Custom Circuits," Proc of IEEE International Symposium on Circuits and Systems, ISCAS'03, Vol. 5, pp197-200, May 2003.
8. R. H. J. M Otten and, R.K. Brayton, "Planning for Performance," Proc of the 35th Design Automation Conference, pp121-126, June 1998.
9. H.C. Chen, D.H.C. Du and L.R. Liu, "Critical Path Selection for Performance Optimization", IEEE trans. On CAD of Integrated Circuits and Systems, vol. 12, $n°2$, pp. 185-195, February 1995
10. A.Verle, X. Michel, N. Azémard, P. Maurine, D. Auvergne, "Low Power Oriented CMOS Circuit *Optimization Protocol*", DATE'05, Munich, Germany, 7-11 March 2005, pp 640-645.
11. I.E. Sutherland, B. Sproull, D. Harris, "Logical effort, designing fast cmos circuits", Morgan Kaufmann Publishers, Inc., 1999.

Synthesis of Hybrid CBL/CMOS Cell Using Multiobjective Evolutionary Algorithms[*]

Francisco de Toro[1], Raúl Jiménez[2], Manuel Sánchez[2], and Julio Ortega[3]

[1] Dept. Teoría de la Señal, Telemática y Comunicaciones, Universidad de Granada
ftoro@ugr.es
[2] Dept. Ing. Electrónica de Sist. Informáticos y Automática, Universidad de Huelva
{naharro,msraya}@diesia.uhu.es
[3] Dept. Arquitectura y Tecnología de Computadores, Universidad de Granada
julio@atc.ugr.es

Abstract. In this work, the optimization of circuits design by using multiobjective evolutionary algorithm is addressed. This methodology enable to deal with circuit specifications -formulated as objective functions- that can be conflicting and want to be optimize at the same time. After the optimization process, a set of different trade-off solutions for the design of the circuit is obtained. This way, SPEA (*Strength Pareto Evolutionary Algorithm*) has been tested as optimizer of an hybrid CBL/CMOS configurable cell. As a result, some conclusions about the optimized values of the transistor sizes of this cell in order to minimized some power comsumption and delay timing specifications are obtained.

1 Introduction

Many real world optimization problems deal with several (and normally confliction) objectives functions, which need to be accomplished at the same time. A multiobjective optimization problem (MOP) can be defined ([ ]) as that of finding a *vector of decision variables* belonging to a given input search space $x \in \Theta \subseteq R^n$, which meets a series of constraints and optimizes a *vector of objective functions*:

$$f(x) = [f_1(x), f_2(x), ..., f_k(x)] \tag{1}$$

where the k elements represent the objectives. The meaning of optimum is not well defined in this context, so if the objectives are conflicting an unique solution that optimize all the objectives can not be found. Therefore, the concept of *Pareto Optimality* is used. Considering that all the components of the vector of objective functions want to be *maximized*, a solution $x* \in \Theta \subseteq R^n$ is defined as Pareto optimal if the following condition is satisfied:

$$\forall x \in \Theta, \exists i \in 1, .., k \quad | \quad f_i(x\star) > f_i(x) \quad and \quad \forall j \neq i \in 1, ..., k \quad f_j(x\star) \geq f_j(x) \tag{2}$$

---

[*] This work has been partially sponsored by UHU2004-06 project

V. Paliouras, J. Vounckx, and D. Verkest (Eds.): PATMOS 2005, LNCS 3728, pp. 629–637, 2005.
© Springer-Verlag Berlin Heidelberg 2005

This means that x* is Pareto optimal, if no feasible vector x exists that would increase one criterion without causing a simultaneous decrease in, at least, one of the others. The notion of Pareto optimum always gives not just a single solution, but rather a set of *Pareto Optimal Solutions*. If $x \in \Theta$ is not a Pareto optimal point, then it is referred to as *dominated solution*. When a subset of the input search space is considered, Pareto Optimal Solutions are often refered as *non-dominated solutions*.

Evolutionary Algorithms [ ] are stochastic optimization procedures which apply a transformation process (crossover and mutation operators), inspired by the species natural selection process, to a set (*population*) of coded solutions (*individuals*) to the problem. These procedures are specially suitable for solving multiobjective optimization problems because they are able to capture multiple solutions in a single run. Furthermore, *diversity maintaining techniques* can be easily incorporated to these procedures in order to encourage dissimilarity between the found solutions. Multiobjective Optimization Evolutionary Algorithms (MOEAs) is a very active area of research [ ]. A good summary of real-world optimization problems addressed with MOEAs can be found in [ ].

Microelectronic design has the final goal to obtain a circuit implementing a functionality with certain characteristics. In order to achieve this goal, a design must follow a set of stages, all of them included in a design flow. A *top-down* design flow begins with an oral specifications and arrives to a circuit following several stages that can be optimized separately with MOEAs. In most cases, the objective functions are usually characterization parameters related with circuit specifications, such as power consumption and operation frequency, while the input space will be compounded by the employed algorithm, the functional blocks, the logic families and the floor-planning to the stage of algorithm, architecture, logic family and technology mapping respectively. This paper addresses the use of evolutionary multiobjective optimization techniques in microelectronic design at logic family level. More specifically, the algorithm SPEA (*Strength Pareto Evolutionary Algorithm*) [ ] is used to find the sizes of the transistors -ratio W/L- on an hybrid CBL/CMOS cell in order to optimize some circuit specifications -formulated as objective functions to be minimize- related with time delay, power consumption and noise. The obtained Pareto-optimal solutions (transistor sizes) provides an insight into the nature of the input space, so some conclusions can be extracted for the design of circuits using this cell.

This paper is divided as follows. Section 2 analyzes the hybrid CBL/CMOS cell; section 3 addresses the optimization methodology used in the design of the cell and reviews SPEA. Results of the optimization of the CBL/CMOS cell using SPEA are shown in Section 4; Finally, section 5 exposes the conclusions drawn from this work and point out some future lines to improve the optimization methodology presented here.

2   Design of an Hybrid CBL/CMOS Cell

The logic family used in this paper is a configurable CBL/CMOS family [6]. Depending on a signal, the behaviour of the cell is the same than a CMOS or a CBL cell, and hence, the behaviour of a low power or a low noise cell. The schematic of a hybrid inverter is shown in figure 1.

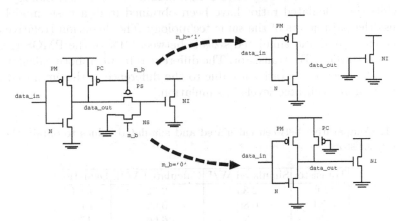

Fig. 1. Schematic of an hybrid inverter

The behaviour of this cell is as following. When the signal m_b is high, the cell is configured as CMOS cell. In this case, the transistors PC and PS are off. The transistor NS is on, and so it puts a low level in the gate of transistor NI, cutting it off. With this configuration, the signal $data\_out$ is the complementary value of the signal $data\_in$ due to the transistors PM and N, that is, a CMOS inverter.

When the signal m_b is low, the cell is configured as CBL cell. In this case, the function of the transistor PM is eliminated because its drain and source terminals are always connected through the transistor PC that is always on. Besides, the transistor PS is on connecting the output signal to the gate of the transistor NI, while the transistor NS is off. With this configuration, the signal $data\_out$ will be low when the signal $data\_in$ is high; in this case, there exists a path between supply and ground with a constant supply current (so the transistor PC must be weaker than transistor N). The signal $data\_out$ will be high when the signal $data\_in$ is low, due to the transistor PC (typical of CBL cell) and PM (typical of MOS cell). The supply current is due to the transistor NI that is on.

In the design of this cell, we must obtain the sizes of each transistor. In order to obtain them, we can use the methods of each family. In the case of CBL cell, a possible method consists to fixe the voltage of low level and the supply current, as we can see in [7, 8], while in CMOS cell, the most usual method consists to match both NMOS and PMOS trees.

The transistors PS and NS have the function of pass-transistors, and hence their effects over parameters are low; then their sizes will be the minimum sizes to switches, due to their low influence. Though these sizes can be calculated by a model based on analitical expresions (equations), that it is not an accurate way to do it because most of the times, these models don't take under consideration minor order effects like channel length, so electrical simulators are a better option. In table 1 we show the simulated and calculated ratios W/L considering a low level signal (voltage=0.162 v), and a supply current of 0.47 mA. The simulation has been performed by ELDO in a standard CMOS technology of 0.35 μm; while the calculated ratios have been obtained from a basic model. Both cases use the parameters of the same technology. The deviation (relative error) between calculated and simulated sizes is between 21% of the PMOS tree, and 0.4% of the PMOS CBL transistor. The differences between both calculated and simulated transistor size ratios are due to the difference of the employed model level, with more accurated levels in simulation.

Table 1. Comparative between calculated and simulated values to obtain the ratios W/L of transistors

Transistor	Simulated W/L	Calculated W/L	Deviation
PC	2.33	2.32	0.4%
NI	0.80	0.69	16%
N	5	6.06	-17%
PM	15	18.87	-21%

3   Multiobjective Genetic Algorithm and Optimization Methodology

SPEA (*Strength Pareto Evolutionary Algorithm*), used as multiobjective optimizer (see Figure 2, uses an external archive (E) containing the non-dominated solutions found so far [ ] to be updated (step 04 and step 05 in figure 2) in each iteration . The non-dominated solutions stored in E are intended to approximate the Pareto-optimal solutions of the problem after a convenient number of iterations. A *clustering* algorithm [ ](step 06 in figure 2) is applied to the external archive in order to achieve two main goals: (1) to avoid that the size of the external archive exceeds an user defined maximum size (*maxESize*); (2) to distribute uniformly the found non-dominated solutions along the objective space by encouraging the dissimilarities between the stored solutions. The transformation operators (step 09) are a single point crossover operator and uniform mutation operator for real-coded individuals [ ]. Each individual (coded solution) is composed by a vector of real numbers containing the sizes (width and length) of all transistors involved in the cell. The overall employed optimization methodology is depicted in figure 3. In order to reduce the input search space size, the following restrictions are imposed:

$$3\mu m < W_i < 0.3\mu m \quad \forall \quad transistors$$
$$2\mu m < L_i < 0.3\mu m \quad \forall \quad transistors$$
$$\frac{W}{L}_{PC} < \frac{W}{L}_{N}$$

```
01  Create randomly Population P of Candidate Solutions (individuals) of
    Size Popsize
02  Create external archive E
03  While (stop_condition) FALSE
04      Copy non dominated individuals form P to E
05      Delete dominated individuals form E
06      if Size(E) > maxESize then reduce Size of E with clustering algorithm
07      Assign fitness to each individuals in P and E
08      Binary Torunament Selection of Popsize individuals from P+E
09      Crossover and Mutation of individuals in P
10      Evaluate the Performance of each individual in P using ELDO Simulator
11  EndWhile
```

Fig. 2. SPEA pseudocode

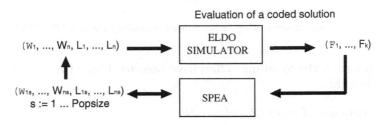

Evaluation of a coded solution

Fig. 3. Block diagram of optimization methodology

Simulation has been performed in the enviroment shown in figure 4(a). In it, the studied inverter is the center one; while the other inverters generate a more realistic enviroment, that is, a *fan_in* and a *fan_out* of one inverter, and hence, a realistic waveform in all signals of studied inverter. Also, the supply source is not ideal due to the inductor.

Both CMOS and CBL configurations are considered in the same simulation. As example, the supply voltage is shown in figure 4(b). In it, we can see the behaviour of both configurations: firstly CBL and secondly CMOS configurations. In a normal operation, the cell configuration will not change, and so, the initialization of both configurations is not considered in order to obtain the different parameters (because the behaviour is different from the normal one).

4   Results

We have focused for this work in the followings objective functions (cost factors) to be minimized simultaneously: supply current (I_{dd}) peak, supply voltage (V_{dd}) peak, the RMS value of supply current, average power consumption (Pow), propagation delays (T_{delay}, T_{fall}, T_{rise}), power-delay product (PDP) and the low level (V_{OL}) in CBL configuration; all these parameters will be obtained for both configurations, except the last parameter (typical of CBL configuration).

All these cost factors are measured for each coded solution using ELDO simulator as indicated in figure 3. The input signal is a pulses train to 100 KHz and the supply voltage will be maintained to 3.3 v. Regarding the adjustment

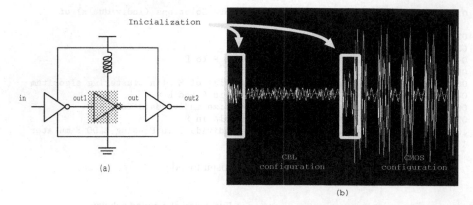

Fig. 4. Schematic of simulation environment including the CMOS/CBL inverter

of the optimizer, the following values have been used for SPEA in all the runs
of this algorithm:

- Maximun size of external archive=80
- population size=30
- Crossover probability=0.8
- mutation probability=0.01

We can be interested in studying the dependency of a given transistor size
over a certain objective function. In this context, the ploting of the non-dominated ratio W/L solutions of each transistor with respect to a given objective
function show us which are the areas of the input search space related with a
"good design" for that objective when the *simultaneous* minimization of all cost
factors is considered. As an example, the solutions for two different cost factors
are shown in figure 5. In this figure, the peak of supply voltage and the average
propagation delay versus W/L of each transistor are depicted (W/L in y-axis
and the cost factor in x-axis) for the sum of CBL and CMOS configuration (both
configurations are contributing to the output). Firstly, we can see the different
behaviour to minimize both factors in some transistors. For example, the N
transistor shows a better behaviour with a W/L ratio near to one and to five to
power consumption and propagation delay respectively. We can see that in this
case the cost factors are in confliction and the use of multiobjective evolutionary
optimization is plenary justified then. In Table 2 values minimizing every cost
factor are shown.

Secondly, we can extract some recommendations for the design of this cell.
The transistors with a special influence in peak of supply voltage are the transistors N, PM and PC; the optimum ratio is near to one in all these transistors.
The others transistors do not show excesive influence in this parameter. In the
case of propagation delay, the transistors with a special influence are the transistors N, PM and PC; the optimum ratio is near to three, five and four to the
transistors PC, N and PM respectively.

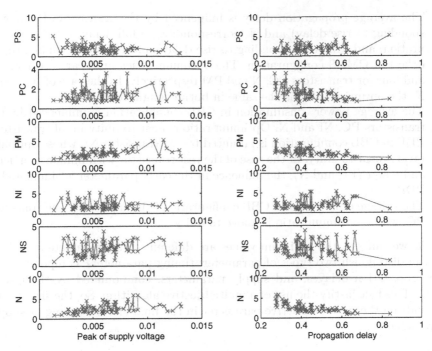

Fig. 5. (V_{dd}) and propagation delay for the sum of CBL and CMOS configurations versus W/L for every transistor of the cell

From the study of non-dominated solutions for this cell, we extract the following conclusions:

- (I_{dd}) peak shows dependency with the transistors N, NI and PC in CBL configuration, and N and PM in CMOS configuration. In all of them, optimum W/L ratios are near to unity.
- (V_{dd}) peak is influenced by the transistors N and NI in CBL configuration, and PM in CMOS configuration. In all of them, the optimum ratios are near to unity.
- (I_{dd}) RMS value is influenced by the transistors PC, NI and N in CBL configuration, but none influence is appreciated in CMOS configuration. The optimum ratios are near to unity.

Table 2. Transistor sizes minimizing *separately* cost factors

Minimized factor	PM	PC	N	PS	NS	I
Sum of CBL and CMOS Configuration						
Peak $I_{dd} = 0.099+0.092$mA.	0.61/0.98	0.86/1.00	0.77/0.81	1.99/0.38	0.52/0.88	0.89/0.97
RMS $I_{dd} = 0.422+0.009$m.	1.28/0.46	0.89/0.73	1.14/0.93	0.99/0.92	1.73/0.52	0.65/0.89
Peak $V_{dd} = 0.300+1.100$mV.	0.61/0.98	0.86/1.00	0.77/0.81	1.99/0.38	0.52/0.88	0.89/0.97
$T_{rise} = 0.073+0.169$ns.	1.86/0.38	1.52/0.60	1.60/0.59	0.59/0.38	1.02/0.33	1.06/0.50
$T_{fall} = 0.092+0.051$ns.	1.47/0.65	0.90/0.90	1.99/0.34	0.86/0.58	1.54/0.84	1.21/0.49
$T_{delay} = 0.113+0.159$ns.	1.28/0.67	1.52/0.43	1.60/0.33	0.59/0.56	1.04/0.33	0.91/0.67

– The average propagation delay is influenced by the transistors PC (corresponding to rise delay) and N (corresponding to fall delay) in CBL configuration, and PM (corresponding to rise delay) and N (corresponding to fall delay) in CMOS configuration. The optimum ratios are near to three, five and four for transistors PC, N and PM respectively. In the case of transistor N, the optimum value is the same in both configuration.
– The average power consumption in CBL configuration is influenced by the transistors PC, NI and N. Optimum ratio is near to unity in all transistors.
– PDP in CBL configuration is influnced by the transistor NI, whose optimum ratio is near to unity. In the case of the rest transistor with influence in delay and power (PC and N), the influence in the early parameter is eliminated in PDP.
– The voltage of low level in CBL configuration is influenced by the transistor PC, whose optimum ratio is above to unity.

As we can see, the optimum values are different to all parameters. In fact, we can distinguish two types of parameter: timing parameters, compounded by the propagation delayes; and supply parameters, compounded by the rest of them. This fact justify the use of a multiobjective algorithm. So, the sizes to use depends on the most restrictive parameter in the final circuit: timing or supply parameters.

5   Conclusions

In this work, the optimized design of a CBL/CMOS cell by using a multiobjective evolutionary algorithm has been addressed. Then, the following conclusions from the encountered solutions have been obtained:

– The pass transistors (PS and NS) do not show a clear dependency with none of parameters.
– There exists a clear difference between the best sizes to optimize timing parameters and the rest of them. In the cases of timing parameters, the ratio W/L tends to high values (between three and five); while in the rest of parameters, the ratio W/L tends to values near to one.
– The transistor NI only shows a clear tendency to all current parameters in CBL configuration (peak, RMS and average value of supply current, and power delay product). Always, this tendency is to one. In the other parameters, there does not exist a tendency to a value of the ratio W/L.
– The transistor PM shows a tendency to one in the ratio W/L to current and voltage peak in CMOS configuration. In the case of the delay in CMOS configuration, the tendency of ratio W/L is to four.
– The transistor PC shows a tendency similar to the transistor NI in current parameters. In the case of the delay, the tendency is to three.
– The transistor N shows two clear behaviours. In the cases of current parameters and peak of supply voltage, the tendency of ratio W/L is to one in both configurations. But in the case of the delay, the tendency is to higher values, four in CBL configurations and five in CMOS.

Table 3. Optimum ratios of each transistor depending on the more restrictive parameter

Transistor	Timing parameter	Supply parameter	Optimum value
PC	3	1	Depending on parameter
PM	4	1	Depending on parameter
PS	-	-	-
N	4-5	1	Depending on parameter
NI	-	1	1
NS	-	-	-

Then, the optimum ratios are shown in table 3, depending on the more restrictive parameter.

6 Future Work

This work is the init of a optimization process whose main objective is to achieve a cells library addressed to mixed signal applications. With this library, a new optimization process will be done in order to obtain the optimum configurations of each cell in a general digital system.

References

1. J.P. Cohoon and D.H. Marks, " A review and Evaluation of Multiobjective Programming Techniques", *Water Resources Research*, Vol 11, No. 2, pp. 208-220, 1975
2. A.E.Eiben and J.E. Smith,*Introduction to Evolutionary Computing*, Natural Computing Series: Springer, 2003.
3. K.Deb, *Multiobjective Optimization using Evolutionary Algorithms*, John Wiley & Sons, 2002
4. C.A. Coello and G.B. Lamont, *Applications of Multiobjective Evolutionary Algorithms*, World Scientific, 2004.
5. E.Zitzler and L.Thiele, "Multiobjective Evolutionary Algorithms: A Comparative Case Study and the Strengh Pareto Approach", *IEEE Transactions on Evolutionary Computation*, Vol. 3 No. 4, pp. 257-271, 1999.
6. R. Jiménez, P. Parra, P. Sanmartín, and A.J. Acosta, "Optimum current/voltage mode circuit partitioning for low noise applications", *in Proc. XVIII Conference of Design of Circuits and Integrated Systems*, 2003, pp. 63-68.
7. Albuquerque, E., Fernandes, J. and Silva, M.: "NMOS Current-Balanced Logic" *Electronics Letters*, vol. 32, pp. 997-998, May 1996.
8. Albuquerque, E. and Silva M.: "A new low-noise logic family for mixed-signal IC's" *IEEE Trans. Circuits Systems I*, vol. 46, pp. 1498-1500, 1999.
9. J. Morse, "Reducing the size of the nondominated set:Pruning by clustering", *Computers and Operations Research*, Vol. 7, No. 2, pp. 55-66, 1980.
10. Z. Michalewicz, *Genetic Algorithms Data Structures = Evolution Programs*, Springer, 1999.

Power-Clock Gating in Adiabatic Logic Circuits

Philip Teichmann[1], Jürgen Fischer[1], Stephan Henzler[1],
Ettore Amirante[2], and Doris Schmitt-Landsiedel[1]

[1] Institute for Technical Electronics, Technical University Munich,
Theresienstrasse 90, D-80290 Munich, Germany
teichmann@tum.de
http://www.lte.ei.tum.de
[2] Now with Infineon Technologies, Munich

Abstract. For static CMOS Clock-Gating is a well-known method to decrease dynamic losses. In order to reduce the static power consumption caused by leakage currents, Power-Gating has been introduced. This paper presents for the first time Clock-Gating and Power-Gating in Adiabatic Logic. As the oscillator signal is both the power and the clock in Adiabatic Logic, a Power-Clock Gating is implemented using a switch to detach the adiabatic logic block from the oscillator. Depending on the technology the optimum switch topology and dimension is discussed. This paper shows that a boosted n-channel MOSFET as well as a transmission gate are good choices as a switch. Adiabatic losses are reduced greatly by shutting down idle adiabatic circuit blocks with Power-Clock Gating.

1 Introduction

The evolution in modern semiconductor technologies and the increasing demand for more complex systems leads to a rising number of gates per chip. Dynamic and leakage losses are a major concern in static CMOS circuits and have been adressed by many proposals in the past ([1],[2],[3]). Adiabatic circuits are a way to dissipate less than the fundamental limit for the dynamic energy loss per logic operation ($E_{CMOS} = \frac{1}{2}CV_{DD}^2$) in static CMOS. Therefore a constant charging current is used. The most promising adiabatic logic circuits like the Efficient Charge Recovery Logic (ECRL) [4] and the Positive Feedback Adiabatic Logic (PFAL) [5] are dual-rail encoded and use a four-phase power supply. At each cycle charge is transferred to one of the output nodes and recycled to the supply again, leading to losses even for constant input signals. Power-Clock Gating (PCG) avoids these losses by detaching the Power-Clock from idle circuit blocks.

This work treats PCG for the ECRL family, which provides a large saving against static CMOS. Simulations are performed with a BSIM3 model and parameters of an industrial 130nm process. The operating frequency for the investigated circuits is 100MHz, this is a suitable frequency for many digital signal processing tasks. After a short description of the adiabatic Power-Clock and the so-called adiabatic losses, basic considerations of PCG are shown. A major decision is the choice of a suitable switch. Section 3 deals with the optimization

V. Paliouras, J. Vounckx, and D. Verkest (Eds.): PATMOS 2005, LNCS 3728, pp. 638–646, 2005.

of the switch. In Section 4 basic PCG switching theory is discussed, followed by the verification of the theoretical results in the simulation section.

2 Adiabatic Losses and Power-Clock Gating

A four-phase Power-Clock Φ is used for the investigated ECRL familiy. Each cycle consists of four states, as can be seen in Fig. 1 a). Adjacent phases are shifted by a quarter of the phase period T.

During the Evaluation state **E**, the Power-Clock voltage rises and one of the output capacitances is loaded depending on the input signals. The Hold state **H** provides a stable output for the Evaluation state of the following adiabatic gate. In the Recover state **R** the charge from the output node is recycled to the oscillator. For symmetry reasons a Wait state **W** is inserted.

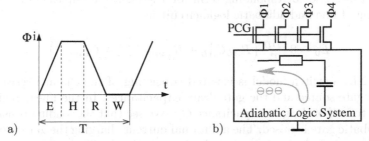

Fig. 1. a) One phase of the adiabatic four phase clock. The cycle consists of four states. b) The general ideal behind Power-Clock Gating: The circuit is detached from the oscillator via a switch

In static CMOS, no dynamic losses occur as long as the logic state at the input does not change. As the output node is charged and discharged during every cycle in Adiabatic Logic, energy dissipation occurs even for a steady output state. Therefore Power-Clock Gating is introduced, which enables us to detach the oscillator from our circuit. Fig. 1 b) shows the principle of PCG: Switches are used to disconnect the adiabatic circuit from the oscillator.

The dissipated energy per cycle of an adiabatic gate can be expressed by

$$E_{diss} = 8\frac{R_{AL}C_{AL}^2}{T}V_{DD}^2 \ . \tag{1}$$

where R_{AL} is the path resistance and C_{AL} is the output node capacitance. These losses are called adiabatic losses and Equation (1) is true as long as

$$\frac{T}{4} \gg max(RC) \tag{2}$$

where for Equation (1) $max(RC) = R_{AL}C_{AL}$.

An equivalent circuit is presented in Fig. 2 a) including a model for the line connecting the circuit to the oscillator.

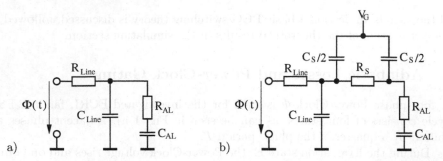

Fig. 2. a) The equivalent circuit consists of the model for the line and an adiabatic circuit model. b) An equivalent model for the switch is inserted between the line and the adiabatic circuit model

If the condition of Equation (2) holds, the currents loading the capacitances are constant. The corresponding term for the energy dissipation of the system consisting of line and adiabatic logic circuit is

$$E_0 = 8 \frac{V_{DD}^2}{T} \left(R_{AL} C_{AL}^2 + R_{Line} \left(C_{AL} + C_{Line} \right)^2 \right) . \tag{3}$$

In Fig. 2 b) a switch model is inserted consisting of the channel resistance R_S and the gate-source and the gate-drain capacitances. In the linear region both amount to half of the gate capacitance C_S. We see that the loading resistance for the adiabatic gate is raised. The additional current charging the gate capacitance C_S of the switch device causes adiabatic losses in the line resistance R_{Line} and in R_S. If the switch is in on-state, Equation (3) is extended to

$$E_{on} = 8 \frac{V_{DD}^2}{T} \left(R_{AL} C_{AL}^2 + R_S \left(\frac{C_S}{2} + C_{AL} \right)^2 + \right.$$

$$\left. + R_{Line} \left(C_{AL} + C_S + C_{Line} \right)^2 \right) . \tag{4}$$

So the switch is adding additional losses to the system. To reduce the influence of the switch, we are looking for a low-resistance, low-capacitance switch. Taking a MOSFET as switching device a trade-off between resistance and capacitance has to be made. If we enlarge the width of the switch on the one hand the resistance is decreased but on the other hand the gate capacitance is increased. So an optimum width can be found that generates a minimum energy overhead. Turning off the switch the overall losses are reduced to the losses of the supply line and the switch.

$$E_{off} = 8 \frac{V_{DD}^2}{T} \left(R_{Line} \left(\frac{C_S}{2} + C_{Line} \right)^2 \right) . \tag{5}$$

The stand-by energy dissipation E_{off} is mainly dependent on the Power-Clock line length.

3   Switch Topologies

The choice of the best PCG switch topology is a major concern. Preferably, we are looking for a switch with a low on-resistance and a small threshold drop. Three topologies (Fig. 3) were investigated in aspect of their suitability as switch. The n-channel and the p-channel MOSFETs are boosted by a voltage V_{OV} to provide full swing operation and reduce the on-resistance. The on-resistance in the linear region is equal to

$$R_S = \frac{1}{\mu C_{OX} \frac{W}{L} |V_{DD} + V_{OV} - \Phi - V_{th}|} \tag{6}$$

where μ is the mobility and C_{OX} is the specific oxide capacitance. W and L are the channel width and length.

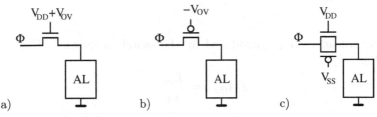

a)   b)   c)

Fig. 3. Three topologies have been investigated. For a) n- and b) p-channel devices a boost voltage V_{OV} has been applied in order to achieve full swing operation. c) For the transmission gate the width of the p-channel W_p is twice W_n

The p-channel width W_p of the transmission gate is twice the width W_n of the n-channel to compensate for the differing mobilities of n-channel and p-channel transistors. The transmission gate's inverse on-resistance is

$$R_S^{-1} = \mu_n C_{OX} \frac{W_n}{L} (V_{DD} - \Phi - V_{th,n} + |-\Phi - V_{th,p}|) \ . \tag{7}$$

Looking at Equation (4) we want a small C_S. As C_S is directly proportional to the width, a small PCG switch is desired. So for a MOSFET switch a trade-off between the on-resistance R_S and the capacitance C_S has to be made.

A MATLAB simulation was performed, calculating the resistance R_S and the energy from Equation (4). A line length of $100\mu m$ has been chosen. The adiabatic load consists of 16 ECRL inverters. For the 16 inverters a model using R_{AL} and C_{AL} has been chosen, producing the same energy dissipation like 16 ECRL inverters. For the n-channel and the p-channel switch an overdrive voltage of $V_{OV} = 400mV$ has been applied. The results are presented in Fig. 4. In Fig. 4 a) the transmission gate's resistance is almost constant over the whole Power-Clock voltage range. The n-channel has its lowest resistance for low voltage of the Power-Clock Φ and the p-channel for a voltage close to $V_{DD} = 1.2V$. The relative

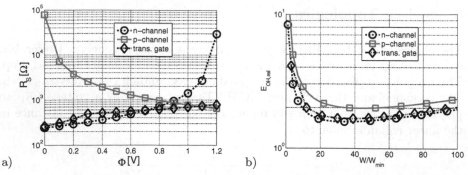

Fig. 4. a) The resistance of n-channel and p-channel MOSFET is dependent on Φ whereas the transmission gate shows an almost constant resistance over the whole Power-Clock voltage range. The n-channel and p-channel devices are boosted by a voltage $V_{OV} = 400mV$. b) Looking at the relative energy overhead, we see that the boosted n-channel device adds the lowest overhead in on-state, closely followed by the transmission gate

energy overhead $E_{OH,rel}$ introduced by the switch is shown in Fig. 4 b). It is defined as

$$E_{OH,rel} = \frac{E_{on}}{E_0} - 1 \ . \tag{8}$$

On the x-axis the allover width is specified. For the transmission gate this is $W = W_p + W_n = 3W_n$. The boosted n-channel switch introduces the lowest overhead $E_{OH,rel}$. The transmission gate differs slightly, the boosted p-channel doubles the dissipated energy. The optimum width W_{opt} is found at the minimum of $E_{OH,rel}$.

Summarizing the properties of the switch topologies, we see that the boosted n-channel MOSFET seems to be the best choice at first sight. But the close results for n-channel and transmission gate in respect of energy overhead $E_{OH,rel}$ and area consumption $A(W_{opt})$ show that the transmission gate is a good choice as well, as it is not boosted and the boost circuit itself will rise the area and the energy overhead for the boosted n-channel solution. Furthermore, the n-channel switch will reach its highest R_S for values of the Power Clock close to V_{DD}. This leads to a misshaped Power-Clock signal after the switch, as the full V_{DD} level cannot be reached (see Fig. 5). This limit can be reduced boosting the n-channel with higher voltages but it will lead to a higher gate voltage and this will compromise the reliability of a system. On the other hand, voltage scaling is limited due to the reduced voltage level at the adiabatic block caused by the voltage drop over the n-channel switch.

Simulations show that the transmission gate and the non-boosted n-channel switch (Fig. 5) cause an energy penalty $E_{OH,rel}$ of 18%, if driving 16 Inverters with an area $A = 15A_{min}$ for the switches. Full swing operation with the n-channel needs a V_{OV} of 400mV. By increasing V_{OV} to 400mV, $E_{OH,rel}$ can be reduced to 14% for the n-channel.

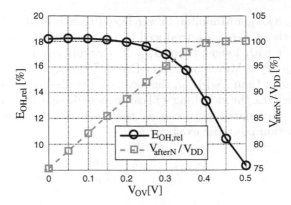

Fig. 5. For the n-channel device a high boost voltage V_{OV} has to be applied in order to allow full swing operation of the adiabatic circuit. V_{afterN} is the voltage after the n-channel switch. A V_{OV} of 400mV needs to be applied to reach V_{DD}. $E_{OH,rel}$ decreases according to the reduced R_S for higher boost voltages

4   PCG Switching Theory

In a system consisting of different functional units, some units are not used permanently. In other applications the whole system can be switched off for certain times. During idle times a shut-off via PCG is performed. As PCG not only introduces an overhead due to losses through R_S and C_S, but also an switching overhead E_{SOH}, a minimum time can be found, where applying PCG for a circuit pays. This minimum of T_{off} is called Minimum Power Down Time T_{MPD}. The switching overhead E_{SOH} is caused during turn-off and turn-on in the adiabatic circuit. Additional energy is needed to charge the gates of the switching devices, but this overhead is not taken into account in the considerations in this paper.

First the mean energy dissipation using PCG $(\overline{E_{PCG}})$ is calculated by integrating the energy dissipation over $T_{ges} = T_{off} + T_{on}$.

$$\overline{E_{PCG}} = \frac{1}{T_{ges}} \int_{0}^{T_{ges}} E(t)dt \tag{9}$$

If we assume that the switching overhead appears within one cycle of the Power-Clock T_Φ Equation (9) can be written as

$$\overline{E_{PCG}} = \frac{1}{T_{ges}} \left(T_{on}E_{on} + T_{off}E_{off} + T_\Phi E_{SOH} \right) . \tag{10}$$

The Minimum Power-Down Time $T_{MPD,0}$ gives us the time for the power-down, where $\overline{E_{PCG}}$ is equal to the energy dissipation per cycle for a system with no switch E_0. In Equation (11) E_{off} is assumed to be much smaller than E_{on}.

$$T_{MPD,0} = E_{OH,rel}T_{on} + T_\Phi \frac{E_{SOH}}{E_0} \tag{11}$$

For a longer T_{on} a longer Minimum Power-Down Time is required, to compensate the relative overhead $E_{OH,rel}$ introduced by the switch. The switching overhead looses its impact with longer times T_{on}. If a circuit is already equipped with PCG we can specify a Minimum Power-Down $T_{MPD,on}$. When the circuit is in Power-Down longer than $T_{MPD,on}$, the mean energy dissipation $\overline{E_{PCG}}$ is pushed below E_{on}. Otherwise the switching will cause more dissipation than can be saved by shutting off. Summarizing we can see that each switching process should be rated, if a reduction potential exists or not.

$$T_{MPD,on} = T_\Phi \frac{E_{SOH}}{E_0} \tag{12}$$

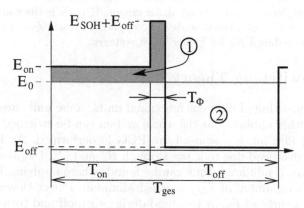

Fig. 6. PCG introduces overhead (1) in the on-state and with each switching event. In order to gain from a Power-Down, the overhead has to be compensated by the savings during a Power-Down (2)

Now looking at the switches, a boosted n-channel is adding little relative overhead $E_{OH,rel}$. For a long T_{on} a n-channel device combined with a high boost voltage V_{OV} allows the minimum product $T_{on}E_{OH,rel}$, but as mentioned before, the boost circuit itself will cause an energy overhead as well, leading to the conclusion that the transmission gate is a suiting switching device as well.

5   Simulation Setup and Results

A 16bit Carry Lookahead Adder (CLA) structure assembled with ECRL gates [ ] is simulated using a BSIM3 model with 130nm CMOS technology parameters. The Power-Clock phases are detached from the CLA in a way, such that each phase is connected and disconnected in the **W** state. Charge stored in the circuit is recovered into the oscillator in the **R** phase. Virtually no charge is remaining in the system, and the shut-off can be performed with best energy efficiency in the **W** state.

To provide realistic input signals, two ECRL inverters are chained in front of each input. At the outputs, a chain of two ECRL inverters is connected to have a realistic load. The simulation arrangement is sketched in Fig. 7 a).

For the simulation we choose the relation $T_{on} = T_{off}$. Equation (10) simplifies to

$$\overline{E_{PCG}} = \frac{1}{2}\left(E_{on} + E_{off} + \frac{T_{\Phi}}{T_{off}}E_{SOH}\right) . \tag{13}$$

For long T_{off} the switching overhead E_{SOH} loses its impact and $\overline{E_{PCG}}$ converges to $\frac{1}{2}\left(E_{on} + E_{off}\right) \approx \frac{1}{2}E_{on}$. Both devices, transmission gate and boosted

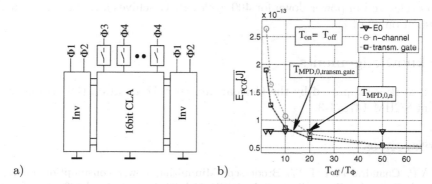

a)   b)

Fig. 7. a) The input signals for the simulated 16bit CLA are conditioned with two chained inverters. An output load is provided by an inverter chain at the outputs. N-channel and transmission gate switches are investigated. b) Simulation results for the CLA show that $T_{MPD,0}$ for the transmission gate is lower than that for the n-channel, using the relation $T_{on} = T_{off}$. This leads to the assumption that the transmission gate produces less switching overhead

n-channel, where sized to $15A_{min}$, this is an area penalty of approximately 12% for the CLA. The simulation results are presented in Fig. 7 b) for the boosted n-channel device and the transmission gate. $T_{MPD,0}$ is $13T_{\Phi}$ for the transmission gate and $19T_{\Phi}$ for the n-channel switch. Knowing that E_{on} is less for the n-channel device, we can draw the conclusion from Equation (13), that it creates a higher switching overhead E_{SOH}. Equation (12) shows a dependency on T_{on}. Thus for long T_{on}, $T_{MPD,0}$ of n-channel and transmission gate will cross. Each application has to be considered in aspect of T_{off} and T_{on}, to choose the right switch. For a $T_{on} \neq T_{off}$ simulations show, that the circuit is gaining large savings from PCG. Running the CLA for $20T_{\Phi}$ and shutting it down for $60T_{\Phi}$ saves 48% for the boosted n-channel and 53% for the transmission gate switch. If the CLA is in on-state for $100T_{\Phi}$ and in off-state for $400T_{\Phi}$ we achieve 73.5% reduction with the n-channel and 73.2% with the transmission gate.

6 Conclusion

Power-Clock Gating for Adiabatic Logic is proposed that incorporates Power Gating as well as Clock Gating in one switch. A theoretical approach to PCG has been presented, that gives us a first advice for the choice of a suitable switch topology.

It was shown that both, boosted n-channel MOSFET and transmission gate are reasonable switch topologies for PCG. The n-channel is producing less static overhead but the boost circuit itself will diminish this advantage.

The Minimum Power-Down Time T_{MPD} has been presented as a Figure of Merit. It tells us the minimum off-time for which we gain savings from the introduction of PCG for a given T_{on}. Power Clock Gating was applied at a 16Bit Carry Lookahead Adder (CLA) structure, proving that PCG can reduce losses greatly with an acceptable area penalty. For an application that is in on-state for 100 cycles and in power-down for 400 cycles PCG achives a reduction in energy dissipation of 73%.

Acknowledgements

This work is supported by the German Research Foundation (DFG) under the grant SCHM 1478/1-3.

References

1. A.P. Chandrakasan, R.W. Brodersen: Minimizing power consumption in digital CMOS circuits. Proceedings of the IEEE, Vol. 83, Issue 4, April 1995, pp 498-523
2. James W. Tschanz, Siva G. Narendra, Yibin Ye, Bradley A. Bloechel, Shekhar Borkar, Vivek De: Dynamic Sleep Transistor and Body Bias for Active Leakage Power Control of Microporcessors. IEEE Journal of Solid-State Circuits, Vol. 38, No. 11, pp. 1838-1845, 2003.
3. Stephan Henzler, Thomas Nirschl, Stylianos Skiathitis, Joerg Berthold, Juregen Fischer, Philip Teichmann, Florian Bauer, Georg Georgakos, Doris Schmitt-Landsiedel: Sleep Transistor Circuits for Fine-Grained Power Switch-Off with Short Power-Down Times. IEEE International Solid-State Circuits Conference, p. 13, 2005.
4. Y. Moon, D. Jeong: An Efficient Charge Recovery Logic Circuit. IEEE Journal of Solid-State Circuits, Vol. 31, No. 4, pp. 514-522, 1996.
5. A. Vetuli, S. Di Pascoli, L. M. Reyneri: Positive feedback in adiabatic logic. Electronics Letters, Vol. 32, No. 20, pp. 1867-1869, 1996.
6. E. Amirante: Adiabatic Logic in Sub-Quartermicron CMOS Technologies. Selected Topics of Electronics and Micromechanics, Volume 13, Shaker, Aachen, 2004.
7. J. Fischer, E. Amirante, F. Randazzo, G. Iannaccone, D. Schmitt-Landsiedel: Reduction of the Energy Consumption in Adiabatic Gates by Optimal Transistor Sizing. Proceedings of the 13th International Workshop on Power And Timing Modeling, Optimization and Simulation, PATMOS'03, Turin, Italy, September 2003, pp. 309-318
8. A. Bargagli-Stoffi, G. Iannaccone, S. Di Pascoli, E. Amirante, D. Schmitt-Landsiedel: Four-phase power clock generator for adiabatic logic ciruits. Electronics Letters, 4th July 2002, Vol. 38, No. 14, pp. 689-690

The Design of an Asynchronous Carry-Lookahead Adder Based on Data Characteristics

Yijun Liu and Steve Furber

APT Group, School of Computer Science
University of Manchester, Manchester M13 9PL, UK
yijun.liu@cs.manchester.ac.uk, steve.furber@manchester.ac.uk

Abstract. Addition is the most important operation in data processing and its speed has a significant impact on the overall performance of digital circuits. Therefore, many techniques have been proposed for fast adder design. An asynchronous ripple-carry adder is claimed to use a simple circuit implementation to gain a fast average performance as long as the worst cases input patterns rarely happen. However, based on the input vectors from a number of benchmarks, we observe that the worst cases are not exceptional but commonly exist. A simple carry-lookahead scheme is proposed in the paper to speed up the worst-case delay of a ripple-carry adder. The experiment result shows the proposed adder is about 25% faster than an asynchronous ripple-carry adder with only small area and power overheads.

1 Introduction

As a basic computation function of data processing, integer addition is the most commonly used and important operation in digital circuit design. Therefore the speed and power consumption of adders have great impact on the overall system speed and power consumption. A ripple-carry adder [ ] is implemented by using multiple copies of a 1-bit full adder, where the carry output of the $(i-1)$-th full adder is fed into the carry input of the i-th full adder and the lowest order full adder has a carry input of 0. Because the carry input of the i-th adder depends on the carry output bit of the $(i-1)$-th adder, the carry input of the $(i-1)$-th adder depends on the carry output bit of the $(i-2)$-th adder, and so on, the carries must ripple from the least-significant bit to the most-significant bit, resulting in an addition time of O(n) (n is the word length of the adder). Although a ripple-carry scheme is very slow for building a wide adder, small ripple-carry adders are often used as building blocks in larger adders because the constant factor of a full adder's delay is very small.

To speed up adders, a wide variety of techniques have been proposed. These techniques include: parallelizing adder segments by using redundant hardware, like carry-select adders [ ]; using dedicated carry propagation circuits to increase the speed of carry propagation, like carry-lookahead (or Manchester) adders [ ];

V. Paliouras, J. Vounckx, and D. Verkest (Eds.): PATMOS 2005, LNCS 3728, pp. 647–656, 2005.

or totally avoiding carry propagation, like carry-skip adders [ ]. These schemes can be used to satisfy different speed and hardware requirements. Apart from these synchronous adders, asynchronous adders [ ][ ][ ] are claimed to achieve high performance by gaining 'average performance'.

The remainder of the paper is organized as follows: Section 2 presents the principle of asynchronous ripple-carry adders and tests the 'average latency' of ripple-carry adders by using several benchmarks; Section 3 proposes a new scheme to speed up the worst-case latency of asynchronous adders. It also gives the circuit implementation of the proposed adder; Section 4 gives the experimental results; and Section 5 concludes the paper.

2   Average Latency of Asynchronous Ripple-Carry Adders

Asynchronous logic [ ] is claimed to have speed and power advantages over synchronous logic design because an asynchronous circuit can use small hardware to achieve 'average latency' which is smaller than the worst-case delay of its synchronous counterparts. Asynchronous ripple-carry adders are very commonly used examples to demonstrate the average latency of asynchronous design. The schematic of a precharge asynchronous ripple-carry adder is shown in Figure 1 [ ].

This is a hybrid asynchronous circuit, which has single-rail inputs and outputs, but its carry chain uses dual-rail 4-phase signalling. The carry bits of this full adder are weakly indicating. Therefore, if the two inputs (a and b) of the

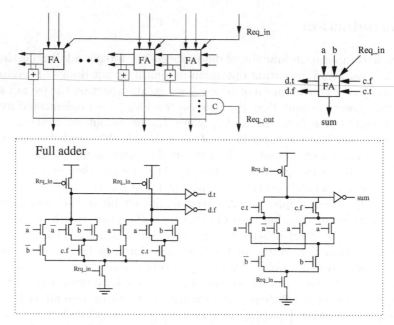

Fig. 1. A hybrid asynchronous adder that displays average-case latency

full adder are equal to each other, the carry-generating circuit can generate the valid carry and propagate it to the high-order full adder without waiting for the valid carry input from the low-order full adder. Only when $a \neq b$ does the full adder need to wait for the carry from the low-order full adder. So there is only a 50% probability that the full adder needs to propagate carries. The carry chain of the asynchronous ripple-carry adder is shown in Figure 2. The weak indication of the asynchronous full adders makes the carry chain not very 'solid' — every node has 50 percent chance to break. So the worst-case latency of an addition depends on the longest carry chain segment in the addition. Figure 3 plots the average latency of an weakly-indicating asynchronous adder as a function of its word-length when it is fed with randomly generated numbers. For a 32-bit ripple-carry adder, the average latency is only 5.34 full adder delays, which is smaller than the worst-case delay of a much more complex synchronous adder.

Fig. 2. A weakly indicating asynchronous carry chain

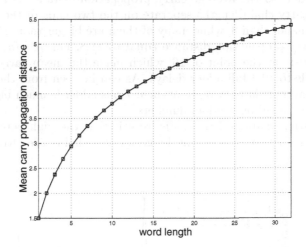

Fig. 3. Average size of longest carry chain for different word lengths assuming random data distribution[ ]

Unfortunately, the low average latency of asynchronous adders is based on the assumption that all input vectors are random numbers. However, in real applications, the input operands fed to an asynchronous adder are not random numbers and the 'average latency' of the asynchronous adder is not generally achieved. Garside demonstrated unbalanced distribution of both data processing and address calculation using the Dhrystone benchmark [ ]. Ten sets of input vectors are also used in this work to test the practical average latency of asynchronous adders. The characteristics of the input vectors are shown in Table1.

Table 1. Input vectors

Input vectors	Description
Rand	Randomly generated inputs
Gauss	Gaussian samples having $\mu = 0$ and $\sigma^2 = 1000$
adp(a)	Branch calculations of a audio/video encoding/decoding program
adp(d)	Data processing operations of a audio/video program
espr(a)	Branch calculations of a logic minimization algorithm
espr(d)	Data processing operations of a logic minimization algorithm
jpeg(a)	Branch calculations of a JPEG image processing program
jpeg(d)	Data processing operations of a JPEG image processing program
qsort(a)	Branch calculations of a quick sort program
qsort(d)	Data processing operations of a quick sort program

The vectors are taken from the ALU and address incrementer (discarding sequential calculations) of an ARM microprocessor when it runs several benchmarks. Figure 4 and Figure 5 show the different distributions of the longest carry propagation distance. For *rand*, the carry chain lengths congregate in the area between 2 and 7, so the average carry propagation distance is about 5.4. For *jpeg(d)*, the carry chain lengths separate on the two ends of the x-axis. Many of them are smaller than 8, while many of them are bigger than 24. This results in a mean carry propagation distance equal to 11.8. For *jpeg(a)*, a significant extra peak exists at the point of 16, which make the mean carry propagation distance equals to 10.4 full adder delays. As can be seen from the comparison, the real-time average latency of an asynchronous adder is much bigger than the average latency based on random numbers.

An interesting characteristic of Figure 4 is that the high percentage distributions gather on two ends — the longest carry chains are either very short

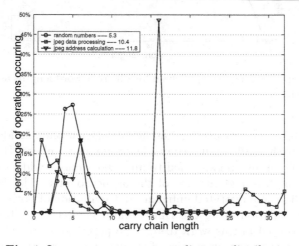

Fig. 4. Longest carry propagate distance distribution

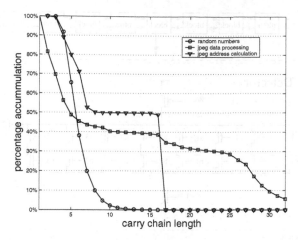

Fig. 5. Proportion of longest carry chains exceeding given length

or very long. Through further analysis of data processing operations, the short carry chains represent the additions of two inputs having the same signs (they are both negative or positive). The long carry chains represent the additions of one small negative number and one positive number. For example, $0 - 1$ (0xFFFFFFFF+0x00000000) has a longest carry chain that contains 32 full adder delays. Since there is a 50% chance to add positive numbers and negative numbers, the average latency of an n-bit asynchronous ripple-carry adder is about $n/2$ full adder delays, which is a big latency.

The reason why a big proportion of address calculations have a longest carry propagation distance equal to half the word length is due to compilers and the specific CPU architectures. In the tests, the ARM programs are loaded at the address 0x8000. As is well known, branches dominate the actual trace of a program. Among branches, jump back instructions having a small jump distance are most common. The specific loaded addresses and small negative offset result in a longest carry chain segment. For example, 0x8010 - 9 (0x00008010+0xFFFFFFF7) has a carry chain that contains 16 full adder delays. This is the reason why the delays of more than 50% address calculations gather in the area of half word length.

3   Proposed Asynchronous Carry-Lookahead Adder

Because the long delays usually happen when calculating the addition of a negative and a positive number, for the applications where most operands are positive numbers, asynchronous adders still have the low average latency advantage. However, if the average latency of an asynchronous adder can not meet a given performance requirement, hardware additions are needed to speed up the asynchronous adder. Those high performance techniques used in synchronous adders can also be used to increase the performance of asynchronous adders. However, the easiest way may be deduced from the observation mentioned above. If there

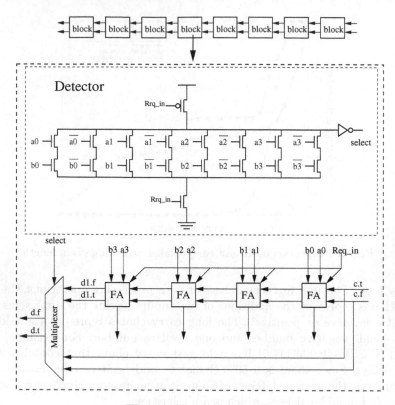

Fig. 6. The proposed adder

is a pair of same-order input bits equal to each other, the carry chain is broken at this node.

A fast asynchronous adder is shown in Figure 6. A 32-bit adder is subdivided into 8 blocks. Each block contains a 4-bit asynchronous ripple-carry adder and a detector. The detector detects if there exists a bit pair that has two bits equal to each other. If not, the detector passes the carry in from the lower-order block directly to the high-order block; otherwise, the detector propagates the normal carry bit. This adder is similar to a synchronous carry-lookahead adder. This scheme is very efficient and low overhead — it reduces the worst-case delay by approximately 8 times (ignoring the multiplexer delay). The hardware overhead includes only 7 detectors and multiplexers. Using a precharge logic style, the detector is very small and fast. The hardware overhead is also very small (25%). The delay of the detector and the multiplexer is 0.8 of the half adder delay. The worst-case delay of the 4-4-4-4-4-4-4-4 (8,4) scheme (8 blocks and each block has 4 full adders) is $7 \times 1 + 7 \times 0.8 = 12.6$ full adder delays. The worst-case delay can be minimized by reorganizing full adders. A 5-5-5-5-5-7 scheme is also tested for comparison. Figure 7 shows the evaluation speeds (without including the delay of the completion detector) of different schemes. As can be seen, the (8,4) scheme is good in terms of speed. Another advantage of the (8,4) scheme

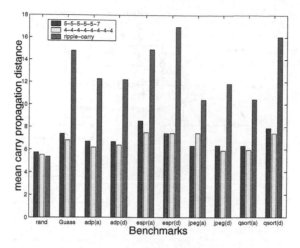

Fig. 7. Delay comparison of three asynchronous adders

is that it keeps the detectors a reasonable size. Using the proposed technique, the average latency of address calculation is sped up by 44% and the average latency of data processing is sped up by 53% with 25% hardware overhead.

3.1   Completion Detector Design

Figure 7 does not include the delay of the asynchronous completion detector which indicates the completion of additions. A completion detector can greatly affect the overall speed because it has a C-gate with a fan-in of n (see Figure 1). Nielsen and Sparsø [7] combined strongly indicating and weakly indicating full adders to minimize the number of nodes needing to be checked. However, this scheme also increases the possibility of carry dependence, which affects the average latency of asynchronous adders (24.6% based on the benchmarks).

A tree-style detector is used in the proposed adder as shown in Figure 8 (a). The delay of the circuit can be reduced by several inverter delays by interlacing n-pass-transistor gates and p-pass-transistor gates as shown in Figure 8 (b). This tree detector is very fast because it detects the completion of carries in parallel, so for n-bit adders, it has a $log_2 n$-level logic delay. Moreover, the completion signal is propagated along with the carries propagating, so delays mostly overlap. The detector's contribution to the overall delay is very small.

The completion detector is very fast because it uses high performance pass-transistors and drivers for long wires. However, the whole completion path needs to be reset after every addition. The pass-transistors are closed after initialization, so the internal nodes after the pass-transistors need to be precharged before the next addition. When using n-pass-transistor gates and p-pass-transistor gates together, the nodes after the n-transistors are precharged to 1 and the nodes after the p-transistors are precharged to 0.

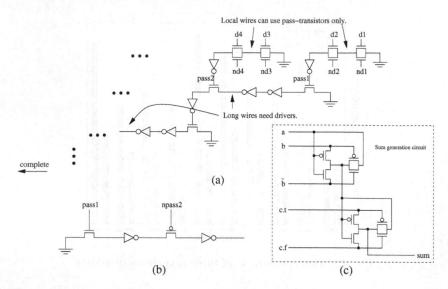

Fig. 8. The pass-transistor tree completion detector and sum generation circuit

4 Experimental Results

The proposed adder has been implemented using a $0.18\mu m$ ST CMOS technology. An asynchronous ripple-carry adder and a synchronous carry-select adder are also implemented for comparison. The comparison of the three adders is shown in Table 2. The simulation is schematic-based and wire capacitances are ignored. The supply voltage is 1.8 volts.

The use of a dual-rail internal carry chain not only results in a low average latency but also reduces the unnecessary switching activities caused by glitches from the carry chain, because carries are not propagated until they are valid. However, the dual-rail carry chain seems not to be good at power consumption although it minimizes glitches, because after each operation, the carry chain needs to return to zero. This introduces extra transitions. The same thing happens with the completion detector. This is the reason why the asynchronous ripple-carry adder is not two times better in power consumption than the synchronous carry-select adder as expected, although it uses less than half the hardware. For this reason, we decided to not precharge the sum generation circuit every time. A pass-transistor logic style is used to build the sum generation circuit as shown in Figure 8 (c). The double pass-transistor [ ] sum circuit contributes to the reductions of both area and power consumption.

As can be seen from the table, the proposed adder is 27% faster than the asynchronous ripple-carry adder in data processing and 19% faster than the ripple-carry adder in address calculation at the cost of 25% hardware overhead and 15% power overhead when running these benchmarks. For the worst-case latency, the proposed adder is more than two times faster than the ripple carry adder. The proposed adder is also faster than a synchronous carry-select adder in

Table 2. The comparison of different adders

Adders		proposed precharged sum	proposed pass-transistor sum	asynchronous ripple-carry	synchronous carry-select
worst-case delay		1.68	1.68	3.48	1.38
transistor number		1,730	1,690	1,332	3,144
area (ratio)		0.5	0.5	0.4	1
rand	delay	1.18	1.18	1.11	1.38
	power	0.51	0.45	0.40	0.64
Gauss	delay	1.23	1.23	1.69	1.38
	power	0.46	0.43	0.37	0.56
data processing	delay	1.28	1.28	1.75	1.38
	power	0.48	0.42	0.37	0.45
address calculation	delay	1.18	1.18	1.45	1.38
	power	0.45	0.42	0.35	0.45

both data processing and address calculation. For power efficiency, this proposed adder using pass-transistor sum circuits is only slightly better than the carry-select adder. The reason is due to the return-to-zero operation of asynchronous adder chain and completion detector. The synchronous adder also saves transitions when the current operand pair is similar to the one it has just finished. However, since the synchronous adder has a bigger die size and more complex wire layout, we estimate that the proposed adder has a better real chip measurement than the carry-select adder.

5   Conclusion

Asynchronous logic is claimed to have advantages in high speed and low power consumption because asynchronous circuits can display average-case performance. Therefore, if the worst cases of an asynchronous circuit rarely happen, there is no need to use expensive dedicated circuits specially for speeding up the worst-case delay and the asynchronous circuit can still achieve a reasonable performance. Asynchronous ripple-carry adders are very commonly used examples to demonstrate the average-case performance of asynchronous circuits.

In the paper, we use 10 sets of input vectors taken from a number of benchmarks to test the average latency of an asynchronous ripple-carry adder. The results show that the average performance of an asynchronous ripple-carry adder is much slower than estimated by using random numbers. Through analysis, we find the reason is due to the additions of small positive numbers and negative numbers. Under such a circumstance, the longest carry chain propagates from the least-significant bit to the most-significant bit. Since the chance of adding a small positive number and a negative number is very big, even asynchronous adders have to decrease their worst-case delay. We propose a low-overhead carry-lookahead scheme which decreases the worst-case delay of an asynchronous ripple-carry adder by half. Based on the comparison, the proposed adder is 25% faster than an asynchronous ripple-carry adder. The proposed adder is also 11% faster and 7% more power efficient than a synchronous carry-select adder.

In this paper, we also propose a tree completion detector which has a $O(log_2 n)$ delay. The tree completion detector is a general design that can be used in any asynchronous circuits using a precharge logic style.

References

1. D. Goldberg, "Computer arithmetic", Computer Architecture: A Quantitative Approach, Morgan Kaufmann Publishers, 1990.
2. O. J. Bedrij, "Carry-select adder", IRE Transactions on Electronic Computers, vol. EC-11, Jun. 1962, pp. 340-346.
3. O. L. MacSorley, "High-speed arithmetic in binary computers", IRE proceedings, vol. 49, 1961, pp. 67-91.
4. S. Tuttini, "Optimal group distribution in carry-skip adders", Proceedings of the 9th Symposium on Computer Arithmetic, Sep.1989, pp. 96-103.
5. A. J. Martin, "Asynchronous datapaths and the design of an asynchronous adder", Formal Methods in System Design, 1(1):119-137, July. 1992.
6. J. D. Garside, "A CMOS VLSI Implementation of an Asynchronous ALU", Proceedings of the IFIP Working Conference on Asynchronous Design Methodologies, Manchester, England (1993).
7. L. S. Nielsen and J. Sparsø, "A low-power asynchronous data-path of a FIR filter bank", Proceeding of Async1996, 1996, pp. 197-207
8. J. Sparsø, S. Furber (eds). "Principles of Asynchronous Circuit Design: A systems perspective". Kluwer Academic Publishers, 2001.
9. R. Zimmermann and W. Fichtner, "Low-Power Logic Styles: CMOS versus Pass-Transistor Logic", IEEE Journal Of Solid State Circuits, 32(7):1079-1090, July 1997.

Efficient Clock Distribution Scheme
for VLSI RNS-Enabled Controllers

Daniel González, Luis Parrilla, Antonio García,
Encarnación Castillo, and Antonio Lloris

Departament of Electronics and Computers Technology,
University of Granada, 18071- Granada, Spain
{dgonzal,lparrilla,grios,encas,lloris}@ditec.ugr.es

Abstract. Clock distribution has become an increasingly challenging problem
for VLSI designs because of the increase in die size and integration levels,
along with stronger requirements for integrated circuit speed and reliability.
Additionally, the great amount of synchronous hardware in integrated circuits
makes current requirements to be very large at very precise instants. This paper
presents a new approach for clock distribution in PID controllers based on
RNS, where channel independence removes clock timing restrictions. This approach generates several clock signals with non-overlapping edges from a
global clock. The resulting VLSI RNS-enabled PID controller, shows a significant decrease in current requirements (the maximum current spike is reduced to
a 14% of single clock distribution one at 125 Mhz) and a homogeneous time
distribution of current supply to the chip, while keeping extra hardware and
power to a minimum.

1 Introduction

The Proportional-Integral-Derivate (PID) controller is used to solve about 90-95% of
control applications, including special applications requiring ultra-high precision
control [1]. The controllers for the above applications are usually implemented as
discrete controllers. The Residue Number System (RNS) [2] has shown itself as a
valid alternative for supporting high-performance arithmetic with limited resources.
In this way, RNS enables high-precision arithmetic-intensive applications using limited and modular building blocks. Concretely, when applied to PID controllers over
Field Programmable Logic Devices (FPLDs), RNS has shown an improvement of
331% in speed over the binary solution [3]. On the other hand, implementation of this
controllers on VLSI leads to difficulties in the proper synchronization of such systems. As clock distribution represents a challenge problem for VLSI designs, consuming an increasing fraction of resources such as wiring, power, and design time; an
efficient strategy for clock distribution is needed.

For True Single Phase Clock (TSPC), which is a special dynamic logic clocking
technique and should not be used as a general example, clock lines must be distributed all over the chip, as well as being distributed within each operating block. More
complex clocking schemes may require the distribution of two or four non-
overlapping clock signals [4], thus increasing the resources required for circuit synchronization. Moreover, for clock frequencies over 500 MHz, phase differences between the clock signal at different locations of the chip (skew) are presenting serious

V. Paliouras, J. Vounckx, and D. Verkest (Eds.): PATMOS 2005, LNCS 3728, pp. 657–665, 2005.
© Springer-Verlag Berlin Heidelberg 2005

problems [5]. An added problem with increasing chip complexity and density is that the length of clock distribution lines increases along with the number of devices the clock signal has to supply, thus leading to substantial delays that limit system speed. A number of techniques exists for overriding clock skew, with the most common being RC tree analysis. This method represents the circuit as a tree, modeling every line through a resistor and a capacitor, and modeling every block as a terminal capacitance [6]. Thus, delay associated with distribution lines can be evaluated and elements to compensate clock skew can be subsequently added. Minimizing skew has negative sides, especially as simultaneous triggering of so many devices leads to short but large current demands. Because of this, a meticulous design of power supply lines and device sizes is required, with this large current demand resulting in area penalties. If this is not the case, parts of the chip may not receive as much energy as required for working properly. This approximation to the problems related to fully synchronous circuits and clock skew has been previously discussed [7]; this paper will present an alternative for efficiently synchronizing RNS-based circuits while keeping current demand to a minimum. The underlying idea is to generate out-of-phase clock signals, each controlling an RNS channel, thus taking advantage of the non-communicating channel structure that characterizes RNS architectures in order to reduce the clock synchronization requirements for high-performance digital signal processing systems. In the present work, this synchronization strategy is applied to RNS-PID controllers [3] implemented on VLSI, achieving important reductions in current demand and current change rate.

2 Digital PID Controllers

The PID controller is described by the following equation [1]:

$$y(t) = K_p \left(x(t) + \frac{1}{T_i} \int_0^t x(\tau)d\tau + T_d \frac{dx(t)}{dt} \right) \tag{1}$$

where K_p is the proportional constant, T_i is the integral time and T_d is the derivative time. The discrete implementation of the controller is usually derived from the following approximations:

$$\int_0^t x(\tau)d\tau \approx \sum_{i=0}^{n-1} x[i]h$$

$$\frac{dx(t)}{dt} \approx \frac{x[n]-x[n-1]}{h} \tag{2}$$

where $x[j]$ is the jth sample and h is the sampling period. Using (2) in equation (1), the discrete version of (1) is:

$$y[n] = y[n-1] + K_p \left(x[n] - x[n-1] \right) + \frac{K_p \cdot h}{T_i} x[n-1] +$$

$$+ \frac{K_p \cdot T_d}{h} \left(x[n] - 2x[n-1] + x[n-2] \right) \tag{3}$$

Equation (3) may be rewritten more conveniently just defining the constants C_0, C_1 and C_2:

$$C_0 = K_p \left(1 + \frac{T_d}{h} \right)$$

$$C_1 = K_p \left(\frac{h}{T_i} - 1 - 2\frac{T_d}{h} \right) \tag{4}$$

$$C_2 = \frac{K_p T_d}{h}$$

Thus, the discrete version of the PID controller is:

$$y[n] = y[n-1] + C_0 x[n] + C_1 x[n-1] + C_2 x[n-2] \tag{5}$$

For a typical high-precision application, a 10-bit input may be considered, as well as 12-bit representations of the coefficients K_P, K_I, K_d and a 26-bit output.

3   Clock Skew

Clock skew occurs when the clock signal has different values at different nodes within the chip at the same time. It is caused by differences in the length of clock paths, as well as by active elements that are present in these paths, such as buffers. Clock skew lowers system throughput compared to that obtainable from individual blocks of the system, since it is necessary to guarantee the proper function of the chip with a reduced speed clock. Skew will cause clock distribution problems if the following inequality holds [8]:

$$\frac{D}{v} > \frac{k}{f_{app}} \tag{6}$$

where $k<0.20$ (typical value) is a constant, D is the typical size of the system, v is the propagation speed for the clock signal and f_{app} is the applied clock frequency. Existing solutions for clock skew provide two different approaches to the problem:

1. Equalize the length of clock paths to processing elements using buffer and delay elements or through H-tree, mesh or X-tree topologies [6, 9-12].
2. Eliminate or minimize the skew caused by variations during chip fabrication [13-14].

Typically, synchronous systems consist of a chain of registers separated by combinational logic that performs data processing. The maximum clock frequency is derived from:

$$\frac{1}{f_{max}} = T_{min} \geq T_{PD} + T_{skew} \tag{7}$$

where T_{PD} is the time between the arrival of the clock signal at the i-th register and stable processed data at the output of the $(i+1)$-th register. T_{skew} is the time between the arrival of the clock signal at the i-th register and the arrival of the same signal at the $(i+1)$-th register.

Clock skew can be considered as either positive or negative, although the sign criteria is not standardized. Hatamian [11] considers the skew to be positive when the clock signal arrives at the i-th register before that to the $(i+1)$-th register, as illustrated by Fig. 1. If positive, then from equation (7), the minimum system clock period is increased, while if negative, T_{min} decreases. An excessive positive skew results in a decrease in system performance, but if the skew is negative race-related problems may arise if data processing time is lower than the skew.

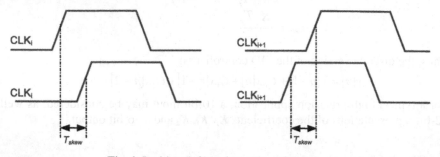

Fig. 1. Positive (left) and negative (right) skew

4 Efficient Synchronization Scheme

The proposed synchronization scheme for RNS-based systems, introduces the generation of several signal clocks from the master clock. These clocks are slightly out-of-phase, thus with non-overlapping edges. Each one of these clock signals synchronizes one of the RNS channels, while global data synchronization (mainly at the global inputs and outputs of the system) is carried out by the global master clock. Thus, each channel computes at different time instants and, consequently, the current demand is distributed over the whole clock cycle. This has the effect of reducing current spikes on the power supply lines approximately by a factor that is the number of generated clock signals. The phase difference between the generated clocks has to satisfy some specifications. First of all, the number of clock signals with overlapping active cycles has to be minimized, as well as the time two or more active cycles overlap. Moreover, clock edges must not coincide. Finally, data coherency has to be respected at both the input and output of the system. Clear advantages are obtained when these requisites are satisfied, since current spikes are reduced and power dissipation is distributed over the master clock cycle, rather than concentrated around the master clock edges. Moreover, not only absolute current values are reduced, but also its temporal variation. Also, as a side effect, power supply lines may be scaled and clock distribution resources reduced, thus simplifying the chip design task.

At first sight, the synchronization scheme described above may seem to be impractical because of the presence of several clock signals within the chip, with associated synchronization problems. However, the nature of RNS [2,15], with non–communicating channels, perfectly suits this clocking scheme. Thus, a generated clock signal is applied to each independent channel, while the master clock signal used to generate these other clocks can also synchronize the global input and output. Moreover, it will be shown that the resources required for implementing this new strategy are minimum and a few transistors, basically three inverters, are required for each channel. More specifically, the master clock signal is routed through an inverter chain, thus being delayed at every point of the chain. Meanwhile, the generated clock signals are extracted at adequate points of this chain and conditioned to be used as clock signal for a complete RNS channel. This scheme requires the inverter chain to alternate large and small input capacitances and low driving capabilities, so appropriate delays can be generated. This has the effect of generating low-quality clock signals within the inverter chain, so additional buffers are required in order to obtain adequate clock signals. Fig. 2 illustrates the hardware required to generate the proposed synchronization scheme, where dCLK stands for generated delayed clocks, while Fig. 3 shows the detailed scheme for the so-called dCLK_cell cell, which consists of three inverters.It can be deduced from Fig. 3 that three design parameters, L_d, W_b and L_b, are available in order to obtain the system specifications, while L_{min} represents the feature size of the fabrication process and W_{min} the minimum usual width for pMOS transistors. Connecting CHout pads to CHin pads, the inverter chain described above is built, while the master global clock is used as input to this chain. Fig. 3 illustrates how large capacitance inverters are alternated with minimum-size devices. Thus, the low driving capabilities of the latter allow modeling of the required delay using the L_d parameter. Meanwhile, the generated clock signal dCLK is regenerated by a third inverter that includes the parameters W_b and L_b. These allow matching of the timing specifications for a proper clock signal for a given system, also allowing the adaptation of the cell to the overall capacitance to be driven by the generated clock. However, these three design parameters L_d, W_b and L_b are not fully independent, and their relation needs a careful study of the final system to be synchronized in order to select their optimum values. Fig. 4 shows the resulting generated clocks in a simple design example for a 300 MHz master global clock, with 0.1 pF loads for every dCLK signal. It can be noted that the requirements enumerated above about non-overlapping edges and active cycles are matched, with every dCLK signal being to be used as clock for a given RNS channel.

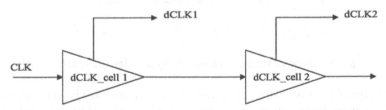

Fig. 2. dCLK_cell chain for out-of-phase clock generation

5 Design Example

A real RNS-based processing application [2] was considered for the evaluation of the proposed synchronization technique. Specifically, a fast PID Controller with 26-bit dynamic range was designed at the transistor level simulated using PSpice for both a single global clock and the proposed technique. For these simulations, a public domain MOSIS CMOS 0.6 μm process [16] was used. This is a three-metal, one-poly, 3.3V CMOS process that is available to MOSIS costumers through Agilent Technologies. RNS [15] have been shown to be a useful alternative for binary implementations for a variety of digital processing applications. Concretely, high performance PID controllers can be designed taken advantage of RNS properties [3].

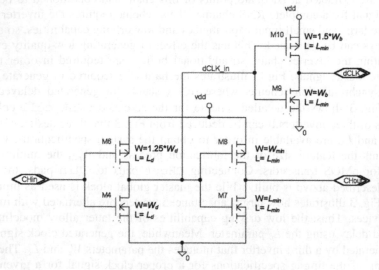

Fig. 3. dCLK_cell schematic

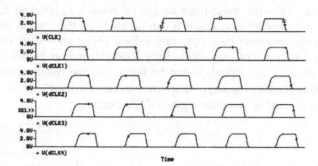

Fig. 4. Resulting generated clock signals for a design example (125 MHz)

For a typical high-precision application, a 12-bit input may be considered, as well as 16-bit representations of the coefficients C_0, C_1, C_2 (5). Thus, it is possible to obtain a 10-bit output without round errors. This RNS-enabled system requires four channels with moduli {256, 63, 61, 59}. Each one of this channels includes LUTs for fixed coefficient multiplications, adders, and a compensated modulo accumulator for synchronization of datasets. Fig. 5 shows the structure of this output acumulator. Since the system is composed just of adders, tables and registers, the well-known two-stage modulo adder [3] was used, while registers were implemented using negative edge triggered D flip-flops (nETDFF) based on TSPC logic [1]. TSPC was selected because it only requires a single-phase clock, thus minimizing synchronization resources and simplifying the implementation of the proposed alternative. Because of the great connection locality for the example system, load driving is kept to a minimum and device sizes can be fixed to the process minimum for most of the transistor involved. Only transistors involved in clock management will have larger sizes since they have to drive large loads. The systems under simulation include around 50.000 transistors.

In order to get illustrative comparison results, the RNS-enabled PID controller was simulated under two different clocking strategies: first of all, a single global clock used to synchronize the whole circuit, using a train pulse voltage source; and second, the proposed strategy.

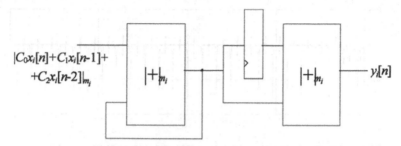

Fig. 5. Corrected modulo m_i accumulator for PID applications

Since the RNS-enabled PID controller requires four channels, as mentioned above, the proposed design example was synchronized using four dCLK_cell cells and four generated dCLK signals, each one synchronizing an RNS channel. The design parameters for the dCLK_cell cells, after careful selection, were fixed at L_d=1 µm, W_d=2 µm and W_b=9 µm. These two alternatives have been simulated for two different clock frequencies, 125 MHz and 300 MHz. A comparison between the proposed strategy and the buffered clock simulation will illustrate the affordable power penalty introduced by this new clocking strategy.

Fig. 6 shows the current on power supply lines for the PID full system working with a 125 MHz clock for both a single clock and the proposed synchronization scheme. Fig. 7 shows the corresponding currents when a 300 MHz clock is applied. Clearly evident is the considerable decrease in the magnitude of the current spikes. In this way, current supply to the chip is distributed over time when the new strategy is considered, while for a global clock current spikes are around four times larger. This indicates that the expected benefits derived from the proposed synchronization scheme are confirmed through simulation. Table 1 summarizes the results obtained for the different simulations and both clock frequencies. We note that the maximum current spike is clearly reduced when this new clocking strategy is considered, as well as the maximum value of the current change rate (di/dt). Finally, if power dissipation is considered, the comparison between the single clock and the proposed strategy shows that the latter introduce an affordable increase in power.

6 Conclusions

This paper has presented a new alternative for synchronizing RNS-based systems and reducing current demand. The proposed strategy was tested using an RNS-Enabled PID controller consisting of around 50.000 transistors. Simulation results demonstrate the effectiveness of this new clocking strategy in reducing the maximum current spike as well as reducing the maximum time derivative of the current spike. Concretely, the maximum current spike is reduced to 14-23% of the single clock strategy one at 125-300Mhz, and the time derivate to 3-1%. On the other hand, the power penalty introduced by the new scheme is clearly affordable. Thus, the use of this synchronization scheme may lead to reduced skew-related problems as well as to reducing chip area through the reduction of the size of power supply lines, caused by the reduction in current and current change rate requirements.

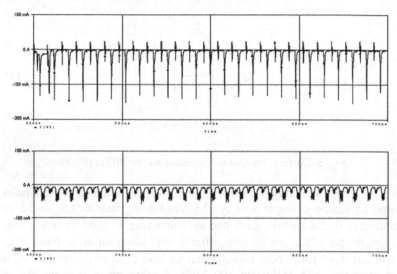

Fig. 6. Current from power supply line for a single clock (above) and the proposed alternative (below) for a 125 MHz frequency

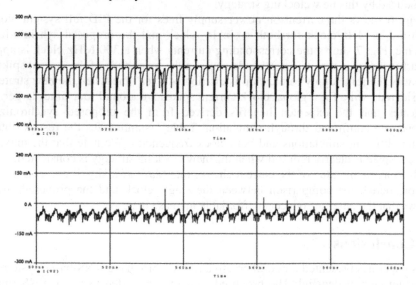

Fig. 7. Current from power supply line for a single clock (above) and the proposed alternative (below) for a 300 MHz frequency

Table 1. Simulation results for RNS-Enabled PID controller using different synchronization approaches

	Single clock		Proposed strategy	
	125 MHz	300 MHz	125 MHz	300 MHz
Max Spike	275.06 mA	348.4 mA	38.84 mA	82.143 mA
Max di/dt	45.11 A/ns	261.75 A/ns	1.67 A/ns	2.47 A/ns
Power	55.37 mW	119.67 mW	66.95 mW	146.11mW

Acknowledgements. Authors wish to acknowledge financial support from Investigation General Direction (Spain) under project TIC2002-02227. The CAD tools were provided by Mentor Graphics Inc, trough their university program.

References

1. Aström, K.J., Hägglund, T.: PID Control. Theory, Design and Tunning, 2ended., Instrument Society of America. Research Triangle Par., NC (1995).
2. Szabo, N.S., Tanaka, R.I.: Residue Arithmetic and Its Applications to Computer Technology, McGraw-Hill, NY (1967).
3. Parrilla, L., García, A., Lloris, A.: Implementation of High Performance PID Controllers using RNS and Field-Programmable Devices, Proc. of 2000 IFAC Workshop on Digital Control PID'00 (Terrassa, Apr. 5-7 2000), (2000) 628-631.
4. Yuan, J., Svensson, C.: High-speed CMOS Circuit Technique, IEEE Journal of Solid State Circuits, Vol 24, No. 1, (1989) 62-70.
5. Bailey D.W., Benchsneider, B.J.: Clocking Design and Analysis for a 600-MHz Alpha Microprocessor, IEEE Journal of Solid State Circuits, Vol 33, (1998) 1627-1633.
6. Ramanathan, P., Dupont A.J., Shin, K.G.:Clock Distribution in General VLSI Circuits. IEEE Transactions on Circuits and Systems I: Fundamental Theory and Applications, Vol 41, No. 5, (1994) 395-404.
7. Yoo, J., Gopalakrishnan G., Smith, K.F.: Timing Constraints for High-speed Counterflow-clocked Pipelining, IEEE Transactions on VLSI Systems, Vol. 7, No. 2, (1999) 167-173.
8. Grover, W.D.: A New Method for Clock Distribution. IEEE Transactions on Circuits and Systems I: Fundamental Theory and Applications, Vol 41, No. 2, (1994) 149-160.
9. Jackson, M.A.B., Srinivasan, A., Kuh, E.S.: Clock Routing for High Performance IC's. 27th ACM/IEEE Design Automation Conference (1990).
10. Wann, D.F., Franklin, N.A.: Asynchronous and Clocked Control Structures for VLSI Based Interconnect Networks. IEEE Transactions on Computers, Vol 32, No.5, (1983) 284-293.
11. Hatamian, M.: Chapter 6, Understanding clock skew in synchronous systems. In Concurrent Computations (Algorithms, Architecture, and Technology). S.K. Tewksbury, B.W. Dickinson, and S.C. Schwartz (Eds.), Plenum Publishing, New York, (1988) 87-96.
12. Friedman, E.G.: Clock Distribution Networks in Synchronous Digital Integrated Circuits. Proceedings of the IEEE, Vol. 89, No. 5, (2001).
13. Friedman, E.G., Powell, S.: Design and Analysis of an Hierarchical Clock Distribution System for Synchronous Cell/macrocell VLSI. IEEE Journal of Solid State Circuits, Vol. 21, No. 2, (1986) 240-246
14. Shoji, M.: Elimination of Process-dependent Clock skew in CMOS VLSI. IEEE Journal of Solid State Circuits, vol. 21, (1986) 869-880.
15. Soderstrand, M.A., Jenkins, W.K., Jullien G.A., Taylor, F.J.: Residue Number System Arithmetic: Modern Applications in Digital Signal Processing. IEEE Press (1986).
16. MOSIS Process Information: Hewlett Packard AMOS14TB, http://www.mosis.org/technical/processes/proc-hp-amos14tb.html

Power Dissipation Impact of the Technology Mapping Synthesis on Look-Up Table Architectures

Francisco-Javier Veredas[1,2] and Jordi Carrabina[3]

[1] Infineon Technologies AG, Munich, D-81699, Germany
[2] Microelectronics Department, University of Ulm, D-89081, Germany
[3] Microelectronics Department, University Autonoma of Barcelona, E-08193, Spain

Abstract. This paper presents a study of the power dissipation repercussion on the logic and physical synthesis using LUT architectures. It is observed that the same function with different mappings on a LUT show different power dissipation. In concrete, the study reveals that the difference depends on the number of inputs of the Boolean function mapped. A power model based on this concept is developed. The power model is used to analyze the efficiency of the synthesis concerning power dissipation in LUTs. Also, a study of the fan-out and the function mapped is done. A power cost model is created to associate the fan-out and our power model. A set of circuits have been synthesized with an academic FPGA synthesis tool. The synthesis tool is used with the options of optimal delay, optimal area and delay-area trade-off. Our study shows that, in this case, synthesizing for area or delay does not affect the power dissipation. Three different LUT architectures have been studied. Results show that a four input LUT is a good choice concerning power dissipation.

1 Introduction

A k-input Look-up table (LUT) can map any k-input Boolean function. LUTs are extensively used in architectures such as Field-Programmable Gate Arrays (FPGAs) [1] or Mask-Programmable Gate-Arrays (MPGA) [2] for both flexibility and unitary cost in low and mid volume productions, even with low power requirements. Therefore, power dissipation becomes a critical factor for LUT-based architectures. LUT-based architectures are power hungry in comparison with ASICs. Power dissipation in FPGA has been extensively studied. In [3,4,5,6] several techniques are proposed at CAD level. Power aware LUT-based architectures can be found in [7,8,9]. Different contributions [10,11,12] describe circuit level techniques.

In this paper we study the relation between power dissipation and the Boolean mapping. We know that in configurable logic not all the k inputs of a LUT are used after synthesis. Power dissipation in a k-input LUT depends of the input signal used. Therefore, it is possible to create a model with the synthesis results of both maximum power dissipation and the lower power dissipation.

In Section 2 the power model is developed. Section 3 describes the evaluation framework used. Section 4 shows the experimental results. Finally, section 5 concludes the paper.

2 Power Dissipation in Look-Up Tables

Power dissipation in CMOS technology has two main contributions: static and dynamic [13]. Static power dissipation is mainly due to leakage current, therefore being

V. Paliouras, J. Vounckx, and D. Verkest (Eds.): PATMOS 2005, LNCS 3728, pp. 666–673, 2005.
© Springer-Verlag Berlin Heidelberg 2005

a common contribution difficult to reduce using design techniques other than managing power supply activation. The study of this paper is focused on dynamic power dissipation.

Dynamic power dissipation is the major contribution of the total power dissipation in CMOS circuits. It is well know that the dynamic power dissipation in a CMOS technology is,

$$P_{dyn} = \alpha \cdot C_{load} \cdot V_{DD}^2 \cdot f \tag{1}$$

where α is the switching activity of the circuit, C_{load} is the load capacity, V_{DD} is the supply voltage and f is the circuit operating frequency. The switching activity is the number of toggles of a node in a clock period. We assume that a clock has an activity of 1.

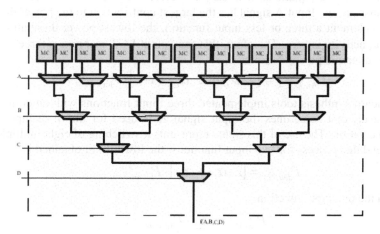

Fig. 1. Four- input LUT architecture

There are three sources of power dissipation in a LUT from an architectural point of view: configuration circuitry, memory cells and multiplexer power dissipation.

The configuration circuitry has dynamic power dissipation when the configuration is loaded but not dynamic power dissipation after configuration since in our study we do not consider dynamic reconfiguration.

Moreover, memory cells have low switching activity, since they are clocked devices. Therefore, the dynamic power dissipation of the memory cells is relatively small. The major contribution in the configuration memory is static power dissipation.

2.1 Input Permutation Considerations

For simplicity reasons, we consider that activity value is equal for all input signals of the LUT. Applying this assumption to all the multiplexers of a LUT in the equation (1) let to,

$$P_{dyn,LUT} = N \cdot \alpha_{inLUT} \cdot C_{load} \cdot V_{DD}^2 \cdot f_{mux} \tag{2}$$

where N is the total number of multiplexors, i.e. 15 in the four input LUT.

In this section a four-input LUT is used as example for the dynamic power dissipation analysis. Its extension for LUTs with more that four inputs can be performed directly. As shown in Fig. 1, each input of the LUT has different loading capacitance, i.e. each input signal drives a different number of multiplexers. Therefore, this lets to a different load capacitance to be driven by each input signal, so that the total dynamic power dissipation for internal multiplexers in a four-input LUT is,

$$P_{dyn,LUT4} = [8 \cdot \alpha_A + 4 \cdot \alpha_B + 2 \cdot \alpha_C + \alpha_D] \cdot P_{int,mux} \tag{3}$$

where α_A, α_B, α_C and α_D are the activities of the A, B, C and D input signals respectively. $P_{int,mux}$ is,

$$P_{int,mux} = C_{in,mux} \cdot V_{DD}^2 \cdot f_{mux} \tag{4}$$

Looking at the equations, it is easy to derive the upper and lower limits of the power dissipation. Input A signal has the largest contribution to the power dissipation. If we implement a three or less input function, the lowest power dissipation appears when α_A activity is zero. Therefore, this four-input LUT mapping a three input function has dynamic power dissipation,

$$P_{dyn,LUT4} = [4 \cdot \alpha_B + 2 \cdot \alpha_C + \alpha_D] \cdot P_{int,mux} \tag{5}$$

Standard synthesis tools implemented three input functions with random mapping. This means that sometimes the input signal A is used for the three input function implementation. The use of this signal represents a switching of eight multiplexors.

Extend this process for two input functions, the lower limit obtained is,

$$P_{dyn,2in} = [2 \cdot \alpha_C + \alpha_D] \cdot P_{int,mux} \tag{6}$$

And the one input function,

$$P_{dyn,1in} = \alpha_D \cdot P_{int,mux} \tag{7}$$

Power factor for the extreme best/worst cases are shown in the Table 1. Results shows the cases of a three input LUT, four-input LUT and five-input LUT.

Table 1. Power dissipation differences (worst/best) with the input function implemented

Implemented Function	3-input LUT	4-input LUT	5-input LUT
five-input	-	-	x1
four-input	-	x1	x2
three-input	x1	x2	x4
two-input	x2	x4	x8
one-input	x4	x8	x16

3   Evaluation Framework

The MCNC benchmark set [17] is used to get experimental resultsas it is extensively used in FPGA research. We selected 10 of the largest circuits. Circuits have been synthesized within the Berkeley SIS framework [20]. The script used in SIS is the *sript.rugged*. The *dmig* package is used to create a network of two-bonded functions.

The FPGA technology mapping tool used is RASP [21]. RASP is an academic FPGA synthesis tool. As traditional FPGA synthesis tools, the tool is quite technology independent. RASP integrates different algorithms for delay or area optimization, but not for power. In this paper three different algorithms are used. CutMap [22]   does technology mapping with simultaneous area delay minimization. FlowMap [23] is for depth optimal technology mapping. FlowMap-r [24] does an area-delay tradeoff. We use two different options for FlowMap-r. These options are selected authomatically with the provided RASP script.

Perl-scripts have been developed for the analysis of the results. The main focus of these scripts is to show statistics (like number of LUTs, fan-out,...).

4   Experimental Results

4.1   Technology Mapping Results: Function Distribution

Technology mapping results are shown in Fig. 2 for the ten circuits. The architecture used for the synthesis is a four-input LUT. The algorithm used is the *FlowMap-r*. Synthesis results show the LUT utilization classified according to by the function fan-in. The use of one input functions is practically non-existent, due to the fact that in the FPGA synthesis all the inverters are packet into the LUT, except when there are multiple fan-out between the positive and the negative functions of for buffering high fan-out signals. We obtained that around 65 % of the functions mapped are four input functions.

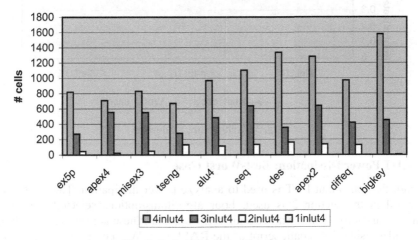

Fig. 2. Technology Mapping Results

4.2   Architecture Exploration

Three different architectures are used with the *FlowMap-r* synthesis algorithm. The average power dissipation is shown in Fig. 3. It is important to note that there are not major differences also in the best-case scenario for the three-input LUT and the four-input LUT. A four-input LUT is 7% better than a three-input LUT. This conducts to the conclusion that the synthesis tool is a bit more efficient for four-input LUT than for three-input LUT.

4.3   Fan-Out and Cost Power Model

A Perl-script has been written to count the fan-out of each netlist. The case used is a four-input LUT with *FlowMap-r* synthesis. Results (Fig. 4) show that the fan-out increases when raising the number of inputs,.

So, a simple model is developed to study the fan-out with the previous results. First we take the interconnect- fanout relation of [25],

$$W \_ seg = 3.68 \cdot Fan \_ out + 3.5 \qquad (8)$$

The interconnect segmentation can be considered proportional to the load capacitance of the equation (1). Then, the power cost function on accumulating current fan-out contribution multiplied by the interconnect load factor to the previous power results. The graphic obtained is displayed in Fig 5. We see that power cost function grows exponential with the number of inputs.

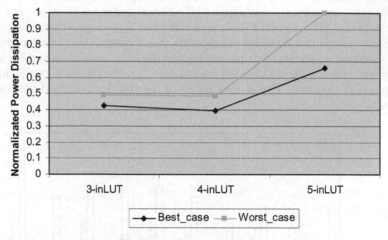

Fig. 3. Architecture Evaluation

4.4   LUT Power Dissipation: Best-Worst Case

As said, the four-input LUT is used to analyze power dissipation. The power model described in the section 2 is used. Four algorithms/options selected are *CutMap*, *FlowMap* and *FlowMap-r* with two different options. These algorithms are automatically called using the default script of the RASP tool. We see in Fig. 6 that there are not many differences between the results obtained from these algorithms. All of them are in the same order for the best/worst power dissipation case.

5   Conclusions

Power dissipation of LUT-based design is studied with special care in technology mapping aspects. First, we observed that different mapped functions in a LUT have different power dissipation. Our study reveals that the difference depends of the number of inputs of the Boolean function mapped and their assignment to the physical inputs to the LUT. A power model with these concepts has been developed. The power model is used to analyze the dependence of the power dissipation in LUT with synthesis tools. The study of the fan-out concerning the function mapped has carried out. A power cost model has been created to associate the fan-out and our power model. A set of circuits has been synthesized with an academic FPGA synthesis tool. The synthesis tool has been used with the options of optimal delay, optimal area and a trade-off delay-area. The study shows that synthesis for area or delay is not affecting power dissipation. Three different LUT architectures have been studied. Results show that for power dissipation a four input LUT is a good choice.

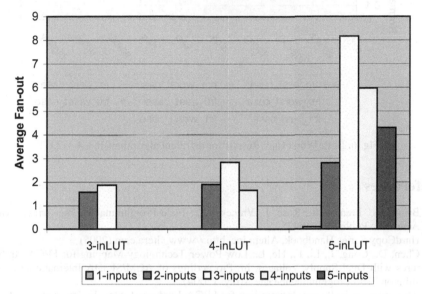

Fig. 4. Fan-out depending of the function implemented and LUT used

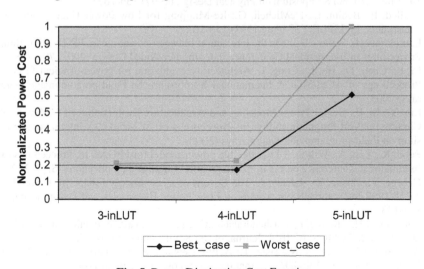

Fig. 5. Power Dissipation Cost Function

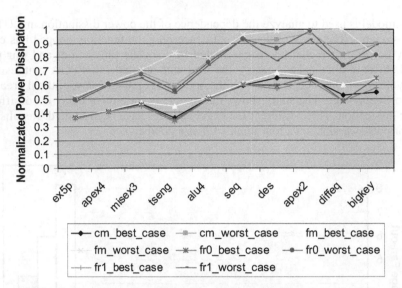

Fig. 6. Best/ Worst Case Results for different algorithms in a 4-in LUT

References

1. Brown, S., Francis, R., Rose, J., Vranesic, Z.: Field-Programmable Gate Arrays. Kluwer Academic Publishers (1992)
2. HardCopy Series Handbook, Altera Inc. http://www.altera.com (2005)
3. Chen, D., Cong, J., Li, F., He, L.: Low-Power Technology Mapping for FPGA Architectures with Dual Supply Voltages. In: Proceedings of ACM/SIGDA International Symposium on Field-programmable Gate Arrays (2004)
4. Alexander, M.J.: Power Optimization for FPGA Look-up Tables. In: Proceedings of ACM 1997 International Symposium on Physical Design (1997) 156-162
5. Vuilled, P., Benini, L., DeMicheli, G.: Re-Mapping for Low Power Under Tight Timing Constraints. In: ISPLED'97 (Aug.1997) 156-162
6. Lamoureux, J., Wilton, S.J.: On the interaction between power-aware FPGA algorithms. In: Proceedings of International Conference in Computer-Aided Design (2003) 701-708
7. Li, F., Chen, D., He, L., Cong, J.: Architecture evaluation for power-efficient FPGAs. In: Proceedings of ACM/SIGDA International Symposium on Field-programmable Gate Arrays (2003)
8. Lin, Y., Li, F., He, L.: Power Modeling and Architecture Evaluation for FPGA with Novel Circuits for Vdd Programmability. In: Proceedings of ACM/SIGDA International Symposium on Field-programmable Gate Arrays (2005)
9. Poon, K., Yan, A., Wilton, S.J.: A flexible power model for FPGAs. In: Proceedings of 12th International Conference on Field-Programmable Logic and Applications (Sept. 2002)
10. George, V., Zhang, Z., Rabaey, J.: The Design of a low Energy FPGA. In: Proceedings of ACM 1999 International Symposium on Low Power Electronics and Design (Aug. 1999) 188-193
11. Calhoun, B.H., Honore, F.A., Chandrakassan, A.P.: A Leakage Reduction Methodology for Distributed MTCMOS. In: IEEE Journal of the Solid-State Circuits (May 2004) 818-826
12. Anderson, J.H., Najm, F.N., Tuan, T.: Active Leakage Power Optimization for FPGA. In: Proceedings of ACM/SIGDA International Symposium on Field-programmable Gate Arrays (2004) 33-41

13. Weste, N.H.E., Eshraghian, K.: Principles of CMOS VLSI Design. A System Perspective (Second ed.) Addison Wesley Longman (1993)
14. Kusse E.A., Rabaey J.: Low-Energy FPGA Structures. In: Proceedings of ACM 1998 International Symposium on Low Power Electronics and Design (Aug. 1998) 155-160
15. Shang, L., Kaviani, A.S., Bathala, K.: Dynamic Power Consumption in Virtex-II FPGA Family. In: Proceedings of ACM/SIGDA International Symposium on Field-programmable Gate Arrays (2002) 157-164
16. Gayasen, A., Tsai, Y., Vijaykrishnan, Kandemir, M., Irwin M.J.: Reducing Leakage Energy in FPGAs Using Region-Constrained Placement. In: Proceedings of ACM/SIGDA International Symposium on Field-programmable Gate Arrays (2004)
17. Yang, S.: Logic Synthesis and Optimization Benchmarks, Version 3.0. Tech. Report. Microelectronics Centre of North Carolina (1991)
18. Betz, V., Rose, J., Marquardt, A.: Architecture and CAD for Deep-Submicron FPGAs. Kluwer Academic Publishers (1999)
19. Veredas-Ramirez, F. J., Scheppler, M. and Pfleiderer, H.-J.: A Survey on Reconfigurable Computing Systems: Taxonomy and Metrics. In: Proceedings of the IV Workshop on Reconfigurable Computing and Applications (Sept. 2004) 25-36
20. Sentovich, E.M.: SIS: A System for Sequential Circuit Analysis. Tech. Report No. UCB/ERL M92/41, Univeristy of California, Berkeley (1992)
21. Cong, J., Peck, J., Ding, Y.: RASP: A General Logic Synthesis System for SRAM-based FPGAs. In: Proceedings of ACM/SIGDA International Symposium on Field-programmable Gate Arrays (1996) 137-143
22. Cong, J., Hwang, Y.: Simultatious Depth and Area Minimization in LUT-Based FPGA Mapping. In: Proceedings of ACM/SIGDA International Symposium on Field-programmable Gate Arrays (1995) 68-74
23. Cong, J., Ding, Y.: FlowMap: An Optimal Technology Mapping Algorithm for Delay Optimization in Lookup-Table based FPGA Designs. In: IEEE Trans. On CAD (Jan.1994) 1-12
24. Cong, J., Ding, Y.: On Area/Depth Trade-off in LUT-Based FPGA Technology Mapping. In: IEEE Trans. On VLSI Systems (June 1994) 137-148
25. Chen, D., Cong, J., Li, F., He, L.: Low-Power Technology Mapping for FPGA Architectures with Dual Supply Voltages. In: Proceedings of ACM/SIGDA International Symposium on Field-programmable Gate Arrays (2004)
26. Betz, V., Rose, J., Marquardt, A.: Architecture and CAD for Deep-Submicron FPGAs. Kluwer Academic Publishers (1999)
27. Anderson, J.H., Najm, F.N.: Power Estimation Techinques for FPGAs. In: IEEE Transactions on VLSI Systems (October 2004) 1015-1027
28. George, V., Rabaey, J.: Low-Energy FPGAs: Architecture and Design. Kluwer (2001)

The Optimal Wire Order for Low Power CMOS

Paul Zuber[1], Peter Gritzmann[2], Michael Ritter[2], and Walter Stechele[1]

[1] Institute for Integrated Systems, TU München, Germany
paul.zuber@tum.de
lis.ei.tum.de/people/pz.html
[2] Institute for Combinatorial Geometry, TU München, Germany

Abstract. If adjacent wires are brought into a simple specific order of their switching activities, the effect of power optimal wire spacing can be increased. In this paper we will present this order along with a prove of this observation. For this purpose, it is shown how to derive the new power optimal wire positions by solving a *geometric program*. Due to their simplicity in implementation, both principles reported substantially differ from previous approaches. We also quantify the power optimization potential for wires based on a representative circuit model, with promising results.

1 Introduction

Today it is widely accepted that one of the new key issues in designing CMOS circuits at 130, 90 and 65nm technologies is power. We have to work under power constraints that stem from heat removal, reliability or battery lifetime limitations. It is not possible to benefit from either the integration complexity or performance features of a new technology node without optimizing for a low power consumption at all levels of the design.

Structure of This Document. This section continues with a short overview of existing work in the field of wire ordering and wire spacing and of our approach. Also to the Introduction belongs an illustration of CMOS power basics and the current situation for on-chip wires to an extent we will need in subsequent sections. The next section is the description of power optimal wire spacing and ordering. We create a rule for power optimal wire ordering. The rule is mathematically proved in Section 3. All of Section 4 deals with experiments to quantify the optimization potential of wire spacing and ordering. Section 5 will conclude the article together with some future remarks.

Related Work and Presented Approach. Wire ordering and spacing both have a long history in Electronic Design Automation. People attempt to space and order bus wires for different objectives like power [ ], crosstalk [ ] [ ], area [ ], or timing [ ]. The latter work contains a more complete list.

Our approach reveals two basic phenomenons which have not been previously published. First, the wire spacing problem is written as a geometric program rather than developing a heuristic or using exhaustive searching. Taking a step

V. Paliouras, J. Vounckx, and D. Verkest (Eds.): PATMOS 2005, LNCS 3728, pp. 674–683, 2005.
© Springer-Verlag Berlin Heidelberg 2005

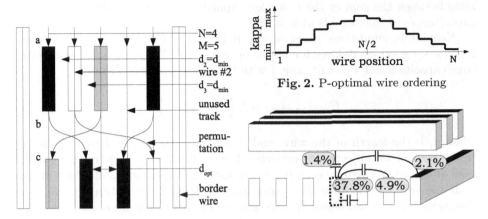

Fig. 2. P-optimal wire ordering

Fig. 1. Wires: the darker the more active   **Fig. 3.** Capacitances in a 0.13μm process

forward, a mathematical formulation for a new globally power optimal wire order is formulated. Figure 1 shows an unoptimized bus (a) with an un-populated routing track. The idea is to place the wires off-grid (c), so that the unused space can be exploited: An individual distance is assigned to each wire pair based on the wire activities. A lower capacitance for highly active wires and vice versa is the result. If the wires are put into the order (b) indicated by Figure 2 beforehand, the power savings increase.

1.1 CMOS Power Basics

The power consumption of a CMOS gate is usually decomposed into a static and a dynamic component. The static one is about to reach orders of magnitudes similar to the dynamic one [ ]. However, this paper solely deals with part of the dynamic component. More precisely, we try to reduce the capacitive power caused by the edge capacitances to be described shortly. For our purpose we will reduce the capacitive power formula to $\kappa \cdot C$ and provide expressions for C and κ in this subsection.

Capacitance. One widely known fact in integrated system design is that the average wire capacitances tend to increase when compared to the average gate capacitances [ ][ ]. The first address for capacitance minimization should therefore be the interconnects.

There has been another important trend in the physics of wires: the demand for ever higher integration densities and yet acceptable sheet resistances required a typical on-chip wire to become physically more thick than wide. This implied that now the edge-to-edge (other terms found in literature: x-coupling, lateral, sidewall) capacitances within one layer dominate [ ], as opposed to the past.

Figure 3 shows a capacitance break down simulated with FastCap [ ]. The test setup was a wire (dotted) on metal2 of a typical 0.13μm process embedded into a fully crowded proximity. Note that the bottom layer is not shown. The ratio between the sum of the two edge capacitances of that wire and all other capacitances of that wire is almost 78%.

Keeping in mind these observations, it is clear that the edge-to-edge capacitances are good candidates for power optimization. We approach the edge-to-edge capacitance of a circuit node i with the plate capacitor formula:

$$C_{\text{edge-to-edge},i} \propto l_i \left(\frac{1}{d_i} + \frac{1}{d_{i+1}} \right), \tag{1}$$

where l_i is the length of the wire, and d_i and d_{i+1} the left and right distances to the next metal objects, respectively, cf. wire 2 in Figure 1. Note that for the sake of brevity we assume each net to consist of only one segment. We will also omit the wire length since this paper only deals with the optimization of the distances between two wires. A more general disquisition can be found in [ ].

Switching Activity et al. The switching activity is a major factor that separates high- from low-power nets. It can represent a probability based on assumptions. Or, it is derived by a simulation on gate level and is related to the simulation time interval. It then represents the actual number of toggles of a node. Since there are more factors which influence the capacitive power, we want to define a variable κ that captures all these factors. We call it κ because all factors are invariant after synthesis.

Definition 1. *Let* $\alpha_{01,i}$, f_i, *and* $V_{DD,i}$ *of a node* i *be the switching activity or toggle rate, the frequency of the associated clock domain, and the supply voltage of the driving gate, respectively. We define a power weighting factor* κ_i *for a node* i *as*

$$\kappa_i := \alpha_{01,i} f_i V_{DD,i}^2 \tag{2}$$

2   Power Optimal Wire Spacing and Ordering

Let us now consider N parallel wires of width w routed on M tracks as in Figure 1. By combining (1) and (2) into the well-known formula for the capacitive power dissipation, $P_C = C\alpha_{01} f V_{dd}^2$, we get the objective function of a geometric program:

2.1   Wire Spacing

$$P_{\text{edge-to-edge},i} \propto \sum_{n=1}^{N+1} \frac{(\kappa_n + \kappa_{n-1})}{d_n} = \min! \tag{3}$$

$$d_n \geq d_{min} \quad \forall n = 1 \ldots N + 1 \tag{4}$$

$$\sum_{n=1}^{N+1} d_n \leq \beta \tag{5}$$

At this point it should be noted that we model the whole scenario to be enclosed in between two static wires with the numbers 0 and $N + 1$ to avoid edge effects which could influence the results. The analogon on a chip could be power or shield wires.

Between the two border wires, now arbitrary wire distances can occur. The only exception is that no wire pair must get closer than a minimum distance d_{min}, cf. (4). The value of this minimum spacing is technology dependent. In the second constraint (5), $\beta := (M+1)D_{pitch} - N \cdot w$ was introduced to reflect the chip boundaries. We assume $w + d_{min} = d_{pitch}$ which is the case for many processes. Obviously, if there are no more than N tracks availiable between the above mentioned border wires routed on tracks 0 and $M+1$, there will be no freedom to optimize the distances and thus the power objective. Hence, $M > N$ is the prerequisite to do wire spacing and shall therefore be true for the remainder of the document.

2.2 Wire Ordering

Wire ordering is the deliberate assignment between wires and available tracks in the effort to optimize some objective function. For our case, the non-linearity induced by $1/d$ in (3) lets anticipate an influence of the arrangement of (κ_n) on the effect of wire spacing for low power.

Definition 2. *Given is a set $(\kappa_0, \kappa_1, \ldots, \kappa_N, \kappa_{N+1})$ of κ-factors of the N nets to be routed as defined in (2). For a permutation π of the numbers $\{1 \ldots, N\}$ we call $(\kappa_0, \kappa_{\pi(1)}, \ldots, \kappa_{\pi(N)}, \kappa_{N+1})$ a wire ordering. We denote the set of all wire orderings by $\mathbb{K}$. A particular wire ordering is called a* power optimal wire ordering *if the solution for (3) - (5) is minimal over all wire orderings in $\mathbb{K}$.*

An investigation of the problem with different wire orderings revealed the following observation, cf. Figure 2.

Theorem 1. *A wire ordering $(q_n) \in \mathbb{K}$ is a power optimal wire ordering if and only if it is constructed in the following way:*

1. *Start with $q_0^{(0)} := q_{N+1}^{(0)} := 0$, $K^{(0)} := \bigcup_{i=1}^{N}\{\kappa_i\}$.*
2. *For $s = 1, \ldots, N$:*
   (a) Let $q_a^{(s-1)} \leq q_{a+1}^{(s-1)}$ be the two greatest elements of $(q_i^{(s-1)})$ (in theorem 4 we will prove that these have to be adjacent).
   (b) Let $c := \min K^{(s-1)}$ and define $K^{(s)} := K^{(s-1)} \setminus \{c\}$.
   (c) Define $(q_i^{(s)})$ by inserting c between $q_a^{(s-1)}$ and $q_{a+1}^{(s-1)}$, i.e.

$$q_i^{(s)} := \begin{cases} q_i^{(s-1)}, & \text{for } i \leq a \\ c, & \text{for } i = a+1 \\ q_{i-1}^{(s-1)}, & \text{for } i \geq a+2. \end{cases}$$

3 Proof of Theorem 1

3.1 Known Results from Convex Programming

We first state a basic results from convex programming that we shall need in our subsequent considerations. In what is to follow we denote by u_n the n-th unit vector and by e the all-ones vector $(1, \ldots, 1)^T$.

Theorem 2. *Let $P \subset \mathbb{R}^n$ be a closed convex subset of the open convex set C and let $f : C \rightarrow \mathbb{R}$ be a convex differentiable function. Then $x^* \in P$ minimizes f over P if and only if*

$$-\nabla f(x^*) \in N_P(x^*),$$

where $N_P(x^*)$ is the cone of the outer normals in x^*, defined by

$$N_P(x^*) := \{c \in \mathbb{R}^n : \max_{x \in P} c^T x = c^T x^*\}.$$

For the proof of this theorem and some background on convex programming, we refer the reader to [10], theorem 27.4. The next lemma is a simple inequality of the square root function, which will be essential for the proof of our results.

Lemma 1. *Let $a \geq b \geq 0$ and let $c > 0$. Then $\sqrt{a} - \sqrt{b} \geq \sqrt{a+c} - \sqrt{b+c}$ with equality if and only if $a = b$.*

Proof. This follows straightforward from the fact that the function $x \mapsto \sqrt{x}$ is differentiable in $(0, \infty)$ and has strictly decreasing slope. $\square$

3.2   Characterization of the Optimal Distances

For ease of notation let us now denote $\kappa_n + \kappa_{n-1}$ by γ_n. We first consider the problem of characterizing the optimal d-vector of (3)-(5) for a given wire ordering.

Theorem 3. *Let $P := \{d \in \mathbb{R}^{N+1} : d \geq d_{min}e \wedge \sum_{n=1}^{N+1} d_n \leq \beta\}$. Then $d^* \in P$ is optimal for (3)-(5) if and only if there exists $\tilde{\gamma} > 0$ that satisfies both*

$$d_n^* = \max\{\tilde{\gamma}\sqrt{\gamma_n}, d_{min}\}   \forall n = 1, \ldots N+1$$

$$and \quad \sum_{n=1}^{N+1} d_n^* = \beta.$$

Proof. Let $d^* \in P$ and $\tilde{\gamma}$ as required by the theorem. We have to show that $-\nabla f(d^*) \in N_P(d^*)$. For this purpose let I denote the index set $I := \{n : d_n^* = d_{min}\}$. We need to find $\lambda_0, \lambda_1, \ldots, \lambda_{N+1} \geq 0$ such that

$$-\nabla f(d^*) = \begin{pmatrix} \frac{\gamma_1}{d_1^{*2}} \\ \vdots \\ \frac{\gamma_{N+1}}{d_{N+1}^{*2}} \end{pmatrix} = \sum_{n \in I} \lambda_n(-u_n) + \lambda_0 e.$$

One can easily calcuate $\lambda_0 = \frac{1}{\tilde{\gamma}^2}$ and $\lambda_n = \frac{1}{\tilde{\gamma}^2} - \frac{\gamma_n}{d_{min}^2}$ for $n \in I$. As $\tilde{\gamma} \leq \frac{d_{min}}{\sqrt{\gamma_n}}$ for $n \in I$, the λ_n are all nonnegative, so $-\nabla f(d^*) \in N_P(d^*)$ and d^* is optimal by theorem 2.

For the converse, suppose $d^* \in P$ is an optimum for (3)-(5). Let $m := \max\{\frac{\gamma_n}{d_n^{*2}} : n = 1, \ldots, N+1\}$ and $M := \{n : \frac{\gamma_n}{d_n^{*2}} = m\}$. According to theorem 2 we have $-\nabla f(d^*) \in N_P(d^*)$, hence there are $\lambda_0, \lambda_1, \ldots, \lambda_{N+1} \geq 0$

such that $-\nabla f(d^*) = \sum_{n=1}^{N+1} \lambda_n(-u_n) + \lambda_0 e$. Of course $\nabla f(d^*) \neq 0$ (because $d_n^* \geq d_{min} \ \forall n$), so λ_0 must attain some value $\geq m$ (note this implies that d^* is on the hyperplane $e^T d = \beta$, hence $\sum_{n=1}^{N+1} d_n^* = \beta$) and $\lambda_n = \lambda_0 - \frac{\gamma_n}{d_n^{*2}}$ for $n = 1, \ldots, N+1$. In case $\lambda_n > 0$ for all $n = 1, \ldots, N+1$, the vector d^* would be determined by the intersection of $N+2$ hyperplanes with normal vectors $-u_1, \ldots, -u_{N+1}$ and e, which is clearly impossible as $M > N$. So at least one λ_n must be 0, and this can only be the case if $\lambda_0 = m$, which means $\lambda_n = 0 \iff n \in M$. So for $n \in M$ we have $d_n^* = \frac{\sqrt{\gamma_n}}{\sqrt{m}} = \tilde{\gamma}\sqrt{\gamma_n}$ with $\tilde{\gamma} = \frac{1}{\sqrt{m}}$, whereas for $n \notin M$ the value of d_n^* is determined by the intersection of the hyperplanes $-u_n^T d = d_{min}$ with $e^T d = \beta$, therefore $d_n^* = d_{min}$ for all $n \notin M$. Of course, for $i \notin M$ and $j \in M$ the inequality $\frac{\gamma_i}{d_{min}^2} < \frac{\gamma_j}{(d_j^*)^2} = \frac{1}{\tilde{\gamma}^2}$ holds, so d^* is of the form stated above. $\qquad\square$

For the following consideration we assume that the optimal wire spacing d^* is of the form $d_n^* = \tilde{\gamma}\sqrt{\kappa_n + \kappa_{n-1}}$. If one or more distances are at their lower bound, things get a bit more technical, but the result is basically the same. So the objective function (3) is reduced to

$$\tilde{\gamma} \sum_{n=1}^{N+1} \sqrt{\kappa_n + \kappa_{n-1}} = \text{min!}$$

3.3   Power Optimal Wire Ordering

Before we prove our main result, we first provide the "building blocks". The basic idea is to make use of the inductive nature of the proposed algorithm. The next theorem provides us with the key ideas for the proof of theorem 1, but first let us formalize the notion of a *unimodal wire ordering*.

Definition 3. *Let* $(q_n)_{n=0,\ldots,N+1} \in \mathbb{K}$ *be a wire ordering of the* (κ_i). *If there exists an index* $t, 1 < t < N+1$, *such that* $q_{n-1} \leq q_n \ \forall \ n \leq t$ *and* $q_n \geq q_{n+1}$ *$\forall \ n \geq t$, *the wire ordering* (q_n) *is called a* unimodal *wire ordering* with mode t.

Theorem 4. *Let* $(q_n)_{n=0,\ldots,N+1} \in \mathbb{K}$ *be a power optimal wire ordering. Then* (q_n) *is unimodal. Furthermore, if we denote by* $q_t \geq q_s \geq q_r$ *the three greatest elements of* (q_n), *then these can (and in case they are uniquely determined must) be chosen such that one of them is adjacent to both of the others; if* $q_t > q_s, q_r$, *then* q_t *is located between* q_r *and* q_s, *i.e. either* $r = t-1 \wedge s = t+1$ *or* $r = t+1 \wedge s = t-1$.

Proof. To avoid some technical details we only prove the case where all elements of (q_n) are pairwise distinct. Similar arguments can be applied for the general case, but some special instances must be taken care of. We first prove that (q_n) has to be unimodal. To see this, let us assume the existence of a wire ordering (q_n) minimizing (3)-(5) that is not unimodal. Then there exists an index $1 < t < n$ such that $q_{t-1} > q_t < q_{t+1}$, and we choose t to be minimal with that property; we may w.l.o.g. assume $q_{t-1} \leq q_{t+1}$. Let (p_n) be the sequence defined by

$$p_n := \begin{cases} q_n, & \text{for } n \neq t-1, t, t+1 \\ q_t, & \text{for } n = t-1 \\ q_{t-1}, & \text{for } n = t \\ q_{t+1}, & \text{for } n = t+1. \end{cases}$$

Then the objective function for (p_n) differs from that of (q_n) by

$$\sqrt{q_t + q_{t-2}} + \sqrt{q_{t-1} + q_t} + \sqrt{q_{t+1} + q_{t-1}}$$
$$-\sqrt{q_{t-1} + q_{t-2}} - \sqrt{q_t + q_{t-1}} - \sqrt{q_{t+1} + q_t}$$
$$= \sqrt{q_{t-1} + q_{t-2} - (q_{t-1} - q_t)} - \sqrt{q_{t+1} + q_{t-1} - (q_{t-1} - q_t)}$$
$$-\left(\sqrt{q_{t-1} + q_{t-2}} - \sqrt{q_{t+1} + q_{t-1}}\right)$$

As $q_{t+1} + q_{t-1} \geq q_{t-1} + q_{t-2} > 0$ and $q_{t-1} - q_t > 0$ we can apply lemma 1 to see that the objective for (p_n) is less than for (q_n), an obvious contradiction.

For the second claim of our theorem assume again that (q_i) is optimal with maximal element q_t and q_s, q_r as defined in the theorem. Again, to avoid some technicalities we assume all elements of (q_n) to be pairwise distinct. Now suppose $q_t > q_r$, q_s is not located between q_s and q_r, then due to unimodality both q_s and q_r have to be on the same side of q_t, we assume w.l.o.g. that $s, r > t$, therefore $q_{t-1} \leq q_r \leq q_s \leq q_t$. Also due to unimodality, $s = t+1$, $r = t+2$ (there can be no smaller element between them, because q_t is the unique mode of the sequence). Now we can reorder the sequence by changing the places of q_t and q_s without destroying unimodality, hence we define (p_n) by

$$p_n := \begin{cases} q_n, & \text{for } n \neq t, t+1 \\ q_{t+1}, & \text{for } n = t \\ q_t, & \text{for } n = t+1. \end{cases}$$

Then

$$\sum_{n=1}^{N+1} \sqrt{q_n + q_{n-1}} \leq \sum_{n=1}^{N+1} \sqrt{p_n + p_{n-1}}$$
$$\Longleftrightarrow \sqrt{q_t + q_{t-1}} + \sqrt{q_{t+1} + q_t} + \sqrt{q_{t+2} + q_{t+1}}$$
$$\leq \sqrt{p_t + p_{t-1}} + \sqrt{p_{t+1} + p_t} + \sqrt{p_{t+2} + p_{t+1}}$$
$$\Longleftrightarrow \sqrt{q_{t+2} + q_t - (q_t - q_{t+1})} - \sqrt{q_t + q_{t-1} - (q_t - q_{t+1})}$$
$$\leq \sqrt{q_{t+2} + q_t} - \sqrt{q_t + q_{t-1}},$$

and we use the same argument as above to obtain a contradiction. Consequently, q_s and q_r both have to be adjacent to the maximal element q_t.   □

From the two statements of theorem 4 we may now deduce our central conclusions. We will prove the induction step separately to make things more concise.

Theorem 5. *A wire ordering* (q_n) *for a problem of size* $N+1$ *is optimal if and only if the sequence* (q'_n) *defined by removing a maximal element from* (q_n) *is an optimal wire ordering for the reduced problem of size* N.

Proof. First, let (q_n) minimize the sum $v := \sum_{n=1}^{N+1} \sqrt{q_n + q_{n-1}}$ and let c be the maximal element of the wire ordering, a and b the two elements next in size which are both adjacent to c by theorem 4. Then by removing c we define a wire ordering (q'_n) with objective value $v' = v - \sqrt{a+c} - \sqrt{c+b} + \sqrt{a+b}$. Now suppose there is a wire ordering (p'_n) with objective value $w' < v'$. The elements a and b are the two greatest elements of (p'_n), therefore they have to be adjacent and we can define a sequence (p_i) by inserting c between a and b. The objective value of (p_n) is $w = w' - \sqrt{a+b} + \sqrt{a+c} + \sqrt{c+b}$, so $w < v$, contradicting the optimality of (q_n).

To see the other direction, let (q_n) be some wire ordering of length $N+1$ with objective value v, greatest element c and adjacent elements a and b, such that (q'_n) defined by removing c from (q_n) minimizes $v' = \sum_{n=1}^{N} \sqrt{q'_n + q'_{n-1}}$. Suppose (q_n) is not the optimal wire ordering for length $N+1$, then there exists a sequence (p_n) with objective value $w < v$ and we can define (p'_n) with objective value w' by removing c from (p_i). Again, we know $w = w' + \sqrt{a+c} + \sqrt{c+b} - \sqrt{a+b}$ and $v = v' + \sqrt{a+c} + \sqrt{c+b} - \sqrt{a+b}$, so $w' < v'$, contradicting the optimality of (q'_n). $\quad\square$

It is now easy to see that the construction provided in theorem 1 simply formalizes the induction step given in theorem 5. We are finally ready to prove theorem 1.

Proof (of theorem 1). We proceed by induction. The construction given in the theorem mimics exactly the statement of theorem 5, so the induction step is clear. For the induction basis, let us examine the case of $n = 3$. Here we have real numbers $0 < d \le e \le f$ and the ordering arising from the construction is either $(0, d, f, e, 0)$ or $(0, e, f, d, 0)$, depending on whether we insert e on the right or on the left of d. From theorem 4 we know that the optimal solution has to be unimodal with mode f and d and e have to be adjacent to f, so apart from the constructed sequences there are no possible solutions. Furthermore, the objective values of the two possible solutions are equal, so both are optimal. $\quad\square$

4   Experiments

Without proof we propose that there also exists a permutation for which the power is worse than for any other permutation. This will shed light on the benefits expected from power optimal wire ordering. In other words, one who does not consider the actual wire order could abandon power savings anywhere between 0 and the maximum optimization potential. Note that the effect of wire spacing [ ] alone is not subject of this document.

In our experiments we are considering (κ_n) to be similar to an industrial μprocessor [ ]. For several values of N, we randomly selected a set of N parallel wires. Each of these sets were permuted three times: for power optimal, for

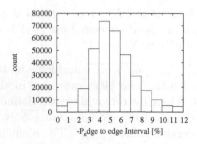

Fig. 4. Histogram for N=64, M=65

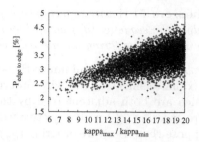

Fig. 5. $\kappa_{max} : \kappa_{min}$ vs. $-P$ scatter plot

power worst, and a for a typical ordering found for buses. The latter distribution is simply ordered by κ. Buses are typically arranged that way, assuming a descending toggle rate from the least to the most significant bit. A geometric program solver [ ] is used to find the optimal power optimal values for various Ms. An outer loop repeated the test some 300,000 times. We related the resulting average worst-case power and bus power values to the best-case power for each N, M combination.

Tables 1 and 2 display the optimization potentials for the worst possible scenario and the bus scenario, respectively. For example, The edge-to-edge power for $N = 64$ wires routed on $M = 65$ tracks could be up to 3.4% worse if wire ordering was not cared for. It is interesting to note that this number is almost the same for any value of N, if $M = N + 1$. We further remark that most of the optimization potential can be exploited by adding only limited extra space.

Figure 4 shows the experimental results for $N = 64$ and $M = 65$ as histogram in 1% intervals. For the same parameters we scatter plotted the optimization potential as a function of the fraction of the highest and lowest κ appearing in the design, cf. Figure 5. This is an interesting source of information for the system designer. One can make out an upper optimization limit depending on only two circuit properties. The CPU-time to optimize an $N = 256$ (512, 768, 1024) case is 1.2s (15.1s, 41.2s, 113.9s) on a 3GHz PC.

5   Conclusion

Future Remarks. The model does not respect the effect of fringe capacitances and capacitances to other wires on the same layer. However, the applicability of the simpler $1/d$ model for C is shown in [ ]. If a more detailed model is desired, capacitance extractions can be done to find fitting functions for C. Presuming these fitting functions remain posynomial, a globally optimal solution

Table 1. Max. Reduction potential [%]

$_N\backslash^M$	N+1	1.25N	1.5N	1.75N
8	3.5	5.8	8.3	9.6
16	3.6	9.3	12.4	13.7
64	3.4	16.5	19.2	19.9
256	3.0	20.3	22.5	22.7

Table 2. Red. potential [%] for buses

$_N\backslash^M$	N+1	1.25N	1.5N	1.75N
8	2.6	4.3	6.2	7.2
16	2.7	7.4	9.9	10.9
64	2.6	14.4	17.0	17.6
256	2.3	19.0	21.1	21.3

exists for the newly created problem [ ]. Miller-capacitances and hence crosstalk power and signal integrity issues are not considered in this paper. Furthermore, the effect on timing of the proposed methodology has not been in the focus of this contribution. However, with little modifications in the objective function, targeting the timing problem with the same notion becomes possible.

Summary. In this paper a significant step forward was taken from power optimal wire spacing through *geometric optimization* alone. A proof was given for the presence of a power optimal order of wires that increases the effect of spacing. The order can be very simply arranged given the sorted power weighting factors of the involved wires. After ordering, *geometric optimization* delivers the globally best possible result without the use of heuristics.

Extensive investigations show the potential of power optimal ordering. On broad buses, the power values for optimally ordered wires and those for an unoptimized order can differ by a two-digit percentage. Interesting results are further the saturating optimization potential for increased space and the dependency of the expected savings only on the highest and lowest value of κ.

References

1. S. Boyd, S. J. Kim, and S. S. Mohan. *Geometric Programming and its Applications to EDA Problems.* Date 05 Tutorial Notes, 2005.
2. J. Cong, C. Koh, and Z. Pan. Interconnect Sizing and Spacing with Consideration of Coupling Capacitance. *IEEE Transactions on Computer-Aided Design of Integrated Circuits and Systems*, 6:1164–1169, 2001.
3. W. Embacher. *Analysis of Automated Power Saving Techniques using Power Compiler (TM).* LIS Diploma Thesis, TU München, Germany, May 2004.
4. P. Groeneveld. Wire ordering for detailed routing. *Design & Test of Computers*, 6:6–17, 1989.
5. R. Ho, K. W. Mai, and M. A. Horowitz. The future of wires. *Proceedings Of The IEEE*, 89(4):490–504, April 2001.
6. Computational Optimization Laboratory. *A geometric programming solver, COPL_GP.* Internet: http://www.stanford.edu/~yyye/Col.html, 2000.
7. E. Macii, M. Poncino, and S. Salerno. Combining wire swapping and spacing for low-power deep-submicron buses. *Proceedings of the 13th ACM Great Lakes symposium on VLSI*, pages 198–202, 2003.
8. K. Moiseev. *Net-Ordering for Optimal Circuit Timing in Nanometer Interconnect Design.* CCIT Report #506, Haifa, Israel, October 2004.
9. K. Nabors. *FastCap.* MIT, 1992, 2005.
10. R. Rockafellar. *Constrained Global Optimization: Algorithms and Applications.* Springer Verlag, 1987.
11. SIA. *International Technology Roadmap for Semiconductors.* Internet: http://public.itrs.net, 2005.
12. A. Windschiegl and W. Stechele. Exploiting metal layer characteristics for low-power routing. *Power and Timing Modeling Workshop PATMOS*, 2002.
13. P. Zuber, F. Müller, and W. Stechele. Optimization Potential of CMOS Power by Wire Spacing. *Lecture Notes in Informatics*, 2005.
14. P. Zuber, A. Windschiegl, and W. Stechele. Reduction of CMOS Power Consumption and Signal Integrity Issues by Routing Optimization. *Design, Automation & Test in Europe DATE*, March 2005.

Effect of Post-oxidation Annealing on the Electrical Properties of Anodic Oxidized Films in Pure Water

Bécharia Nadji

Laboratoire d'Eectrification des Entreprises Industrielles
Faculté des Hydrocarbures et de la chimie, Université M'hamed Bougara Boumerdès
bnadji@umbb.dZ, b_nadji@yahoo.com

Abstract. The work presented here consists of investigating and studying the electronic properties of anodic oxide film (SiO2). This study deals to the determination of interface states density Si/SiO2 and the study of electronic conduction. MOS capacitors with anodic oxides (9nm) were elaborated. The anodic silica films were produced by anodization of monocristalline silicon wafers in pure water in an electrolysis cell (P.T.F.E) at room temperature, with a constant current density of 20 µA/cm2. Film thickness increases linearly as a function of total charge during oxidation Using C(V),G(ω),I(V) measurements, we have determined the interface states density, fixed charges density and conduction mechanism which is of Fowler - Nordheim type for annealed oxides at various temperatures.

Keywords: Anodic oxidation,Pure water, Fixed Charges, interface States density, Si/SiO2,Electrical characterisation, MOS Structures, Fowler-Nordheim tunnelling.

I Introduction

The most significant application of SiO2 layers is undoubtedly their use as a gate dielectric in MOS transistors, an injection dielectric in non-volatile memories (EEPROM) and as dielectric in storage capacitors of DRAM memories. The increasing miniaturization and the higher integration density in CMOS technology and the capacitive structures for memories require the fabrication of ultra thin gate oxides (lower than 10 nm). These oxides must be perfectly homogeneous, be deprived of defects such as pin-holes, with high field breakdown and present very good Si/SiO2 interface in particular, both the interface states density and the fixed charges density. must be low. In addition, the reproducibility and the homogeneity of these properties are essential factors for the reliability and the reproducibility of device performances. Moreover, the thickness of the oxide films must be precisely controlled during the fabrication process. The conventional method of high temperature oxidation hardly meets these requirements. With the progressive shrinking of LSI device size, device and circuit is complexity increasing. The fabrication process is also becoming increasingly difficult. Moreover, there is a great demand to reduce energy consumption on the grounds of preventing global warming and to environmental pollution. An oxidation process in VLSI fabrication requires a high temperature and thermal stress damages the silicon wafer inconspicuously. From this viewpoint, low temperature processes have been researched. Many low temperature processes have been proposed to fabricate insulation films on silicon such as plasma deposition (13), chemical vapour deposition (CVD) , photo-CVD, jet vapour deposition JVD, anodic oxidation. How-

V. Paliouras, J. Vounckx, and D. Verkest (Eds.): PATMOS 2005, LNCS 3728, pp. 684–692, 2005.
© Springer-Verlag Berlin Heidelberg 2005

ever, the pollution which systematically accompanied these oxides formation had made, so far, their use unthinkable in technology. To obtain oxides free from any contamination, the anodic oxidation of silicon in pure ultra water [3, 4] can be used. This is an oxidation process which is carried out at room temperature. On the other hand, the electrochemical oxidation of silicon in pure water permits to obtain very thin homogeneous oxide layers whose thickness can be easily controlled by simple coulometry. Introducing this oxidation step into semiconductor technology would help to reduce the total thermal budget during processing .This is very important regarding dopant diffusion for example. The anodic oxide can be applied for the following purposes:

1. Low-temperature VLSI processes.
2. Bonded and Etched SOI.
3. Oxide film for micro-machining.

Here, we report on the experimental method and the characteristics of anodic oxidized silicon dioxide films grown in pure water at room temperature as a possible oxidation process at a low temperature. For the characterization of the grown films, we systematically used the electrical methods (static, quasi-static, C (V) as well as I (V) measurements).

II Experimental Procedure

MOS structure were fabricated on a p -type <100> oriented silicon substrate , doped with boron in the range $10^{15} - 10^{16}$cm-3. After a conventional chemical cleaning, a thermal oxidation under dry oxygen is carried out in order to obtain layers from 2000 to 3000 Angstroms .Windows are opened by lithography and The anodic oxides are grown in an electrolysis cell (P.T.F.E) with pure water under constant current density of 20μA/cm2. All anodizations are performed at room temperature. A given oxide thickness is obtained by stopping anodization as soon as an charge quantity defined by the calibration curve of figure (1)[3] has flown in the circuit. After post anodization annealing under nitrogen atmosphere at various temperatures (600°C, 800°C, 1050°C), an aluminium film is deposited by evaporation and annealed at 450°C during 30 mn. The obtained C (V), G (ω) characteristics were used to determine interface states densities, oxide fixed charges. The I(V) characteristics are obtained either by applying voltage ramp to the sample, or by applying fixed voltage which is incremented step by step, and used to determine the conduction type.

III Experimental Results and Discussion

A. Fixed Charges
The presence of electric charges in the oxide results in a simple C (V) ideal curve shifted along the voltage axis. The fixed charges quantity Q$_f$ contained in the oxide is measured by comparing the experimental value of flat band voltage with the computed value starting from the relation [7, 9]:

$$V_{FB}^{C} = \left[\phi_m - \chi_{sc} - \frac{E_g}{2q} - KT \, Ln\left(\frac{N_A}{n_i} \right) \right] \tag{1}$$

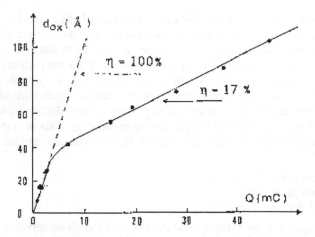

Fig. 1. Calibration curve: oxide thickness formed according to the electricity quantity in the electrochemical cell; current density $=10\mu A/cm^2$, surface $= 1cm^2$

Where $q(\phi_m - \lambda_{sc})$ is the barrier height between the metal and the semiconductor

The experimental value of the flat band voltage $\left(V_{FB}^e\right)$ is deduced from the gate voltage value for which the measured capacitance is equal to the structure theoretical capacitance at flat band voltage.

The density N of fixed charges is obtained using the following expression

$$N = \frac{Q_f}{q} = \frac{\Delta V_{FB} C_{ox}}{q} \tag{2}$$

Where $\Delta V_{FB} = V_{FB}^C - V_{FB}^e$

The experimental C(V) characteristic of figure (2) is practically overlapping on the ideal characteristic.

By taking $q(\phi_m - \lambda_{sc}) = 0.11eV$, the theoretical value of flat band voltage is calculated from relation (1).

We can reasonably state that the oxide charge density in these anodic films is lower than $10^{11}cm-2$, which confirms that there is no appreciable pollution occurring during the anodic oxidation of silicon in pure water. It is noticed besides that the lack of sensitivity on the determination of charges is related to the low oxide thickness.

B. Interface States

a) Quasi-static method: The $C_{LF}\left(V_g\right)$ curve is obtained by subjecting MOS structure to a voltage ramp of small slope. This characteristic is deduced from the displacement current flowing in the circuit. Voltages were ramped from 20 to 50 mV/s The determination of interface states density can be obtained by comparing such a curve either with a high frequency experimental curve (C_{LF} - C_{HF} method) or with a theoretical curve calculated by supposing that the states do not give any contribution to the MOS structure capacitance [1].

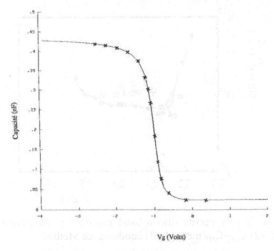

Fig. 2. Experimental capacitance (x) and theoretical curve; temperature annealing of 980°C, surface (400x400) µm², oxide thickness: 9nm

The density of Si/SiO2 interface states is given by:

$$D_{it} = \frac{C_{ox}}{qs} \left[\frac{\frac{C_{LF}}{C_{ox}}}{1 - \frac{C_{LF}}{C_{ox}}} - \frac{\frac{C_{TH}}{Cox}}{1 - \frac{C_{TH}}{C_{ox}}} \right] \tag{3}$$

b) Conductance method: This method requires the conductance measurement of the G_m and the capacitance C_m associated with the structure as a function of ω for various gate bias values Vg. According to the Nicollian model [7], this method allows the determination, at the same time, interface states density Dit in terms of energy and the principal parameters associated with these states (the standard deviation of the surface potential Ψ_s and the effective capture section σ

$$D_{it} = \frac{1}{qf_N} \left(\frac{G_p}{\omega} \right)_{max} \tag{4}$$

By using the Nicollian model, $\frac{G_p}{\omega} = f(\omega)$ curves for various values of Vg (-0.56 V and -0.67 V) have been exploited.

Figure (3) summarizes the interface states density, determined by the C(V) and G(ω) methods for a 600°C anneal. Even if, the conductance method leads to states densities slightly lower than that measured by the quasi-static methods, the values remain of the same order of magnitude;$2 * 10^{11} eV^{-1} cm^{-2}$ for energies close to silicon mid gap .

For the annealing temperature of 800°C (figure (4)), the number of surface states is sufficiently high to allow their determination by the various techniques. On the other hand, for very high temperature annealing 1050°C only the conductance method proved sufficiently sensitive to allow the determination of the states density.

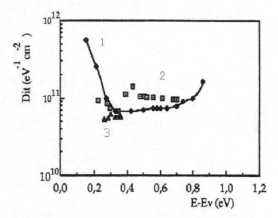

Fig. 3. Interface state density versus silicon band gap energy, temperature 600°C (30 mn). (1) C_{LF}–C_{TH} method; (2) C_{LF}–C_{HF} method, (3) conductance Method

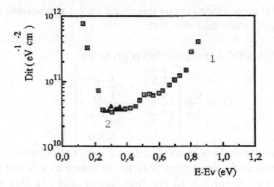

Fig. 4. Interface state density versus silicon band gap energy at a temperature of 800°C (30 mn). (1) C_{LF} – C_{TH} method; (2) conductance method

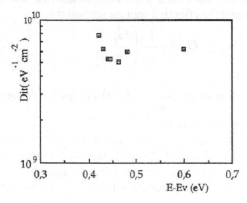

Fig. 5. Interface State density versus silicon band gap energy at a temperature of 1050°C (30 mn) with conductance Method

In spite of the measurements difficulties, it seems well that high annealing temperatures really lead to states densities lower than $10^{10} eV^{-1} cm^{-2}$.

C. Conduction Current in the Oxide

Figure (6) and figure (7) show the conduction characteristics plotted in a Fowler-Nordheim representation for two annealing temperature 1050°C, 600°C. The curves forms obtained are classic. At strong fields: the conduction current is dominated by tunnel effect through a triangular barrier at the oxide [5]; this field domain is repre-

sented by the linear part of the curves $Log\left(\dfrac{J}{E_{ox}^2}\right) = f\left(\dfrac{1}{E_{ox}}\right)$

On the other hand, at the weak fields, the F-N effect is negligible and the current is an indication of imperfections present in the oxide [11].

The presence of a strong field in an oxide results in a thinning of the energy barrier which allows the flow of the electrons by tunnel effect. The expression of the F-N electronic current is given by [10]:

$$A = \frac{q^3 m}{8 \pi h_{ox} \phi_B^{\frac{3}{2}}} \qquad (5)$$

$$\text{with} \quad A = \frac{q^3 m}{8 \pi h m_{ox} \phi_B^{3/2}} \qquad (6)$$

$$B = \frac{4}{3} \frac{2\pi}{qh} \sqrt{2m_{ox}} \, \phi_B^{3/2} \qquad (7)$$

and where Φ_B is the barrier height at the Al/SiO2 interface in eV, m is the electron mass, m_{ox} it's effective mass and E_{ox} is the electric field in the oxide.

The constant B of the line obtained under strong field gives access to the barrier height Φ_B between aluminium and anodic silica.

After calculation of B and by taking $m_{ox} = 0.5$ m [10], One finds thus, the value for Φ_B to be 2.8 eV. This value corresponds to the effective barrier height between aluminium and anodic silica, which depends on the applied electric field E_{ox}, because of the Schottky effect. According to [9], one has indeed:

$$\phi_B = \phi_{B0} - \sqrt{\frac{qE_{ox}}{4\pi\varepsilon o}} \qquad (8)$$

Φ_{B0} is the barrier height under no field, Φ_B is the barrier height in the presence of an electric field Eox.

In order to determine Φ_{B0}, we draw a network of Fowler-Nordheim theoretical curves with Φ_{B0} as a parameter, (Figure(7)). The experimental points shown on this figure indicate a value of 3.1 eV [8,2], which corresponds precisely to the commonly know value for the Al/SiO2 interface. Therefore, these results show that for annealing temperatures higher than 800°C, the conduction under strong field is dominated by the Fowler-Nordheim effect and that the defects giving rise to states are sufficiently very few so that their influence is negligible. The conduction current under weak field is more signify-cant for the low temperatures annealing.

The breakdown occurs at very high electric fields (13-14 MV/cm).

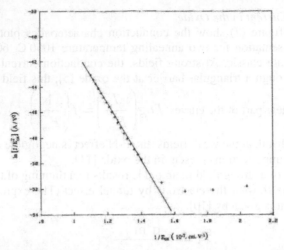

Fig. 6. Fowler-Nordeim Characteristic ,Vg < 0 Annealing temperature:1050°C Surface (500 * 500)μm², thickness 9nm

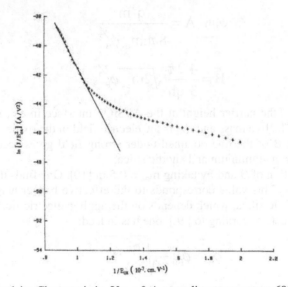

Fig. 7. Fowler-Nordeim Characteristic ,Vg < 0 A annealing temperature 600°C Surface 100 * 100μm², thickness 9nm

IV Conclusion

Our experiments showed, first of all, that the anodic oxidation of silicon carried out in pure water and in the absence of any electrolyte support, led indeed to oxides free from any contamination. The obtained results showed that the surface states densities decreased regularly with the increase in annealing temperature, down to measurable (lower than 10^{10}eV −1cm−2) when the oxide was reheated with 1050°C. The density of states is of the order of $10^{11} eV - 1 cm - 2$ for the annealing temperature of $600°C$. It thus appears that in all cases the density of states is maintained at a very low levels, comparable with those obtained with the best thermal oxides.

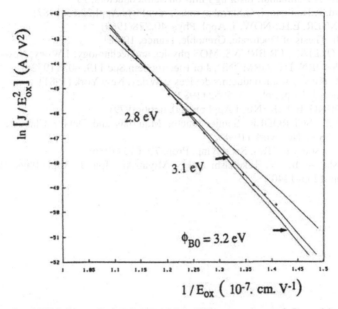

Fig. 8. Network of F-N theoretical calculated for different values of Φ_{B0} with correction of Shotctky effect and experimental curve: anneal $1050°C$; surface $= 500 * 500\mu m2$; thickness 9nm

The study of the conduction currents carried out showed that, anodic oxidation constitutes a method to, particularly grown ultra thin oxides free from short circuit currents. The currents obtained under weak and average field still depend, on the annealing temperature and show that the annealing temperature is an important factor. However these conduction currents values are acceptable. With the strong fields, conduction is dominated by the Fowler-Nordheim tunnel mechanism. The analysis of F-N conduction characteristics enabled us to measure the barrier height of the Al/SiO2 interface. The obtained value of 3.1 eV after correction of the effect Schottky, confirms the absence of any appreciable conduction related to defects in the oxide. The annealing temperature of $800°C$ must be regarded as minimal value giving rise to oxides in all respects comparable with the best thermal oxides. One will note however, that an annealing with $800°C$ constitutes, in micro-electronic technology, a low temperature process, insofar as the redistribution phenomena of doping agent, remains extremely limited at this temperature. If the anodic oxide can be applied to an oxidized substrate to fabricate a Bonded and Etched -SOI wafer, an inexpensive SOI substrate can be obtained. We consider that the most realistic application is the use of the anodic oxide for micromachining. We attempted to apply the anodic oxidation to aid the suppression of global warming and environmental pollution, We hope this study will contribute to positive changes in the use of energy

References

1. C.N.Berglund, IEEE.trans.Elect.Dev, ED-13701
2. J.Capilla, G.sarrabayrouse, Rev.Phys.Appl. 19,343(1984).
3. F.Gaspard, A.Halimaoui, Insuling Films on Semicoductors, J.J.
4. F.Gaspard, A.Halimaoui, G.Sarrabayrouse, rev. Phys.Appl.22.65(1987)
5. M.LEZINGER, E.H.SNOW, J. Appl. Phys. 40,278(1969)
6. B.NADJI. Thesis of Doctorate, Grenoble, France, 1990.
7. E.H.NICOLLIAN, J.R.BREWS, MOS physics and technology, J.Wiley, New-York(1982)
8. C. M, OSBURN, DW.ORMOND, J of Electrochem.Soc 119, 603 (1972).
9. S.M.SZE, phys of semiconductors devices, J.Wiley, New York (1981).
10. Z.a Weinberg, J.Appl.phy.53, 5052 (1982)
11. D.R. YOUNG, E.A.IRENE, J.Appl.phy.50, 6366(1979)
12. DIETER K. SCHRODER, Semiconductor Materials and Device Characterization, John Wiley and Son, New york (1998)
13. M.J. Thompson, Mat. Res. Soc. Symp. Prop. 70, 613 (1986)
14. K. Ohnishi, A. Ito, Y. Takahashi and S. Miyazaki , Jpn. J. Appl. Phys. Part1 ,Vol. 41 (2002) pp. 1235–1240.

Temperature Dependency in UDSM Process

B. Lasbouygues[1], Robin Wilson[1], Nadine Azémard[2], and Philippe Maurine[2]

[1] STMicroelectronics Design Department 850 rue J. Monnet, 38926, Crolles, France
[2] LIRMM, Univ. Montpellier II, (C5506), 161 rue Ada, 34392 Montpellier, France

Abstract. In low power UDSM process the use of reduced supply voltage with high threshold voltages may reverse the temperature dependence of designs. In this paper we propose a model to define the true worst Process, Voltage and Temperature conditions to be used to verify a design. This model will provide an accurate worst case definition for high performance designs where standard design margins are not applicable. This model is validated at either cell level or path level on two different 130nm process.

I Introduction

With the scaling of technologies, the leakage current contribution to the power consumption has dramatically increased. One solution to reduce the static to dynamic power ratio is to use higher threshold voltage (V_T) devices. However due to the opposite temperature sensitivities of the mobility and the threshold voltage, the current delivered by the devices exhibits, for such biasing conditions, a complex behavior with temperature [1, 2] that may completely modify the temperature at which a design reaches its critical corner. This temperature effect, called temperature inversion phenomenon, becomes particularly critical when the threshold voltage value approaches half V_{DD}. Unfortunately this configuration may occur in some variants of low power UDSM processes (130nm and 90nm), in particular for high V_T process options.

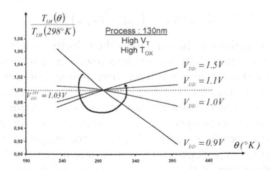

Fig. 1. Temperature evolution of the simulated inverter delay for different V_{DD} values

To validate most digital synchronous designs, it is usually sufficient to verify the design behavior for the worst and the best case timing conditions. In actual low power design, due to the combined use of high threshold voltage devices and reduced supply voltages, the worst case timing conditions becomes less predictable and can occur at different temperatures. In the majority of cases this effect is accounted for through design margins however for high performance designs where more accurate timing

V. Paliouras, J. Vounckx, and D. Verkest (Eds.): PATMOS 2005, LNCS 3728, pp. 693–703, 2005.

validation is required, it may be necessary to perform validations in more than 2 PVT conditions in order to guarantee the correct behavior of the design.

Even if the sensitivity of design performances to V_{DD} and V_T values has already been identified as one of the major limitations in V_{DD} scaling [11], little attention has been given to characterize their temperature sensitivity in the low voltage domain. To illustrate the problem we give in Fig.1 the evolution of the normalized propagation delay of a simple inverter. As shown, for high values of the V_{DD} the critical case appears, at high temperature, while for low V_{DD} values the temperature sensitivity is completely reversed. For this reason, this temperature effect is often called the temperature inversion phenomenon.

The main contribution of this paper is to propose a representation of the timing performance of a CMOS structure [4] that facilitates in finding the true worst case condition during the STA. It is organized as follows. In section II, we characterize the transistor current temperature inversion phenomenon. Section III introduces the analytical model from which the timing performance representation is deduced. In section IV we apply these models to characterize the temperature sensitivity of the process, the cell and a data path of a circuit in two different 130nm technologies to demonstrate the influence of the process conditions to the temperature inversion phenomenon. Section IV concludes this work.

II Temperature Inversion: Process Characterization

The threshold voltage and the carrier mobility values are temperature dependent [1, 2]. If both threshold voltage and carrier mobility values monotonically decrease when the temperature increases, the resulting impact on the gate switching current and thus on its timing performance depends on the considered range of V_{DD} values. To analyze this impact let us consider the Sakurai's drain source current (I_{DS}) representation [6]:

$$I_{N,P}^{Fast} = K_{N,P} \cdot W_{N,P} \cdot \left(V_{DD} - V_{TN,P}\right)^{\alpha_{N/P}} \qquad (1)$$

where $K_{N/P}$ is a conduction coefficient, and α the velocity saturation index. From (1), it is obvious that a decrease of the $V_{TN/P}$ values results in an increase of the saturation transistor currents while a decrease of the carrier mobility induces an opposite variation of the timing performances. As a result the temperature coefficient of the transistor current and the associated performance parameters may exhibit supply voltage and temperature dependent behaviours. As an example, it has been experimentally observed [8] that for a specific range of V_{DD} value the temperature coefficient of the transition time becomes V_{DD} dependent and may have negative values for $V_{DD} < 2 \cdot V_T$. This is a direct illustration of the temperature inversion phenomenon that could appear in nowadays processes with high V_T values and either regular or reduced V_{DD} values.

To characterize the temperature inversion phenomenon, we analyzed the I_{DS} temperature sensitivity of a transistor designed with two 130nm processes A and B, differing one from the other by their threshold value. More precisely, we extracted from eldo simulations, the evolution with temperature of the maximum drain source courant ($V_{GS}=V_{DD}$) delivered by NMOS transistors for different values of V_{DD}. As Fig.2

reports the obtained I_{DS} evolution of an NMOS transistor designed with the process B for three different temperature values.

As shown, for the process B the I_{DS} evolution is non monotonic. Indeed considering the crossing point of the three I_{DS} curves ($V_{DD}=V_{GS}=1.08V$), it is possible to distinguish two different V_{DD} domains. In the domain I, the NMOS current has a greater value at low temperature. In the domain II, the situation is reversed, the current value is smaller at 233°K. This gives evidence that depending on the V_{DD} value the worst case configuration can be obtained for a completely different temperature value. We can thus conclude that the temperature coefficient of the I_{DS} current is V_{DD} dependent and is equal to zero for the crossing point value $V_{DD}=1.05V$.

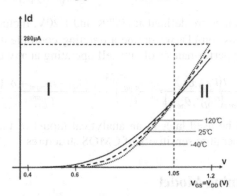

Fig. 2. Temperature evolution of the NMOS transistor current for the Process B

As discussed in [1,2] the evolution of the worst case PVT condition from high to low temperatures can be justified considering the relative variation of the V_{TN} and K_N values with temperature. Although V_T and K have both a negative temperature coefficient, depending on the supply voltage value they have opposite influence on the timing performance. At high V_{DD} value, the temperature induced variations of V_T are small with respect to V_{DD}. As a consequence the temperature induced variations of I_{DS} are mainly due to the K term. At the contrary, at low V_{DD} value, the relative variation of V_T, with respect to V_{DD}, becomes significant enough to counterbalance and even reverse the temperature sensitivity of the I_{DS} courant. In others words, at high V_{DD} values the K term dominates while for lower V_{DD} values the variation of the V_{DD}-V_T term becomes preponderant, reversing the I_{DS} temperature sensitivity.

Table 1. Inversion voltage and nominal threshold voltages for A and B processes

	Process A		Process B	
	NMOS	PMOS	NMOS	PMOS
Inversion point	0.92 V	1.08 V	1.05 V	1.38 V
Vtnom.	0.56 V	0.47 V	0.64 V	0.56 V

One interesting point is that at a V_{DD} value equal to the crossing point value the effect of the V_T variation on the current is exactly balanced by the effect of the K one. Thus imposing this operating point may results in quasi temperature insensitive circuit. Therefore the inversion voltage value of N and P transistors are key metrics to characterize the temperature sensitivity of a design, or to predict unexpected temperature inversion. Table 1 gives the inversion voltage values extracted from the simulations of the transistor current sensitivity for both A and B processes.

Considering the V_{DD} corners (1.08V and 1.32V) of these processes, we can deduce for the process A that both N and P transistors are never subject to temperature inversion, but for the process B, if the NMOS device is not subject to T inversion, the PMOS is always in inversion. To characterize a process with respect to the temperature inversion phenomenon, two coefficients have to be considered δ and Xk [9, 10] that give respectively the sensitivities of V_T and K to the temperature

$$Vt = Vtnom - \delta \cdot (\theta - \theta nom) \quad K = Knom \cdot \left(\frac{\theta nom}{\theta} \right)^{Xk} \tag{2}$$

where the nominal values are defined at 298°K and 1.20V. Including these parameters into the current expression (1), we define a derating coefficient Der_I(V_{DD},θ) which allows estimating the performances of any cell operating at any PVT condition.

$$Der\_I(V_{DD}, \theta) = \frac{I(V_{DD}, \theta)}{I_{nom}(V_{DD}^{nom}, \theta_{nom})} = \left(\frac{\theta_{nom}}{\theta} \right)^{Xk} \cdot \left(\frac{V_{DD} - V_{Tnom} + \delta \cdot (\theta - \theta_{nom})}{V_{DD}^{nom} - V_{Tnom}} \right)^{\alpha} \tag{3}$$

This coefficient will be used later in an analytical model of timing performances to characterize the temperature sensitivity of CMOS structures.

III Physical Timing Model

It has been shown in [4, 5] that the delay performance of CMOS circuits can be modeled with an analytical representation of the cell output transition time with an explicit identification of the design and process parameters. We summarize in this part the main points of this model considering only the case of falling output edges.

A. Transition Time Modeling

Considering the transistor as a current generator [5], the output transition time of CMOS cell can be directly obtained from the modelling of the (dis)charging current that flows during the switching process of the structure and from the amount of charge ($C_L \cdot V_{DD}$) to be exchanged with the output node as

$$\tau_{outHL} = C_L \cdot V_{DD} \Big/ I_{NMax} \tag{4}$$

where C_L represents the total output load, I_{NMax} is the maximum current available in the structure. The key point here is to evaluate this maximum current which depends on the input controlling condition. For that, two domains have to be considered: the fast input and the slow input range. In the fast input range, the driving condition imposes a constant and maximum current value in the structure. The current expression can then be directly obtained from eq.1 [6]. Combining equations 1 and 4 finally leads to the output transition time expression in the fast input range

$$\tau_{outHL}^{Fast} = \tau_{N/P} \cdot DW_{HL/LH} \cdot \frac{Cl}{C_{N/P}} = \frac{DW_{HL/LH} \cdot Cl}{K_{N/P} \cdot W_{N/P} \cdot (V_{DD} - V_{TN/P})^{\alpha_{N/P}}} \tag{5}$$

where $DW_{HL/LH}$ are the logical effort of the considered CMOS structure, $C_{N/P}$ are N/P gate capacitance and finally

$$\tau_N = \frac{Cox \cdot L \cdot V_{DD}}{K_N \cdot (V_{DD} - V_{TN})^{\alpha_N}} \qquad \tau_P = \frac{Cox \cdot L \cdot V_{DD}}{K_P \cdot (V_{DD} - V_{TP})^{\alpha_P}} \qquad (6)$$

are process metrics since these parameters capture the sensitivity of the output transition time to both the V_{DD} and the V_T values. In the slow input range, the maximum switching current decreases with the input ramp duration. Extending the results of [5] to general value of the velocity saturation index, the maximum switching current flowing in a CMOS structure is

$$I_{N/PMax}^{Slow} = \left\{ \frac{\left(\alpha_{N/P} \cdot K_{N/P} \cdot W_{N/P} \right)^{1/\alpha_{N/P}} \cdot Cl \cdot V_{DD}^2}{\tau_{in}} \right\}^{\frac{\alpha_{N/P}}{\alpha_{N/P}+1}} \qquad (7)$$

Combining (4) and (5) with (7), we finally obtain a manageable transition time expression for a falling output edge

$$\tau_{outHL}^{Slow} = \left(\frac{V_{DD} - V_{TN}}{\alpha^{1/\alpha_N} \cdot V_{DD}} \right)^{\frac{\alpha_N}{1+\alpha_N}} \cdot \left(\tau_{outHL}^{Fast} \cdot \tau_{IN} \right)^{1/\alpha_N + 1} = \left(\frac{Cl \cdot \tau_{IN} \cdot V_{DD}^{1-\alpha_N}}{\alpha^{1/\alpha_N} \cdot K_N \cdot W_N} \right)^{\frac{1}{1+\alpha_N}} \qquad (8)$$

with an equivalent expression for the rising edge. To conclude with the modeling of the output transition time, one can observe that in the fast input range the transition time only depends on the output load while in the slow input range, it also depends on the input transition time duration but is threshold voltage independent.

B. Propagation Delay Modeling

The delay of a CMOS gate is load, gate size and input slew dependent. Following [7], the input slope and the I/O coupling can be introduced in the propagation delay as

$$t_{HL/LH} = \frac{\tau_{in}}{\alpha_{N/P}+1} \left(\frac{\alpha_{N/P}-1}{2} + v_{TN/P} \right) + \left(1 + \frac{2C_M}{C_M + C_L} \right) \frac{\tau_{outHL/LH}}{2} \qquad (9)$$

This demonstrates the full delay sensitivity to the switching environment (τ_{IN}, τ_{out}). Considering the transition time dependency to the current (4), eq.9 will be of great importance in characterizing the temperature sensitivity of designs.

IV   Temperature Inversion: Cell Characterization

This section details how it is possible to include the temperature derating coefficient, defined in section II, in the timing performance representation summarized below. In the section III, we show that the τ parameter is a direct representation of the process performance, and that the transition time and the delay expressions directly characterize the cell and circuit level with respect to τ parameter. Thus the modeling of the τ parameter temperature sensitivity must allow characterizing the temperature sensitivities of the timing performances of CMOS structures. This section highlights this point.

A. Process Level

First, we study the impact of the temperature inversion phenomenon on the performance metric. As shown in (4) the temperature dependency of the τ parameter is completely defined from that of the current. From (3) we can also easily define a derating
factor $Der\_\tau(V_{DD}, \theta)$ of the process parameter τ as

$$\frac{\tau(V_{DD},\theta)}{\tau_{nom}\left(V_{DD}^{nom},\theta_{nom}\right)} = Der\_\tau(V_{DD},\theta) = \left(\frac{\theta}{\theta_{nom}}\right)^{Xk}\left(\frac{V_{DD}}{V_{DD}^{nom}}\right)\cdot\left(\frac{V_{DD}^{nom}-V_{Tnom}}{V_{DD}-V_{Tnom}+\delta\cdot(\theta-\theta_{nom})}\right)^{\alpha} \quad (10)$$

To characterize the temperature sensitivity of the process it is just necessary to apply
this coefficient to the τ metric of each transistor, since this parameter captures the
output transition time and propagation delay sensitivity to both the supply and threshold voltage values.

Table 2. Simulated and calculated $\tau_{N/P}$ values for each PVT corner

Process B	τ_P (ps) Simulation			Error (%) Model vs. Sim.			τ_N (ps) Simulation			Error (%) Model vs. Sim.		
V_{DD} (V)	1.08	1.20	1.32	1.08	1.20	1.32	1.08	1.20	1.32	1.08	1.20	1.32
θ=233K	36.6	26.9	21.9	-0.7	-2.2	-6.1	13.9	9.6	7.4	0.1	-4.4	-7.2
θ=298K	33.2	26.2	21.8	0.3	0	-1.1	14.2	10.2	8.0	-1.0	0	0.1
θ=398K	31.2	25.6	21.9	-3.2	-0.7	0.9	14.5	11.1	9.1	-2.5	1.4	3.1

To validate this approach, we applied the derating coefficient (10) to the PMOS and
NMOS transistors of the process B that exhibits (Fig.1) a temperature inversion point
above the operating V_{DD} value. This has been done in three steps. First step has been
devoted to the extraction of the $\tau_{N/P}$ values from simulations for all PVT corners. In a
second step, we calibrated the model on the nominal V_{DD} and θ values to obtain the
values of $\delta_{N/P}$ $Xk_{N/P}$ coefficients. Finally, we applied the derating coefficient (10) to
the nominal $\tau_{N/P}$ values to obtain the $\tau_{N/P}$ values for all others PVT conditions. Results
are reported in Table 2. As expected, the PMOS transistor of the process B is inverted
on the whole V_{DD} range and is temperature independent at V_{DD}=1.32V. On the other
hand, Table 2 shows that the NMOS transistor is not impacted by the temperature inversion phenomenon. This could be guessed from Table 1 hat shows that the inversion voltage (1.05V) is smaller than the worst case supply voltage. Finally, according
to the expressions (5, 8, 9), we may conclude that the timing performance parameters
(rising edge only) will be impacted by the temperature inversion. The worst case PVT
occurs at the worst case process defined by small V_{DD} and low temperature values.
For this process the main result obtained on the τ sensitivity is that the falling edge is
not impacted by the temperature inversion. At 1.08V the sensitivity to the temperature
is very small but the worst case operating mode corresponds to the standard definition
with high temperature and low V_{DD} values. As a summary, for the B process, both
edges vary in an opposite ways. The temperature coefficient of the transition time is
negative (-0.15ps/°K) for the rising edges, while the falling edges exhibit a small but
positive temperature coefficient (+0.04ps/°K). In that condition the critical operating
mode will be defined by the rising edge.

The conclusion of this paragraph is that the $\tau_{N/P}$ coefficients are usefull to characterize the I_{DS} variations with respect to all the PVT parameters. We have shown that

the model differentiates both edges to account of the N/P transistor dissymmetry. With respect to the V_{DD}, the threshold voltage appears to be the most temperature sensitive parameter, a 5% variation of V_T, can completely reverse the I_{DS} sensitivity.

B. Cell Level

At cell level we must consider the evolution of the transition time and the propagation delay values. As explain in [5] and sum up in part 3, the output transition time can exhibit two different evolutions depending on the controlling input conditions. Thus two different expressions have been developed: one for the fast input range (5) and the other for the slow input range (8). Considering the transition time, a brief analysis of these equations shows that, depending on the input controlling conditions, the temperature variation may affect differently the output transition time. Indeed an analysis of the switching current shows that (5) does depend on the V_T value while (8) does not. This means that in the slow input range the effect of temperature variations can affect the output transition time only through the parameter K. Consequently, the temperature coefficient of the output transition time can never be negative in the slow input range while it may be reversed in the fast input range. This result can be verified in Fig.3 and Fig. 4 that illustrates the evolution of the output transition of an inverter with the input ramp duration. As shown the output transition time exhibits a positive temperature coefficient in the slow input range, while in the fast input range, this coefficient is positive for the A process, and negative for the B process.

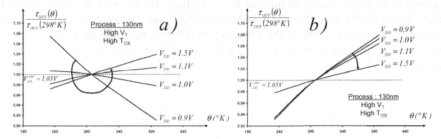

Fig. 3. Inverter output transition time (process B) vs temperature for different V_{DD} and input ramp duration a) a fast input ramp (τ_{IN}=10ps) and b) a slow input ramp (τ_{IN} =1780ps)

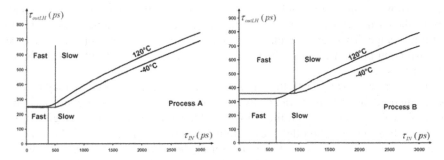

Fig. 4. Inverter output transition time vs input ramp duration for two different θ values

As for the output transition time, the analysis of temperature variation effects on the propagation delay must also distinguish between the fast and slow input ramp domains. As shown by eq.9, the propagation delay is the sum of two terms. Since the second term of the propagation delay is a fraction of the output transition time, all the results obtained in the preceding paragraph remain true for this fraction of the propagation delay. However these results are not sufficient to capture all the temperature variation effects on the propagation delay. Indeed, the first term of the propagation delay, which models the input slope effect, is directly proportional to the input ramp duration and to the threshold voltage value as

$$\frac{1}{\alpha_N + 1} \left(\frac{(\alpha_N - 1)}{2} + v_{TN} \right) \cdot \tau_{IN} \tag{11}$$

Thus the evolution of the propagation delay with the temperature is strongly influenced by the decrease of v_T value with the increase of temperature. More precisely for any structure there is always an input ramp value, $\tau_{IN}^{Inversion}$, after which the temperature coefficient of the propagation delay is necessarily negative. This limit value can either occur in the fast or the slow input design range. In order to evaluate, the value of $\tau_{IN}^{Inversion}$, we compute in the fast input range the partial derivative of the propagation delay expression with respect to the temperature and cancelled it to obtain the input ramp limit:

$$\tau_{IN}^{Inversion} = \frac{\alpha_N + 1}{\delta_N} \cdot \left(1 + \frac{2C_M}{C_M + C_L} \right) \cdot \tau_{outHL}^{Fast} (\theta_{nom}, V_{DD}) \cdot \left[\frac{X_{kN}}{\theta} - \frac{\alpha_N \cdot \delta_N}{V_{DD} - V_{Tnom} + \delta(\theta - \theta_{nom})} \right] \tag{12}$$

As shown the input transition time value controlling the temperature inversion of the delay is load dependent. One can note that this limit can be negative. This means that the temperature coefficient of the propagation delay is negative on all the fast input range. In the slow input range, the preceding procedure leads to the following input ramp duration limit:

$$\tau_{IN}^{Inversion} = \left(\frac{\alpha_N + 1}{\delta_N} \cdot \frac{\chi_{kN}}{\theta} \right)^{\frac{\alpha_N + 1}{\alpha_N}} \cdot \left(1 + \frac{2C_M}{C_M + C_L} \right) \left(\frac{V_{DD} - V_{Tnom}}{\alpha^{\frac{1}{\alpha}} \cdot V_{DD}} \right) \cdot \left(\tau_{outHL}^{Fast} (\theta_{nom}, V_{DD}) \right)^{\frac{1}{\alpha_N}} \tag{13}$$

As previously mentioned expressions (12) and (13) predict that there is always an input ramp value after which the temperature coefficient of the propagation delay is necessarily negative due to the reduction of the V_T value with the temperature increase. However in all the studied processes, the reduction with increasing temperature of the term (11) was enough important to place the input ramp limit value in the fast input range. This can be understood since the only process subjected to temperature inversion phenomenon is a high threshold voltage process used to limit the leakage currents.

Moreover, as shown in (9), the temperature coefficient of the delay may be less or more imposed by the input ramp duration dependency, that is proportional to the threshold voltage value. As an illustration of this temperature effect, we give in Fig.5 the normalized propagation delay evolutions with θ for a fast input and a slow input ramp. Fig.6 gives respectively the evolution of the propagation delay with the input ramp duration for two different temperature and output load values. As shown for the process B, the temperature dependency of the delay is as expected: the critical configuration is observed at low temperature. On the contrary for process A, a strong input ramp duration dependency of the delay temperature coefficient is obtained. For quite slow input ramps the temperature coefficient is reversed and the critical case is observed at the lowest temperature.

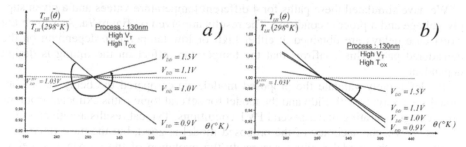

Fig. 5. Inverter propagation delay (process A) vs θ: a) for a fast input ramp (τ_{IN}= 10ps), b) for a slow input ramp (τ_{IN} = 1780ps)

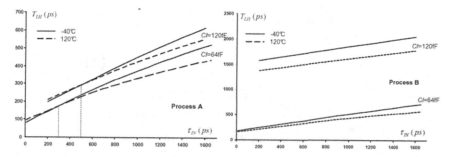

Fig. 6. Inverter propagation delay vs input ramp duration for different θ and C_L conditions

In Table 3, we validate (12). For each loading condition (process A) we impose an input ramp corresponding to the limit defined in (12), and simulate the delay of an inverter for the corresponding load and different temperature conditions. As shown we obtain a quasi temperature insensitivity of the delay, as predicted by (12).

Table 3. Delay for several temperature and C_L values for the input slope calculated with (12)

Temp.	Load	Delay (ps) at τ_{IN} = τ_{IN} limit			
		4 fF	64fF	120fF	640fF
-40 C		42.8	176.5	301.5	1474.6
25 C		42.3	175.3	299.6	1466.2
125 C		42.2	175.2	299.8	1467.9

Independently of the considered input range, expressions (12) and (13) allow to conclude that the $\tau_{IN}^{Inversion}$ value depends on design parameters such as the load and the transistor widths. It thus appears difficult to characterize the temperature effects on performances using few and constant derating factors, since the temperature effects are radically different and depending on the operating design conditions.

C. Path / Circuit Level

For a real design both edges have to be taken into account during the STA and the timing behavior with respect to the temperature is not easy to predict. To illustrate this situation let us consider the four critical paths extracted from a real design.

We have simulated these paths for 4 different temperature values and a given supply voltage and a process condition. The results are given on figure 7a. As shown the worst case delays are obtained at either high or low temperature depending on the considered path. This confirms that the temperature effect on the timing is design dependent.

To definitively validate the proposed model, we compared the delay values obtained successively with Eldo and the model for several logic paths extracted from the same design operating under several PVT conditions. Typical results are given in fig. 7b. As shown, the accuracy of the proposed model is good, but the most interesting result is that the model predicts accurately the evolution of the propagation delays with temperature.

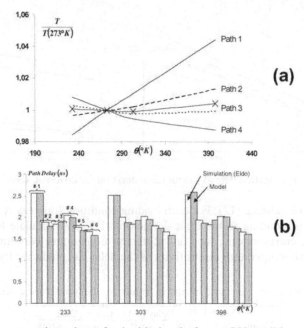

Fig. 7. (a) Temperature dependence for 4 critical paths for one PV condition, (b) Simulated and calculated Paths delay evolution with the temperature for the process B

V Conclusion

Considering the physical parameters that affect the temperature sensitivity of transistor currents, we have defined a model allowing capturing the non linear effects of temperature variations on performances. We have demonstrated the possibility to predict the occurrence of the temperature inversion phenomenon on design performance. It has been shown that critical corner for design characterization may evolve from high to low temperature. Then, with the model properties, design guidelines and methodologies have been defined to take into account the temperature inversion phenomenon for design validation of real circuits.

References

1. S.M. Sze,"Physics of semiconductor devices", Wiley ed. 1983.
2. Changhae Park et al, "Reversal of temperature dependence of integrated circuits operating at very low voltages", Proc. IEDM conference, pp.71-74, 1995.
3. Synopsys Inc.,"Scalable Polynomial Delay And Power Model", Rev 5, October 2002.
4. P. Maurine and al, "Transition time modeling in deep submicron CMOS" IEEE Trans. on Computer Aided Design, vol.21, n11, pp.1352-1363, nov.2002.
5. B. Lasbouygues and al "Continuous representation of the performance of a CMOS library", European Solid-State Circuits, ESSIRC'03 Conf. pp.595-598, 16-18 Sept 2003.
6. T. Sakurai and A.R. Newton,"Alpha-power model, and its application to CMOS inverter delay and other formulas", J. Solid State Circuits vol. 25,pp. 584-594, April 1990.
7. K.O. Jeppson, "Modeling the Influence of the Transistor Gain Ratio and the Input-to-Output Coupling Capacitance on the CMOS Inverter Delay", IEEE JSSC, Vol. 29, pp. 646-654, 1994.
8. J.M. Daga, E. Ottaviano, D. Auvergne, "Temperature effect on delay for low voltage applications", Design Automation and Test in Europe, pp 680-685, 23-26 Feb. 1998, Paris.
9. J.A. Power and all, "An Investigation of MOSFET Statistical and Temperature Effects", IEEE Int. Conf. on Microelectronic & Test Structures, Vol. 5, pp.202-207, March 1992.
10. A. Osman and al, "An Extended Tanh Law MOSFET Model for High Temperature Circuit Simulation", IEEE JSSC, Vol. 30, No2, pp.147-150, Feb. 1995.
11. Shih-Wei Sun, P.G.Y.Tsui, "Limitations of CMOS Supply- Voltage scaling by MOSFET threshold voltage variation", IEEE J. of Solid State Circuits, Vol.30, N°8, pp.947-949, August 1995.

Circuit Design Techniques for On-Chip Power Supply Noise Monitoring System

Howard Chen[1] and Louis Hsu[2]

[1] IBM Research Division, Yorktown Heights, New York 10598, USA
[2] IBM Systems & Technology Group, Hopewell Junction, New York 12533, USA

Abstract. This paper describes the novel design of an on-chip noise-monitoring device, which can measure the power supply noise of each individual macro and determine the noise interference between different macros. A hierarchical noise-monitoring system is proposed to monitor and store the power supply noise information for core-based design, as part of the built-in-self-test system. The method can be further extended from system-on-chip to system-on-package to provide a complete coverage of noise testing methodology. The circuit of a properly-sized noise monitor has been simulated with the actual VDD and GND noise waveforms of a 4-GHz benchmark microprocessor to verify its functionality.

1 Introduction

Power supply noise is the switching noise that causes power supply voltage fluctuation, which can be subsequently coupled onto the evaluation nodes of a circuit. For an under-damped low-loss network, the power supply noise problem can manifest in the form of a slowly decaying transient noise, or a potentially more dangerous resonant noise. As the power supply voltage and threshold voltage continue to scale down in nanometer technology, the noise margin will become very small, and the control of power supply noise will be critical in determining the performance and reliability of VLSI circuits.

Power supply noise can be simulated by modeling the inductance, resistance, and capacitance of the power distribution network. However, it is often difficult to verify the accuracy of simulation results without the actual hardware measurement data. Furthermore, for high-performance system-on-chip design, the analog circuits, which are more susceptible to noise, may have multiple supply voltages that must be isolated from the digital circuits and analyzed separately [ ]. To calibrate the simulation model and provide a better estimate of the power supply noise, hardware measurement can be performed by using an amplification circuit to send the analog noise waveform off chip to an external tester [ ]. This method is difficult to implement, however, due to the resolution required to measure high-frequency noise, and the large number of noise sources that need to be monitored. To minimize the possible noise interference in an analog apparatus, a sampling method and multiple voltage comparators can be used to send the output to a digital tester [ ]. Unfortunately, the use of clocks in the sampling circuit will limit the time resolution of noise measurement. The placement

V. Paliouras, J. Vounckx, and D. Verkest (Eds.): PATMOS 2005, LNCS 3728, pp. 704–713, 2005.
© Springer-Verlag Berlin Heidelberg 2005

of voltage comparators may also introduce uncertainty on the reference voltage due to additional voltage drops on the interconnect. Since sub-sampled waveforms may not capture the supply noise behavior during normal chip operation, an autocorrelation technique [ ] that treats supply noise as a random process and uses its statistical properties has been developed to measure the dynamics of cyclostationary noise process.

In this paper, we will first describe an innovative circuit design technique to implement an on-chip noise-monitoring device, and the methodology to measure the noise on the chip. Then we will illustrate the design of a noise-monitoring system that comprises multiple on-chip noise-monitoring devices distributed strategically across the chip, and collects the noise data for further analysis. Finally, the simulation results of a hierarchical noise-monitoring system are presented to demonstrate the accuracy of noise measurement data, which can be stored in the memory and scanned out to an external tester.

2   On-Chip Noise-Monitoring Device

The proposed hierarchical power supply noise monitoring system is based on a novel design and implementation of an individual on-chip noise analyzer unit that comprises a reference voltage generator, a noise-monitoring device, a noise data latch, and an optional power supply regulator. Controlled by a higher level built-in-self-test unit or external tester, each noise analyzer unit is capable of measuring the noises in signal or power bus lines. The fluctuating supply (VDD) and ground (GND) voltages are monitored by the noise-monitoring device and compared to the reference voltages. As the reference voltages are adjusted upward or downward, the noise data are recorded by the sampling latch. Alternatively, a noise analyzer unit can be implemented with an externally shared reference voltage generator to minimize the circuit area and power. In addition to measuring the VDD and GND noise directly, the noise analyzer unit can also be designed to measure the voltage differential between VDD and GND.

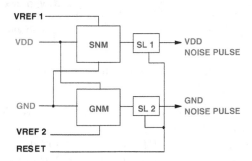

Fig. 1. Integrated noise-monitoring device

Figure 1 shows an integrated noise-monitoring device, comprising one supply noise monitor (SNM), one ground noise monitor (GNM), one control signal, two voltage reference levels ($VREF1$ and $VREF2$), and two sampling latches (SL_1 and SL_2). The power supply voltage VDD is monitored by SNM and measured against the reference voltage $VREF1$. The ground voltage GND is monitored by GNM and measured against the reference voltage $VREF2$. The supply noise pulses generated by SNM are sampled by latch SL_1, and the ground noise pulses generated by GNM are sampled by latch SL_2 in the noise-monitoring system.

2.1 VDD Supply Noise Monitor

Figure 2 depicts the circuit schematic of the VDD supply noise monitor SNM, which is designed with a strong PMOS device and a weak NMOS device that are connected as an inverter. The source and the body of the PMOS device are connected to $VREF1$, and the source and the body of the NMOS device are connected to GND. If an SOI device, which has a floating body, instead of a bulk device is used, it is desirable to provide the SOI device with a body contact so as to minimize the variability of threshold voltage due to body effect. The gates of the PMOS and NMOS devices are connected to the varying VDD to be measured, and the drains of the PMOS and NMOS devices are connected to the noise pulse output NP. In order to provide a strong driving power that pulls up the output when both PMOS and NMOS devices are on, the PMOS device must be designed with a wider channel, lower threshold voltage, or thinner gate oxide than the NMOS device.

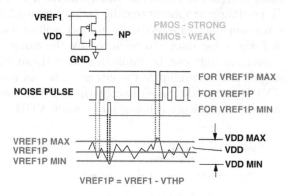

Fig. 2. VDD supply noise monitor

To determine if the PMOS device is turned on, we define $VREF1P = VREF1 - V_{THP}$, where V_{THP} is the threshold voltage of the PMOS device. If the VDD input is higher than $VREF1P$, the PMOS device will be turned off, while the NMOS device will be turned on, and the NP output will be GND. On the other hand, if VDD drops below $VREF1P$, the PMOS device will be turned on by switching from weak inversion to strong inversion, and the NP output will change to $VREF1$ because the PMOS device has been carefully designed to overpower the NMOS device.

As we adjust the reference voltage $VREF1$, the supply noise monitor SNM can generate the noise pulse output NP in one of 3 distinct regions, where NP is always 1 (high) in region A, always 0 (low) in region C, and toggles between 0 and 1 in region B (Figure 3). When $VREF1P$, which equals $VREF1 - V_{THP}$, is greater than the maximum supply voltage VDD_{MAX}, the output NP remains in region A. When $VREF1P$ is less than the minimum supply voltage VDD_{MIN}, the output NP remains in region C. Therefore, we can identify VDD_{MAX} by sweeping $VREF1$ through the boundary between region A and region B, and identify VDD_{MIN} by sweeping $VREF1$ through the boundary between region B and region C. The width of region B, which equals $VDD_{MAX} - VDD_{MIN}$, thus defines the worst-case range of VDD noise fluctuations.

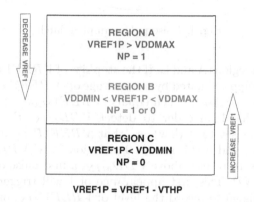

Fig. 3. VDD scanning based on noise pulse output characteristics

The noise pulse signal NP from the supply noise monitor SNM is sent to the sampling latch SL_1, which generates a write-enabling signal when the boundary between region B and region C is reached. The inverted noise pulse output signal $\overline{NP}$ from the supply noise monitor SNM is sent to the sampling latch SL_2, which in turn generates a write-enabling signal when the boundary between region A and region B is reached. These write-enabling signals will facilitate the recording of VDD_{MIN} and VDD_{MAX} values into memory, based on the corresponding $VREF1$ levels at the boundaries. Figure 4 shows the circuit diagram of a set-reset (SR) sampling latch to detect and record the boundary between regions A and B, as well as the boundary between regions B and C. Each SR latch, comprising 2 NOR gates, is triggered on the positive edge of the SET signal.

Two reference voltage scanning mechanisms are implemented in the sampling latch design to detect the maximum and minimum VDD noise. In order to detect VDD_{MAX}, the reference voltage $VREF1$ is initially set to 1 (high), so that $VREF1P$ is much greater than estimated VDD_{MAX} and NP is always 1 in region A. As $VREF1P$ is adjusted downward step by step to just below VDD_{MAX}, a first 0 pulse will appear on the output node NP. This first appearance of 0 will trigger SR_2 to generate a write-enabling signal to record the level of $VREF1_{MAX}$, or $VREF1P_{MAX} + V_{THP}$, which corresponds to the

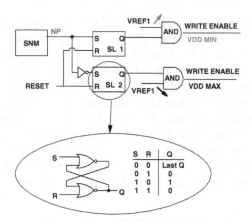

Fig. 4. Noise pulse sampling latch

boundary between regions A and B. If the step size of $VREF1$ adjustment is ΔV, VDD_{MAX} can be approximated by the average of $(VREF1_{MAX} - V_{THP} + \Delta V)$ and $(VREF1_{MAX} - V_{THP})$, which equals $(VREF1_{MAX} - V_{THP} + \Delta V/2)$.

On the other hand, in order to detect VDD_{MIN}, the reference voltage $VREF1$ is initially set to 0 (low), so that $VREF1P$ is much less than the expected VDD_{MIN} and NP is always 0 in region C. As $VREF1P$ is incrementally adjusted upward to just above VDD_{MIN}, a first pulse of 1 will appear on the output node NP. This first appearance of 1 will trigger SR_1 to generate a write-enabling signal to record the level of $VREF1_{MIN}$, or $VREF1P_{MIN} + V_{THP}$, which corresponds to the boundary between regions B and C. If the step size of $VREF1$ adjustment is ΔV, VDD_{MIN} can be approximated by the average of $(VREF1_{MIN} - V_{THP} - \Delta V)$ and $(VREF1_{MIN} - V_{THP})$, which equals $(VREF1_{MIN} - V_{THP} - \Delta V/2)$. The resulting range of VDD noise fluctuations, also known as the peak-to-peak VDD noise, can be calculated from $(VDD_{MAX} - VDD_{MIN})$, which equals $(VREF1_{MAX} - VREF1_{MIN} + \Delta V)$.

2.2   GND Noise Monitor

Similarly, Figure 5 depicts the circuit schematic of the ground noise monitor GNM, which is designed with a strong NMOS device and a weak PMOS device that are connected as an inverter. The source and the body of the PMOS device are connected to VDD, and the source and the body of the NMOS device are connected to $VREF2$. The gates of the PMOS and NMOS devices are connected to the varying GND to be measured, and the drains of the PMOS and NMOS devices are connected to the noise pulse output NP. In order to pull down the output when both PMOS and NMOS devices are on, the NMOS device is designed with stronger driving power than the PMOS device.

In order to determine if the NMOS device is turned on, we define $VREF2N = VREF2 + V_{THN}$, where V_{THN} is the threshold voltage of the NMOS device. If the GND input is lower than $VREF2N$, the NMOS device will be turned off,

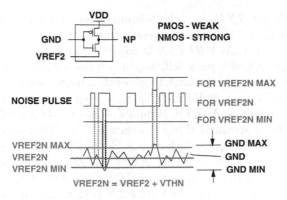

Fig. 5. GND noise monitor

the PMOS device will be turned on, and the NP output will be VDD. On the other hand, if GND rises above $VREF2N$, the NMOS device will be turned on by switching from weak inversion to strong inversion, and the NP output will change to $VREF2$ because the NMOS device has been carefully designed to overpower the PMOS device.

As we adjust the levels of reference voltage $VREF2$, the GND noise monitor GNM can generate the noise output NP in one of three distinct regions as shown in Figure 6, where NP is always 1 (high) in region A, always 0 (low) in region C, and toggles between 0 and 1 in region B. When $VREF2N = VREF2 + V_{THN}$ is greater than the maximum ground voltage GND_{MAX}, the output NP remains 1 in region A. When $VREF2N$ is less than the minimum supply voltage GND_{MIN}, the output NP remains 0 in region C. Therefore, we can identify GND_{MAX} by sweeping $VREF2$ through the boundary between region A and region B, and identify GND_{MIN} by sweeping $VREF2$ through the boundary between region B and region C. The width of region B, which equals $GND_{MAX} - GND_{MIN}$, thus defines the worst-case range of GND noise fluctuations.

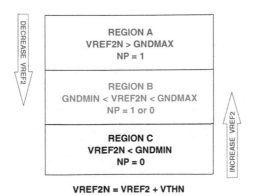

Fig. 6. GND scanning based on noise pulse output characteristics

In order to detect GND_{MAX}, the reference voltage $VREF2$ is initially set to 1 (high), so that $VREF2N$ is much greater than estimated GND_{MAX}, and NP is always 1 in region A. As $VREF2N$ is adjusted downward step by step to just below GND_{MAX}, a first 0 pulse will appear on the output node NP. This first appearance of 0 will trigger the latch to generate a write-enabling signal to record the level of $VREF2_{MAX}$, or $VREF2N_{MAX} - V_{THN}$, which corresponds to the boundary between regions A and B. If the step size of $VREF2$ adjustment is ΔV, GND_{MAX} can be approximated by the average of $(VREF2_{MAX} + V_{THN} + \Delta V)$ and $(VREF2_{MAX} + V_{THN})$, which equals $(VREF2_{MAX} + V_{THN} + \Delta V/2)$.

On the other hand, in order to detect GND_{MIN}, the reference voltage $VREF2$ is initially set to 0 (low), so that $VREF2N$ is much less than GND_{MIN} and NP is always 0 in region C. As $VREF2N$ is incrementally adjusted upward to just above GND_{MIN}, a first 1 pulse will appear on the output node NP. This first appearance of 1 will trigger the latch to generate a write-enabling signal to record the level of $VREF2_{MIN}$, or $VREF2N_{MIN} - V_{THN}$, which corresponds to the boundary between regions B and C. If the step size of $VREF2$ adjustment is ΔV, GND_{MIN} can be approximated by the average of $(VREF2_{MIN} + V_{THN} - \Delta V)$ and $(VREF2_{MIN} + V_{THN})$, which equals $(VREF2_{MIN} + V_{THN} - \Delta V/2)$. The resulting range of GND noise fluctuations, also known as the peak to peak GND noise, can be calculated from $(GND_{MAX} - GND_{MIN})$, which equals $(VREF2_{MAX} - VREF2_{MIN} + \Delta V)$.

3   Hierarchical Noise-Monitoring System

Figure 7 shows a noise-monitoring system, comprising a noise monitor controller, a reference voltage generator, multiple VDD and GND noise monitors and sampling latches, and a memory storage unit. The noise monitor controller receives signals from the external tester and sends control signals to the reference voltage generator and the noise monitors. The reference voltage generator provides a set of reference voltages for the noise monitor to determine the noise level of VDD and GND. The noise data are then latched and stored in the memory to be scanned out. Depending on the switching condition and the periodicity of

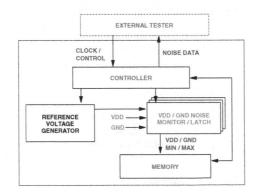

Fig. 7. Power supply noise monitoring system

the noise, the measurement period can cover multiple clock cycles to collect not only the high-frequency switching noise data, but also the mid-frequency and low-frequency ΔI noise data.

The reference voltage generator can be implemented by using a positive charge pump to generate a reference voltage that is higher than nominal Vdd, and a negative charge pump to generate a reference voltage that is lower than nominal GND. A band-gap reference circuit, which is insensitive to temperature and supply voltage variations, can then be used to adjust the reference voltage level in finer steps.

The noise-monitoring system can be further implemented in a hierarchical manner for system-on-chip design, where the in-core noise-monitoring system for each functional unit is composed of multiple noise analyzer units to measure the local power supply noise. Control and data lines are routed to each noise monitor in the core, similar to those of a scan chain. The measured noise data are then latched in the local noise analyzer unit and scanned out sequentially. The reference voltages can be generated by the local built-in-self-test unit LBIST to reduce the circuit size and power consumption of the noise analyzer unit. On the chip level, a global noise analyzer can be implemented with multiple in-core noise monitoring units and a global built-in-self-test unit GBIST. The GBIST unit sends control signals through control wires to the local built-in-self-test unit LBIST of each core. The noise data are then scanned out from the data wires and stored in the memory buffer inside the GBIST unit. Alternatively, an existing memory core on the chip can be utilized to store the noise data. Measured with a comprehensive built-in self test procedure, the worst-case range of power supply voltage fluctuation can be recorded for each test pattern and workload.

4 Noise Monitor Simulation Results

The implementation of a multi-stage power supply noise monitor has been simulated with the SOI device models under the 65-nm technology. The VDD noise monitor, which consists of a 3-inverter buffer chain with properly sized devices shown in Table 1, is used to measure the VDD waveform of a 4-GHz benchmark microprocessor, shown on the top of Figure 8. The GND noise monitor, which also consists of a 3-inverter buffer chain with properly sized devices shown in Table 1, is used to measure the GND waveform of a 4-GHz benchmark microprocessor, shown on the top of Figure 9. Both the VDD and GND waveforms exhibit a high-frequency noise at 4 GHz and a mid-frequency noise near the resonant frequency of 250 MHz. The worst-case crossbar current is limited under 1 mA when both $PFET1$ and $NFET1$ are turned on.

Table 1. Device sizes of a 3-inverter noise monitor

Width (um)	PFET1	NFET1	PFET2	NFET2	PFET3	NFET3
VDD monitor	20	1	2	1	2	1
GND monitor	1	10	2	1	2	1

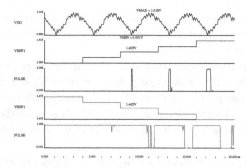

Fig. 8. $VREF1$ sweep to detect VDD_{MIN} and VDD_{MAX}

Figure 8 shows the circuit simulation results of VDD supply noise monitor, where the equivalent threshold voltage of the PMOS device is about 0.450V. As the reference voltage $VREF1$ is adjusted from low to high at $\Delta V = 5mV$ intervals, the first pulse of 1 is detected when $VREF1$ reaches 1.405V and VDD_{MIN} turns on the PMOS device. With the adjustment of $\Delta V/2 = 3mV$ resolution which equals half of the $5mV$ measurement interval, the calculated VDD_{MIN} is 1.405 - 0.450 - 0.003 = 0.952V, which correlates well with the actual VDD_{MIN} of 0.951V. Similarly, as the reference voltage $VREF1$ is adjusted from high to low at $\Delta V = 5mV$ intervals, the first pulse of 0 is detected when $VREF1$ reaches 1.465V and VDD_{MAX} turns off the PMOS device. With the adjustment of $\Delta V/2 = 3mV$ resolution which equals half of the $5mV$ measurement interval, the calculated VDD_{MAX} is 1.465 - 0.450 + 0.003 = 1.018V, which correlates well with the actual VDD_{MIN} of 1.018V.

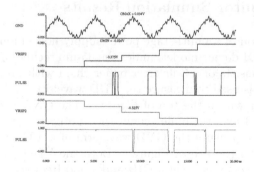

Fig. 9. $VREF2$ sweep to detect GND_{MIN} and GND_{MAX}

Figure 9 shows the circuit simulation results of a GND noise monitor, where the equivalent threshold voltage of the NMOS device is about 0.354V. As the reference voltage $VREF2$ is adjusted from low to high at $\Delta V = 5mV$ intervals, the first pulse of 1 is detected when $VREF2$ reaches -0.375V and GND_{MIN} turns off the NMOS device. With the adjustment of $\Delta V/2 = 3mV$ resolution which equals half of the $5mV$ measurement interval, the calculated GND_{MIN} is

-0.375 + 0.354 - 0.003 = -0.024V, which correlates well with the actual GND_{MIN} of -0.026V. Similarly, as the reference voltage $VREF2$ is adjusted from high to low at $\Delta V = 5mV$ intervals, the first pulse of 0 is detected when $VREF2$ reaches -0.325V and GND_{MAX} turns on the NMOS device. With the adjustment of $\Delta V/2 = 3mV$ resolution which equals half of the $5mV$ measurement interval, the calculated GND_{MAX} is -0.325 + 0.354 + 0.003 = 0.032V, which correlates well with the actual GND_{MAX} of 0.034V.

In addition to detecting VDD_{MIN}, VDD_{MAX}, GND_{MIN}, and GND_{MAX}, which determine the range of VDD and GND noise, the noise monitors can also be used to measure the period during which VDD or GND voltage stays below or above a certain critical voltage such as $\pm10\%$ of nominal Vdd. Since the width of each noise pulse directly corresponds to the time window that VDD or GND voltage stays above or below the critical voltage, the sweep of reference voltage $VREF1$ or $VREF2$ can continue after it detects the first pulses, until the critical voltage is reached, to measure the amount of time that the VDD or GND noise exceeds its limit, based on power, timing and other performance specifications.

5   Conclusions

A new circuit design technique for an on-chip noise analyzer unit has been developed and demonstrated to achieve a measurement accuracy of $\Delta V/2$, where ΔV is the adjustment interval of reference voltage. To minimize the uncertainty of process variation, the threshold voltage of the PMOS and NMOS devices can be measured by using a current criterion of $100nA \times W/L$, and calibrated by setting VDD and GND to a known reference voltage during the initial voltage scan. The subsequent voltage scanning mechanism then exploits the device characteristics of inverters and sampling latches, and uses the VDD and GND noise monitoring devices in each noise analyzer unit to measure the power supply noise and calibrate the simulation model of a power supply network. Based on the proposed in-core noise monitoring system, a hierarchical power supply noise monitoring methodology can be developed for system-on-chip design to improve system performance and reliability with an ubiquitous noise feedback mechanism.

References

1. E. Alon, et al., "Circuit and techniques for high-resolution measurement of on-chip power supply noise," International Symposium on VLSI Circuits, June 2004.
2. H. Aoki, et al., "On-chip voltage noise monitor for measuring voltage bounce in power supply lines using a digital tester," International Conference on Microelectronic Test Structures, March 2000, pp. 112-117.
3. M. Badaroglu, et al., "Methodology and experimental verification for substrate noise reduction in CMOS mixed-signal ICs with synchronous digital circuits," International Solid-State Circuits Conference, February 2002, pp. 274-275.
4. H. Partovi and A. Barber, "Noise-free analog islands in digital integrated circuits," U.S. Patent No. 5453713.

A Novel Approach to the Design
of a Linearized Widely Tunable Very Low Power
and Low Noise Differential Transconductor

Hamid Reza Sadr M.N.

Islamic Azad University, South Tehran Branch
+98-21-6556926, +98-21-6516959, +98-912-2503048
hsmn1@yahoo.com

Abstract. In this paper, a novel economically linearized widely tunable very
low power and low noise differential transconductor is proposed and compared
with the conventional differential pair. The application of the resulted
transconductor as a negative resistor is presented. The linearity, power, noise,
speed, and tunability performances are simulated, discussed, and compared with
the conventional differential pair's.

Keywords: Differential transconductors, negative resistors, analog circuits and
filters, low noise, low power, widely tunable circuits, GHz range frequencies,
continuous-time filters, Q-enhanced active filters, Gm-C filters, RF, VCO.

1 Introduction

Generally speaking, as well as tunability, due to their simple structure and hence the
superior speed and noise performances, transconductors have occupied a significant
role in the design of analog circuits especially continuous-time filters. These cells also
offer the excellent features such as high input and output impedances, ability to work
in GHz range frequencies, ability to replace OP-Amps in math operations, ability to
model the positive and negative resistors, and small chip area.

Most of the time, the linearity of these cells in their conventional shape, which due
to the least possible parasitics and active parts enjoy superior speed and noise performances does not satisfy the applications at hand. On the other hand, due to technology non-idealities, as soon as the simple topology of the conventional differential
pair is modified to improve the linearity of these cells, the speed and noise performances degrade. Many different linearization techniques have been discussed in design
of differential transconductors [1][2][3][5][7][8][9][10]. The target essentially is to
present linearization techniques, which, while being low power, do not degrade the
noise performance and keep the speed of the linearized transconductor as close as
possible to its equivalent conventional differential pair's. The speed of interest nowadays is in the range of GHz, while the linearized topologies capable of working in this
range and at the same time satisfy the RF requirements are rare.

The new linarized transconductor proposed in this paper, while being simple and
economic, has the advantage of being very low power and low noise, widely tunable,
and as fast as the simple conventional differential pair to work in GHz range frequencies. One important application of these cells is in Silicon based Q-enhancement tech-

V. Paliouras, J. Vounckx, and D. Verkest (Eds.): PATMOS 2005, LNCS 3728, pp. 714–723, 2005.
© Springer-Verlag Berlin Heidelberg 2005

niques; in which transconductors are used as tunable negative resistors to eliminate the losses of on-chip reactive elements (especially spiral inductors) [3][4]. Other important applications of these cells are in VCOs and Gm-C filters, where they are used as voltage to current converters, positive and negative resistors, and the gyrator-C, simulating the inductor's behavior [4].

Since the differential transconductors have the advantage over the single ended ones that they can omit the even order harmonics [5][7][8], here the discussion is made on these configurations. As well, since the analysis of a differential transconductor and its resulted negative resistor are similar, here, in order to show the application, it is chosen to analyze and simulate the proposed transconductor in its negative resistor configuration.

In section 2, the new differential transconductor named GGD3 (God Given Design 3) is analyzed. Its transconductance is calculated and compared with the conventional differential pair's. GGD3, while securing wider tuning range, collects the advantages of its two ancestors i.e. GGD [1] and GGD2 [1,2] transconductors.

In sections 3, 4, and 5 the linearity, power, and noise performances of GGD3 are evaluated respectively, and compared with conventional diff. pair's. Finally, in section 6, the Hspice simulation results of the two approaches are compared.

2   GGD3 Transconductor

Figure1 shows the GGD3 transconductor connected to act as a negative resistor. This is a modification to the GGD3's ancestor, i.e. GGD transconductor shown in figure 2. As can be seen, the only difference between the two topologies is that, in GGD, $M_{3,4}$ are diode connected and hence their output impedances can not be practically very high. Also, there, despite the superior advantages, we have the disadvantage of decreased value to the degeneration resistor caused by the GGD's topology. To overcome these disadvantages and improve the performance of GGD's topology while maintaining its simplicity and hence advantages, the easiest solution is to use a cross connection as in GGD3 in figure1. If matching is assumed, due to the symmetry of both topologies, the DC voltages at the sources of $M_{1,2}$ are essentially the same, and hence GGD and GGD3 are expected to have the same power performances. Also, the same as GGD, in GGD3 no DC current passes through R_{deg} and it is always isolated from the DC current path. $M_{3,4}$ work in saturation and hence are ideally true current sources with high output impedances which act as the paths for the DC currents.

As $M_{3,4}$ configure a negative resistor in parallel with R_{deg}, considering R^- as the negative real part of this resistor, and $R_{eq} = R_{deg} \parallel R^-$ seen at the sources of $M_{1,2}$, we are dealing with three major work regions for GGD3; $R^- \gg R_{deg}$: in this case $R_{eq} \cong R_{deg}$, $R^- \geq R_{deg}$: in this case $R_{eq} =$ a positive value, and $R^- \leq R_{deg}$: in this case $R_{eq} =$ a negative value. As it is clear, an analytical solution to the measurement of R_{eq} and hence the output impedance of GGD3 is not easy. On the other hand, for GGD3 in its negative resistor shape, the region of our interest is where we can have a negative resistance at its output. As well, if we can measure the current passing through R_{deg} directly, then, using the KVL we can ignore the effect of the negative resistance of $M_{3,4}$ on R_{eq} and the analysis of GGD3 becomes much easier. Using this idea, since $M_{1,2}$ are in saturation, the transconductance (Gm) of GGD3 may be calculated as:

$$Gm = \frac{\sqrt{KI}}{1 + k \cdot R_{deg} \cdot \sqrt{KI}} \quad \text{where} \quad K = \frac{\mu C_0 W}{2L}$$

$$, \ k = \frac{i'}{i} \geq 1 \ , \quad \text{and} \quad \sqrt{KI} = \frac{I}{V_{GS\,1,2} - V_{T\,1,2}} = G_{meff} \qquad (1)$$

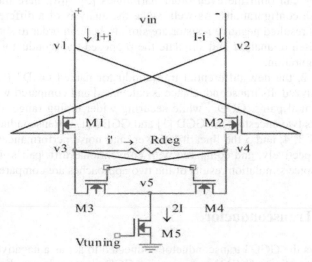

Fig. 1. GGD3 transconductor connected to act as a negative resistor

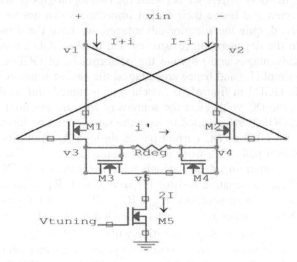

Fig. 2. GGD transconductor connected to act as a negative resistor [1]

The idea of including the coefficient k in equation (1) will be cleared in section 3. If k=1, then equation (1) is the same as what is calculated for the conventional differential pair in figure3, designed using N-MOSFETs whose sources are degenerated by the degeneration resistors ($R_{deg}/2$). However, in GGD3, since no DC current passes through R_{deg}, Gm can be changed by changing I and R_{deg} independently, if R⁻ is considered properly. This fact offers a degree of freedom to the designer and makes GGD3's Gm more control friendly. On the other hand, if k > 1, GGD3 gives less Gm than the conventional differential pair. Though, this may seem to be a negative point for GGD3, later in section 3 it will be shown that this helps GGD3 to be more linearized (source-degenerated) than its conventional counterpart. The same effect is caused by R⁻, as, in the work region of our interest, this negative resistor causes R_{eq} to be more than R_{deg}. This means more degeneration and hence linearization of the circuit.

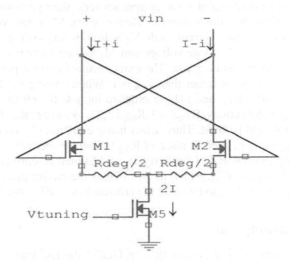

Fig. 3. Conventional differential pair transconductor connected to act as a negative resistor

3 Linearity

In figure1, the 2nd and 3rd harmonic distortions of GGD3 may be calculated as:

$$HD_3 = \frac{KI}{32\, I^2\, (1 + k.R_{deg}\, .\sqrt{KI}\,)^3}\, V_m^2 \quad \text{and} \quad HD_2 = 0 \tag{2}$$

Where V_m is the peak value of the sinusoidal assumed input voltage vin, and the other parameters are the same as in equation (1). Again, if k=1, then equation (2) is the same as what is calculated for the conventional differential pair in figure3. In equation (2), HD_2 is zero (as expected) and HD_3 is a function of the DC current I, k.R_{deg}, and K. K is affected by the secondary effects, the analysis of which can be found elsewhere [5]. Here, just the effects of k.R_{deg} and I are discussed. Since in GGD3 no DC current passes through R_{deg}, if the DC current I in equation (2) changes, the DC voltage across R_{deg} does not change, and hence $\sqrt{KI}$ is not affected. Also, $\sqrt{KI}$ does not change, if R_{deg} changes. Thus, HD_3 can be adjusted by changing R_{deg} and $\sqrt{KI}$ while the DC voltage across the former and the magnitude of the latter are independent of each other. Again, this fact offers a degree of freedom to the designer and makes GGD3's HD_3 more control friendly than the conventional diff. pair's in figure3. Another significant advantage of GGD3 is that, due to the special connection of $M_{3,4}$, the ac current passing through R_{deg} (i.e., i' = i $_{rdeg}$) is more than input/output ac current (i.e. i $_{vin}$ = i). This fact provides an added value to the degeneration resistor (i.e. R_{deg}) in GGD3, and hence, compared to the conventional differential pair, using the same value for the degeneration resistor, an extra linearization (source-degeneration) is achieved. This phenomenon may be explained as follows:

$M_{3,4}$'s gate-source connections receive ac voltages opposite the ac voltages applied to the $M_{1,2}$'s gate-source connections respectively. Hence, when M_1's gate voltage (v_{G1}) increases, M_3's drain voltage increases, but its gate voltage (v_{G3}) decreases. As $M_{3,4}$ are always in saturation and act as current sources, their gate-source voltages are dominant in changing their drain currents. Hence, when M_1's gate voltage increases, its source wants to give more current while M_3's drain wants to draw less current. The same analysis is true for M_2's gate voltage and M_4's drain current with polarities opposite to their discussed counterparts. The extra resulted current passes through R_{deg} and hence i_{rdeg} = i' becomes larger than i $_{vin}$ = i. When writing the KVL around the loop including v_{in}, DS_1, R_{deg}, and SD_2, in order to include the effect of this extra current the following substitutions ($R_{deg}.i'$ = k.$R_{deg}.i$ $_{vin}$, k>1) or ($R_{deg}.i'$ = R'$_{deg}.i$, R'$_{deg}$ = k.R_{deg} , k>1) should be used. Thus, when using equation (2) to calculate HD$_3$ for GGD3, R'$_{deg}$>R_{deg} must be used in place of R_{deg} to include the effect of the additional ac current passing through R_{deg}. Obviously, this additional ac current may be considered as an added value to R_{deg}, which in turn makes the circuit more linear than normally expected by using R_{deg} to linearize the conventional differential pair.

4 Power Consumption

If figures1,3 are compared, it is seen that, in GGD3 the DC current passes through $M_{3,4}$ in place of the two $R_{deg}/2$ in diff. pair. From the DC viewpoint, $V_{DS3\ or\ 4}$ in figure1 replaces the ($R_{deg}/2$).I in figure3. Here, keeping the other parts in figures1,3 the same, based on the sizes chosen for $M_{3,4}$, GGD3 may be either more or less power consuming than the conventional differential pair. On the other hand, Regardless of the magnitude of R_{deg}, if the two extra transistors are chosen properly, GGD3 can consume much less power than its conventional counterpart. Because, in conventional diff. pair, when R_{deg} is changed to linearize the transconductor, both the linearity and the power consumption are affected. But, in GGD3, the magnitude of R_{deg} and the linearity obtained from it does not affect the power consumption. Hence, the power consumption and the linearization may be treated individually, if R$^-$ is considered properly. This can be considered as a significant advantage for GGD3.

5 Noise Analysis

With a short look at the two circuits in figures1,3, GGD3, due to the presence of $M_{3,4}$, seems to be noisier than the conventional differential pair. However, later we see that as the simulations show, GGD3 is much less noisy than the coventional differential pair, if the noise is measured around a small resistor in series with the output current path instead of around R_{deg} as in [1,2]. One major reason to this is the very lower power that GGD3 consumes. It seems that, regarding the noise performance, nothing does work better than simulations. Here, the opportunity is taken to say that in GGD2

[1,2] the same phenomenon is true as simulations show. Hence, hereby it is declared that GGD2 [1,2] is also less noisy than the conventional pair, as there in [1,2] the noise has been mesured around R_{deg}, and there, because of the added value to the R_{deg}, the measured noise has been evaluated more than its real value at the output current.

6 Simulations

The n-channel MOSFET model has been taken from the TSMC 0.35micron technology. Due to the complexity of GGD3 because of the performance of $M_{3,4}$ as a negative resistor, in order to study GGD3's work regions, it is chosen to keep $M_{1,2}$'s and M_5's aspect ratios fixed at 280/1 and 40/1 respectively, and change $M_{3,4}$'s aspect ratios as 40/1, 80/1, 128.57/1, and 157.14/1 for GGD3a, GGD3b, GGD3c, and GGD3d respectively. For conventional diff. pair the aspect ratios of $M_{1,2}$ and M_5 are chosen 160/1 and 20/1 respectively. Connecting all substrates to ground, and keeping v1 and v2 at 1.5Vdc (2.5Vdc supply is assumed), the curves resulted from the Hspice simulations of GGD3 and diff. pair as the negative resistors in figures1,3 are drawn in figures 4 to 9. In these curves (with V_m=0.2v at 1.8GHz) the features of GGD3 may be analyzed as follows:

In figure 4, clearly in all cases GGD3 provides an added value to R_{deg} by passing an extra ac current through it. In figures 5 and 6, it seems that with R_{deg}=300Ohm, for low values of the aspect ratio of $M_{3,4}$, as in GGD3a curve, GGd3 produces less

$$Gm = \left| \frac{1}{Zreal} \right|$$ than the diff. pair, while its Gm is somewhat constant (it increases very

slightly) as Vtuning increases. This is due to the fact that, in this case, R⁻ is so large that in parallel with R_{deg} its effect is negligible. On the other hand,

in $$R_{eq} = \frac{R^- . R_{deg}}{R^- + R_{deg}}$$, since R⁻ is always negative, the numerator is always negative,

and the denominator decreases as R⁻ approaches R_{deg} which in turn increases R_{eq} and hence decreases the Gm of GGD3. This can be observed at curves GGD3b-d. Another characteristic of GGD3 is that, as in GGD3b-d is observed, when the aspect ratio of $M_{3,4}$ increses, the Gm starts increasing as Vtuning and hence the power increase. This phenomenon, in contrast with diff. pair and other transconductor topologies, is again due to the effect of R⁻. Because, R⁻ decreases as the DC current I increases. In GGD3a this decrement in not dominant, while in GGD3b-d it is dominant enough to overcome the increment of Gm caused by the increment of the DC current I. This means that, in this work region, GGD3, in the lower extreme of Vtuning, while giving its highest Gm, dissipates its lowest power. Obviously, this is contrary to all the other transconductor topologies. In figure 7, it is absolutely clear that GGD3, in all cases, is much less power consuming than the diff. pair. In figure 8, again it is clear that GGD3 is essentially very-low-noise compared to the conventional diff. pair. In figure 9, GGD3a seems much less linear than the diff. pair. With a short look at figure 7 the reason becomes clear. As a matter of fact, in this case, GGD3, compared to the diff. pair, consumes such a lower power than the diff. pair, that the effect of its provided added value to R_{deg} can not be observed, since the linearity increase with DC current I and k.R_{deg} both (see equation 2). Clearly, in GGD3b-d, in which the DC current I and the added value coefficient k are increased noticeably compared to the GGD3a,

GGD3 has become more linear than the conventional diff. pair. The last simulation compares the speed of the two topologies as in figure 10. In order to compare the speeds, the same power consumption and DC voltages at the drains of $M_{1,2}$ has been considered for both transconductors. This helps evaluate the effect of the added $M_{3,4}$ in GGD3 on the speed of this topology. Obviously, GGD3 has essentially the same speed as the diff. pair's. This is a very significant advantage for GGD3, since it uses extra elements and gains lots of features, but does not degrade the target speed. To obtain the same power and voltage at the drains of M1,2, for both GGD3 and diff. pair, it is chosen for GGD3: (M1,2=280/1, M3,4=157.14/1, M5=40/1, M6,7=100/1, Vtuning=2.35v, Rdeg=80Ohm), and for diff. pair: (M1,2=280/1, M5=40/1, Rdeg/2=50Ohm, Vtuning=1.001v and M6,7=100/1). Where M6,7 are the diode connected loads.

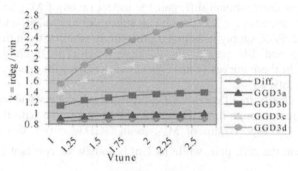

Fig. 4. The added value coefficient (k) that GGD3 provides for the degeneration resistor (i.e. R_{deg}) used to linearize the transconductor (R_{deg} = 300ohm)

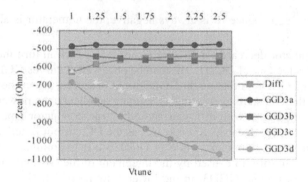

Fig. 5. GGD3 and conventional differential pair negative resistances (R_{deg} = 300ohm)

7 Conclusions

A novel approach to the design of a linearized widely tunable very low power and low noise differential transconductor, named GGD3 has been proposed.

- GGD3 is a simple, tunable differential circuit. Having different work regions, in either case, GGD3 is essentially a linear, very low power and low noise transconductor compared to the conventional diff. pair.

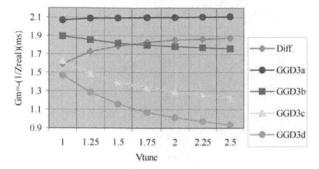

Fig. 6. GGD3 and conventional differential pair Gm (R$_{deg}$ = 300ohm)

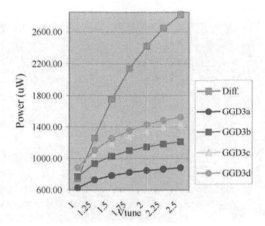

Fig. 7. GGD3 and conventional differential pair power consumptions (R$_{deg}$ = 300ohm)

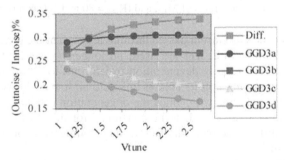

Fig. 8. GGD3 and conventional differential pair input to output noise transfer measured around a 1.0 ohm resistor in series with output (R$_{deg}$ = 300ohm)

- GGD3 is Gm and HD3 control friendly (i.e., because the degeneration resistor R$_{deg}$ and the DC current path are isolated from each other, in contrast with the conventional diff. pair, GGD3's Gm and HD$_3$ may be controlled more easily). Also, considering CMRR, the tuning range of GGD3 is much better than the conventional diff. pair. This is due to way that M$_{3,4}$ are connected to work as current sources with large output impedances seen at M$_{3,4}$'s drains individually.

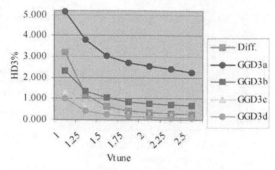

Fig. 9. GGD3 and conventional differential pair Harmonic Distortions (R_{deg} = 300ohm)

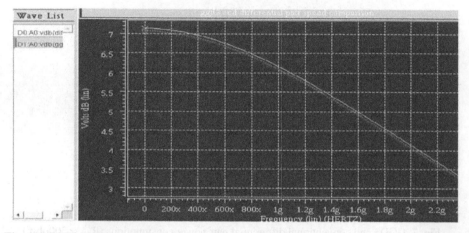

Fig. 10. GGD3 and conventional diff. pair's speeds compared under the same power consumption, the same DC voltages at Drains of $M_{1,2}$, and (R_{deg} = 80 and 100ohm respectively)

- Since in GGD3 no DC current passes through R_{deg}, when the technologies with the smaller minimum feature sizes are used, no longer do we need to be concerned about the heat that the poly degeneration resistor has to withstand because of the DC current passing through it. This means that, no longer do we need to increase the width of the poly line with which the degeneration resistors are made, and as a result no extra chip area needs to be consumed for this reason. Compared to the conventional differential pair, this feature may compensate for the extra chip area that $M_{3,4}$ consume.
- Since GGD3 gives an additional value to R_{deg}, compared to the conventional diff. pair, with the same magnitude of the degeneration resistor, more linearization can be achieved. This means less chip area for the linearization purpose. Hence, GGD3 may be considered as an economic approach to the design of a linearized differential transconductor and/or negative resistor.
- In contrast with the diff. pair, in GGD3, the magnitude of R_{deg} (and hence the degree of the linearization) does not affect the power consumption. The same is true when Gm is changed by changing R_{deg}. These are two other important advantages, obtained from the isolation of R_{deg} from the DC current path.

- As GGD3 improves linearity while is less noisy, given the same power and Gm as the diff. pair, it is very promissing to yield much better dynamic range.

References

1. Hamid Reza Sadr M. N., "A Novel Linear, Low Noise, Low Power Differential Transconductor & A Novel Linearization Technique", 10th IEEE International Conference on Electronics, Circuits and Systems, ICECS2003, vol. II, pp. 412-415, Dec. 2003
2. Hamid Reza Sadr M. N., "A Novel Approach to the Linearization Of the Differential Transconductors", IEEE International Symposium on Circuits and Systems, ISCAS2004, vol. I, pp. 1020-1023
3. Dandan Li, "Theory and design of Active LC filters on silicon", Ph.D. thesis, Columbia University, 2000
4. William B. Kuhn, "Design of Integrated, Low power, Radio Receivers in BiCMOS Technologies", Ph.D. thesis, faculty of Virginia Polytechnic Institute and State University, 1996
5. Darwin Cheung, 10-MHz 60-dB Dynamic-Range 4th-order Butterworth low pass filter, M.Sc. thesis, Department of Electrical and Electronics Engineering, Hong Kong University of Science and Technology, 1996
6. R. Schaumann, "Continuous-time Integrated Filters - A Tutorial, "in Integrated Continuous-Time Filters, Tsividis, Y. P., and Voorman, J.O., Ed., New York, IEEE Press, pp. 3-14
7. Wu, P. and Schaumann, R. and Latham, P. "Design considerations for common-mode feedback circuits in fully-differential operational transconductance amplifiers with tuning, "IEEE International Symposium on circuits and Systems, vol. 5, pp. Xlviii+3177, Jun 1991
8. Czarnul, Z. and Takagi, S., "Design of linear tunable CMOS differential transconductor cells," Electronics Letters, vol. 26, no. 21, pp. 1809-11, Oct 1990
9. S. Szczepanski, J. Jakusz, and Rolf Schaumann, "A Linear Fully Balanced CMOS OTA for VHF Filtering Applications", IEEE Trans. Circuits Syst. II, vol. 44, no. 3, pp. 174-187, March 1997
10. Y. P. Tsividis, "Integrated Continuous-time Filter Design", IEEE 1993 custom integrated circuits conference

A New Model for Timing Jitter Caused by Device Noise in Current-Mode Logic Frequency Dividers

Marko Aleksic, Nikola Nedovic*, K. Wayne Current, and Vojin G. Oklobdzija

Department of Electrical and Computer Engineering,
University of California, Davis, CA 95616
maleksic@ucdavis.edu

Abstract. A new method for predicting timing jitter caused by device noise in current-mode logic (CML) frequency dividers is presented. Device noise transformation into jitter is modeled as a linear time-varying (LTV) process, as opposed to a previously published method, which models jitter generation as a linear time-invariant (LTI) process. Predictions obtained using the LTV method match jitter values obtained through exhaustive simulation with an error of up to 7.7 %, whereas errors of the jitter predicted by the LTI method exceed 57 %.

1 Introduction

Timing jitter (or phase noise, if observed in the frequency domain) is a major constraint in modern high-speed communication systems. Jitter imposes limitations to the maximum signaling rate for which the bit error rate (BER) does not exceed its maximum acceptable level, the minimum spacing between channels in order to avoid inter-channel interference, etc. Unfortunately, jitter does not scale down with the signal period. As the signaling rates increase and are now in the range of tens of Gbps, even small amounts of jitter can severely impair the performance of these systems.

It would be very desirable to know the amount of jitter that will be caused by device noise of circuits in a system before the system is actually fabricated. This way, it would be possible to determine whether the system meets the requirements in the pre-fabrication phase of the design process, and hence, reduce the cost of the design. For that reason, a lot of research has been done on jitter and phase noise analysis of different circuits constituting precise frequency synthesizers, namely phase-locked loops (PLLs). Most of this research targeted voltage-controlled oscillators (VCOs) and several different theories emerged for predicting VCO jitter and phase noise, [1], [2], [3]. However, oscillators are not the only source of jitter in a PLL. As shown in [4], phase noise caused by the frequency divider, which is an integral part of every high-frequency PLL, can be amplified by the loop and occur at the PLL output, degrading performance of the subsequent circuitry. Beside PLLs, frequency dividers can also be found in multi-phase clock generators, multiplexers and demultiplexers of high-speed front-ends, etc. Therefore, there is an evident need for studying the jitter of frequency dividers.

Before we start analyzing jitter, it needs to be defined: jitter is a variation of the signal period from its nominal value. It is caused by various factors such as power

---

* The author is currently with Fujitsu Laboratories of America, 1240 E Arques Ave., M/S 345, Sunnyvale, CA 9485, USA

V. Paliouras, J. Vounckx, and D. Verkest (Eds.): PATMOS 2005, LNCS 3728, pp. 724–732, 2005.
© Springer-Verlag Berlin Heidelberg 2005

supply noise, substrate noise, cross-talk etc. In the analysis presented here, only jitter caused by the circuit device noise will be studied. The effects of other sources of jitter can be minimized through the circuit design (e.g. differential circuits are resistant to power supply noise) or technological process (e.g. circuits fabricated in a triple-well process are resistant to substrate noise). Since it is caused by random processes, jitter itself is a random process and is described by its variance or root-mean-square (RMS) value, rather than its instantaneous value.

Until recently, the only reliable way to predict jitter of frequency dividers was through exhaustive simulation. Several models for frequency divider jitter and phase noise were proposed, [5], [6], but these were based on experimental results and could not be applied to circuits other than those described in the studies. Even though empirical, model described in [5] revealed some properties of frequency divider jitter that can be used to simplify the analysis. Since multi-stage frequency dividers are often implemented as asynchronous counters, total jitter at the output of a multi-stage divider is the sum of jitters of each stage. Therefore, stages can be analyzed only one at a time. The first analytical model was proposed in [7]. This model was an application of a VCO jitter model described in [1] to a current-mode logic (CML) frequency divider, and it models generation of jitter as a linear time-invariant (LTI) process.

The work presented in this paper proposes a new way of modeling the process of device noise transformation into jitter in CML frequency dividers. The focus is on CML circuits since that is the usual design technique of choice in high-speed systems. Jitter generation is modeled as a linear time-varying (LTV) process and the model achieves more accurate predictions than the LTI model.

2   CML Frequency Divider Circuit Design

As will be shown shortly, jitter generation of a circuit is closely related to the circuit topology and output waveforms. For that reason, Fig. 1 shows a block-diagram of a CML frequency divider-by-two. It consists of a master-slave latch connected as a T-flip-flop and a level-shifter. Transistor-level schematics of the latch and the level-shifter are given in Fig. 2. The circuit was implemented in a triple-well process by Fujitsu, with the minimum channel length of 110 nm and a power supply voltage $V_{DD} = 1.2$ V. Maximum frequency of the input signal for which the circuit still operates properly is 10 GHz. CML design technique employs low-swing differential-voltage signaling and in this case, circuits were designed for a nominal output swing of 300 mV. Waveforms of the master-slave latch and the level-shifter output voltages around the switching time are given in Fig. 3. Active (rising) edge of the input signal (v_{in} in Fig. 1) arrives at $t = 0$ in Fig. 3. For all the following considerations, it will be assumed that the input signal is perfect and does not contain any jitter.

3   The LTI Frequency Divider Jitter Model [7]

According to the model proposed in [7], noise can affect the output crossing time only around the nominal crossing time. Then, variance of the output jitter is equal to the variance of the output-referred voltage device noise, multiplied by the inverse of the squared derivative of the differential output voltage at the nominal crossing time:

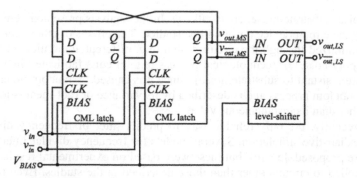

Fig. 1. Block-diagram of a CML frequency divider-by-two

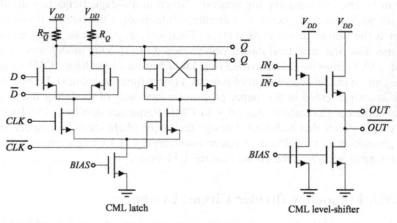

Fig. 2. Transistor-level schematics of the CML latch and the level –shifter

$$\sigma_T^2 = \left[\frac{\partial (v_{out} - v_{\overline{out}})}{\partial t} \right]_{@t_x}^{-2} \times \sigma_{v_n}^2 \quad (1)$$

In case of a CML frequency divider, variance of jitter at the output will be equal to the sum of variances of the master-slave latch jitter and the level-shifter jitter, assuming that noise sources in the latch and the level-shifter are uncorrelated:

$$\sigma_{T,tot}^2 = \left[\frac{\partial (v_{out,MS} - v_{\overline{out},MS})}{\partial t} \right]_{@t_{x,MS}}^{-2} \times \sigma_{v_n,MS}^2 + \left[\frac{\partial (v_{out,LS} - v_{\overline{out},LS})}{\partial t} \right]_{@t_{x,LS}}^{-2} \times \sigma_{v_n,LS}^2 \quad (2)$$

Since the circuits are differential, there are noise sources at both outputs. Assuming that these noise sources are mutually uncorrelated, total jitter variance will be double the value given in (2):

$$\sigma_{T,tot}^2 = 2 \times \left\{ \left[\frac{\partial (v_{out,MS} - v_{\overline{out},MS})}{\partial t} \right]_{@t_{x,MS}}^{-2} \times \sigma_{v_n,MS}^2 + \left[\frac{\partial (v_{out,LS} - v_{\overline{out},LS})}{\partial t} \right]_{@t_{x,LS}}^{-2} \times \sigma_{v_n,LS}^2 \right\} \quad (3)$$

Jitter RMS value is found as the square-root of the variance given in (3), i.e. $\sigma_{T,tot}$.

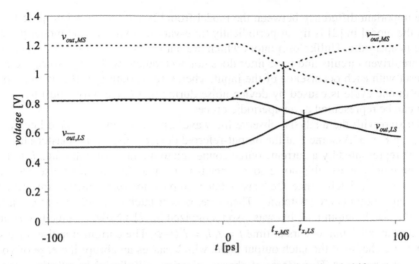

Fig. 3. Output voltages of the master-slave latch and the level-shifter

According to (1), impulse response function of an abstract system that converts device noise into jitter can be defined as:

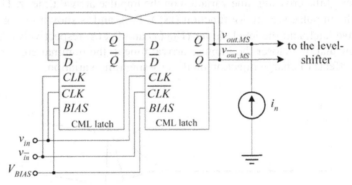

Fig. 4. Setup for determining the impulse sensitivity function (ISF) of the master-slave latch

$$h(t) = \left[\frac{\partial (v_{out} - v_{\overline{out}})}{\partial t} \right]^{-1}_{@t_x} \times \delta(t) \; , \tag{4}$$

where $\delta(t)$ is the Dirac impulse. The system defined by (4) is a linear time-invariant (LTI) system, since its response does not depend on the occurrence time of the input (i.e. noise, in this case).

4 The Proposed LTV Jitter Model

Frequency divider jitter model proposed here was inspired by the oscillator phase noise model presented recently in [2], even though representation of electric oscillators as linear time-varying systems for noise was known even in the late 1960s [8].

One important difference between the model from [2] and the model proposed here is that the model in [2] is linear periodically time-varying, whereas the model presented here is aperiodic. Unlike oscillators which are autonomous circuits, frequency dividers are driven circuits and their jitter does not accumulate with time, since the timing is reset with each occurrence of the input. Therefore, variation of the output crossing time in one cycle is caused by device noise during that cycle only, and jitter generation can be represented as an aperiodic process.

To start with the analysis, observe the waveform of the master-slave latch output, $v_{out,MS}$ in Fig. 3. Assume that the output-referred device noise of the master-slave latch can be represented by a current source connected to the latch output node, i_n in Fig. 4. (To be more precise, this noise source needs to include only device noise originating from the slave latch, since the master latch outputs are stable at the time when the slave latch outputs are switching. Therefore, master latch noise cannot affect jitter of the slave latch output and this was also recognized in [7].) Now, assume that source i_n is a unit impulse that occurs at time $t = \tau$, $i_n = \delta(t-\tau)$. The current impulse injects 1 C of electric charge at the latch output node, which causes an abrupt increase of voltage $v_{out,MS}$ at time $t = \tau$. The effects of charge injection will diminish with time, but will also affect the crossing time of the master-slave latch outputs. Naturally, if the impulse occurs after the nominal crossing time (i.e. $\tau > t_{x,MS}$) charge injection will not cause any crossing time variation. It is possible to define a function that shows the dependence of the crossing time variation on the impulse arrival time, τ. This function is called the impulse sensitivity function (ISF), $\Gamma(\tau)$, and is shown in Fig. 5 for both master-slave latch and the level shifter ($\Gamma_{MS}(\tau)$ and $\Gamma_{LS}(\tau)$ respectively). The ISFs in Fig. 5 show that, the closer the impulse arrival time to the nominal crossing time, the larger the effects of charge injection on the crossing time variation.

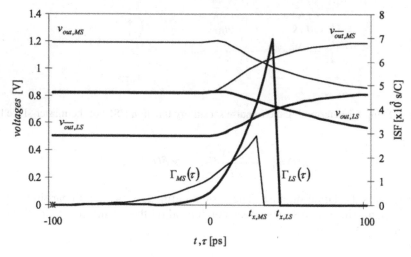

Fig. 5. ISFs of the master-slave latch and the level-shifter ($\Gamma_{MS}(\tau)$ and $\Gamma_{LS}(\tau)$, respectively)

Once $\Gamma(\tau)$ is known, we can define the impulse response function of the noise-to-jitter conversion system, $h(t, \tau)$:

$$h(t,\tau) = \Gamma(\tau) \times u(t-\tau) \; , \tag{5}$$

where $u(t)$ is the unit step function. Now it is obvious that this system is an LTV system, since its response depends on the impulse arrival time.

The response of an LTV system to an input signal $x(t)$ is given by the convolution integral:

$$y(t) = \int_{-\infty}^{+\infty} h(t,\tau) x(\tau) d\tau \; . \tag{6}$$

If the input to the system is a random process whose autocorrelation function is given by $R_X(t,\xi)$, autocorrelation of the resulting random process, at the output of the LTV system, $R_Y(t,\xi)$, will be [9]:

$$R_Y(t,\xi) = \int_{-\infty}^{+\infty} \int_{-\infty}^{+\infty} h(t,r) h(\xi,s) R_X(r,s) dr ds \; . \tag{7}$$

To demonstrate the proposed LTV model for the noise-to-jitter conversion, we will assume that the output-referred noise of the slave latch (i_n in Fig. 4) is white and has a RMS value $\sigma_{n,MS}$. Then, the autocorrelation function, $R_{n,MS}(t,\xi)$, of that noise source is:

$$R_{n,MS}(t,\xi) = \sigma_{n,MS}^2 \times \delta(t-\xi) \; . \tag{8}$$

Combining (5), (7) and (8), with $\Gamma(\tau) = \Gamma_{MS}(\tau)$ and $R_X(t,\xi) = R_{n,MS}(t,\xi)$, it can be shown that the autocorrelation of the output random process (i.e. master-slave latch jitter), $R_{T,MS}(t,\xi)$, is given by the following expression:

$$R_{T,MS}(t,\xi) = \sigma_{n,MS}^2 \times \int_{-\infty}^{\min\{t,\xi\}} \Gamma_{MS}^2(s) ds \; . \tag{9}$$

Since the output random process is jitter (i.e. crossing time variation), it would be natural to evaluate (9) at the nominal crossing time, $t = t_{x,MS}$. In addition, variance of a random process is given as the autocorrelation function $R_X(t,\xi)$ for $t = \xi$. Hence, variance of the master-slave latch jitter will be equal to $R_{T,MS}(t_{x,MS}, t_{x,MS})$:

$$\sigma_{T,MS}^2 = R_{T,MS}(t_{x,MS}, t_{x,MS}) = \sigma_{n,MS}^2 \times \int_{-\infty}^{t_{x,MS}} \Gamma_{MS}^2(s) ds \; . \tag{10}$$

Note that evaluating (10) at a time instant $t > t_{x,MS}$ will not change the result, since $\Gamma_{MS}(s) = 0$ for $s > t_{x,MS}$, which is shown in Fig. 5.

Equation (10) does not give the total variance of the master-slave latch jitter. Remember that another noise source exists at the other output of the master-slave latch, which will also affect the output crossing time. We will assume that the RMS value of this noise source is also $\sigma_{n,MS}$ and that the cross-correlation between the noise sources at the two outputs is zero. In addition we will assume that the rising and falling edges of the master-slave output signal are symmetrical, hence, ISFs of both outputs of the master-slave latch are the same. Under these assumptions, total variance of the master-slave output jitter is double the value given in (10).

Similar analysis can be conducted for jitter caused by the level-shifter device noise. In this case, a relationship similar to (10) can be derived, but using the level-shifter

ISF, $\Gamma_{LS}(\tau)$, and the RMS value of the level-shifter output-referred device noise, $\sigma_{n,MS}$. (Again, it will be assumed that the noise is white.) Also, in this case, jitter is evaluated at the nominal crossing time of the level-shifter outputs, $t_{x,LS}$.

Finally, if noise sources in the master-slave latch and the level-shifter are not correlated, the total jitter variance of the CML frequency divider-by-two is given by the following expression:

$$\sigma_{T,tot}^2 = 2 \times \left[\left(\sigma_{n,MS}^2 \times \int_{-\infty}^{t_{x,MS}} \Gamma_{MS}^2(s)ds \right) + \left(\sigma_{n,LS}^2 \times \int_{-\infty}^{t_{x,LS}} \Gamma_{LS}^2(s)ds \right) \right] \tag{11}$$

Total jitter RMS value is given as the square-root of the variance in (11).

5   Comparison of the Two Jitter Models

The LTI and the LTV jitter models were used to predict jitter of the one-stage CML frequency divider from Fig. 1 and 2. The predictions were compared against jitter results obtained through exhaustive simulation in HSPICE. In all three cases, output-referred noise was white, and the RMS values of noise at each output of the master-slave latch and the level-shifter were $\sigma_{n,MS} = 1.61 \times 10^{-12}$ A and $\sigma_{n,LS} = 1.73 \times 10^{-12}$ A, respectively. These values were determined using the NOISE analysis in HSPICE, with DC conditions that are equal to the instantaneous voltages and currents at the moments when the master-slave latch and the level-shifter outputs are crossing, $t_{x,MS}$ and $t_{x,LS}$.

For the LTI model, derivatives of the output voltages were obtained through simulation in HSPICE, using the DERIVATIVE function in the MEASURE statement. Variances of the output-referred voltage noise sources were calculated using the following relationships:

$$\sigma_{Vn,MS}^2 = \int \sigma_{n,MS}^2 |Z_{MS}(f)|^2 df$$
$$\sigma_{Vn,LS}^2 = \int \sigma_{n,LS}^2 |Z_{LS}(f)|^2 df \tag{12}$$

where $Z_{MS}(f)$ and $Z_{LS}(f)$ are output impedances of the master-slave latch and the level-shifter, which were obtained in HSPICE. Finally, jitter was found using (3).

For the LTV model, ISFs of the master-slave latch and the level-shifter were found in HSPICE in the following manner: a charge was injected via a short current impulse, while changes of the output crossing times were measured for different impulse arrival times. (The ISFs shown in Fig. 5 were obtained this way.) Nominal crossing times $t_{x,MS}$ and $t_{x,LS}$ were also found in HSPICE. Finally, jitter was calculated using (11).

To compare the accuracy of the LTI and the LTV jitter models, jitter values obtained through exhaustive transient simulation in HSPICE were used. However, in the HSPICE transient analysis, it is not possible to specify noise sources. Therefore, device noise from the master-slave latch and the level-shifter was simulated by four piecewise linear (PWL) current sources connected to the outputs of the circuits. Instantaneous values of these sources were determined using a random number generator in MATLAB, while keeping their RMS values equal to $\sigma_{n,MS}$ for the sources at the master-slave latch outputs, and $\sigma_{n,LS}$ for the sources at the level-shifter outputs. Time step of the PWL sources was small enough so that the power spectral density of the sources was white over the bandwidth of interest. The four PWL sources were mutually uncorrelated, which was verified in MATLAB. Finally, the transient analysis was run for 100 periods of the frequency divider output signal, and variations of the crossing times of the outputs were recorded. Simulation time was over 24 hours. In contrast, time needed to obtain data for the LTI and the LTV model from HSPICE was in the range from a few minutes for NOISE analyses to an hour for determining the ISFs for the LTV model.

The results are summarized in Table 1. Jitter values are given for three cases of the biasing current: case 1 is the nominal value, case 2 is a value 20 % less than nominal, and case 3 is a value 20 % greater than nominal. Last two columns in Table 1 show relative errors of the jitter values predicted by the LTV and the LTI models, with respect to the values obtained through the exhaustive HSPICE simulation.

It can be seen from Table 1 that the new LTV model gives jitter predictions which are much closer to the values obtained through exhaustive simulation. The reason for this is that the LTV model integrates effects of noise prior and at the crossing time, while the LTI model evaluates effects of noise only at the nominal crossing time and hence predicts jitter values which are smaller than the actual ones.

6 Conclusions

A new method for predicting jitter caused by device noise in current-mode logic (CML) frequency dividers is presented. As opposed to a previously published method, the new method models device noise transformation into jitter as a linear-time-varying (LTV) process. Predictions of the LTV model match jitter values obtained through exhaustive simulation with an error of up to 7.7 %. In contrast, error of the predictions obtained by the previously published method exceeds 57 %.

In this work, the new jitter model was demonstrated for CML frequency dividers and white noise only. However, the method is not limited to this case. As long as the impulse sensitivity function (ISF) of a circuit and the autocorrelation function of the output-referred noise are known, the LTV jitter model formulated in (11) can be used to estimate jitter.

Table 1. Comparison of the LTI and the LTV jitter models with exhaustive simulation

	Exhaustive simulation $\sigma_{T,tot}$ [fs]	LTV model (this work) $\sigma_{T,tot}$ [fs]	LTI model [7] $\sigma_{T,tot}$ [fs]	Error LTV [%]	Error LTI [%]
case 1	62.2	65.9	26.6	5.9	57.3
case 2	82.8	89.2	36.0	7.7	56.5
case 3	48.8	52.5	21.6	7.7	55.7

Acknowledgements

This work was funded, in part, by SRC contract 01TJ923, NSF grant CCR01-19778 and Fujitsu Laboratories of America. The authors would also like to thank the staff of the Advanced LSI Technology Group of Fujitsu Laboratories of America and its director, William W. Walker, for valuable suggestions and comments.

References

1. J. A. McNeill, "Jitter in Ring Oscillators", IEEE Journal of Solid-State Circuits, Vol. 32, No. 6, pp. 870-879, June 1997.
2. A. Hajimiri and T. H. Lee, "A General Theory of Phase Noise in Electrical Oscillators", IEEE Journal of Solid-State Circuits, Vol. 33, No. 2, February 1998.
3. A. Demir, A. Mehrotra and J. Roychowdhury, "Phase Noise in Oscillators: A Unifying Theory and Numerical Methods for Characterization", IEEE Transactions on Curcuits and Systems-II, Vol. 47, pp. 655-674, May 2000.
4. A. Hajimiri, "Noise in Phase-Locked Loops", Invited Paper, Proc. of IEEE Southwest Symposium on Mixed-Signal Circuits, pp. 1-6, Feb. 2001.
5. W. F. Egan, "Modeling Phase Noise in Frequency Dividers", IEEE Transactions on Ultrasonics, Ferroelectrics and Frequency Control, Vol. 37, No. 4, pp. 307-315, July 1990.
6. V. F. Kroupa, "Jitter and Phase Noise in Frequency Dividers", IEEE Transactions on Instrumentation and Measurement, Vol. 50, No. 5, pp. 1241-1243, October 2001.
7. S. Levantino, L. Romano, S. Pellerano, C. Samori, A. L. Lacaita, "Phase Noise in Frequency Dividers", IEEE Journal of Solid State Circuits, Vol. 39, No. 5, pp. 775-784, May 2004.
8. K. Kurokawa, "Noise in Synchronized Oscillators", IEEE Transactions on Microwave Theory and Techniques, Vol. MTT-16, No. 4, April 1968.
9. W. A. Gardner, "Introduction to Random Processes: with applications to signals and systems", Second edition, McGraw-Hill Publishing Company, New York, 1990.

Digital Hearing Aids: Challenges and Solutions for Ultra Low Power

Wolfgang Nebel[1], Bärbel Mertsching[2], and Birger Kollmeier[1]

[1] Oldenburg University, Germany
[2] Paderborn University, Germany

Abstract. Demands for Digital hearing aids and other portable digital audio devices are becoming more and more challenging: On the one hand more computational performance is needed for new algorithms including noise reduction, improved speech intelligibility, beam forming, etc. On the other hand flexibility through programmability is needed to allow for product differentiation and longer lifetime of hardware designs. Both requirements have to be met by extremely low power solutions eing operated out of very small batteries. Hence, new hearing aids are examples of ultra low power designs and technologies.

The session comprises a tutorial introduction into modern signal processing algorithms of hearing aids and respective quality metrics. Next recent developments in industrial hearing aid architectures are presented. Finally new research results in power optimization of signal processing algorithms and circuit implementations suitable for hearing aids are shown.

V. Paliouras, J. Vounckx, and D. Verkest (Eds.): PATMOS 2005, LNCS 3728, p. 733, 2005.
© Springer-Verlag Berlin Heidelberg 2005

Tutorial Hearing Aid Algorithms

Thomas Rohdenburg, Volker Hohmann, and Birger Kollmeier

CvO University of Oldenburg, Faculty V – Medical Physics Group,
26111 Oldenburg, Germany
Thomas.Rohdenburg@uni-oldenburg.de
http://www.medi-ol.de/

Abstract. The normal-hearing system extracts monaural and binaural features from the signals at the left and right ears in order to separate and classify sound sources. Robustness of source extraction is achieved by exploiting redundancies in the source signals (auditory scene analysis, ASA). ASA is closely related to the "Cocktail Party Effect", i.e., the ability of normal-hearing listeners to perceive speech in adverse conditions at low signal-to-noise ratios. Hearing-impaired people show a reduced ability to understand speech in noisy environments, stressing the necessity to incorporate noise reduction schemes into hearing aids. Several algorithms for monaural, binaural and multichannel noise reduction have been proposed, which aim at increasing speech intelligibility in adverse conditions. A summary of recent algorithms including directional microphones, beamformers, monaural noise reduction and perceptual model-based binaural schemes will be given. In practice, these schemes were shown to be much less efficient than the normal-hearing system in acoustically complex environments characterized by diffuse noise and reverberation. One reason might be that redundancies in the source signals exploited by the hearing system are not used so far by noise reduction algorithms. Novel multidimensional statistical filtering algorithms are introduced that might fill this gap in the future.

Noise reduction schemes often require high computational load which results in a high power consumption. To reduce the computational expense one promising approach could be to reduce the numerical precision in specific parts of the algorithm or to replace costly parts by computationally simpler functions. This might lead to additional distortion in the signal that reduces the perceived audio signal quality. Quality Assessment is needed to control the negative effects of power optimization in the algorithms. However, subjective listening tests are time-consuming and cost-intensive and therefore inappropriate for tracking small changes in the algorithm. Objective quality measures based on auditory models that can predict subjective ratings are needed. We introduce a quality test-bench for noise reduction schemes that helps the developer to objectively assess the effects of power optimization on audio quality. Hence, a compromise between audio quality degradation and power consumption can be obtained in a fast and cost-efficient procedure that is based on auditory models.

For more information and related literature please refer to:
"http://www.physik.uni-oldenburg.de/Docs/medi/publhtml/
publdb.byyeardoctype.html"

V. Paliouras, J. Vounckx, and D. Verkest (Eds.): PATMOS 2005, LNCS 3728, p. 734, 2005.
© Springer-Verlag Berlin Heidelberg 2005

Optimization of Digital Audio Processing Algorithms Suitable for Hearing Aids

Arne Schulz and Wolfgang Nebel

CvO University of Oldenburg, Faculty II – Department of Computing Science,
26111 Oldenburg, Germany
Arne.Schulz@Uni-Oldenburg.DE

Abstract. In the current embedded systems market the hearing aid technology is one very special target. These very small devices suffer - more than usual mobile devices - on the one hand from insufficient power supplies and thus very limited operating times. On the other hand the need for improved signal processing is obvious: noise reduction, speech intelligibility, auditory scene analysis are promising enhancements for these devices to enable the hearing impaired user a more comfortable and normal life.

These facts make the design gap seen in normal embedded systems design even worse. Therefore early power estimation and optimization become more important. Nowadays hearing aids are often based on specially optimized DSPs and ?Ps primarily for commercial reasons like time-to-market, the ability of last minute changes and during lifetime adaptations of software to new research results or market demands. However, a number of computation intensive functions need to be performed by almost any digital hearing aid, e.g. Fourier Transforms, band pass filters, Inverse Fourier Transform. These can be realized much more power efficiently by dedicated HW-coprocessors, which may be parameterized. Hence Low Power design for hearing aids includes: algorithm design and optimization, HW-/SW-partitioning, processor selection, SW-optimization and co-processor design and optimization.

To meet this need, power estimation even on the algorithmic level has become an important step in the design flow. This helps the designer to choose the right algorithm right from the start and much optimisation potential can be used due to the focus on the crucial parts. Here we focus on the co-processor design and power optimization.

The power consumption of a co-processor is primarily determined by the dynamic switching power during calculations and the static leakage power. An estimation of both contributors requires analyzing the architecture and its activity during the expected operation of the device. In early stages of the design neither information is available yet. A new analysis and optimization tool has been developed to generate this information from an executable specification and a typical application scenario. The tool ORINOCO, which was developed at the OFFIS research institute and now is being commercialized by ChipVision Design Systems, analyzes a design specification given in C or SystemC and automatically generates the necessary information of a power optimized architecture implementing that specification and the activity within that

V. Paliouras, J. Vounckx, and D. Verkest (Eds.): PATMOS 2005, LNCS 3728, pp. 735–736, 2005.
© Springer-Verlag Berlin Heidelberg 2005

architecture. It further includes power macro models of the components of that architecture which allow making an estimate of the to-be-expected static and dynamic power consumption of a co-processor. It thus enables power optimization in the earliest and thus most efficient and effective phases of a design. The talk gives an overview of the actual approach of designing hardware implementations of digital audio signal processing algorithms under strong power constraints. The main focus of the talk is on the analysis and optimization of algorithms in the early stage of design and on the modelling of components needed for high-level power estimation. The different optimization steps will be illustrated by a co-processor design of a hearing aid.

References

W. Nebel. System-Level Power Optimization. Invited Keynote in: Euromicro Symposium on Digital System Design (DSD 2004), Rennes, France,2004, pp 27-34

Optimization of Modules
for Digital Audio Processing

Thomas Eisenbach, Bärbel Mertsching, Nikolaus Voß, and Frank Schmidtmeier

GET Lab, University of Paderborn
eisenbach@upb.de

Abstract. Today the demands of power optimized strategies have become one of the biggest challenges for developers. The sooner the aspect of low power is considered in the design process the greater the effect is on power consumption of the system design later on. Therefore the developer should have access to power information at an early stage of the design process.

Our approach is to develop a design framework which allows a high level specification of a given algorithm while allowing a consistent hardware implementation of a digital signal processing system across all steps in the tool-chain, from the behavioral specification up to the placed and routed netlist. In this paper we present how to use this framework for the optimization of digital audio processing modules and show some results.

1 Introduction

We begin by describing the background of our research project. In the second section we give an overview of our design framework. Then the example DSP system will be described, followed by an overview on the hardware design workflow with our framework for an algorithm which is part of this system in the fourth section and a description of the power estimation in section five. We close with some results we achieved with this framework supported workflow.

1.1 The Context: The PRO-DASP Project

The project PRO-DASP[1] (Power Reduction for Digital Audio Signal Processing) is carried out together with the MEDI group and the EHS group of the University of Oldenburg. In this co-operation the GET Lab is developing low power architectures for different signal processing algorithms applicable for instruments for the hearing impaired.

The DSP algorithms are provided by the MEDI group. As hearing aid systems are wearable devices, they should be small and lightweight. A DSP system inside a hearing aid device is operating out of a small battery and should consume

---

[1] We gratefully acknowledge partial funding of this work by the Deutsche Forschungsgemeinschaft under grant Me 1289/6 "PRO-DASP"

V. Paliouras, J. Vounckx, and D. Verkest (Eds.): PATMOS 2005, LNCS 3728, pp. 737–746, 2005.

as little power as possible to provide a long battery lifetime. Hence implementing hardware architectures for hearing aid algorithms is a challenging task for developers providing a huge demand for support tools as e.g. a development framework.

In this paper we present the implementation of a low power architecture for the so called gammatone resynthesis algorithm. This work is an example for the typical workflow with the framework.

2   Framework Overview

Today's trends in semiconductor technology cause two main problems which require a new design methodology. These trends are the availability of cheap and powerful silicon and the demand for complex applications which led to large designs. On the one hand these large designs consume a lot of power and on the other hand their development including design and verification increases the time to market. The design gap expands.

Our approach to address both problems is to develop a framework with an integrated low power design methodology which assists a design flow over all levels of abstraction starting from a high level specification. Our framework contains a hierarchical macro module library providing modules of often used DSP and arithmetic functions.

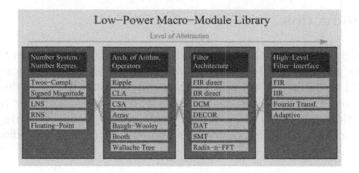

Fig. 1. Hierarchical Macro Module Library

The designer is able to construct the algorithm to be implemented within the framework by putting together several modules out of the library depicted in Fig. 1. Depending on the application context, there can be chosen high level modules like filters or different Fourier transformations or low level modules, which implement basic arithmetic functions like addition and multiplication for various number systems and number representations.

As, depending on the statistic properties of the input signal and the data flow of a given algorithm, the right choice of the number representation can significantly reduce the power consumption in terms of bit activity.

To connect the system level to the behavioral level each of these framework modules consists of a C++ part and a behavioral VHDL part which behave identical. By this, we met one of our goals, the closure of the design gap, as the workflow can be started on the system level with the implementation of a C++ prototype of a given algorithm.

These framework supported implementation steps are described in more detail in the subsection "Preliminary Inspections" later in this paper. By starting with a C++ implementation all inspections can be done very flexible and faster as on RT-level and with regard to an efficient power optimization this is also important, because design decisions on an early stage of the design process have a greater effect on the design's power consumption as on later stages.

After the high level specification and the preliminary inspections of the DSP algorithm, the VHDL part of each chosen macro module can be used to explore different architectures, which is described in more detail in subsection "Exploration of Architectures". At the architectural level the developer is able to chose special low power architectures, provided by the framework, activate power management functionality like gated clocks or glitch minimization or use the propagation of constants or an isolation of operands both provided by our framework.

As there are different power minimizing capabilities on all abstraction layers of the design process and intercommunicable effects between these abstraction layers, it is important to have an enclosing and continuous design methodology. To ensure this also for levels below the behavioral one, we adapt our framework to a tool-chain of commercial synthesis tools.

At all stages, a functional verification of the implemented DSP system can be done by evaluating its audio quality with an integrated audio quality testbench.

The main emphasis in this paper is put on a typical design workflow with our framework on the basis of an exemplary algorithm and show some power estimation results we achieved with our power estimation flow. As the chosen example is a simple one, only parts of the available features of our framework were necessary for its implementation. A more detailed description of the framework can be found in [ ].

3 Digital Signal Processing System

The core audio processing system of the digital hearing aid of our co-operation partner MEDI is based on a frequency analysis of the audio signal (called gammatone filter bank), a subsequent processing of each of the frequency bands (e.g. a level adaptation, noise reduction or dynamic compression), and a final reintegration of the split-up signal into a single audio stream. This system is depicted in Fig. 2.

3.1 Example Algorithm: The Resynthesis

An inverse gammatone filter bank is used to reintegrate the frequency bands (resynthesis). (see [ ] [ ]). In this paper we focus on this resynthesis algorithm.

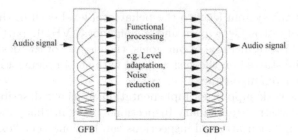

Fig. 2. Overview of the core system [ ]

As the different bandpass filters of the gammatone filter bank cause different group delays for each frequency band, a simple summation of all band signals can cause destructive interferences which can result in poor signal quality. Thus the output signal should not be calculated by a simple summation of all band signals, rather before a summation, all different group delays have to be compensated. This can be achieved by delaying the band signals of higher frequencies more than the lower frequency signals.

According to [ ] the signal of the highest band has to be delayed with a maximal delay of 14 ms to compensate the difference to the longest delay time of the lowest band. As a hearing aid user will be irritated if a delay between auditory and visual impression is greater than 4 ms, this is unacceptable.

Therefore each of the gammatone filter output signals $x_i(k)$ is delayed with a frequency channel-dependent delay time $D_i \leq 4$ ms and multiplied with an also frequency channel-dependent complex phase factor C_i.

$$y(k) := \sum_{i=1}^{30} C_i \cdot x_i(k - D_i) \qquad (1)$$

In this way all channels have their envelope maximum (by delaying) and their fine structure maximum (by multiplying with a complex factor) at the same instant of time.

4   Hardware Design Workflow

This section contains an overview over all necessary steps from the algorithm specification to a synthesizeable VHDL description using our framework.

The workflow steps can be separated into two main parts. The first part is a preliminary inspection of all system requirements such as bit-widths, etc. This work proceeds in the upper area depicted in Fig. 3. Here the whole DSP system is implemented in software (block C-simulation) using C++ modules. This software implementation acts as a reference for all further steps. The second part, which mainly proceeds in the lower area of Fig. 3 is an exploration of different low power architectures.

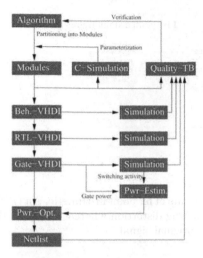

Fig. 3. Design flow and tool-chain

4.1 Preliminary Inspections

The DSP algorithms are originally implemented in Matlab using 32-bit floating point numbers. To implement such a system with our framework, the following steps have to be done:

1. Partition the algorithm into modular blocks (e.g. analysis, processing, resynthesis)
2. Create a C++ implementation of the system as reference implementation
3. Replace the reference implementation step by step with modules from the macro-library (e.g. arithmetic operators, filters)
4. Constrain the modules (e.g. specify a bit-width for the operators)
5. Validate the system by comparing the processing result with the result of the reference implementation.

Once we have built the C++ reference implementation we replace as many parts of it as possible with modules of the framework's macro module library. In case of our example dsp-algorithm, the resynthesis, only low level modules like number representations are chosen out of the macro module library. After this the developer is able to simulate the bit-accurate behavior of a later hardware system simply be running the executable of the C++ implementation which is faster and more flexible as a hardware simulation.

As all of these bit-accurate modules such as arithmetic operators of different number representations and filters can be parameterized, the developer can easy analyze the effect of its parameter choice to quantization errors. To ensure that the signal processing does not distort the signal beyond a threshold, our framework embodies an audio quality testbench. With this audio quality testbench, developed by the MEDI group, a psycho acoustically motivated measurement of the output signal quality can be performed ([ ]). Fig. 4 shows results of a fix point implementation of the resynthesis algorithm.

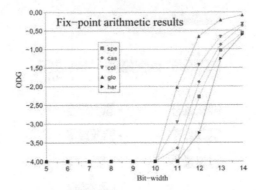

Fig. 4. Quality measure results of fix point arithmetic. An ODG (Objective Differential Grade) of -4 means very strong distortion whereas -1 means that most persons hardly hear the difference to the original signal

By iterating the above mentioned steps the developer can find the minimal necessary bit-width of all operator parameters. In our case study a satisfactory output signal quality ($ODG \geq -1$) for the resynthesis algorithm can be achieved with fixed point arithmetic of 14 bits.

4.2 Exploration of Architectures

The next step is the implementation of the VHDL hardware description of the DSP system. Due to the fact that all bit-accurate C++ modules of the framework are combined with VHDL-module generators, the developer is able to easily generate VHDL modules with the chosen parameters. Now it is possible to explore different architectures of the given algorithm. In this paper we will concentrate on another aspect of the architecture exploration.

In the case of the resynthesis system, we build an adjustable architecture, with that it is possible to adjust a grade of parallelism. We used this system to analyze how different architecture parameters and design constraints effect the power consumption of the DSP system. Therefore we developed a power estimation flow which is supported by our framework. This will be shown in the next section.

5 Power Estimation

In our framework we used a tool-chain consisting of Synopsys behavioral compiler for the step from the behavioral level to the Register Transfer (RT) level and design compiler for the step from RT level to the gate level. For an evaluation of low power architectures we developed a power estimation flow for which we extended the tool-chain with Synopsys' Power Compiler, Physical Compiler and PrimePower. The target technology has been a standard cell low power ASIC process such as UMCL 0.18 μ or Mietec 0.35 μ.

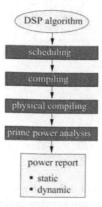

Fig. 5. Power estimation workflow

Our tool-chain starts on the behavioral level. The ready parameterized VHDL model of the DSP algorithm available in different grades of parallelism will be scheduled by the behavioral compiler. Depending on the grade of parallelism different state machines are generated. Next the scheduled hardware description will be synthesized by the design compiler to gate level using design ware components. Now the Physical Compiler creates a floor plan based on a fill factor of 70 % to allow the calculation of interconnect capacities. Finally a gate level simulation using Cadence's ncsim digital simulator combined with Prime Power is started. For the simulation the DSP system is feed with the 30 frequency bands of a typical audio sample. As a result, Prime Power provides a list of the average power consumption of the whole system and all parts separated in dynamic power and static power.

6 Power Estimation Results

In this section we will show results of our architectural explorations. For the simulation of the resynthesis, the DSP system is fed with the 30 frequency bands of a 100 ms speech sample. We simulated different architectures with a grade of parallelism of p=0 (sequential design, processes only one of all 30 frequency channels) up to p=3 (processes three frequency channels in parallel) and p=8 (hand optimized, processes all 30 frequency channels in parallel).

The results shown in the following figures only depict the power consumption of the resynthesis part of the simulated chip design.

6.1 Impact of Clock Gating

With our power estimation flow we were able to easily analyze the effects of different tool supported power minimizing design methodologies e.g. clock gating, on the power consumption of our example DSP design.

Fig. 6 depicts the average overall power consumption. It illustrates how clock gating on the RT level effects the power consumption.

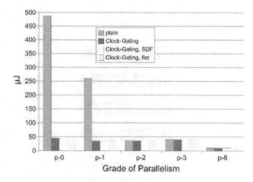

Fig. 6. Overall power consumption of different resynthesis architectures: The resynthesis architectures labelled as "plain" are synthesized without clock gating. The variants labelled "flat" were created by a removal of the design's hierarchy during compiling which enables hierarchy independent design optimizations

As one can see, the improvement of lower power consumption due to clock gating decreases for the more parallel architectures. For the sequential architecture (p=0) the difference between power consumption of architectures with clock gating and those without is much greater than for grade of parallelism of p=2 or higher. This is due to higher clock frequencies for the serial architectures than for parallel ones. For maintaining the same data rate for all architectures it is necessary to increase the clock frequencies with a lower grade of parallelism.

6.2   Impact of Parallelism

A more detailed view of the overall power consumption of the resynthesis system without the plain architectures can be seen in Fig. 7. As we were interested in highly accurate power estimates, we have annotated information about the delay time due to interconnect capacity to the simulator. Therefore we let the physical compiler generate a SDF file which contains these delay time informations. Architectures with an annotation of these SDF files are labelled SDF.

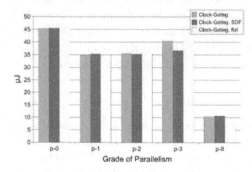

Fig. 7. Overall power consumption of different resynthesis architectures in more detailed view

First it can be seen that there is no significant difference between estimation results with SDF annotation and those without.

Furthermore the influence of parallelism to the overall power consumption can be noticed. The second architecture (p=1) consumes less power than the sequential one (p=0). As one can see, there is nearly no difference between p=1 and p=2 and at p=3 there is even a higher power consumption even though it processes more in parallel. The reason for this can be found in the scheduling step. Behavioral Compiler's search space increases for these more parallel architectures. As a result, it creates a suboptimal state machine of the design's controller which consumes more power compared to our simple hand done implementation (p=8).

6.3   Leakage Power Versus Area Consumption

The last result presented in this paper shows the leakage power consumption of our resynthesis design versus the chip area. In the left diagram of Fig. 8 it can be seen that the leakage power consumption increases with the grade of parallelism.

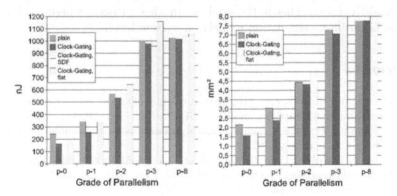

Fig. 8. Leakage power of different resynthesis architectures versus area consumption of the complete chip

In the right diagram the area consumption of the different architectures of the resynthesis DSP system is shown. Comparing the area consumption of the design depending to grades of parallelism and leakage power consumption of the design one can find a similar dependency.

For the chosen UMC 0.18 μm 1.8 V process the impact of leakage currents is proportional to the chip area. As it can be seen comparing the left diagram of Fig. 8 with the diagram in Fig. 7 the leakage power of the parallel architecture (p=8) is with 1 μJ nearly 10% of its overall power. With decreasing power supply voltages the leakage power becomes more relevant.

7 Summary and Outlook

In this paper we presented a design framework which assists the hardware designer in implementing power optimized designs for audio signal processing systems. It provides fast and efficient design and validation capabilities and implements a low power design flow with commercial tools for synthesis, power estimation and simulation. It additionally includes a custom made psycho-acoustically valid assessment for correct noise evaluation. The designer's workflow with this framework was shown, considering as an example the implementation of the resynthesis algorithm. We described the preliminary inspections of different bitwidths and arithmetic operands with a C++ reference implementation and the following architectural explorations. Furthermore we presented results created with our power estimation flow. By this the use of the framework was illustrated.

To enable a fast functional verification and to demonstrate the use of our low power DSP architectures we currently implement the core system on a rapid prototyping system, which, based on a PCI card containing onboard RAM, a Virtex-II FPGA and several communication ports, can be connected to an external audio codec device to ensure a closed audio processing loop.

Furthermore we develop currently a link between the low power framework and Matlab which is used by the majority of the algorithm developers in the hearing aid context. This is the goal of another project, where we intend to provide bit-accurate simulation models on system level.

References

1. M. Hansen and B. Kollmeier. Objective modeling of speech quality with a psychoacoustically validated auditory model. *J. Audio Eng. Soc.*, 2000.
2. V. Hohmann. Frequency analysis and synthesis using a gammatone filterbank. *Acustica / acta acustica*, 2002.
3. N. Voss, T. Eisenbach, and B. Mertsching. A Rapid Prototyping Framework for Audio Signal Processing Algorithms. *2004 IEEE International Conference on Field-Programmable Technology, December 6-8, 2004, Brisbane, Australia*, 2004.
4. N. Voss and B. Mertsching. A framework for low power audio design. *Proc. VLSI-2004*, 2004.

Traveling the Wild Frontier
of Ultra Low-Power Design

Jan Rabaey

University of California, Berkeley

Abstract. Power concerns have been at the forefront for the last decade, yet were always considered a second order citizen with respect to other design metrics. Today however, few will dispute that CMOS has entered the "power-limited scaling regime," with power dissipation becoming the limiting factor on what can be integrated on a chip and how fast it can run. Many approaches have been proposed over to address the concerns regarding both active and standby power. Yet, none of these provides a persistent answer enabling technology scaling may go on in the foreseeable future. Fortunately, a number of researchers are currently engaging in ultra-low power design (ULP), providing a glimpse on potential innovative solutions as well as clear showstoppers. In this talk, we first will present a perspective on power roadmaps and challenges. The second part of the presentation will present some of the solutions currently being considered in the ULP community. The talk will conclude with some long-term perspectives.

V. Paliouras, J. Vounckx, and D. Verkest (Eds.): PATMOS 2005, LNCS 3728, p. 747, 2005.
© Springer-Verlag Berlin Heidelberg 2005

DLV (Deep Low Voltage): Circuits and Devices

Sung Bae Park

Samsung Advanced Institute of Technology (SAIT)

Abstract. There will be a lot of different approaches to achieve the low power on a given system. Multiple modest speed processors will provide the better power-performance product compared to that of a high speed single one. Application dependent memory partitioning and non-blocking memory subsystem will make it possible to reduce unoptimized bus switching power including low power cache design. One of the most efficient way to reduce the power is going to implement the lowest operating voltage circuits as low as possible. The power equation below 130nm process is as follows: Power $= afCV^2 + bV^3 + cV^5$. The a term can be decided from architectural features such as swithing efficiency. The b term is from subthreshold leakage and the c term from gate leakage. If we operate 1GHz processor at 0.1V compared to that of current 1V, then power reduction can be 1/100 for a term, 1/1,000 for b term, and 1/100,000 for c term. We will explore how low the operating voltage would be possible in the CMOS circuits and devices, and we will discuss the barriers and challenges to achieve the DLV for ultra low power design.

V. Paliouras, J. Vounckx, and D. Verkest (Eds.): PATMOS 2005, LNCS 3728, p. 748, 2005.
© Springer-Verlag Berlin Heidelberg 2005

Wireless Sensor Networks: A New Life Paradigm

Magdy Bayoumi

Center of Advanced Computer Studies
University of Louisiana, USA

Abstract. Computers, Communication, and sensing technologies are converging to change the way we live, interact, and conduct business. Wireless Sensor networks reflect such convergence. These networks are based on collaborative efforts of a large number of sensor nodes. They should be low-cost, low-power, and multifunction. These nodes have the capabilities of sensing, data processing, and communicating. Sensor networks have wide range of applications, from monitoring sensors in industrial facilities to control and management of energy applications to military and security fields.

Because of the special features of these networks, new network technologies are needed for cost effective, low power, and reliable communication. These network protocols and architectures should take into consideration the special features of sensor networks such as: the large number of nodes, their failure rate, limited power, high desity..etc.

In this talk the impact of wireless sensor networks will be addressed, several of the design and communication issues will be discussed, and a case study of a current project of using such networks in drilling and management off-shore oil will be given.

V. Paliouras, J. Vounckx, and D. Verkest (Eds.): PATMOS 2005, LNCS 3728, p. 749, 2005.

Cryptography: Circuits and Systems Approach

O. Koufopavlou, George N. Selimis, N. Sklavos, and P. Kitsos

Electrical and Computer Engineering Department,
University of Patras, Patras, Greece

Abstract. Wireless Communications have become a very attractive and interesting sector for the provision of electronic services. Mobile networks are available almost anytime, anywhere and the user's acceptance of wireless hand-held devices is high. The services, are offered, are strongly increasing due to the different large range of the users' needs. In our days, the wireless communication protocols have specified security layers, which support security with high level strength. These wireless protocols security layers use encryption algorithms, which in many cases have been proved unsuitable and outdated for hardware implementations. The software and especially the hardware implementations of these layers are proved hard process for a developer/implementer. The performance results of those implementations are not often acceptable for the wireless communication standards and demands. Especially in the hand held devices and mobile communications applications with high speed and performance specifications such implementation solutions are not acceptable. In this talk, first, the mobile communication protocols are introduced and their security layers are described. The software implementations results of the used cryptographic algorithms are given for a fair comparison with hardware. The available hardware devices and latest technologies are described. The VLSI implementation approaches are also demonstrated in details with emphasis on the problems that a hardware implementer has to solve. Flexible solutions are proposed in order the implementation problems to be faced successfully with today's needs of the mobile data transfers. Finally, recommendations and observations for the wireless security engines are discussed.

V. Paliouras, J. Vounckx, and D. Verkest (Eds.): PATMOS 2005, LNCS 3728, p. 750, 2005.
© Springer-Verlag Berlin Heidelberg 2005

Author Index

Lecture Notes in Computer Science

For information about Vols. 1–3592

please contact your bookseller or Springer